The Physics of Phonons

This fully updated second edition of **The Physics of Phonons** remains the most comprehensive theoretical discussion devoted to the study of phonons, a major area of condensed matter physics.

It contains exciting new sections on phonon-related properties of solid surfaces, atomically thin materials (such as graphene and monolayer transition metal chalcogenides), in addition to nanostructures and nanocomposites, thermoelectric nanomaterials, and topological nanomaterials, with an entirely new chapter dedicated to topological nanophononics and chiralphononics. Although primarily theoretical in approach, the author refers to experimental results wherever possible, ensuring an ideal book for both experimental and theoretical researchers.

The author begins with an introduction to crystal symmetry and continues with a discussion of lattice dynamics in the harmonic approximation, including the traditional phenomenological approach and the more recent ab initio approach, detailed for the first time in this book. A discussion of anharmonicity is followed by the theory of lattice thermal conductivity, presented at a level far beyond that available in any other book. The chapter on phonon interactions is likewise more comprehensive than any similar discussion elsewhere. The sections on phonons in superlattices, impure and mixed crystals, quasicrystals, phonon spectroscopy, Kapitza resistance, and quantum evaporation also contain material appearing in book form for the first time. The book is complemented by numerous diagrams that aid understanding and is comprehensively referenced for further study. With its unprecedented wide coverage of the field, **The Physics of Phonons** is an indispensable guide for advanced undergraduates, postgraduates, and researchers working in condensed matter physics and materials science.

Features

- Fully updated throughout, with exciting new coverage on graphene, nanostructures and nanocomposites, thermoelectric nanomaterials, and topological nanomaterials.
- Authored by an authority on phonons.
- Interdisciplinary, with broad applications through condensed matter physics, nanoscience, and materials science.

W0235009

The Physics of Phonons
Second Edition

Gyaneshwar P. Srivastava

CRC Press
Taylor & Francis Group
Boca Raton London New York

CRC Press is an imprint of the
Taylor & Francis Group, an **informa** business

Second edition published 2023
by CRC Press
6000 Broken Sound Parkway NW, Suite 300, Boca Raton, FL 33487-2742

and by CRC Press
4 Park Square, Milton Park, Abingdon, Oxon, OX14 4RN

© 2023 Gyaneshwar P. Srivastava

First edition published by CRC Press 1990

CRC Press is an imprint of Taylor & Francis Group, LLC

British Library Cataloguing-in-Publication Data
A catalogue record for this book is available from the British Library

ISBN: 978-0-367-68526-3 (hbk)
ISBN: 978-0-367-69316-9 (pbk)
ISBN: 978-1-003-14127-3 (ebk)

DOI: 10.1201/9781003141273

Typeset in Nimbus Roman
by KnowledgeWorks Global Ltd.

Publisher's note: This book has been prepared from camera-ready copy provided by the author.

Vidyām cā'vidyām ca yastadvedobhayam saha.
Avidyayā mṛtyum tīrtvā vidyayā'mṛtamaśnute.
                                        (Yajurveda, XL.14).

(He who knows the nature of knowledge and nescience simultaneously passes over death by means of nescience and attains salvation by the help of true knowledge.)

Dedicated to the memories of my parents, and aunty and uncle Harish

Contents

Preface (Second Edition)

Significant advances have been made in developing concepts, techniques and applications of the physics of phonons since the publication of the first edition of this book in early 1990. In this edition I have attempted to include some aspects of new developments in theoretical and computational studies of phonon physics. These include topics, among others, such as (1) lattice dynamics within the harmonic approximation using the density functional perturbation theory (DFPT), (2) anharmonic phonon interactions (three-phonon and four-phonon processes) using fully *ab initio* treatments and a semi-*ab intio* treatment, (3) *ab initio* treatment of phonon-electron interaction, (4) nanophononics, (5) topological phonons, (6) topological nanophononics, (7) topological phonon chirality and (8) anisotropic effective medium theory for lattice thermal conductivity of nanocomposites.

Chapter-wise, the main additional material in edition 2 is as follows. Chapter 2 deals with lattice dynamics of two-dimensional mesh and provides a discussion on empirical many-body interatomic potential. Chapter 3 has been substantialy rewritten, detailing different approaches for *ab initio* treatment of lattice dynamics at harmonic level. Chapter 4 presents semi-*ab initio* and *ab initio* formalisms for cubic and quartic anharmonicity. The semi-*ab initio* approach has allowed for the presentation of explicit expressions for cubic and quartic anharmonic potential terms. Chapter 6 describes methods of evaluation of phonon-phonon and phonon-electron interaction rates using *ab inito* and semi-*ab initio* treatments. A simplified treatment of phonon relaxation arising from phonon-spin interaction has been added. Chapter 7 presents a spectrum of three-phonon relaxation rates and provides a description of thermal conductivity calculation using different levels of relaxation time theory (*e.g.* single-mode, Callaway, iterative) and different schemes for Brillouin zone integration (*e.g.* Debye's isotropic continuum scheme, special wave-vector scheme using phonon eigensolutions from *ab initio* methods). Chapter 8 discusses *ab intio* results for phonons on semicondcutor surfaces and for mono-layer thin 2D systems. Chapter 9 presents theory and results for thermal transport in periodically structured nanocomposite. This is done at two levels. For ultrathin nanocomposites (periodcity typically less than 10 nm) the theories described in chapters 3–7 have been used. A fully anisotropic effective medium theory (incorporating anisotropy of each of insert, host and insert-host parts) has been developed and applied for calculation of thermal conductivity of nanocomposites with periodicy larger than the lower end of the spectrum for phonon mean free path (typically larger than 10 nm). Modern developments of the topics of topological nanophononics and topological phonon chirality have been presented in chapter 10. Three new appendices have been added, and a few minor corrections and additions have also been made throughout the book. On the whole, more than 35% new material has been added to edition 1 to prepare edition 2.

Several national and international colleagues have indicated that they would like me to produce a second edition of this book. Also, my family members (wife Kusum, daughter Jyoti, son Deepak, daughter-in-law Marina and granddaughter Elizabeth) have been encouraging me to undertake this project. I must say that the work on this edition would not have progressed smoothly during the present COVID-19 pandemic period had it not been for the huge support from my wife Kusum. For this, I am very grateful to her.

I thank Mrs Rebecca Hodges-Davies, Commissioning Editor for Physics, CRC Press, Taylor & Francis Group, for inviting and supporting me to prepare this edition of the book. I also thank her team members and the desk editor for their help with the technical side of the publication.

It is pleasing that since the publication of the first edition I have been receiving encouraging emails and letters from graduate students and academics from different countries. Also, several good and encouraging reviews of the first edition were published in 1990s. Similarly, there have been helpful reviews of the proposal for the inclusion of additional topics in the second edition. I thank all the reviewers of the first edition and the proposed second edition of the book and hope that my effort in preparing this edition will meet readers' expectations. In particular, I hope that this edition will prove useful to undergraduates, postgraduates, postdoctoral fellows, experienced researchers as well as academics interested in condensed matter physics, material science and computational condensed matter physics.

Several postgraduates, postdoctoral fellows and collaborators, particularly Professor Hüseyin Tütüncü, Dr Steve Hepplestone and Dr Iorwerth Thomas, have contributed to my knowledge of the subject matter added to this edition. I am truly thankful to all those. I especially thank Dr Iorweth Thomas for his collaboration in the past decade on thermal transport and for providing useful feedback on a few chapters added to this edition. I would also like to thank Dr S. Raj Vatsya of NRC Canada for useful discussions in 2006 on properties of linear operators, in particular the anharmonic phonon collison operator used in this book. Any misconceptions and mistakes in this edition are failings on my part, and I would welcome suggestions for improvement from readers. It is important to add that many important works and citations are inadvertently missing in this limited-sized book, for which I express my sincere apology. I also apologise for inadvertently using a few symbols to represent more than one variable or physical quantity, but hope that this does not cause much inconvenience to readers.

Gyaneshwar Srivastava, D. Sc.
Professor of Theoretical Condensed Matter Physics
School of Physics, University of Exeter
Exeter, UK
February 2022

Acknowledgments (Second Edition)

I am grateful to the following for granting permission to reproduce figures added to the second edition of this book.

    The authors of all figures not originated by myself.
    Americal Physical Society for figures 3.2, 3.10, 3.11, 3.12, 3.13, 3.14, 7.3, 7.11, 7.12, 7.13, 7.17, 7.24, 8.6, 8.16, 8.17, 8.18, 9.14, 10.7, 10.8, 10.11–10.22, 11.9, 13.12.
    Institute of Physics for figures 7.7, 7.8, 8.10, 9.15, 10.3, 10.4, 10.9.
    American Institute of Physics for figures 7.18, 8.12, 8.13, 8.14.
    Springer-Verlag for figures 7.2, 9.19, 10.6.
    Elsevier Science Publishers for figures 3.9, 8.8, 9.16(b).
    MDPI for figures 9.16(a), 9.17, 9.18, 9.21.
    EDP Sciences for figure 3.8.
    Taylor and Francis Ltd for figure 9.20.
    Nature Communications for figure 5.2
    Oxford University Press for figure 10.10

Preface (First Edition)

This book is an outcome of my research interest and teaching at the undergraduate and postgraduate levels. I hope that it will be useful to final-year undergraduates, postgraduates and researchers who seek to study the physics of phonons beyond the level presented in most textbooks currently available on the subject.

I have covered three basic topics of phonon physics in non-metallic solids and liquid helium: (i) vibrational properties, (ii) phonon interactions and (iii) thermal properties. A brief description of the various chapters is as follows.

Chapter 1 presents a brief discussion of the elements of crystal symmetry. Chapter 2 describes the *semiclassical* concepts of lattice dynamics of a linear chain and of three-dimensional crystals. Calculations of the density of normal modes, specific heat and velocity of elastic waves in crystals are also discussed. In chapter 3 the *ab initio* treatment of lattice dynamics in the harmonic approximation is presented at three different levels of sophistication. Chapter 4 introduces the second-quantised notation and derives expressions for the phonon Hamiltonian up to the third-order anharmonic term for both a crystalline solid and an elastic continuum. Chapter 5 describes the theory of lattice thermal conductivity based on three different methods: (i) relaxation-time approaches, (ii) complementary variational approaches and (iii) linear-response approaches. The kinetic theory expression for phonon conductivity, based on the single-mode relaxation-time result, is shown to arise from the diagonal part of the anharmonic phonon collision operator. It is also shown that the classical work of Callaway on thermal conductivity can be understood in terms of an effective phonon relaxation time which includes a contribution from the off-diagonal three-phonon N-processes operator. Furthermore, Callaway's scheme for an effective phonon relaxation time is extended to include the effect of the off-diagonal part of the three-phonon U-processes operator. The theory of complementary variational principles derives a sequence of lower bounds and a sequence of upper bounds, both converging monotonically towards the exact thermal conductivity. This follows among others the work of Arthurs, Benin and the present author. The Green–Kubo linear response expression for thermal conductivity is evaluated by using two different approaches: the double-time Green's function method and the Zwanzig–Mori projection operator method. Although these two approaches are only used here to evaluate the heat current–heat current correlation function, it is hoped that the details of the schemes given in this section will prove useful when applying the linear-response method to other relevant problems. Chapter 6 discusses phonon scattering mechanisms in insulators and doped semiconductors. Results for phonon relaxation times and matrix elements of the phonon scattering operator are derived by using the linear-response and time-dependent first- and second-order perturbation theories, as appropriate. Particular attention has been paid to three-phonon processes involving both acoustic and optical phonons. A brief discussion of the theory of phonon–photon interaction, leading to infrared absorption and Raman scattering processes, is also presented.

Chapter 7 presents numerical results, and their comparison with experiment, for phonon relaxation rates and thermal conductivity of insulators and semiconductors. The theory of phonons on surfaces and in superlattices, using both the continuum model and the crystal model, is presented in chapter 8. This chapter also discusses the interaction of phonons with a two-dimensional gas. Chapter 9 describes the theory of Elliot and Dawber for localised vibrational modes in semiconductors, and the modified-random-element-isodisplacement model for long-wavelength optical phonons in mixed crystals. Also, this chapter discusses infrared results for localised modes in single crystals and Raman scattering results for mode behaviour in mixed crystals. Chapter 10 discusses the physics of phonons in quasicrystals and in amorphous solids. The concept of quasicrystals, their structure and phonon dispersion in the Fibonacci monatomic linear chain is discussed. Recent views on the

nature of vibrational excitations in amorphous solids, namely, phonons, tunnelling states and fractons, and their contribution to the specific heat and thermal conductivity are discussed. Chapter 11
attempts to describe some of the modern phonon spectroscopic techniques for generating and detecting phonons in crystalline semiconductors and in liquid helium. Finally, chapter 12 provides a
brief discussion on phonons and rotons in superfluid helium, interaction between these excitations,
Kapitza resistance and quantum evaporation.

Each chapter includes references to many original works and reviews which can be consulted
to give the reader in-depth knowledge. The list of references is not meant to be exhaustive and I
apologise if I have left out any important works.

Some knowledge of undergraduate solid state physics, including electronic band structure, and
group theory will be useful in following the subject matter presented in the book.

This book uses a number of unit systems, e.g. CGS and AU; I have not converted these to SI units.
Hopefully this will not cause too much confusion to readers.

A few acknowledgments are in order. I am grateful to Professor G S Verma who introduced me
to the field of phonon interactions in solids. My sincere thanks are due to all those with whom
I have had discussions and collaborations on the subject matter over the years. I am particularly
grateful to Professor J E Parrott, Professor L J Challis, Professor D L Weaire, Professor J C Inkson,
Professor A F G Wyatt, Dr R Jones, Dr R T Phillips, Dr C Lambert, Dr D Kechrakos, Dr Z Ikonić,
Mr P Briddon and Mr P Mulheran who kindly read various chapters of this book and made very
valuable suggestions for improvement. However, all remaining misconceptions and mistakes in the
book are my responsibility, and I would appreciate hearing from readers about any corrections and
suggestions for improvement that could be included in any future edition.

My special thanks go to Mr C S G Cousins whose initial expert advice helped me successfully
prepare this manuscript using TEX.

I would also like to express my sincere thanks to Mr J A Revill, Commissioning Editor for IOP
Publishing, and his staff for their interest and support in the publication of this book.

Finally, it could not have been possible to complete this book had I not received the continuous
and full support of my wife Kusum and our children Jyoti and Deepak. To them I can only give my
love and best wishes in life.

G P Srivastava
Exeter
November 1989

Acknowledgments (First Edition)

I am grateful to the following for granting permission to reproduce figures included in this book.

The authors of all figures not originated by myself.

The American Physical Society for figures 2.7(a), 2.9, 2.12(b), 3.4, 4.4, 5.1, 7.7–7.9, 7.12–7.14, 8.3, 8.6–8.8, 8.12, 8.13, 9.1, 9.4(a)–(d), (h)–(j), 9.5(a), (c), 9.7, 9.9, 10.2, 10.4, 10.9(d), 10.10–10.12, 10.16, 11.2 and 11.8.

International Atomic Energy Agency for figure 2.7(b).

Pergamon Press plc for figures 2.10, 7.1–7.4, 9.4(e), (f) and 9.6.

John Wiley and Sons Inc for figure 2.13.

Akademie-Verlag for figures 4.2, 9.5(e), 9.10 and 9.11.

Springer-Verlag for figures 8.9, 8.11, 10.7 and 10.17.

Pion, London for figures 7.5, 7.6 and 7.10.

Les Editions de Physique for figure 7.11.

Plenum Publishing Corp for figures 9.5(b) and 12.2.

American Institute of Physics for figure 9.5(d).

Academic Press Inc for figures 10.3, 10.13 and 10.15.

Gordon and Breach Science Publishers Inc for figure 10.6.

Flammarion for figure 10.8.

Taylor and Francis Ltd for figures 10.9(a)–(c).

Friedr. Vieweg & Sohn for figure 10.14.

Elsevier Science Publishers for figures 12.3–12.5.

1 Elements of Crystal Symmetry

In this chapter an introduction to crystal symmetry is presented. Some of the results will be required for the rest of the book. The discussion is intended to be brief, and the reader is recommended to consult further literature for more details. References used in preparing the presentation in this book are, among others, Heine (1960), Bhagavantam (1966), Morgan (1969), Mariot (1972), Boardman *et al* (1973), Callaway (1974), Joshi (1982), Burns (1985) and Kittel (1986).

1.1 DIRECT LATTICE

A *lattice* is defined as an infinite regular array of *points* in which all points have identical surroundings. As a point is a mere realisation, a lattice is just a mathematical construction. From a chosen point, the position vector of another point of a three-dimensional lattice is given by a *translation vector*

$$\boldsymbol{T} = n_1\boldsymbol{a}_1 + n_2\boldsymbol{a}_2 + n_3\boldsymbol{a}_3, \tag{1.1}$$

where n_1, n_2 and n_3 are any integers, and $\boldsymbol{a}_1, \boldsymbol{a}_2$ and $\boldsymbol{a}_3$ are three independent *primitive translation vectors*. The parallelepiped formed by $\boldsymbol{a}_1, \boldsymbol{a}_2$ and $\boldsymbol{a}_3$, which contains one lattice point, is called a *primitive unit cell*. As all points have identical surroundings, the whole space can be generated by suitable translations of the primitive unit cell. Therefore, the essential features of a lattice can be extracted from its primitive unit cell.

A cell or parallelepiped which contains more than one lattice point is called a *non-primitive unit cell*.

A primitive unit cell in the shape of a parallelepiped defined by $\boldsymbol{a}_1, \boldsymbol{a}_2$ and $\boldsymbol{a}_3$ has a volume equal to the *Wigner–Seitz cell*, or *proximity cell,* which is constructed as follows. Choose a lattice point and draw lines joining all neighbouring points. Construct perpendicular planes bisecting these planes. The smallest volume thus enclosed is the Wigner–Seitz cell.

A total of 14 distinct types of three-dimensional lattices can be realised. These are known as the 14 *Bravais lattices.* For convenience, these are grouped into seven *lattice systems* according to their unit cell lengths and interaxial angles. The seven systems, with the number of their lattices given in parentheses, are triclinic (1), monoclinic (2), orthorhombic (4), tetragonal (2), cubic (3), trigonal (1) and hexagonal (1).

Here we will only consider one of the seven systems, namely, the *cubic system,* which is characterised by $a_1 = a_2 = a_3$ and $\hat{\boldsymbol{a}}_1 \cdot \hat{\boldsymbol{a}}_2 = \hat{\boldsymbol{a}}_2 \cdot \hat{\boldsymbol{a}}_3 = \hat{\boldsymbol{a}}_3 \cdot \hat{\boldsymbol{a}}_1 = 0$. The three different lattice types of this system are simple cubic (SC), face-centred cubic (FCC) and body-centred cubic (BCC). These are shown in figure 1.1, where the cube is known as the *conventional cell* and a is called the cubic lattice constant.

SC

The primitive unit cell for the SC lattice is its conventional cell. The primitive translational vectors are

$$
\begin{aligned}
\boldsymbol{a}_1 &= a(1,0,0)\\
\boldsymbol{a}_2 &= a(0,1,0)\\
\boldsymbol{a}_3 &= a(0,0,1),
\end{aligned}
\tag{1.2}
$$

DOI: 10.1201/9781003141273-1

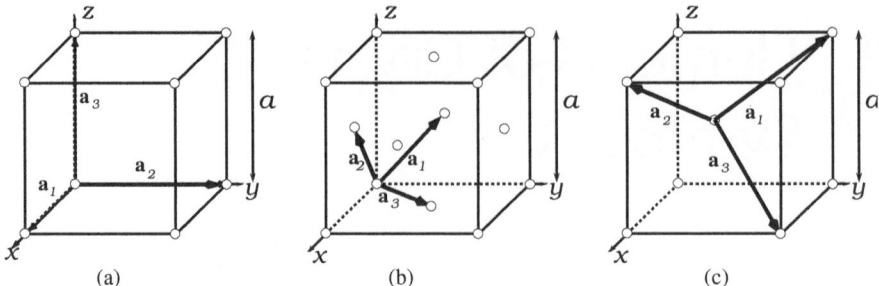

Figure 1.1 The cubic lattices: (*a*) SC, (*b*) FCC and (*c*) BCC. The cell shown is the conventional cell and *a* represents the cubic lattice constant. The primitive translation vectors a_1, a_2 and a_3 are also shown.

where

$$(m_1, m_2, m_3) = m_1 \hat{x} + m_2 \hat{y} + m_3 \hat{z}, \tag{1.3}$$

and $\hat{x}, \hat{y}, \hat{z}$ are unit vectors along the x, y and z directions in the cartesian system. The volume of the primitive cell, and hence volume per lattice point, is $\Omega = |a_1 \cdot (a_2 \times a_3)| = a^3$.

FCC

The primitive unit cell for the FCC lattice is defined by the three primitive translation vectors

$$
\begin{aligned}
a_1 &= \frac{a}{2}(0,1,1) \\
a_2 &= \frac{a}{2}(1,0,1) \\
a_3 &= \frac{a}{2}(1,1,0).
\end{aligned} \tag{1.4}
$$

The volume of the unit cell is $\Omega = |a_1 \cdot (a_2 \times a_3)| = a^3/4$.

BCC

The primitive unit cell for the BCC lattice is defined by the three primitive translation vectors

$$
\begin{aligned}
a_1 &= \frac{a}{2}(-1,1,1) \\
a_2 &= \frac{a}{2}(1,-1,1) \\
a_3 &= \frac{a}{2}(1,1,-1).
\end{aligned} \tag{1.5}
$$

The volume of the cell is $\Omega = |a_1 \cdot (a_2 \times a_3)| = a^3/2$.

1.2   RECIPROCAL LATTICE

The mathematical construction of a three-dimensional direct space lattice can be translated into an equivalent three-dimensional reciprocal space lattice. Such a mapping is achieved by using the relationship

$$\exp(iG \cdot T) = 1. \tag{1.6}$$

Corresponding to a direct translation vector T, this relationship defines a *reciprocal translation vector* G. The points generated by the vectors G form the *reciprocal lattice*. As direct lattice vectors T have the dimensions of [length], the reciprocal lattice vectors G have the dimensions of [1/length].

The reciprocal lattice is an essential concept for the study of crystalline solids and their diffraction properties. The space generated by reciprocal lattice points is called the *reciprocal space* (various other names are also commonly used, e.g. *G*-space, *q*-space, Fourier space, or momentum space).

The reciprocal lattice can be constructed, in conformity with the relationship in equation (1.6), both geometrically and mathematically. Geometrically, the reciprocal lattice is constructed as follows. Consider the lattice as a network of interpenetrating planes. Choose a lattice point as the origin and consider drawing normals to all possible planes in the direct lattice. If d is the distance of a plane from the origin, then represent this plane by a point at a distance $1/d$ on its normal. A collection of all such points produces the reciprocal lattice. Mathematically, the reciprocal lattice is defined by a set of vectors G:

$$G = m_1 b_1 + m_2 b_2 + m_3 b_3, \tag{1.7}$$

where m_1, m_2 and m_3 are any integers and b_1, b_2 and b_3 are the primitive translation vectors of the reciprocal lattice which are defined in terms of the primitive translation vectors a_1, a_2 and a_3 of the direct lattice as follows:

$$
\begin{aligned}
b_1 &= \frac{2\pi}{\Omega}(a_2 \times a_3) \\
b_2 &= \frac{2\pi}{\Omega}(a_3 \times a_1) \\
b_3 &= \frac{2\pi}{\Omega}(a_1 \times a_3),
\end{aligned}
\tag{1.8}
$$

where $\Omega = |a_1 \cdot (a_2 \times a_3)|$ is the volume of the primitive unit cell of the direct lattice. It can be easily shown that $G \cdot T = 2\pi N$, where N is an integer, so that equation (1.6) is verified. The volume of the parallelepiped defined by b_1, b_2 and b_3 is $\Omega_{rl} = |b_1 \cdot (b_2 \times b_3)| = (2\pi)^3/\Omega$. It can be easily verified that a reciprocal lattice vector $G_{hkl} = hb_1 + kb_2 + lb_3$ is perpendicular to a plane in the direct lattice with Miller indices (hkl).

SC

Using equations (1.2) and (1.8), the primitive translation vectors of the reciprocal lattice of the SC lattice are

$$
\begin{aligned}
b_1 &= \frac{2\pi}{a}(1,0,0) \\
b_2 &= \frac{2\pi}{a}(0,1,0) \\
b_3 &= \frac{2\pi}{a}(0,0,1).
\end{aligned}
\tag{1.9}
$$

The volume of the parallelepiped defined by b_1, b_2 and b_3 is $\Omega_{rl} = (2\pi/a)^3$.

FCC

The primitive translation vectors of the reciprocal lattice of the FCC direct space lattice are, using equations (1.4) and (1.8),

$$
\begin{aligned}
b_1 &= \frac{2\pi}{a}(-1,1,1) \\
b_2 &= \frac{2\pi}{a}(1,-1,1) \\
b_3 &= \frac{2\pi}{a}(1,1,-1).
\end{aligned}
\tag{1.10}
$$

The volume of the parallelepiped defined by b_1, b_2 and b_3 is $\Omega_{rl} = 4(2\pi/a)^3$.

BCC

The primitive translation vectors of the reciprocal lattice of the BCC direct space lattice are

$$
\begin{aligned}
\boldsymbol{b}_1 &= \frac{2\pi}{a}(0,1,1) \\
\boldsymbol{b}_2 &= \frac{2\pi}{a}(1,0,1) \\
\boldsymbol{b}_3 &= \frac{2\pi}{a}(1,1,0).
\end{aligned}
\tag{1.11}
$$

The volume of the primitive cell of the reciprocal lattice is $\Omega_{rl} = 2(2\pi/a)^3$.

Notice that the reciprocal lattice of the FCC direct space lattice is BCC, and vice versa. The reciprocal lattice of the SC direct space lattice remains SC. It can be verified that the reciprocal lattice of the reciprocal lattice is the original direct space lattice.

1.3 THE BRILLOUIN ZONE

It is customary in solid-state physics to choose the primitive unit cell of the reciprocal lattice as its central Wigner–Seitz cell, and not the parallelepiped defined by the primitive translation vectors $\boldsymbol{b}_1, \boldsymbol{b}_2$ and $\boldsymbol{b}_3$. The central Wigner–Seitz cell of a reciprocal lattice is known as its *first Brillouin zone* (or simply the Brillouin zone). The Wigner–Seitz cells of a reciprocal lattice outer to the central Wigner–Seitz cell generate the second Brillouin zone, third Brillouin zone, etc. For example, the second Brillouin zone is the volume enclosed between the central and the next outer Wigner–Seitz cells, and so on. Each Brillouin zone has the volume equal to $\Omega_{BZ} = \Omega_{rl} = |\boldsymbol{b}_1 \cdot (\boldsymbol{b}_2 \times \boldsymbol{b}_3)|$.

One of the most significant features of the Brillouin zone concept is that, for specular reflection from a set of parallel planes perpendicular to a reciprocal lattice vector $\boldsymbol{G}$, an elastic scattering $\boldsymbol{q}' = \boldsymbol{q} + \Delta\boldsymbol{q}$ ($|\boldsymbol{q}'| = |\boldsymbol{q}|$) can be expressed as

$$
\begin{aligned}
\text{or} \qquad \Delta\boldsymbol{q} &= \boldsymbol{G} \\
q^2 &= (\boldsymbol{q} - \boldsymbol{G})^2 \\
\text{or} \qquad G^2 - 2\boldsymbol{q}\cdot\boldsymbol{G} &= 0 \\
\text{or} \qquad \boldsymbol{q}\cdot\hat{\boldsymbol{G}} &= \frac{1}{2}|\boldsymbol{G}|.
\end{aligned}
\tag{1.12}
$$

In other words, the projection of $\boldsymbol{q}$ along a shortest reciprocal lattice vector $\boldsymbol{G}$ lies on the Brillouin zone face. This relation provides an alternative definition of the Bragg reflection, or diffraction, condition $2d_{hkl}\sin\theta = n\lambda$ for a wave of wavelength $\lambda = 2\pi/q$ propagating at an angle θ with respect to a set of parallel planes whose interplanar spacing is given by $d_{hkl} = 2\pi/G_{hkl}$, and where n is an integer.

SC

A general reciprocal lattice vector of the SC lattice is

$$
\begin{aligned}
\boldsymbol{G} &= m_1\boldsymbol{b}_1 + m_2\boldsymbol{b}_2 + m_3\boldsymbol{b}_3 \\
&= \frac{2\pi}{a}(m_1, m_2, m_3),
\end{aligned}
\tag{1.13}
$$

where m_i are any integers.

The shortest $\boldsymbol{G}$s are $\pm(2\pi/a)\hat{\boldsymbol{x}}, \pm(2\pi/a)\hat{\boldsymbol{y}}, \pm(2\pi/a)\hat{\boldsymbol{z}}$. Therefore, there are six perpendicular bisector planes at distances, from the origin, $\pm(\pi/a)\hat{\boldsymbol{x}}, \pm(\pi/a)\hat{\boldsymbol{y}}, \pm(\pi/a)\hat{\boldsymbol{z}}$. These planes enclose a cube of volume $(2\pi/a)^3$, which is the first Brillouin zone shown in figure 1.2(*a*).

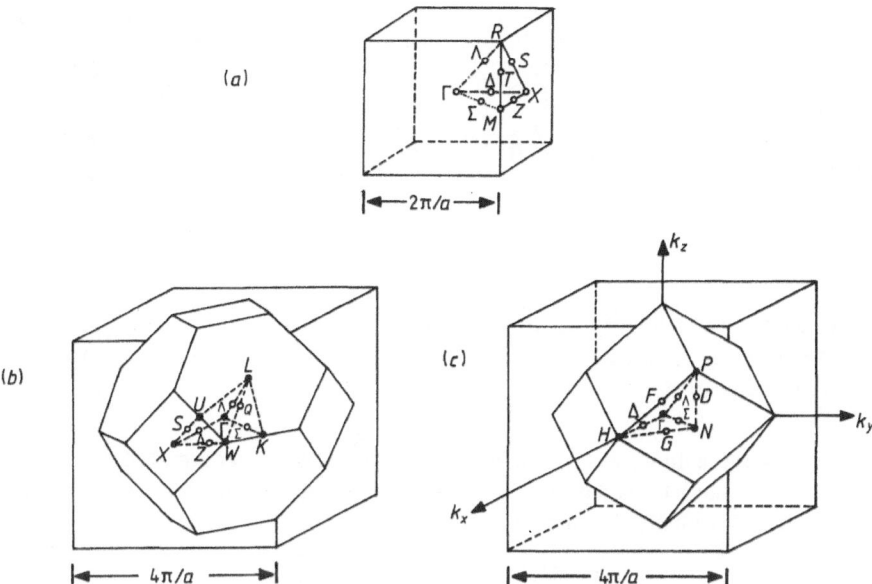

Figure 1.2 The Brillouin zone for cubic lattices: (a) SC, (b) FCC and (c) BCC. The irreducible part of the zone is also shown. (The shape of the Brillouin zone of the BCC lattice is identical to the shape of the Jones zone for the diamond/zincblende structure crystals (see section 1.8).

BCC

A general reciprocal lattice vector of the FCC lattice is

$$
\begin{aligned}
\boldsymbol{G} &= m_1\boldsymbol{b}_1 + m_2\boldsymbol{b}_2 + m_3\boldsymbol{b}_3 \\
&= \frac{2\pi}{a}(-m_1 + m_2 + m_3, m_1 - m_2 + m_3, m_1 + m_2 - m_3),
\end{aligned}
\tag{1.14}
$$

where m_i are any integers.

The shortest $\boldsymbol{G}$s are the eight vectors $(2\pi/a)(\pm1,\pm1,\pm1)$. The next shortest $\boldsymbol{G}$s are the six vectors $\pm(2\pi/a)(2,0,0), \pm(2\pi/a)(0,2,0), \pm(2\pi/a)(0,0,2)$. The Brillouin zone enclosed by the bisector planes perpendicular to these 14 vectors is a *truncated octahedron*, shown in figure 1.2(b).

BCC

A general reciprocal lattice vector of the BCC lattice is

$$
\boldsymbol{G} = \frac{2\pi}{a}(m_2 + m_3, m_3 + m_1, m_1 + m_2),
\tag{1.15}
$$

where m_i are any integers.

The shortest $\boldsymbol{G}$s are the 12 vectors $(2\pi/a)(0,\pm1,\pm1), (2\pi/a)(\pm1,0,\pm1), (2\pi/a)(\pm1,\pm1,0)$. The Brillouin zone defined by the bisector planes perpendicular to these planes is a regular 12-faced figure, *a rhombic dodecahedron*, as shown in figure 1.2(c).

1.4   CRYSTAL STRUCTURE

A crystal, or crystalline solid, is a regular three-dimensional arrangement of *atoms*. The concept of a lattice is useful in describing the arrangement of atoms in a crystal. For this purpose, we identify a

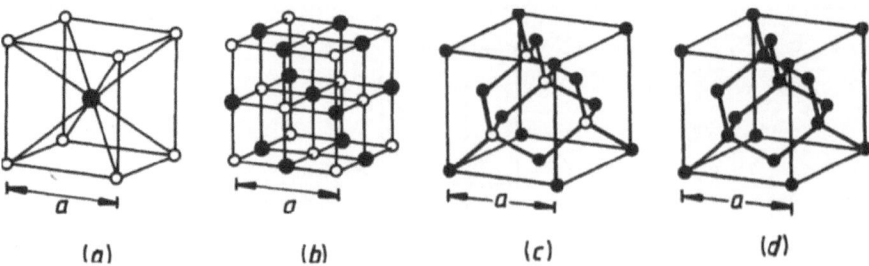

Figure 1.3 Some examples of crystal structure with cubic lattices: (*a*) CsCl structure with SC lattice, (*b*) NaCl structure with FCC lattice, (*c*) zincblende structure with FCC lattice, (*d*) diamond structure with FCC lattice. *a* denotes the cubic lattice constant.

particular lattice corresponding to a crystal, and assign one atom or a group of atoms, called *a basis*, to every lattice point. Thus a crystal has the repeat motif of its underlying lattice. A *crystal structure* is defined by the details of its lattice and basis:

$$\text{crystal structure} = \text{lattice} + \text{basis}$$

i.e. by the details of the parallelepiped of the primitive unit cell of its lattice and the distribution of atoms inside the unit cell.

Examples of crystal structures

We will consider a few examples of crystal structures based on the cubic lattice system.

Crystal structure with the SC lattice
The most common crystal structure with the SC lattice is the CsCl structure. In this structure, the basis consists of two atoms (one molecule of CsCl) with Cs at $(0,0,0)$ and Cl at $a(1/2,1/2,1/2)$ (or vice versa), where a is the cubic lattice constant. This structure is shown in figure 1.3(*a*). The number of nearest neighbours of each atom (the coordination number) of this structure is 8.

Crystal structures with the FCC lattice
We consider four structures with the FCC lattice.

The NaCl (or rocksalt) structure In this structure, which is also known as the alkali halide or rocksalt structure, each point of the FCC lattice is assigned a basis of one unit of NaCl: Na (group I atom) at $(0,0,0)$ and Cl (group VII atom) at $a(1/2,1/2,1/2)$ (or vice versa), as shown in figure 1.3(*b*). The coordination number of this structure is 6.

The CaF$_2$ structure
In this structure, the basis consists of Ca at $(0,0,0)$ and 2F at $\pm a(1/4,1/4,1/4)$.

The zincblende (sphalerite, or cubic ZnS) structure This is another type of crystal structure with the FCC lattice. As shown in figure 1.3(*c*), the basis consists of one atom at $(0,0,0)$ and another at $a(1/4,1/4,1/4)$. Semiconductors composed of atoms from group III and group V of the periodic table (known as III–V semiconductors) and many II–VI semiconductors crystallise in the zincblende structure. Each atom in this structure is tetrahedrally bonded to four atoms of the other type.

The diamond (or adamantine) structure The zincblende structure turns into the diamond crystal structure when the two basis atoms are identical. This crystal structure, therefore, is characterised by a centre-of-inversion at the midpoint between the two basis atoms. Crystals of group IV elements such as C (diamond), Si, Ge and α-Sn (grey tin) exhibit the diamond structure. In this structure, each atom is tetrahedrally bonded to four neighbours (of the same type of course), as shown in figure 1.3(*d*).

The diamond and zincblende structures are relatively open as the coordination number is only 4.

Many IV–IV, III–V and II–VI compounds can also be crystallised with the wurtzite structure (based on the hexagonal lattice), maintaining tetrahedral bonding. Some examples are SiC, GaN, ZnS, CdS, CdSe and ZnO.

Examples of crystal structures with the BCC lattice

An example of a crystal structure with the BCC lattice is provided by alkali halide metals such as Na and K. In this case, the basis consists of just one alkali atom, so that the structure of both the lattice and the crystal is the same: BCC. Another example is the SiF_4 crystal which also has the BCC lattice.

1.5 POINT GROUPS

Apart from its lattice translational symmetry, a crystal also shows two other types of symmetries: *point group* and *space group* symmetries.

An operation carried out about an axis passing through a point in space which brings the crystal structure into itself while leaving the point fixed is called a *point symmetry operation*. The collection of all point symmetry operations of a crystal forms a group which is called the *point symmetry group* (or simply *point group*). An allowed point symmetry operation can be one of the following types:

(i) rotation about an n-fold axis: rotations through $(2\pi/n)$, with only $n=2,3,4$, or 6 are possible;

(ii) reflection: reflection in a plane, called a *mirror plane*;

(iii) inversion: an inversion through a point such that every point r is brought into $-r$;

(iv) rotation–reflection: rotation through $(2\pi/n)$ followed by a reflection in a plane perpendicular to the rotation axis;

(v) rotation–inversion: rotation through $(2\pi/n)$ followed by an inversion about a point lying on the rotation axis.

Operations of type (i) are called proper rotations, while operations of types (ii)–(v) are called improper rotations. Thus a point group is a group of allowed proper and improper rotations of a lattice or crystal.

The total number of distinct point groups, corresponding to the seven lattice systems, is 32. These are generally known as the 32 *crystallographic point groups*. The total number of point groups for the cubic system is 5. Of particular interest to our discussion in this book are the diamond and zincblende cubic structures. The point group of the diamond structure is the full octahedral group O_h (or $m3m$) with 48 operations. The point group of the zincblende structure is the tetrahedral group T_d (or $\bar{4}3m$) with 24 operations; as discussed earlier, the zincblende structure lacks the inversion symmetry present in the diamond structure. (Here we have used two commonly used notations: e.g. O_h is the *Schoenflies notation* (Schoenflies 1923) and $m3m$ is the *International* (or *Hermann–Mauguin*) *notation* (Hermann 1930, Mauguin 1931).)

1.6 SPACE GROUPS

An operation about an axis passing through a point in space which leaves a crystal invariant, but which may involve moving the point within the primitive unit cell, is called a *space symmetry operation*. Such an operation can be represented by $\hat{O} = \{\mathscr{R} \mid t\}$, where $\mathscr{R}$ represents a point symmetry operation and t is a *fractional translation* associated with $\mathscr{R}$,

$$t = t(\mathscr{R}) = u_1 a_1 + u_2 a_2 + u_3 a_3, \tag{1.16}$$

with $|u_i| < 1$. The collection of all space symmetry operations of a crystal forms a group which is called the *space symmetry group* (or simply *space group*). The group of operations $\{\mathscr{R} \mid 0\}$, called the *point group of the space group*, corresponds to one of the 32 crystallographic point groups.

Symmetry operations involving fractional translations are as follows.

(i) *Screw axes*: a screw axis operation is identified as a rotation followed by a fractional translation parallel to the rotation axis.

(ii) *Glide planes*: a glide plane operation is a fractional translation followed by a reflection in a plane containing the translation vector.

A space group operator $\{\mathscr{R} \mid t\}$ corresponds to a coordinate transformation of the form

$$r' = \mathscr{R}r + t, \tag{1.17}$$

where $\mathscr{R}$ can be represented by a 3×3 real orthogonal matrix in cartesian coordinates.

There are two types of space groups: *symmorphic* (or simple) and *non-symmorphic*. A symmorphic space group involves symmetry operations which do not involve any fractional translations. A non-symmorphic space group is specified by at least one operation which involves a fractional translation. There are a total of 230 (non magnetic) space groups, out of which 73 are symmorphic.

The space group of the diamond structure is the non-symmorphic group O_h^7 (or $Fd3m$), involving 48 operations. The space group of the zincblende structure is the symmorphic group T_d^2 (or $F\bar{4}3m$), involving 24 operations. The operations of these space groups are listed in Table 1.1.

It should be mentioned that the fractional translations of a space group are dependent on the choice of origin. This is shown in Table 1.1 for two different choices of origin in the diamond and zincblende structure crystals. In particular, it can be noticed that the space group of the zincblende structure changes from symmorphic to non-symmorphic (i.e. non-zero fractional translations are introduced) when the origin is changed from one of the basis atoms to the midpoint between the two basis atoms.

1.7 SYMMETRY OF THE BRILLOUIN ZONE

If $\mathscr{R}$ is an operation in the point group of a lattice, then $\mathscr{R}^{-1}$ is also an operation of the same point group. Also, $R^{-1}T$ is a lattice translation if T is, so that

$$\mathscr{R}^{-1}T \cdot G = 2\pi N, \tag{1.18}$$

where N is some integer. As $\mathscr{R}$ represents an orthogonal transformation, i.e. $\mathscr{R}^T = \mathscr{R}^{-1}$, we can express

$$
\begin{aligned}
T \cdot \mathscr{R}G &= \sum_{ij} T_i \mathscr{R}_{ij} G_j \\
&= \sum_{ij} T_i \mathscr{R}_{ji}^{-1} G_j \\
&= \mathscr{R}^{-1}T \cdot G = 2\pi N,
\end{aligned}
\tag{1.19}
$$

so that $\mathscr{R}G$ must be a reciprocal lattice vector if G is. Therefore, it can be concluded that a reciprocal lattice has the same point group as its direct lattice. As a consequence of this result, it follows that the reciprocal lattice belongs to the same *system* as the direct lattice, although not necessarily the same *type*.

Although a primitive unit cell in the form of the parallelepiped defined by the vectors a_1, a_2 and a_3 does not automatically display the point symmetry of its lattice, the symmetrical unit cell in the form of the Wigner–Seitz cell does. Therefore, the Brillouin zone of a lattice displays the point symmetry of the reciprocal lattice and hence the direct lattice. This means that, apart from the translational symmetry $f(q + G) = f(q)$, a function or property in the reciprocal space will also obey the point group symmetry of the direct lattice when q lies inside the Brillouin zone. Thus if the number of point group operations is M, then unique values of $f(q)$ exist only for q-values lying in a $(1/M)$th part of the Brillouin zone. Such a part of the Brillouin zone is called its *irreducible part*.

Table 1.1

Space group operations of the diamond structure (O_h^7 group). The first 24 operations correspond to the zincblende structure (T_d^2 group).

j	$\{\mathscr{R} \mid t\}^{-1}r$			$\{\mathscr{R} \mid t\}^{-1}r$		
	origin at atomic site			origin midway between the two atoms		
1	x	y	z	x	y	z
2	$\bar{x}$	$\bar{y}$	z	$\bar{x}-\frac{a}{4}$	$\bar{y}-\frac{a}{4}$	z
3	x	$\bar{y}$	$\bar{z}$	x	$\bar{y}-\frac{a}{4}$	$\bar{z}-\frac{a}{4}$
4	$\bar{x}$	y	$\bar{z}$	$\bar{x}-\frac{a}{4}$	y	$\bar{z}-\frac{a}{4}$
5	y	$\bar{x}$	$\bar{z}$	y	$\bar{x}-\frac{a}{4}$	$\bar{z}-\frac{a}{4}$
6	$\bar{y}$	x	$\bar{z}$	$\bar{y}-\frac{a}{4}$	x	$\bar{z}-\frac{a}{4}$
7	$\bar{x}$	z	$\bar{y}$	$\bar{x}-\frac{a}{4}$	z	$\bar{y}-\frac{a}{4}$
8	$\bar{x}$	$\bar{z}$	y	$\bar{x}-\frac{a}{4}$	$\bar{z}-\frac{a}{4}$	y
9	$\bar{z}$	$\bar{y}$	x	$\bar{z}-\frac{a}{4}$	$\bar{y}-\frac{a}{4}$	x
10	z	$\bar{y}$	$\bar{x}$	z	$\bar{y}-\frac{a}{4}$	$\bar{x}-\frac{a}{4}$
11	$\bar{y}$	$\bar{x}$	z	$\bar{y}-\frac{a}{4}$	$\bar{x}-\frac{a}{4}$	z
12	$\bar{z}$	y	$\bar{x}$	$\bar{z}-\frac{a}{4}$	y	$\bar{x}-\frac{a}{4}$
13	x	$\bar{z}$	$\bar{y}$	x	$\bar{z}-\frac{a}{4}$	$\bar{y}-\frac{a}{4}$
14	y	x	z	y	x	z
15	z	y	x	z	y	x
16	x	z	y	x	z	y
17	z	x	y	z	x	y
18	y	z	x	y	z	x
19	z	$\bar{x}$	$\bar{y}$	z	$\bar{x}-\frac{a}{4}$	$\bar{y}-\frac{a}{4}$
20	$\bar{y}$	$\bar{z}$	x	$\bar{y}-\frac{a}{4}$	$\bar{z}-\frac{a}{4}$	x
21	$\bar{z}$	$\bar{x}$	y	$\bar{z}-\frac{a}{4}$	$\bar{x}-\frac{a}{4}$	y
22	$\bar{y}$	z	$\bar{x}$	$\bar{y}-\frac{a}{4}$	z	$\bar{x}-\frac{a}{4}$
23	$\bar{z}$	x	$\bar{y}$	$\bar{z}-\frac{a}{4}$	x	$\bar{y}-\frac{a}{4}$
24	y	$\bar{z}$	$\bar{x}$	y	$\bar{z}-\frac{a}{4}$	$\bar{x}-\frac{a}{4}$
25	$\bar{x}+\frac{a}{4}$	$\bar{y}+\frac{a}{4}$	$\bar{z}+\frac{a}{4}$	$\bar{x}$	$\bar{y}$	$\bar{z}$
26	$x+\frac{a}{4}$	$y+\frac{a}{4}$	$\bar{z}+\frac{a}{4}$	$x+\frac{a}{4}$	$y+\frac{a}{4}$	$\bar{z}$
27	$\bar{x}+\frac{a}{4}$	$y+\frac{a}{4}$	$z+\frac{a}{4}$	$\bar{x}$	$y+\frac{a}{4}$	$z+\frac{a}{4}$
28	$x+\frac{a}{4}$	$\bar{y}+\frac{a}{4}$	$z+\frac{a}{4}$	$x+\frac{a}{4}$	$\bar{y}$	$z+\frac{a}{4}$
29	$\bar{y}+\frac{a}{4}$	$x+\frac{a}{4}$	$z+\frac{a}{4}$	$\bar{y}$	$x+\frac{a}{4}$	$z+\frac{a}{4}$
30	$y+\frac{a}{4}$	$\bar{x}+\frac{a}{4}$	$z+\frac{a}{4}$	$y+\frac{a}{4}$	$\bar{x}$	$z+\frac{a}{4}$
31	$x+\frac{a}{4}$	$\bar{z}+\frac{a}{4}$	$y+\frac{a}{4}$	$x+\frac{a}{4}$	$\bar{z}$	$y+\frac{a}{4}$
32	$x+\frac{a}{4}$	$z+\frac{a}{4}$	$\bar{y}+\frac{a}{4}$	$x+\frac{a}{4}$	$z+\frac{a}{4}$	$\bar{y}$
33	$z+\frac{a}{4}$	$y+\frac{a}{4}$	$\bar{x}+\frac{a}{4}$	$z+\frac{a}{4}$	$y+\frac{a}{4}$	$\bar{x}$
34	$\bar{z}+\frac{a}{4}$	$y+\frac{a}{4}$	$x+\frac{a}{4}$	$\bar{z}$	$y+\frac{a}{4}$	$x+\frac{a}{4}$
35	$y+\frac{a}{4}$	$x+\frac{a}{4}$	$\bar{z}+\frac{a}{4}$	$y+\frac{a}{4}$	$x+\frac{a}{4}$	$\bar{z}$
36	$z+\frac{a}{4}$	$\bar{y}+\frac{a}{4}$	$x+\frac{a}{4}$	$z+\frac{a}{4}$	$\bar{y}$	$x+\frac{a}{4}$
37	$\bar{x}+\frac{a}{4}$	$z+\frac{a}{4}$	$y+\frac{a}{4}$	$\bar{x}$	$z+\frac{a}{4}$	$y+\frac{a}{4}$
38	$\bar{y}+\frac{a}{4}$	$\bar{x}+\frac{a}{4}$	$\bar{z}+\frac{a}{4}$	$\bar{y}$	$\bar{x}$	$\bar{z}$
39	$\bar{z}+\frac{a}{4}$	$\bar{y}+\frac{a}{4}$	$\bar{x}+\frac{a}{4}$	$\bar{z}$	$\bar{y}$	$\bar{x}$
40	$\bar{x}+\frac{a}{4}$	$z+\frac{a}{4}$	$\bar{y}+\frac{a}{4}$	$\bar{x}$	z	$\bar{y}$
41	$\bar{z}+\frac{a}{4}$	$\bar{x}+\frac{a}{4}$	$\bar{y}+\frac{a}{4}$	$\bar{z}$	$\bar{x}$	$\bar{y}$
42	$\bar{y}+\frac{a}{4}$	$\bar{z}+\frac{a}{4}$	$\bar{x}+\frac{a}{4}$	$\bar{y}$	$\bar{z}$	$\bar{x}$
43	$\bar{z}+\frac{a}{4}$	$x+\frac{a}{4}$	$y+\frac{a}{4}$	$\bar{z}$	$x+\frac{a}{4}$	$y+\frac{a}{4}$
44	$y+\frac{a}{4}$	$z+\frac{a}{4}$	$\bar{x}+\frac{a}{4}$	$y+\frac{a}{4}$	$z+\frac{a}{4}$	$\bar{x}$
45	$z+\frac{a}{4}$	$x+\frac{a}{4}$	$\bar{y}+\frac{a}{4}$	$z+\frac{a}{4}$	$x+\frac{a}{4}$	$\bar{y}$
46	$y+\frac{a}{4}$	$\bar{z}+\frac{a}{4}$	$x+\frac{a}{4}$	$y+\frac{a}{4}$	$\bar{z}$	$x+\frac{a}{4}$
47	$z+\frac{a}{4}$	$\bar{x}+\frac{a}{4}$	$y+\frac{a}{4}$	$z+\frac{a}{4}$	$\bar{x}$	$y+\frac{a}{4}$
48	$\bar{y}+\frac{a}{4}$	$z+\frac{a}{4}$	$x+\frac{a}{4}$	$\bar{y}$	$z+\frac{a}{4}$	$x+\frac{a}{4}$

Consider the diamond and zincblende structures. For both these structures, the Bravais lattice is FCC, and the point group is the octahedral group O_h with 48 operations. (Note that this is same as the point group of the diamond crystal.) Thus the irreducible part of the Brillouin zone of these crystal structures has the volume $\Omega_{BZ}/48$. One such irreducible segment is defined by the relations

$$0 \leq q_x \leq q_y \leq q_z \qquad \leq \frac{2\pi}{a}$$
$$q_x + q_y + q_z \qquad \leq \frac{3}{2}\frac{2\pi}{a}. \tag{1.20}$$

This is the *pentahedral* $\Gamma X U L K W$ shown in figure 1.2(b) with five faces and six vertices at

$$\begin{aligned}
\Gamma &= \frac{2\pi}{a}(0,0,0) & X &= \frac{2\pi}{a}(1,0,0) \\
U &= \frac{2\pi}{a}\left(1,\frac{1}{4},\frac{1}{4}\right) & L &= \frac{2\pi}{a}\left(\frac{1}{2},\frac{1}{2},\frac{1}{2}\right) \\
K &= \frac{2\pi}{a}\left(\frac{3}{4},\frac{3}{4},0\right) & W &= \frac{2\pi}{a}\left(1,\frac{1}{2},0\right).
\end{aligned} \tag{1.21}$$

The irreducible part of the Brillouin zone of the BCC lattice is the *tetrahedron* $\Gamma H N P$ shown in figure 1.2(c) which has the volume $\Omega_{BZ}/48$, and where

$$\begin{aligned}
\Gamma &= \frac{2\pi}{a}(0,0,0) & H &= \frac{2\pi}{a}(1,0,0) \\
N &= \frac{2\pi}{a}\left(\frac{1}{2},\frac{1}{2},0\right) & P &= \frac{2\pi}{a}\left(\frac{1}{2},\frac{1}{2},\frac{1}{2}\right).
\end{aligned} \tag{1.22}$$

The irreducible part of the Brillouin zone for the SC lattice is the *tetrahedron* $\Gamma X M R$ of volume $\Omega_{BZ}/48$, shown in figure 1.2(a), with its vertices at

$$\begin{aligned}
\Gamma &= \frac{2\pi}{a}(0,0,0) & X &= \frac{2\pi}{a}\left(\frac{1}{2},0,0\right) \\
M &= \frac{2\pi}{a}\left(\frac{1}{2},\frac{1}{2},0\right) & R &= \frac{2\pi}{a}\left(\frac{1}{2},\frac{1}{2},\frac{1}{2}\right).
\end{aligned} \tag{1.23}$$

1.8   JONES ZONE

The first Brillouin zone, which is mathematically defined from the translational symmetry of the lattice, can accommodate in one energy band up to two valence electrons per primitive cell of the crystal. The usual concept of Brillouin zone boundaries is that they act as Bragg reflection planes for waves travelling inside the crystal. However, for a crystal with more than two valence electrons per primitive unit cell, the highest occupied electron energy surface (the Fermi surface) extends beyond the boundaries of the first Brillouin zone and, within the nearly-free-electron model, no splitting of the electron levels at the boundaries of the zone is expected. In other words, the boundaries of the first Brillouin zone do not act as Bragg reflection planes. In such a case, it is therefore useful to construct a larger zone, known as the *Jones zone*, with a volume just sufficient to accommodate all the valence electrons in the primitive unit cell. The faces of such a zone act as Bragg reflection planes, leading to discontinuities in electron energy levels.

For the diamond and zincblende structures, the Jones zone is constructed by the planes bisecting the 12 reciprocal lattice vectors of type (220): $(2\pi/a)(\pm1,\pm1,0)$, $(2\pi/a)(0,\pm1,\pm1)$, $(2\pi/a)(\pm1,0,\pm1)$. The shape of this zone is identical to the first Brillouin zone of the BCC lattice (shown in figure 1.2(c)): a rhombic dodecahedron.

We will use the concept of the Jones zone of the diamond structure in section 3.3.2.3.

1.9 SURFACE BRILLOUIN ZONE

Surfaces of solids are the subject of a great deal of interest to physicists and chemists. The importance of studying semiconductor surfaces lies in their technological usage. To study electronic, optical or vibrational properties of a solid surface, it is essential to develop the concepts of surface periodicity and surface Brillouin zone.

Mathematically, we may construct the surface Brillouin zone by identifying the periodicity in the surface plane and introducing an artificial periodicity in the direction perpendicular to the surface. For example, for the case of the ideal zincblende (110) surface whose unit mesh is shown in figure 1.4(a), we choose

$$
\begin{aligned}
\boldsymbol{a}_1 &= \frac{a}{2}(-1,1,0) \\
\boldsymbol{a}_2 &= a(0,0,1) \\
\boldsymbol{a}_3 &= c(1,1,0),
\end{aligned}
\tag{1.24}
$$

where $\boldsymbol{a}_1$ and $\boldsymbol{a}_2$ are the primitive translation vectors in the surface plane, and the vector perpendicular to the surface $\boldsymbol{a}_3$ is constructed with c arbitrarily related to the lattice constant a. In this particular case, the basis vectors define a parallelepiped of dimensions $(a/\sqrt{2}) \times a \times c\sqrt{2}$. As now

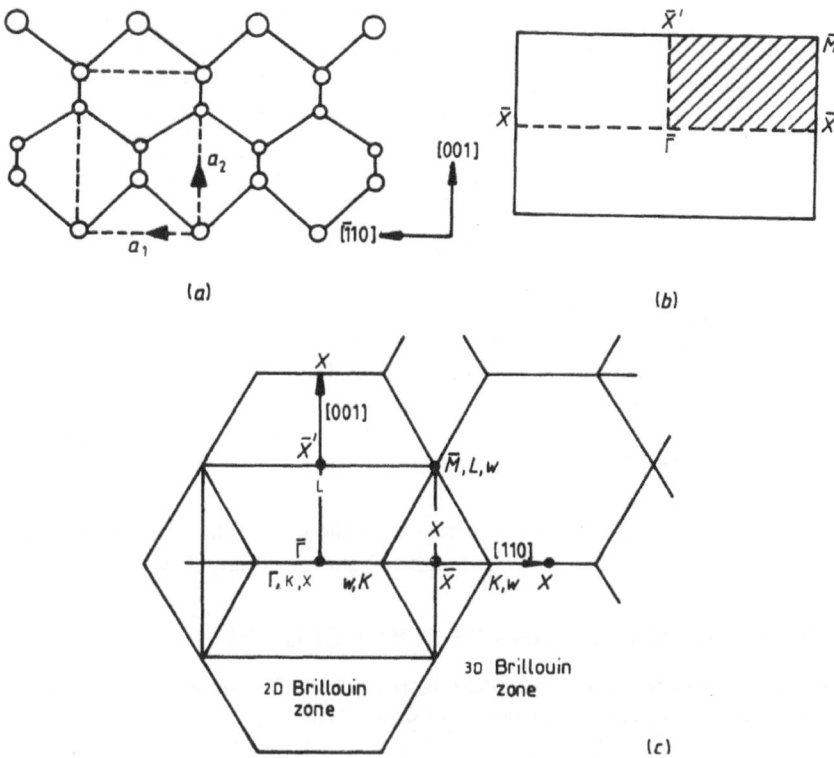

Figure 1.4 (a) The (110) unit mesh for the zincblende structure. Smaller dots represent atoms in the subsurface layer. (b) The corresponding surface Brillouin zone. The symmetry points are labelled and the irreducible segment is shown by the hatched area. (c) Projection of the zincblende bulk Brillouin zone onto the (110) surface Brillouin zone. The back projection of bulk symmetry points is represented by small capital letters and points not projected are labelled by larger capital letters.

we are dealing with a periodic system, the procedure described in section 1.2 can be used to construct the reciprocal lattice. The primitive translation vectors of the reciprocal lattice in the present example are

$$
\begin{aligned}
\boldsymbol{b}_1 &= \frac{2\pi}{\Omega}(\boldsymbol{a}_2 \times \boldsymbol{a}_3) = \frac{2\pi}{|\boldsymbol{a}_1|}\hat{\boldsymbol{a}}_1 \\
\boldsymbol{b}_2 &= \frac{2\pi}{\Omega}(\boldsymbol{a}_3 \times \boldsymbol{a}_1) = \frac{2\pi}{|\boldsymbol{a}_2|}\hat{\boldsymbol{a}}_2 \\
\boldsymbol{b}_3 &= \frac{2\pi}{\Omega}(\boldsymbol{a}_1 \times \boldsymbol{a}_2) = \frac{2\pi}{|\boldsymbol{a}_3|}\hat{\boldsymbol{a}}_3,
\end{aligned}
\tag{1.25}
$$

where $\Omega = |\boldsymbol{a}_1 \cdot (\boldsymbol{a}_2 \times \boldsymbol{a}_3)|$ is used. It should be mentioned that the vector $\boldsymbol{a}_3$ is only introduced to facilitate the construction of the reciprocal lattice using the concept described in section 1.2. The surface reciprocal lattice is defined by setting the vector $\boldsymbol{b}_3$ to zero. Obviously the primitive unit cell of the reciprocal lattice of the surface is the area defined by $\boldsymbol{b}_1$ and $\boldsymbol{b}_2$.

A general vector of the surface reciprocal lattice is

$$
\boldsymbol{G}_\| = m_1\boldsymbol{b}_1 + m_2\boldsymbol{b}_2 = \{m_1, m_2\},
\tag{1.26}
$$

where m_i are any integers. For the present case, the first two sets of shortest $\boldsymbol{G}_\|$s are simply $\pm\boldsymbol{b}_1$ and $\pm\boldsymbol{b}_2$. Therefore, the corresponding surface Brillouin zone is the area determined by the bisector planes perpendicular to these vectors. figure 1.4(*b*) shows the surface Brillouin zone for the (110) surface of the zincblende structure. This surface Brillouin zone exhibits the C_{2v} point group symmetry of the (110) surface. As this point group has four operations, the irreducible part of the surface Brillouin zone has a volume $(\frac{1}{4})$th of the zone. This is the area $\bar{\Gamma}\bar{X}\bar{M}\bar{X}'$ shown in figure 1.4(*b*), where

$$
\begin{aligned}
\bar{\Gamma} &= \{0,0\} & \bar{X}' &= \{0,\tfrac{1}{2}\} \\
\bar{X} &= \{\tfrac{1}{2},0\} & \bar{M} &= \{\tfrac{1}{2},\tfrac{1}{2}\},
\end{aligned}
\tag{1.27}
$$

with $\{m_1, m_2\}$ defined in equation (1.26).

Geometrically, the surface Brillouin zone can be obtained by projecting the bulk Brillouin zone along the direction perpendicular to the surface. Figure 1.4(*c*) shows the projection of the zincblende bulk Brillouin zone onto the (110) surface Brillouin zone. The following projections of the symmetry points of the bulk Brillouin zone onto the surface Brillouin zone can be identified: Γ, κ, $x \to \bar{\Gamma}$; $x \to \bar{X}$; $L \to \bar{X}'$; $L, w \to \bar{M}$.

The surface Brillouin zones for two different periodicities in the (111) plane of the diamond structure, together with their irreducible segments, are shown in figure 1.5.

1.10 MATRIX REPRESENTATIONS OF POINT GROUPS

The symmetry operations of a point group can be represented by square matrices (real or complex) which obey the multiplication table of the point group. The number of rows or columns in such a matrix is called the *dimension* of the representation. As can be realised, an infinite number of matrix representations of a point group is possible. However, for a given point group, there exists a set of fundamental representations, called *inequivalent irreducible representations*, from which *all* other representations can be obtained. Stated mathematically, if a representation for a point symmetry operation $\mathscr{R}$ could be expressed in the block form

$$
D(\mathscr{R}) = \begin{pmatrix} D_1(\mathscr{R}) & 0 \\ 0 & D_2(\mathscr{R}) \end{pmatrix}
\tag{1.28}
$$

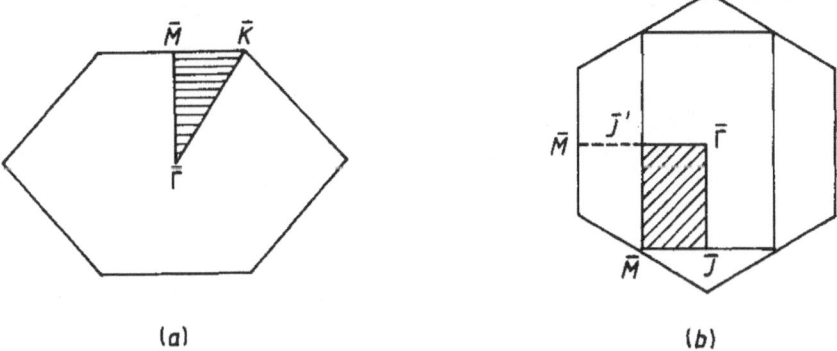

Figure 1.5 Brillouin zone for the (111) surface of the diamond/zincblende structure: (a) surface periodicity-1×1; (b) surface periodicity-2×1.

then $D(\mathscr{R})$ is said to be *reducible*. If $D_i(\mathscr{R})$ are diagonal matrices so that their further reduction is not possible, then these are called *irreducible representations*.

An important quantity for an irreducible representation is its character which is the trace of its matrix

$$\chi_i(\mathscr{R}) = \sum_\mu D_{\mu\mu}^{(i)}(\mathscr{R}). \tag{1.29}$$

The characters of two different irreducible representations satisfy the orthogonality relation

$$\sum_{\mathscr{R}} \chi_i^*(\mathscr{R})\chi_j(\mathscr{R}) = h\delta_{ij}, \tag{1.30}$$

where h is the number of elements in the group. Another important relation is

$$\sum_i l_i^2 = h, \tag{1.31}$$

where l_i is the dimensionality of the ith irreducible representation. Any two representations which have the same character are equivalent within a similarity transformation and are said to belong to the same *class*. Thus it follows that *the number of inequivalent irreducible representations of a group is equal to the number of classes in the group.*

There are three widely used notations for matrix representations of point groups: Bethe's notation $\Gamma_1, \Gamma_2, ..., \Gamma_5$ (Bethe 1929); the BSW notation $\Gamma_1, \Gamma_2, \Gamma_{12}, \Gamma_{15}, \Gamma_{25}, ...$ (Bouckaert *et al* 1936); and the EWK notation A, B, E, T (Eyring *et al* 1940). In this book, we will use all these notations. To make the reader familiar with these notations, we list here the irreducible representations of the point groups O_h, T_d, D_{2d} and C_{2v}. The point group $O_h(m3m)$ has 10 irreducible representations. In the BSW (EWK) notation, there are 4 one-dimensional irreducible representations: $\Gamma_1(A_{1g})$, $\Gamma_{1'}(A_{1u})$, $\Gamma_2(A_{2g})$, $\Gamma_{2'}(A_{2u})$; 2 two-dimensional irreducible representations: $\Gamma_{12}(E_g)$, $\Gamma_{12'}(E_u)$; and 4 three-dimensional irreducible representations: $\Gamma_{15'}(T_{1g})$, $\Gamma_{15}(T_{1u})$, $\Gamma_{25'}(T_{2g})$, $\Gamma_{25}(T_{2u})$. The point group $T_d(\bar{4}3m)$ has five irreducible representations: in the (BSW, EWK, Bethe) notation, these are $(\Gamma_1, A_1, \Gamma_1)$, $(\Gamma_2, A_2, \Gamma_2)$, $(\Gamma_{12}, E, \Gamma_3)$, $(\Gamma_{25}, T_1, \Gamma_4)$ and $(\Gamma_{15}, T_2, \Gamma_4)$. In the Bethe (EWK) notation, the five irreducible representations of the point group D_{2d} $(\bar{4}2m)$ are $\Gamma_1(A_1)$, $\Gamma_2(A_2)$, $\Gamma_3(B_1)$, $\Gamma_4(B_2)$ and $\Gamma_5(E)$. The four irreducible representations of the point group C_{2v} in the Bethe (EWK) notation are $\Gamma_1(A_1)$, $\Gamma_3(A_2)$, $\Gamma_2(B_1)$ and $\Gamma_4(B_2)$. In the above list, A_i are one-dimensional representations with character $+1$, B_i are one-dimensional representations with character -1, and the subscripts g and u represent the parity of the representation under inversion : $g = gerade = even$, $u = ungerade = odd$. The irreducible representations of other point groups can be found in many standard textbooks on group theory.

1.11   EFFECT OF SPACE GROUP OPERATIONS ON PLANE WAVES

It was discussed earlier that the space group of a crystal is its full symmetry group. A space group operator is expressed as $\{\mathscr{R} \mid t\}$, where $\mathscr{R}$ is either a proper or an improper rotation operator and t represents a fractional translation associated with $\mathscr{R}$. (The group of lattice translation vectors, which has operations of the type $\{0|T\}$, is also a part of space group but we have excluded this from our discussion.) Using equation (1.17), it can be verified that the effect of two successive operations is given by

$$\{\mathscr{R}' \mid t'\}\{\mathscr{R} \mid t\} = \{\mathscr{R}'\mathscr{R} \mid \mathscr{R}'t + t'\}. \tag{1.32}$$

The identity element of the space group is $\{E \mid 0\}$. Furthermore, the inverse of the operator $\{\mathscr{R} \mid t\}$ is

$$\{\mathscr{R} \mid t\}^{-1} = \{\mathscr{R}^{-1} \mid -\mathscr{R}^{-1}t\}. \tag{1.33}$$

This can be verified by showing, with the help of equation (1.32), that

$$\{\mathscr{R} \mid t\}\{\mathscr{R} \mid t\}^{-1} = \{E \mid 0\}. \tag{1.34}$$

The effect of an operation $\hat{O} = \{\mathscr{R} \mid t\}$ on a function or property of a crystal at a point in its unit cell is given as

$$\begin{aligned} \hat{O}f(r) &= f(\{\mathscr{R} \mid t\}^{-1}r) & (1.35) \\ &= f(\mathscr{R}^{-1}r - \mathscr{R}^{-1}t). & (1.36) \end{aligned}$$

Let us consider $f(r)$ as a plane wave $\exp(iq \cdot r)$. Then

$$\begin{aligned} \hat{O}\exp(iq \cdot r) &= \exp(iq \cdot \{\mathscr{R} \mid t\}^{-1}r) \\ &= \exp[iq \cdot (\mathscr{R}^{-1}r - \mathscr{R}^{-1}t)] \\ &= \exp(-iq \cdot \mathscr{R}^{-1}t)\exp(iq \cdot \mathscr{R}^{-1}r) \\ &= \exp(-i\mathscr{R}q \cdot t)\exp(i\mathscr{R}q \cdot r) \\ &= \gamma\exp(i\mathscr{R}q \cdot r), & (1.37) \end{aligned}$$

where we have used the orthogonal nature of the matrix representing $\mathscr{R}$. Thus upon the application of a space group operation, a plane wave characterised by a wave vector q is transformed, to within a phase factor $\gamma \equiv \exp(-i\mathscr{R}q \cdot t)$, into another plane wave which is characterised by a wave vector $q' = \mathscr{R}q$. The phase factor γ is $+1$ for all operations of a symmorphic space group. In general, however, for the cubic system γ takes values ± 1, or $\pm i$.

All operations $\mathscr{R}$ which leave q invariant or carry it into an equivalent point, i.e. when

$$q' = \mathscr{R}q = q + G \tag{1.38}$$

form a group $C(q)$, which is called the *group of the wave vector q*. This group is a subgroup of the full space group of the crystal under consideration. The set of distinct q' (taking q to new positions) generated by the application of all the group operations is called the 'star' of q. In general, the more symmetric the point q, the fewer the number of vectors in its star. The product of the order of the group of q and the number of vectors in its star equals the order of the full point group.

The irreducible representations of the group $C(q)$ can be used to obtain all the irreducible representations of the full space group. The reader is referred to the excellent review article by Koster (1957) for details on this topic.

2 Lattice Dynamics in the Harmonic Approximation – Semiclassical Treatment

2.1 INTRODUCTION

It is convenient to visualise an atom in a solid in terms of its ion core and valence electrons. The ion cores of a solid vibrate about their equilibrium positions and the (valence) electrons move about their (in some cases many more) ion cores. In a crystalline solid the description of the motion of all the ion cores and electrons can be simplified by taking advantage of its lattice translational periodicity $f(r + T) = f(r)$ as discussed in Chapter 1. This essentially reduces the problem to the dynamics of the atoms in the primitive unit cell, containing p ion cores (say) and the electrons associated with them. In general, a detailed description of even this 'small' system is a formidable task. In order to simplify the problem, we make two approximations.

Adiabatic approximation

As the ion cores are much heavier than the electrons, their motion can be treated separately. This is called the *adiabatic approximation*. As a consequence of this approximation (a) while dealing with the motion of the electrons, it is assumed that the ion cores are in their equilibrium positions and (b) the motion of the ions cores is determined in a potential field generated by the average motion of the electrons. It should be mentioned here that it is part (b) above which is one of the formidable ingredients for a first-principles study of lattice dynamics. We will discuss this point in some detail in chapter 3.

Harmonic approximation

Let us consider the total potential energy of a crystal in terms of interatomic potentials. While for simple metals and rare-gas solids interatomic potentials can be reasonably well expressed in terms of pair (or two-body) interactions, for strongly bonded crystals (such as covalently bonded semiconductors) consideration of at least three-body interactions becomes very important. Multi-body interactions may, however, be simplified to contributions involving two-body sums (Biswas and Hamann 1987). For the present discussion, we therefore restrict ourselves to two-body interactions. Let us represent a two-body term as $\mathscr{U}(R)$, where R is the interatomic separation between a pair of atoms. In general the shape of an interatomic potential $\mathscr{U}(R)$ is complicated, as shown in figure 2.1. However, $\mathscr{U}(R)$ can be expanded in a Taylor series in powers of a small displacement $x = R - R_0$ around its minimum at R_0:

$$\mathscr{U} = \mathscr{U}(R_0) + \left(\frac{\partial \mathscr{U}}{\partial R}\right)_0 \cdot x + \frac{1}{2}\left(\frac{\partial^2 \mathscr{U}}{\partial R^2}\right)_0 x^2 + \dots. \tag{2.1}$$

The first term is a constant and is unimportant for dynamical problems, and the second term produces a force and must vanish in the equilibrium configuration. Therefore, the first important term in the above expansion is quadratic in the displacement x. The consideration of only the quadratic term in equation (2.1) is known as the *harmonic approximation*. In this picture an atom in a crystal can be described as a three-dimensional simple harmonic oscillator, and the term $(\partial^2 \mathscr{U}/\partial R^2)_0$ represents an interatomic force constant.

DOI: 10.1201/9781003141273-2

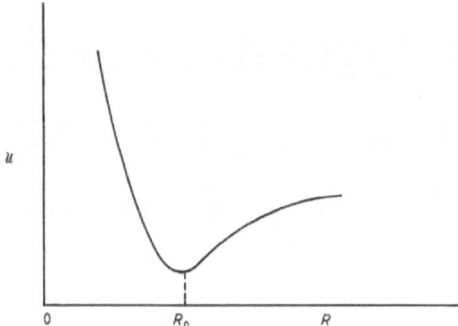

Figure 2.1 A typical two-body interatomic potential curve in a crystal. R_0 is the equilibrium interatomic distance. The shape of $\mathscr{U}$ near its minimum can be regarded as parabolic.

2.1.1 THE PHONON AS AN EXCITATION IN A CRYSTAL

In the classical picture within the harmonic approximation, the atoms (say, N in number) of a crystal are visualised as joined by harmonic springs and the crystal dynamics is analysed in terms of a linear combination of $3N$ *normal modes of vibration*. A normal mode of vibration is expressed as a travelling wave of the form $A\exp[i(q \cdot r - \omega t)]$, where q shows the direction of wave propagation, ω is the circular frequency of the wave and A is the amplitude of vibration. The energies of the normal modes of a crystal are quantised: for the qth mode the energies are $(n_q + \frac{1}{2})\hbar\omega(q), n_q = 0, 1, 2, \ldots$ (see chapter 4 for more details). The quantum of energy $\hbar\omega(q)$ is associated with an *elementary excitation* called a *phonon*. Thus a phonon is a quantum of crystal vibrational energy. The concept of a phonon is similar to that of a photon which is a quantum of energy in an electromagnetic field. Clearly in the harmonic approximation, we have the physical picture of *non-interacting phonons* in a crystal.

As the concept of a phonon originates from relative motion of the atoms, rather than the motion of their centre of mass, a phonon in a crystal does not carry a momentum. However, for practical purposes, we assign a momentum $\hbar q$ to a phonon in the qth mode. For this reason, a phonon is called a *quasi-particle*.

2.1.2 STATEMENT OF THE PROBLEM OF LATTICE DYNAMICS

The problem of lattice dynamics in the harmonic approximation is to find the normal modes of a crystal. In other words, we seek to calculate the energies (or frequencies) of the non-interacting phonons as a function of their wave vectors q. The relationship between ω and q, namely $\omega = \omega(q)$, is called *phonon dispersion*.

In order to calculate the phonon dispersion in a crystal, we need to know the interatomic force constants. This can be done at two levels: the semiclassical and phenomenological level or the quantum mechanical and *ab initio* level. In this chapter we discuss the basic ingredients of the semiclassical treatment. In Chapter 3, we will describe an *ab initio* treatment.

2.1.3 PHONON STATISTICS

Phonons in a crystal are in thermal equilibrium with each other. The average number of phonons in the qth mode, quasi-particles of zero spin, in thermal equilibrium at temperature T is given by the Bose–Einstein distribution function

$$\bar{n}_q = \frac{1}{\exp(\hbar\omega(q)/k_B T) - 1},\tag{2.2}$$

where k_B is Boltzmann's constant. From this expression, it is clearly seen that at absolute zero there are no phonons in a crystal. At low temperatures $\hbar\omega \gg k_B T$, $\bar{n} \simeq \exp(-\hbar\omega/k_B T)$ and there is an exponentially small probability for a phonon to be present. At high temperatures $k_B T \gg \hbar\omega$, $\bar{n} \simeq k_B T/\hbar\omega$ and the number of phonons increases linearly with temperature. As the total number of phonons is not conserved, but is determined by the temperature, the distribution function in equation (2.2) corresponds to zero chemical potential.

2.2 LATTICE DYNAMICS OF A LINEAR CHAIN

To describe the problem of lattice dynamics in the harmonic approximation, we first consider a linear chain of atoms. We will extend the idea to two- and three-dimensional cases in sections 2.3 and 2.4.

2.2.1 MONATOMIC LINEAR CHAIN

Consider a monatomic linear chain of an infinitely large number, N, of atoms separated a apart. For all the atoms to have an identical environment, let us assume that it is a closed chain so that the $(N+1)$th atom is the 1st atom. This is known as the *periodic, cyclic, or Born–von Kármán boundary condition* and provides a mathematical scheme to avoid unimportant end effects on the dynamical problem of an infinitely long chain. Further, assume that only nearest-neighbour forces are significant.

Suppose that at a particular time the nth atom in the chain has a displacement u_n from its equilibrium (figure 2.2). Then from Newton's second law and Hooke's law the equation of motion of the nth atom is

$$m\frac{d^2 u_n}{dt^2} = \Lambda[(u_{n+1} - u_n) + (u_{n-1} - u_n)]. \tag{2.3}$$

Here m is the mass of an atom and Λ is the nearest-neighbour force constant.

We try a solution

$$\begin{aligned} u_n &= A\exp[i(qx - \omega t)] \\ &= A\exp[i(qna - \omega t)] \end{aligned} \tag{2.4}$$

for $x = na$. Here A denotes the amplitude of the motion of the nth atom. Also, the direction of atomic motion, namely that of u_n, is given by that of A. Note that we only have one allowed mode of vibration here – either longitudinal in which case atoms vibrate along the chain which is the line of wave propagation, or transverse in which case atoms vibrate perpendicular to the chain. For this reason, we have not represented u_n or A as vectors. Further, in figure 2.2 and in the text we have only discussed the longitudinal case. With equation (2.4), equation (2.3) becomes the following dynamical equation (an eigenvalue equation):

$$\omega^2 A = \frac{2\Lambda}{m}(1 - \cos qa)A \equiv DA. \tag{2.5}$$

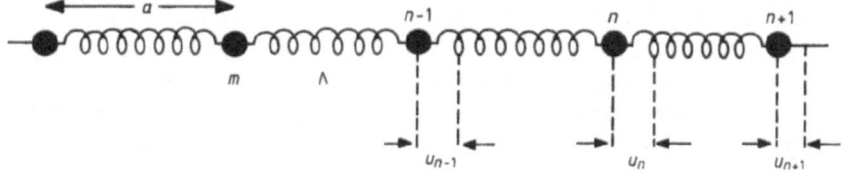

Figure 2.2 A monatomic linear chain showing displacements u_{n-1}, u_n, u_{n+1} for the $(n-1)$th, nth, $(n+1)$th atoms from equilibrium, respectively.

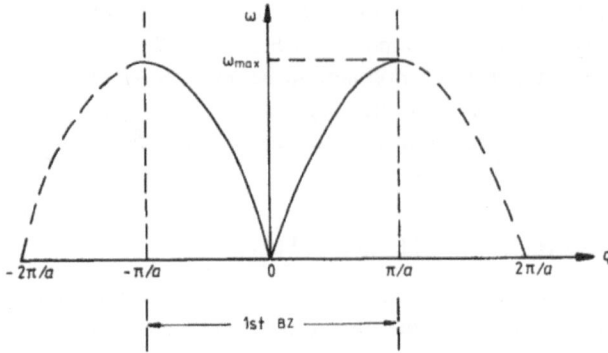

Figure 2.3 The phonon dispersion curve for the monatomic linear chain shown in figure 2.2. The Brillouin zone is the line segment between $-\pi/a$ and π/a.

D is calle1d the dynamical matrix and in this case is just a (1×1) matrix. For a non-trivial solution $(A \neq 0)$, equation (2.5) gives

$$\omega = 2\sqrt{\frac{\Lambda}{m}}|\sin qa/2|. \tag{2.6}$$

It is clearly seen from figure 2.3 that (i) the dispersion curve shows the translational symmetry in q space: $\omega(q + G_n) = \omega(q)$, where $G_n = \pm 2n\pi/a$, (n = integer), is the magnitude of a reciprocal lattice vector corresponding to the chain, and (ii) $\omega(q)$ is a symmetric function between q and $-q$.

Using the *periodic boundary condition* $u_{N+1} = u_1$, or $u(x = 0) = u(x = Na)$ one can easily verify that there are N allowed values of q in the Brillouin zone of the linear chain. As each q-value represents a normal mode, we say that there are N distinct normal modes of the linear chain.

It is interesting to define two types of phonon velocity. For a phonon characterised by frequency ω and wave vector $\boldsymbol{q}$ its phase velocity $\boldsymbol{c}_p$ is defined by the relation $\omega = \boldsymbol{q} \cdot \boldsymbol{c}_p$. For a group of phonons having frequencies around ω, a group velocity is defined by $c_g = \boldsymbol{\nabla}_q \omega$. For the case of a linear chain, we obtain

$$c_p \quad = v\left(\frac{\sin(qa/2)}{(qa/2)}\right), \tag{2.7}$$

$$c_g \quad = v\cos(qa/2), \tag{2.8}$$

with

$$v = a\sqrt{\frac{\Lambda}{m}} = \frac{a\omega_{max}}{2}, \tag{2.9}$$

where

$$\omega_{max} = 2\sqrt{\frac{\Lambda}{m}} \tag{2.10}$$

is the maximum frequency for the normal modes of the chain. Two points are noteworthy.

(i) $c_g \to 0$ when $q = \pm\pi/a$, i.e. when $\lambda = 2a$. From equation (2.4) we note that for $q = \pm\pi/a$ (i.e. at the Brillouin zone boundary) the phases of vibrations of two neighbouring atoms differ by π, so that a lattice wave becomes a standing wave. This can also be interpreted in terms of the first-order Bragg reflection condition $\lambda = 2a\sin\theta$. In the present example of longitudinal waves, the Bragg angle θ is 90°, so that $\lambda = 2a$. Therefore, we can state that when a lattice wave with wavelength twice the interatomic distance, $\lambda = 2a$, is incident normally to a Brillouin zone face, it is reflected back and hence becomes a standing wave.

Figure 2.4 A diatomic linear chain with alternating masses m and M ($m < M$). The unit cell has length $2a$.

(ii) In the long-wavelength limit ($qa << 1$) $c_p = c_g = v$, the sound velocity. In this limit, the chain looks like a continuum and the lattice wave becomes a sound wave (or acoustic wave) given by the equation of motion

$$\frac{1}{v^2}\frac{\partial^2 u}{\partial t^2} = \frac{\partial^2 u}{\partial x^2}. \tag{2.11}$$

This equation can be formally obtained from equation (2.3) by regarding u as a continuous function of x and expanding it in a Taylor series:

$$u_{n+1} = u(x+a) \simeq u(x) + a\frac{\partial u(x)}{\partial x} + \frac{1}{2}a^2\frac{\partial^2 u(x)}{\partial x^2}. \tag{2.12}$$

2.2.2 DIATOMIC LINEAR CHAIN

The monatomic linear chain is the simplest system we can study and yet its dispersion curve provides some fundamental knowledge of what should be expected for the normal modes of a simple (monatomic) crystal. The essential features of the lattice dynamics of crystals with a basis of more than one atom can be understood most simply by studying the dynamics of a diatomic linear chain, which can be viewed as a linear chain with a basis of two atoms.

Consider an infinitely long chain of $2N$ atoms forming N unit cells, each of length $2a$, as shown in figure 2.4. Assume it to be a closed chain so that the Born–von Kármán periodic boundary condition is applicable. Let m and M ($m < M$) be the two masses of the basis, and let only nearest neighbour forces be significant. Further, for simplicity assume that the force constant is the same for both types of atoms.

Then, as for the monatomic linear chain, the equations of motion for the two types of atom are

$$m\frac{d^2 u_{2n}}{dt^2} = \Lambda[u'_{2n+1} + u'_{2n-1} - 2u_{2n}], \tag{2.13}$$

$$M\frac{d^2 u'_{2n+1}}{dt^2} = \Lambda[u_{2n+2} + u_{2n} - 2u'_{2n+1}]. \tag{2.14}$$

Let us try solutions

$$u_{2n} = A_1\exp[i(2nqa - \omega t)], \tag{2.15}$$

$$u'_{2n+1} = A_2\exp\{i[(2n+1)qa - \omega t]\}. \tag{2.16}$$

With equations (2.15) and (2.16), equations (2.13) and (2.14) turn into the coupled eigenvalue equations

$$-\omega^2 mA_1 = \Lambda[A_2 e^{iqa} + A_2 e^{-iqa} - 2A_1], \tag{2.17}$$

$$-\omega^2 MA_2 = \Lambda[A_1 e^{-iqa} + A_1 e^{iqa} - 2A_2]. \tag{2.18}$$

Equations (2.17) and (2.18) can be combined into the form

$$\omega^2 A_i = \sum_{j=1}^{2} D_{ij}A_j \qquad i = 1,2, \tag{2.19}$$

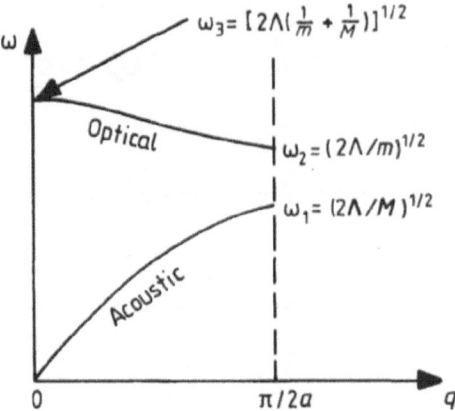

Figure 2.5 The phonon dispersion spectrum of the diatomic linear chain plotted in the first Brillouin zone. The acoustic branch extends from 0 to ω_1 and the optical branch extends from ω_2 to $\omega_3 = \omega_{max}$. The gap $(\omega_2 - \omega_1)$ in the spectrum is due to different atomic masses $(m \neq M)$ in the unit cell.

where D is a (2×2) dynamical matrix given by

$$D = \begin{pmatrix} \frac{2\Lambda}{m} & -\frac{2\Lambda}{m}\cos qa \\ -\frac{2\Lambda}{M}\cos qa & \frac{2\Lambda}{M} \end{pmatrix}. \tag{2.20}$$

Non-trivial solutions of equations (2.19) are given by solving the secular equation

$$\left| D_{ij} - \omega^2 \delta_{ij} \right| = 0, \tag{2.21}$$

where δ_{ij} is the Kronecker delta. The solutions are

$$\omega^2 = \Lambda\left(\frac{1}{m} + \frac{1}{M}\right) \pm \Lambda\left[\left(\frac{1}{m} + \frac{1}{M}\right)^2 - \frac{4}{mM}\sin^2 qa\right]^{1/2}. \tag{2.22}$$

Also, from equation (2.17),

$$\frac{A_1}{A_2} = \frac{2\Lambda\cos qa}{2\Lambda - \omega^2 m} = \frac{2\Lambda - M\omega^2}{2\Lambda\cos qa}. \tag{2.23}$$

Corresponding to the two signs in equation (2.22), there are two branches of the phonon dispersion curve of the diatomic linear chain.

Let us examine the two solutions in the small-q (or long-wavelength) region ($qa \ll 1$). For the upper branch in figure 2.5 A_1 and A_2 have opposite signs, meaning that the two atoms in the unit cell move in opposite directions. If the two atoms had opposite charges on them, such a mode of vibration could be excited by an electric field of the appropriate frequency. In ionic crystals, this corresponds to the electric field associated with the infrared part of the visible light. For this reason, the upper branch is called the *optical branch*. For the lower branch $A_1/A_2 = 1$, meaning that both atoms in the unit cell move in phase with each other. This is characteristic of a sound wave. Hence the lower branch is called the *acoustic branch*.

For finite values of q (i.e. away from the zone centre), there is no simple distinction between the acoustic and optical branches. It is interesting to note that, at the zone boundary ($q = \pi/2a$), the group velocity of both types of wave is zero. In other words, the waves become stationary at the

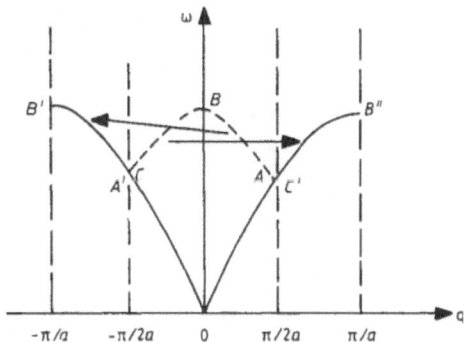

Figure 2.6 The special case of a diatomic linear chain. When $m = M$ the optical branch is just extended over the acoustic branch. The portion AB of the curve is translated through a reciprocal lattice vector to take the portion $A'B'$, and the portion CB is similarly unfolded out to the portion $C'B''$.

zone boundary. Also, from equation (2.23),

$$\frac{A_1}{A_2} = \frac{2\Lambda - M\omega^2}{2\Lambda \cos qa} = \infty \qquad \text{as} \qquad \cos\frac{\pi}{2} = 0, \tag{2.24}$$

$$\frac{A_1}{A_2} = \frac{2\Lambda \cos qa}{2\Lambda - m\omega^2} = 0. \tag{2.25}$$

From equation (2.25), it is apparent that at $\omega = \omega_2 = (2\Lambda/m)^{1/2}$ atoms with the heavier mass are stationary: $A_2 = 0$. Further, from equation (2.24), at $\omega = \omega_1 = (2\Lambda/M)^{1/2}$ atoms with the lighter mass are stationary: $A_1 = 0$. As expected, when $m \to M$ the dispersion spectrum of the diatomic chain becomes similar to that of the monatomic chain. The gap at $q = \pi/2a$ disappears and the top branch can be unfolded to cover the regions $\pi/2a$ to π/a and $-\pi/2a$ to $-\pi/a$, as is shown in figure 2.6.

2.3 LATTICE DYNAMICS OF MONATOMIC TWO-DIMENSIONAL MESH

We will consider two types of monatomic two-dimensional mesh: a square lattice and a hexagonal lattice. Atoms will be considered to interact with their nearest and next nearest neighbours for the square lattice and only with their nearest neighbours for the hexagonal lattice. The following concept will be useful in developing the lattice dynamics of such systems (see de Launay (1956) for further description). Consider interaction between atom 0 with displacement u_0 and an atom n with displacement u_n. Let R_n be the vector separating atom n from atom 0 and $\varepsilon_n = R_n/R_n$ be a unit vector parallel to R_n (see figure 2.7). The central force experienced by atom 0 due to atom n is

$$F_n = \Lambda_n \left[\varepsilon_n \cdot (u_n - u_0) \right] \varepsilon_n, \tag{2.26}$$

where Λ_n is the force constant between the two atoms. The equation of motion for atom 0 is then

$$m\frac{\mathrm{d}^2 u_0}{\mathrm{d}t^2} = \sum_n F_n, \tag{2.27}$$

where the sum over n includes neighbours under consideration (nearest and/or next nearest).

2.3.1 MONATOMIC SQUARE MESH

Figure 2.8 shows a ball and spring model of a monatomic two-dimensional square lattice, with m as atomic mass, a nearest neighbour distance, Λ_1 the nearest neighbour force constant and Λ_2 the

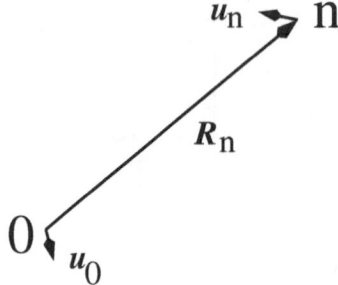

Figure 2.7 Schematic illustration of atomic positions and displacement vectors of two neighbouring atoms.

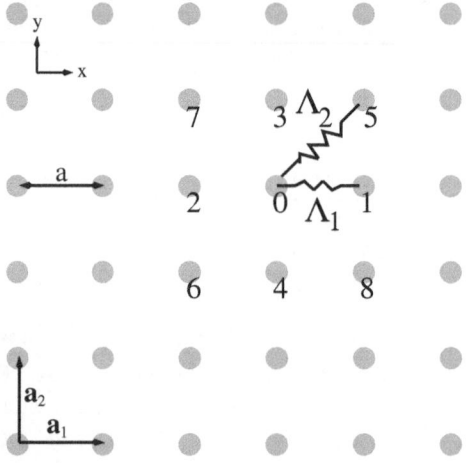

Figure 2.8 Ball and spring model of a two-dimensional monatomic square lattice. Λ_1 and Λ_2 are the nearest neibhbour and next-nearest neighbour force constants, respectively.

next nearest neighbour force constant. The unit cell is spanned by the primitive translation vectors a_1 and a_2. The Brillouin zone for this system is the horizontal square centred the Γ point in figure 1.2. Using equations (2.26) and (2.27), we can write down the equation of motion for atom 0 as

$$m\frac{\mathrm{d}^2 u_0}{\mathrm{d}t^2} \;=\; \Lambda_1 \sum_{n=1}^{4}(u_n - u_0) + \Lambda_2 \sum_{n=5}^{8}(u_n - u_0). \tag{2.28}$$

Following equation (2.4), we express the displacement vector of the nth atom at R_n as

$$u_n = A\exp(\mathrm{i}R_n \cdot q - \omega t) \tag{2.29}$$

where A is the amplitude vector and q is the phonon wave vector. Note that $R_n = (x_n, y_n)$, $A = (A_x, A_y)$ and $q = (q_x, q_y)$ are two-dimensional vectors. Following figure 2.8 and R_0 taken as the origin, for $n > 0$ the vectors R_n and ε_n can readily be determined. As an example, we will show how to derive the expression for the force on atom labelled 0 by atom labelled 8. From figure 2.8, $R_8 = a(1, -1)$ and $\varepsilon_8 = (\frac{1}{\sqrt{2}}, -\frac{1}{\sqrt{2}})$. Using equation (2.26),

$$\begin{aligned} F_8 &= \Lambda_2\big[\varepsilon_8 \cdot (u_8 - u_0)\big]\varepsilon_8 \\ &= \Lambda_2\{\exp(\mathrm{i}q \cdot R_8) - 1\}(\varepsilon_8 \cdot u_0)\varepsilon_8 \end{aligned}$$

$$= \Lambda_2\{\exp(i\mathbf{q}\cdot\mathbf{R}_8) - 1\}\left(\frac{A_x}{\sqrt{2}} - \frac{A_y}{\sqrt{2}}\right)\left(\frac{1}{\sqrt{2}}, -\frac{1}{\sqrt{2}}\right)\exp(-i\omega t). \tag{2.30}$$

With the total force on atom labelled 0 worked out and using equation (2.29), equation (2.28) can be expressed as the matrix eigenvalue equation

$$\omega^2 A_\alpha = \sum_\beta{}' D_{\alpha\beta}A_\beta, \quad \alpha, \beta = x, y. \tag{2.31}$$

For a chosen phonon vector $\mathbf{q}$, non-trivial eigenvalues (*i.e.* phonon frequencies $\omega(\mathbf{q}s), s = 1, 2$) can be obtained by solving the secular equation

$$\left|D_{ij} - \omega^2\delta_{ij}\right| = 0, \tag{2.32}$$

where the elements of the two-dimensional dynamical matrix D are

$$
\begin{aligned}
D_{xx} &= \frac{2}{m}[\Lambda_1(1 - \cos q_x a) + \Lambda_2(1 - \cos q_x a \cos q_y a)], \\
D_{xy} &= \frac{2\Lambda_2}{m}\sin q_x a \sin q_y a, \\
D_{yx} &= D_{xy}, \\
D_{yy} &= \frac{2}{m}[\Lambda_1(1 - \cos q_y a) + \Lambda_2(1 - \cos q_x a \cos q_y a)].
\end{aligned}
\tag{2.33}
$$

Eigenvector components $A_x(\mathbf{q}s)$ and $A_y(\mathbf{q}s)$ corresponding to an eigenvalue $\omega(\mathbf{q}s)$ can then be obtained from equation (2.31).

It is important to compare and contrast the discussion in this section with that in sections 2.2.1 and 2.2.2. For the monatomic one-dimensional case in section 2.2.1, we have a single *acoustic phonon branch*. For the monatomic two-dimensional case in this section, we have two *acoustic phonon branches*: these are labelled as the lower or *transverse acoustic* (TA) and the upper or *longitudinal acoustic* (LA). Strictly correct descriptions of the TA and LA branches require the eigenvector $\mathbf{A}$ to be respectively perpendicular and parallel to the phonon wavevector $\mathbf{q}$. In contrast, for the diatomic one-dimensional case in section 2.2.2 we have two phonon branches: *acoustic* and *optical*. These concepts of phonon *polarisations* and *branches* will be extended further when discussing the three-dimensional case in section 2.4.

The phonon dispersion curves along the Δ and Σ symmetry directions in the Brillouin zone are plotted in figure 2.9. It is instructive to discuss the solutions along these directions.

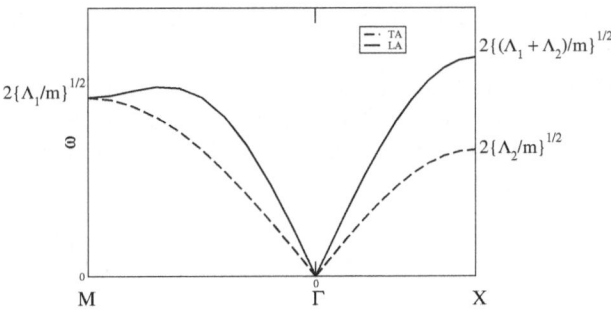

Figure 2.9 Phonon dispersion curves for a two-dimensional monatomic square lattice, with $\Lambda_2 = \Lambda_1/2$

The Δ direction: We can choose $q_x = q$ and $q_y = 0$, making $D_{xy} = D_{yx} = 0$. The eigenvalues are then

$$\omega^2(\text{TA}) = \frac{2}{m}\Lambda_2(1 - \cos qa), \tag{2.34}$$

$$\omega^2(\text{LA}) = \frac{2}{m}(\Lambda_1 + \Lambda_2)(1 - \cos qa) \tag{2.35}$$

and the normalised eigenvectors $e = A/A$ are

$$e(\text{TA}) = (0, 1) \quad \text{and} \quad e(\text{LA}) = (1, 0). \tag{2.36}$$

The Σ direction: Along this line $q_x = q_y = q/\sqrt{2}$. The eigenvalues are

$$\omega^2(\text{TA}) = \frac{2}{m}\Lambda_1(1 - \cos qa/\sqrt{2}), \tag{2.37}$$

$$\omega^2(\text{LA}) = \frac{2}{m}[\Lambda_1(1 - \cos qa/\sqrt{2}) + \Lambda_2(1 - \cos\sqrt{2}qa)] \tag{2.38}$$

and the normalised eigenvectors are

$$e(\text{TA}) = (-\frac{1}{\sqrt{2}}, \frac{1}{\sqrt{2}}) \quad \text{and} \quad e(\text{LA}) = (\frac{1}{\sqrt{2}}, \frac{1}{\sqrt{2}}). \tag{2.39}$$

Note that at the Γ and M points the directions of the eigenvectors become indeterminate and any pair of orthogonal unit vectors will suffice.

When the next-nearest neighbour interaction is not taken into account ($\Lambda_2 = 0$), then only the LA branch exists along the Δ direction. In other words, the monatomic two-dimensional mesh then vibrates like two independent monatomic linear chains (one along the x-axis and the other along the y-axis). And along the Σ direction the two branches become degenerate, characterised by the nearest neighbour force constant Λ_1.

2.3.2   MONATOMIC HEXAGONAL CLOSE PACKED MESH

As a second example of two-dimensional system, we consider the hexagonal close packed mesh of atomic mass m with only nearest neighbour interactions, as shown in figure 2.10. The unit cell for this hexagonal lattice may be taken as the parallelogram spanned by the primitive translation vectors $a_1 = a(1, 0)$ and $a_2 = a(\frac{1}{2}, \frac{\sqrt{3}}{2})$, where a is the interatomic distance. The corresponding primitive reciprocal translation vectors are $b_1 = \frac{2\pi}{a}(1, -\frac{1}{\sqrt{3}})$ and $b_2 = \frac{2\pi}{a}(0, \frac{2}{\sqrt{3}})$. The Brillouin zone for this hexagonal lattice is identical to that shown in figure 1.5(a). The symmetry points can be chosen as $M = \{0, \frac{1}{2}\} = \frac{2\pi}{a}(0, \frac{1}{\sqrt{3}})$ and $K = \{\frac{1}{3}, \frac{2}{3}\} = \frac{2\pi}{a}(\frac{1}{3}, \frac{1}{\sqrt{3}})$.

With atom labelled 0 at the origin, the atomic position vector R_n and the unit vector $\varepsilon_n = R_n/R_n$ for each of the six neighbours interacting with force constant Λ can be easily worked out. Following the steps described in the previous sub-section, we can derive the following expressions for the elements of the two-dimensional dynamical matrix:

$$
\begin{aligned}
D_{xx} &= \frac{\Lambda}{m}\left(3 - 2\cos q_x a - \cos\frac{q_x a}{2}\cos\frac{\sqrt{3}q_y a}{2}\right), \\
D_{xy} &= -\frac{\sqrt{3}\Lambda}{m}\sin\frac{q_x a}{2}\sin\frac{\sqrt{3}q_y a}{2}, \\
D_{yx} &= D_{xy}, \\
D_{yy} &= \frac{3\Lambda}{m}\left(1 - \cos\frac{q_x a}{2}\cos\frac{\sqrt{3}q_y a}{2}\right).
\end{aligned} \tag{2.40}
$$

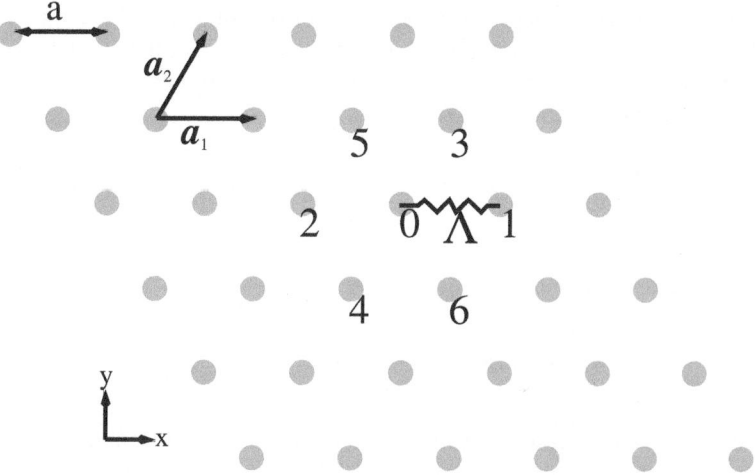

Figure 2.10 Ball and spring model of a two-dimensional monatomic square lattice. Λ is the next-nearest neighbour force constant.

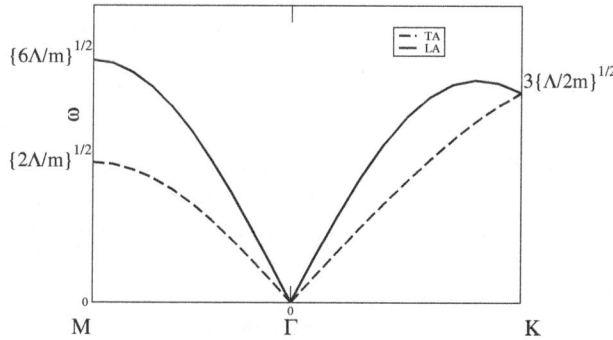

Figure 2.11 Phonon dispersion curves for a two-dimensional monatomic hexagonal lattice.

The phonon dispersion curves along the symmetry directions $M - \Gamma$ and $\Gamma - K$ are shown in figure 2.11. It is easy to see that the off-diagonal terms of the dynamical matrix, D_{xy} and D_{yx}, vanish at the M and K points. The frequencies at the M point are $\omega(\text{TA}) = \sqrt{\frac{2\Lambda}{m}}$ and $\omega(\text{LA}) = \sqrt{\frac{6\Lambda}{m}}$. The TA and LA modes become degenerate at the K point with frequency $\omega = 3\sqrt{\frac{\Lambda}{2m}}$.

2.4 LATTICE DYNAMICS OF THREE-DIMENSIONAL CRYSTALS – PHENOMENO-LOGICAL MODELS

The interatomic force constant method described for the one- and two-dimensional cases can be extended to calculate the normal modes of a general three-dimensional crystal. In this section we will first derive the equations of motion and discuss their solutions. Then we will present a brief discussion of various phenomenological models for interatomic force constants which have been used in lattice dynamical studies of semiconductors. A full derivation of the crystal Hamiltonian will be presented in section 4.2.

Consider a crystal with p atoms per unit cell, and let $\boldsymbol{u}(\boldsymbol{lb})$ represent the displacement of the bth atom in the lth unit cell. Assume that the total potential energy $\mathscr{V}$ of the crystal is a function of the instantaneous position of all atoms. Following equation (2.1), we now expand $\mathscr{V}$ in a Taylor series in powers of the atomic displacements $\boldsymbol{u}(\boldsymbol{lb})$:

$$
\begin{aligned}
\mathscr{V} &= \mathscr{V}_0 + \sum_{lb\alpha} \left.\frac{\partial \mathscr{V}}{\partial u_\alpha(\boldsymbol{lb})}\right|_0 u_\alpha(\boldsymbol{lb}) + \frac{1}{2}\sum_{lb,l'b'}\sum_{\alpha\beta}\Phi_{\alpha\beta}(\boldsymbol{lb};\boldsymbol{l'b'})u_\alpha(\boldsymbol{lb})u_\beta(\boldsymbol{l'b'}) + \cdots \\
&= \mathscr{V}_0 + \mathscr{V}_1 + \mathscr{V}_2 + \cdots,
\end{aligned}
\tag{2.41}
$$

where $\mathscr{V}_0$ is the equilibrium value, and

$$
\Phi_{\alpha\beta}(\boldsymbol{lb};\boldsymbol{l'b'}) = \left.\frac{\partial^2 \mathscr{V}}{\partial u_\alpha(\boldsymbol{lb})\partial u_\beta(\boldsymbol{l'b'})}\right|_0
\tag{2.42}
$$

is an element of the harmonic force constant between the atoms located at $(\boldsymbol{lb})$ and $(\boldsymbol{l'b'})$. As discussed in section 2.1, the first term in equation (2.41) is unimportant for the dynamical problem and can be set to zero, and the second term vanishes in the equilibrium configuration, so that in the harmonic approximation we have

$$
\mathscr{V}_{harm} \equiv \mathscr{V}_2 = \frac{1}{2}\sum_{lb,l'b'}\sum_{\alpha\beta}\Phi_{\alpha\beta}(\boldsymbol{lb};\boldsymbol{l'b'})u_\alpha(\boldsymbol{lb})u_\beta(\boldsymbol{l'b'}).
\tag{2.43}
$$

The equations of motion now become a generalisation of equation (2.3)

$$
m_b\ddot{u}_\alpha(\boldsymbol{lb}) = -\sum_{l'b'\beta}\Phi_{\alpha\beta}(\boldsymbol{lb};\boldsymbol{l'b'})u_\beta(\boldsymbol{l'b'}),
\tag{2.44}
$$

where m_b is the mass of the bth atom, $\alpha = 1,2,3$, and Φ is the interatomic harmonic force constant matrix defined in equation (2.42). In fact, $\Phi_{\alpha\beta}(\boldsymbol{lb};\boldsymbol{l'b'})$ represents the negative of the linear force on atom $(\boldsymbol{lb})$ along the α direction due to a unit displacement of atom $(\boldsymbol{l'b'})$ along the β direction.

The force constant matrix Φ obeys two important symmetry relations. From the *lattice translational symmetry*, we have

$$
\Phi_{\alpha\beta}(\boldsymbol{lb};\boldsymbol{l'b'}) = \Phi_{\alpha\beta}(\boldsymbol{0b};(\boldsymbol{l'}-\boldsymbol{l})\boldsymbol{b'})
\tag{2.45}
$$

and the *infinitesimal translational invariance* of the crystal (i.e. when all the atoms are equally displaced, there is no force on any atom) leads to

$$
\Phi_{\alpha\beta}(\boldsymbol{lb};\boldsymbol{lb}) = -\sum_{l'b'\neq lb}\Phi_{\alpha\beta}(\boldsymbol{lb};\boldsymbol{l'b'}).
\tag{2.46}
$$

The force constant $\Phi_{\alpha\beta}(\boldsymbol{lb};\boldsymbol{l'b'})$ is called a 'self-term'. Using the lattice translational symmetry, equation (2.29) can be expressed as

$$
m_b\ddot{u}_\alpha(\boldsymbol{0b}) = -\sum_{l'b'\beta}\Phi_{\alpha\beta}(\boldsymbol{0b};\boldsymbol{l'b'})\mathbf{u}_\beta(\boldsymbol{l'b'}).
\tag{2.47}
$$

To solve equation (2.47), we try a solution of the form

$$
u_\alpha(\boldsymbol{lb}) = \frac{1}{\sqrt{m_b}}\sum_q U_\alpha(\boldsymbol{q};b)\exp\left[i(\boldsymbol{q}\cdot\boldsymbol{x}(\boldsymbol{l})-\omega t)\right],
\tag{2.48}
$$

where $\boldsymbol{x}(\boldsymbol{l})$ is the equilibrium position vector of the lth unit cell and $U_\alpha(\boldsymbol{q};b)$ is independent of $\boldsymbol{l}$. By substituting this into equation (2.47) we get

$$
\omega^2 U_\alpha(\boldsymbol{q};b) = \sum_{b'\beta}D_{\alpha\beta}(bb'|\boldsymbol{q})U_\beta(\boldsymbol{q};b')
\tag{2.49}
$$

a non-trivial solution of which is obtained by solving the following determinantal equation:

$$\left| D_{\alpha\beta}(bb'|q) - \omega^2 \delta_{\alpha\beta} \delta_{bb'} \right| = 0. \tag{2.50}$$

The expression for the dynamical matrix reads

$$D_{\alpha\beta}(bb'|q) = \frac{1}{\sqrt{m_b m_{b'}}} \sum_{l'} \Phi_{\alpha\beta}(0b;l'b') \exp\left(iq \cdot x(l')\right). \tag{2.51}$$

This definition of the dynamical matrix in Fourier space is known as the 'D-type' (cf Maradudin *et al* 1971). Equations (2.49) and (2.50) are the three-dimensional analogues of equations (2.19) and (2.21).

If we consider the pairs (α, b) and (β, b') as indices then D is a $3p \times 3p$ Hermitian matrix:

$$D^*_{\alpha\beta}(bb'|q) = D_{\beta\alpha}(b'b|q). \tag{2.52}$$

Also, it follows from equation (2.51) that

$$D^*_{\alpha\beta}(bb'|q) = D_{\alpha\beta}(bb'|-q). \tag{2.53}$$

Equation (2.50) will produce $3p$ eigenvalues $\omega^2(qs)$, where $s = 1, 2, \ldots, 3p$. Since D is Hermitian, its eigenvalues $\omega^2(qs)$ are real and we consider $\omega(qs)$ as real to meet the condition of the stability of the crystal. From time reversal symmetry, it follows that

$$\omega^2(qs) = \omega^2(-qs) \tag{2.54}$$

and hence

$$\omega(qs) = \omega(-qs). \tag{2.55}$$

For each of the $3p$ eigenvalues $\omega^2(qs)$ corresponding to a given q, there exists an eigenvector $e(b;qs)$, so that equation (2.49) can be rewritten as

$$\omega^2(qs)e_\alpha(b;qs) = \sum_{b'\beta} D_{\alpha\beta}(bb'|q)e_\beta(b';qs). \tag{2.56}$$

The components of $e(b;qs)$ satisfy the following orthonormality and completeness relations:

$$\sum_{b\alpha} e^*_\alpha(b;qs)e_\alpha(b;qs') = \delta_{ss'}, \tag{2.57}$$

$$\sum_s e^*_\beta(b';qs)e_\alpha(b;qs) = \delta_{\alpha\beta}\delta_{bb'}. \tag{2.58}$$

The eigenvectors $e(b;qs)$ can be complex only for crystals with more than one atom per unit cell. From equations (2.53) and (2.56), it can be noted that

$$e^*(b;-qs) = e^{i\xi}e(b;qs) \tag{2.59}$$

and it is usual to make the choice $e^{i\xi} = 1$ (Born and Huang 1954, Maradudin *et al* 1971).

Sometimes it is convenient to express equation (2.48) in the form

$$u_\alpha(lb) = \frac{1}{\sqrt{m_b}} \sum_q U'_\alpha(q;b) \exp\left[i(q \cdot x(lb) - \omega t)\right], \tag{2.60}$$

where $x(lb) \equiv x(l) + x(b)$ is the equilibrium position vector of the bth atom in the lth unit cell. The equations of motion in (2.47) then become

$$\omega^2(qs)e'_\alpha(b;qs) = \sum_{b'\beta} C_{\alpha\beta}(bb'|q)e'_\beta(b';qs), \tag{2.61}$$

where the eigenvector $e'(b;qs)$ and the 'C-type' dynamical matrix are related to $e(b;qs)$ and the 'D-type' matrix by the relations

$$e'(b;qs) = \exp(-iq \cdot x(b))e(b;qs),\qquad\qquad (2.62)$$

$$C_{\alpha\beta}(bb'|q) = \frac{1}{\sqrt{m_b m_{b'}}}\sum_{l'}\Phi_{\alpha\beta}(0b;l'b')\exp[-iq\cdot(x(0b)-x(l'b'))]$$

$$= \exp(-iq\cdot x(b))D_{\alpha\beta}(bb'|q)\exp(iq\cdot x(b')). \qquad (2.63)$$

The eigenvectors $e'(b;qs)$ satisfy the orthonormality and closure relations similar to those given in equations (2.57) and (2.58) for $e(b;qs)$.

A complete solution of the eigenproblem in equation (2.56) or (2.61) is sought in terms of $\omega = \omega(qs)$ (called the phonon dispersion relation for eigenvalues) and $e = e_s(q)$ (the dispersion relation for eigenvectors). The index s is a branch index. For a crystal with N_0 unit cells and p atoms per unit cell $s = 1, 2, \ldots, 3p$, so that there will be $3p$ phonon branches, each with N_0 distinct q-vectors (or normal modes). Out of these there will be three acoustic branches such that $\omega(q) \to 0$ as $q \to 0$ and $3p - 3$ optical branches such that $\omega(q) \to$ constant as $q \to 0$. Atomic vibrations corresponding to any of the branches, acoustic or optical, can either be longitudinal such that $e \parallel q$ or transverse such that $e \perp q$, or a mixture of longitudinal and transverse. In an isotropic crystal, and in an isotropic elastic continuum, it is always possible to construct three mutually independent polarisation modes for a given q: $e \parallel q$ and $e_{T_1} \perp e_{T_2} \perp q$ (with degenerate eigenvalues for the two transverse polarisations). In an anisotropic crystal a clear relationship between e and q does not exist, except when q is along a high symmetry direction. For example, in cubic crystals, the concept of pure longitudinal and transverse polarisation modes is only defined when q is along the symmetry directions $[100], [110]$ and $[111]$.

Consider two crystals : Ne (FCC structure) and Si (diamond structure). As the underlying lattice in both cases is the same (FCC), the Brillouin zone for both the structures is identical in shape (shown in figure 1.2). In both cases, we can speak of pure longitudinal and transverse phonon modes along the symmetry directions ΓX (or $[100]$), ΓK (or $[110]$) and ΓL (or $[111]$). In the case of Ne (one atom per unit cell), we only have acoustic branches which can be labelled LA (longitudinal acoustic), and $T_1 A$, $T_2 A$ (two transverse acoustic branches), so that the three values of the index s can be expressed as $s =$ LA, $T_1 A$, $T_2 A$. In the case of Si (two atoms per unit cell), we have three acoustic branches (LA, $T_1 A$, $T_2 A$), and three optical branches (LO, $T_1 O$, $T_2 O$), so that the index s now stands as $s =$ LA, $T_1 A$, $T_2 A$; LO, $T_1 O$, $T_2 O$. Notice, from figures 2.12(a), 2.12(b), 2.15 and 2.16, that for crystals based on the FCC lattice the transverse modes (in acoustic or optical branch) are degenerate along $[100]$ and $[111]$ but not along $[110]$.

The subject of lattice dynamics of metals and ionic solids has been discussed in numerous articles and books (see, e.g., Born and Huang 1954, Maradudin $et\ al$ 1971, Joshi and Rajagopal 1968). We concentrate on a brief description of the theory of the lattice dynamics of semiconductors. The theoretical models which have been employed for constructing the dynamical matrix for tetrahedrally bonded crystals with diamond and zincblende structures can be broadly classified in three catagories: (i) force constant models, (ii) the rigid ion model and (iii) dipole approximation models.

2.4.1 MODELS OF INTERATOMIC FORCES

Some of the phenomenological force models which have been employed for crystals with the diamond/zincblende structure are (a) Born model, (b) Born–von Kármán model (c) valence force field model and (d) empirical many-body interatomic potential. Some of these have been compiled in a handbook by Torres and Stoneham (1985).

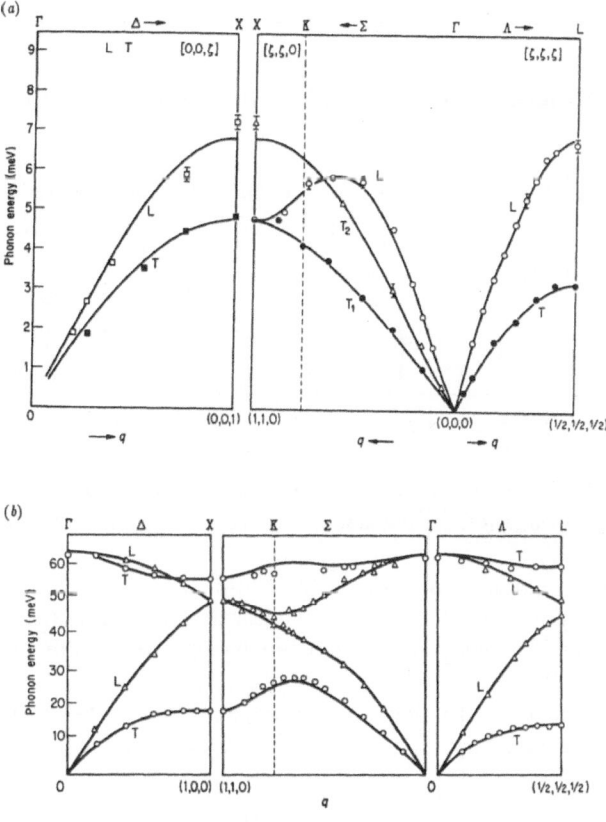

Figure 2.12 (a) Measured phonon dispersion in Ne (after Leake *et al* (1969); reproduced from Elliott and Gibson (1982)). (b) Measured phonon dispersion in Si (after Dolling (1963); reproduced from Elliott and Gibson (1982)). Note: $33.3563 \text{ cm}^{-1} = 1 \text{ THz} = 4.1357 \text{ meV}$.

2.4.1.1 The Born model

Born (1914) proposed a model for the interatomic potential of diamond-type crystals with nearest-neighbour central and non-central force constants α_B and β_B, respectively. The potential energy for an atom i is given by

$$\mathscr{V}_i = \frac{1}{4}\sum_j (\alpha_B - \beta_B)\left|(\boldsymbol{u}(i) - \boldsymbol{u}(j))\cdot\hat{\boldsymbol{r}}_{ij}\right|^2 + \frac{1}{4}\sum_j \beta_B\left|\boldsymbol{u}(i) - \boldsymbol{u}(j)\right|^2, \qquad (2.64)$$

where j refers to the nearest-neighbours of the atom i, $\hat{\boldsymbol{r}}_{ij}$ is a unit vector along the bond between i and j, $\boldsymbol{u}(i)$ is the displacement of atom i, and the summations are over the bonds to site i.

This simple model is generally found to be inadequate for explaining the frequencies and dispersion of TA phonons away from zone centre. This is primarily due to neglect of long-range interatomic interactions in this model.

2.4.1.2 The Born–von Karman model

In the Born–von Kármán model (see Born and Huang 1954), the effective crystal potential energy is expressed as

$$\mathscr{V} = \frac{1}{2}\sum_i \sum_j \sum_\alpha \sum_\beta \Phi_{\alpha\beta}(i,j) u_\alpha(i) u_\beta(j), \qquad (2.65)$$

Table 2.1

The Born–von Karman force constant matrices for diamond-type and zincblende-type crystals.

Order of neighbour	Force constant matrix: general model	Force constant matrix: central model
First neighbour	$\begin{pmatrix} A & B & B \\ B & A & B \\ B & B & A \end{pmatrix}$	$\begin{pmatrix} A_1 & A_1 & A_1 \\ A_1 & A_1 & A_1 \\ A_1 & A_1 & A_1 \end{pmatrix}$
Second neighbour	$\begin{pmatrix} C & D & E \\ D & C & E \\ -E & -E & F \end{pmatrix}$	$\begin{pmatrix} A_2 & A_2 & 0 \\ A_2 & A_2 & 0 \\ 0 & 0 & 0 \end{pmatrix}$

where $\Phi_{\alpha\beta}(i,j)$ are the force constants between atoms i and j. In the general Born–von Kármán model the number of independent force constants is reduced, for a given crystal structure, by noting that:

(a) the order of differentiation in equation (2.27) is immaterial;

(b) the potential energy $\mathscr{V}$ is invariant under rigid translations and rotations of the crystal, and

(c) the potential energy $\mathscr{V}$ is invariant to all symmetry operations belonging to the space group of the crystal.

When condition (a) is not used, the model is called an extended Born–von Kármán model (Powell 1970). In the central Born–von Kármán model the interaction between two atoms i and j is purely central, of the form

$$\frac{1}{2}\alpha_C \left| (u(i) - u(j)) \cdot \hat{r}_{ij} \right|^2, \tag{2.66}$$

where $\hat{r}_{ij}$ is the unit vector along the distance of atoms i and j.

The general and central force constant matrices for the diamond structure up to the sixth nearest neighbours have been worked out (Herman 1959). From table 2.1, it is clear that for interactions up to the second nearest-neighbours two (six) force constants are required for the central (general) Born–von Kármán model.

As described in section 1.4, in diamond structure there are two atoms per FCC unit cell located at $(0,0,0)$ and $a(1/4,1/4,1/4)$, where a is the cubic lattice constant. High symmetry points and directions in the FCC Brillouin zone are presented in section 1.7. Using the two first-neighbour general force constants A and B, and the three-dimensional extension of the procedures described in sections 2.2.2 and 2.3, Parrott (1969a) has set up the dynamical matrix and obtained analytical expressions for the phonon eigenvalues and eigenvectors along the Δ line joining the Γ and X points in the Brillouin zone. We will not detail the steps but simply quote the expressions for the dispersion relations:

$$\omega_{LA}^2 = \frac{4A}{m}\left(1 - \cos\frac{q_x a}{4}\right), \quad \omega_{LO}^2 = \frac{4A}{m}\left(1 + \cos\frac{q_x a}{4}\right),$$

$$\omega_{TA}^2 = \frac{4A}{m} - \left[\left(\frac{4A\cos\frac{q_x a}{4}}{m}\right)^2 + \left(\frac{4B\sin\frac{q_x a}{4}}{m}\right)^2\right]^{1/2},$$

$$\omega_{TO}^2 = \frac{4A}{m} + \left[\left(\frac{4A\cos\frac{q_x a}{4}}{m}\right)^2 + \left(\frac{4B\sin\frac{q_x a}{4}}{m}\right)^2\right]^{1/2}. \tag{2.67}$$

These results are plotted in figure 2.13. Both TA and TO branches are doubly degenerate, and the LA and LO modes become degenerate at X. Note, however, that the TO branch lies higher than the LO branch for finite q_x values along this symmetry line. It can be judged as a limitation of this

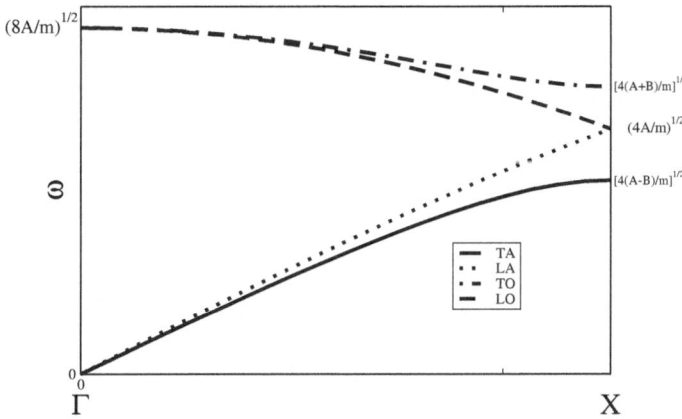

Figure 2.13 Phonon dispersion curves for diamond structure materials using the two nearest-neighbour general force constants A and B (see table 2.1). Calculations were performed by setting $B = 3A/8$.

simple model, as in diamond structure materials the TO and LO branches cross each other along the Δ line (see, *e.g.* figure 2.6(*b*)).

2.4.1.3 The valence force field model

One can express the interatomic potential energy in terms of internal displacements, namely changes in bond lengths and bond angles. Since valence coordinates are involved, the potential is called the valence force potential.

A general form of the rotationally invariant valence force potential for the unit cell in the diamond/zincblende type structure is given as (Musgrave and Pople 1962, Martin 1970, Torres and Stoneham 1985)

$$
\begin{aligned}
\mathscr{V} \;=\; & \frac{1}{2}\sum_{i=1}^{4} k_r (\Delta r_{1i})^2 + \sum_{n=1}^{2}\left(\frac{1}{2}\sum_{i,j>i} k_\theta^n r_0^2 (\Delta\theta_{inj})^2 + \sum_{i,j\neq i} r_0 k_{r\theta}^n (\Delta r_{ni})(\Delta\theta_{inj}) \right. \\
& + \sum_{i,j>i} k_{rr}^n (\Delta r_{ni})(\Delta r_{nj}) + \sum_{i,j\neq i, k\neq j} r_0^2 k_{\theta\theta}^n (\Delta\theta_{inj})(\Delta\theta_{jnk}) \\
& \left. + \frac{1}{2}\sum_{i,j>i,k} r_0^2 k_{\theta\theta}^{*n}(\Delta\theta_{jni})(\Delta\theta_{nik}) \right),
\end{aligned}
\tag{2.68}
$$

where n (=1,2) represents the two atoms in the unit cell, r_0 is the equilibrium bond length, r_{ni} is the bond vector between atoms n and i, the sums over i and j include the first neighbour of atom n, θ_{inj} is the angle between the bonds in and nj, and in the last term k represents a second nearest neighbour of the atom n. Notice that the above potential includes three-body interactions.

Keating (1966) simplified the valence force field energy in equation (2.68) by retaining only the squares of the scalar variations $\Delta(r_{ni}\cdot r_{nj})$ about the atom n. The resulting potential energy per unit cell in the diamond/zincblende structure is Torres and Stoneham 1985)

$$
\begin{aligned}
\mathscr{V} \;=\; & \frac{1}{2}\alpha_K \sum_{i=1}^{4}[\Delta(r_{1i}\cdot r_{1i})]^2 + \frac{1}{2}\sum_{n=1}^{2}\beta_K^n\left(\frac{3}{4r_0^2}\right)\sum_{i,j>i}[\Delta(r_{ni}\cdot r_{nj})]^2 \\
& + \sum_{n=1}^{2}\sigma_K^n\left(\frac{3}{4r_0^2}\right)\sum_{i,j>i}\Delta(r_{ni}\cdot r_{ni})\Delta(r_{nj}\cdot r_{nj}).
\end{aligned}
\tag{2.69}
$$

The parameters α_K, β_K and σ_K are called the central, non-central, and stretch–stretch Keating force constants, respectively. The potential energy in equation (2.69) is an invariant function of bond lengths and bond angles.

The Keating potential becomes equivalent to the full valence force potential in the limit of small displacements, and the two sets of parameters are related as follows:

$$
\begin{aligned}
k_r &= 3\alpha_K + \frac{1}{2}\beta_K, \\
k_\theta &= \frac{2}{3}\beta_K, \\
k_{rr} &= \frac{1}{12}\beta_K + 3\sigma_K, \\
k_{r\theta} &= \frac{1}{3\sqrt{2}}\beta_K, \\
k_{\theta\theta} &= k_{\theta\theta}^* = 0,
\end{aligned}
\tag{2.70}
$$

where we have used force constants averaged over the two atoms, e.g., $\beta_K = \frac{1}{2}(\beta_K^1 + \beta_K^2)$, etc.

2.4.1.4 Empirical many-body interatomic potentials

The valence force field model described in the previous subsection is reasonably good for describing elastic properties and phonon spectra of crystalline diamond/zincblende materials. In order to satisfactorily describe properties of materials made of the same elements in a wider configuration space requiring larger displacements from the ideal tetrahedral geometry (e.g. to describe other crystalline forms or the liquid phase), it becomes desirable to seek more general forms of interatomic potential. Stillinger and Weber (1985) and Tersoff (1988, 1989) developed empirical many-body interatomic potentials which are more general in nature and are found to be excellent choices for explaining properties of covalently bonded materials such as Si, Ge and C in different crystalline forms and in the liquid phase.

Stillinger and Weber (1985) expressed the interatomic potential by using a combination of the pair (two-body, central) v_2 term and a triplet (three-body) v_3 term that possesses full translational and rotational symmetry. The two-body term is expressed as

$$
v_2(r) = \begin{cases} A\left(Br^{-p} - r^{-q}\right)\exp[(r-a)^{-1}], & r < a \\ 0, & r \geq a, \end{cases}
\tag{2.71}
$$

where A, B, p and a are parameters with positive value. The three-body term is expressed as

$$
v_3(\mathbf{r}_i, \mathbf{r}_j, \mathbf{r}_k) = h(r_{ij}, r_{ik}, \theta_{jik}) + h(r_{ji}, r_{jk}, \theta_{ijk}) + h(r_{ki}, r_{kj}, \theta_{ikj}),
\tag{2.72}
$$

where θ_{jik} is the angle between $\mathbf{r}_j$ and $\mathbf{r}_k$ subtended at vertex i, etc. The function h is expressed as

$$
h(r_{ij}, r_{ik}, \theta_{jik}) = \lambda \exp\left[\gamma(r_{ij}-a)^{-1} + \gamma(r_{ik}-a)^{-1}\right]\left(\cos\theta_{jik} + \frac{1}{3}\right)^2,
\tag{2.73}
$$

where λ and γ are two further parameters. Note that the function h is considered only when both r_{ij} and r_{ik} are less than the cut-off a introduced earlier.

Several improved versions of three-body potential have been attempted. Here we refer the reader to the work of Tersoff (1988, 1989). He relied upon the central idea that in real systems the bond order (i.e. the strength of each bond) depends upon the local environment. Including environment-dependent bond order explicitly, he expressed the interatomic potential in the following form

$$
V_{ij} = f_c(r_{ij})\left[f_R(r_{ij}) + b_{ij}f_A(r_{ij})\right]
\tag{2.74}
$$

where f_R presents a repulsive pair potential, which includes the orthogonalization energy when atomic wave functions overlap, f_A represents an attractive pair potential associated with bonding, and f_C is a smooth cutoff function to limit the range of the potential. The function b_{ij} represents a measure of the bond order. For a multicomponent system, the functions f_R, f_A and f_C are expressed as follows (Tersoff 1989):

$$f_R(r_{ij}) = A_{ij}\exp(-\lambda_{ij}r_{ij}), \tag{2.75}$$

$$f_A(r_{ij}) = -B_{ij}\exp(-\mu_{ij}r_{ij}), \tag{2.76}$$

$$f_C(r_{ij}) = \begin{cases} 1, & r_{ij} < R_{ij} \\ \frac{1}{2}+\frac{1}{2}\cos[\pi(r_{ij}-R_{ij})/(S_{ij}-R_{ij})], & R_{ij} < r_{ij} < S_{ij} \\ 0, & r_{ij} > S_{ij} \end{cases} \tag{2.77}$$

where

$$b_{ij} = \chi_{ij}(1+\beta_i^{n_i}\zeta_{ij}^{n_i})^{-1/2n_i}, \quad \zeta_{ij} = \sum_{k\neq i,j} f_c(r_{ik})g(\theta_{ijk}),$$

$$g(\theta_{ijk}) = 1+c_i^2/d_i^2 - c_i^2/[d_i^2+(h_i-\cos\theta_{ijk})^2],$$

$$\lambda_{ij} = (\lambda_i+\lambda_j)/2, \quad \mu_{ij}=(\mu_i+\mu_j)/2, \quad A_{ij}=(A_iA_j)^{1/2},$$

$$B_{ij} = (B_iB_j)^{1/2}, \quad R_{ij}=(R_iR_j)^{1/2}, \quad S_{ij}=(S_iS_j)^{1/2}. \tag{2.78}$$

Here i, j and k label atoms, r_{ij} is the length of the ij bond and θ_{oijk} is the angle between bonds ij and ik. The parameter χ_{ij} indicates weakness or strength of the heteropolar bond ij, so that $\chi_{ii}=1$. In total there are 11 parameters $(A, B, \lambda, \mu, \beta, n, c, d, h, R$ and $S)$ for a single species system, and two additional parameters (χ_{ij}) for a two-species system.

2.4.2 THE RIGID ION MODEL

The rigid ion model, originally developed for the study of lattice dynamics of ionic crystals such as alkali halides, involves central pair interactions between the ions which are regarded as point charges (Kellermann 1940). The application of this model to semiconductors, in particular partially ionic III–V and II–VI compounds, has been made (Banerjee and Varshni 1969, Vetelino and Mitra 1969) by assuming that the forces in the crystal arise from two contributions: (a) long range or Coulomb interactions between 'effective' charges $\pm Z^\star e$ on the ions and (b) short-range central and non-central interactions, as described in the Born–von Kármán model. Thus this model for zincblende crystals combines the essential features of an alkali halide with the diamond lattice. For a good fit of experimental phonon dispersion curves, a large number of adjustable parameters (usually 11) are needed (Kunc 1973–74).

2.4.3 DIPOLE APPROXIMATION MODELS

The Born–von Kármán and the rigid ion models have been extended to two relatively more successful dipole models for diamond/zincblende type crystals: (a) shell model and (b) bond charge model.

2.4.3.1 The shell model

The shell model was developed by Dick and Overhauser (1958) for ionic crystals and by Cochran (1959a) for covalent crystals. Essentially, an atom is represented by a non-polarisable ion core with charge X (consisting of the nucleus and tightly bound electrons) and a shell of valence electrons with charge Y. The displacement of the shell relative to the core produces a dipole. An ion core and its shell are regarded as point charges when dealing with the Coulombic interaction with other

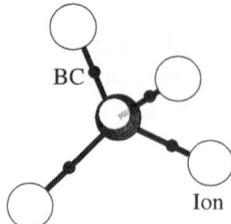

Figure 2.14   The concept of the bond charge model for a tetrahedrally bonded covalent solid.

ion cores and shells. In the simplest shell model, a minimum of five independent parameters are needed to describe core–core, core–shell and shell–shell forces extending only to nearest neigh-bours. However, for a successful explanation of phonon dispersion in many diamond/zincblende type semiconductors it is necessary to use a 14-parameter model (Waugh and Dolling 1963). An extension of the shell model is the *deformable shell model* or breathing shell model, proposed by Schröder (1966) and Kress (1972).

2.4.3.2   The bond charge model

Semiconductors with the diamond and zincblende structures are characterised by covalent bonds between neighbouring atoms. The maximum of the electronic charge density in semiconductors with the diamond structure is at the centre of such a bond. In semiconductors with the zincblende structure the maximum of the charge density is shifted slightly towards the anion taking part in a bond. This gives rise to the concept of *bond charges* in tetrahedrally coordinated semiconductors.

The bond charge model can be regarded as an extension of the shell model. Phillips (1968) and Martin (1969) proposed a simple bond charge model, and Weber (1974, 1977) and Rustagi and Weber (1976) developed an adiabatic bond charge model for semiconductors with the diamond structure. In the adiabatic bond charge model, one considers point ion cores and point bond charges of zero mass located somewhere in the ion–ion bonds (see figure 2.14). Four types of interactions are considered: (i) Coulombic interaction between ions and bond charges; (ii) nearest-neighbour ion–ion central interaction; (iii) ion–bond charge central interaction; and (iv) bond charge–bond charge non-central (or bond-bending) interaction of Keating type.

Interaction (i) is long ranged, while interactions (ii)–(iv) are short ranged and are considered only between first neighbours. This model has been successfully applied to the study of the lattice dynamics of bulk diamond and zincblende type semiconductors, and semiconductor superlattices (Yip and Chang 1984). Figure 2.15 shows the phonon dispersion curve in Ge calculated from the bond charge model, together with a comparison with the experimental results of Nilsson and Nelin (1971) and Nelin and Nilsson (1972). With the adiabatic motion of the bond charges, Weber (1977) has shown that the typical flattening of the TA branch away from the zone centre is achieved when the effective ion–bond charge coupling (i)+(iii) is weak compared to the bond–bond interaction (iv). A detailed description of this model for calculations of phonon modes can be found in Srivastava (1999).

The phonon dispersion curve for GaAs is shown in figure 2.16. The results from the bond charge model of Rustagi and Weber (1976) agree very well with experiment.

2.4.4   LONG-WAVELENGTH OPTICAL PHONONS IN IONIC CRYSTALS

When comparing the phonon dispersion curves in figures 2.15 and 2.16, it becomes apparent that, while the general shapes in the two figures are similar, there is one important difference. At $q = 0$ the

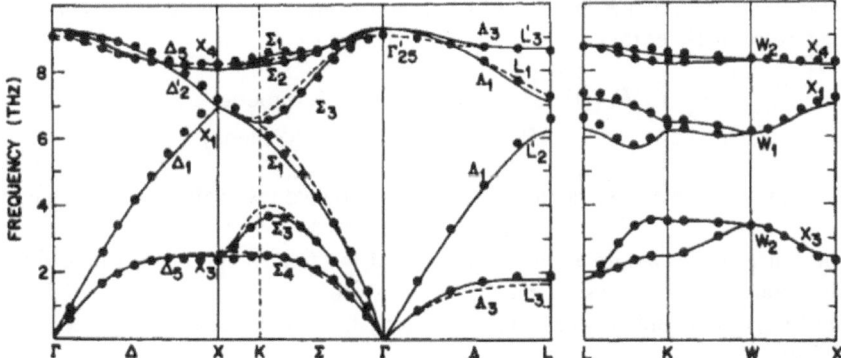

Figure 2.15 Phonon dispersion curves for Ge: solid lines show the calculations from the adiabatic bond charge model of Weber (1977) and the experimental values are from neutron scattering experiments (Nilsson and Nelin 1971, Nelin and Nilsson 1972).

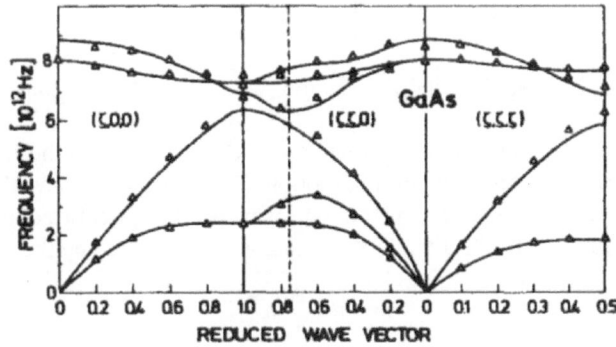

Figure 2.16 Phonon dispersion curves for GaAs. The solid curves represent the calculations of Rustagi and Weber (1976) based on the adiabatic bond charge model. The experimental results are from Dolling and Waugh (1965).

transverse and longitudinal optical branches are degenerate in Ge, while $\omega(\text{LO}) > \omega(\text{TO})$ in GaAs. In fact it is found that the splitting of the LO and TO branches for long wavelengths occurs in almost all crystals which are heteropolar (partially ionic such as GaAs) or ionic (such as NaCl). The origin of this splitting is the electrostatic field created by the long wavelength modes of vibrations in such crystals, as explained below.

From section 2.2.2 we note that for a long-wavelength optical mode, the two atoms in the unit cell vibrate against each other. In ionic crystals such a vibrational mode will generate a finite polarisation density P, which in turn will give rise to a macroscopic electric field E and an electric displacement D. These quantities are related as

$$D = \varepsilon E = E + 4\pi P, \tag{2.79}$$

where ε is the dielectric constant of the medium (here we consider only cubic crystals, so that ε is not a tensor). Further, D and E satisfy the electrostatic equations

$$\nabla \cdot D = 0, \tag{2.80}$$

$$\nabla \times E = 0. \tag{2.81}$$

In a cubic crystal $D \parallel E \parallel P$ and it can be shown from equations (2.80) and (2.81) (see, e.g., Ashcroft and Mermin 1976) that

$$D = 0 \quad E \;=\; -4\pi P \qquad \text{(LO mode)} \tag{2.82}$$

$$\qquad\quad E \;=\; 0 \qquad\quad \text{(TO mode)} \tag{2.83}$$

i.e. there is a non-vanishing long-range electric field associated with the long wavelength LO vibrations.

In the presence of the polarisation density P and the long-range electrostatic field E a long-wavelength vibration experiences an electrostatic restoring field (called the *local field* E_{loc} or the *effective field* E_{eff}). In a cubic crystal, this is given by the Lorentz relation (see, e.g., Ashcroft and Mermin 1976)

$$E_{\text{eff}} = E + \frac{4\pi P}{3}. \tag{2.84}$$

Using equations (2.82)–(2.84) we get

$$E_{\text{eff}} \;=\; -\frac{8\pi P}{3} \qquad \text{(LO modes)} \tag{2.85}$$

$$E_{\text{eff}} \;=\; \frac{4\pi P}{3} \qquad \text{(TO modes)}. \tag{2.86}$$

Thus for LO vibrations E_{eff} reduces the polarisation, while for TO vibrations it supports the polarisation.

If $Q_b^\star$ represents an effective charge associated with the bth ion then the force $Q_b^\star E_{\text{eff}}$ should be added to the right-hand side of the equation of motion, (2.44). This will have the effect of producing $\omega_{\text{TO}} < \bar{\omega}_{\text{op}}$ and $\omega_{\text{LO}} > \bar{\omega}_{\text{op}}$, where $\bar{\omega}_{\text{op}}$ is the long-wavelength optical frequency in the absence of the ionicity (i.e. when $E_{\text{eff}} = 0$). The LO and TO frequencies are related by the *Lyddane–Sachs–Teller relation*

$$\omega_{\text{LO}}^2 = \frac{\varepsilon_0}{\varepsilon_\infty}\omega_{\text{TO}}^2, \tag{2.87}$$

where ε_0 and ε_∞ are, respectively, the static and optical dielectric constants of the crystal. Thus in an ionic crystal the long-wavelength LO and TO frequencies will be split by the finite amount $\omega_{\text{LO}} - \omega_{\text{TO}}$.

A theoretical model for studying the long-wavelength optical phonons in mixed crystals will be discussed in section 11.4.

2.4.5   SOFT PHONON MODES

In section 2.4 we considered $\omega \geq 0, \omega^2 \geq 0$ for phonons in a crystal. For negative ω^2 the phonon frequencies ω become imaginary and the atomic motion increases exponentially with time, i.e. the crystal structure becomes unstable.

Consider a phonon mode with a finite frequency ω (i.e. other than a long-wavelength acoustic mode). If due to some effect (or effects) the frequency of this mode is reduced and eventually brought down to zero, then such a mode is called a *soft mode*. The presence of a soft mode deforms the original crystal structure in favour of a more stable structure.

An example of a soft TO phonon mode is found in some ferroelectric crystals. As the ferroelectric transition temperature T_c is approached from above, the frequency of the lowest TO branch in the paraelectric phase decreases as $\omega_{\text{TO}}^2 \propto (T - T_c)$. At T_c the restoring force of this TO mode becomes zero so that the crystal distorts into another structure. More discussion on this point can be found in many text books on solid state physics (see, e.g., Burns 1985).

An example of a displacive phase transition in GaAs due to a soft TA mode will be discussed in section 4.7.3

2.5   DENSITY OF NORMAL MODES

One of the most important quantities in lattice dynamical studies is the density of normal modes. This is defined as the number of normal modes between frequencies ω and $\omega + d\omega$, or, equivalently, between wave vectors q and $q + dq$. Assuming that the dispersion relation (for phonons or for electrons) in a crystal is known, a general expression for the density of states (of normal modes or of electrons) can be obtained by using the periodic boundary condition. In this section we specifically deal with phonons but the general result derived here is valid for electrons as well.

2.5.1   GENERAL EXPRESSION FOR A THREE-DIMENSIONAL CRYSTAL

For obtaining a general expression for the density of states consider an infinitely large cube of dimensions $L \times L \times L$ containing N^3 primitive unit cells of the crystal (N along each of the x, y, z directions). The technique of chain-folding used in section 2.2.1 can be applied along each of the x, y and z directions, so that now we have the following *cyclic boundary conditions*:

$$u(r) \equiv u(r + L\hat{x}) \equiv u(r + L\hat{y}) \equiv u(r + L\hat{z}) \tag{2.88}$$

(see equation (4.8) for a generalised case). Thus for a trial solution of the form given in equation (2.4) we have

$$\exp\left[i(q_x x + q_y y + q_z z)\right] = \exp\left\{i[q_x(x + L) + q_y y + q_z z]\right\} \tag{2.89}$$

or

$$e^{iq_x L} = 1 \equiv e^{i2\pi n} \qquad n = 0, \pm 1, \pm 2, \ldots. \tag{2.90}$$

Similarly along the y and z directions. Thus

$$q_x, q_y, q_z = 0; \pm \frac{2\pi}{L}; \pm \frac{4\pi}{L}; \ldots; \pm \frac{N\pi}{L} \tag{2.91}$$

i.e. the allowed q values form a cubic mesh in q-space, with one q confined to a volume $(2\pi/L)^3$. Therefore, for each phonon polarisation index (or for each spin orientation and for each band of an electron), a unit volume in q-space contains $(L/2\pi)^3 = N_0\Omega/8\pi^3$ values of q. Here $N_0\Omega$ is the volume of the solid containing N_0 unit cells.

To derive an expression for the density of states, we need to find the number of q-values, $g(q)$, in a range dq (or correspondingly the number of frequencies, $g(\omega)$, in a frequency interval $d\omega$). To derive an expression for $g(\omega)$ consider two surfaces of constant phonon frequencies ω and $\omega + d\omega$. Consider an elementary area dS_ω on the $\omega = $ constant surface and construct a right cylinder between the two constant frequency surfaces. The volume of this right cylinder of height $dq_\perp$ is

$$\int_{cylinder} d^3q = \int dS_\omega dq_\perp = \int dS_\omega \frac{d\omega}{|\nabla_q \omega|}, \tag{2.92}$$

where we have used the fact that $\nabla_q \omega$ is perpendicular to the $\omega = $ constant surface. Since the density of momentum space is $N_0\Omega/8\pi^3$, the number of q-values in this elementary volume is

$$
\begin{aligned}
g(\omega)d\omega &= \frac{N_0\Omega}{8\pi^3} \times \text{volume of cylinder} \\
&= \frac{N_0\Omega}{8\pi^3} \int dS_\omega \frac{d\omega}{|\nabla_q \omega|}.
\end{aligned}
$$

Therefore, the density of states is given by the expression

$$g(\omega) = \frac{N_0\Omega}{8\pi^3} \int \frac{dS_\omega}{|c_g|}, \tag{2.93}$$

where $c_g = \nabla_q \omega$ is the phonon group velocity.

Equation (2.93) can be expressed more generally as

$$g(\omega) = \frac{N_0 \Omega}{8\pi^3} \sum_s \int \frac{\mathrm{d}S_\omega}{|\boldsymbol{\nabla}_q \omega_s|}, \tag{2.94}$$

where s denotes a phonon polarisation index and the integration is over a constant frequency surface $\omega_s(\boldsymbol{q}) = \omega$. (When dealing with electrons s should stand for band index, and the expression in equation (2.94) should be multiplied by 2 to account for spin degeneracy.) In general the expression in equation (2.94) must be calculated numerically. This point is discussed in the next section.

For some $\boldsymbol{q}$-values the group velocity is (or approaches) zero ($c_g \to 0$) which leads to $g(\omega) \to \infty$. Such points in $\boldsymbol{q}$-space are called critical points and the singularities in the density of states are known as the von Hove singularities.

At a critical point $\boldsymbol{q} = \boldsymbol{q}_c$ the curvature of the dispersion curve is zero, $\boldsymbol{\nabla}_q \omega = 0$, so that near $\boldsymbol{q}_c$ one can approximate

$$\omega(\boldsymbol{q}s) = \omega(\boldsymbol{q}_c s) + \sum_{i=1}^{3} a_{is}(q_i - q_{ci})^2, \tag{2.95}$$

where q_i is a component of $\boldsymbol{q}$. The characteristics of the singularity at $\boldsymbol{q}_c$ can be studied by substituting equation (2.95) in equation (2.94). The critical point is a *maximum* if all $a_i < 0$, a *minimum* if all $a_i > 0$, and a *saddle point* if two of the a_is are positive and the third is negative, or vice versa. Phillips (1956) has presented a detailed group-theoretical and topological discussion of lattice vibrations and critical points.

2.5.2 TWO-DIMENSIONAL CASE

When dealing with a two dimensional system (such as a surface of area $A = L^2$), the density of $\boldsymbol{q}$-space will be $A/(2\pi)^2$. Thus the phonon density of states will be given by

$$g(\omega) = \frac{A}{4\pi^2} \sum_s \int \frac{\mathrm{d}l_\omega}{\boldsymbol{\nabla}_q \omega_s} \tag{2.96}$$

where $\mathrm{d}l_\omega$ is an elementary length of constant frequency in the two-dimensional region.

2.5.3 ONE-DIMENSIONAL CASE

Let us apply the periodic boundary condition to the monatomic linear chain of length $L = Na$ as shown in figure 2.2. It is easy to show that the density of the q-space is $L/2\pi$ for $-\pi/a \le q \le \pi/a$. As both positive and negative ranges of q are included, the density of normal modes is given by

$$
\begin{aligned}
g(\omega) &= \frac{L}{\pi} \frac{1}{\boldsymbol{\nabla}_q \omega} \tag{2.97}\\[2mm]
&= \frac{2N}{\pi \omega_{max}} [\cos(qa/2)]^{-1}\\[2mm]
&= \frac{2N}{\pi} \frac{1}{\sqrt{\omega_{max}^2 - \omega^2}}. \tag{2.98}
\end{aligned}
$$

Similar arguments can be used to calculate the density of normal modes of a diatomic linear chain. However, the resulting expression will not be as simple as in the case of a monatomic linear chain. Figure 2.17 shows the main features of the density of normal modes of a monatomic linear chain and a diatomic linear chain. There is a critical point at $\omega = \omega_{max}$ for the monatomic linear chain. For the diatomic linear chain in figure 2.4, there is a critical point at $\omega_3 = \sqrt{\Lambda/m + \Lambda/M}$. In addition, because of two different masses ($m < M$) there are two extra singularities at $\omega_1 = \sqrt{2\Lambda/M}$ and $\omega_2 = \sqrt{2\Lambda/m}$.

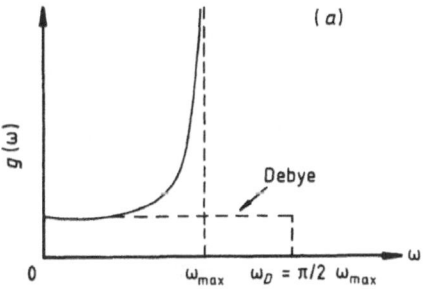 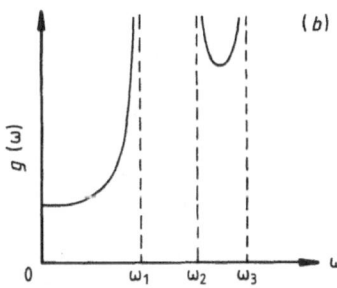

Figure 2.17 The density of normal modes in a linear chain of atoms: (a) the monatomic chain in figure 2.2 and (b) the diatomic chain in figure 2.4. The density of states in the Debye model which extends up to the Debye frequency $\omega_D = \frac{1}{2}\pi\omega_{max}$, as discussed in section 2.6.4, is also shown in (a).

2.6 NUMERICAL CALCULATION OF DENSITY OF STATES

A calculation of $g(\omega)$ using the general expression in equation (2.94) requires a realistic integration over the Brillouin zone of the crystal under consideration. In general, such an integration requires a knowledge of phonon frequencies for wave vectors q in the entire Brillouin zone. In practice, advantage is taken of the point group symmetry associated with the Brillouin zone so that phonon frequency calculations are needed for wave vectors q lying only within the irreducible part of the zone. Although symmetry arguments reduce the amount of work, the problem of performing the integration in equation (2.94) still remains. In this section we describe two schemes of different sophistication for a numerical calculation of $g(\omega)$.

A useful collection of phonon dispersion and density-of-states curves for many insulators and semiconductors obtained from theoretical as well as experimental studies, with further discussion of force models, can be found in the book by Bilz and Kress (1979).

2.6.1 ROOT SAMPLING METHOD

This is a simple scheme of calculating density of states based on expressing equation (2.94) in the form [1]

$$g(\omega) = \frac{N_0\Omega}{8\pi^3} \sum_{qs} \delta(\omega - \omega(qs)). \tag{2.99}$$

The dynamical problem is solved for phonon frequencies at a large number of q_v-points distributed throughout the irreducible part of the zone. With these frequencies one computes $g(\omega)$ by replacing the Dirac delta function in equation (2.75) by a Heaviside function as explained below:

$$g(\omega) = \text{constant} \times \sum_{q_v,s}^{IBZ} \Theta(\omega - \omega(q_v s)), \tag{2.100}$$

where

$$\Theta = \begin{cases} 1 & \text{for } |\omega - \omega(q_v s)| \leq \frac{\Delta\omega}{2} \\ 0 & \text{otherwise.} \end{cases}$$

---

[1] This follows from the following property of the Dirac delta function:

$$\delta(f(x)) = \sum_i \frac{\delta(x - x_i)}{|f'(x_i)|}$$

where x_i are simple zeros of $f(x)$. (See equation (4.5.18) in Maradudin *et al* (1971).)

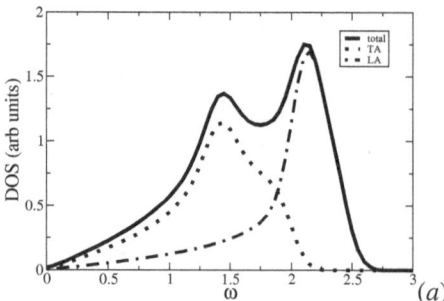
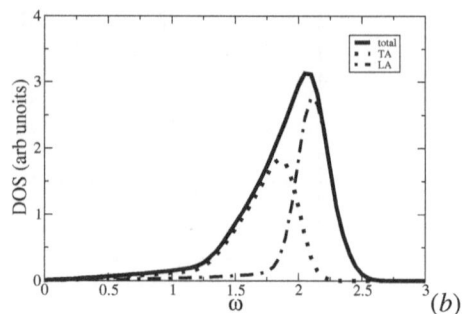

Figure 2.18 Phonon density of states for two-dimensionl systems: (a) monatomic square mesh, with the next nearest-neighbour force constant taken as half of the nearest-neighbour force constant: $\Lambda_2 = \Lambda_1/2$, and frequencies expressed in units of $\sqrt{\Lambda_1/m}$, and (b) monatomic hexagonal mesh, with frequencies expressed in units of $\sqrt{\Lambda/m}$.

with an appropriately small frequency width (say, $\Delta\omega \sim 0.005$ THz). This method of calculating density of states can be very time-consuming. This is because, to obtain detailed features in $g(\omega)$ with a reasonable accuracy, one requires a first-hand knowledge of $\omega(qs)$ at a very large number of q-points (typically 2000 or more) in the irreducible part of the zone.

Following this procedure we present in figure 2.18 the phonon density of states curves for the two two-dimensional systems studied in sections 2.3.1 and 2.3.2.

Phonon eigenvalues were computed by solving the secular equation in (2.32) and using the dynamical matrices as described in sections 2.3.1 and 2.3.2. Density of states calculations were performed numerically using 325 q-points in the IBZ region $\Gamma - X - M$ for the square lattice [cf. figure 1.2(a)] and 529 q-points in the IBZ region $\Gamma - M - K$ for the hexagonal lattice [cf. figure 1.5(a)]. For the square lattice with nearest and next nearest neighbour force constants Λ_1 and Λ_2, there are two peaks (van Hove singularities) in the total density of states curve, corresponding to the transverse acoustic (TA) and longitudinal acoustic (LA) dispersion curves. For the hexagonal lattice with the consideration only nearest neighbour interactions, there is a single peak (van Hove sigularity) in the density of states. Also presented are the density of states curves for the individual polarisations TA and LA.

2.6.2   LINEAR ANALYTICAL APPROACHES

Efficient Brillouin zone integration schemes which can provide detailed features in $g(\omega)$ have been developed by Gilat and Raubenheimer (1966) and Lehman and Taut (1972). Several modifications of these schemes also exist. According to these schemes, the dynamical problem is solved for a relatively small number of wave vectors q_c within the irreducible part of the Brillouin zone (IBZ) and linear extrapolation or interpolation is used to perform zone integration analytically.

2.6.2.1   The method of Gilat and Raubenheimer

The idea behind the method of Gilat and Raubenheimer (1966) is to calculate phonon frequencies $\omega(q_c, s)$ over a crude mesh of wave vectors q_c inside the irreducible part of the Brillouin zone. Then cells are formed around the points q_c. The question of efficient shapes of these cells depends on the type of lattice under study, as discussed by Gilat (1976) in some detail. For the sake of brevity, we only discuss here the case of cubic cells. This, however, may require considering more than one IBZ so that the whole Brillouin zone can be filled exactly and in an exhaustive manner.

Having formed the cubic cells, the next step is to define constant frequency surfaces $\omega(qs) = \omega$ inside every cell. This is done by defining a set of parallel planes inside every cell. The direction cosines of these planes are given by the direction cosines l_i of $\nabla\omega(q_c s)$. The gradient of frequencies inside a cell can be approximated by the linear relation

$$\nabla\omega \cdot (q_i - q_c) = \omega(q_i) - \omega(q_c) \tag{2.101}$$

for any q_i inside the cell. If w is the distance of a particular frequency plane from the centre q_c of the cell, then

$$g(\omega) = \sum_{q_c,s} g_s(q_c;\omega) \tag{2.102}$$

with

$$
\begin{aligned}
g_s(q_c;\omega)d\omega &= CW(q_c)S_s(l_1 l_2 l_3;w)dw \\
&= CW(q_c)S_s(l_i;w)\frac{d\omega}{|\nabla\omega(q_c s)|},
\end{aligned} \tag{2.103}
$$

where ω lies in the range

$$\omega(q_c) - w_{max}|\nabla\omega(q_c s)| \le \omega \le \omega(q_c s) + w_{max}|\nabla\omega(q_c s)|$$

and

$$w = (\omega - \omega(q_c s))/|\nabla\omega(q_c s)|. \tag{2.104}$$

Here C is a normalisation constant, $W(q_c)$ is the statistical weight of the point q_c for its occurrence in the irreducible part of the Brillouin zone, and w_{max} is the maximum distance w confined within each cell.

For a cube of side $2b$, and the direction cosines ordered in a decreasing sequence $l_1 \ge l_2 \ge l_3 \ge 0$, we can distinguish between four different ranges for $w_i > 0$, namely

$$
\begin{aligned}
w_1 &= b|l_1 - l_2 - l_3| \\
w_2 &= b(l_1 - l_2 + l_3) \\
w_3 &= b(l_1 + l_2 - l_3) \\
w_4 &= b(l_1 + l_2 + l_3).
\end{aligned} \tag{2.105}
$$

For any interval (w_i, w_{i+1}), there could be more than one type of cross-sectional area $S_s(l_i,w)$. We quote here the results of Gilat and Raubenheimer (1966):

(i) $0 \le w \le w_1$

$$
\begin{aligned}
S_s(l_1 l_2 l_3;w) &= \frac{4b^2}{l_1} \quad\quad \text{for} \quad\quad l_1 \ge l_2 + l_3 \tag{2.106}
\end{aligned}
$$

$$
= \frac{2b^2(l_1 l_2 + l_2 l_3 + l_3 l_1) - (w^2 + b^2)}{l_1 l_2 l_3}
$$
$$
\text{for} \quad l_1 < l_2 + l_3. \tag{2.107}
$$

(ii) $w_1 \le w \le w_2$

$$S_s(l_1 l_2 l_3;w) = \frac{2b^2(3l_2 l_3 + l_1 l_2 + l_3 l_1) + bw(l_1 - l_2 - l_3) - \frac{1}{2}(w^2 + wb^2)}{l_1 l_2 l_3}. \tag{2.108}$$

(iii) $w_2 \le w \le w_3$

$$S_s(l_1 l_2 l_3;w) = \frac{2b[b(l_1 + l_2) - w]}{l_1 l_2}. \tag{2.109}$$

(iv) $w_3 \leq w \leq w_4$

$$S_s(l_1 l_2 l_3; w) = \frac{[b(l_1 + l_2 + l_3) - w]^2}{2 l_1 l_2 l_3}. \tag{2.110}$$

All these expressions, as well as their derivatives, are continuous at their respective range boundaries. Also, these expressions are even functions of w, so that

$$S_s(l_i; -w) = S_s(l_i; w) \tag{2.111}$$

from which expressions can be obtained for the range $w < 0$.

It should be remembered that this method for calculating density of states requires the evaluation of the gradient of frequency (see equation (2.104)). Gilat (1976) has proposed an *extrapolation* method of gradient calculation. Other methods may also be used for gradient calculation (e.g. using the $k \cdot p$ method in electron band structure calculations).

2.6.2.2   The tetrahedron method

There are two practical difficulties with the method of Gilat and Raubenheimer: (i) it requires an explicit evaluation of the gradient $\nabla_q \omega$, and (ii) as cubes do not usually fit the Brillouin zone exactly, it is necessary to determine the statistical weights of the cubes for the irreducible part of the zone. The *analytical tetrahedron method* developed by Jepsen and Anderson (1971), and independently by Lehman and Taut (1972), overcomes these difficulties. In this method, the irreducible Brillouin zone is exactly filled with a finite number of non-overlapping tetrahedra. In cubic crystals, this is most efficiently done by using a cubic grid of q_c points in the Brillouin zone and then dividing portions of the cubes inside the irreducible part into a number of tetrahedra of equal volume. Within each tetrahedron, the frequency $\omega(qs)$ is linearly *interpolated* to the form

$$\omega(q) = \omega(q_c) + a \cdot (q - q_0). \tag{2.112}$$

Here $\omega(q_0)$ are the frequencies at the corners q_0 of the tetrahedron. The coefficient $a \equiv \nabla \omega(q_0)$ is approximated from $\omega(q_{0i})$ at the four corners of the tetrahedron q_i, $i = 1, 2, 3, 4$.

The density of states is then evaluated as the sum of contributions from all tetrahedra within the irreducible Brillouin zone

$$g(\omega) \propto \sum_s \sum_j g_{s,j}(\omega), \tag{2.113}$$

where j indicates a tetrahedron and s represents a phonon polarisation index. Analytical expressions for $g_{s,j}(\omega)$, as a function of the volume V_i of the ith tetrahedron and frequencies at its four corners, are given in the book by Skriver (1984) and will not be reproduced here.

Several improvements of the linear tetrahedron method have been presented. Blöchl *et al* (1994) developed a simple correction formula that extends the expansion in equation (2.112) up to the quadratic term. This is found to improve the results for systems with many band crossings. More recently, Kawamura *et al* (2014) have introduced further improvement to the method based on the third-order interpolation and the least-squares method that reduces the number of q points required to obtain converged results for Brillouin zone integrations. Readers are referred to these research articles for further details.

2.6.3   THE ISOTROPIC CONTINUUM APPROXIMATION

A simple and useful approximation in the calculation of density of states is the long-wavelength or continuum approximation. This is based on the reasoning that at low temperatures most of the phonons excited in a solid are confined to low-q or long-wavelength acoustic branches. Within this approximation, details of crystal structure are ignored and the isotropic dispersion relation $\omega(qs) = c_s q$ is used for all the normal modes lying within a sphere of radius q.

For an isotropic continuum, the polarisation index s can be considered to take values $s = \text{L}, \text{T}$ (degenerate transverse modes, and both longitudinal and transverse modes of only acoustic type). Futhermore, in this case a constant frequency surface is the surface of a sphere of radius q. An area dS_ω at $\omega = \text{constant}$ surface is an element of area on the surface of this sphere, which in spherical polar coordinates is given by

$$dS_\omega = q^2 \sin\theta\, d\theta\, d\phi. \qquad (2.114)$$

Further, in the continuum approximation the group velocity is the same as the phase velocity: $c_g = c$. Therefore, equation (2.94) becomes

$$
\begin{aligned}
g(\omega) &= \frac{N_0 \Omega q^2}{8\pi^3} \sum_s \frac{1}{c_s} \int\int d\phi\, d\theta\, \sin\theta \\
&= \frac{N_0 \Omega q^2}{2\pi^2} \sum_s \frac{1}{c_s} \\
&= \frac{N_0 \Omega}{2\pi^2} \sum_s \frac{\omega^2}{c_s^3}.
\end{aligned}
\qquad (2.115)
$$

Obviously this expression is very simple to use in any calculation and in fact does not require a solution of the dynamical problem.

2.6.4 THE DEBYE APPROXIMATION

As discussed in the previous subsection, the continuum model uses the long-wavelength approximation and deals with only the acoustic phonon branches. However, Debye (1912) used the continuum approximation in the whole q-space and determined a cut-off q_D (called the Debye radius) or equivalently an energy cut-off ω_D (called the Debye frequency) such that the volume of the sphere of radius q_D (the Debye sphere) contains the correct number $(3N)$ of all phonon modes in a crystal with N atoms:

$$
\begin{aligned}
3N &= \int_0^{\omega_D} d\omega\, g(\omega) \\
&= \frac{N_0 \Omega}{2\pi^2} \sum_s \frac{1}{c_s^3} \int_0^{\omega_D} d\omega\, \omega^2 \\
&= \frac{N_0 \Omega}{2\pi^2} \frac{\omega_D^3}{\bar{c}^3}
\end{aligned}
\qquad (2.116)
$$

with $\bar{c}$ defined as

$$\frac{3}{\bar{c}^3} = \sum_s \frac{1}{c_s^3}. \qquad (2.117)$$

Equation (2.116) defines ω_D and q_D:

$$\omega_D = \bar{c} q_D = \bar{c} \left(\frac{6\pi^2 N}{N_0 \Omega} \right)^{1/3}. \qquad (2.118)$$

It is also useful to define the Debye temperature Θ_D for the solid by the relation $k_B \Theta_D = \hbar \omega_D$, where k_B is Boltzmann's constant.

It should be borne in mind that in the continuum approximation one can only talk of the acoustic phonon branches. This means that the Debye theory accommodates only acoustic phonon modes. This readily is achieved for Bravais crystals, *i.e.* for crystals with only one atom per unit cell. In this case, $N = N_0$ and the radius of Debye sphere can be expressed as $q_D^{\text{Bravais}} = \left(6\pi^2 / \Omega \right)^{1/3}$. Suppose that the unit cell of the crystal contains two atoms of masses m_1 and m_2. If the mass ratio m_1 / m_2 is close to unity, then the slope of the optical branch can be comparable to that of the acoustic branch

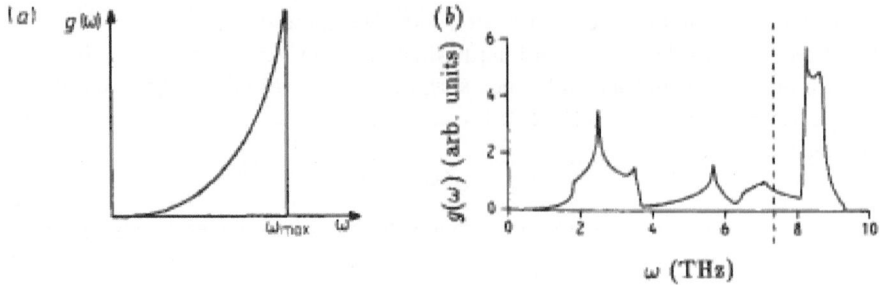

Figure 2.19 The density of normal modes in a three-dimensional crystal. (*a*) The Debye model, (*b*) The density of states for Ge, as calculated with the adiabatic bond charge model (Weber 1977).

and the role of optical phonons can be quite important. The original Debye scheme can be extended to include the contribution of optical phonons in $g(\omega)$, in an approximate way. This can be done by defining a ω_{max}, bigger than $\omega_D^{Bravais}$, so that the corresponding sphere of radius q_{max} includes both acoustic and optical phonons. However, this scheme may not be reasonable if the mass ratio is significantly bigger than unity, but in that case the group velocity of optical phonons is quite small and their contribution to transport properties is negligibly small (see Chapters 6 and 7 for further discussion on this point).

The density of states in the Debye model is a smooth function of the phonon frequency ($g(\omega) \propto \omega^2$) and has just one van Hove singularity at the Debye frequency ω_D. The density of states for an actual crystal structure in general shows a few singular points. Figure 2.19 gives an schematic illustration of the density of states in the Debye model and its comparison with the density of states for a realistic crystal.

The Debye model can also be used to obtain analytical expressions for the density of states for a linear chain and for a two-dimensional lattice. From equation (2.97), we get for a monatomic linear chain

$$g_{1D}(\omega) = L/\pi v = 2N/\pi w_{max} = \text{constant}, \tag{2.119}$$

where the terms are defined in section 2.2.1. For a two-dimensional lattice we can replace $l_\omega = 2\pi q$ in equation (2.96), so that

$$g_{2D}(\omega) = \frac{A}{2\pi} \sum_s \frac{\omega_s}{c_s^2}. \tag{2.120}$$

Thus g_{1D} is a constant (shown in figure 2.17 for the monatomic linear chain), while $g_{2D} \propto \omega$.

2.7 LATTICE SPECIFIC HEAT

The heat capacity at constant volume is a fundamental quantity from the point of view of thermal properties of solids. It is defined as

$$C_V = \left(\frac{\partial E}{\partial T}\right)\bigg|_{N_0\Omega}, \tag{2.121}$$

where E is the thermal energy of the crystal with volume $N_0\Omega$ at temperature T. The heat capacity at constant pressure C_p is related to C_v by the following thermodynamic relation:

$$C_p = C_v + \beta^2 B N_0 \Omega T, \tag{2.122}$$

where β is the volume thermal expansion coefficient, and B is the bulk modulus of the solid. In general $C_p \geq C_v$ but, as $T \to 0$, $C_p \to C_v$ so that at low temperatures the two quantities are nearly identical. Experimental observations suggest the following temperature dependence for C_v:

$$
\begin{aligned}
&\text{at high temperatures} &&C_v &&\simeq 3Nk_B &&\text{for all solids} \\
&\text{at low temperatures} &&C_v &&\propto T^3 &&\text{in insulators} \\
& && &&\propto T &&\text{in metals.}
\end{aligned}
$$

Here N is the number of atoms in the solid. In this section, we only discuss the behaviour in insulators. (The $C_v \propto T$ behaviour in metals is due to conduction electrons.)

The molar heat capacity of a solid at high temperature is $C_{vm} = 3N_a k_B \equiv 3R$, where N_a is Avogadro's number and R is the gas constant. This is known as the law of Dulong and Petit.

The specific heat (or specific heat capacity) of a solid is defined as the heat capacity per unit volume,

$$
C_v^{sp} = \frac{1}{N_0 \Omega} C_v. \tag{2.123}
$$

2.7.1 CLASSICAL THEORY OF C_V

In the simple classical theory, atoms of a solid are regarded as independent three-dimensional simple harmonic oscillators. The average thermal energy associated with the vibrations of an atom is $3k_B T$. For a solid with N atoms, therefore, $E = 3Nk_B T$ and hence $C_v = 3Nk_B$ in agreement with the Dulong–Petit law at high temperatures. Clearly this simple picture completely fails to explain the low temperature behaviour of C_v.

2.7.2 QUANTUM THEORY OF C_V

In the quantum theory we consider the crystal thermal energy as a sum of energies of phonons in thermal equilibrium with each other. From statistical mechanics the average energy for a phonon in polarisation mode s is (also see equation (4.42))

$$
\bar{\varepsilon}_s = \bar{n}_s \hbar \omega_s \tag{2.124}
$$

with $\bar{n}_s$ given by equation (2.2).

2.7.2.1 Einstein's model

In 1907 Einstein formulated the first quantum theory of C_v (Einstein 1907). He assumed that all atoms in a solid vibrate independently and with the same frequency ω_E. This means that each phonon in a solid has an average energy $\bar{\varepsilon}_E = 3\bar{n}(\omega_E)\hbar\omega_E$, where the factor 3 accounts for the three polarisation modes. With this the crystal thermal energy is

$$
\begin{aligned}
E &= N\bar{\varepsilon}_E \\
&= 3N\hbar\omega_E / [\exp(\hbar\omega_E / k_B T) - 1]
\end{aligned} \tag{2.125}
$$

and therefore the heat capacity is

$$
C_v = 3Nk_B \left(\frac{\hbar\omega_E}{k_B T} \right)^2 \bar{n}_E (\bar{n}_E + 1). \tag{2.126}
$$

At high temperatures $k_B T \gg \hbar\omega_E$ and we can approximate $\bar{n}_E \simeq k_B T / \hbar\omega_E$, which gives $C_v = 3Nk_B$ in agreement with the Dulong–Petit law. At low temperatures $k_B T \ll \hbar\omega_E$ and we can approximate $\bar{n}_E \simeq \exp(-\hbar\omega_E / k_B T)$ and $\bar{n}_E + 1 \simeq 1$, which gives

$$
C_v \big|_{LT} = 3Nk_B \left(\frac{\hbar\omega_E}{k_B T} \right)^2 \exp(-\hbar\omega_E / k_B T)
$$

$$\propto T^{-2}e^{-B/T}. \tag{2.127}$$

Thus Einstein's model predicts a temperature dependent behaviour of C_v at low temperatures, which is an improvement over the classical theory. However, although it predicts the correct behaviour $C_v \to 0$ as $T \to 0$, the actual temperature variation is wrong.

2.7.2.2 Debye's model

A satisfactory explanation for both high- and low-temperature variations of C_v did not emerge until the Debye model for the density of normal modes was incorporated in the quantum picture of lattice vibrations. Debye in 1912 used the isotropic continuum approximation for phonon dispersion (sections 2.6.3 and 2.6.4) and considered contributions from phonons in all acoustic polarisation branches and with wave vectors throughout the Debye sphere (Debye 1912). With these considerations, the crystal thermal energy is expressed as follows.

We first write

$$E = \sum_q \sum_s \hbar\omega_s(q)\bar{n}_s(q), \tag{2.128}$$

where $\bar{n}_s(q)$ is given by equation (2.2), s is the polarisation index, and the sum over q includes all independent phonon states in the Brillouin zone (whch is considered as the Debye sphere). In practice, the crystal volume $N_0\Omega$ is very large and the q-values are densely spaced, so that the sum over q can be replaced by an integral

$$\begin{aligned}
\sum_q \to \int \mathrm{d}q &\equiv \frac{N_0\Omega}{8\pi^3}\int \mathrm{d}^3q \\
&= \frac{N_0\Omega}{8\pi^3}\int \mathrm{d}S_\omega \frac{\mathrm{d}\omega}{|\nabla_q\omega|} \\
&= \int g(\omega)\mathrm{d}\omega. \tag{2.129}
\end{aligned}$$

Equation (2.128) then becomes

$$E = \sum_s \int_0^{\omega_D} \mathrm{d}\omega_s g_s(\omega)\hbar\omega_s(q)\bar{n}_s(q) \tag{2.130}$$

ω_D being the Debye frequency. Inserting the expression for $g_s(\omega)$ from equation (2.115) we get

$$E = \frac{N_0\Omega\hbar}{2\pi^2}\sum_s \frac{1}{c_s^3}\int_0^{\omega_D} \mathrm{d}\omega_s \omega_s^3 \bar{n}_s, \tag{2.131}$$

where the q dependence of ω and $\bar{n}$ is not shown for simplicity. From here the expression for the lattice, heat capacity comes as

$$C_v = \frac{N_0\Omega\hbar^2}{2\pi^2 k_B T^2}\sum_s \frac{1}{c_s^3}\int_0^{\omega_D} \mathrm{d}\omega_s \omega_s^4 \bar{n}_s(\bar{n}_s+1). \tag{2.132}$$

By introducing dimensionless quantities $z = \hbar\omega_s/k_B T$ and $z_D = \Theta_D/T$, and by using equations (2.94) and (2.97), we can rewrite equations (2.131) and (2.132) as

$$E = 9Nk_B T\left(\frac{T}{\Theta_D}\right)^3\int_0^{z_D} \mathrm{d}z\frac{z^3}{e^z-1} \tag{2.133}$$

$$C_v = 9Nk_B\left(\frac{T}{\Theta_D}\right)^3\int_0^{z_D} \mathrm{d}z\frac{z^4 e^z}{(e^z-1)^2}. \tag{2.134}$$

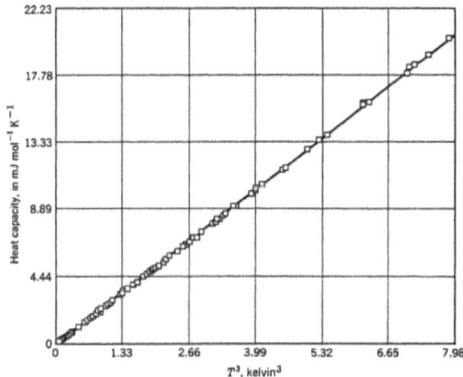

Figure 2.20   The low-temperature heat capacity of solid argon compared with the Debye T^3 prediction with $\Theta_D = 92$ K (solid line). (Reproduced from Kittel (1986).)

At high temperatures $z \ll 1$ and $e^z - 1 \simeq z$, so it can be easily seen that $E = 3Nk_BT$ and $C_v = 3Nk_B$, in agreement with the classical result. At low temperatures $z \gg 1$ so that z_D can be taken as ∞ and

$$
\begin{aligned}
E\big|_{LT} &= 9Nk_BT \left(\frac{T}{\Theta_D}\right)^3 \int_0^\infty dz \frac{z^3}{e^z - 1} \\
&= 9Nk_BT \left(\frac{T}{\Theta_D}\right)^3 \Gamma(4)\zeta(4) \\
&= \frac{3\pi^4}{5}Nk_BT \left(\frac{T}{\Theta_D}\right)^3,
\end{aligned}
\tag{2.135}
$$

where ζ is the Riemann zeta function: $\zeta(4) = \pi^4/90$. Differentiating equation (2.135) with respect to temperature we get

$$
\begin{aligned}
C_v\big|_{LT} &= \frac{12\pi^4}{5}Nk_B \left(\frac{T}{\Theta_D}\right)^3 \\
&\propto T^3.
\end{aligned}
\tag{2.136}
$$

Thus the Debye theory successfully explains both the low- and high-temperature dependence of the lattice heat capacity of insulating crystals. The result in equation (2.136) also indicates that the smaller the Debye temperature, the larger the low-temperature heat capacity of a solid would be. An excellent fit of the expression in equation (2.134) to the low-temperature heat capacity for solid argon is shown in figure 2.20.

It should be emphasized that real crystals cannot be adequately treated as an elastic continuum and their atomic nature must be considered. But then an accurate calculation of C_v becomes quite involved as it necessitates computation of $g(\omega)$ using a realistic lattice dynamical model. If instead the Debye formula for C_v is preferred, then it becomes necessary to treat Θ_D as a single empirical parameter in equation (2.134). A reasonable way to choose the Debye temperature as a temperature-dependent parameter $\Theta_D(T)$ is to fit equation (2.134) with the observed heat capacity. A list of Debye temperatures for a number of elements, determined by fitting the observed heat capacity to the Debye formula in equation (2.134) at the temperature where the heat capacity is half the Dulong–Petit value ($\frac{3}{2}Nk_B$), has been presented by de Launay (1956).

2.8   ELASTIC WAVES IN CUBIC CRYSTALS

The theory of elastic waves in crystals is very well described in many books and articles (see, e.g., de Launay 1956, Drabble 1966, Brüesch 1982). We will use a standard notation for elastic constants to express the Hamiltonian of an anharmonic elastic continuum in section 4.5. Here we simply mention that the equations of elastic waves in a three-dimensional crystal can be derived by appropriately extending the ideas in equation (2.11). It can be shown that in the long-wavelength limit equation (2.44) can be expressed as

$$\rho \ddot{u}_\alpha = \sum_{\beta\gamma\delta} C_{\alpha\beta,\gamma\delta} \frac{\partial^2 u_\gamma}{\partial r_\delta \partial r_\beta}, \tag{2.137}$$

where ρ is the mass density and $C_{\alpha\beta,\gamma\delta}$ are the second-order elastic constants of the crystal.

Using the trial function

$$\boldsymbol{u} = \boldsymbol{U} \exp[i(\boldsymbol{q}\cdot\boldsymbol{r} - \omega t)] \tag{2.138}$$

we can express equation (2.137) as

$$\sum_\beta \left(\sum_{\gamma\delta} C_{\alpha\gamma,\beta\delta} q_\gamma q_\delta - \delta_{\alpha\beta}\rho\omega^2 \right) u_\beta = 0. \tag{2.139}$$

These are known as the *Green–Christoffel equations*, whose non-trivial solutions are obtained by solving the secular determinant

$$\left| \sum_{\gamma\delta} C_{\alpha\gamma,\beta\delta} q_\gamma q_\delta - \delta_{\alpha\beta}\rho\omega^2 \right| = 0. \tag{2.140}$$

A comparison of the solutions of equation (2.140) with the solutions of equation (2.50) in the long-wavelength limit can be made to relate the atomic force constants to the elastic constants.

The elastic tensor $C_{\alpha\beta,\gamma\delta}$ is symmetric under the transpositions $\alpha \leftrightarrow \beta$, $\gamma \leftrightarrow \delta$ and $\alpha\beta \leftrightarrow \gamma\delta$, so that there are only 21 independent components. It is usual to use a simplified notation for the elastic constants by contracting the indices as follows

$$11 \rightarrow 1, 22 \rightarrow 2, 33 \rightarrow 3, 23, 32 \rightarrow 4, 31, 13 \rightarrow 5, 12, 21 \rightarrow 6. \tag{2.141}$$

In the contracted notation, the following relations between the elastic constants of *cubic crystals* hold: $c_{11} = c_{22} = c_{33}$, $c_{12} = c_{13} = c_{23} = c_{31} = c_{32}$, $c_{44} = c_{55} = c_{66}$, and others are zero. Out of these c_{11}, c_{12} and c_{44} are generally considered as the independent second-order elastic constants.

If u_x, u_y, u_z represent the x,y,z components of atomic displacement in a *cubic crystal*, then the equations of the elastic waves (equation (2.137)) become

$$\rho \frac{d^2 u_x}{dt^2} = c_{11} \frac{\partial^2 u_x}{\partial x^2} + c_{44}\left(\frac{\partial^2 u_x}{\partial y^2} + \frac{\partial^2 u_x}{\partial z^2} \right) + (c_{12}+c_{44})\left(\frac{\partial^2 u_y}{\partial x \partial y} + \frac{\partial^2 u_z}{\partial x \partial z} \right) \tag{2.142}$$

and similar equations for u_y and u_z obtained by cyclic permutation of (u_x, u_y, u_z) and (x, y, z). The secular determinant in equation (2.140) now reads

$$\begin{vmatrix} c_{11}q_x^2 + c_{44}(q_y^2+q_z^2) - \rho\omega^2 & (c_{12}+c_{44})q_xq_y & (c_{12}+c_{44})q_xq_z \\ (c_{12}+c_{44})q_xq_y & c_{11}q_y^2 + c_{44}(q_z^2+q_x^2) - \rho\omega^2 & (c_{12}+c_{44})q_yq_z \\ (c_{12}+c_{44})q_xq_z & (c_{12}+c_{44})q_yq_z & c_{11}q_z^2 + c_{44}(q_x^2+q_y^2) - \rho\omega^2 \end{vmatrix} = 0. \tag{2.143}$$

The solutions of this equation give the frequencies ω and hence the sound velocity v in terms of the elastic constants c_{PQ}.

Now let us obtain the results for the velocities of pure longitudinal and pure transverse elastic waves along the symmetry directions [100], [110] and [111] in cubic crystals.

For a wave along [100], i.e. $\boldsymbol{q} \parallel [100]$, we can try a solution $u_x = U_x \exp[i(q_x x - \omega t)], u_y = u_z = 0$. From equation (2.143) the speed of the longitudinal wave is

$$v_L[100] = \frac{\omega}{q} = \sqrt{\frac{c_{11}}{\rho}}. \tag{2.144}$$

To obtain the speed of a transverse wave, we may try a solution $u_y = U_y \exp[i(q_y y - \omega t)], u_x = u_z = 0$, or we may equally well try $u_z = U_z \exp[i(q_z z - \omega t)], u_x = u_y = 0$. In both cases equation (2.143) yields the same speed of transverse wave

$$v_T[100] = \sqrt{\frac{c_{44}}{\rho}}. \tag{2.145}$$

Thus for a wave propagation along the x axis, we have a longitudinal wave with particle displacement along the x axis, and two identical but independent transverse waves with particle displacements along the y and z axes.

For a wave along [110], $q_x = q_y = q/\sqrt{2}$. In this case a little bit of algebra is required to determine the direction of atomic displacements. The longitudinal wave with atomic displacement along [110] has speed

$$v_L[110] = \sqrt{\frac{c_{11} + c_{12} + 2c_{44}}{2\rho}}. \tag{2.146}$$

However, there are two non-degenerate transverse waves with speeds

$$\text{atomic displacement along } [001]: \quad v_{T_1}[110] = \sqrt{\frac{c_{44}}{\rho}} \tag{2.147}$$

$$\text{atomic displacement along } [1\bar{1}0]: \quad v_{T_2}[110] = \sqrt{\frac{c_{11} - c_{12}}{2\rho}}. \tag{2.148}$$

For a wave along [111], $q_x = q_y = q_z = q/\sqrt{3}$. In this case the longitudinal wave has speed

$$v_L[111] = \sqrt{\frac{c_{11} + 2c_{12} + 4c_{44}}{3\rho}}. \tag{2.149}$$

The two transverse waves have identical speed:

$$v_T[111] = \sqrt{\frac{c_{11} - c_{12} + c_{44}}{3\rho}} \tag{2.150}$$

and correspond to atomic displacements along $[1\bar{1}0]$ and $[\bar{1}\bar{1}2]$.

In directions other than [100], [110] and [111] the waves in a cubic crystal crystal cannot all be interpreted as purely longitudinal and purely transverse unless the material is isotropic. The isotropy of a material requires that the transverse velocity be the same in all directions. In the case of cubic crystals we get, from equations (2.145), (2.147)–(2.148) and (2.150),

$$c_{11} - c_{12} = 2c_{44}. \tag{2.151}$$

If each atom of an unstrained crystal is at a centre of inversion (such as in a monatomic crystal) and if the interatomic forces are central, then there exist certain relations between the elastic constants of the crystal. These are known as the *Cauchy relations*. For cubic crystals there is only one such relation:

$$c_{12} = c_{44}. \tag{2.152}$$

Thus a monatomic cubic crystal is isotropic if

$$c_{11} = 3c_{44} \tag{2.153}$$

and then

$$v_L = \sqrt{3} v_T = \sqrt{\frac{3c_{44}}{\rho}}. \tag{2.154}$$

The elastic constants of crystals with diamond and zincblende structures, which are characterised by the presence of non-central forces due to the formation of directional bonds, do not in general meet the isotropy condition in equation (2.151) and do not satisfy the Cauchy relation.

3 Lattice Dynamics in the Harmonic Approximation – *Ab initio* Treatment

3.1 INTRODUCTION

All phenomenological theories of lattice dynamics suffer from at least two problems: they are not necessarily applicable to all types of solid and their parameters do not necessarily contain conceptual simplicity. It was felt in the early 1970s that the number of phenomenological models had gone to the extent that one wanted to see the development of an *a priori* quantum mechanical calculation of phonon spectra. This is especially true for semiconductors and insulators. In 1974 Sham described several semi-phenomenological models (Sham 1974) which give a flavour of what should be involved in a first-principles calculation of phonon frequencies in non-metals.

In a calculation of the phonon spectra of a solid, the direct ion–ion interaction (described by a central force model) must be corrected by including the effect of indirect ion–ion interaction (also known as ion–electron–ion interaction, or electronic polarisation). In phenomenological models, this is done in an *ad hoc* way. An *ab initio* calculation requires an accurate and parameter less knowledge of the microscopic electronic response to frozen-in lattice vibrations. Such a calculation is relatively easier for simple metals with only s and p screening electrons and with rather unimportant local-field effects. However, for transition metals, semiconductors and insulators local-field effects are very important and an *ab initio* calculation of phonon spectra in such cases must accurately account for these effects.

The basic idea is to determine force constants via the total energy of the crystal under investigation. (Here the term crystal energy refers to the total energy of a system of frozen nuclei.) Within the adiabatic approximation, valence electrons respond instantaneously to nuclear motion and the total energy of the system can therefore be considered as a function only of the positions $\boldsymbol{R}$ of the ions,

$$E_{tot}(\boldsymbol{R}) = E_{ion-ion}(\boldsymbol{R}) + E_{el}(\boldsymbol{R}). \tag{3.1}$$

Here $E_{ion-ion}$ is the direct ion–ion interaction energy, and E_{el} is the total energy of the (valence) electrons moving in the potential field of ions, including quantum mechanical kinetic, exchange and correlation energies. The equilibrium internuclear distance R_0 is determined by minimisation of the total energy, for which the force on every atom is zero. Phonon energies are determined from $E_{tot}(\boldsymbol{R})$ for $\boldsymbol{R}$ away from equilibrium.

There are two commonly used approaches for the calculation of $E_{tot}(\boldsymbol{R})$: (i) direct methods and (ii) linear response methods. In this chapter we present the underlying theory for both the direct and the linear response approaches. The coverage of the subject is as follows. In section 3.2 we will describe total energy and force calculations based on the application of *density functional theory* (DFT) and the plane wave pseudopotential formalism. In section 3.3 we will discuss evaluation of the harmonic force constants and phonon eigensolutions using a few variants of both the direct and linear response methods, including the *density functional perturbation theory* (DFPT).

3.2 TOTAL ENERGY AND FORCES

In the direct method, calculation of the electronic part of the total energy, E_{el}, is sought from a solution of the Schrödinger equation for electrons. This is a very difficult problem, and indeed numerous

DOI: 10.1201/9781003141273-3

attempts have been made to solve it since the early days of solid state physics: first attempts were made by Wigner and Seitz (1933, 1934) and by Fuchs (1935). We can categorise the direct method into two approaches: (a) real space calculations and (b) momentum space formalism. Real space calculations for the total energy of solids include direct solutions of the Schrödinger equation in the Hartree–Fock, density functional or model potential approximations. A major difficulty in these calculations is in dealing with divergent terms in the expression for the total energy. This has led to a number of 'shape' approximations for the electronic charge density $\rho(r)$; for example, muffin-tin or cellular approximations. A simple muffin-tin approximation produces unsatisfactory results for open structure solids like tetrahedrally bonded semiconductors and insulators, but non-muffin-tin corrections are quite complicated.

A direct calculation of the total energy from a self-consistent solution of the Schrödinger equation within a pseudopotential scheme has been presented by Ihm *et al* (1979). This method uses the momentum space formalism and does not use any shape approximation to the electronic charge density $\rho(r)$. More importantly, the formalism results into a drastic simplification of the total energy expression. Although their formalism is particularly designed to be applicable within a pseudopotential scheme with a plane wave basis set, it can be extended to mixed basis sets as well (e.g. plane waves plus Gaussians). In the following we describe the underlying theory of this method.

3.2.1 GENERAL CONSIDERATION

We begin by explicitly writing the crystal Hamiltonian $\mathscr{H}$ as

$$\begin{aligned} \mathscr{H} &= T+V \\ &= T_{\text{ion}} + T_{\text{el}} + V_{\text{ion--ion}} + V_{\text{el--el}} + V_{\text{ion--el}}, \end{aligned} \tag{3.2}$$

where ionic and electronic contributions to the kinetic energy T and potential energy V are indicated by subscripts. The total crystal energy E_{tot} is the expectation value of the Hamiltonian operator $\mathscr{H}$ which satisfies the many-body Schrödinger equation

$$\mathscr{H}\eta(\boldsymbol{R},\boldsymbol{r}) = E_{\text{tot}}\eta(\boldsymbol{R},\boldsymbol{r}), \tag{3.3}$$

where $\eta(\boldsymbol{R},\boldsymbol{r})$ is the wavefunction of the crystal with ions at $\{\boldsymbol{R}\}$ and electrons at $\{\boldsymbol{r}\}$. Assuming that electrons adiabatically follow ions, we use the Born-Oppenheimer approximation and express

$$\eta(\boldsymbol{R},\boldsymbol{r}) = \chi(\boldsymbol{R})\Psi(\boldsymbol{R},\boldsymbol{r}), \tag{3.4}$$

where χ and Ψ are ionic and electronic wavefunctions, respectively. This allows us to express

$$T_{\text{ion}}\eta(\boldsymbol{R},\boldsymbol{r}) \simeq \Psi(\boldsymbol{R},\boldsymbol{r})T_{\text{ion}}\chi(\boldsymbol{R}) \tag{3.5}$$

and thus split equation (3.3) in the form of two inter-linked equations

$$[T_{\text{ion}} + V_{\text{ion--ion}} + E_{\text{el}}(\boldsymbol{R})]\chi(\boldsymbol{R}) = E_{\text{ion}}\chi(\boldsymbol{R}) \tag{3.6}$$

and

$$[T_{\text{el}} + V_{\text{el--ion}} + V_{\text{el--el}}]\Psi(\boldsymbol{R},\boldsymbol{r}) = E_{\text{el}}(\boldsymbol{R})\Psi(\boldsymbol{R},\boldsymbol{r}). \tag{3.7}$$

While the terms T_{ion} and $V_{\text{ion--ion}}$ can be considered classically and if $V_{\text{el--ion}}$ is taken in the Coulombic electron-nuclear form, then the many-body Schrödinger equation in (3.7) must be solved carefully for obtaining the remaining terms. If, as discussed later, $V_{\text{el--ion}}$ is considered in some other form, then it also may be required to be evaluated by solving a many-body Schrödinger equation. Different choices for $V_{\text{el--ion}}$ and $\Psi(\boldsymbol{R},\boldsymbol{r})$ have led to different schemes for solving the electronic many-body Schrödinger equation in (3.7). Two broad categories of formalisms have been adopted for the consideration of $\Psi(\boldsymbol{R},\boldsymbol{r})$: *wavefunction formalisms* such as the Hartree-Fock method, and *density functional formalisms*.

An example of the application of the Hartree–Fock based wavefunction approach can be found in a recent publication (Dovesi *et al* 2020). The concept of density functional started in the late 1920s and early 1930s, with the development of the Thomas–Fermi–Dirac–Wigner scheme (see Appendix A). In the Thomas–Fermi model it is assumed that electrons move independently in an effective electrostatic potential, obtained from solving Poisson's equation and requiring that chemical potential is constant (Thomas 1927, Fermi 1928). The kinetic energy of the electrons is approximated as the kinetic energy of a system of non-interacting electrons with a spatially slowly varying charge density $n(\boldsymbol{r})$. Dirac (1930) added the concept of exchange to the Thomas–Fermi theory. Later, the electron correlation contribution, *i.e.* Coulomb repulsion between antiparallel spin electrons, was added by Wigner (1934). A fundamentally new and robust density functional scheme was developed by Kohn and co-workers in mid 1960s. The resultant Hohenberg–Kohn–Sham formalism of the density functional scheme (Hohenberg and Kohn 1964, Kohn and Sham 1965) is found to be far more successful in describing electronic, optical and vibrational properties of solids. It has gained unparallel recognition and is simply known as the *density functional theory* (DFT). Here we will describe how the Hohenberg–Kohn–Sham DFT is applied to solve equation (3.7) for the electronic part of the crystal total energy.

3.2.2 DENSITY FUNCTIONAL THEORY

The fundamentals of the DFT and derivations of important quantities are described in Appendix A. In essence, this theory replaces the consideration of an N-electron wavefunction $\Psi(\boldsymbol{r}_1,\boldsymbol{r}_2,...,\boldsymbol{r}_N)$ in equation (3.7) with that of the electronic charge density $n(\boldsymbol{r})$ defined as

$$n(\boldsymbol{r}) = N \int d\boldsymbol{r}_2... \int d\boldsymbol{r}_N \Psi^*(\boldsymbol{r},\boldsymbol{r}_2,...,\boldsymbol{r}_N)\Psi(\boldsymbol{r},\boldsymbol{r}_2,...,\boldsymbol{r}_N) \tag{3.8}$$

for a chosen ionic configuration $\{\boldsymbol{R}\}$. In other words, $\Psi(\boldsymbol{r}_1,\boldsymbol{r}_2,...,\boldsymbol{r}_N)$ is a unique functional of $n(\boldsymbol{r})$:

$$\Psi(\boldsymbol{r}_1,\boldsymbol{r}_2,...,\boldsymbol{r}_N) \to \Psi[n(\boldsymbol{r})], \tag{3.9}$$

which, however, is not known. The total energy of the electronic system takes the lowest value for the correct ground-state value $n(\boldsymbol{r}) = \rho(\boldsymbol{r})$ of the charge density. As explained in Appendix A, the ground-state electronic energy, when the system is subjected to an external potential $V_{\text{ext}}(\boldsymbol{r})$, can be expressed as

$$E_{\text{el}} = T_0[\rho] + \int d\boldsymbol{r} V_{\text{ext}}(\boldsymbol{r})\rho(\boldsymbol{r}) + \int d\boldsymbol{r} V_{\text{H}}(\boldsymbol{r})\rho(\boldsymbol{r}) + \int d\boldsymbol{r} V_{\text{xc}}(\boldsymbol{r})\rho(\boldsymbol{r}) \tag{3.10}$$

with the various terms explained below.

The ground-state electronic charge density is obtained as

$$\rho(\boldsymbol{r}) = 2\sum_j^{\text{occ}} |\psi_j(\boldsymbol{r})|^2, \tag{3.11}$$

where the factor 2 accounts for the spin degeneracy of the electron states and $\psi_j(\boldsymbol{r})$ is the jth occupied wave function of a system of *fictitious non-interacting particles* which satisfies the following Schrödinger-like equation, called the Kohn–Sham equation

$$\left(-\frac{\hbar^2}{2m}\nabla^2 + V_{\text{KS}}(\boldsymbol{r})\right)\psi_j(\boldsymbol{r}) = \varepsilon_j\psi_j(\boldsymbol{r}), \tag{3.12}$$

with ε_j representing the corresponding eigenvalue. The Kohn–Sham potential $V_{KS}(r)$ in equation (3.12) is

$$V_{KS}(r) = V_{ext}(r) + V_H(r) + V_{xc}(r), \qquad (3.13)$$

where V_H is the Hartree potential and V_{xc} is the many-body exchange-correlation potential. V_{xc} is usually unknown and various levels of approximation have been made to express it as a functional of ρ, some of which are discussed in Appendix A. The term $T_0[\rho]$ in equation (3.10) is the kinetic energy of the system of *fictitious non-interacting particles*

$$T_0[\rho] = \sum_j^{occ} \varepsilon_j - \int dr V_{KS}(r)\rho(r). \qquad (3.14)$$

Equation (3.12) must be solved self-consistently by making a plausible choice for ψ_j and utilizing an acceptable functional form of V_{xc}.

Before proceeding, it should be noted that the Kohn–Sham eigensolutions (ψ_j and ε_j) should not not be directly identified as the electronic eigensolutions of the system under study. The only meaningful interpretation of Kohn–Sham eigenfunctions $\{\psi_j\}$ comes via equation (3.11) which defines the electronic charge density of the system. The Kohn–Sham eigenvalues $\{\varepsilon_j\}$ are merely Lagrange multipliers for the application of the Kohn–Sham variational principle (see Appendix A), satisfying the requirement that the total number of states with $\varepsilon_j \leq \mu$ equals the number of electrons, subject to the orthonormality of ψ_j. Only the highest occupied ε_j of a large system is meaningful: being equal to the chemical potential μ it represents the ionisation energy of the system (or the work function of a metal).

3.2.3 MOMENTUM SPACE FORMULATION OF ELECTRONIC BAND STRUCTURE

The external potential V_{ext} is the potential electrons in the system are subjected to. For a system in its ground state in absence of any external perturbation, V_{ext} is the potential exerted by the ions(nuclei) on the valence(any) electron. For most purposes it is sufficient to consider an atom as an ion plus valence electrons. This helps us replace V_{ext} by a weak pseudo-potential V_{ps} and the highly oscillatory wavefunction ψ by a smooth function ϕ, known as pseudo wavefunction. These considerations are detailed in Appendix B.

3.2.3.1 Kohn–Sham equations using plane wave pseudopotential method

A pseudo wavefunction ϕ can be built using either an atomic orbital basis (such as Slater orbitals or Gaussian orbitals) or plane waves, the latter being more common. Using a plane wave basis, we express $\phi_j(r)$ for an electron of wave vector k as

$$\phi_{j,k}(r) = \frac{1}{\sqrt{N_0\Omega}} \sum_G A_j(k+G) e^{i(k+G)\cdot r}, \qquad (3.15)$$

where $N_0\Omega$ is the volume of the crystal containing N_0 unit cells, G is a reciprocal lattice vector and $A_j(k+G)$ is a Fourier component of the pseudo wavefunction.

A *non-local* ionic pseudopotential is energy dependent and shows space non-locality, *i.e.* it is expressed as $v(r,r';E)$. As explained in Appendix B, in a small energy range we can express it in a *semi-local* form

$$\hat{v}_{ps}(r) = \sum_\ell \mathscr{P}_\ell^* v_\ell(|r|) \mathscr{P}_\ell, \qquad (3.16)$$

where $\mathscr{P}_\ell$ is a projection operator which projects out the ℓth angular momentum component of the core wave function. The total *non-local* pseudopotential for a crystal can be constructed as a sum of non-overlapping ionic pseudopotentials,

$$V_{ps}(\boldsymbol{r}) = \sum_{\boldsymbol{p}} \sum_{b} \hat{v}_b(\boldsymbol{r} - \boldsymbol{p} - \boldsymbol{\tau}_b), \tag{3.17}$$

where $\boldsymbol{p}$ is a Bravais lattice vector and $\boldsymbol{\tau}_b$ is the position of the bth ion in the unit cell. A Fourier component of $V_{ps}(\boldsymbol{r})$ can be expressed as (see Appendix B)

$$V_{ps}(\boldsymbol{k} + \boldsymbol{G}, \boldsymbol{k} + \boldsymbol{G}') = \frac{1}{M} \sum_{b} S_b(\boldsymbol{G} - \boldsymbol{G}') \sum_{\ell} v_{b,\ell}(\boldsymbol{k} + \boldsymbol{G}, \boldsymbol{k} + \boldsymbol{G}'), \tag{3.18}$$

where M is the number of ions (atoms) in the unit cell,

$$S_b(\boldsymbol{G} - \boldsymbol{G}') = e^{i(\boldsymbol{G} - \boldsymbol{G}') \cdot \boldsymbol{\tau}_b} \tag{3.19}$$

is the structure factor for the bth atom at position $\boldsymbol{\tau}_b$, and

$$v_{b,\ell}(\boldsymbol{k} + \boldsymbol{G}, \boldsymbol{k} + \boldsymbol{G}') = \frac{1}{\Omega_{at}} \int d\boldsymbol{r} \, e^{-i(\boldsymbol{k} + \boldsymbol{G}') \cdot \boldsymbol{r}} \hat{v}_b(\boldsymbol{r}) e^{i(\boldsymbol{k} + \boldsymbol{G}) \cdot \boldsymbol{r}} \tag{3.20}$$

is the ℓth component of the form factor for the bth atom of volume Ω_{at}.

For a *local* pseudopotential $v_b(\boldsymbol{r}) = v_b(|\boldsymbol{r}|)$ and equation (3.18) reduces to

$$V_{ps}^{L}(\boldsymbol{G}) = \frac{1}{M} \sum_{b} S_b(\boldsymbol{G}) v_b(|\boldsymbol{G}|) \tag{3.21}$$

with $S_b(\boldsymbol{G})$ given by equation (3.19) and

$$v_b(|\boldsymbol{G}|) = \frac{1}{\Omega_{at}} \int d\boldsymbol{r} \, v_b(|\boldsymbol{r}|) e^{i\boldsymbol{G} \cdot \boldsymbol{r}}. \tag{3.22}$$

Using the plane-wave pseudopotential scheme described above, we can easily express the Kohn–Sham equation in (3.12) in momentum space (see, e.g., Appendix B)

$$\sum_{\boldsymbol{G}'} \left[\left(\frac{\hbar^2}{2m} (\boldsymbol{k} + \boldsymbol{G})^2 - \varepsilon_j \right) \delta_{\boldsymbol{G}, \boldsymbol{G}'} + V_{ps}(\boldsymbol{k} + \boldsymbol{G}, \boldsymbol{k} + \boldsymbol{G}') \right.$$
$$\left. + V_H(\boldsymbol{G}' - \boldsymbol{G}) + V_{xc}(\boldsymbol{G}' - \boldsymbol{G}) \right] A_j(\boldsymbol{k} + \boldsymbol{G}') = 0. \tag{3.23}$$

This is now a matrix eigenvalue equation which must be solved self-consistently for the eigenvalues ε_j and the eigenvectors $A_j(\boldsymbol{k} + \boldsymbol{G})$. From the eigenvectors, a Fourier component of the charge density $\rho(\boldsymbol{G})$ can be calculated:

$$\rho(\boldsymbol{G}) = \sum_{\boldsymbol{k}}^{\text{BZ}} \rho_{\boldsymbol{k}}(\boldsymbol{G}) \tag{3.24}$$

with

$$\rho_{\boldsymbol{k}}(\boldsymbol{G}) = 2 \sum_{j}^{\text{occ}} \sum_{\boldsymbol{G}_1} \sum_{\boldsymbol{G}_2} A_j^*(\boldsymbol{k} + \boldsymbol{G}_2) A_j(\boldsymbol{k} + \boldsymbol{G}_1) \delta_{\boldsymbol{G}_1 - \boldsymbol{G}_2, \boldsymbol{G}}, \tag{3.25}$$

where the factor 2 accounts for the spin degeneracy of the electronic states, and $\rho(\boldsymbol{G} = 0)$ is normalised to the number of valence electrons in the unit cell. A Fourier component of the Coulomb (Hartree) potential $H_H(\boldsymbol{G})$ is obtained by solving the necessary Poisson equation:

$$V_H(\boldsymbol{G}) = \frac{4\pi e^2 \rho(\boldsymbol{G})}{G^2}. \tag{3.26}$$

The term $V_{xc}(G)$ is the Fourier transform of an effective one-particle exchange-correlation potential $V_{xc}(r)$ defined within the DFT as the functional derivative of the exchange-correlation energy $E_{xc}[\rho[r]]$ (Kohn and Sham 1965)

$$V_{xc}(r) = \frac{\delta E_{xc}[\rho(r)]}{\delta \rho(r)}. \tag{3.27}$$

While there is no simple exact expression for $E_{xc}[\rho(r)]$, for slowly varying $\rho(r)$ one can write (Hohenberg and Kohn 1964)

$$E_{xc} = \int dr \rho(r) \varepsilon_{xc}[\rho(r)], \tag{3.28}$$

where $\varepsilon_{xc}[\rho(r)]$ is the energy per particle of of homogeneous electron gas with density $\rho(r)$ in the neighbourhood of r. Within this *local density approximation* (LDA), the expression for the exchange-correlation potential becomes

$$V_{xc} = \frac{d(\rho \varepsilon_{xc})}{d\rho} = \varepsilon_{xc} + \rho \frac{d\varepsilon_{xc}}{d\rho}. \tag{3.29}$$

In order to account for the inhomogeneous density distribution in real solids, modifications to the LDA are required. Several improvements over the LDA have been proposed. The most commonly used is the so-called *generalised gradient approximation* (GGA). A brief discussion of $E_{xc}[\rho(r)]$ and $V_{xc}(r)$ within the LDA and GGA has been presented in Appendix A. Here we will continue within the LDA.

3.2.3.2 Special k-points method for Brillouin zone summation

The amount of labour required to do the Brillouin zone summation in equation (3.25) can be reduced by using a few special k-points in the irreducible part of the Brillouin zone of the system. The special k-points scheme uses a concept which is based on the use of group symmetry considerations.

Baldereschi (1973), Chadi and Cohen (1973), Cunningham (1974) and Monkhorst and Pack (1976) have developed a method for determining 'special' sets of k-points in the irreducible segment of the Brillouin zone which are 'most efficient' in averaging periodic functions. By 'most efficient' one means determining the average to a high degree of accuracy with a modest number of sampled k-points. It is found that Baldereschi's mean-value k-point, or Chadi and Cohen's two special k-points, are adequate for a reasonably good representation of the electronic charge density, total energy and forces in semiconductors with diamond and zincblende structures. A good description of special k-points for crystals of different symmetries and their use in the calculation of electronic structure and related physical properties is given in Evarestov and Smirnov (1983) where references to other works can also be found.

To understand the use of the special k-points scheme for Brillouin zone summation let us consider evaluation of electronic charge density $\rho(r)$ at a point r in the system. For a periodic system, we express

$$\begin{aligned} \rho(r) &= \sum_G \rho(G) e^{iG \cdot r} \\ &= \sum_G \left[\sum_k^{BZ} \rho(k, G) \right] e^{iG \cdot r}, \end{aligned} \tag{3.30}$$

with the Fourier component $\rho(k, G)$ for a k point inside the Brillouin zone given in equation (3.24). Consideration of group theory can be made to reduce the Brillouin zone integration to a summation involving a few special k points inside the irreducible part of the zone. We will explain this below. [1]

---

[1] An alternative procedure for obtaining symmetrized electronic charge is described in section 3.2.5.

We first consider replacing the infinite k-sum (or integration) in equation (3.30) by a finite sum using N_k k points inside the Brillouin zone.

$$\rho(r) \simeq \frac{1}{N_k} \sum_{k(\text{BZ})}^{N_k} \sum_G \rho(k,G) e^{iG\cdot r}. \tag{3.31}$$

Then, using equations (1.35) – (1.37) for the effect of a space group operation $\hat{O} = \{\mathscr{R} \mid t\}$ on a plane wave, we can express

$$
\begin{aligned}
\rho(r) &\simeq \frac{1}{M} \sum_{\{\mathscr{O}_j\}}^{M} \frac{M}{N_k} \mathscr{O}_j \sum_{k(\text{IBZ})}^{N_k/M} \sum_G \rho(k,G) e^{iG\cdot r} \\
&= \frac{1}{M} \sum_m^{M} \frac{1}{L} \sum_{k(IBZ)}^{L} \sum_G \rho(k,G) e^{iG\cdot\mathscr{R}_m^{-1}r - iG\cdot\mathscr{R}_m^{-1}t_m} \\
&= \frac{1}{M} \sum_m^{M} \frac{1}{L} \sum_{k(IBZ)}^{L} \sum_G \rho(k,G) e^{i\mathscr{R}_m G\cdot r} e^{-i\mathscr{R}_m G\cdot t_m} \\
&= \frac{1}{L} \sum_G \sum_{k(IBZ)}^{L} \left[\frac{1}{M} \sum_m^{M} \rho(k,\mathscr{R}_m^{-1}G) e^{-iG\cdot t_m} \right] e^{iG\cdot r} \\
&= \frac{1}{L} \sum_G \sum_{k(IBZ)}^{L} \rho_{\text{sym}}(k,G) e^{iG\cdot r},
\end{aligned}
\tag{3.32}
$$

where M is the order of the symmetry group and $L = N_k/M$ is the number of k points in the irreducible part of the Brillouin zone (IBZ). Also, in the above we have made use of the fact that the sets $\{G\}$ and $\{\mathscr{R}_m^{-1}G\}$ are the same. The Fourier component $\rho_{\text{symm}}(k,G)$ is the symmetrized charge density for the k point in the IBZ:

$$\rho_{\text{sym}}(k,G) = \frac{1}{M} \sum_m^{M} \rho(k,\mathscr{R}_m^{-1}G) e^{-iG\cdot t_m}. \tag{3.33}$$

Note that for symmorphoic space groups there are no fractional translations ($t_m = 0$) and hence $\exp(-iG.t_m) = 1$.

The average $\frac{1}{L}\sum_{k(\text{IBZ})}^{L} \rho_{\text{sym}}(k,G)$ over the IBZ can be obtained by choosing a finite set of special k points $\{k_i\}$:

$$\frac{1}{L} \sum_{k(\text{IBZ})}^{L} \rho_{\text{sym}}(k,G) = \sum_{k_i} w(k_i) \rho_{\text{sym}}(k_i,G), \tag{3.34}$$

where $\{w(k_i)\}$ are non-negative weight factors associated with $\{k_i\}$ satisfying the normalisation condition

$$\sum_i w(k_i) = 1. \tag{3.35}$$

In the (extended) Monkhorst and Pack scheme (Monkhorst and Pack 1976, Fehlner and Vosko 1977) special k points are considered in the form

$$k = u_1 b_1 + u_2 b_2 + u_3 b_3 + k_0, \tag{3.36}$$

where b_i are the primitive translation vectors of the reciprocal lattice and k_0 is an arbitrary choice. The coefficients u_i are chosen as follows:

for cubic Bravis lattices

$$u_i = \frac{2p_i - l_0 - 1}{2l_0}; \qquad p_i = 1,2,3,...,l_0; \qquad i = 1,2,3. \tag{3.37}$$

for hexagonal Bravais lattices

$$u_i = \frac{p_i - 1}{l_0}; \qquad p_i = 1, 2, 3, ..., l_0; \qquad i = 1, 2$$

$$u_3 = \frac{2p_3 - l_0 - 1}{2l_0}; \qquad p_3 = 1, 2, 3, ..., l_0. \tag{3.38}$$

The integer l_0 in the above description determines the number of special points in the set. The weight $w(\mathbf{k}_i)$ associated with $\mathbf{k}_i$ is simply the ratio of the order of the entire point group to the order of the group of the wave vector $\mathbf{k}_i$, and is used following the normalization condition in equation (3.35). When using this scheme, it is usual practice to speak of the $(u_1 \times u_2 \times u_3)$ $\mathbf{k}$-points mesh.

3.2.3.3 Self-consistent solutions of Kohn–Sham equations

As the Kohn–Sham potential $V_{KS}(\mathbf{r})$ is obtained from a knowledge of the Kohn–Sham orbitals $\{\psi(\mathbf{r})\}$, it is imperative to seek a self-consistent solution of the Kohn–Sham equation. Indeed, the same consideration miust be made when solving any many-body Schrödiner equation. In the present context we shall consider the momentum space version of the Kohn–Sham equation in (3.23). Several iterative schemes have been adopted for obtaining self-consistency between the charge density ρ and the screening potential $V_{scr} = V_H + V_{xc}$.

Let, at the mth iteration, input and output $V_{scr}(\mathbf{G})$ components be denoted as vectors $V_{in}^{(m)}$ and $V_{out}^{(m)}$, respectively. In the simple linear mixing scheme the screening input to $(m+1)$th iteration is taken as a linear mix of the input and output at the mth iteration:

$$V_{in}^{(m+1)} = (1 - \alpha)V_{in}^{(m)} + \alpha V_{out}^{(m)}, \tag{3.39}$$

with a judicious choice for the parameter α in the range $0 < \alpha < 1$. A choice close to $\alpha = 0.5$ works reasonably well for systems with small-size unit cells for which the smallest reciprocal vector $\mathbf{G}$ entering the screening potential is not too small. For system with large unit cells (e.g. in surface calculations based on the use of an artificially large unit cell) for which the screening potential, particularly the Hartree term for small $\mathbf{G}$ vectors, change rapidly from one iteration to the next, a much smaller choice of α is found helpful in achieving convergence.

Several schemes have been attempted to obtain covergence acceleration. One successful scheme is based on Broyden's updated version of the quasi Newton–Raphson iterative procedure (Broyden 1965). Srivastava (1984) adopted Broyden's second method which avoids storage and multiplication of $N \times N$ matrices produced by equation (3.23) in favour of only storage of m vectors of length N, where m is the number of iterations. In this scheme a rank-one inverse-Jacobian updating procedure is followed and, for iterations $m > 1$, equation (3.39) is updated to the form

$$V_{in}^{(m+1)} = V_{in}^{(m)} + \alpha F^{(m)} - \sum_{i=2}^{m} B_i^T F^{(m)} A_i, \tag{3.40}$$

where

$$F^{(m)} = V_{out}^{(m)} - V_{in}^{(m)}$$

$$A^{(i)} = \alpha(F^{(i)} - F^{(i-1)}) + V_{in}^{(i)} - V_{in}^{(i-1)} - \sum_{j=2}^{i-1} a_{ij} V_{in}^{(j)}$$

$$a_{ij} = V_{out}^{T\,(m)}(F^{(i)} - F^{(i-1)})$$

$$B^{(i)} = \frac{(F^{(i)} - F^{(i-1)})^T}{(F^{(i)} - F^{(i-1)})^T (F^{(i)} - F^{(i-1)})}. \tag{3.41}$$

Vanderbilt and Louie (1984) used a modified version of Broyden's direct-Jacobian updating procedure in which they incorporated information from all previous iterations. However, in their approach the storing and the multiplication of $N \times N$ matrices are required. Johnson (1988) used the modifications of Vanderbilt and Louie but used Srivastava's computational scheme to avoid the storage and multiplications of $N \times N$ matrices. Johnson's scheme is now routinely employed in DFT calculations.

3.2.3.4 Numerical solution of Kohn–Sham equation

For setting up the Kohn–Sham matrix eigenvalue equation in (3.23), the number of planewaves (*i.e.* G vectors) are decided by setting a kinetic energy cutoff:

$$E_{cut} = \frac{\hbar^2}{2m}[(k+G)^2]_{max}. \tag{3.42}$$

Clearly, different number of plane waves may be generated for different k points. Use of E_{cut} to choose the number of plane waves is preferable over the straightforward option of fixing the number of plane waves. It can be shown that the number of plane waves varies as $E_{cut}^{3/2}$. In general, higher the value of E_{cut} better the convergence of the eigensolutions. But in practice judicious practical considerations for E_{cut} are made depending on the availability of computational resources and the quality of the pseudopotentials employed.

For matrix diagonalization using standard techniques, such as the Choleski–Householder method, per k point computer memory requirement scales as $O(N^2)$ and the CPU requirement scales as $O(N^3)$. It is therefore useful to use iterative schemes for diagonalization of huge matrices. Two routinely used iterative scheme are (i) residual minimization by direct inversion in iterative space (RM-DIIS) and (ii) conjugate gradient (CG) method. These techniques have been discussed in several places (see, e.g. Srivastava 1999, Martin 2004) and thus will not be presented here.

3.2.4 MOMENTUM SPACE FORMULATION OF TOTAL ENERGY

Using equations (3.1), (3.10), (3.28) and (3.29), the ground-state total energy of the system can now be expressed as

$$
\begin{aligned}
E_{tot} &= N_0 \frac{e^2}{2} \sum_{p,b,b'}^{\prime} \frac{z_b z_{b'}}{|p + \tau_b - \tau_{b'}|} + T_0[\rho] + \int dr V_{ps}(r)\rho(r) \\
&\quad + \frac{e^2}{2} \int \int dr\, dr' \frac{\rho(r)\rho(r')}{|r - r'|} + \int dr\rho(r)\varepsilon_{xc}[\rho(r)] \tag{3.43} \\
&= E_{ion-ion} + E_{kin}^0 + E_{el-ion} + E_H + E_{xc}, \tag{3.44}
\end{aligned}
$$

where N_0 is the number of unit cells, p is a Bravais lattice vector, and $z_b e$ and τ_b are the ionic charge and the position vector, respectively, for the bth atom in the unit cell. $E_{kin}^0 \equiv T_0[\rho]$ is the kinetic energy of non-interacting electrons. The prime in equation (3.43) indicates that the term $p + \tau_b - \tau_{b'} = 0$ is omitted from the summation. Using the momentum-space formalism of the previous subsection, we can express the total energy per unit cell as

$$
\begin{aligned}
E_{tot} &= \frac{e^2}{2} \sum_{p,b,b'}^{\prime} \frac{z_b z_{b'}}{|p + \tau_b - \tau_{b'}|} + \frac{\hbar^2}{2m} \sum_k \sum_G \sum_j^{occ} w(k)|A_j(k+G)|^2(k+G)^2 \\
&\quad + \frac{1}{M} \sum_k \sum_j^{occ} w(k) \sum_{G,G',b} A_j^*(k+G')A_j(k+G)e^{i(G-G')\cdot\tau_b} \\
&\quad \times \sum_\ell v_{b,\ell}(k+G, k+G')
\end{aligned}
$$

$$+\frac{4\pi e^2}{2}\sum_G \frac{|\rho(G)|^2}{G^2}+\sum_G \rho^*(G)\varepsilon_{xc}(G) \tag{3.45}$$

$$= E_{ion-ion}+E_{kin}^0+E_{el-ion}+E_H+E_{xc}, \tag{3.46}$$

where $v_{b,\ell}$ is the contribution towards the form factor v_b from the ℓth angular momentum (see equation (3.20) and Appendix B).

If the ionic pseudopotentials $v_{b,\ell}$ are known, then E_{tot} can in principle be evaluated provided the coefficients $\rho(G)$ are obtained from a self-consistent solution of the Kohn–Sham equations (3.23). In practice, however, one must be careful in dealing with the divergences in the first, third and fourth terms, and in dealing with the rate of convergence of the lattice sum in the first term. Ihm *et al* (1979) showed that with some mathematical manipulations the three individually divergent terms can be added together to produce a non-divergent result: we can express

$$E_{ion-ion}+E_{el-ion}+E_H = \gamma_E+E'_{el-ion}+E'_H+E_1, \tag{3.47}$$

where the prime indicates that the divergent terms in E_H (with $G=0$) and E_{el-ion} (with $G-G'=0$) are omitted,

$$\begin{aligned}
\gamma_E &\equiv E'_{ion-ion} \\
&= E_{ion-ion}-\frac{1}{2}\frac{1}{\Omega_{at}}\sum_b \int dr \frac{z_b e^2}{r} \\
&= E_{ion-ion}-\frac{1}{2}\frac{1}{\Omega_{at}}\lim_{G\to 0}\sum_b \frac{4\pi z_b^2 e^2}{G^2}
\end{aligned} \tag{3.48}$$

is the electrostatic energy of point ions in a uniform gas of valence electrons (called the Ewald energy), and

$$E_1 = \bar{z}\frac{1}{\Omega_{at}}\sum_b \int dr \left(v_b(|r|)+\frac{z_b e^2}{r}\right) \tag{3.49}$$

is the correction arising from the 'pseudo' nature of the potential. With this consideration the eigenvalue problem in equation (3.23) is solved with $V_{ps}(0)$ and $V_H(0)$ set to zero. Equation (3.45) then reads

$$E_{tot} = \gamma_E+E_{kin}^0+E'_{el-ion}+E'_H+E_{xc}+E_1. \tag{3.50}$$

Thus, an alternative expression for the total crystal energy per unit cell in the momentum-space representation becomes

$$\begin{aligned}
E_{tot} &= \gamma_E+\sum_j^{occ}\varepsilon_j-\frac{1}{2}\sum_{G\neq 0}\frac{4\pi e^2}{|G|^2}|\rho(G)|^2 \\
&\quad +\sum_G \rho^*(G)[\varepsilon_{xc}(G)-V_{xc}(G)]+E_1
\end{aligned} \tag{3.51}$$

$$= \gamma_E+\sum_j^{occ}\varepsilon_j-E'_H+\Delta E_{xc}+E_1, \tag{3.52}$$

where ε_j are the eigenvalues obtained with $V_H(0)$ and $V_{ps}(0)$ set equal to zero in equation (3.23), and $V_{xc}(G)$ is a Fourier component of the exchange-correlation potential defined in equations (3.27) and (3.29).

The Ewald summation method (Ewald 1921) can be used to efficiently evaluate the electrostatic energy term γ_E. This has been discussed in many books and review articles (see e.g. Born and Huang 1954, Maradudin *et al* 1971, Martin 2004). The result is

$$
\gamma_E = \frac{e^2}{2} \sum_{bb'} z_b z_{b'} \left(\frac{4\pi}{\Omega} \sum_{G \neq 0} \cos[G \cdot (\tau_b - \tau_{b'})] \frac{\exp(-G^2/4\eta^2)}{|G|^2} \right.
$$
$$
\left. - \frac{\pi}{\eta^2 \Omega} + \sum_p{}' \frac{\operatorname{erfc}(\eta x)}{x} \Big|_{x=|p+\tau_b-\tau_{b'}|} - \frac{2\eta}{\sqrt{\pi}} \delta_{b,b'} \right), \tag{3.53}
$$

where η is a parameter that controls the convergence of the two summations in the above result, and $\operatorname{erfc}(y)$ is the complementary error function. The choice $\eta \simeq 1/r_0^2$, where r_0 is the nearest-neighbour distance, seems to produce rapid convergence of both sums (the direct space sum and the reciprocal space sum) in the above equation.

3.2.5 HELLMANN–FEYNMAN THEOREM FOR FORCE CALCULATION

The pseudopotential–DFT formalism for the total crystal energy can be easily extended to calculate the related forces. On grounds of crystal symmetry, in undisplaced configuration the force on each atom in a unit cell will be zero. This will be true even beyond the equilibrium lattice constant (e.g. under hydrostatic pressure). However, if one of the basis atoms in the unit cell, say the bth atom, is displaced from its equilibrium crystal structural site, then there will be a force F^b acting on the nucleus at τ_b. A quantum mechanical expression for this force can be derived by taking the analytic gradient of the total energy expression in equation (3.44)

$$
\begin{aligned}
F^b &= -\nabla_b E_{\text{tot}} \\
&= e^2 \sum_{p,b' \neq b} \frac{z_b z_{b'} (p + \tau_b - \tau_{b'})}{|p + \tau_b - \tau_{b'}|^3} - \int dr \, \nabla_b [V_{\text{ext}}(r)\rho(r)] \\
&= F^b_{\text{ion}} + F^b_{\text{el}},
\end{aligned} \tag{3.54}
$$

where for time being we have used V_{ext}, a general external potential, in place of the pseudopotential V_{ps}. The first term on the right-hand side can be evaluated in a straightforward manner using Ewald's method. The calculation of the second term, however, requires some careful consideration, as explained below.

We first realise that $V_{ext}(r)$ depends directly on nuclear positions $\tau = \{\tau_b\}$, so that a given crystal geometry represents a unique external potential $V_{ext}(r, \tau)$. This in turn gives rise to a unique electronic charge density $\rho(r, \tau)$ with an implicit dependence on τ. This means that we can express the second term on the right-hand side in equation (3.54) as

$$
\begin{aligned}
F^b_{\text{el}} &= -\int dr \rho(r, \tau) \, \nabla_b [V_{\text{ext}}(r, \tau)] - \int dr V_{ext}(r, \tau) \nabla_b [\rho(r, \tau)] \\
&= -\int dr \rho(r, \tau) \, \nabla_b [V_{\text{ext}}(r, \tau)] - \int dr \frac{\delta E_{el}}{\delta \rho} \, \nabla_b [\rho(r, \tau)] \\
&= F^b_{\text{el}(1)} + F^b_{\text{el}(2)}.
\end{aligned} \tag{3.55}
$$

Combining equations (3.54) and (3.55) we have

$$
\begin{aligned}
F^b &= F^b_{\text{ion}} + F^b_{\text{el}(1)} + F^b_{\text{el}(2)} \\
&= F^b_{\text{HF}} + F^b_{\text{el}(2)},
\end{aligned} \tag{3.56}
$$

where F^b_{HF} is the Hellmann–Feynman force (Feynman 1939), which is equal to the negative gradient of the classical electrostatic potential arising from all positive charged nuclei (ions) and the quantum mechanical electronic charge density.

The term $F^b_{\text{el}(2)}$ is due to inaccuracies in the calculated charge density, i.e. due to inaccuracies in solving the Kohn–Sham equations in (3.11) and (3.12), or in (3.23) and (3.24). Therefore $F^b_{\text{el}(2)}$ may be referred to as the variational force. We can express this term as follows (Bendt and Zunger 1983, Scheffler *et al* 1985, Srivastava and Weaire 1987):

$$
\begin{aligned}
\boldsymbol{F}^b_{\mathrm{el}(2)} &= -2\Re \sum_j \int \mathrm{d}\boldsymbol{r} \frac{\partial \psi_j^*}{\partial \boldsymbol{\tau}_b} \left(\frac{\delta E}{\delta \rho} \right) \psi_j \\
&= -2\Re \sum_j \int \mathrm{d}\boldsymbol{r} \frac{\partial \psi_j^*}{\partial \boldsymbol{\tau}_b} \left(-\frac{\hbar^2}{2m} \nabla^2 + V_{\mathrm{KS}} - \varepsilon_n \right) \psi_n \\
&= -2\Re \sum_j \int \mathrm{d}\boldsymbol{r} \frac{\partial \psi_j^*}{\partial \boldsymbol{\tau}_b} \left(-\frac{\hbar^2}{2m} \nabla^2 + \tilde{V}_{\mathrm{KS}} - \varepsilon_j \right) \psi_j \\
&\quad - \int \mathrm{d}\boldsymbol{r} \frac{\partial \rho}{\partial \boldsymbol{\tau}_b} [V_{\mathrm{KS}} - \tilde{V}_{\mathrm{KS}}] \\
&= \boldsymbol{F}^b_{\mathrm{IBS}} + \boldsymbol{F}^b_{\mathrm{NSC}}.
\end{aligned}
\tag{3.57}
$$

In the above $\tilde{V}_{\mathrm{KS}}$ is an effective 'input' Kohn–Sham potential, calculated after a finite number of interations in a self-consistent solution of equations (3.11) and (3.12).

Thus, there are two separate origins of the variational force. The first term, $\boldsymbol{F}^b_{\mathrm{IBS}}$, reflects the correction due to an incomplete basis set. In molecular Hartree–Fock theory, such a force is called the Pulay force (Pulay 1969). If ψ is an exact eigenstate of the Kohn–Sham equation with an effective 'input' potential, then $\boldsymbol{F}_{\mathrm{IBS}} = 0$. A weaker statement would be that if the wavefunction ψ comprised originless orbitals, i.e. did not depend on ionic positions (e.g. plane waves), then $\boldsymbol{F}_{\mathrm{IBS}} = 0$. Thus for a plane-wave basis set representation of ψ this term is identically zero, while in a linear-combination-of-atomic-orbitals approach one has to worry about it. The second term, $\boldsymbol{F}_{\mathrm{NSC}}$, is a measure of non-self-consistency (or lack of self-consistency) in the solution of the Kohn–Sham equation: if the effective potential $\tilde{V}_{\mathrm{KS}}$ is equal to the exact potential V_{KS}, then $\boldsymbol{F}_{\mathrm{NSC}} = 0$.

In a fully self-consistent pseudopotential calculation, based on a plane-wave basis set as in section 3.2.4, we have $\boldsymbol{F}_{\mathrm{IBS}} = \boldsymbol{F}_{\mathrm{NSC}} = 0$, and we recover the Hellmann–Feynman theorem. Thus in this approach the force on the bth atom is

$$
\boldsymbol{F}^2 \equiv \boldsymbol{F}^b_{\mathrm{HF}} = \boldsymbol{F}^b_{\mathrm{ion}} + \boldsymbol{F}^b_{\mathrm{el}(1)}.
\tag{3.58}
$$

The force $\boldsymbol{F}^b_{\mathrm{ion}}$ is obtained by taking the derivative of equation (3.53) with respect to the basis vector $\boldsymbol{\tau}_b$:

$$
\begin{aligned}
\boldsymbol{F}^b_{\mathrm{ion}} &= e^2 z_b \sum_{b' \neq b} z_{b'} \left[\frac{4\pi}{\Omega} \sum_{\boldsymbol{G} \neq 0} \frac{\boldsymbol{G}}{G^2} \sin(\boldsymbol{G} \cdot (\boldsymbol{\tau}_b - \boldsymbol{\tau}_{b'})) \exp(-G^2/4\eta^2) \right. \\
&\quad \left. + \sum_p' \left(\mathrm{erfc}(\eta x) \frac{\boldsymbol{x}}{|\boldsymbol{x}|^3} + \frac{2\eta \boldsymbol{x}}{\sqrt{\pi} x^2} \exp(-|\eta x|^2) \right) \Big|_{\boldsymbol{x} = \boldsymbol{p} + \boldsymbol{\tau}_b - \boldsymbol{\tau}_{b'}} \right].
\end{aligned}
\tag{3.59}
$$

The term $\boldsymbol{F}^b_{\mathrm{el}(1)}$ can be obtained by differentiating the expression for E_{el-ion} in equation (3.45),

$$
\begin{aligned}
\boldsymbol{F}^b_{\mathrm{el}(1)} &= -\Re \left(\frac{\mathrm{i}}{M} \sum_{\boldsymbol{k}} \sum_j^{occ} w(\boldsymbol{k}) \sum_{\boldsymbol{G},\boldsymbol{G}'} A_j^*(\boldsymbol{k} + \boldsymbol{G}') A_j(\boldsymbol{k} + \boldsymbol{G})(\boldsymbol{G} - \boldsymbol{G}') \right. \\
&\quad \left. \times \exp[\mathrm{i}(\boldsymbol{G} - \boldsymbol{G}') \cdot \boldsymbol{\tau}_b] \sum_\ell v_{b,\ell}(\boldsymbol{k} + \boldsymbol{G}, \boldsymbol{k} + \boldsymbol{G}') \right).
\end{aligned}
\tag{3.60}
$$

If a local pseudopotential is used, then equation (3.60) is reduced to

$$
\boldsymbol{F}^b_{\mathrm{el}(1)}(\text{local potential}) = \Re \left[\frac{\mathrm{i}}{M} \sum_{\boldsymbol{G}} \rho^*(\boldsymbol{G}) \boldsymbol{G} \exp(-\mathrm{i}\boldsymbol{G} \cdot \boldsymbol{\tau}_b) v_b(|\boldsymbol{G}|) \right]
\tag{3.61}
$$

with $v_b(|\boldsymbol{G}|)$ as the form factor of a local ionic potential $v_b(|\boldsymbol{r}|)$.

It is worth noting that as a convenient alternative to using equation (3.25), a complex fast Fourier transform (CFFT) [see, e.g. Press *et al* (1986)] of equation (3.15) can be made to produce the pseudo wavefunction $\phi_{j,k}(r)$ at a point r in real space. From this the charge density in phase space is obtained as

$$\rho(k,r) = 2\sum_{j}^{\text{occ}} \phi_{j,k}^{*}(r)\phi_{j,k}(r). \tag{3.62}$$

An inverse CFFT of this then produces the Fourier component $\rho(k,G)$. This would be in the k–r phase space. A non-symmetrized charge density in direct space r can be calculated

$$\rho^{(\text{non-sym})}(r) = \sum_{k_i}^{\text{IBZ}} w(k_i)\rho(k_i,r), \tag{3.63}$$

where the k_i vectors sample the IBZ. Then the symmetrized charge density can be obtained by using equation (1.35). The symmetrized charge density expression is (Giannozzi *et al* 2009)

$$\rho^{(\text{sym})}(r) = \frac{1}{N_{\text{sym}}} \sum_{\mathscr{R}} \rho^{(\text{non-sym})}(\mathscr{R}^{-1}r - \mathscr{R}^{-1}t), \tag{3.64}$$

where $\mathscr{R}$ and t are the rotational and fractional translation parts of the N_{symm} symmetry operations.

The force symmetrisation is done in a similar manner (Giannozzi *et al* 2009). The symmetrized Hellmann–Feynman force on bth atom is

$$F^b = \frac{1}{N_{\text{sym}}} \sum_{\hat{O}} \hat{O} F^{\text{non-sym}}(\tau_b) = \frac{1}{N_{\text{sym}}} \sum_{\hat{O}} \mathscr{R} F^{\text{non-sym}}(\mathscr{R}^{-1}\tau_b), \tag{3.65}$$

where the sum is over N_{sym} space group symmetry operations $\hat{O} = \{\mathscr{R} \mid t\}$ and $\mathscr{R}^{-1}(\tau_b)$ labels the atom into which the bth atom transforms (modulo a lattice translation vector) after the application of $\mathscr{R}^{-1}$.

3.3 HARMONIC FORCE CONSTANTS AND PHONON EIGENSOLUTIONS

With total energy and Hellmann–Feynman force results being available, we can calculate harmonic force constants to construct the dynamical matrix and solve it for phonon eigensolutions. There are several routes for doing this, depending to some extent on the requirement level. We will describe some of the methods. In general, we can group such methods in two categories: *direct methods* and *perturbative methods*.

3.3.1 DIRECT METHODS

Direct methods use the total energy and/or forces directly to construct harmonic force constant(s) and thus obtain phonon eigensolutions. These have been used at different levels, as described below.

3.3.1.1 Frozen phonon method

The frozen-phonon method (Yin and Cohen 1982, Kunc and Martin 1983) is a 'direct' approach in which a distorted crystal is treated as a crystal in a new structure with a lower symmetry than the undistorted crystal. Exactly the same method is used for the calculation of the total energy of both the undistorted and the distorted crystals.

Consider a wave (defined by a fixed phonon wave vector q) propagating through a crystal. This causes crystal ion cores to vibrate with a definite displacement pattern. Within the adiabatic approximation the electron cloud quickly responds to such a displacement pattern. If we imagine taking a snapshot, then it is possible to consider the system in a new crystal structure corresponding to the 'frozen' vibrational mode. Such a new crystal structure will correspond to a lower symmetry and a higher total energy than the original (undistorted) structure.

If $E_{tot}(0)$ and $E_{tot}(u)$ are, respectively, the total energy of the undistorted and distorted structures, then the frequency ω of the 'frozen phonon' is defined by

$$\frac{1}{2}\omega^2 \sum_b m_b |u(b)|^2 = E_{tot}^{harm}(u) - E_{tot}(0), \qquad (3.66)$$

where E_{tot}^{harm} is the harmonic part of the total energy, u is the distortion amplitude and $u(b)$ is the displacement vector of the bth atom with mass m_b in the unit cell. For small distortions, only cubic and quartic anharmonicities are expected to be contained in E_{tot} around the equilibrium with $u = 0$. Calculations with u, $-u$ and $2u$ are needed to extract E_{tot}^{harm} after getting rid of cubic and quartic anharmonicities.

The phonon frequency defined in equation (3.66) can, in principle, also be obtained from the force equation

$$F_{HF}^b \Big|^{harm} = -\sum_{b'} \Phi(b;b')u(b'), \qquad (3.67)$$

where the left-hand side represents the harmonic contribution to the Hellmann–Feynman force on atom b, and Φ is the interatomic force constant matrix. Once the force constant matrix Φ, or its Fourier transform (the dynamical matrix) is determined, it is straightforward to calculate the vibrational frequencies of normal modes as explained in section 2.4.

Consider an example of a highly symmetric distortion produced by stretching or compressing the bond along [111] between atoms A and B (figure 3.1) in the primitive unit cell for the diamond or zincblende structure. The translational symmetry of this new crystal structure is the same as that of the undisplaced structure, but the point-group symmetry has been reduced to C_{3v}. Such a vibrational mode corresponds to the zone-centre phonon $LTO(\Gamma)$ in the diamond structure. In the zincblende structure such a mode is restricted to the $TO(\Gamma)$ mode, as the longitudinal optical mode $LO(\Gamma)$ cannot be treated by this frozen phonon approach because of the long-range Coulomb interactions (see section 2.4.4). If we consider ξ as the relative displacement between atoms A and B, and μ as the reduced mass of the unit cell, then for the $TO(\Gamma)$ mode equation (3.67) can be expressed by the simple equation

$$F = -\Lambda\xi \qquad (3.68)$$

and the phonon frequency is given by

$$\omega = \sqrt{\frac{\Lambda}{\mu}}, \qquad v = \frac{\omega}{2\pi}. \qquad (3.69)$$

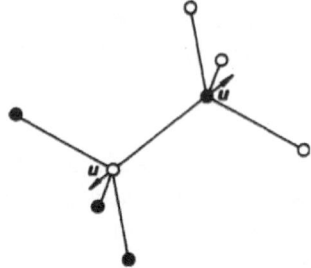

Figure 3.1 The displacement pattern of atoms in the zincblende structure for the evaluation of the frequency of the $TO(\Gamma)$ mode using the frozen-phonon method.

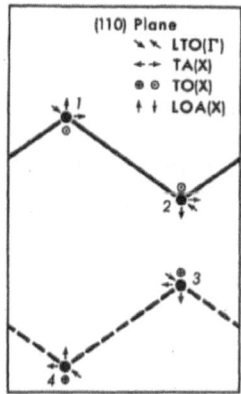

Figure 3.2 The displacement pattern of atoms in the diamond structure for the evaluation of the frequency of the LTO(Γ), TA(X), TO(X) and LOA(X) modes using the frozen-phonon method. Reproduced from Yin and Cohen (1982).

Both the total energy method (equation (3.66)) and the force method (equations (3.67) and (3.69)) produce results for ω_Γ(TO) which are identical within numerical accuracy.

In order to calculate phonon frequencies at the symmetry point X, a four atom unit cell is required. The projection of such a unit cell on the (110) plane is shown in figure 3.2. Also shown in that figure are the atomic displacement patterns, determined by group-theoretic method, for phonon polarizations at the Γ and X symmetry points (Yin and Cohen 1982). Calculated phonon frequencies in Si, Ge, GaAs, NaCl and Al for TO(Γ), as well as zone-boundary phonon modes such as LO(X), LA(X), TO(X) and TA(X) are compiled in Table 3.1. As can be seen, the first-principles results agree very well with experiment.

Although in principle a frozen phonon mode can be calculated either by using the total energy method or the force method, it should be mentioned that the force method has some advantages over the energy method. Firstly, if N represents the degree of freedom, then from one self-consistent calculation one obtains $3N_f$ force constants instead of just one value of total energy. This helps one consider a relatively small number of distorted geometries ($\propto N_f$) when calculating Hellmann–Feynman forces against a large number ($\propto N_f^2$) of distorted geometries when determining the dynamical matrix using the finite energy difference method. Secondly, in $\Delta E = E(u) - E(0)$ the energy difference is usually a very small fraction of the total energy, and thus high numerical accuracy in $E(u)$ and $E(0)$ is required for phonon calculation. On the other hand, as changes in forces are of the same magnitude as the forces themselves, high numerical accuracy is attainable in the gradient method.

3.3.1.2 Finite displacement method for zone-centre phonons

Investigation of zone-centre optical phonon modes of a small group of atoms in a solid, or adsorbate atoms on a solid surface, can be made by exerting finite displacements of atoms away from their equilibrium positions and extracting the linear part of the resultant Hellmann–Feynman forces. Then the dynamical matrix can be built up by using the second-order force constants $\Phi(b; b')$ in equation (3.67). With the dynamical matrix specified, zone-centre optical modes and frequencies of the considered 'small molecular unit' can be obtained by solving the relevant form of the determinantal equation in (2.50). We will present a few numerical results in section 8.4.3.

Table 3.1

Calculated vibrational frequencies obtained from the frozen-phonon approach, using the pseudopotential–LDA scheme. Experimental results, shown in parentheses, are cited in the references for the theoretical works. Units are THz.

Material	Reference	$v_\Gamma(TO)$	$v_X(LO)$	$v_X(TO)$	$v_X(LA)$	$v_X(TA)$
Si	1	15.16	12.16	13.48		4.45
		(15.53)	(12.32)	(13.90)		(4.49)
Ge	1	8.90	7.01	7.75		2.44
		(9.12)	(7.21)	(8.26)		(2.40)
GaAs	2	8.29	7.55	7.94	7.20	1.87
		(8.19)	(7.22)	(7.56)	(6.89)	(2.41)
NaCl	3	3.09	3.48	3.26	3.58	1.53
		(3.25)	(3.26)	(3.39)	(2.67)	
Al	4				6.11	3.72
					(6.08)	(3.65)

References: (1) Yin and Cohen (1982); (2) Kunc (1985); (3) Froyen and Cohen (1984); (4) Lam and Cohen (1982).

3.3.1.3 Planar force constant method

An extension of the frozen-phonon method described in section 3.3.1.1 is the planar force constant method developed by Kunc and Martin (1982). The concept in this *ab initio* method consists in selecting one direction of the wave vector q and in representing the lattice vibrations as the motion of rigid planes of atoms perpendicular to q. The rigid planes are assumed to be connected by harmonic *interplanar forces*, so that the situation can be exactly described by the equation of motion written for a linear chain of atoms (Kittel 1986) and a complete phonon dispersion along q can be obtained.

To explain the method we consider crystals of diamond and zincblende type. The concept of the method is simple and efficient for q along the symmetry directions [100], [110] and [111] and we restrict our discussion to these directions. Of these three directions, [110] has the most elaborate concept of interplanar forces. Figure 3.3 shows the system of the (110) sublayers schematically. Each of the (110) atomic planes is composed of two types of atomic site, denoted c (for cation) and a (for anion), which can vibrate independently. For phonons propagating along [110], we need to consider interplanar force constants Λ_n^{ca}, Λ_n^{cc}, Λ_n^{ac} and Λ_n^{aa} as shown in figure 3.3. Suitable supercell atomic displacement patterns are shown in figure 3.4. This particular choice of unit cell allows determination of *interplanar force constants* $\{\Lambda_n\}$ up to the fifth nearest planar distance (or planar neighbours).

Referring to figure 3.4, group theoretical analysis (Maradudin and Vosko 1968, Warren 1968) shows that while the transverse vibrations T_2 with displacement $u(\text{T}_2) = (u_1, -u_1, 0)$ are eigenmodes, the longitudinal vibrations L with $u(\text{L}) = (u_1, u_1, 0)$ and the transverse vibrations T_1 with $u(\text{T}_1) = (0, 0, u)$ are not independent eigenmodes when $q \parallel [110]$. Let us first consider the T_2 mode. If u_0^b represents the displacement of the sublayer b at the plane 0, then the forces on atoms in the plane n are given by the equation of motion

$$-F_n^{b'} = \Lambda_n^{b'b}(\text{T}_2)u_0^b, \tag{3.70}$$

where $b = c, a$ and $b' = c, a$. The occurrence of Λ_0^{ca} and Λ_0^{ac} (the force constants connecting two different sublayers of the same plane) is somewhat reminiscent of the inclusion of the (local) core–shell interactions in the formalism of the shell model. The 'self-terms' Λ_0^{cc} and Λ_0^{aa} can be determined through the translational symmetry (cf equation (2.46)) from all the other force constants.

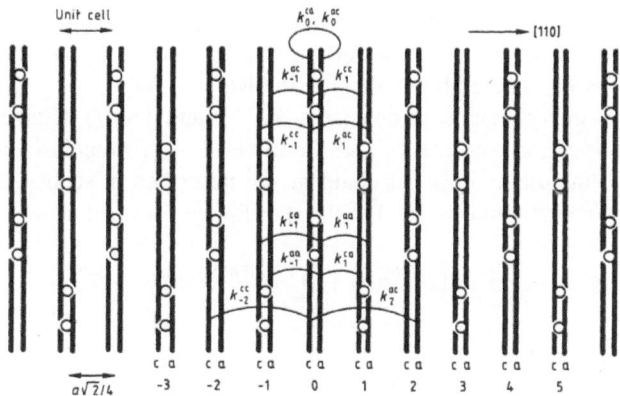

Figure 3.3 The (110) atomic planes are composed of two types of atomic sites denoted c and a, which for vibrations with $q \parallel [110]$, move in phase as compact sets forming 'sublayers' c and a. A phonon propagating along the [110] direction can be described as vibrations of a linear chain of planar sublattices c and a which are connected by interplanar force constants k_n^{ca}, k_n^{cc}, etc. Reproduced from Srivastava and Kunc (1988).

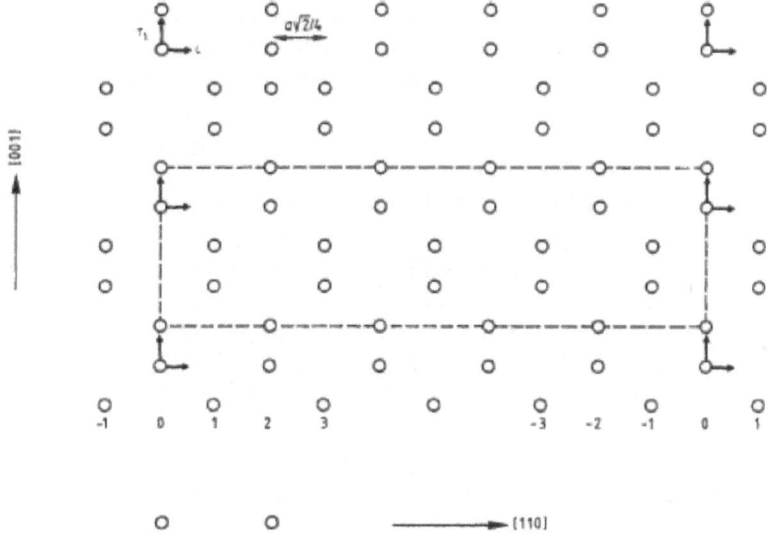

Figure 3.4 The supercell and displacement patterns used for *ab initio* determination of the force constants connecting the atomic planes in figure 3.3. The T_2 direction is perpendicular to the figure and is not shown. This size of the supercell allows force constants up to fifth planar neighbours to be determined. Reproduced from Srivastava and Kunc (1988).

For the 'mixed' case $L - T_1$, we must consider a linear combination of the L and T_1 displacement patterns

$$u = u(L) + u(T_1),\tag{3.71}$$

with the ratio u_{T_1}/u_L to be calculated as a part of the dynamical problem. The equations of motion are now given by

$$-\begin{bmatrix} F_n(L) & F_n(T_1)\end{bmatrix} = \begin{bmatrix} \Lambda_n(LL) & \Lambda_n(LT_1) \\ \Lambda_n(T_1L) & \Lambda_n(T_1T_1) \end{bmatrix} \begin{bmatrix} u_0(L) & u_0(T_1)\end{bmatrix},\tag{3.72}$$

where we have used $\Lambda_n(ij) = \{\Lambda_n^{bb'}(ij)\}$ and $u_0(i) = \{u_0^b(i)\}$ for simplicity. There exist certain symmetry relations for the interplanar force constants, both in the T_2 case and in the $L - T_1$ case. These have been discussed in detail by Srivastava and Kunc (1988).

Once the harmonic interplanar force constants $\Lambda_n(T_2)$ and $\{\Lambda_n(ij)\}$ have been calculated, the phonon dispersion curves for both the T_2 and the mixed $L - T_1$ branches can be calculated from suitable extensions of the diatomic linear chain model described in section 2.2.2. Using equation (2.63), the 'C-type' dynamical matrix for a vector $q = (q_1, q_1, 0)$ is set up as follows:

$$C_{ij}(bb'|q) = \frac{1}{\sqrt{m_b m_{b'}}} \left(\Lambda_0^{b'b}(ji) + \sum_{m>0} [\Lambda_m^{b'b}(ji)e^{imq_1} + \Lambda_{-m}^{b'b}(ji)e^{-imq_1}] \right) \tag{3.73}$$

for the $L - T_1$ mixed case, and

$$C_{ij}(bb'|q) = \frac{1}{\sqrt{m_b m_{b'}}} \left(\Lambda_0^{b'b}(ji) + 2 \sum_{m>0} \Lambda_m^{b'b}(ji)\cos(mq_1) \right) \tag{3.74}$$

for the T_2 case. Equations (3.73) and (3.74) are valid for both the diamond and zincblende crystal structures.

Kunc and co-workers (see Srivastava and Kunc 1988, and references therein) have used the planar force constant method to calculate the phonon dispersion in Ge and GaAs along [100], [111] and [110]. For each case a suitable unit cell is chosen and the planar force constants are evaluated for longitudinal and transverse modes of distortion in the unit cell. Writing $\Lambda = \Lambda_{\text{ion}-\text{ion}} + \Lambda_{\text{el}-\text{ion}}$, the ion–ion and electron–ion parts of the force constants are calculated, respectively, from the derivative of the Hellmann–Feynman forces $F_{\text{ion}-\text{ion}}$ (equation (3.59)) and $F_{\text{el}(1)}$ (equation (3.60)). The electronic part of the force $F_{\text{el}-\text{ion}}(= F_{\text{el}(1)})$ is calculated from a self-consistent pseudopotential method, within the LDA (equation 3.61).

If we disregard anharmonicity, then all the planar force constants for $q \parallel [110]$ can be determined from just three self-consistent calculations for the Hellmann–Feynman forces: with a small displacement u for each of the L, T_1 and T_2 directions. However, in reality even with a fairly small displacement u, some anharmonic effects are usually present which must be carefully eliminated from the forces before the harmonic force constants $\{\Lambda\}$ can be calculated. The anharmonicity can arise (i) by making equations (3.70) and (3.72) non-linear, and (ii) by making the forces F_n non-parallel to the displacement u_0 which caused it. In addition, for $q \parallel [110]$ effect (ii) and the $L - T_1$ coupling must be carefully separated.

For the T_2 mode, the cubic anharmonicity is absent, and the quartic anharmonicity can be eliminated by calculating forces with displacements $u = (u_1, -u_1, 0)$ and $2u$. For the mixed $L - T_1$ modes, the lowest order of anharmonicity is cubic for the longitudinal and transverse response to $u(L)$, and quartic for both the longitudinal and transverse response to $u(T_1)$. These anharmonicities can be eliminated by considering displacements $u(L) = \pm(u_1, u_1, 0)$, $2u(L)$, $u(T_1) = \pm(0, 0, \sqrt{2}u_1)$ and $2u(T_1)$. In all these displacements the quantity u_1 can be taken to be of the order of 1% of the lattice constant.

Figure 3.5 shows the phonon dispersion in Ge along the [100], [110] and [111] directions calculated from the planar force constant method. The input to these calculations is a local version of the pseudopotential for Ge^{+4} which is obtained by averaging the pseudopotentials of Pickett et al (1978) for Ga^{+3} and As^{+5}. The Schrödinger equation in (3.24) was solved by employing Slater's X_α method (Slater 1974) with $\alpha = 0.8$, and plane waves up to kinetic energy $E_1 = 2.55$ Ryd were treated exactly, and plane waves between kinetic energies E_1 and $E_2 = 9.15$ Ryd were treated by Löwdin's perturbation method (Löwdin 1951).) Also shown are the experimental data of Nilsson and Nelin (1971). The T_2 modes have symmetry $\Sigma_2 + \Sigma_4$, and the mixed $L - T_1$ modes have symmetry $\Sigma_1 + \Sigma_3$. In general the agreement between theory and experiment is very encouraging. There is some disagreement in the transverse acoustic branches. The source of this disagreement is considered to lie in the quality of the *local* pseudopotential and the Slater's X_α exchange method used in

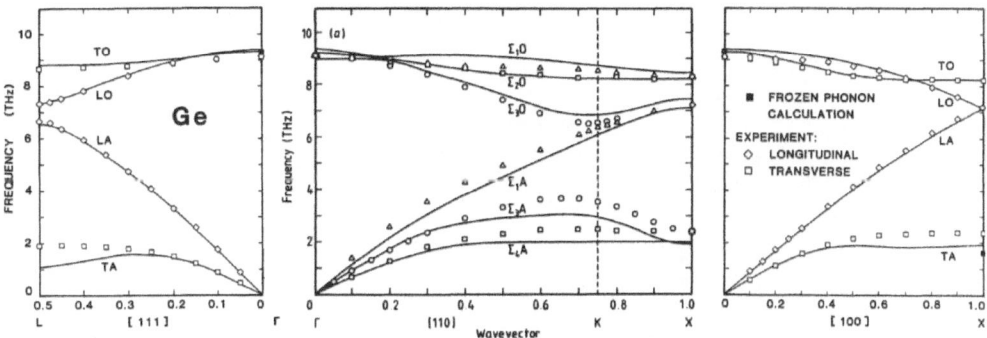

Figure 3.5   Phonon dispersion in Ge along the [100],[110] and [111] directions calculated from the planar force constants which are determined *ab initio using* the local pseudopotential method within the local density approximation. For the [110] direction, interactions were considered up to the third interplanar neighbours for the T_2 modes, and up to the fourth neighbours for the $L - T_1$ modes. Experimental data are taken from Nilsson and Nelin (1971): the branches Σ_2 and $\Sigma_4 (\lozenge)$ are the pure T_2 modes, $\Sigma_1 (\triangle)$ and $\Sigma_3 (\circ)$ label the mixed $L - T_1$ modes. (Taken from Kunc and Gomes Dacosta 1985, Srivastava and Kunc 1988.)

the calculations. However, the main message of this approach is that the lattice dynamics of semi-conductors can be studied *ab initio*, without using any adjustable parameters. Another important conclusion to draw from this *ab initio* work is that for an adequate study of lattice vibrations in Ge forces extending at least up to a distance equal to three (110) interplanar distance must be taken into consideration.

Information on phonon frequencies alone is not sufficient to obtain unique information on the normal modes of a crystal. Full eigenvector information must also be obtained. From equation (2.60) the normalised displacement of atom lb in the mode qs can be expressed as

$$u_\alpha(\boldsymbol{lb}|\boldsymbol{q}s) = \frac{1}{\sqrt{m_b}} \big|e_\alpha(\boldsymbol{b};\boldsymbol{q}s)\big| \exp[(i\varphi_\alpha(\boldsymbol{b};\boldsymbol{q}s)) + i\boldsymbol{q}\cdot\boldsymbol{x}(\boldsymbol{lb}) - i\omega t], \qquad (3.75)$$

where the normalised complex eigenvector components $e_\alpha(\boldsymbol{b};\boldsymbol{q}s)$ have been assigned the real amplitudes $\big|e_\alpha(\boldsymbol{b};\boldsymbol{q}s)\big|$ and phases $\exp(i\varphi_\alpha(\boldsymbol{b};\boldsymbol{q}s))$. The calculated dispersions of the eigenvectors e in terms of the amplitudes and phases for $\boldsymbol{q} \parallel [110]$ in Ge are shown in figures 3.6 and 3.7. Clearly for the pure T_2 modes both the acoutic $(\Sigma_4 A)$ and $(\Sigma_2 O)$ branches have identical amplitude for all $\boldsymbol{q}$-values along the direction [110]. For the mixed $L - T_1$ modes, the acoustic $(\Sigma_1 A, \Sigma_3 A)$ and optic $(\Sigma_1 O, \Sigma_3 O)$ branches show dispersive behaviour along [110] for both the longitudinal (L) and transverse (T_1) components. The dispersion of the amplitudes of the mixed $L - T_1$ modes is the most interesting information provided by the *ab initio* calculation: such information cannot be predicted by symmetry, nor is it available from experiment.

The method discussed above for calculations of phonon dispersion results can be applied for $\boldsymbol{q}$ along any symmetry direction for non-polar crystals such as Si and Ge. However, it is not applicable to treat LO phonons in polar crystals close to the BZ (*i.e.* when $\boldsymbol{q} \to 0$) due to the presence of nonanalytic $\boldsymbol{q}$ dependence of the microscopic electric fields developed by such vibrations (see Born and Huang (1954) and section 2.4.4). It was shown by Kunc and Martin (1982) that the depolarizing field can be calculated from the slope of the average electrostatic potential in the central part of the supercell. The forces resulting from the depolarizing field can be subtracted from the total forces to obtain the relavant force constant. This procedure works for any $\boldsymbol{q} \neq 0$. Kunc and Martin (1982) applied this method successfully to calculate the phonon dispersion curves for GaAs along the [100] symmetry direction.

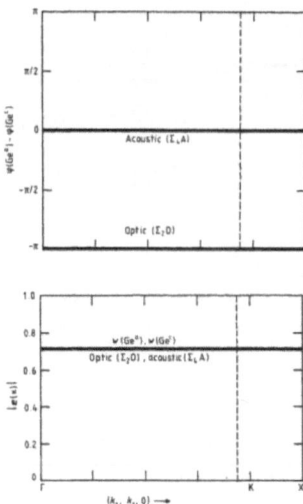

Figure 3.6 Dispersion of the normalised amplitudes of the transverse eigenmodes $T_2(\Sigma_2 + \Sigma_4)$ in Ge for $q \parallel [110]$: $|e|$ represents the amplitude and ϕ the phase factor. $\varphi(Ge^c)$ and $\varphi(Ge^a)$ are the $\varphi(b)$ for the two basis subplanes. Reproduced from Srivastava and Kunc (1988).

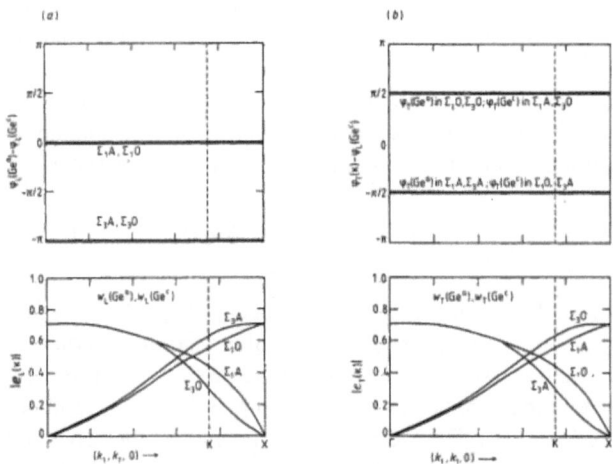

Figure 3.7 Same as in figure 3.6 for the mixed $L - T_1$ modes $(\Sigma_1 + \Sigma_3)$ in Ge: (a) longitudinal components, (b) transverse components. The phase factors are displayed relative to the longitudinal of the c-type sublayer. Reproduced from Srivastava and Kunc (1988).

3.3.1.4 Real-space method for an arbitrary q in BZ

The planar force constant method for obtaining phonon eigensolutions along a symmetry direction can be generalized for obtaining solutions for any phonon wave vector q in the Brillouin zone of a periodic system. This would require two considerations. Firstly, an adequate sized supercell should be chosen to ensure that up to four or five nearest neighbour interactions can be evaluated in order to obtain well-converged phonon frequencies, especially for TA modes. Secondly, a method must be adopted for dealing with long wave (short q) LO modes in polar systems.

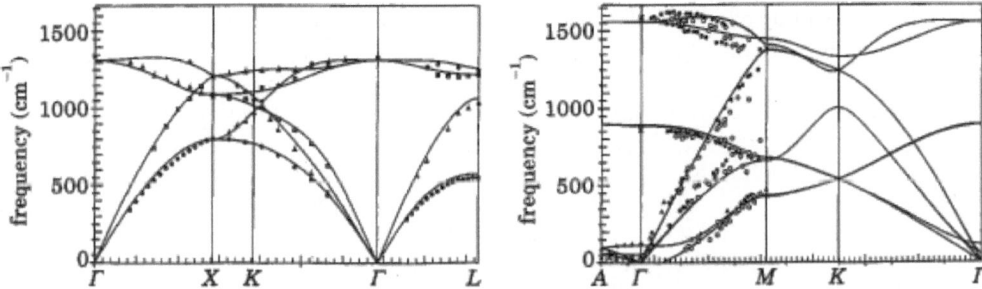

Figure 3.8 Phonon dispersion curves for diamond (left) and graphite (right). Calculations were performed using the plane-wave pseudopotential method and the DFT-LDA. Dots show the inelastic neutron scattering data. Note: $33.3563 \text{ cm}^{-1} = 1$ THz. Reproduced from Kress *et al* (1995).

Kress *et al* (1995) developed this method for non-polar systems using the following procedure. A single atom (lb) in the supercell (where l labels the elementary cells and b represents the atoms in the basis) is displaced by a small vector $u(lb)$ and the Hellmann–Feynman forces $F(l'b')$ acting on the surrounding atoms ($l'b'$) are calculated. Space group symmetry operations $\mathscr{O}$ are applied to the displacemnt vector $u(lb)$ and the forces $F(l'b')$. Linearly independent vectors $\tilde{u}(lb)$ and $\tilde{F}(l'b')$ are stored. A linear transformation is applied to generate linearly independent Cartesian components $\tilde{u}_\alpha(lb)$ and $\tilde{F}_\alpha(l'b')$. Harmonic force constants are calculated as

$$\Phi_{\alpha\beta}(lb;l'b') = -\frac{\tilde{F}_\alpha(l'b)}{\tilde{u}_\alpha(lb)}. \tag{3.76}$$

The calculated force constant matrix is again symmetrized with respect to the space group operations $\mathscr{O}$. Finally, the 'C-type' dynamical matrix is generated

$$C_{\alpha\beta}(bb'|q) = \frac{1}{\sqrt{m_b m_{b'}}} \sum_{l'} \Phi_{\alpha\beta}(lb;l'b') \exp\left[-iq \cdot (x(lb) - x(l'b'))\right], \tag{3.77}$$

where m_b is the mass of the bth atom in the lth cell with its equilibrium position vector $x(lb)$. Phonon eigensolutions are obtained by solving the secular equation (see equation (2.61)). Kress *et al* (1995) performed calculations on a $(2 \times 2 \times 2)$ cubic supercell containing 64-atom for diamond and on a $(3 \times 3 \times 2)$ hexagonal supercell containing 72 atoms for graphite. The phonon dispersion curves, shown in figure 3.8, are in very good agreement with neutron scattering data.

As mentioned earlier, for polar systems calculations of the LO mode close to the BZ centre requires additional effort to deal with the macrscopic electric field arising from the ionic displacements. In the previous sub-section we cited the work of Kunc and Martin (1982) which subtracts the depolarizing forces from the total forces to obtain the relavant force constant. In principle, this method can be adopted within the supercell scheme discussed in this sub-section. An alternative scheme was adopted by Parlinski *et al* (2000). In this scheme, LO phonon frequencies obtained for a few finite q points from calculations using an elongated supercell are extrapolated to obtain the result in the limit $q \to 0$.

3.3.2 LINEAR RESPONSE METHODS

In contrast to using a periodically repeated supercell in direct methods, linear response methods use the primitive or the smallest repeated unit cell for evaluating the harmonic force constant. For a given lattice distortion u, the total electronic energy is expressed as

$$E_{el}(u) = E_{el}(0) + \delta E_{el}(u, \delta\rho), \tag{3.78}$$

where $\Delta\rho$ is rhe resulting change in the electronic charge density. Writing $E_{tot} = E_{ion-ion} + E_{el}$ (see equation (3.43)), the harmonic force constant matrix elements are expressed as

$$\Phi_{\alpha,\beta} = \frac{\partial^2 E_{ion-ion}}{\partial\tau_\alpha\partial\tau_\beta} + \int dr\rho(r)\frac{\partial^2 V_{ext}}{\partial\tau_\alpha\partial\tau_\beta} + \int dr\frac{\partial\rho(r)}{\partial\tau_\alpha}\frac{\partial V_{ext}(r)}{\partial\tau_\beta} \tag{3.79}$$

$$= -\frac{\partial F_\beta^{ion}}{\partial\tau_\alpha} - \int dr\rho(r)\frac{\partial F_\beta^{el}(r)}{\partial\tau_\alpha} - \int dr\frac{\partial\rho(r)}{\partial\tau_\alpha}F_\beta^{el}(r) \tag{3.80}$$

$$= \Phi_{\alpha,\beta}^{ion} + \Phi_{\alpha,\beta}^{el}. \tag{3.81}$$

From the above it is clear that while the ionic contribution Φ^{ion} and the first term in the electronic contribution Φ^{el} involve only the unperturbed electronic charge density $\rho(r)$, the second term in Φ^{el} requires knowledge of the change in $\rho(r)$ with respect to the displacement in the atomic coordinate τ_b.

In the linear response theory the induced charge density change $\delta\rho$ is related to the change in the externally applied potential $\delta V_{ext}(= \delta V_{ps})$

$$\delta\rho = \chi\delta V_{ps} \tag{3.82}$$

with χ as the *density response matrix*. With this the last term in equation (3.79) can be expressed as

$$\int dr\frac{\partial\rho(r)}{\partial\tau_\alpha}\frac{\partial V_{ps}(r)}{\partial\tau_\beta}$$

$$= \int\int drdr'\frac{\partial V_{ps}(r')}{\partial\tau_\alpha}\frac{\partial\rho(r)}{\partial V_{ps}(r')}\frac{\partial V_{ps}(r)}{\partial\tau_\beta} \tag{3.83}$$

$$= \int\int drdr'\frac{\partial V_{ps}(r')}{\partial\tau_\alpha}\chi(r,r')\frac{\partial V_{ps}(r)}{\partial\tau_\beta}. \tag{3.84}$$

Another quantity of interest is the *polarisability matrix* $\tilde{\chi}$ defined as

$$\delta\rho = \tilde{\chi}\delta V^{ind}, \tag{3.85}$$

where δV^{ind} is the potential induced by the applied potential δV_{ps}

$$\delta V^{ind} = \varepsilon^{-1}\delta V_{ps}. \tag{3.86}$$

The inverse dielectric matrix ε^{-1} is related to χ and $\tilde{\chi}$ through equations (3.82)–(3.85)

$$\chi = \tilde{\chi}\varepsilon^{-1}. \tag{3.87}$$

The above discussion outlines two schemes for implementing the linear response theory for phonon calculations: the DFPT using $\{\frac{\partial\rho}{\partial\tau_\alpha}\}$ as in equation (3.79) or the *dielectric matrix theory* using $\{\frac{\partial V_{KS}}{\partial\tau_\alpha}\}$ and χ as in equation (3.84). We will describe both schemes.

3.3.2.1 Density functional perturbation theory (DFPT)

We will discuss the DFPT using the plane-wave pseudopotential approach, as developed by Giannozzi *et al* (1991), de Gironcoli (1995) and Baroni *et al* (2001). As noted in equation (3.79) determination of the electronic part of the harmonic force constant $\Phi_{\alpha,\beta}^{el}$ requires numerical evaluation of changes to the electronic charge density and the Kohn–Sham potential energy with respect to lattice distortion. An atomic displacement u of periodicity q from its equilibrium position τ in a periodic crystal will impose a perturbation ΔV_{ps}, which in turn will generate a change ΔV_{KS} in the self-consistent

Kohn–Sham potential energy and the corresponding change $\Delta\rho$ in the ground state electronic charge density. These changes are inter-related following the equation below

$$\Delta V_{\text{KS}}(r) = \Delta V_{\text{ps}}(r) + e^2 \int dr' \frac{\Delta\rho(r')}{|r-r'|} + \Delta\rho(r) \frac{dV_{\text{KS}}}{d\rho}. \tag{3.88}$$

Using first-order perturbation theory, the linear variation $\Delta\rho$ in the charge density (see equation (3.62)) can be expressed as .

$$\Delta\rho(r) = 4\Re \sum_j^{\text{occ}} \phi_j^*(r) \Delta\phi_j(r) \tag{3.89}$$

and obtained by solving the equations

$$(H_{\text{scf}} - \varepsilon_j)|\Delta\phi_j> = -(\Delta V_{\text{KS}} - \Delta\varepsilon_j))|\phi_j> \tag{3.90}$$

and

$$\Delta\varepsilon_j = <\phi_j|\Delta V_{\text{KS}}|\phi_j>, \tag{3.91}$$

where H_{scf} is the self-consistent Hamiltonian. Using equations (3.89)-(3.91) the expression for $\Delta\rho$ reads

$$\Delta\rho(r) = 4 \sum_j^{\text{occ}} \sum_{m\neq j} \phi_j^*(r) \phi_m(r) \frac{<\phi_m|\Delta V_{\text{KS}}|\phi_j>}{\varepsilon_j - \varepsilon_m}. \tag{3.92}$$

The sum over unoccupied states (conduction bands) in equation (3.92) can be avoided by projecting onto the unoccupied states of the correction to the occupied state orbitals (see Giannozzi *et al* 1991 and Baroni *et al* 2001 for details). A Fourier component of $\Delta\rho$ is then expressed as

$$\Delta\rho(q+G) = \frac{4}{N_0\Omega} \sum_k \sum_j^{\text{occ}} <\phi_{j,k}|e^{-i(q+G)\cdot r} P_c|\Delta\phi_{j,k+q}>, \tag{3.93}$$

where k is the electron wavevector, q is the phonon wavevector, $\Delta\phi_{j,k+G}$ is the solution of the linear system

$$[\varepsilon_{j,k} - H_{\text{scf}}]|\Delta\phi_{j,k+q}> = P_c \Delta V_{\text{KS}}(q)|\phi_{j,k}> \tag{3.94}$$

and P_c is the projector over the electronic unoccupied-state manifold. Expressions for Fourier components of relevant terms in V_{KS} can easily be obtained by following the development in sub-section 3.2.3.1.

It should be pointed that in the discussion of the application of the DFT and DFPT methods, we have tacitly assumed non-metallic systems (*i.e.* systems with finite gap in electronic band structure). A slight extension is required for application of the method to metallic systems. Essentially the BZ integration needs to be carried out by smearing electronic bands to account for occupied states up to Fermi energy. Many kinds of smearing techniques can be used: Fermi-Dirac broadening, Lorentzian, Gaussian, or Gaussian combined with polynomials, etc. Moreover, we have assumed spin degenerate electronic states. Consideration of spin-polarised states is straightforward. The DFPT is also routinely applied with the inclusion of spin-orbit interaction in DFT electronic calculations. These developments have been covered in many publications, notably in the book by Martin (2004) and in the publication by Verstraete *et al* (2008). Furthermore, a real-space formalism of DFPT using localized, atom-centred basis set has also been recently developed Shang *et al* (2017).

The momentum-space expression for the electronic part of the harmonic force constant is (Baroni *et al* 2001)

$$\Phi_{\alpha\beta}^{\text{el}}(bb'|q) = \sum_p e^{-iq\cdot p} \Phi_{\alpha\beta}^{\text{el}}(p) = \frac{1}{N_0} \frac{\partial^2 E_{\text{tot}}}{\partial X_\alpha^\dagger(q,b) \partial X_\beta(q,b')}$$

$$
= \frac{1}{N_0} \left[\int d\boldsymbol{r} \left(\frac{\partial \rho(\boldsymbol{r})}{\partial X_\alpha(\boldsymbol{q}, \boldsymbol{b})} \right)^\dagger \frac{\partial V_{\mathrm{ps}}(\boldsymbol{r})}{\partial X_\beta(\boldsymbol{q}, \boldsymbol{b}')} \right.
$$
$$
\left. + \int d\boldsymbol{r}\, \rho(\boldsymbol{r}) \frac{\partial^2 V_{\mathrm{ps}}(\boldsymbol{r})}{\partial X_\alpha^\dagger(\boldsymbol{q}, \boldsymbol{b}) \partial X_\beta(\boldsymbol{q}, \boldsymbol{b}')} \right], \tag{3.95}
$$

where $\boldsymbol{p} = \boldsymbol{l} - \boldsymbol{l}'$ is a difference vector between two unit cells located at $\boldsymbol{l}$ and $\boldsymbol{l}'$, and $\boldsymbol{X}(\boldsymbol{q}, \boldsymbol{b})$ is a Fourier component of the displacement vector for the bth atom (see, section 4.2).

The ionic contribution to the harmonic force constant is obtained as the second derivative of the energy term $E_{\mathrm{ion-ion}}$ (cf. equations (3.46) and (3.53)) or the first derivative of the force term F_{ion} (cf. equation (3.54) and (3.59)). The result is (see, Maradudin *et al* 1971 and Baroni *et al* 2001)

$$
\begin{aligned}
\Phi_{\alpha\beta}^{\mathrm{ion}}(bb'|\boldsymbol{q}) &= \frac{4\pi e^2}{\Omega} \sum_G \frac{e^{-(\boldsymbol{q}+\boldsymbol{G})^2/4\eta}}{(\boldsymbol{q}+\boldsymbol{G})^2} (q_\alpha + G_\alpha)(q_\beta + G_\beta) e^{i(\boldsymbol{q}+\boldsymbol{G})\cdot(\tau_b - \tau_{b'})} \\
&\quad - \frac{2\pi e^2}{\Omega} \sum_{G \neq 0} \frac{e^{-G^2/4\eta}}{G^2} \left[z_b \sum_{b''} z_{b''} G_\alpha G_\beta e^{iG\cdot(\tau_b - \tau_{b''})} + c.c \right] \delta_{bb'} \\
&\quad + e^2 \sum_l z_b \left\{ z_{b'} e^{i\boldsymbol{q}\cdot\boldsymbol{l}} \left[f_2(x)\delta_{\alpha\beta} + f_1(x) x_\alpha x_\beta \right]_{x=\tau_b - \tau_{b'} - l} \right. \\
&\quad \left. - \delta_{bb'} \sum_{b''} z_{b''} \left[f_2(x)\delta_{\alpha\beta} + f_1(x) x_\alpha x_\beta \right]_{x=\tau_b - \tau_{b''} - l} \right\}, \tag{3.96}
\end{aligned}
$$

where the sum over $\boldsymbol{G}$ excludes $\boldsymbol{q} + \boldsymbol{G} = 0$, the sum over the direct space translation vector $\boldsymbol{l}$ excludes $\tau_b - \tau_{b''} - \boldsymbol{l} = 0$ and the functions f_1 and f_2 are defined as

$$
f_1(x) = \frac{3\,\mathrm{erfc}(\sqrt{\eta}x) + 2\sqrt{\frac{\eta}{\pi}} x (3 + 2\eta x^2) e^{-\eta x^2}}{x^5} \tag{3.97}
$$

$$
f_2(x) = \frac{-\,\mathrm{erfc}(\sqrt{\eta}x) - 2\sqrt{\frac{\eta}{\pi}} x e^{-\eta x^2}}{x^3}. \tag{3.98}
$$

The discussion above for $\Phi^{\mathrm{ion-ion}}$ is complete for all phonon modes, except for LO modes with $\boldsymbol{q} \to 0$ in polar materials. As discussed in section 2.4.4, the long-range character of the Coulomb forces produces a macroscopic field to which long-wavelength LO phonons are coupled. The corresponding electronic potential is not lattice periodic, which results in a non-analytic part of the force constant $\Phi^{\mathrm{ion-ion}}$ for zone-centre LO phonon modes. This contribution has the form (Cochran and Cowley 1962)

$$
{}^{\mathrm{non-analytic}}\Phi_{\alpha\beta}^{\mathrm{ion}}(bb'|\boldsymbol{q} \to 0) = \frac{4\pi}{\Omega} e^2 \frac{(\boldsymbol{q} \cdot \boldsymbol{Q}_b^\star)_\alpha (\boldsymbol{q} \cdot \boldsymbol{Q}_b^\star)_\beta}{(\boldsymbol{q} \cdot \varepsilon_\infty \cdot \boldsymbol{q})}, \tag{3.99}
$$

where $\boldsymbol{Q}_b^\star$ is the *Born effective charge* of the bth ion and ε_∞ is the optical dielectric constant.

The DFPT has emerged as popular for determining phonon eigensolutions as the DFT is for electronic eigensolutions. Indeed, the DFPT scheme has been very successfully applied to accurately reproduce measured phonon dispersion relation in bulk materials, and in predicting results for low-dimensional materials. Results for a few low-dimensional systems will be presented in later chapters. Here we present results for a few materials, extending compilation of results presented earlier for other materials obtained from other methods (e.g. for Ne and Si in figure 2.12, for Ge in figures 2.15 and 3.5, for GaAs in figure 2.16 and for diamond and graphite in figure 3.8).

Figure 3.9 shows the phonon dispersion curves for graphene, obtained from the application of a real-space formalism of DFPT (Shang *et al* 2017). The DFPT results are validated by being in excellent agreement with the results obtained with force constants determined with finite-difference

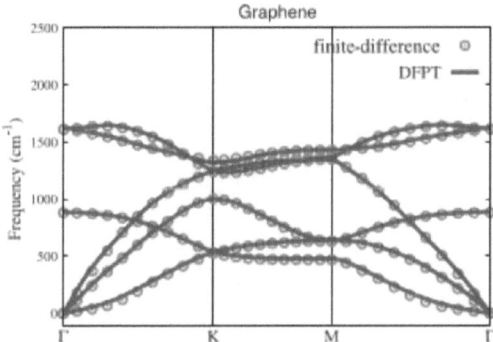

Figure 3.9 Phonon dispersion curves for graphene, obtained from the application of a real-space formalism of DFPT using the all-electron potential, an atomic-orbital basis set and the LDA. Reproduced from Shang *et al* (2017).

calculations. Comparison with figure 3.8 shows that the phonon dispersion curves of graphene (two-dimensional or monolayer structure) and graphite (quasi-three-dimensional structure) are very similar. This is not surprising, as the van der Waals type inter-layer interaction in graphite is very weak compared to the strong covalent inter-layer bonding. In contrast, the phonon dispersion spectrum of the three-dimensional diamond structure is quite different. The maximum frequency for diamond is much smaller than for graphite or graphene. This is largely due to the inter-layer bonding in graphite and graphene being much stronger than the covalent bonding in diamond. While the long wavelength TA and LA modes show linear dispersion for graphene, graphite and diamond, the long wavelength ZA branch in graphene and graphite shows a quadratic dispersion. This is typical of layered materials. The ZO branch in graphite, the optical counterpart of the out-of-plane acoustic ZA branch, has a considerable lower frequency than the other optical branches.

Figure 3.10 shows the phonon dispersion curves for the fcc simple metals Al and Pb and for the bcc transition metal Nb, obtained from the application of the plane-wave pseudopotential formalism of DFPT (de Gironcoli 1995). The theoretical results for Al and Nb agree very well with experimental data. For Pb, however, the theoretical results are at variance with experimental data along all the three high symmetry directions. This has been identified by Verstraete *et al* (2008) as a consequence of the lack of spin-orbit coupling (SOC) in the theoretical calculations reported by de Gironcoli (1995). Panel (b) in figure 3.10 shows the phonon dispersion curves of Pb from the DFPT calculations by Verstraete *et al* (2008) with and without the inclusion of SOC. They find that the inclusion of SOC changes the phonon modes at many q points appreciably. In general, the TA modes are softened and the LA modes are slightly hardened. The softening of the lower TA mode is strongest at the X point, where its frequency is almost halved.

Phonon dispersion curves for complex crystal structures

The DFPT technique has also been successfully applied to map out phonon dispersion curves and density of states for complex crystal structures. Here we present two examples.

In the first example we consider skutterudite crystal structure. In figures 3.11 and 3.12 we present the crystal structure and phonon dispersion curves for the lanthanide skutterudite $LaRu_4As_{12}$ which crystallizes in the $CoAs_3$-type skutterudite filled by La atoms and is a superconductor with critical temperature of 10.45 K. The BCC primitive unit cell contains 17 atoms, resulting in 3 acoustic branches and 48 optical branches. The superconductivity in this material is intrinsic to the $[Ru_4As_{12}]$ polyanion and La electronically stabilizes the structure. A trademark feature of this and similar filled-skutterudite compounds is the existence of a flat electronic band very close to the Fermi level, which generates a peak in the electronic density of states and gives rise to an enhancement in their

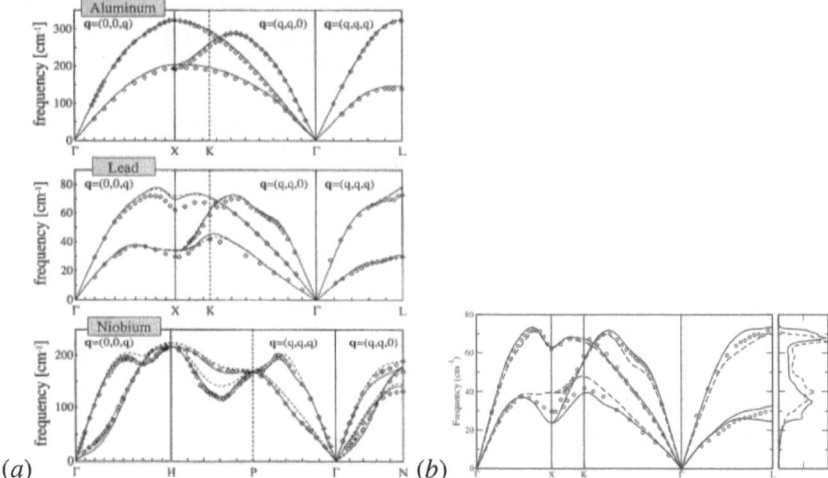

Figure 3.10   (a) Phonon dispersion curves for the fcc simple metals Al and Pb and for the bcc transition metal Nb, obtained from the application of the plane-wave pseudopotential formalism for DFPT–LDA. Solid and dashed curves are obatined by using different smearing widths for BZ summations for metallic systems. Also shown by diamonds are experimental data. Reproduced from de Gironcoli (1995). (b) Effect of spin-orbit coupling on phonon modes in Pb, calculated by Verstraete *et al* (2008) using the plane-wave pseudopotential formalism for DFPT-GGA. Reproduced from Verstraete *et al* (2008).

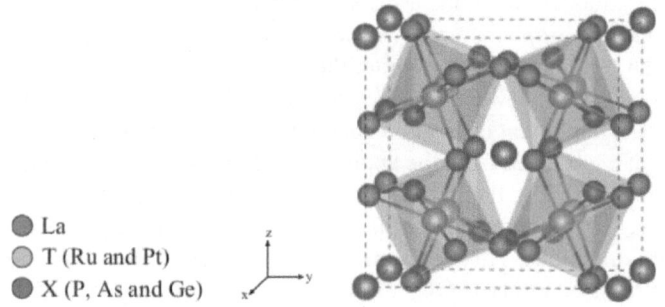

Figure 3.11   The BCC structure of filled skutterudite with the general formula LnT_4X_{12} (Ln = lanthanide, T = transition metal, X = pnictogen). (Original diagram in colour) Reproduced from Tütüncü *et al* (2017).

superconducting properties. Absence of imaginary phonon frequencies in the vibrational spectrum indicates that $LaRu_4As_{12}$ is dynamically stable in its BCC structure. Partial phonon density of states results show that the heaviest element La dominates the low-frequency region up to 3.0 THz and strong hybridization between Ru-related and As-related vibrations is present in the frequency region above 6.5 THz.

In the second example, we consider semiconducting clatherate structures. These are open-structured compounds consisting of three-dimensional network of atoms. Clatherate frameworks formed by group IV elements are of the type X_mE_n, where E represents a group IV element and X represents a 'guest' or 'filler' atom in the voids, or cages, formed by E type atoms. Pristine clathrate frameworks, however, do not contain any guest atoms or molecules in the voids.

Clatherates are classified into various types. The crystal structure for semiconducting type-II and type-VIII pristine Si clathrates are shown in figure 3.13. The Bravais lattice for the type-II structure

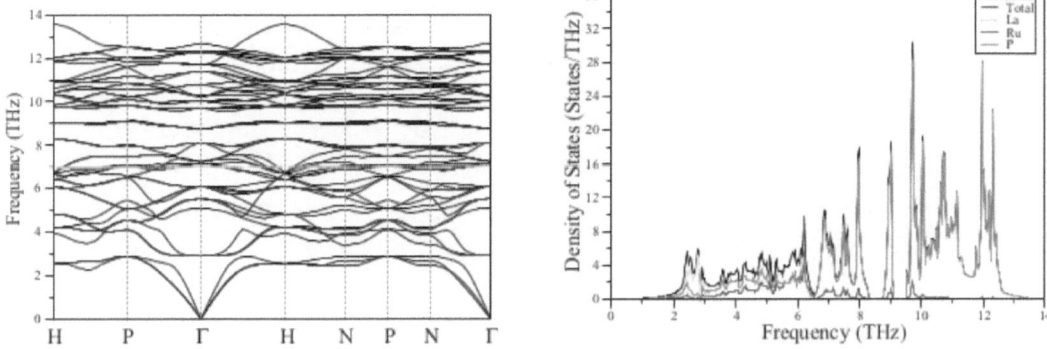

Figure 3.12 Phonon dispersion curves (left panel) and total and partial density of states (right panel) for the BCC filled skutterudite LaRu$_4$As$_{12}$, obtained from the application of the plane-wave pseudopotential formalism of DFPT-GGA. (Original diagram in colour) Reproduced from Tütüncü *et al* (2017).

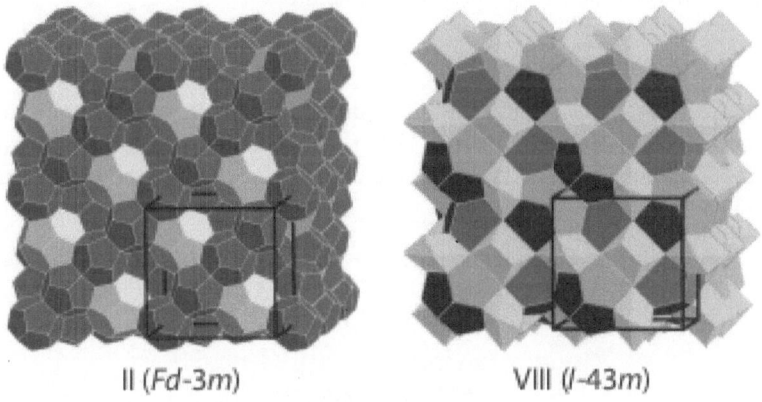

II (*Fd*-3*m*)   VIII (*I*-43*m*)

Figure 3.13 Crystal structure of pristine Si clatherate structures II and VIII. (Original diagram in colour) Reproduced from Härkönen and Karttunen (2016).

is FCC and there are 34 Si atoms per unit cell (denoted as Si$_{34}$), with space group Fd$\bar{3}$m. Sometimes this is denoted as Si$_{136}$ when referring to 136 Si atoms per cubic unit cell. The building blocks are 20-atom and 28-atom cages. The Si$_{20}$ cage is composed of 12 five-membered rings, and the Si$_{28}$ cage is composed of 12 five-membered rings and 2 six-membered rings (Karttunen *et al* 2011). The Bravais lattice for the VIII framework is BCC and there are 23 Si atoms per unit cell (denoted as Si$_{23}$), with space group I$\bar{4}$3m. Sometimes this is denoted as Si$_{46}$ when referring to 46 Si atoms per cubic unit cell. The basic framework contains three-, four- and five-membered rings (Karttunen *et al* 2011). In both type-II and type-VIII structures, the atoms are tetrahedrally coordinated. However, Si-Si interatomic distances are slightly longer than in diamond-type Si and the tetrahedral arrangement of neighbouring atoms is distorted, with some distribution of bond angles around the perfect tetrahedral angle 109.5^o. The electronic band gap in these structures is larger than in diamond-Si.

The phonon dispersion curves and density of states for the type-II and type-VIII Si clatherates, obtained from the application of the DFPT described in section 3.3.2.1 (Harkonen and Karttunen, 2016), are presented in figure 3.14. For both structures, clear span of acoustic branches is visible only upto approximately 100 cm^{-1} (3 THz). Between 100 and 200 cm^{-1} (3–6 THz) there are several flat-band (molecular like) branches. Branches above 350 cm^{-1} (10.5 THz) are flat (molecular like).

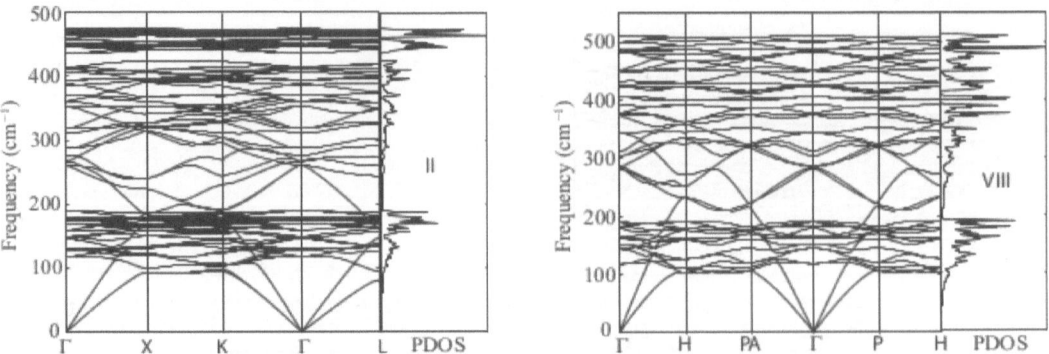

Figure 3.14 Phonon dispersion curves for pristine Si clatherate structures II and VIII. Reproduced from Härkönen and Karttunen (2016).

There are up to two true gaps in the spectrum above 400 cm^{-1} (12 THz). The gap slightly above 400 cm^{-1} is comparatively broader for the type-II structure.

3.3.2.2 The dielectric matrix method

As discussed in equations (3.82)–(3.87), an alternative to the DFPT is to treat the linear response theory via the inverse of the dielectric matrix. Using the Hohenberg–Kohn theorem (Appendix A) one can derive the following expression for the dielectric matrix within the LDA (Sham and Kohn 1966, Martin and Kunc 1983):

$$\varepsilon = 1 - \tilde{\chi}V_{\mathrm{H}} \tag{3.100}$$
$$= 1 - \tilde{\chi}_0(1 - V_{\mathrm{xc}}\tilde{\chi}_0)^{-1}V_{\mathrm{H}} \tag{3.101}$$

and

$$\varepsilon^{-1} = 1 + \chi V_{\mathrm{H}}, \tag{3.102}$$

where V_{H} and V_{xc} are the Coulomb (Hartree) and exchange-correlation potentials, respectively. $\tilde{\chi}_0$ is the polarisability matrix in the random-phase approximation (RPA) (time-dependent Hartree form) whose elements are given by (Wiser 1963)

$$\begin{aligned}
\tilde{\chi}_0(q+G, q+G') &= \frac{1}{N_0\Omega} \sum_{\substack{n,n' \\ (n\neq n')}} \sum_k^{BZ} \frac{\bar{f}_{n'}(k+q) - \bar{f}_n(q)}{\varepsilon_{n'}(k+q) - \varepsilon_n(k)} \\
&\quad \times \langle n,k| \exp\left[-\mathrm{i}(q+G)\cdot r\right]|n',k+q\rangle \\
&\quad \times \langle n',k+q| \exp\left[\mathrm{i}(q+G')\cdot r\right]|n\cdot k\rangle.
\end{aligned} \tag{3.103}$$

Here $|n,k\rangle$ represents the single-particle Bloch state with wave vector k, band index n and energy $\varepsilon_n(k)$, and the occupation number $\bar{f}_n(k)$ is the Fermi–Dirac distribution function. The band index n includes all valence and conduction bands, G and G' are reciprocal lattice vectors and q is a phonon wave vector.

From equations (3.100)–(3.103), we see that the construction of the matrices ε and χ requires computation of $\tilde{\chi}_0, V_{\mathrm{H}}$ and V_{xc} from the output of an electronic band structure study for k-points throughout the Brillouin zone. Clearly, evaluation of ε and χ is extremely time-consuming. The labour in the evaluation of these matrices can be reduced by using (i) group symmetry considerations to restrict the sum over k to within the irreducible part of the zone, and (ii) a special k-points scheme for Brillouin zone summation. Furthermore, the slow convergence of the sum in equation (3.60) which includes *all* conduction bands can be speeded up by using the moment expansion method developed by van Camp *et al* (1983). Let us write equation (3.103) as

$$\tilde{\chi}_0 = \frac{1}{N_0\Omega} \sum_{\substack{m,m' \\ (m\neq m')}} \frac{\bar{f}_m - \bar{f}_{m'}}{\varepsilon_m - \varepsilon_{m'}} \langle m|z|m'\rangle \langle m'|z'|m\rangle, \tag{3.104}$$

where

$$|m\rangle = |n,\mathbf{k}\rangle, \qquad |m'\rangle = |n',\mathbf{k}+\mathbf{q}\rangle$$
$$z = \exp[-\mathrm{i}(\mathbf{q}+\mathbf{G})\cdot\mathbf{r}], \qquad z' = \exp[\mathrm{i}(\mathbf{q}+\mathbf{G}')\cdot\mathbf{r}]. \tag{3.105}$$

Expand $(\varepsilon_m - \varepsilon_{m'})^{-1}$ in a finite geometrical series with $p+1$ terms and define moments as follows:

$$\mu_s = \sum_{m'} \varepsilon_{m'}^s \langle m|z|m'\rangle \langle m'|z'|m\rangle. \tag{3.106}$$

Then we can express

$$\tilde{\chi}_0 = \frac{2}{N_0\Omega} \left[\sum_m \bar{f}_m \left(\frac{\mu_0}{\varepsilon_m} + \frac{\mu_1}{\varepsilon_m^2} + \dots + \sum_{m'} \frac{\varepsilon_m^p \langle m|z|m'\rangle \langle m'|z'|m\rangle}{\varepsilon_m^{p+1}(\varepsilon_m - \varepsilon_{m'})} \right) \right]. \tag{3.107}$$

Letting $p \to \infty$ the series in equation (3.107) can be transformed into a continued fraction

$$\tilde{\chi}_0 = \frac{2}{N_0\Omega} \sum_m \bar{f}_m \cfrac{\alpha_1}{\varepsilon_m + \cfrac{\alpha_2}{\varepsilon_m + \cfrac{\alpha_3}{\varepsilon_m + \dots}}} \tag{3.108}$$

where

$$\alpha_1 = \mu_0, \quad \alpha_2 = -\frac{\mu_1}{\mu_0}, \quad \alpha_3 = \frac{\mu_1^2 - \mu_0\mu_2}{\mu_0\mu_1}, \quad \text{etc} \dots . \tag{3.109}$$

In practice equation (3.108) can be truncated after a few terms. The advantage of the moment expansion scheme is that only valence states are required. van Camp *et al* (1983) have presented a discussion of the polarisability matrix calculation using the moment expansion method and the direct summation over the conduction bands. The question of the convergence of the moment expansion method must be checked carefully before proceeding with phonon calculations.

The normal modes of lattice vibration with wave vector $\mathbf{q}$ are determined by the 'D-type' dynamical matrix

$$D_{\alpha\beta}(\mathbf{bb'}|\mathbf{q}) = \sum_l \Phi_{\alpha\beta}(\mathbf{0b};\mathbf{lb'})\exp(\mathrm{i}\mathbf{q}\cdot\mathbf{l}), \tag{3.110}$$

where Φ is the force constant matrix and $\mathbf{l}$ is a direct lattice vector. The matrix D has contributions from ion–ion and el–ion parts. The ion–ion part of the matrix can be straightforwardly evaluated using Ewald's method (see Kellermann 1940 and section 3.3.2.1). The electron–ion part can be expressed as (Sham 1969)

$$D_{\alpha\beta}^{ei}(\mathbf{bb'}|\mathbf{q}) = \frac{1}{\sqrt{m_b m_{b'}}} \left(X_{\alpha\beta}(\mathbf{bb'}|\mathbf{q}) - \delta_{bb'} \sum_{b''} X_{\alpha\beta}(\mathbf{bb''}|0) \right), \tag{3.111}$$

where

$$\begin{aligned}
X_{\alpha\beta}(\mathbf{bb'}|\mathbf{q}) = {} & \frac{1}{\Omega} \sum_{G,G'} (\mathbf{q}+\mathbf{G})_\alpha \exp[\mathrm{i}(\mathbf{q}+\mathbf{G})\cdot\boldsymbol{\tau}_b] w_b(|\mathbf{q}+\mathbf{G}|) \\
& \times \chi(\mathbf{q},\mathbf{G},\mathbf{G}') w_{b'}(|\mathbf{q}+\mathbf{G}'|)
\end{aligned}$$

$$\times \exp\left[-\mathrm{i}(\boldsymbol{q}+\boldsymbol{G}')\cdot\boldsymbol{\tau}_{b'}\right](\boldsymbol{q}+\boldsymbol{G}')_\beta. \tag{3.112}$$

Here m_b is the mass of the bth ion in the unit cell.

Although the moment expansion method is fairly simple, it is important to test the convergence of the series in equation (3.107). In a calculation of phonon spectrum of Si, van Camp *et al* (1983) only considered the zeroth and first moments, but their results are not particularly good.

3.3.2.3 A simple perturbative method

A simple perturbative method for lattice dynamics of covalent crystals, based on a local pseudopotential scheme, has been presented by Morita *et al* (1972), Soma and Morita (1972), and Soma (1976, 1978a). In this method crystal properties are calculated without doing the band structure calculation.

In the nearly free electron picture the total energy of the crystal can be expressed as

$$E_{\mathrm{tot}} = E'_{\mathrm{ion-ion}} + E'_{\mathrm{el}} \equiv \gamma_{\mathrm{E}} + E'_{\mathrm{el}} \tag{3.113}$$

with

$$E'_{\mathrm{el}} = E_0^0 + E_1 + E_2 + \gamma_c E_{\mathrm{cov}}. \tag{3.114}$$

Here γ_{E} is the Ewald energy as in equations (3.48) and (3.53), E_0^0 is the total of the kinetic, exchange and correlation energies of a uniform gas of electrons, E_1 and E_2, are, respectively, first- and second-order perturbation energies of the electron gas due to the pseudopotential and E_{cov} is the covalent bond correction to the crystal energy corresponding to higher-order perturbations. γ_c is an adjustable parameter which modifies the contribution of E_{cov}.

The energy E_0^0, per atom, is given by

$$\frac{E_0^0}{z} = \frac{2.21}{r_\mathrm{s}^2} - \frac{0.916}{r_\mathrm{s}} + E_{\mathrm{corr}}, \tag{3.115}$$

where r_s is the average interelectronic distance and E_{corr} is the free electron correlation energy. The energy E_1, per atom, is obtained from equation (3.49):

$$\frac{E_1}{z} = \frac{1}{M}\left[\lim_{G\to 0}\left(\sum_b^M V_b(|\boldsymbol{G}|) + \frac{4\pi z e^2}{\Omega_{\mathrm{at}} G^2}\right)\right], \tag{3.116}$$

where Ω_{at} is the atomic volume, M is the number of atoms in the unit cell and $z = \sum_b z_b$. The second-order perturbation energy of the electron gas E_2 is also called the band structure energy, and is given by (Harrison 1966)

$$E_2 = \frac{\Omega_{\mathrm{at}}}{8\pi e^2}\sum_{G\neq 0}\frac{G^2}{(1-f(G))}\left(\frac{1}{\varepsilon_g^*(G)}-1\right)|V(G)|^2. \tag{3.117}$$

where $V(\boldsymbol{G}) = \sum_b S_b(\boldsymbol{G})v_b(|\boldsymbol{G}|)$ is the crystal ionic (local) pseudopotential, $S_b(\boldsymbol{G})$ is the structure factor for the bth atom, and $\varepsilon_g^*(G)$ is the dielectric function for a uniform gas of free electrons, given by

$$\varepsilon_g^*(G) = 1 + \frac{4\pi e^2 z}{\Omega_{\mathrm{at}} G^2}\times(1-f(G))\frac{1}{\frac{2}{3}E_F}\left(\frac{1}{2}+\frac{1-y^2}{4y}\ln\left|\frac{1+y}{1-y}\right|\right) \tag{3.118}$$

with $y = G/2k_F$, k_F being the Fermi wave vector. $f(G)$ is the Hubbard exchange factor

$$f(G) = \frac{\frac{1}{2}G^2}{G^2 + k_F^2 + 2k_F/\pi}. \tag{3.119}$$

With $f = 0$, equation (3.118) becomes the Lindehart dielectric function.

It can be noticed that the present expression for the dielectric function represents a simplified version of the diagonal approximation of the dielectric matrix given in equation (3.101). Although this approximation is found to be quite good in the theory of lattice dynamics of simple metals (Joshi and Rajagopal 1968), it is found to produce incorrect results for crystals with localised valence electrons, e.g. semiconductors and insulators. In covalent semiconductors, this effect is accounted for by the additional term E_{cov}. For diamond and zincblende structure semiconductors, Morita *et al* (1972) have used the isotropic energy-gap model of Heine and Jones (1969) in the calculation of E_{cov}. The Jones zone of these crystals is nearly spherical (see section 1.8 and figure 1.2(*c*)) and its zone faces can be matched to the Fermi surface. Heine and Jones showed that the band gap at the point $k_X = (1,1,0)$ on the Jones zone faces is approximately given by

$$E_g^X = 2|U_{\text{eff}}(220)|, \tag{3.120}$$

where $U_{\text{eff}}(220)$ is an effective screened pseudopotential which includes higher order terms beyond the first-order result $U(220)$:

$$U_{\text{eff}}(220) = U(220) + \sum_Q \frac{\langle -k_X|U|k_X - Q\rangle \langle k_X - Q|U|k_X\rangle}{(\hbar^2/2m)[k_X^2 - (k_X - Q)^2]}. \tag{3.121}$$

Here Q is restricted to $\{111\}$ plane waves, such that $k_X - Q = (0,0,\pm 1)$, and the Fourier transform of U in the summation is of the type $U(111)$. The screened potential U is constructed using the linear screening scheme

$$U(G) = \frac{1}{\varepsilon_g^*(G)} \frac{1}{M} \sum_b S_b(G) u_b(|G|) = \frac{1}{M} \sum_b S_b(G) u_b(|G|), \tag{3.122}$$

where, for the bth atom, $S_b(G) = \exp(iG \cdot \tau_b)$ is the structure factor and $u_b(|G|)$ is the screened atomic pseudopotential. The covalent bond correction E_{cov} to the crystal energy is given by (Morita *et al* 1972)

$$E_{cov} = -g(E_F)\{|U_{\text{eff}}(220)|^2 - |U(220)|^2\}, \tag{3.123}$$

where $g(E_F) = z/\frac{2}{3}E_F$ is the density of states per atom of the uniform gas of valence electrons at the Fermi surface.

The force constants for the dynamical matrix in equation (3.110) are given by

$$\Phi_{\alpha\beta}(0b; hb') = \Phi_{\alpha\beta}^0(0b; hb') + \gamma_c\Delta\Phi_{\alpha\beta}(0b; hb'). \tag{3.124}$$

Here Φ^0 corresponds to paired two-body forces as in second-order perturbation theory of simple metals, and is given by

$$\begin{aligned}
\Phi_{\alpha\beta}^0(0b; hb') &= \frac{1}{M} \sum_{q \neq 0} \left[\frac{4\pi z^2 e^2}{\Omega_{at}q^2} - |v_b(q)|^2 \left(1 - \frac{1}{\varepsilon_g^*(q)}\right) \right. \\
&\quad \left. \times \frac{\Omega_{at}q^2}{(1 - f(q))4\pi z e^2} \right] q_\alpha q_\beta \exp(iq \cdot h) \delta_{bb'},
\end{aligned} \tag{3.125}$$

where the first term represents the direct Coulomb force constant, and the second term represents the indirect two-body force constant. $\Delta\Phi$ represents unpaired three- and four-body forces corresponding to third- and fourth-order perturbation energy (E_{cov}) and is given by

$$\Delta\Phi_{\alpha\beta}(0b; hb') = \frac{1}{ME_F} \sum_Q^{\{220\}} U_{eff}^b(Q)$$

$$\times \sum_{q} \frac{u_b(|\boldsymbol{Q}-\boldsymbol{q}|)u_b(|\boldsymbol{q}|)}{(\hbar^2/2m)[(\tfrac{1}{2}\boldsymbol{Q})^2-(\tfrac{1}{2}\boldsymbol{Q}-\boldsymbol{q})^2]}$$

$$\times [(\boldsymbol{Q}-\boldsymbol{q})_\alpha q_\beta + q_\alpha(\boldsymbol{Q}-\boldsymbol{q})_\beta]\exp{(i\boldsymbol{q}\cdot\boldsymbol{h})}\delta_{bb'}. \tag{3.126}$$

Soma (1978a) and Soma *et al* (1981) have applied this approach to study the lattice dynamics of Si, Ge and α-Sn. Soma's group has also extended this approach to the study of lattice vibrations in partially ionic semiconductors, namely, III–V and II–VI compounds (Soma and Kagaya 1983a,b).

4 Anharmonicity

4.1 INTRODUCTION

In the harmonic approximation we analyse a crystal with N atoms as a set of 3N harmonic quantum oscillators or normal modes. Since a normal mode of vibration of a crystal is treated as a quantum of energy, and hence a quasi-particle – a phonon, we also associate with it a (quasi) momentum $\hbar q$. The use of the term 'quasi-momentum' is appropriate, since unlike a particle momentum the phonon momentum cannot increase indefinitely. When the phonon momentum increases by $\hbar G$ (where G is a reciprocal lattice vector), Bragg reflection of the normal mode takes place from atomic planes in the crystal and momentum $\hbar G$ is 'transferred' to the lattice as a whole.

In the harmonic approximation phonons are independent of each other. The vibrations of a real crystal, however, are not purely harmonic and the meaning of independent phonons breaks down. Anharmonicity leads to coupling between phonons of the harmonic crystal, which becomes more important as the temperature of the crystal increases. Phonon–phonon interaction is a typical many-body problem.

In this chapter we review the phonon Hamiltonian of a general three-dimensional crystal, with a view to analysing the lowest order anharmonicity. We then review the Hamiltonian of an elastic continuum, again with a view to analysing the lowest order anharmonicity. We simplify the expression for the Fourier transform of the anharmonic force tensor and derive an expression for the Grüneisen constant in the isotropic continuum approximation. Finally, we discuss the role of Grüneisen's constant in the crystal phase transition. Towards the end, two recent developments have been added to express the cubic and quartic anharmonic potentials beyond the elastic continuum level. One of these is a semi-*ab initio* approach, which derives expressions for the potentials within the elastic continuum approximation but can make use of phonon dispersion relations obtained at *ab initio* level. The other approach is fully at *ab initio* levels, based either on the density functional theory (DFT) or the density functional perturbation theory (DFPT).

4.2 HAMILTONIAN OF A GENERAL THREE-DIMENSIONAL CRYSTAL

Consider a general three-dimensional crystal. Consider a unit cell situated at a position vector l and let $b\{\equiv \tau_b\}$ be an atomic position in it. Let $x(lb)$ be the actual coordinates of the bth atom at a particular time t, so that

$$u(lb) = x(lb) - (l+b) \tag{4.1}$$

is the deviation of the atomic position from the equilibrium position $(l+b)$. We now expand the crystal potential energy $\mathscr{V}$ in a Taylor series in powers of the displacements $u(lb)$:

$$
\begin{aligned}
\mathscr{V} &= \mathscr{V}_0 + \sum_{lb\alpha} \frac{\partial \mathscr{V}}{\partial u_\alpha(lb)}\bigg|_0 u_\alpha(lb) \\
&\quad + \frac{1}{2} \sum_{lb,l'b'} \sum_{\alpha\beta} \frac{\partial^2 \mathscr{V}}{\partial u_\alpha(lb)\partial u_\beta(l'b')}\bigg|_0 u_\alpha(lb)u_\beta(l'b') \\
&\quad + \frac{1}{3!} \sum_{lb,l'b',l''b''} \sum_{\alpha\beta\gamma} \frac{\partial^3 \mathscr{V}}{\partial u_\alpha(lb)\partial u_\beta(l'b')\partial u_\gamma(l''b'')}\bigg|_0 \\
&\quad \times u_\alpha(lb)u_\beta(l'b')u_\gamma(l''b'') + \dots \\
&= \mathscr{V}_0 + \mathscr{V}_1 + \mathscr{V}_2 + \mathscr{V}_3 + \dots. \tag{4.2}
\end{aligned}
$$

DOI: 10.1201/9781003141273-4

For the equilibrium state to be a minimum in energy

$$\frac{\partial \mathcal{V}}{\partial x(lb)}\bigg|_0 = 0. \tag{4.3}$$

Also, the constant term $\mathcal{V}_0$ which fixes the zero of the potential can be set to zero. Equation (4.2) then reads

$$
\begin{aligned}
\mathcal{V} \;=\; & \frac{1}{2}\sum_{lb,l'b'}\sum_{\alpha\beta}\Phi_{\alpha\beta}(lb,l'b')u_\alpha(lb)u_\beta(l'b') \\
& + \frac{1}{3!}\sum_{lb,l'b',l''b''}\sum_{\alpha\beta\gamma}\Psi_{\alpha\beta\gamma}(lb,l'b',l''b'')u_\alpha(lb)u_\beta(l'b')u_\gamma(l''b'') \\
& + \ldots,
\end{aligned}
\tag{4.4}
$$

where Φ is a matrix (a cartesian tensor of second rank with 3^2 elements) and Ψ is a cartesian tensor of third rank with 3^3 elements, defined as

$$\Phi_{\alpha\beta}(lb,l'b') \;=\; \frac{\partial^2 \mathcal{V}}{\partial u_\alpha(lb)\partial u_\beta(l'b')}\bigg|_0 \tag{4.5}$$

$$\Psi_{\alpha\beta\gamma}(lb,l'b',l''b'') \;=\; \frac{\partial^3 \mathcal{V}}{\partial u_\alpha(lb)\partial u_\beta(l'b')\partial u_\gamma(l''b'')}\bigg|_0. \tag{4.6}$$

The matrix Φ has already been defined in equation (2.42) but we have repeated it for the sake of completeness of equation (4.4).

Let $p(lb)$ be the momentum operator of the atom located at $l+b$, with mass m_b. Then the crystal Hamiltonian can be written as (retaining terms only up to the cubic power of u)

$$
\begin{aligned}
H \;=\; & \sum_{lb}\frac{p(lb)\cdot p(lb)}{2m} + \frac{1}{2}\sum_{lb,l'b'}\sum_{\alpha\beta}\Phi_{\alpha\beta}(lb,l'b')u_\alpha(lb)u_\beta(l'b') \\
& + \frac{1}{3!}\sum_{lb,l'b',l''b''}\sum_{\alpha\beta\gamma}\Psi_{\alpha\beta\gamma}(lb,l'b',l''b'')u_\alpha(lb)u_\beta(l'b')u_\gamma(l''b'').
\end{aligned}
\tag{4.7}
$$

The *cyclic boundary condition* in equation (2.88) can be generalised to a crystal in the form of a parallelepiped of dimensions $L_1 \times L_2 \times L_3$ ($\equiv N_1 a_1 \times N_2 a_2 \times N_3 a_3$), containing $N_1 \times N_2 \times N_3$ unit cells with lattice constants a_1, a_2 and a_3:

$$u_b(l) = u_b(l+N_1 a_1) = u_b(l+N_2 a_2) = u_b(l+N_3 a_3). \tag{4.8}$$

It is evident that the form of the Hamiltonian in equation (4.7), involving arguments lb and $l'b'$ for $\mathcal{V}_2$ and $lb, l'b'$ and $l''b''$ for $\mathcal{V}_3$, is quite complicated. However, it is possible to simplify the form of the Hamiltonian by introducing new coordinates. (The art of simplifying a crystal Hamiltonian within the harmonic approximation is some times called *finding a diagonal representation of the Hamiltonian*.) There are two steps involved.

In the first step of Hamiltonian diagonalisation, we make a Fourier analysis of the coordinate (u) and momentum (p) variables

$$u(lb) \;=\; \frac{1}{\sqrt{N_0\Omega}}\sum_q X(q,b)e^{iq\cdot l} \tag{4.9}$$

$$p(lb) \;=\; \frac{1}{\sqrt{N_0\Omega}}\sum_q P(q,b)e^{-iq\cdot l}, \tag{4.10}$$

where $N_0\Omega$ is the volume of the crystal with $N_0 = N_1 N_2 N_3$ number of unit cells. Thus we have transformed the variables u and p into *normal coordinate operators* X and P. Since $u_{l,b}$ and $p_{l,b}$ are Hermitian, we must have

$$X^\dagger(q,b) \;=\; X(-q,b) = \frac{1}{\sqrt{N_0\Omega}} \sum_l u_{l,b} \exp\left(iq \cdot l\right) \tag{4.11}$$

$$P^\dagger(q,b) \;=\; P(-q,b) = \frac{1}{\sqrt{N_0\Omega}} \sum_l p(l,b) \exp\left(-iq \cdot l\right) \tag{4.12}$$

i.e. the new coordinate operators are non-Hermitian. These operators satisfy the following commutation relations:

$$
\begin{aligned}
[X(q,b), P(q',b')] &= \frac{1}{N_0\Omega} \sum_{l,l'} \exp\left[-i(q \cdot l - q' \cdot l')\right][u(l,b), p(l',b')] \\
&= \frac{1}{N_0\Omega} \sum_{l,l'} \exp\left[-i(q \cdot l - q' \cdot l')\right]\hat{\mathbf{I}}\, i\hbar \delta_{ll'} \delta_{bb'} \\
&= \hat{\mathbf{I}}\, i\hbar \delta_{qq'} \delta_{bb'}.
\end{aligned}
\tag{4.13}
$$

In other words, the new 'displacements' and 'momenta' are canonically conjugate and non-commuting if they correspond to the same wave vector and basis vector; otherwise they are dynamically independent variables.

Substitution of equations (4.9) and (4.10) into equation (4.7) gives

$$
\begin{aligned}
H \;=\;\; & \frac{1}{N_0\Omega} \sum_{qq'lb} \frac{P(q,b) \cdot P(q',b)}{2m_b} \exp\left[-i(q+q') \cdot l\right] \\
& + \frac{1}{2} \frac{1}{N_0\Omega} \sum_{\substack{q,q' \\ lb,l'b'}} \sum_{\alpha\beta} \Phi_{\alpha\beta}(lb, l'b') X_\alpha(q,b) X_\beta(q',b') \exp\left[i(q \cdot l + q' \cdot l')\right] \\
& + \frac{1}{3!} \frac{1}{(N_0\Omega)^{3/2}} \sum_{\substack{qq'q'' \\ lb,l'b',l''b''}} \sum_{\alpha\beta\gamma} \Psi_{\alpha\beta\gamma}(lb, l'b', l''b'') \\
& \times X_\alpha(qb) X_\beta(q'b') X_\gamma(q''b'') \exp\left[i(q \cdot l + q' \cdot l' + q'' \cdot l'')\right].
\end{aligned}
\tag{4.14}
$$

The first term in this equation can be simplified by performing the summation over l,

$$
\begin{aligned}
\text{First term} \;=\;\; & \sum_{q,q'b} \frac{P(qb) \cdot P(q'b)}{2m_b} \frac{1}{N_0\Omega} \sum_l \exp\left[-i(q+q') \cdot l\right] \\
&= \sum_{qq'b} \frac{P(qb) \cdot P(q'b)}{2m_b} \delta_{q+q',0} \\
&= \sum_{qb} \frac{P(qb) \cdot P^\dagger(qb)}{2m_b}
\end{aligned}
\tag{4.15}
$$

where the last step is obtained by using equation (4.12).

In order to simplify the second term in equation (4.14), we first note that because of lattice translational symmetry the harmonic force constant matrix Φ can be expressed as (see equation (2.45))

$$\Phi_{\alpha\beta}(lb, l'b') = \Phi_{\alpha\beta}(0b, (l'-l)b'). \tag{4.16}$$

Introducing $h = l' - l$ and defining

$$\Phi_{\alpha\beta}(bb' \,|\, q) \;=\; \sqrt{m_b m_{b'}} D_{\alpha\beta}(bb' \,|\, -q)$$

$$= \sum_{h} \Phi_{\alpha\beta}(\mathbf{0}b, \mathbf{h}b') \exp(-i\mathbf{q} \cdot \mathbf{h}) \tag{4.17}$$

we can express the second term in equation (4.14) as

$$\text{Second term} = \frac{1}{2} \sum_{\substack{q,b,b' \\ \alpha\beta}} \Phi_{\alpha\beta}(bb' \mid \mathbf{q}) X_{\alpha}(\mathbf{q}b) X_{\beta}^{\dagger}(\mathbf{q}b'). \tag{4.18}$$

To simplify the third term, we change the summations over $\mathbf{l}'$ and $\mathbf{l}''$ to those over new variables $\mathbf{h}'$ and $\mathbf{h}''$, defined by $\mathbf{h}' = \mathbf{l}' - \mathbf{l}, \mathbf{h}'' = \mathbf{l}'' - \mathbf{l}$. Then, following the scheme used in equations (4.16)–(4.18),

$$\mathcal{V}_3 = \frac{1}{3!} \frac{1}{(N_0 \Omega)^{3/2}} \sum_{\mathbf{q}b, \mathbf{q}'b', \mathbf{q}''b''} \sum_{\alpha\beta\gamma} \sum_{\mathbf{l}} \exp[i(\mathbf{q} + \mathbf{q}' + \mathbf{q}'') \cdot \mathbf{l}]$$
$$\times \Psi_{\alpha\beta\gamma}(\mathbf{q}b, \mathbf{q}'b', \mathbf{q}''b'') X_{\alpha}(\mathbf{q}b) X_{\beta}(\mathbf{q}'b') X_{\gamma}(\mathbf{q}''b''), \tag{4.19}$$

where

$$\Psi_{\alpha\beta\gamma}(\mathbf{q}b, \mathbf{q}'b', \mathbf{q}''b'') \equiv \sum_{\mathbf{h}', \mathbf{h}''} \Psi_{\alpha\beta\gamma}(\mathbf{0}b, \mathbf{h}'b', \mathbf{h}''b'') e^{i\mathbf{q}' \cdot \mathbf{h}'} e^{i\mathbf{q}'' \cdot \mathbf{h}''}. \tag{4.20}$$

Summing over $\mathbf{l}$ we get

$$\mathcal{V}_3 = \frac{1}{3!} \frac{1}{\sqrt{N_0 \Omega}} \sum_{\substack{\mathbf{q}b, \mathbf{q}'b' \\ \mathbf{q}''b''}} \delta_{\mathbf{G}, \mathbf{q}+\mathbf{q}'+\mathbf{q}''} \sum_{\alpha\beta\gamma}$$
$$\times \Psi_{\alpha\beta\gamma}(\mathbf{q}b, \mathbf{q}'b', \mathbf{q}''b'') X_{\alpha}(\mathbf{q}b) X_{\beta}(\mathbf{q}'b') X_{\gamma}(\mathbf{q}''b''), \tag{4.21}$$

where $\mathbf{G}$ is a reciprocal lattice vector.

Collecting all the three terms, the crystal Hamiltonian is expressed as

$$H = \sum_{\mathbf{q}b} \frac{\mathbf{P}(\mathbf{q}b) \cdot \mathbf{P}^{\dagger}(\mathbf{q}b)}{2m_b} + \frac{1}{2} \sum_{\substack{\mathbf{q}bb' \\ \alpha\beta}} \Phi_{\alpha\beta}(bb' \mid \mathbf{q}) X_{\alpha}(\mathbf{q}b) X_{\beta}^{\dagger}(\mathbf{q}b')$$
$$+ \frac{1}{3!} \frac{1}{\sqrt{N_0 \Omega}} \sum_{\substack{\mathbf{q}b, \mathbf{q}'b'\mathbf{q}''b'' \\ \alpha\beta\gamma}} \delta_{\mathbf{G}, \mathbf{q}+\mathbf{q}'+\mathbf{q}''}$$
$$\times \Psi_{\alpha\beta\gamma}(\mathbf{q}b, \mathbf{q}'b', \mathbf{q}''b'') X_{\alpha}(\mathbf{q}b) X_{\beta}(\mathbf{q}'b') X_{\gamma}(\mathbf{q}''b''). \tag{4.22}$$

At this stage of the first step of simplification the crystal Hamiltonian is viewed in terms of the coordinates $X(\mathbf{q}b)$ and momenta $P(\mathbf{q}b)$ of pN_0 atoms, coupled by a set of harmonic force constants $\Phi(bb' \mid \mathbf{q})$ and a set of anharmonic force constants $\Psi(\mathbf{q}b, \mathbf{q}'b', \mathbf{q}''b'')$. (Here p is the number of atoms per unit cell and N_0 is the number of unit cells in the crystal.) For each value of $\mathbf{q}$, the problem of finding the normal modes of the system is equivalent to finding the eigenstates of this Hamiltonian.

The discussion of the crystal Hamiltonian based on the variables $X(\mathbf{q}b)$ and $P(\mathbf{q}b)$ can be completed by introducing eigenvectors of the normal modes of the system. Following section 2.3, we introduce the polarisation vector $e(b \mid \mathbf{q}s)$ to represent the magnitude and direction of vibration of the atom b in the vibrational mode $(\mathbf{q}s)$, where s denotes the polarisation branch. $e(b \mid \mathbf{q}s)$ obey the orthogonality relation

$$\sum_{b} e^*(b \mid \mathbf{q}s) \cdot e(b \mid \mathbf{q}s') = \delta_{ss'}. \tag{4.23}$$

With the introduction of the eigenvectors $e(b|qs)$, we make another set of normal coordinate transformations

$$X(qs) = \sum_b \sqrt{m_b} e^*(b|qs) \cdot X(qb) \tag{4.24}$$

$$P(qs) = \sum_b \frac{1}{\sqrt{m_b}} e(b|qs) \cdot P(qb) \tag{4.25}$$

with $P(qs)$ as canonically conjugate to $X(qs)$. If desired, equations (4.24) and (4.25) can be inverted by using the closure relation

$$\sum_s e_\alpha^*(b|qs) e_\beta(b'|qs) = \delta_{\alpha\beta} \delta_{bb'}. \tag{4.26}$$

The normal coordinate operators $X(qs)$ and $P(qs)$ defined by equations (4.24) and (4.25) can be used to re-express the Hamiltonian in equation (4.22). However, it is more convenient to proceed now to the second step of the proposed simplification of the crystal Hamiltonian.

In the second step of simplifying the crystal Hamiltonian, we make yet another set of transformations:

$$u_{qs} = \frac{1}{\sqrt{2\hbar\omega(qs)}} P(qs) - i\sqrt{\frac{\omega(qs)}{2\hbar}} X^\dagger(qs) \tag{4.27}$$

$$a_{qs}^\dagger = \frac{1}{\sqrt{2\hbar\omega(qs)}} P^\dagger(qs) + i\sqrt{\frac{\omega(qs)}{2\hbar}} X(qs). \tag{4.28}$$

The operators a_{qs} and $a_{qs}^\dagger$ are known as phonon annihilation and creation operators, respectively. It can be verified that these operators obey the following commutation relations:

$$[a_{qs}, a_{q's'}^\dagger] = \delta_{q,q'} \delta_{s,s'} \hat{\mathbf{I}}. \tag{4.29}$$

From equations (4.27) and (4.28), we can express

$$X(qs) = -i\sqrt{\frac{\hbar}{2\omega(qs)}} (a_{qs}^\dagger - a_{-qs}) \tag{4.30}$$

$$P(qs) = \sqrt{\frac{\hbar\omega(qs)}{2}} (a_{qs} + a_{-qs}^\dagger), \tag{4.31}$$

where we have used $\omega(-qs) = \omega(qs)$, $X^\dagger(qs) = X(-qs)$, and $P^\dagger(qs) = P(-qs)$. Therefore, from equations (4.24) and (4.25), and (4.30) and (4.31) we have

$$\begin{aligned} X(qb) &= \frac{1}{\sqrt{m_b}} \sum_s e(b|qs) X(qs) \\ &= -i \sum_s \sqrt{\frac{\hbar}{2m_b\omega(qs)}} e(b|qs) (a_{qs}^\dagger - a_{-qs}) \end{aligned} \tag{4.32}$$

$$\begin{aligned} P(qb) &= \sqrt{m_b} \sum_s e^*(b|qs) P(qs) \\ &= \sum_s \sqrt{\frac{m_b\hbar\omega(qs)}{2}} e^*(b|qs) (a_{qs} + a_{-qs}^\dagger). \end{aligned} \tag{4.33}$$

Thus we now have transformations which express the coordinate and momenta vectors $X(qb)$ and $P(qb)$ in terms of the phonon creation and annihilation operators and the polarisation vectors.

We now substitute equations (4.32) and (4.33) in equation (4.22) to simplify the terms in the Hamiltonian:

$$\text{First term} = \frac{1}{2}\sum_{qb}\frac{1}{m_b}P(qb) \cdot P^{\dagger}(qb)$$

$$= \frac{1}{4}\sum_{qs}\hbar\omega(qs)(a_{qs} + a^{\dagger}_{-qs})(a^{\dagger}_{qs} + a_{-qs}), \tag{4.34}$$

where equation (4.23) is used.

$$\text{Second term} = \frac{1}{2}\sum_{\substack{q,b,b' \\ \alpha\beta}}\Phi_{\alpha\beta}(bb'|q)X_{\alpha}(qb)X^{\dagger}_{\beta}(qb')$$

$$= \frac{1}{2}\sum_{\substack{q,b,b' \\ s\alpha\beta}}\Phi_{\alpha\beta}(bb'|q)\frac{\hbar}{2\omega(qs)}\frac{1}{\sqrt{m_b m_{b'}}}e_{\alpha}(b|qs)e^{*}_{\beta}(b'|qs)$$

$$\times(a^{\dagger}_{qs} - a_{-qs})(a_{qs} - a^{\dagger}_{-qs})$$

$$= \frac{1}{4}\sum_{\substack{q,b \\ s\alpha}}\hbar\omega(qs)e_{\alpha}(b|qs)e^{*}_{\alpha}(b|qs)(a^{\dagger}_{qs} - a_{-qs})(a_{qs} - a^{\dagger}_{-qs})$$

$$= \frac{1}{4}\sum_{qs}\hbar\omega(qs)(a^{\dagger}_{qs} - a_{-qs})(a_{qs} - a^{\dagger}_{-qs}), \tag{4.35}$$

where we have made use of equations (4.17), (4.23), (2.53), (2.54), (2.56) and (2.59).

Adding terms 1 and 2, we get

$$H_{\text{harm}} = \text{term } 1 + \text{term } 2$$

$$= \frac{1}{4}\sum_{qs}\hbar\omega(qs)\big[(a_{qs} + a^{\dagger}_{-qs})(a^{\dagger}_{qs} + a_{-qs})$$

$$+ (a^{\dagger}_{qs} - a_{-qs})(a_{qs} - a^{\dagger}_{-qs})\big]$$

$$= \frac{1}{4}\sum_{qs}\hbar\omega(qs)(a_{qs}a^{\dagger}_{qs} + a^{\dagger}_{qs}a_{qs} + a_{-qs}a^{\dagger}_{-qs} + a^{\dagger}_{-qs}a_{-qs})$$

$$= \frac{1}{2}\sum_{qs}\hbar\omega(qs)(a_{qs}a^{\dagger}_{qs} + a^{\dagger}_{qs}a_{qs}), \tag{4.36}$$

where we have used the fact that a summation over allowed values of $-q$ merely duplicates the sum over q. It is easy to simplify the above result a little further. Using the commutation relations in equation (4.29), we can finally write the crystal Hamiltonian in the harmonic approximation (i.e. including terms up to $\frac{1}{2}$) as

$$H_{\text{harm}} = \sum_{qs}\hbar\omega(qs)(a^{\dagger}_{qs}a_{qs} + \frac{1}{2}). \tag{4.37}$$

The expression in equation (4.37) is in the required (*diagonal*) form, as can be appreciated below. Before proceeding further, it is worth recapitulating that this form of the Hamiltonian has been obtained by making the two-step coordinate transformation as described above. In the first step we changed the picture from the classical coordinates and momenta (*particle picture*) $X(lb), P(lb)$ to the first quantisation variables (*wave picture*) $X(qs), P(qs)$. In the second step we introduced the second quantisation variables (*quasi-particle picture*) $a_{qs}, a^{\dagger}_{qs}$. To appreciate the diagonal nature of the expression in equation (4.37) in the second quantised notation, let us calculate the crystal eigenvalues. Let us denote by $|n_{qs}\rangle$ a state which has n phonons of wavevector q and polarisation s. The effects of the operators $a^{\dagger}_{qs}, a_{qs}$ and $a^{\dagger}_{qs}a_{qs}$ on the state $|n_{qs}\rangle$ are given as follows:

$$
\begin{aligned}
a_{qs}^{\dagger}|n_{qs}\rangle &= \sqrt{n_{qs}+1}\,|n_{qs}+1\rangle \\
a_{qs}|n_{qs}\rangle &= \sqrt{n_{qs}}\,|n_{qs}-1\rangle \\
a_{qs}^{\dagger}a_{qs}|n_{qs}\rangle &= n_{qs}|n_{qs}\rangle.
\end{aligned}
\tag{4.38}
$$

In other words, while $a_{qs}^{\dagger}$ and a_{qs} are, respectively, phonon creation and destruction operators, $a_{qs}^{\dagger}a_{qs}$ is a phonon number operator. Thus using equations (4.37) and (4.38) we get

$$
\begin{aligned}
H_{\text{harm}}|n_{qs}\rangle &= \sum_{qs}\hbar\omega(qs)(n_{qs}+\tfrac{1}{2})|n_{qs}\rangle \\
&= \sum_{qs}\varepsilon_{qs}|n_{qs}\rangle
\end{aligned}
\tag{4.39}
$$

so that the eigenvalues of a three-dimensional simple harmonic oscillator with the Hamiltonian

$$
h_{\text{harm}} = \hbar\omega(qs)(a_{qs}^{\dagger}a_{qs}+\tfrac{1}{2})
\tag{4.40}
$$

are

$$
\varepsilon_{qs} = \hbar\omega(qs)(n_{qs}+\tfrac{1}{2}).
\tag{4.41}
$$

From here it is clearly seen that the average energy of phonons in mode (qs) is

$$
\bar{\varepsilon}_{qs} = \hbar\omega(qs)\bar{n}_{qs}
\tag{4.42}
$$

where the thermal average $\bar{n}_{qs}$ is the Bose–Einstein distribution function given in equation (2.2). The second term in equation (4.42) is the zero-point energy.

Having discussed the harmonic part of the crystal Hamiltonian in the second quantised formulation, we now proceed to express the cubic anharmonic term $\mathcal{V}_3$ in this notation. In the first quantised notation, this is the third term in equation (4.22). Using the transformation in equation (4.32), we re-write equation (4.22) as

$$
\begin{aligned}
\mathcal{V}_3 &= \frac{1}{3!}\frac{1}{\sqrt{N_0}}(i)\sum_{\substack{qb,q'b',q''b'' \\ ss's''\alpha\beta\gamma}}\left(\frac{\hbar^3}{8m_b m_{b'} m_{b''}\,\omega(qs)\omega(q's')\omega(q''s'')}\right)^{1/2} \\
&\quad \times \delta_{G,q+q'+q''}\,e_\alpha(b|qs)e_\beta(b'|q's')e_\gamma(b''|q''s'')\Psi_{\alpha\beta\gamma}(qb,q'b',q''b'') \\
&\quad \times (a_{qs}^{\dagger}-a_{-qs})(a_{q's'}^{\dagger}-a_{-q's'})(a_{q''s''}^{\dagger}-a_{-q''s''}) \\
&= \frac{1}{3!}\sum_{\substack{qs,q's' \\ q''s''}}\delta_{G,q+q'+q''}\Psi(qs,q's',q''s'') \\
&\quad \times (a_{qs}^{\dagger}-a_{-qs})(a_{q's'}^{\dagger}-a_{-q's'})(a_{q''s''}^{\dagger}-a_{-q''s''}),
\end{aligned}
\tag{4.43}
$$

where

$$
\begin{aligned}
\Psi(qs,q's',q''s'') &= \frac{i}{\sqrt{N_0\Omega}}\sum_{\substack{bb'b'' \\ \alpha\beta\gamma}}\left(\frac{\hbar^3}{8m_b m_{b'} m_{b''}\,\omega(qs)\omega(q's')\omega(q''s'')}\right)^{1/2} \\
&\quad \times e_\alpha(b|qs)e_\beta(b'|q's')e_\gamma(b''|q''s'') \\
&\quad \times \Psi_{\alpha\beta\gamma}(qb,q'b',q''b'').
\end{aligned}
\tag{4.44}
$$

The factor $\Psi(qs,q's',q''s'')$ in the cubic anharmonic term in the Hamiltonian is proportional to an average of the Fourier transformed tensor $\Psi(qb,q'b',q''b'')$, projected upon the directions of the polarisation vectors $e(b|qs), e(b'|q's')$ and $e(b''|q''s'')$.

This completes our transformation of both harmonic and cubic anharmonic terms in the crystal Hamiltonian using the second quantisation scheme for coordinate transformation. Once again, notice that while in this representation the harmonic term is diagonal, the anharmonic term is not. This coordinate transformation scheme can also be used to express higher order terms in the Hamiltonian, if desired.

4.3 EFFECT OF ANHARMONICITY ON PHONON STATES

From equation (4.39), it is clear that the harmonic term in the Hamiltonian gives the picture of non-interacting phonons in a crystal. In reality at finite temperatures all crystals contain anharmonic lattice forces, a fact which is borne out from measurements of thermal expansion and lattice thermal conductivity, and from widths of phonon peaks in neutron scattering from crystals. However, neutron scattering experiments clearly indicate one-phonon peaks, which suggest that anharmonicity can be viewed as a perturbation on the non-interacting phonon states of a crystal. This is what is generally accepted in theoretical treatments of anharmonicity. The effect of anharmonicity is thus to introduce interactions among the independent phonons of a crystal. For example, the effect of the cubic term $\mathcal{V}_3$ is to cause, in first-order perturbation, interactions involving three phonons and, in second-order, interactions involving four phonons. Similarly the quartic term, $\mathcal{V}_4$, causes, in first-order perturbation, four-phonon interactions, and so on.

In this section we discuss the effect on phonon states of $\mathcal{V}_3$ in first-order perturbation. Details of transition probabilities will be discussed in Chapter 6, but here we investigate types of three-phonon processes and the conservation rules governing them.

From equation (4.43), we see that, due to translational invariance of the crystal potential energy, the phonon wave vectors q, q' and q'' have to satisfy the momentum conservation

$$q + q' + q'' = G, \tag{4.45}$$

where G is a reciprocal lattice vector including zero, and q, q', and q'' are restricted to the first Brillouin zone. Phonon–phonon interaction processes with $G = 0$ and $G \neq 0$ are called *normal* or *N*-processes, and *umklapp* or *U*-processes (Peierls 1929), respectively.

From equation (4.43) it is clear that the effect of $\mathcal{V}_3$ is governed by the operator

$$(a_{qs}^{\dagger} - a_{-qs})(a_{q's'}^{\dagger} - a_{-q's'})(a_{q''s''}^{\dagger} - a_{-q''s''}). \tag{4.46}$$

Expanding this product we get

$$a_{qs}^{\dagger}a_{q's'}^{\dagger}a_{q''s''}^{\dagger} - a_{qs}^{\dagger}a_{q's'}^{\dagger}a_{-q''s''} - a_{qs}^{\dagger}a_{-q's'}a_{q''s''}^{\dagger}$$
$$+a_{qs}^{\dagger}a_{-q's'}a_{-q''s''} - a_{-qs}a_{q's'}^{\dagger}a_{q''s''}^{\dagger} + a_{-qs}a_{q's'}^{\dagger}a_{-q''s''}$$
$$+a_{-qs}a_{-q's'}a_{q''s''}^{\dagger} - a_{-qs}a_{-q's'}a_{-q''s''}. \tag{4.47}$$

Each of these operators acts on a three-phonon state $|n_{qs}n_{q's'}n_{q''s''}\rangle$ according to the rules given in equation (4.38). Therefore, the effect of the operator in term 1 of equation (4.47) is to increase by unity the number of each of the phonons represented by $qs, q's'$ and $q''s''$. In other words, three phonons are simultaneously created. Similarly the effect of other operators in equation (4.47) can be inferred. It can be seen that there are four basic processes of three-phonon type: (i) annihilation of two phonons and creation of a third phonon (we shall refer to this as a *class 1* event in chapter 6); (ii) annihilation of one phonon and creation of two phonons (referred to as a *class 2* event in chapter 6); (iii) simultaneous annihilation of three phonons; and (iv) simultaneous creation of three phonons. However, only (i) and (ii) are real possibilities, as they satisfy energy conservation. Possibilities (iii) and (iv) violate energy conservation, and can, therefore, only be included as possible virtual three-phonon processes when considering higher-order anharmonic processes.

Let us consider creation or annihilation of a phonon in mode qs. The allowed processes are described by the following energy and momentum conservation laws

Class 1 events:

$$\begin{aligned} \omega(qs) + \omega(q's') &= \omega(q''s'') \\ q + q' &= q'' + G \end{aligned} \tag{4.48}$$

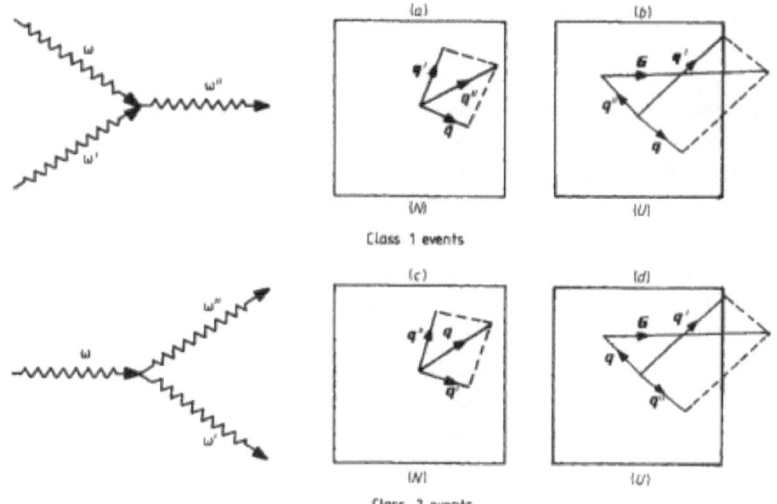

Figure 4.1   A two-dimensional representation of three-phonon N- and U-processes: (a) $q+q' = q''$, (b) $q + q' = q'' + G$, (c) $q = q' + q''$, (d) $q + G = q' + q''$.

Class 2 events:

$$\begin{aligned}
\omega(qs) &= \omega(q's') + \omega(q''s'') \\
q+G &= q' + q''.
\end{aligned} \tag{4.49}$$

In a class 1 event, a phonon $(qs; \omega(qs))$ interacts with another phonon $(q's'; \omega(q's'))$, they both get annihilated, and as a result a third phonon $(q''s''; \omega(q''s''))$ is generated. If the momentum $q+q'$ lies within the first Brillouin zone, the process is called normal (N) process. If $q+q'$ lies outside the first Brillouin zone, then it is *flipped back* into the first zone with the help of an appropriate reciprocal lattice vector G and the phonon interaction process is called an umklapp (U) process (see figure 4.1). Evidently, in a U-process the momentum q'' of the created phonon differs from $q+q'$ by a non-zero reciprocal lattice vector. For this reason we, therefore, sometimes refer to N- and U-processes as momentum conserving and momentum non-conserving processes, respectively. Further, as the direction of q'' in a U-process is opposite to the direction of the resultant $q+q'$, such a process obviously creates a resistance to heat flow by phonons. A similar picture can be presented for a class 2 event in equation (4.49). We will consider class 1 and class 2 events in calculations of phonon lifetimes in Chapter 6.

4.4   EFFECTS OF THE SELECTION RULES ON THREE-PHONON PROCESSES

We can analyse the effects of energy and momentum conservation on both class 1 and class 2 events. Consider the scattering of phonon (qs) in a class 1 event. Each of q' and q'' has three degrees of freedom. Out of these six degrees of freedom, four are fixed by equation (4.48):

$$\begin{aligned}
q_x + q'_x &= q''_x + G_x \\
q_y + q'_y &= q''_y + G_y \\
q_z + q'_z &= q''_z + G_z \\
\omega(qs) + \omega(q's') &= \omega(q''s'').
\end{aligned} \tag{4.50}$$

The remaining two degrees of freedom define some surface $\mathscr{S}'$ in reciprocal space, on which q' must lie for the process in equation (4.48) to take place. The area $\mathscr{S}'$ has in general a very complicated shape depending on the value of q and upon the branches s, s', s'' of the three phonons involved.

To obtain the form of the energy conservation surface $\mathscr{S}'$ for a three-phonon N- or U-process, we should have full knowledge of a realistic phonon dispersion relation throughout the first Brillouin zone. This is a very demanding task and indeed may not be affordable for a study of phonon–phonon interactions in a crystal at various temperatures. When dealing with N-processes, a simplified picture of phonon dispersion relations and of the Brillouin zone (e.g. the Debye approximation) has been widely used. In fact this is what we will do in discussing phonon interactions in this chapter and in Chapters 6 and 7. However, the continuum approximation used in Debye's scheme does not allow for a physical picture of an U-process as there is no concept of a reciprocal lattice vector in a continuum. To overcome this difficulty, Parrott (1963) grafted the following scheme for a *pseudo-reciprocal* lattice vector $\boldsymbol{G}$ when dealing with U-processes within an isotropic continuum model:

$$\text{class 1 events}: \boldsymbol{G} = 2q_D \frac{\boldsymbol{q} + \boldsymbol{q}'}{|\boldsymbol{q} + \boldsymbol{q}'|}, \tag{4.51}$$

where q_D is the Debye radius. We can follow Parrott's scheme to graft a *pseudo*-reciprocal lattice vector for class 2 events (Srivastava 1974, 1976a):

$$\text{class 2 events}: \boldsymbol{G} = 2q_D \frac{\boldsymbol{q} - \boldsymbol{q}'}{|\boldsymbol{q} - \boldsymbol{q}'|}. \tag{4.52}$$

The selection rules in equation (4.50) impose certain restrictions on combinations of polarisation branches of phonons taking part in three-phonon processes. Consider the class 1 N-process $\boldsymbol{q} + \boldsymbol{q}' = \boldsymbol{q}''$. The energy conservation condition requires that the frequency of the created phonon must be higher than that of one of the phonons which are annihilated: $\omega'' > \omega, \omega'$ (where we have used $\omega' \equiv \omega(\boldsymbol{q}'s')$ etc.). Within the isotropic continuum model $\omega = c_s q$ etc. Thus we have

$$c_{s''}q'' = c_s q + c_{s'}q' \tag{4.53}$$
$$c_{s''}q'' > c_s q, c_{s'}q'. \tag{4.54}$$

Also, the momentum requirement $\boldsymbol{q} + \boldsymbol{q}' = \boldsymbol{q}''$ gives

$$q + q' > q''. \tag{4.55}$$

The conditions in equations (4.53) and (4.54) require that $c_{s''} \geq c_s, c_{s'}$. Furthermore, we note that within the isotropic continuum model the strict inequality $c_{s''} > c_s$ holds only if $s'' = \mathsf{L}$ (longitudinal) and $s = \mathsf{T}$ (doubly degenerate transverse). Thus the only distinct types of class 1 N-processes are $\mathsf{T} + \mathsf{T} \to \mathsf{T}$, $\mathsf{T} + \mathsf{T} \to \mathsf{L}$, $\mathsf{T} + \mathsf{L} \to \mathsf{L}$, and $\mathsf{L} + \mathsf{L} \to \mathsf{L}$. The processes $\mathsf{T} + \mathsf{T} \to \mathsf{T}$ and $\mathsf{L} + \mathsf{L} \to \mathsf{L}$ require that all the three participating phonons be collinear. For a non-linear dispersion relation the process $\mathsf{L} + \mathsf{L} \to \mathsf{L}$ is forbidden, but the processes $\mathsf{T}_1 + \mathsf{T}_1 \to \mathsf{T}_2$, $\mathsf{T}_1 + \mathsf{T}_2 \to \mathsf{T}_2 (\omega(\mathsf{T}_2) > \omega(\mathsf{T}_1))$, and $\mathsf{T}_1 + \mathsf{L} \to \mathsf{T}_2 (\omega(\mathsf{T}_2) > \omega(\mathsf{T}_1), \omega(\mathsf{L}))$ may be allowed.

Next consider the class 1 U-process governed by the selection rules $\boldsymbol{q} + \boldsymbol{q}' = \boldsymbol{q}'' + \boldsymbol{G}$ and $\omega + \omega' = \omega''$. It can be appreciated from a configuration of wave vectors for this process (see, e.g., figure 4.1) that each of q, q' and q'' must be small enough to fit inside the first Brillouin zone:

$$q, q', q'' \leq \tfrac{1}{2}G. \tag{4.56}$$

Further, to meet the vector equality $\boldsymbol{q} + \boldsymbol{q}' = \boldsymbol{q}'' + \boldsymbol{G}$ we must have

$$q + q' + q'' \geq G. \tag{4.57}$$

The requirement $\omega + \omega' = \omega''$ together with the constraints in equations (4.56) and (4.57) imposes a lower limit on the value of q''. We will discuss this point in more detail in Chapter 6 where we will consider both class 1 and class 2 processes. Furthermore, an analysis analogous to that presented in equations (4.53)–(4.55) rules out U-processes of the types $\mathsf{T} + \mathsf{T} \to \mathsf{T}$ and $\mathsf{L} + \mathsf{L} \to \mathsf{L}$. However, for a non-linear dispersion relation the processes $T_1 + T_1 \rightarrow T_2$ and $T_1 + L \rightarrow T_2 (\omega(T_2) > \omega(T_1), \omega(L))$ may be allowed.

Therefore, the allowed combinations of phonon polarisation branches within the isotropic continuum model are

$$
\begin{array}{ll}
T + T & \rightleftharpoons L \\
T + L & \rightleftharpoons L
\end{array}
\tag{4.58}
$$

for both N and U processes, and

$$
\begin{array}{ll}
T + T & \rightleftharpoons T \\
L + L & \rightleftharpoons L
\end{array}
\tag{4.59}
$$

for N processes only.

A more detailed discussion of the effects of the selection rules, and of the form of $\mathscr{S}'$ based on crystal symmetry arguments, can be found in Ziman (1960) and Herring (1954).

4.5 HAMILTONIAN OF AN ANHARMONIC ELASTIC CONTINUUM

It is evident from section 4.2 that even the cubic anharmonic term in the crystal Hamiltonian is very complicated in its detail. This is best realised when dealing with anharmonic U-processes. It is, therefore, very useful to have a simplified description of the crystal Hamiltonian. The simplest thing to do is to smear out all the structure and treat the crystal as a continuum. We have some consolation in doing so: the low-lying acoustic modes resemble simple elastic waves.

In the continuum model we replace the lattice vector l, which goes through discrete values, by a continuous position vector r. Sums over all values of l are then replaced by an integral over r. The displacement vector $u(lb)$ is replaced by a vector $u(r)$ which represents the displacement of the continuum crystal at point r. With this definition then $\partial u / \partial r$ is the *elastic strain tensor* and its components represent the rates of change of the displacement in various directions. The second-order and third-order elastic energy terms are given as follows:

$$
\mathcal{V}_2 = \frac{1}{2} \int d^3 r \sum_{\substack{\alpha\beta \\ \gamma\delta}} J_{\alpha\beta\gamma\delta} \frac{\partial u_\alpha}{\partial r_\beta} \frac{\partial u_\gamma}{\partial r_\delta}
\tag{4.60}
$$

$$
\mathcal{V}_3 = \frac{1}{3!} \int d^3 r \sum_{\substack{lmn \\ ijk}} A_{ijk}^{lmn} \frac{\partial u_l}{\partial r_i} \frac{\partial u_m}{\partial r_j} \frac{\partial u_n}{\partial r_k},
\tag{4.61}
$$

where J and A are tensors in cartesian components of fourth and sixth rank, respectively.

From the theory of elasticity we can write equations (4.60) and (4.61) in an expanded form. In general one writes the potential energy expression as follows (see, e.g. Drabble 1966):

$$
\mathcal{V} = \int d^3 r \left[\mathcal{V}_0 + \sum_{mn,pq} \left(\frac{1}{2} c_{mn,pq} \eta_{mn} \eta_{pq} + \frac{1}{3!} \sum_{rs} c_{mn,pq,rs} \eta_{mn} \eta_{pq} \eta_{rs} \right) \right],
\tag{4.62}
$$

where $c_{mn,pq}$ and $c_{mn,pq,rs}$ are components of second-order and third-order elastic constants, respectively. Using the contracted notation as explained in equation (2.141), we can express equation (4.62) as

$$
\mathcal{V} = \int d^3 r \left(\mathcal{V}_0 + \frac{1}{2} \sum_{PQ} c_{PQ} \eta_P \eta_Q + \frac{1}{3!} \sum_{PQR} c_{PQR} \eta_P \eta_Q \eta_R \right)
\tag{4.63}
$$

(sums over subscripts P, Q, R go from 1 to 6). In equation (4.63) we have used

$$
\eta_1 = \eta_{11} \qquad \eta_2 = \eta_{22} \qquad \eta_3 = \eta_{33},
$$

$$\eta_4 = \eta_{23} + \eta_{32} \qquad \eta_5 = \eta_{31} + \eta_{13} \qquad \eta_6 = \eta_{12} + \eta_{21}, \text{ and}$$

$$\eta_{ab} = \frac{(1 + \delta_{ab})}{2} \eta_k. \tag{4.64}$$

In the above equation η_k is defined for $k = 1$ to 6, and η_{ij} are the Lagrangian strain components defined as

$$\eta_{ij} = \frac{1}{2} \left(\sum_k \frac{\partial r_k}{\partial a_i} \frac{\partial r_k}{\partial a_j} - \delta_{ij} \right), \tag{4.65}$$

where

$$r_i = a_i + u_i \tag{4.66}$$

with $\mathbf{r}$ and $\mathbf{a}$ representing the coordinates of the strained and initial (or unstrained) states, respectively. Now, since $r_i = u_i + a_i$, we can write

$$\frac{\partial r_i}{\partial a_j} = \frac{\partial u_i}{\partial a_j} + \frac{\partial a_i}{\partial a_j} = \frac{\partial u_i}{\partial a_j} + \delta_{ij}$$

and hence

$$
\begin{aligned}
\eta_{ij} &= \frac{1}{2} \left[\sum_p \left(\frac{\partial u_p}{\partial a_i} + \delta_{pi} \right) \left(\frac{\partial u_p}{\partial a_j} + \delta_{pj} \right) - \delta_{ij} \right] \\
&= \frac{1}{2} \left(\sum_p \frac{\partial u_p}{\partial a_i} \frac{\partial u_p}{\partial a_j} + \frac{\partial u_j}{\partial a_i} + \frac{\partial u_i}{\partial a_j} \right) \\
&= \frac{1}{2} \left(\frac{\partial u_i}{\partial r_j} + \frac{\partial u_j}{\partial r_i} + \sum_p \frac{\partial u_p}{\partial r_i} \frac{\partial u_p}{\partial r_j} \right) \tag{4.67} \\
&= \frac{1}{2} \left(u_{ij} + u_{ji} + \sum_p u_{pi} u_{pj} \right) \\
&= u_{ij} + \frac{1}{2} \sum_p u_{pi} u_{pj} \qquad p = 1, 2, 3. \tag{4.68}
\end{aligned}
$$

Also, it can be verified that

$$\eta_{ij} = \eta_{ji}. \tag{4.69}$$

For cubic systems equation (4.63) can be expressed in Brugger's notation (1964) as follows:

$$
\begin{aligned}
\mathscr{V} = \int \mathrm{d}^3 r \big[& \mathscr{V}_0 + \frac{1}{2} \{ c_{11}(\eta_1^2 + \eta_2^2 + \eta_3^2) + c_{44}(\eta_4^2 + \eta_5^2 + \eta_6^2) \\
& + 2c_{12}(\eta_1 \eta_2 + \eta_2 \eta_3 + \eta_3 \eta_1) \} + \frac{1}{6} \{ C_{111}(\eta_1^3 + \eta_2^3 + \eta_3^3) \\
& + 3C_{112}[\eta_1^2(\eta_2 + \eta_3) + \eta_2^2(\eta_3 + \eta_1) + \eta_3^2(\eta_1 + \eta_2)] \\
& + 3C_{144}(\eta_1 \eta_4^2 + \eta_2 \eta_5^2 + \eta_3 \eta_6^2) \\
& + 3C_{166}[\eta_4^2(\eta_2 + \eta_3) + \eta_5^2(\eta_3 + \eta_1) + \eta_6^2(\eta_1 + \eta_2)] \\
& + 6C_{123} \eta_1 \eta_2 \eta_3 + 6C_{456} \eta_4 \eta_5 \eta_6 \} \big]. \tag{4.70}
\end{aligned}
$$

From equations (4.60)–(4.61) and (4.70), we can obtain expressions for $\mathscr{V}_2$ and $\mathscr{V}_3$.

Note that equation (4.70) is written in Brugger's notation. There are other notations available as well; for example Birch's notation (Birch 1947). The two sets of elastic constants are related as follows:

$$C_{111}^{Br} = 6C_{111}^B \qquad C_{112}^{Br} = 2C_{112}^B \qquad C_{123}^{Br} = C_{123}^B$$

$$C_{144}^{Br} = \frac{1}{2} C_{144}^B \qquad C_{166}^{Br} = \frac{1}{2} C_{166}^B \qquad C_{456}^{Br} = \frac{1}{4} C_{456}^B. \tag{4.71}$$

Here we have used $C_{ijk} \equiv C_{ijk}^{Br}$ for the elastic constants in Brugger's notation, and C_{ijk}^{B} for the elastic constants in Birch's notation.

Using equations (4.64), (4.67) and (4.69), we can express equation (4.70) as

$$
\begin{aligned}
\mathscr{V} &= \int d^3 r \Big[\mathscr{V}_0 + \frac{1}{2} \{ c_{11}(\eta_{11}^2 + \eta_{22}^2 + \eta_{33}^2) \\
&\quad + 4c_{44}(\eta_{23}^2 + \eta_{31}^2 + \eta_{12}^2 + \eta_{12}^2) + 2c_{12}(\eta_{11}\eta_{22} + \eta_{22}\eta_{33} + \eta_{33}\eta_{11}) \} \\
&\quad + \frac{1}{6} \{ C_{111}(\eta_{11}^3 + \eta_{22}^3 + \eta_{33}^3) \\
&\quad + 3C_{112}[\eta_{11}^2(\eta_{22} + \eta_{33}) + \eta_{22}^2(\eta_{33} + \eta_{11}) + \eta_{33}^2(\eta_{11} + \eta_{22})] \\
&\quad + 12C_{144}(\eta_{11}\eta_{23}^2 + \eta_{22}\eta_{13}^2 + \eta_{33}\eta_{12}^2) \\
&\quad + 12C_{166}[\eta_{23}^2(\eta_{22} + \eta_{33}) + \eta_{13}^2(\eta_{11} + \eta_{33}) + \eta_{12}^2(\eta_{11} + \eta_{22})] \\
&\quad + 6C_{123}\eta_{11}\eta_{22}\eta_{33} + 48C_{456}\eta_{23}\eta_{31}\eta_{12} \} \Big].
\end{aligned}
\tag{4.72}
$$

Then we can use equation (4.67) to expand each term in equation (4.72). For example,

$$
\begin{aligned}
c_{11}(\eta_{11}^2 + \eta_{22}^2 + \eta_{33}^2) &= \frac{1}{4}c_{11}\Bigg[\left(\frac{\partial u_1}{\partial r_1} + \frac{\partial u_1}{\partial r_1} + \sum_p \frac{\partial u_p}{\partial r_1}\frac{\partial u_p}{\partial r_1} \right)^2 \\
&\quad + \left(\frac{\partial u_2}{\partial r_2} + \frac{\partial u_2}{\partial r_2} + \sum_p \frac{\partial u_p}{\partial r_2}\frac{\partial u_p}{\partial r_2} \right)^2 \\
&\quad + \left(\frac{\partial u_3}{\partial r_3} + \frac{\partial u_3}{\partial r_3} + \sum_p \frac{\partial u_p}{\partial r_3}\frac{\partial u_p}{\partial r_3} \right)^2 \Bigg] \\
&= \frac{1}{4}c_{11}\Bigg[\left(2u_{11} + \sum_p u_{p1}u_{p1} \right)^2 + \left(2u_{22} + \sum_p u_{p2}u_{p2} \right)^2 \\
&\quad + \left(2u_{33} + \sum_p u_{p3}u_{p3} \right)^2 \Bigg] \\
&= c_{11}(u_{11}^2 + u_{22}^2 + u_{33}^2) \\
&\quad + c_{11} \sum_p (u_{11}u_{p1}u_{p1} + u_{22}u_{p2}u_{p2} + u_{33}u_{p3}u_{p3}) \\
&\quad + \frac{1}{4}c_{11} \sum_p \{ (u_{p1}u_{p1})^2 + (u_{p2}u_{p2})^2 + (u_{p3}u_{p3})^2 \}.
\end{aligned}
\tag{4.73}
$$

Similarly,

$$
\begin{aligned}
c_{44}(\eta_{23}^2 + \eta_{31}^2 + \eta_{12}^2) &= \frac{1}{4}c_{44}\Bigg[\left(u_{23} + u_{32} + \sum_p u_{p2}u_{p3} \right)^2 \\
&\quad + \left(u_{31} + u_{13} + \sum_p u_{p1}u_{p3} \right)^2 \\
&\quad + \left(u_{12} + u_{21} + \sum_p u_{p1}u_{p2} \right)^2 \Bigg]
\end{aligned}
\tag{4.74}
$$

$$c_{12}(\eta_{11}\eta_{22} + \eta_{22}\eta_{33} + \eta_{33}\eta_{11}) = c_{12}\left[\left(u_{11} + \frac{1}{2}\sum_p u_{p1}u_{p1}\right)\left(u_{22} + \frac{1}{2}\sum_p u_{p2}u_{p2}\right)\right.$$

$$+\left(u_{22} + \frac{1}{2}\sum_p u_{p2}u_{p2}\right)\left(u_{33} + \frac{1}{2}\sum_p u_{p3}u_{p3}\right)$$

$$\left.+\left(u_{33} + \frac{1}{2}\sum_p u_{p3}u_{p3}\right)\left(u_{11} + \frac{1}{2}\sum_p u_{p1}u_{p1}\right)\right] \tag{4.75}$$

$$C_{111}(\eta_{11}^3 + \eta_{22}^3 + \eta_{33}^3) = C_{111}\left[\left(u_{11} + \frac{1}{2}\sum_p u_{p1}u_{p1}\right)^3\right.$$

$$\left.+\left(u_{22} + \frac{1}{2}\sum_p u_{p2}u_{p2}\right)^3 + \left(u_{33} + \frac{1}{2}\sum_p u_{p3}u_{p3}\right)^3\right]$$

$$\tag{4.76}$$

and so on.

In our further discussion, we will need terms of the order of u_{rs}^3 to extract $\mathscr{V}_3$ out of $\mathscr{V}$. Therefore, retaining only terms of third-order in u_{rs}, we get

$$\mathscr{V}_3 = \int d^3r\left(\frac{1}{2}c_{11}\sum_p(u_{11}u_{p1}u_{p1} + u_{22}u_{p2}u_{p2} + u_{33}u_{p3}u_{p3})\right.$$

$$+\frac{1}{2}c_{12}\sum_p\{u_{11}(u_{p2}u_{p2} + u_{p3}u_{p3}) + u_{22}(u_{p1}u_{p1} + u_{p3}u_{p3})$$

$$+u_{33}(u_{p1}u_{p1} + u_{p2}u_{p2})\}$$

$$+c_{44}\sum_p\{(u_{12} + u_{21})u_{p1}u_{p2} + (u_{23} + u_{32})u_{p2}u_{p3}$$

$$+(u_{13} + u_{31})u_{p1}u_{p3}\}$$

$$+\frac{1}{6}C_{111}(u_{11}^3 + u_{22}^3 + u_{33}^3)$$

$$+C_{123}(u_{11}u_{22}u_{33})$$

$$+\frac{1}{2}C_{112}\{u_{11}^2(u_{22} + u_{33}) + u_{22}^2(u_{33} + u_{11}) + u_{33}^2(u_{11} + u_{22})\}$$

$$+\frac{1}{2}C_{144}\{u_{11}(u_{23} + u_{32})^2 + u_{22}(u_{31} + u_{13})^2 + u_{33}(u_{12} + u_{21})^2\}$$

$$+C_{456}\{(u_{12} + u_{21})(u_{23} + u_{32})(u_{13} + u_{31})\}$$

$$+\frac{1}{2}C_{166}\{(u_{11} + u_{22})(u_{12} + u_{21})^2 + (u_{22} + u_{33})(u_{23} + u_{32})^2$$

$$\left.+(u_{33} + u_{11})(u_{13} + u_{31})^2\}\right). \tag{4.77}$$

We can write the above expression in a compact form:

$$
\begin{aligned}
\mathcal{V}_3 &= \int \mathrm{d}^3 r \sum_{i \neq j \neq k} \left(\frac{1}{2} c_{11} u_{ii} \sum_p u_{pi} u_{pi} + \frac{1}{2} c_{12} u_{ii} \sum_p u_{pj} u_{pj} \right. \\
&\quad + c_{44} u_{ij} \sum_p u_{pi} u_{pj} + \frac{1}{6} C_{111} u_{ii}^3 + C_{123} u_{ii} u_{jj} u_{kk} \\
&\quad + \frac{1}{2} C_{112} u_{ii}^2 (u_{jj} + u_{kk}) + \frac{1}{2} C_{144} u_{ii} (u_{jk} + u_{kj})^2 \\
&\quad + C_{456} (u_{ij} + u_{ji})(u_{jk} + u_{kj})(u_{ki} + u_{ik}) \\
&\quad + \left. \frac{1}{2} C_{166} (u_{ii} + u_{jj})(u_{ij} + u_{ji})^2 \right) \qquad i,j,k,p = 1,2,3.
\end{aligned}
\tag{4.78}
$$

This equation can further be expressed in a compact notation as in equation (4.61).

For the continuum, we use equations (4.9) and (4.32) to write down the expression for the displacement vector $u(r)$ in the second quantised notation:

$$
u(r) = \frac{1}{\sqrt{N_0 \Omega}} (-\mathrm{i}) \sqrt{\frac{\hbar}{2\rho}} \sum_{qs} \sqrt{\frac{1}{\omega(qs)}} e_{qs} (a_{qs}^\dagger - a_{-qs}) \exp(\mathrm{i} q \cdot r).
\tag{4.79}
$$

Here ρ is the mass density in the unit cell of volume Ω and $e_{qs} \equiv e(r|qs)$ is the polarisation vector at point r in the medium. There is now no summation over the basis vector b, which is compensated by the use of ρ rather than m_b. Also, now the polarisation branch is only of acoustic type so that the index s stands for $s = \mathsf{L}, \mathsf{T}_1, \mathsf{T}_2$ (with T_1 and T_2 being degenerate). Differentiating equation (4.79) we get

$$
\frac{\partial u_l}{\partial r_i} = \frac{1}{\sqrt{N_0 \Omega}} \sqrt{\frac{\hbar}{2\rho}} \sum_{qs} \frac{1}{\sqrt{\omega(qs)}} e_{qs}^l q_i (a_{qs}^\dagger - a_{-qs}) \exp(\mathrm{i} q \cdot r).
\tag{4.80}
$$

Substituting terms like (4.80) in equation (4.61) we get

$$
\begin{aligned}
\mathcal{V}_3 &= \frac{1}{3!} \int \mathrm{d}^3 r \left(\frac{\hbar}{2 N_0 \Omega \rho} \right)^{3/2} \sum_{\substack{qs, q's', q''s'' \\ lmn, ijk}} A_{ijk}^{lmn} \\
&\quad \times \frac{1}{\sqrt{\omega(qs)\omega(q's')\omega(q''s'')}} e_{qs}^l q_i e_{q's'}^m q_j' e_{q''s''}^n q_k'' \\
&\quad \times (a_{qs}^\dagger - a_{-qs})(a_{q's'}^\dagger - a_{-q's'})(a_{q''s''}^\dagger - a_{-q''s''}) \exp[\mathrm{i}(q + q' + q'') \cdot r].
\end{aligned}
\tag{4.81}
$$

The integation can be performed by using the Fourier theorem

$$
\frac{1}{N_0 \Omega} \int \mathrm{d}^3 r \exp[\mathrm{i}(q + q' + q'') \cdot r] = \delta_{q+q'+q'', G},
\tag{4.82}
$$

where G is a reciprocal lattice vector. (Note that we cannot talk of a reciprocal lattice vector in the continuum, but shall still keep equation (4.82) for dealing with U-processes. This we can do if we use Parrott's grafting scheme, as discussed in the previous section.) Then equation (4.81) becomes

$$
\begin{aligned}
\mathcal{V}_3 &= \frac{1}{3!} \sum_{qs, q's', q''s''} \delta_{q+q'+q'', G} (a_{qs}^\dagger - a_{-qs})(a_{q's'}^\dagger - a_{-q's'}) \\
&\quad \times (a_{q''s''}^\dagger - a_{-q''s''}) \mathscr{F}_{qq'q''}^{ss's''},
\end{aligned}
\tag{4.83}
$$

where

$$\mathscr{F}^{ss's''}_{qq'q''} = \sqrt{\frac{\hbar^3}{8\rho^3 N_0 \Omega}} \sum_{\substack{lmn \\ ijk}} A^{lmn}_{ijk} \frac{e^l_{qs} q_i e^m_{q's'} q'_j e^n_{q''s''} q''_k}{\sqrt{\omega(qs)\omega(q's')\omega(q''s'')}} \tag{4.84}$$

$$= \sqrt{\frac{\hbar^3}{8\rho^3 N_0 \Omega}} \left\{ \frac{qq'q''}{c_s c_{s'} c_{s''}} \right\}^{1/2} A^{ss's''}_{qq'q''}, \tag{4.85}$$

where in obtaining equation (4.85) we have used the continuum dispersion relations $\omega(qs) = c_s q$ etc. and have defined

$$A^{ss's''}_{qq'q''} = \sum_{\substack{lmn \\ ijk}} e^l_{qs} v_i e^m_{q's'} v'_j e^n_{q''s''} v''_k A^{lmn}_{ijk} \tag{4.86}$$

with v, v' and v'' as unit vectors along q, q' and q'', respectively. From a comparison of equations (4.83)–(4.85) with equations (4.43) and (4.44) it is clearly seen that the anharmonic coefficients $\mathscr{F}^{ss's''}_{qq'q''}$ represent the elastic continuum analogue of the coefficients $\Psi(qs, q's', q''s'')$. Similarly, $\{A^{lmn}_{ijk}\}$ represents the tensor $\Psi(qb, q'b', q''b'')$ for an elastic continuum, i.e. the coefficients A^{lmn}_{ijk} are related to the elastic constants. The coefficients $A^{ss's''}_{qq'q''}$ which are components of the tensor $\{A^{lmn}_{ijk}\}$ referred to e_{qs} and q as axes, measure the strength of a three-phonon process. For reasons which will become clear in Chapter 6, we will call $|A^{ss's''}_{qq'q''}|^2$ *three-phonon scattering strengths*.

From a comparison of equations (4.83)–(4.85) with equation (4.61), we can express the components $A^{ss's''}_{qq'q''}$ in terms of the second- and third-order elastic constants

$$\begin{aligned}
A^{ss's''}_{qq'q''} = \sum_{i \neq j \neq k} \Bigg(& 3c_{11} e_i v_i \sum_p e'_p v'_i e''_p v''_i + 3c_{12} e_i v_i \sum_p e'_p v'_j e''_p v''_j \\
& + 6c_{44} e_i v_j \sum_p e'_p v'_i e''_p v''_j + C_{111} e_i v_i e'_i v'_i e''_i v''_i \\
& + 6C_{123} e_i v_i e'_j v'_j e''_k v''_k + 3C_{112} e_i v_i e'_i v'_i (e''_j v''_j + e''_k v''_k) \\
& + 3C_{144} e_i v_i (e'_j v'_k + e'_k v'_j)(e''_j v''_k + e''_k v''_j) \\
& + 3C_{166} (e_i v_i + q_j v_j)(e'_i v'_j + e'_j v'_i)(e''_i v''_j + e''_j v''_i) \\
& + 6C_{456} (e_i v_j + e_j v_i)(e'_j v'_k + e'_k v'_j)(e''_k v''_i + e''_i v''_k) \Bigg) \\
& \qquad i, j, k, p = 1, 2, 3.
\end{aligned} \tag{4.87}$$

Each term in equation (4.87) is contributed by all permutations of non-distinguishable three phonons $(qs), (q's')$ and $(q''s'')$ taking part in a three-phonon process. This means that we must expand a term such as $e_l v_i e'_m v_j e''_n v''_k$ by taking permutations over indistinguishable contributions from the three phonons as follows:

$$\begin{aligned}
\frac{1}{6} \big[& e_l v_i e'_m v'_j e''_n v''_k + e_l v_i e'_n v'_k e''_m v''_j + e_m v_j e'_l v'_i e''_n v''_k \\
& + e_m v_j e'_n v'_k e''_l v''_i + e_n v_k e'_l v'_i e''_m v''_j + e_n v_k e'_m v'_j e''_l v''_i \big].
\end{aligned} \tag{4.88}$$

In this way, we can finally express all the terms in equation (4.87). Here we express the first few terms, and others can be similarly obtained.

$$\text{term 1} = 3c_{11} \sum_{\substack{p \\ i \neq j \neq k}} e_i v_i e'_p v'_i e''_p v''_i$$

$$
= \frac{3c_{11}}{6} \sum_{\substack{p \\ i \neq j \neq k}} \left[e_i v_i e'_p v'_i e''_p v''_i + e_i v_i e'_p v'_i e''_i v''_i + e_p v_i e'_i v'_i e''_p v''_i \right.
$$
$$
\left. + e_p v_i e'_p v'_i e''_i v''_i + e_p v_i e'_i v'_i e''_p v''_i + e_p v_i e'_p v'_i e''_p v''_i \right]
$$
$$
= c_{11} \left[(e \cdot v)(v' \cdot v'')(e' \cdot e'') + (e' \cdot v')(q \cdot v'')(e \cdot e'') \right.
$$
$$
\left. + (e \cdot e')(v \cdot v')(e'' \cdot v'') \right] \tag{4.89}
$$

$$
\text{term}\,2 = 3c_{12} \sum_{\substack{p \\ i \neq j \neq k}} e_i v_i e'_p v'_j e''_p v''_j
$$
$$
= \frac{3}{6} \sum_{\substack{p \\ i \neq j \neq k}} \left[e_i v_i e'_p v'_j e''_p v''_j + e_i v_i e'_p v'_j e''_p v''_j + e_p v_j e'_i v'_i e''_p v''_j \right.
$$
$$
\left. + e_p v_j e'_p v'_j e''_i v''_i + e_p v_j e'_i v'_i e''_p v''_j + e_p v_j e'_p v'_j e''_i v''_i \right]
$$
$$
= c_{12} \left[(e \cdot v)(e' \cdot e'')(v' \cdot v'') + (e \cdot e'')(v \cdot v'')(e' \cdot v') \right.
$$
$$
\left. + (e \cdot e')(v \cdot q')(e'' \cdot v'') \right]. \tag{4.90}
$$

Similarly

$$
\text{term}\,3 = 6c_{44} \sum_{\substack{p \\ i \neq j \neq k}} e_i v_j e'_p v'_i e''_p v''_j
$$
$$
= c_{44} \left[(e \cdot v')(v \cdot v'')(e' \cdot e'') + (e \cdot v'')(v \cdot v')(e' \cdot e'') \right.
$$
$$
+ (e \cdot e'')(v \cdot e')(v' \cdot v'') + (e \cdot e'')(v' \cdot v'')(v \cdot e'')
$$
$$
\left. + (e \cdot e'')(v \cdot v')(e' \cdot v'') + (e \cdot e')(v \cdot v'')(v' \cdot e'') \right] \tag{4.91}
$$

and so on.

4.6 EVALUATION OF THREE-PHONON SCATTERING STRENGTHS

From equations (4.87) and (4.88), it is clear that for the evaluation of the three-phonon scattering strengths $|A_{qq'q''}^{ss's''}|^2$ we need to have a knowledge of the second- and third-order elastic constants of the material. Furthermore, as the coefficients $A_{qq'q''}^{ss's''}$ depend on directions of propagation $(\hat{q}, \hat{q}', \hat{q}'')$ and on polarisation vectors $(e_{qs}, e_{q's'}, e_{q''s''})$ of the participating phonons, it is desirable to solve the dynamical problem (chapter 2) for each participating phonon within the first Brillouin zone. This may be prohibitive in the practical sense. The problem becomes amenable if we assume isotropy of the medium. In that case we use the following isotropic conditions for the elastic constants (in Brugger's notation):

$$
\begin{aligned}
C_{111} &= C_{123} + 6C_{144} + 8C_{456} \\
C_{112} &= C_{123} + 2C_{144} \\
C_{166} &= C_{144} + 2C_{456} \\
2c_{44} &= c_{11} - c_{12}.
\end{aligned} \tag{4.92}
$$

Furthermore, these elastic constants can be related to the isotropic elastic constants following the notation of Landau and Lifshitz (1959):

$$
\begin{aligned}
\mathscr{A} &= 4C_{456} & \mathscr{B} &= C_{144} & \mathscr{C} &= \frac{1}{2}C_{123} \\
\lambda &= c_{12} & \mu &= c_{44} & \lambda + 2\mu &= c_{11}.
\end{aligned} \tag{4.93}
$$

Here λ and μ are the Lamé constants of isotropic elasticity. In terms of the isotropic elastic constants, we can express the cubic anharmonic term in the crystal Hamiltonian as

$$
\mathcal{V}_3 = \int d^3r \sum_{ijk} \left[\frac{\mathcal{C}}{3} \left(\frac{\partial u_i}{\partial r_i} \right)^3 + \frac{\lambda + \mathcal{B}}{2} \frac{\partial u_i}{\partial r_i} \left(\frac{\partial u_j}{\partial r_k} \right)^2 + \frac{\mathcal{B}}{2} \frac{\partial u_i}{\partial r_i} \frac{\partial u_j}{\partial r_k} \frac{\partial u_k}{\partial r_j} \right.
$$
$$
\left. + \left(\mu + \frac{\mathcal{A}}{4} \right) \frac{\partial u_i}{\partial r_j} \frac{\partial u_k}{\partial r_i} \frac{\partial u_k}{\partial r_j} + \frac{\mathcal{A}}{12} \frac{\partial u_i}{\partial r_j} \frac{\partial u_j}{\partial r_k} \frac{\partial u_k}{\partial r_i} \right]. \tag{4.94}
$$

In order to evaluate the three-phonon scattering strengths, Hamilton and Parrott (1969) and Srivastava *et al* (1972) assumed the isotropy of the medium and considered the three phonons to be coplanar. Let us choose $\hat{q}$ along the z axis and consider q, q' and q'' to lie in the $x - z$ plane. Let θ' and θ'' be the angles between q and q', and q' and q'', respectively. Then if θ_1, θ_2 and θ_3 are the (arbitrary) polarisation angles for the transverse phonons, we have

$$
\begin{aligned}
\hat{q} \equiv e_L &= (0, 0, 1) \\
e_T &= (\cos\theta_1, \sin\theta_1, 0) \\
\hat{q}' \equiv e_L' &= (\sin\theta', 0, \cos\theta') \\
e_T' &= (\cos\theta' \cos\theta_2, \sin\theta_2, -\sin\theta' \cos\theta_2) \\
\hat{q}'' \equiv e_L'' &= (\sin\theta'', 0, \cos\theta'') \\
e_T'' &= (\cos\theta'' \cos\theta_3, \sin\theta_3, -\sin\theta'' \cos\theta_3).
\end{aligned} \tag{4.95}
$$

Substituting equation (4.95) into equation (4.87), making use of equation (4.88), and averaging out over 0 to π wherever θ_1, θ_2 and θ_3 occur, we can obtain the following results:

$$
\left| A_{qq'q''}^{TTT} \right|^2 = 0 \tag{4.96}
$$

$$
\left| A_{qq'q''}^{TLL} \right|^2 = \frac{1}{2} \left[\sin(\theta' + \theta'') \{ c_{11} \cos\theta' \cos\theta'' + (2c_{44} + c_{12}) \sin\theta' \sin\theta'' \right.
$$
$$
\left. + (C_{166}^B + c_{44}) \cos(\theta'' - \theta') \} \right]^2 \tag{4.97}
$$

$$
= \frac{1}{2} (\lambda + 3\mu + \mathcal{A} + 2\mathcal{B})^2 \sin^2(\theta' + \theta'') \cos^2(\theta'' - \theta') \tag{4.98}
$$

$$
\left| A_{qq'q''}^{TTL} \right|^2 = \frac{1}{16} \left[\{ C_{166}^B \cos 2\theta' + 2c_{11} \cos\theta' \cos\theta'' \cos(\theta' + \theta'') \right.
$$
$$
- 2c_{44} \sin^2\theta' + (4c_{44} + 2c_{12}) \cos\theta' \sin\theta'' \sin(\theta' + \theta'') \}^2
$$
$$
+ \frac{1}{4} \{ (C_{166}^B + 2c_{11} - 2c_{12} - C_{144}^B) \cos\theta' \cos 2\theta''
$$
$$
+ (4c_{44} + C_{456}^B) \sin\theta' \sin 2\theta''
$$
$$
\left. + (C_{166}^B + 2c_{11} + 2c_{12} + C_{144}^B) \cos\theta'' \}^2 \right] \tag{4.99}
$$

$$
= \frac{1}{16} \left[\{ \lambda + \mu + (\lambda + 3\mu + \mathcal{A} + 2\mathcal{B}) \cos 2\theta' \}^2 \right.
$$
$$
\left. + \{ 2(\lambda + \mathcal{B}) \cos\theta' + (4\mu + \mathcal{A}) \cos\theta'' \cos(\theta'' - \theta') \}^2 \right] \tag{4.100}
$$

$$
\left| A_{qq'q''}^{LLL} \right|^2 = 9(c_{11} + 2C_{111}^B)^2 \tag{4.101}
$$

$$
= \left[3\lambda + 6\mu + 2(\mathcal{A} + 3\mathcal{B} + \mathcal{C}) \right]^2. \tag{4.102}
$$

Note that the results are expressed in terms of both crystal elastic constants (using the Birch notation) and the isotropic elastic constants. Within the isotropic continuum model, the scattering strength for the process $\mathsf{T} + \mathsf{T} \rightleftharpoons \mathsf{T}$ is zero. The scattering strengths of other polarisation combinations depend on the angles θ' (between q and q') and θ'' (between q and q'').

A coplanar geometry for the evaluation of three-phonon scattering strengths is not a restriction. We can consider a general three-dimensional geometry for a three-phonon process and obtain an angularly averaged result for its scattering strength $\overline{|A_{qq'q''}^{ss's''}|^2}$,

$$\overline{\left|A_{qq'q''}^{ss's''}\right|^2} = \frac{1}{(4\pi)^3} \int \int \int \, d\Omega_1 d\Omega_2 d\Omega_3 \left|A_{qq'q''}^{ss's''}\right|^2. \tag{4.103}$$

In the polar coordinate system, we consider for a phonon (qs)

$$\begin{aligned}
\hat{q} \equiv e_\mathsf{L} &= (\sin\theta\cos\phi, \sin\theta\sin\phi, \cos\theta) \\
e_{\mathsf{T}_1} &= (\cos\theta\cos\phi, \cos\theta\sin\phi, -\sin\theta) \\
e_{\mathsf{T}_2} &= (-\sin\phi, \cos\phi, 0)
\end{aligned} \tag{4.104}$$

with similar coordinates for phonons $(q's')$ and $(q''s'')$. Here we have removed the restriction of coplanar phonons and have considered two mutually perpendicular transverse polarisation directions. Numerically obtained results are (Srivastava 1980)

$$\overline{\left|A_{qq'q''}^{\mathsf{TTT}}\right|^2} = 0 \tag{4.105}$$

$$\begin{aligned}
\overline{\left|A_{qq'q''}^{\mathsf{TLL}}\right|^2} &= 0.0255(3c_{44} + 4C_{456})^2 + 0.1333(c_{12} + 2C_{144})^2 \\
&\quad + 0.0593(3c_{44} + 4C_{456})(c_{12} + 2C_{144}) \tag{4.106} \\
&= 0.0255(\mathscr{A} + 3\mu)^2 + 0.1333(\lambda + 2\mathscr{B})^2 \\
&\quad + 0.0593(\mathscr{A} + 3\mu)(\lambda + 2\mathscr{B}) \tag{4.107}
\end{aligned}$$

$$\begin{aligned}
\overline{\left|A_{qq'q''}^{\mathsf{TTL}}\right|^2} &= 0.1333\left[c_{12}^2 + 1.5C_{144}(c_{12} + C_{144})\right] \\
&\quad + 4(c_{44} + C_{456})\left[0.0496(c_{44} + C_{456})\right. \\
&\quad \left. + 0.0222(c_{12} + 2C_{144})\right] \tag{4.108} \\
&= 0.1333\left[\lambda^2 + 1.5\mathscr{B}(\lambda + \mathscr{B})\right] \\
&\quad + (\mathscr{A} + 4\mu)\left[0.0124(\mathscr{A} + 4\mu) + 0.0222(\lambda + 2\mathscr{B})\right] \tag{4.109}
\end{aligned}$$

$$\overline{\left|A_{qq'q''}^{\mathsf{LLL}}\right|^2} = (3c_{11} + C_{111})^2 \tag{4.110}$$

$$= \left[3\lambda + 6\mu + 2(\mathscr{A} + 3\mathscr{B} + \mathscr{C})\right]^2, \tag{4.111}$$

where the third-order elastic constants C_{ijk} are expressed in the Brugger notation.

4.7 THE QUASI-HARMONIC APPROXIMATION AND GRUNEISEN'S CONSTANT

In sections 4.3 and 4.4, we discussed the effect of anharmonicity in terms of phonon–phonon interactions. Anharmonicity has another effect too: it gives rise to a major component of thermal expansion in crystals. Both phonon–phonon interactions and thermal expansion play important roles in the thermal properties of crystals. For many purposes, temperature-dependent properties of anharmonic crystals can be studied at three different levels: (i) quasi-harmonic approximation, (ii) self-consistent harmonic approximation and (iii) pseudoharmonic approximation.

In the quasi-harmonic approximation (Leibfried and Ludwig 1961), the anharmonicity is considered as a weak effect and the atomic force constants and the phonon frequencies are renormalised by taking only thermal expansion into account. The self-consistent harmonic approximation is used to study the vibrational properties of crystals with very strong anharmonicity (e.g. rare earth crystals, crystal He, and crystal H_2). In this approximation, effective atomic force constants are obtained by self-consistently replacing the harmonic force constants by their thermal averages over all possible motions of atoms other than the one under consideration (see, e.g., Horner 1974, Brüesch 1982, and references therein). In the pseudoharmonic approximation (see, e.g., Reissland 1973), the effects of both thermal expansion and phonon–phonon interactions are considered: the thermal expansion effect is considered as in the quasi-harmonic approximation, and phonon–phonon interactions are studied using either a perturbation method or a linear response theory. We will leave the discussion of phonon–phonon interactions for Chapter 6, but will discuss here the effect of thermal expansion in the quasi-harmonic approximation.

4.7.1   THE EQUATION OF STATE IN THE QUASI-HARMONIC APPROXIMATION

We will use the isotropic continuum model for a crystal to derive the equation of state in the quasi-harmonic approximation (Leibfried and Ludwig 1961, Brüesch 1982). In the isotropic continuum model, the (Helmholtz) free energy F of a crystal depends only on its volume $V \equiv N_0 \Omega$ and not on the coordinates of the atoms. The equation of state is then simply given by

$$P = -\left(\frac{\partial F}{\partial V}\right)\bigg|_T. \tag{4.112}$$

In the quasi-harmonic approximation, the free energy F consists of an effective quadratic potential term $\mathscr{V}_2^{eff}$ (including renormalised force constants in the presence of weak anharmonicity) and a vibrational free energy of atoms originating from the uniform strain in the system

$$F = \mathscr{V}_2^{eff} + F_{vib}. \tag{4.113}$$

The vibrational free energy can be expressed as

$$F_{vib} = F_0 + F_{thermal}, \tag{4.114}$$

where

$$F_0 = \frac{1}{2}\sum_{qs} \hbar\omega(qs) \tag{4.115}$$

is the zero-point energy, and the thermal free energy is given by

$$
\begin{aligned}
F_{thermal} &= -k_B T \ln Z_{thermal} \\
&= -k_B T \ln \prod_{qs} \sum_{n(qs)} \exp\left(-n\hbar\omega(qs)/k_B T\right) \\
&= -k_B T \ln \prod_{qs} [1 - \exp\left(-\hbar\omega(qs)/k_B T\right)]^{-1} \\
&= k_B T \sum_{qs} \ln[1 - \exp\left(-\hbar\omega(qs)/k_B T\right)].
\end{aligned}
\tag{4.116}
$$

The equation of state (equation 4.112) is therefore

$$
\begin{aligned}
P &= -\left(\frac{\partial \mathscr{V}_2^{eff}}{\partial V}\right)\bigg|_T - \left(\sum_{qs} \frac{\partial F_{vib}}{\partial \omega(qs)} \frac{\partial \omega(qs)}{\partial V}\right)\bigg|_T \\
&= -\left(\frac{\partial \mathscr{V}_2^{eff}}{\partial V}\right)\bigg|_T + \frac{1}{V}\sum_{qs} \gamma_{qs}\left(\tfrac{1}{2}\hbar\omega(qs) + \hbar\omega(qs)\bar{n}_{qs}\right)\bigg|_T,
\end{aligned}
\tag{4.117}
$$

where $\bar{n}_{qs}$ is the equilibrium phonon distribution function. In the above equation we have defined Grüneisen's parameter $\gamma(qs)$ for phonons in the mode qs by

$$\frac{\partial \omega(qs)}{\partial V} = -\gamma(qs)\frac{\omega(qs)}{V}. \tag{4.118}$$

Assuming that the volume changes due to thermal expansion are small, we can simplify the equation of state in equation (4.117) by expanding $\mathscr{V}_2^{eff}$ and F_{vib} about the equilibrium volume V_0 at $T = 0$. Writing

$$
\begin{aligned}
\mathscr{V}_2^{eff} &\simeq \mathscr{V}_2^{eff}(T=0) + \frac{1}{2}\left(\frac{\partial^2 \mathscr{V}_2^{eff}}{\partial V^2}\right)\Big|_{V_0}(\Delta V)^2 \\
&= \mathscr{V}_2^{eff}(T=0) + \frac{1}{2}\frac{1}{V_0 \kappa_0}(\Delta V)^2
\end{aligned}
\tag{4.119}
$$

and

$$F_{vib} \simeq F_{vib}(V_0) + \left(\frac{\partial F_{vib}}{\partial V}\right)\Big|_{V_0}\Delta V \tag{4.120}$$

we get after neglecting the zero-point energy

$$P(V_0) \simeq -\frac{\Delta V}{\kappa_0} + \sum_{qs}\gamma_{qs}\bar{\varepsilon}(qs). \tag{4.121}$$

Here κ_0 is the compressibility and $\bar{\varepsilon}(qs)$ is the mean energy of phonons in the state qs. Equation (4.121) is the equation of state derived by Mie (1903) and by Grüneisen (1908).

4.7.2 GRUNEISEN'S CONSTANT AND THERMAL EXPANSION IN THE QUASI-HARMONIC APPROXIMATION

Thermal expansion is the fractional volume change at zero pressure. Thus from equation (4.121), after setting $P = 0$,

$$\varpi = \frac{\Delta V}{V_0} = \frac{\kappa_0}{V_0}\sum_{qs}\gamma_{qs}\bar{\varepsilon}(qs). \tag{4.122}$$

The volume thermal expansion coefficient is

$$\beta = \left(\frac{\partial \varpi}{\partial T}\right)_P = \frac{\kappa_0}{V}\sum_{qs}\gamma_{qs}C_v(qs), \tag{4.123}$$

where $C_v(qs)$ is the heat capacity due to mode qs at constant volume and temperature T. In the above equation we have considered κ_0 to be a constant, as it only weakly depends on temperature. If we define a *mode-independent* Grüneisen parameter

$$\gamma = \frac{\sum_{qs}\gamma_{qs}C_v(qs)}{\sum_{qs}C_v(qs)} \tag{4.124}$$

as a weighted average of the γ_{qs}, then equation (4.123) can be expressed as

$$
\begin{aligned}
\beta &= \frac{\kappa_0}{V_0}\gamma C_v \\
&= \kappa_0 \gamma C_v^{sp},
\end{aligned}
\tag{4.125}
$$

where $C_v^{sp} = C_v/V$ is the *specific* heat capacity and $V = N_0\Omega$ is crystal volume. Equation (4.125) is the original Grüneisen relation. Grüneisen considered κ_0 and γ to be independent of temperature and concluded that the thermal expansion coefficient has the same temperature dependence as the specific heat. This is known as Grüneisen's rule or law. Notice that we defined β as the *volume* thermal expansion coefficient. It is common to define the *linear* thermal expansion coefficient α

$$\alpha = \frac{1}{L}\left(\frac{\partial L}{\partial T}\right)_P = \frac{1}{3V}\left(\frac{\partial V}{\partial T}\right)_P = \frac{\beta}{3}. \tag{4.126}$$

Using this definition, we can express equation (4.125) as

$$\alpha = \frac{\gamma C_v^{\text{sp}}}{3B} \tag{4.127}$$

where $B = 1/\kappa_0$ is the bulk modulus.

In the Debye approximation the mode frequencies scale linearly with the cut-off frequency ω_D, $\omega(qs) = (q/q_D)\omega_D, \hbar\omega_D = k_B\Theta_D$, so that the mode-independent Grüneisen parameter can be easily calculated:

$$\begin{aligned} \gamma_D &= -\frac{V}{\omega_D}\frac{\partial\omega_D}{\partial V} = -\frac{V}{\Theta_D}\frac{\partial\Theta_D}{\partial V} \\ &= -\frac{\partial(\ln\Theta_D)}{\partial(\ln V)}. \end{aligned} \tag{4.128}$$

In the Debye theory, if Θ_D is taken as independent of temperature, γ_D should also be independent of temperature, in agreement with Grüneisen's law. Experimentally it is found that Grüneisen's law is approximately true and only at high temperatures, with γ taking values between 1 and 2 for most solids.

A simple generalisation of equation (4.128) is to write the mode Grüneisen parameter as

$$\gamma_{qs} = -\frac{d[\ln\omega(qs)]}{d(\ln V)}. \tag{4.129}$$

Thus the mode Grüneisen parameter can be calculated from a knowledge of the phonon frequencies as a function of crystal volume. In practice some approximation is required to evaluate the expression in equation (4.129). One may express

$$\gamma_{qs} \simeq -\frac{\Delta\ln\omega(qs)}{\Delta\ln V}. \tag{4.130}$$

One may also express (Kunc and Martin 1983)

$$\gamma_{qs} \equiv -\frac{d\omega(qs)/\omega(qs)}{dV/V} \simeq -\frac{1}{3}\frac{a}{\omega(qs)}\frac{d\omega(qs)}{da} \tag{4.131}$$

where a is the lattice constant. Soma (1977) suggested replacing equation (4.129) as follows:

$$\gamma_{qs} = \frac{-a}{6\omega^2(qs)}\frac{d[\omega^2(qs)]}{da}. \tag{4.132}$$

In terms of the Fourier transform of the second- and third-order elastic constants in equations (4.60) and (4.61) γ_{qs} can be defined, assuming cubic symmetry, as (Ziman, 1960)

$$\gamma_{qs} = -\frac{1}{6}\sum_\sigma A_{qq\sigma}^{ss\sigma}/J_{qq}^{ss}, \tag{4.133}$$

where the third-order elastic constant tensor component $A_{qq\sigma}^{ss\sigma}$ is defined by equation (4.86) and J_{qq}^{ss} is a Fourier component of the second-order elastic constant tensor J in equation (4.60). Also, a change

ΔV in volume generates the change ΔJ in the elastic moduli, which can be related to the third-order moduli A in the form (Ziman 1960)

$$\Delta J_{np}^{lm} = \frac{1}{3}\frac{\Delta V}{V}\sum_{\sigma} A_{np\sigma}^{lm\sigma}. \tag{4.134}$$

Using equations (2.140), (4.37), (4.60), (4.80) and (4.134), we write

$$\delta(\omega^2) = \delta(q^2 c^2) = \rho^{-1}\delta\left(\sum_{\substack{lm\\np}} J_{np}^{lm} e_l e_m q_n q_p\right)$$

$$= \frac{\delta V}{3V}\rho^{-1}\sum_{\substack{lm\\npr}} A_{npr}^{lmr} e_l e_m q_n q_p, \tag{4.135}$$

where ρ is the mass density. With this, and using equation (4.118), the Grüneisen constant can be expressed as

$$\gamma_{qs} = -\frac{1}{6\omega^2(qs)\rho}\sum_{\substack{lm\\npr}} A_{npr}^{lmr} e_l q_n e_m q_p. \tag{4.136}$$

This expression was derived in Srivastava (1980).

It is also possible to obtain an expression for the Grüneisen constant for a mode qs taking part in a three-phonon process of the type $(qs)+(q's') \rightleftharpoons (q''s'')$. Srivastava (1980) has obtained the following expressions within the isotropic continuum approximation:

$$\gamma^2(ss's'') = \frac{|A_{qq'q''}^{ss's''}|^2}{4\rho^2 c_s^4} \tag{4.137}$$

$$\gamma^2 = \bar{c}^2\left\{\frac{c_s^2}{c_{s'}^2 c_{s''}^2}\gamma^2(ss's'')\right\}_{av}$$

$$= \frac{1}{12}\left(\frac{\bar{c}}{c_L}\right)^2\left\{\frac{|A_{qq'q''}^{LLL}|^2}{c_{11}^2} + \frac{|A_{qq'q''}^{TLL}|^2}{c_{11}c_{44}} + \frac{|A_{qq'q''}^{TTL}|^2}{c_{44}^2}\right\}, \tag{4.138}$$

where $\bar{c}$ is an average acoustic phonon speed, $\overline{|A_{qq'q''}^{ss's''}|^2}$ has the same meaning as in equations (4.105)–(4.111), and γ is mode averaged. From equations (4.110) and (4.135)–(4.136), it is easy to see that

$$\gamma_{qs}^2(L+L \to L) = \frac{(3c_{11}+C_{111})^2}{4c_{11}^2}. \tag{4.139}$$

Another approximate, but useful, relation is

$$\overline{|A_{qq'q''}^{ss's''}|^2} = \frac{4\rho^2}{\bar{c}^2}\gamma^2 c_s^2 c_{s'}^2 c_{s''}^2. \tag{4.140}$$

The assumption that the Grüneisen constant is temperature independent is not true. It is expected to show low- and high-temperature dependences of the following types (Yates 1972):

$$\gamma_{lowtemp} = \gamma_0(1+bT^2+\ldots) \tag{4.141}$$

$$\gamma_{hightemp} = \gamma_\infty(1-c/T^2+\ldots) \tag{4.142}$$

where γ_0 and γ_∞ are the limiting low- and high-temperature values, respectively, and b and c are constants.

Table 4.1

Mode Gruneisen constants for phonons at a few symmetry points in Si, Ge and GaAs. The theoretical values are obtained by using the *ab initio* frozen-phonon approach described in section 3.3.1.1. For comparison, experimental results are also given.

Material		$LTO(\Gamma)$	$LOA(X)$	$TO(X)$	$TA(X)$	
Si	γ(cal[a])	0.9	1.3	0.9	−1.5	
	γ(exp[b])	0.98	1.5	0.9	−1.4	
Ge	γ(cal[a])	0.9	1.4	1.0	−1.5	
	γ(exp[c])	0.88 ± 0.08				

		$TO(\Gamma)$	$LO(X)$	$LA(X)$	$TO(X)$	$TA(X)$
GaAs	γ(cal[d])	1.42	0.91	1.11	1.56	−3.48
	γ(exp[e])	1.39			1.73	−1.62

[a] Yin and Cohen (1982); [b] Weinstein and Piermarini (1975); [c] Asaumi and Minomura (1978); [d] Kunc and Martin (1981); [e] Trommer *et al* (1980).

The *ab initio* frozen phonon theory described in section 3.3.1.1 has been used to calculate γ_{qs} by Kunc and Martin (1981) and Yin and Cohen (1982) for diamond and zincblende type semiconductors. The perturbative approach described in section 3.3.2.3 has been applied by Soma (1977) to compute the Grüneisen constant of the individual phonon modes and the temperature-dependent mean Grüneisen constant for Si and Ge. Some of the calculated results are presented in Table 4.1 and figure 4.2 where comparison with experiment is also made. A novel first principles method to calculate mode-dependent as well as mode-averaged temperature dependent Grüneisen parameter has recently been presented by Cuffari and Bongiorno (2020). In contrast to the conventional formulation in equations (4.129)–(4.130), their formulation expresses a mode Grüneisen parameter in terms of a "corrected" stress tensor evaluated within a supercell approach. More discussion on mode Grüneisen constants and temperature-dependent mode-average Grüneisen constant will be provided later in section 4.8 and in Chapter 7 where we will deal with computation of lattice thermal conductivity.

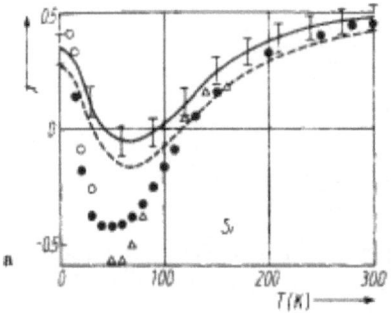 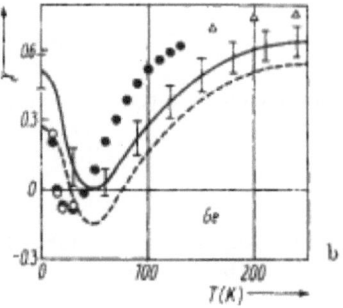

Figure 4.2 The mode-averaged Grüneisen constant γ for Si and Ge. The experimental values (Slack and Bartram 1975) are shown by dots. The theoretical results, obtained by the perturbation theory described in section 3.3.2 (Soma 1977), are shown by the full curve (with only paired two-body forces) and by the broken curve (with the modified perturbation method).

Using the Taylor expansion of the harmonic force constants in terms of the volume or the lattice constant parameter, the mode Grüneisen constant can be expressed in terms of the third-order force constants (Barron and Klein 1974, Fabian and Allen 1997, Ward *et al* 2009, Esfarjani *et al* 2011)

$$
\begin{aligned}
\gamma_{qs} &= -\frac{1}{6\omega^2(qs)} \sum_{b,l'b',l''b''} \sum_{\alpha\beta\delta} \Psi_{\alpha\beta\delta}(0b,l'b',l''b'') \\
&\times \frac{e_\alpha^*(b|qs)e_\beta(b'|qs)}{\sqrt{m_b m_{b'}}} e^{iq\cdot l'} \tau_\delta(l''b''),
\end{aligned}
\tag{4.143}
$$

where $\tau_\delta(lb)$ is the the δth component of the vector describing the position of the bth atom in the lth unit cell.

4.7.3 GRUNEISEN'S CONSTANT AND PHASE TRANSITION

Many crystals show structural changes as the temperature is raised, or when an external pressure is applied. There are many kinds of phase transitions, but here we discuss only the *displacive phase transition* which can be explained by the *soft phonon mode* theory. The concept of a soft mode was briefly discussed in section 2.4.5. The suggestion that certain kinds of solid phase transitions might be triggered by phonon instabilities was made by Cochran (1959b) and by Anderson (1960). Since then, many experimental studies of crystal phase transitions have been made. There are two kinds of phonon instability or lattice softening. One is the softening of the (shear) elastic constants for long wave phonons. The other is the softening of TA phonons near the zone boundaries.

In the quasi-harmonic approximation both the temperature and pressure effects can be treated similarly, as the temperature dependence of thermal expansion is closely related to the pressure dependence of phonon frequencies: these effects change the interatomic potential and renormalise phonon frequencies. A soft mode is characterised by a negative Grüneisen's constant.

Using the perturbative pseudopotential approach described in section 3.3.2.3, Soma (1978b, 1980) studied the role of both kinds of lattice softening in pressure induced phase transition in Si and Ge. His theoretical results indicate that the lattice softening in the long wave phonons does not appear for Si and Ge. By calculating the TA mode Grüneisen constant at X and L points in the Brillouin zone of Si and Ge, Soma roughly correlated the value of fractional lattice constant, $\Delta a/a$, with the phonon instability. He concluded that the effect of the lattice softening at X is prominent, especially in Ge: the phase transition to white-tin type structure is introduced near the volume change $\Delta V/V \simeq 0.1$ under about 100 kbar pressure. His results agree with the effect of pressure on the one- and two-phonon Raman spectra of Weinstein (1977).

Using the first-principles frozen-phonon pseudopotential approach described in section 3.3.1.1 Yin and Cohen (1980), Froyen and Cohen (1982), and Kunc and Martin (1981, 1983) have also studied structural phase transitions in diamond and zincblende type semiconductors. The theoretical predictions agree well with available experimental results. In the present context, consider the displacement pattern S_2 shown in figure 4.3. This displacement can be used to approximate the TA(X) phonon mode in the zincblende structure. In figure 4.4, we show the theoretical results of Kunc and Martin for the energy of the TA(X) mode in GaAs under pressure. These authors find that in GaAs among all $q = X$ modes the TA(X) mode is the most anharmonic and the most sensitive to pressure. This is expected to be true for other zincblende materials as well. Both theoretical and experimental results suggest that the TA(X) is a mode with negative γ. The strong anharmonicity and pressure dependence of the TA(X) mode leads to a fundamental instability of the zincblende structure. The work of Kunc and Martin suggests that if the GaAs crystal is compressed by $\Delta a = -0.144$ Å (a being the cubic lattice constant), then a first-order phase transition to an orthorhombic structure, close to the zincblende structure with a TA(X) displacement, should be expected. The experimental work of Yu *et al* (1978) has reported a transition to an orthorhombic structure (of unspecified nature) corresponding to $\Delta a = -0.28$ Å (about 170 kbar pressure). However, this has not been

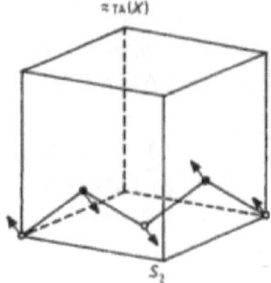

Figure 4.3   The atomic displacement pattern S_2 for predicting the TA(X) vibrational mode in the zincblende structure.

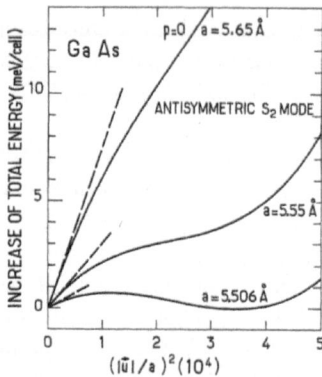

Figure 4.4   The energy of the TA(X) mode in GaAs under pressure. The results of calculation from the harmonic part of the phonon energy are shown by the broken lines. It is clearly seen that with hydrostatic pressure the mode becomes strongly anharmonic, and for certain values of the pressure the variation of the energy difference exhibits a second minimum, with the same energy as the undisplaced crystal. This suggests the possibility of a first-order structural phase change from the zincblende to the orthorhombic structure induced by the soft TA(X) mode. (From Kunc and Martin (1981).)

confirmed so far. Therefore, it is fair to assume that there is no clear experimental evidence for this transition yet. The existence of a similar soft-mode phase transition from the zincblende structure to the orthorhombic structure has been observed in InSb (McWhan and Marezio 1966).

4.8   ANALYTIC EXPRESSIONS FOR CUBIC AND QUARTIC ANHARMONIC POTENTIAL TERMS USING A SEMI-*AB INITIO* SCHEME

A general expression for the cubic anharmonic potential term $\mathscr{V}_3$ was presented in equations (4.43)–(4.45). In equations (4.83)–(4.86) we have presented $\mathscr{V}_3$ of an anharmonic elastic continuum. However, no explicit or analytic expression is possible for the Fourier-transformed third-order force constant tensor $\Psi(qs, q's', q''s'')$ in equation (4.44). Even when regarding a crystal as an elastic continuum, it is not possible to obtain an explicit or analytic expression for the three-phonon scattering strength $A_{qq'q''}^{ss's''}$ in equation (4.86). Whereas the expressions obtained in equations (4.96)–(4.102) assume coplanarity of three-phonon events as stated in equation (4.95), the results in equations (4.103)–(4.111) present angularly averaged results for a general three-dimensional geometry. Accurate numerical and temperature-dependent values of second- and third-order elastic constants are needed for a reliable $\mathcal{V}_3$ expression. Also, it should be reminded that the elastic isotropic continuum model does not treat optical phonon modes.

An approach has been suggested to obtain analytic expressions for both cubic and quartic anharmonic potential terms, $\mathcal{V}_3$ and $\mathcal{V}_4$, considering all phonon modes, acoustic as well as optical, in real crystal structures by Thomas and Srivastava (2017). Their procedure makes use of the continuum approximation in expressing $\mathcal{V}_3$ and $\mathcal{V}_4$ in terms of the mode Grüneisen constants γ_{qs}.

We simplify the problem by restricting the phonon mode (qs) of longitudinal branch so that $e \cdot v = 1$, and take average values $\overline{(e' \cdot v')^2} = \overline{(e'' \cdot v'')^2} = 1/3$. This allows us to approximate the ratio

$$R = \frac{\sum_{\substack{lm \\ npr}} A_{npr}^{lmr} e_l v_n e_m v_p}{A_{qq'q''}^{ss's''}} = \frac{\sum_{\substack{lm \\ npr}} A_{npr}^{lmr} e_l v_n e_m v_p}{\sum_{\substack{lmn \\ ijk}} A_{ijk}^{lmn} e_l v_i e_m' v_j' e_n'' v_k''} \simeq 3, \tag{4.144}$$

where v is a unit vector along q etc., and finally express

$$\frac{|A_{qq'q''}^{ss's''}|}{c_s c_{s'} c_{s''}} = \frac{2\rho|\gamma|}{\bar{c}}, \tag{4.145}$$

where $\bar{c}$ is average acoustic phonon speed. This is the same as equation (4.140). Now we can use equations (4.83), (4.84) and (4.145) to rewrite the cubic anharmonic potential $\mathcal{V}_3$ as

$$\begin{aligned}
\mathcal{V}_3 &= -\frac{1}{3!}\sqrt{\frac{\hbar^3}{2\rho\bar{c}^2 N_0\Omega}} \sum_{qs,q's',q''s''} \gamma_{qs}\sqrt{\omega(qs)\omega(q's')\omega(q''s'')}\delta_{q+q'+q'',G} \\
&\quad (a_{qs}^\dagger - a_{-qs})(a_{q's'}^\dagger - a_{-q's'})(a_{q''s''}^\dagger - a_{-q''s''}).
\end{aligned} \tag{4.146}$$

This expression is preferable over the expression in equations 4.83) and (4.84) as γ_{qs} can be numerically computed using the *ab initio* techniques discussed in section 3.3.2.

For evaluating certain physical quantities, such as anharmonic relaxation of phonons and thermal conductivity, it may be useful to express $\mathcal{V}_3$ using a mode-averaged and temperature dependent Grüneisen constant. This can be done in a manner similar to equation (4.124) within the quasi-harmonic approximation. However, Madsen *et al* (2016) pointed out that it is important to perform the averaging over the squared Grüneisen constant to avoid cancellation between modes with an opposite frequency dependency on volume. So we express

$$\overline{\gamma^2}(T) = \frac{\sum_{qs} \gamma_{qs}^2 C_v(qs)}{\sum_{qs} C_v(qs)}. \tag{4.147}$$

Obviously, the temperature dependence of the root-square-mode-averaged Grüneisen parameter $<\gamma> = \sqrt{\overline{\gamma^2}(T)}$ comes through the specific heat in equation (4.147). With this consideration, the cubic anharmonic term $\mathcal{V}_3$ can be expressed as

$$\begin{aligned}
\mathcal{V}_3 &= \frac{1}{3!}\sqrt{\frac{\hbar^3}{2\rho\bar{c}^2 N_0\Omega}} <\gamma> \sum_{qs,q's',q''s''} \sqrt{\omega(qs)\omega(q's')\omega(q''s'')}\delta_{q+q'+q'',G} \\
&\quad (a_{qs}^\dagger - a_{-qs})(a_{q's'}^\dagger - a_{-q's'})(a_{q''s''}^\dagger - a_{-q''s''}).
\end{aligned} \tag{4.148}$$

From equation (4.143) we know that the mode-dependent Grüneisen constant is related to the anharmonic force constants. Thus, the quasi-harmonic mode-average of the Grüneisen constant represents the temperature dependence of the anharmonic force constants.

The procedure described in section 4.5 to obtain the cubic anharmonic term of an elastic continuum in the form given in equations (4.83)–(4.85) can be extended to express the quartic anharmonic term in the form

$$
\mathcal{V}_4 = \frac{1}{4!} \frac{\hbar^2}{4\rho^2 N_0 \Omega}
$$

$$
\times \sum_{\substack{qs,q's',\\ q''s'',q'''s'''}} \frac{\sqrt{\omega(qs)\,\omega(q's')\,\omega(q''s'')\,\omega(q'''s''')}}{c_s c_{s'} c_{s''} c_{s'''}} A^{ss's''s'''}_{qq'q''q'''} \delta_{q+q'+q''+q''',G}
$$

$$
\times \; (a^\dagger_{qs} - a_{-qs})(a^\dagger_{q's'} - a_{-q's'})(a^\dagger_{q''s''} - a_{-q''s''})(a^\dagger_{q'''s'''} - a_{-q'''s'''}). \tag{4.149}
$$

Here a coupling constant $A^{ss's''s'''}_{qq'q''q'''}$ is a component of the eighth rank tensor $\{A^{lmnl'}_{ijki'}\}$ referred to e_{qs} and q as axes, and is a function of the wave constants:

$$
A^{ss's''s'''}_{qq'q''q'''} = \sum_{\substack{lmnl'\\ ijki'}} e^l_{qs} e^m_{q's'} e^n_{q''s''} e^{l'}_{q'''s'''} v_i v'_j v''_k v'''_{i'} A^{lmnl'}_{ijki'}. \tag{4.150}
$$

The cartesian tensor $\{A^{lmnl'}_{ijki'}\}$ can be related to the elastic constants, and v, v', v'' and v''' are unit vectors parallel to q, q', q'' and q''', respectively. As we will discuss in chapter 6, $|A^{ss's''s'''}_{qq'q''q'''}|^2$ represents *four-phonon scattering strengths*.

Let us re-write equation (4.150) as

$$
A^{ss's''s'''}_{qq'q''q'''} = \frac{1}{R'} \sum_{\substack{lm\\ npr}} e_l e_m v_n v_p A^{lmr}_{npr}, \tag{4.151}
$$

where

$$
R' = \frac{\sum_{\substack{lm\\ npr}} e_l e_m v_n v_p A^{lmr}_{npr}}{\sum_{\substack{lmnl'\\ ijki'}} e_l e'_m e''_n e'''_{l'} v_i v'_j v''_k v'''_{i'} A^{lmnl'}_{ijki'}}. \tag{4.152}
$$

Following the procedure adopted for deriving $\mathcal{V}_3$, we further simplify the problem by restricting ourselves to the relaxing phonon in the mode qs of longitudinal branch so that $e \cdot v = 1$ and take average values $\overline{e' \cdot v'} = \overline{e'' \cdot v''} = \overline{e''' \cdot v'''} = 1/\sqrt{3}$. With these simplifications, and assuming similar magnitudes for third-order and fourth-order elastic constants (consistent with the numerical results of Feng and Ruan (2016)), we finally approximate

$$
R' \approx 3\sqrt{3}, \tag{4.153}
$$

and thus express

$$
A^{ss's''s'''}_{qq'q''q'''} \approx \frac{1}{3\sqrt{3}} \sum_{\substack{lm\\ npr}} e_l e_m v_n v_p A^{lmr}_{npr}. \tag{4.154}
$$

Finally, using equation (4.136), we express the four-phonon coupling coefficient in the following simple from

$$
\frac{A^{ss's''s'''}_{qq'q''q'''}}{c_s c_{s'} c_{s''} c_{s'''}} \approx -\frac{2\rho}{\sqrt{3}} \frac{\rho\gamma}{\bar{c}^2}. \tag{4.155}
$$

With the above approximations, the fourth-order anharmonic potential term can be expressed as

$$
\mathcal{V}_4 = -\frac{1}{4!} \frac{\hbar^2}{2\sqrt{3}\rho^2 \bar{c}^2 N_0 \Omega}
$$

$$
\times \sum_{qs,q's',q''s'',q'''s'''} \gamma_{qs} \sqrt{\omega(qs)\,\omega(q's')\,\omega(q''s'')\,\omega(q'''s''')}\, \delta_{q+q'+q''+q''',G}
$$

$$\times \quad (a_{qs}^{\dagger} - a_{-qs})(a_{q's'}^{\dagger} - a_{-q's'})(a_{q''s''}^{\dagger} - a_{-q''s''})(a_{q'''s'''}^{\dagger} - a_{-q'''s'''}). \tag{4.156}$$

Accepting the earlier argument for adopting the temperature dependent root-square-mode-average Grüneisen parameter $< \gamma >$, we finally express $\mathscr{V}_4$ in the form

$$\begin{aligned}
\mathscr{V}_4 \quad = \quad & \frac{1}{4!} \frac{\hbar^2}{2\sqrt{3}\rho^2 \bar{c}^2 N_0 \Omega} < \gamma > \\
\times \quad & \sum_{qs, q's', q''s'', q'''s'''} \sqrt{\omega(qs)\, \omega(q's')\, \omega(q''s'')\, \omega(q'''s''')} \, \delta_{q+q'+q''+q''',G} \\
\times \quad & (a_{qs}^{\dagger} - a_{-qs})(a_{q's'}^{\dagger} - a_{-q's'})(a_{q''s''}^{\dagger} - a_{-q''s''})(a_{q'''s'''}^{\dagger} - a_{-q'''s'''}). \tag{4.157}
\end{aligned}$$

Although the derivations above are based on several approximations within the concept of elastic anharmonicity, it should be noted that the phonon frequencies $\omega(qs)$ and the mode Grüneisen parameters γ_{qs} appearing in the analytic expressions for $\mathscr{V}_3$ and $\mathscr{V}_4$ in equations (4.148) and (4.157), respectively, can be evaluated numerically by employing the *ab initio* techniques describe in section 3.3.2. Bearing this in mind, we may regard the scheme in the section as semi-*ab initio*.

4.9   *AB INITIO* CALCULATIONS OF CUBIC AND QUARTIC ANHARMONIC POTENTIAL TERMS

In recent years, attempts have been made to determine the *zero-temperature* third- and fourth-order force constants from *ab initio* results of total energy and/or forces described in section 3.3.2. As with the evaluation of harmonic force constants, essentially two distinct routes have been followed to extract cubic force constants: (i) the finite-difference method by employing a real space scheme, and (ii) third-order DFPT, which is an extension of the momentum-space DFPT scheme, based on the so-called $(2n + 1)$ theorem. We will provide a brief discussion of these methods.

4.9.1   FINITE DIFFERENCE METHOD

The finite difference method of evaluating harmonic force constants was described in section 3.3.1. In particular, we detailed the one-dimensional version of the finite difference method in section 3.3.1.3. Extraction of cubic and quartic force constants has been attempted using different levels of generalisation of these schemes.

In the finite difference scheme, harmonic force constants are calculated using the force-displacement relation presented in equation (3.76). That procedure can be extended to extract cubic force constants Ψ and quartic force constants Ξ, as follows:

$$\Psi_{\alpha\beta\gamma}(lb; l'b'; l''b'') \quad \simeq \quad - \frac{F_{\gamma}(l''b'')}{u_{\alpha}(lb) u_{\beta}(l'b')} \tag{4.158}$$

$$\Xi_{\alpha\beta\gamma\delta}(lb; l'b'; l''b''; l'''b''') \quad \simeq \quad - \frac{F_{\delta}(l'''b''')}{u_{\alpha}(lb) u_{\beta}(l'b') u_{\gamma}(l''b'')}. \tag{4.159}$$

Note that for calculating the third-order force constant, $F(l''b'')$ is the force on atom $(l''b'')$ when a pair of atomic displacements $u(lb)$ and $u(l'b')$ are made simultaneously. Similarly, for calculating the fourth-order force constant, $F(l'''b''')$ is the force on atom $(l'''b''')$ when three atomic displacements $u(lb)$, $u(l'b')$ and $u(l''b'')$ are made simultaneously.

For extracting force constants, a three-dimensional supercell of an appropriate size (to allow for consideration of at least up to fourth nearest neighbours) is considered and forces recorded by applying a small atomic displacement. As stated earlier, to account for three-body (four-body) interaction terms, two (three) atoms need to be displaced at a time. Use of symmetry relations is made to help reduce the number of displacements. In section 3.3.1.3 we talked about 'eliminating' cubic and quartic anharmonicities by considering displacements $\pm u$, $\pm 2u$. In the present context the force data with these displacements are used to 'extract' cubic and quartic anharmonicities. The force-displacement data is either Taylor-fitted (Esfarjai and Stokes 2008) or simply a finite difference formula is used (Esfarjani et al 2011, Chaput et al 2011, Togo et al 2015). For polar systems, long-range electrostatic interactions can be included using the scheme described in section 3.3.1.3 (also see Togo et al 2015).

4.9.2 THIRD-ORDER DFPT

The DFPT discussed in section 3.3.2.1 for calculating second-order force constants has been extended as the third-order DFPT by Gonze and Vigneron (1989), Debernardi and Baroni (1994), Debernardi (1998), and Deinzer et al (2003). This formalism avoids the use of supercells and makes use of the so-called $2n+1$ theorem of the perturbation theory. The '$2n+1$ theorem states that in the case of a one-body problem, the first $2n+1$ derivatives of the eigenvalues are directly related to the first n derivatives of the corresponding eigenfunctions (Morse and Feshbach 1953, Epstein 1974). A proof of this theorem for a single particle system is presented in Appendix C. Gonze and Vigneron (1989) showed that the '$2n+1$' theorem is proved to all orders in the DFT of the many-body problem.

Consider the momentum-space displacements $X(q,b)$, $X'(q,b')$, and $X''(q,b'')$ of three atoms labelled b, b' and b''. The third-order force constant in momentum space, involving three phonons of wavevectors q, q' and q'', can be expressed as

$$\Psi^{bb'b''}_{\alpha\beta\gamma}(q,q',q'') = \frac{\partial^3 E_{\text{tot}}}{\partial X_\alpha(q,b) X'_\beta(q',b') X''_\gamma(q'',b'')}. \tag{4.160}$$

Following the discussion presented earlier in this chapter, we can convince ourselves that the coefficients $\Psi^{bb'b''}_{\alpha\beta\gamma}(qq'q'')$ are symmetrical in the three $(i = 1, 2, 3)$ sets of (b_i, α_i, q_i). Using short hand notations $1 = (b,\alpha,q)$, $2 = (b',\beta,q')$ and $3 = (b'',\gamma,q'')$, we express, considering permutations of non-distinguishable three phonons taking part in a three-phonon process (as also done in equation (4.88))

$$\Psi(1,2,3) = \frac{1}{6}\Big[\Psi(1,2,3) + \Psi(2,3,1) + \Psi(3,1,2) \\ + \Psi(1,3,2) + \Psi(3,2,1) + \Psi(2,1,3)\Big]. \tag{4.161}$$

As with the harmonic term, there are ionic and electronic contributions to Ψ

$$\Psi = {}^{\text{ion}}\Psi + {}^{\text{el}}\Psi. \tag{4.162}$$

Let us first consider the electronic term. Using the $(2n+1)$ theorem, the following expression for the electronic contribution is obtained (Deinzer et al 2003)

$$\begin{aligned}
{}^{\text{el}}\Psi^{bb'b''}_{\alpha\beta\gamma}(q,q',q'') &= \frac{1}{3}\delta_{q+q'+q'',G}\delta_{b,b'}\sum_{vk}\Big\langle\phi_{vk}\Big|\frac{\partial^3 V_{\text{ps}}}{\partial X^b_\alpha(0)\partial X^{b'}_\beta(0)\partial X^{b''}_\gamma(0)}\Big|\phi_{vk}\Big\rangle \\
&+ 2\delta_{b,b'}\sum_{vk}\Big\langle\phi_{vk}\Big|\frac{\partial^2 V_{\text{ps}}}{X^{b'}_\beta(-q')\partial X^{b''}_\gamma(0)}P_c\Big|\frac{\partial\phi_{vk}}{\partial X^b_\alpha(0)}\Big\rangle \\
&+ 2\sum_{vk}\Big\langle\frac{\partial\phi_{vk}}{\partial X^b_\alpha(-q)}\Big|P_c\frac{\partial V_{\text{KS}}}{X^{b'}_\beta(q')}P_c\Big|\frac{\partial\phi_{vk}}{\partial X^{b''}_\gamma(-q'')}\Big\rangle \\
&- 2\sum_{vv'k}\Big\langle\frac{\partial\phi_{vk}}{\partial X^b_\alpha(-q)}\Big|P_c\Big|\frac{\partial\phi_{v'k+q''}}{\partial X^{b'}_\beta(-q')}\Big\rangle\Big\langle\phi_{vk+q''}\Big|\frac{\partial V_{\text{KS}}}{X^{b''}_\gamma(q'')}\Big|\phi_{vk}\Big\rangle
\end{aligned}$$

$$+\frac{1}{6}\int d\boldsymbol{r}\int d\boldsymbol{r}'\int d\boldsymbol{r}''E_{\mathrm{xc}}(\boldsymbol{r},\boldsymbol{r}',\boldsymbol{r}'')$$
$$\times\frac{\partial\rho(\boldsymbol{r})}{\partial X_{\alpha}^{b}(\boldsymbol{q})}\frac{\partial\rho(\boldsymbol{r}')}{\partial X_{\beta}^{b'}(\boldsymbol{q}')}\frac{\partial\rho(\boldsymbol{r}'')}{\partial X_{\gamma}^{b''}(\boldsymbol{q}'')},\tag{4.163}$$

where $\phi_{\nu k}$ is the pseudo wavefunction for an electron of wavevector $\boldsymbol{k}$ in the νth valence (occupied) state, P_c is a projector over the conduction band manifold (also see section 3.3.2), and E_{xc} is an exchange-correlation functional.

An expression for the ionic contribution to the third-order anharmonic force constant can be derived by taking the third-derivative of the ion-ion Ewald expression in equation (3.53). Debernardi (1998) has presented the following expression

$$
\begin{aligned}
{}^{\mathrm{ion}}\Psi_{\alpha\beta\gamma}^{bb'b''}(\boldsymbol{q},\boldsymbol{q}',\boldsymbol{q}'') \;=\; & \frac{2\pi e^2 z_b}{3\Omega}\sum_{G\neq0}\frac{e^{-G^2/4\eta}}{G^2}G_\alpha G_\beta G_\gamma \\
& \times\left[\sum_{b'}z_{b'}\sin\{\boldsymbol{G}\cdot(\boldsymbol{\tau}_b-\boldsymbol{\tau}_{b'})\}\right]\delta_{b,b'}\delta_{b',b''} \\
& +\frac{2\pi e^2}{\Omega}z_b z_{b'}\sum_{G\neq0}\frac{e^{-(\boldsymbol{q}'+\boldsymbol{q}''+G)^2/4\eta}}{G^2}\sin\{\boldsymbol{G}\cdot(\boldsymbol{\tau}_b-\boldsymbol{\tau}_{b'})\}\delta_{b',b''} \\
& \times(q'_\alpha+q''_\alpha+G_\alpha)(q'_\beta+q''_\beta+G_\beta)(q'_\gamma+q''_\gamma+G_\gamma),
\end{aligned}\tag{4.164}
$$

where the symbols are the same as used in equations (3.53) and (3.96).

This completes the discussion of *ab initio* determination of the cubic anharmonic force constants in non-polar materials. As discussed in section 2.4.4, in polar crystals, the microscopic electric field, generated by long wavelength optical phonon modes, gives rise to nonanalytic terms in the electronic part of these coefficients. Deinzer *et al* (2004) have shown how these non-analytic terms in the third- and fourth-order force constants can be determined from considerations of other physical quantities including highr-order dipole moments Raman coefficients and nonlinear susceptibilities. Interested readers are referred to their work.

As mentioned before, it should be pointed out that the *ab initio* theory discussed in this sub-section is limited to calculating the *zero-temperature* cubic anharmonic force constants. In contrast, the *semi-ab initio* theory developed in the previous sub-section (section 4.8), although rooted in anharmonic elasticity, has explicitly presented temperature dependent cubic and quartic terms. Further development of the fully *ab initio* theory is required to bring in adequate temperature dependence of anharmonic force constants. This should happen in near future, as accurate approximations for ground-state DFT are being generalised to finite temperature (Pittalis *et al* 2011, Duffy and Trickey 2011, Burke *et al* 2016).

5 Theory of Lattice Thermal Conductivity

5.1 INTRODUCTION

Application of a temperature gradient across a solid excites the elementary excitations such as free electrons, holes and phonons which therefore acquire more energy than the average energy (for electrons or holes) or the zero-point energy (for phonons) and transfer or conduct heat from the hotter to the colder end of the specimen. In metals, both electrons and phonons play their role. In dielectrics or intrinsic semiconductors, almost all heat is conducted by phonons. It was first pointed out by Debye (1914) that if the atomic vibrations in a dielectric remain in normal modes, corresponding to a quadratic potential, then no equilibrium in the phonon distribution is possible and heat transfer would take place at the speed of sound in the solid; i.e. the phonon (or lattice thermal) conductivity would be infinite. However, in reality crystals are of finite size, contain defects and impurities, and become anharmonic at temperatures above the absolute zero. These realities lead to phonon scattering mechanisms. A finite crystal size leads to boundary scattering, while defects and impurities can cause phonon–defect and phonon–electron (hole) scatterings. Anharmonic terms in the crystal potential give rise to phonon–phonon interactions. These phonon scattering mechanisms give rise to thermal resistance (i.e. a finite thermal conductivity) and must be considered in the theory of lattice thermal conductivity.

A microscopic study of phonon conductivity in solids is possible mainly from three approaches: relaxation time approach, variational approach and Green's function approach (or density matrix approach). The relaxation time approach and the variational approach both assume that the thermal transport coefficient (i.e. the lattice thermal conductivity) can be studied from a solution of a linearised Boltzmann equation. The Green's function approach, on the other hand, starts from a more fundamental level in that it deals with the problem in terms of quantum statistics. In this chapter we will present the basic treatment of these approaches.

5.1.1 THE PHONON BOLTZMANN EQUATION

The fundamental assumption in deriving the phonon Boltzmann equation is that there exists a distribution function $n_{qs}(r,t)$ which measures the occupation number of phonons (qs) in the neighbourhood of r at time t. In the presence of a temperature gradient across a dielectric two mechanisms are responsible for the rate of change of this distribution function:

(i) *Diffusion*. The presence of a temperature gradient ∇T involves a spatial dependence of temperature: $T = T(r)$. This causes $n_{qs}(r,t)$ to diffuse (vary from point to point) at the rate

$$\frac{\partial n_{qs}}{\partial t}\bigg|_{diff} = -c_s(q) \cdot \nabla T \frac{\partial n_{qs}}{\partial T}. \tag{5.1}$$

(ii) *Scattering*. Various scattering events also contribute to a rate of change $\partial n_{qs}/\partial t|_{scatt}$ of the distribution function n_{qs} for phonons in mode qs.

The total rate of change of n_{qs} must vanish in the steady state of heat flow through the solid. Thus

$$-c_s(q) \cdot \nabla T \frac{\partial n_{qs}}{\partial T} + \frac{\partial n_{qs}}{\partial t}\bigg|_{scatt} = 0. \tag{5.2}$$

This is the Boltzmann equation for phonons in its general form.

DOI: 10.1201/9781003141273-5

Equation (5.2) is an integro-differential equation whose solution is required for a calculation of phonon conductivity. Unfortunately, this form of the Boltzmann equation is very complicated and in general cannot be solved. This is because equation (5.2) requires a knowledge of the distribution function $n_{q's'}$ of all possible states $q's'$ together with transition rates from $q's'$ to qs.

A simplification of equation (5.2) is possible. We first note that in equilibrium the phonon distribution does not change with time,

$$\frac{\partial \bar{n}_{qs}}{\partial t} = 0. \tag{5.3}$$

With this in mind, we replace in the second term of equation (5.2) n_{qs} by $n_{qs} - \bar{n}_{qs}$, the linear term in a Taylor expansion of n_{qs} about the equilibrium distribution $\bar{n}_{qs}$. Further, we assume that in the *steady state* in the presence of a finite temperature gradient the deviation from equilibrium is small so that we can replace n_{qs} by $\bar{n}_{qs}$ in the first term of equation (5.2) (which is still non-zero as $\partial \bar{n}_{qs}/\partial T \neq 0$). The resulting equation will be called the *linearised Boltzmann equation*.

To derive the linearised form of the Boltzmann equation we consider two cases.

(a) *Elastic scattering case*

Consider an elastic process $(qs) \rightleftharpoons (q's')$. The transition probability for the process $(qs) \to (q's')$ is

$$P_{qs}^{q's'} = n_{qs}(n_{q's'} + 1)\mathcal{Q}_{qs}^{q's'}, \tag{5.4}$$

where $\mathcal{Q}_{qs}^{q's'}$ is the *intrinsic* transition probability (which is independent of phonon distribution) between the states (qs) and $(q's')$, n_{qs} represents the phonon occupancy of the state (qs) and $(n_{q's'} + 1)$ signifies that a phonon has been added to the state $q's'$ with an initial distribution $n_{q's'}$. Similarly, for the reverse process $(q's') \to (qs)$

$$P_{q's'}^{qs} = n_{q's'}(n_{qs} + 1)\mathcal{Q}_{q's'}^{qs}. \tag{5.5}$$

The intrinsic transition rate is the same in either direction, i.e.

$$\mathcal{Q}_{qs}^{q's'} = \mathcal{Q}_{q's'}^{qs} \tag{5.6}$$

which is the principle of microscopic reversibility. The total rate of change of n_{qs} due to the above elastic process is

$$\begin{aligned}
\left.\frac{\partial n_{qs}}{\partial t}\right|_{\text{scatt}} &= \sum_{q's'}(P_{q's'}^{qs} - P_{qs}^{q's'}) \\
&= \sum_{q's'}\left\{n_{q's'}(n_{qs} + 1) - n_{qs}(n_{q's'} + 1)\right\}\mathcal{Q}_{qs}^{q's'}.
\end{aligned} \tag{5.7}$$

(Note that here we have worked out an increase in n_{qs} due to scattering, hence the signs of the two terms in equation (5.7).) We can further express this equation in linearised form

$$\left.\frac{\partial n_{qs}}{\partial t}\right|_{\text{scatt}} = \sum_{q's'}\left\{(n_{q's'} - \bar{n}_{q's'}) - (n_{qs} - \bar{n}_{qs})\right\}\mathcal{Q}_{qs}^{q's'}, \tag{5.8}$$

where we have used $\bar{n}_{qs} = \bar{n}_{q's'}$ which is true for an elastic process with $\hbar\omega(qs) = \hbar\omega(q's')$. Therefore the linearised phonon Boltzmann equation is

$$-c_s(q) \cdot \nabla T \frac{\partial \bar{n}_{qs}}{\partial T} = \sum_{q's'}\left\{(n_{qs} - \bar{n}_{qs}) - (n_{q's'} - \bar{n}_{q's'})\right\}\mathcal{Q}_{qs}^{q's'}. \tag{5.9}$$

Equation (5.9) is relatively easy to solve, particularly if we assume that the energy surfaces in q-space are spherical, so that $c_s(q)$ and q are in the same direction. Let us also assume that the scattering probability depends only on the angle between q and q' and not on their absolute orientations in the crystal. Then we can express the right-hand side of equation (5.9) as

$$(n_{qs} - \bar{n}_{qs}) \int (1 - \cos\theta) \mathscr{Q}(q,\theta) \, d\Pi, \tag{5.10}$$

where $\mathscr{Q}(q,\theta) \, d\Pi$ is a *differential* transition probability for a phonon of momentum $\hbar q$ to scatter through the angle θ into the solid angle $d\Pi$. For conciseness, we write equation (5.10) as $(n_{qs} - \bar{n}_{qs})/\tau_{qs}$, where τ_{qs} is the *relaxation time* for the phonon qs. In other words, for an *elastic* process we write

$$\left. \frac{\partial n_{qs}}{\partial t} \right|_{scatt} = -\frac{(n_{qs} - \bar{n}_{qs})}{\tau_{qs}} \tag{5.11}$$

so that the linearised Boltzmann equation in equation (5.9) takes the form

$$-c_s(q) \cdot \nabla T \frac{\partial \bar{n}_{qs}}{\partial T} = \frac{n_{qs} - \bar{n}_{qs}}{\tau_{qs}}. \tag{5.12}$$

(b) *General inelastic situation*

Let us derive a *general form* for the linearised phonon Boltzmann equation. Let us introduce a new function ψ_{qs} defined by

$$\begin{aligned}
n_{qs} &= \left[\exp(\hbar\omega(qs)/k_B T - \psi_{qs}) - 1 \right]^{-1} \\
&\simeq \bar{n}_{qs} - \psi_{qs} \frac{\partial \bar{n}_{qs}}{\partial(\hbar\omega(qs)/k_B T)} \\
&= \bar{n}_{qs} + \psi_{qs}\bar{n}_{qs}(\bar{n}_{qs} + 1).
\end{aligned} \tag{5.13}$$

Clearly ψ_{qs} is measure of the deviation from equilibrium distribution for phonons in the mode qs. Further, let us consider a general *inelastic* process of the type

$$(qs) + (q's') \rightarrow (q''s'') + (q'''s''').$$

Then, as in equation (5.7), we write

$$\begin{aligned}
\left. \frac{\partial n_{qs}}{\partial t} \right|_{scatt} &= \left[\frac{\partial n_{qs}}{\partial t} \{(q''s'') + (q'''s''') \rightarrow (qs) + (q's')\} \right. \\
&\quad \left. - \frac{\partial n_{qs}}{\partial t} \{(qs) + (q's') \rightarrow (q''s'') + (q'''s''')\} \right] \\
&= \sum_{\substack{q's',q''s'', \\ q'''s'''}} \left(P^{qs,q's'}_{q''s'',q'''s'''} - P^{q''s'',q'''s'''}_{qs,q's'} \right) \\
&= \sum_{\substack{q's',q''s'', \\ q'''s'''}} \{ n_{q''s''} n_{q'''s'''}(n_{qs} + 1)(n_{q's'} + 1) \\
&\quad - n_{qs} n_{q's'}(n_{q''s''} + 1)(n_{q'''s'''} + 1) \} \mathscr{Q}^{q''s'',q'''s'''}_{qs,q's'}.
\end{aligned} \tag{5.14}$$

For an inelastic process the expression on the right-hand side of equation (5.14) cannot be simplified in the manner equation (5.7) was for an elastic process. Let us substitute n_{qs} etc. from equation (5.13) and retain only terms of first order in ψ_{qs}. We then can express

$$-\frac{\partial n_{qs}}{\partial t} = \sum_{\substack{q's',q''s'', \\ q'''s'''}} \left(\psi_{qs} + \psi_{q's'} - \psi_{q''s''} - \psi_{q'''s'''} \right) \bar{P}^{q''s'',q'''s'''}_{qs,q's'}. \tag{5.15}$$

In deriving this result we have used the identity

$$\bar{n}_{qs}\bar{n}_{q's'}(\bar{n}_{q''s''}+1)(\bar{n}_{q'''s'''}+1) = (\bar{n}_{qs}+1)(\bar{n}_{q's'}+1)\bar{n}_{q''s''}\bar{n}_{q'''s'''} \tag{5.16}$$

which maintains detailed balance in the forward and backward directions (and can be verified with the energy conservation condition $\omega(qs)+\omega(q's') = \omega(q''s'')+\omega(q'''s''')$) and

$$\mathscr{Q}^{q''s'',q'''s'''}_{qs,q's'} = \mathscr{Q}^{qs,q's'}_{q''s'',q'''s'''}, \tag{5.17}$$

which is the principle of microscopic reversibility. $\bar{P}$ represents equilibrium transition rate.

An important difference between the general inelastic and the isotropic elastic cases is that while equation (5.11) has been expressed in terms of a relaxation time for a phonon in mode qs, such a simple description is not in general possible for equation (5.15).

(c) *Canonical form of the linearised Boltzmann equation*
From both equation (5.9) (an elastic case) and equation (5.15) (a general inelastic case), it is clear that now we are in a position to express the linearised form of the phonon Boltzmann equation in a standard form:

$$\frac{\partial n_{qs}}{\partial t}\Big|_{diff} + \frac{\partial n_{qs}}{\partial t}\Big|_{scatt} = 0$$

$$-c_s(q)\cdot\nabla T\frac{\partial \bar{n}_{qs}}{\partial T} = -\frac{\partial n_{qs}}{\partial t}\Big|_{scatt}$$

$$X = P\psi$$

$$\text{or} \quad X_{qs} = \sum_{q's'}P^{ss'}_{qq'}\psi^{s'}_{q'}. \tag{5.18}$$

Here X_{qs} is the inhomogeneity created by the temperature gradient and $P^{ss'}_{qq'}$ are the elements of the phonon collision operator and provide a measure of phonon transition probabilities. In the rest of this chapter and in Chapters 6 and 7, we shall refer to this as the *canonical form* of the phonon Boltzmann equation.

5.1.2   EXPRESSION FOR LATTICE THERMAL CONDUCTIVITY

When a finite temperature gradient ∇T is established across a solid, then in steady state the rate of heat energy flow per unit area normal to the gradient is given by the *macroscopic* expression (Fourier law)

$$Q = -\mathscr{K}\nabla T, \tag{5.19}$$

where $\mathscr{K}$ is the thermal conductivity. The *microscopic* expression for the heat current Q in a dielectric solid is obtained by adding contributions from phonons in all possible modes,

$$\begin{aligned}
Q &= \frac{1}{N_0\Omega}\sum_{qs}\hbar\omega(qs)n_{qs}c_s(q) \\
&= \frac{1}{N_0\Omega}\sum_{qs}\hbar\omega(qs)\psi_{qs}\bar{n}_{qs}(\bar{n}_{qs}+1)c_s(q), \tag{5.20}
\end{aligned}$$

where $N_0\Omega$ is the volume of the solid and $c_s(q)$ is the phonon group velocity in the mode qs. The second equality in equation (5.20) is written by noting that it is the deviation of the phonon distribution from equilibrium, $n_{qs}-\bar{n}_{qs}$, which contributes to the heat current. From equations (5.19) and (5.20), we obtain the following expression for the lattice thermal conductivity tensor (see Appendix D for derivation):

$$\mathscr{K}_{ij} = -\frac{Q_i\nabla_j T}{|\nabla T|^2}$$

$$= -\frac{1}{N_0\Omega|\nabla T|^2}\sum_{qs}\hbar\omega(qs)\psi_{qs}\bar{n}_{qs}(\bar{n}_{qs}+1)c_i(qs)\nabla_j T. \quad (5.21)$$

For cubic crystals and for an isotropic medium c is parallel to ∇T, so that the conductivity is a scalar quantity and we can express $\mathscr{K}_{ij} = \mathscr{K}\delta_{ij}$ with

$$\begin{aligned}
\mathscr{K} &= -\frac{1}{N_0\Omega|\nabla T|^2}\sum_{qs}\hbar\omega(qs)\psi_{qs}\bar{n}_{qs}(\bar{n}_{qs}+1)c_s(q)\cdot\nabla T \\
&= \frac{k_B T^2}{N_0\Omega|\nabla T|^2}\sum_{qs}\psi_{qs}X_{qs} \quad (5.22) \\
&= \frac{k_B T^2}{N_0\Omega|\nabla T|^2}(\boldsymbol{\psi},X) \quad (5.23)
\end{aligned}$$

where equation (5.22) defines the inner product $(\boldsymbol{\psi},X)$. Thus we notice that for calculating the lattice thermal conductivity $\mathscr{K}$ we need to solve the phonon Boltzmann equation for the phonon deviation function ψ_{qs}. Throughout this chapter, we will treat $\mathscr{K}$ as a scalar quantity.

In sections 5.2 and 5.3 we will discuss two separate methods of solving the Boltzmann equation for ψ_{qs} and hence for the lattice thermal conductivity, via equation (5.22). In section 5.4 we will develop the linear-response theory for the lattice thermal conductivity.

5.2   RELAXATION-TIME METHODS

The anharmonic interaction of phonons is inelastic in nature and strictly speaking it is not amenable to a realistic relaxation time picture. However, to make the understanding of phonon interaction processes easier, one usually relates the quantity $P\boldsymbol{\psi}$ in equation (5.18) to a relaxation time picture. This can be done at two levels. In section 5.2.1 we will develop the theory of $\mathscr{K}$ using the picture of *single-mode* relaxation time. In sections 5.2.2–5.2.7 we will develop expressions for the conductivity by defining *effective* relaxation times at different levels of sophistication.

5.2.1   SINGLE-MODE RELAXATION-TIME METHOD

The *single-mode relaxation-time (smrt)* method (Srivastava 1974) provides the simplest picture of phonon interaction processes. In this picture one calculates the relaxation rate of phonons in a mode qs, say, on the assumption that all other phonon modes have their equilibrium distribution ($\psi_{q's'} = 0$ for $q's' \neq qs$). In this approximation one therefore expresses

$$\begin{aligned}
(P\psi)_{qs} \simeq P_{qsqs}\psi_{qs} &= \Gamma_{qs}\psi_{qs} \\
&= -\frac{\partial n_{qs}}{\partial t}\Big|_{scatt} \\
&= \frac{n_{qs}-\bar{n}_{qs}}{\tau_{qs}} \\
&= \frac{\bar{n}_{qs}(\bar{n}_{qs}+1)\psi_{qs}}{\tau_{qs}}, \quad (5.24)
\end{aligned}$$

where Γ is the diagonal part of the phonon collision operator $P = \Gamma + \Lambda$. Γ includes the diagonal part of the anharmonic N and U processes operator, as well as all operators representing elastic processes. We shall obtain an expression for Γ in the next chapter. Note that Λ (which will also be used in chapter 6 and 7 represents the off-diagonal part of the anharmonic phonon collision operator, and should not be confused with the same symbol representing force constants in chapters 2, 8, 9, 10 and 11. The single-mode relaxation time τ_{qs} is thus given by

$$\tau_{qs}^{-1} = \frac{\Gamma_{qs}}{\bar{n}_{qs}(\bar{n}_{qs}+1)}. \quad (5.25)$$

The expression for the lattice thermal conductivity is then

$$
\begin{aligned}
\mathscr{K}_{smrt} &= \frac{k_B T^2}{N_0 \Omega |\nabla T|^2} (X\Gamma^{-1}, X) \\
&= \frac{\hbar^2}{N_0 \Omega k_B T^2} \sum_{qs} (c_s(q) \cdot \hat{\nabla}T)^2 \omega_{qs}^2 \tau_{qs} \bar{n}_{qs}(\bar{n}_{qs}+1) \\
&= \frac{\hbar^2}{3N_0 \Omega k_B T^2} \sum_{qs} c_s^2(q) \omega^2(qs) \tau_{qs} \bar{n}_{qs}(\bar{n}_{qs}+1) \\
&= \frac{\hbar^2}{3N_0 \Omega k_B T^2} \langle \tau \omega^2 c^2 \rangle,
\end{aligned}
\tag{5.26}
$$

where the factor $\frac{1}{3}$ is the average of $(\hat{c}_s(q) \cdot \hat{\nabla}T)^2$.

The last line in equation (5.25) uses the following short-hand notation:

$$
\langle f \rangle = \sum_{qs} f_{qs} \bar{n}_{qs}(\bar{n}_{qs}+1).
\tag{5.27}
$$

The *smrt* result for the conductivity will also be referred to as the Debye result: $\mathscr{K}_{smrt} \equiv \mathscr{K}_D$, as it is formally equivalent to the formula derived by Debye (1914).

Limiting our discussion to insulating materials, the inverse of the total *smrt* τ_{qs} for a phonon in mode qs can be expressed as a sum of contributions from boundary scattering $\tau_{qs}(bs)$, mass-defect scattering $\tau_{qs}(md)$, and anharmonic scattering $\tau_{qs}(pp)$:

$$
\tau_{qs}^{-1} = \tau_{qs}^{-1}(bs) + \tau_{qs}^{-1}(md) + \tau_{qs}^{-1}(pp).
\tag{5.28}
$$

In short

$$
\tau_{qs}^{-1} = \tau_r^{-1} + \tau_{pp}^{-1},
\tag{5.29}
$$

where τ_r is the phonon relaxation time due to elastic processes, namely boundary scattering and mass-defect scattering. The phonon–phonon scattering can be expressed as

$$
\tau_{pp}^{-1} = \tau_N^{-1} + \tau_U^{-1},
\tag{5.30}
$$

where the subscripts N and U refer to normal and umklapp processes, respectively. Expressions for τ_r and τ_{pp} will be derived in the next chapter.

Although we have used the additivity of inverse relaxation times for each phonon mode in equation (5.28), it is usually not possible to express the resistivity $W \equiv \mathscr{K}^{-1}$ in the same fashion: i.e. the additivity $W = W(bs) + W(md) + W(pp)$ (a proposition known as Matthiessen's rule in the context of electrical resistivity) is not usually satisfied. This can be understood by noting that the validity of Matthiessen's rule assumes that the resistivity due to one mechanism is independent of the presence of any other, an assumption which is not true if τ is a function of qs: $\tau = \tau_{qs}$.

5.2.2   KLEMENS' MODEL

The Bose–Einstein distribution function in equation (2.2) provides the solution to equation (5.3) in equilibrium for any scattering process which conserves energy. The case of phonon–phonon N scattering processes is, however, special. These processes also conserve momentum, in addition to energy, and equation (5.3) is satisfied for the drifting distribution (Peierls 1935, Klemens 1951, Krumhansl 1965)

$$
\bar{n}_{qs}(u) = \left(\{ \exp[(\hbar\omega(qs) + q \cdot u)/k_B T] - 1 \} \right)^{-1},
\tag{5.31}
$$

where u is an arbitrary vector not dependent on q. Stated differently, in the presence of N processes a non-equilibrium distribution is restored not towards the stationary equilibrium $\bar{n}_{qs}$ but towards a *drifting* distribution $\bar{n}_{qs}(u)$. As N processes conserve momentum, they cannot by themselves contribute to the thermal resistivity. However, when the steady state is maintained in the presence of a temperature gradient, N processes combine with momentum non-conserving processes (elastic or inelastic) and contribute to the thermal resistivity indirectly.

A partial, and approximate, solution to the momentum conserving nature of N processes was given by Klemens (1951, 1958). Consider N processes in the presence of elastic processes which are governed by $\tau_R(\omega)$. Klemens divided three-phonon N processes $q + q' = q''$ into three groups:

$$
\begin{aligned}
&\text{(a)} \quad \omega \sim \omega'' >> \omega'; q \sim q'' \\
&\text{(b)} \quad \omega \sim \omega' \sim \omega''; q \sim q' \sim q'' \\
&\text{(c)} \quad \omega << \omega' \sim \omega''; q \langle\langle\, q''.
\end{aligned}
$$

Processes of type (a) are relatively unimportant for the equilibrium of mode q. Processes of type (b) are important for $\omega > \omega_1 = k_B T/\hbar$. Since $\omega' \sim \omega$, we expect that the total effective relaxation time satisfies $\tau_{eff}(\omega) \sim \tau_{eff}(\omega')$, and $\tau_{eff}^{-1}(\omega) - \tau_R^{-1}(\omega) << \tau_N^{-1}(\omega)$, where τ_N is the single-mode relaxation time for N processes. However, Klemens simply assumed that

$$
\tau_{eff}^{-1}(\omega) - \tau_R^{-1}(\omega) = 0 \qquad \text{if} \quad \omega > \omega', \tag{5.32}
$$

where τ_R^{-1} is the relaxation rate due to momentum non-conserving processes. (This approximation is justified only if $\tau_N(\omega) \rangle \tau_R(\omega)$, and definitely underestimates the effect of N processes.) Processes of type (c) are important only for low-frequency phonons. Klemens argued that one can express

$$
\tau_{eff}^{-1}(\omega) = \tau_R^{-1}(\omega) + \tau_N^{-1}(\omega)\left(1 - \frac{\tau_{eff}(\omega_1)}{\tau_{eff}(\omega)}\right) \qquad \text{if} \quad \omega << \omega_1 = \frac{k_B T}{\hbar}. \tag{5.33}
$$

As in this approximation the relaxation of only low-frequency phonons is modified, Klemens argued that equation (5.33) can apply only to transverse modes. With equations (5.32) and (5.33), we can write, for transverse modes,

$$
\begin{aligned}
\tau_{eff,T}(\omega) &= \tau_{R,T}(\omega)\frac{\tau_{N,T}(\omega) + \tau_{R,T}(\omega_1)}{\tau_{R,T}(\omega) + \tau_{N,T}(\omega)} \qquad \omega \leq \omega_1 \\
&= \tau_{R,T}(\omega) \qquad \omega \geq \omega_1.
\end{aligned} \tag{5.34}
$$

Klemens further argued that longitudinal modes can interact with modes of about the same frequency, so that they undergo processes of type (b), but not of type (c). Assuming that longitudinal modes are coupled to transverse modes of the same frequency, Klemens obtained the following expression for longitudinal modes (i.e. high-frequency phonons):

$$
\tau_{eff,L}(\omega) = \tau_{R,L}(\omega)\frac{\tau_{N,L}(\omega) + r^2\tau_{eff,T}(\omega)}{\tau_{R,T}(\omega) + \tau_{N,T}(\omega)}, \tag{5.35}
$$

where $r = c_T/c_L$ and $\tau_{eff,T}(\omega)$ is given by equation (5.34).

Klemens (1958) also laid down the foundation for an effective relaxation time for U processes. An extension of his basic approach will be discussed in section 5.2.5.

With $\tau_{eff,T}$ as the effective relaxation time for U processes, we can write an overall effective relaxation time τ_K in the Klemens model as $\tau_K^{-1} = \tau_{eff}^{-1} + \tau_{eff,U}^{-1}$, and then the conductivity expression will read

$$
\mathcal{K}_K = \frac{\hbar^2}{3N_0\Omega k_B t^2}\langle\tau_K\omega^2 c^2\rangle. \tag{5.36}
$$

As mentioned earlier, Klemens' model for an effective relaxation time for N processes is based on intuitive arguments and should be regarded as very approximate.

5.2.3 CALLAWAY'S MODEL

Using a more rigorous treatment than Klemens, Callaway (1959) showed that the momentum conserving nature of three-phonon N processes is such that when different phonon scattering processes are combined with N processes there comes about an extra contribution, over and above the Debye term (*i.e.* the *smrt* contribution), to the thermal conductivity expression. Callaway's final expression for lattice thermal conductivity can be written as

$$\mathscr{K}_C = \mathscr{K}_{smrt} + \mathscr{K}_{N-drift}, \tag{5.37}$$

where $\mathscr{K}_{N-drift}$ is a contribution arising from the momentum-conserving nature of N processes. The N-drift contribution may be recognised as the role of the off-diagonal N-processes collision operator. The fundamental principles of Callaway's approach have been investigated and refined by several authors (Nettleton 1963, Krumhansl 1965, Simons 1975, Srivastava 1976a, 1980). The essential steps in the derivation of the above equation are as follows (Parrott 1971, Srivastava 2015).

The rate of change of the phonon distribution function n_{qs} due to normal processes can be written as

$$\frac{\partial n_{qs}}{\partial t}\bigg|_N = \frac{\boldsymbol{q}\cdot\boldsymbol{u} - \psi_{qs}}{\tau_N}\bar{n}_{qs}(\bar{n}_{qs}+1), \tag{5.38}$$

where $\boldsymbol{q}\cdot\boldsymbol{u}$ is the drifting distribution function (equation (5.31)). Although Klemens (1951) suggested that $\boldsymbol{u}$ is some constant vector, making the assumption of isotropy, Callaway considered it as a constant vector parallel to the temperature gradient (*i.e.* $\boldsymbol{u} \parallel \boldsymbol{\nabla}T$). Clearly if $\psi_{qs} = \boldsymbol{q}\cdot\boldsymbol{u}$, then N processes can have no special effect on the phonon distribution. The magnitude of $\boldsymbol{u}$ is calculated by using the fact that the rate of change of total crystal momentum due to N processes must be zero:

$$\begin{aligned}
0 &= \sum_{qs}\left(\frac{\partial n_{qs}}{\partial t}\right)_N \boldsymbol{q} \\
&= \sum_{qs}(\boldsymbol{q}\cdot\boldsymbol{u})\frac{\boldsymbol{q}\cdot\boldsymbol{u} - \psi_{qs}}{\tau_N}\bar{n}_{qs}(\bar{n}_{qs}+1). \tag{5.39}
\end{aligned}$$

To evaluate the magnitude of $\boldsymbol{u}$ from this equation, we need an expression for ψ_{qs} in Callaway's approximation. This is achieved by expressing the Boltzmann equation (equation (5.18)) in the assumed isotropic continuum case (*i.e.* $\omega = cq$, which makes $\boldsymbol{c} \parallel \boldsymbol{q}$),

$$\begin{aligned}
-\frac{\hbar\omega(qs)}{k_B T^2}c_s(q)\hat{\boldsymbol{q}}\cdot\boldsymbol{\nabla}T &= -\left[\frac{\partial n_{qs}}{\partial t}\bigg|_N + \frac{\partial n_{qs}}{\partial t}\bigg|_R\right][\bar{n}_{qs}(\bar{n}_{qs}+1)]^{-1} \\
&= -\frac{\boldsymbol{q}\cdot\boldsymbol{u}}{\tau_N} + \psi_{qs}\left(\frac{1}{\tau_N}+\frac{1}{\tau_R}\right), \tag{5.40}
\end{aligned}$$

where τ_R^{-1} is the relaxation rate due to momentum non-conserving processes. Writing the total relaxation time τ as

$$\tau^{-1} = \tau_N^{-1} + \tau_R^{-1} \tag{5.41}$$

and substituting equation (5.40) in equation (5.39) we get

$$|\boldsymbol{u}| = -\frac{\hbar|\boldsymbol{\nabla}T|}{k_B T^2}\frac{\sum_{qs}\omega(qs)c_s(q)q(\hat{\boldsymbol{q}}\cdot\hat{\boldsymbol{u}})(\hat{\boldsymbol{c}}_s\cdot\widehat{\boldsymbol{\nabla}T})\tau\tau_N^{-1}n_{qs}(\bar{n}_{qs}+1)}{\sum_{qs}q^2(\hat{\boldsymbol{q}}\cdot\hat{\boldsymbol{u}})^2\tau\tau_N^{-1}n_{qs}(\bar{n}_{qs}+1)} \tag{5.42}$$

$$= -\frac{\hbar|\boldsymbol{\nabla}T|}{k_B T^2}\frac{\langle\omega cq\tau\tau_N^{-1}\rangle}{\langle q^2\tau_N^{-1}(1-\tau\tau_N^{-1})\rangle}. \tag{5.43}$$

In writing equation (5.43) we use used the notation $\langle...\rangle$ as defined in equation (5.27) and have used the factor $\frac{1}{3}$ for the averages of $(\hat{\boldsymbol{q}}\cdot\hat{\boldsymbol{u}})^2$ and $(\hat{\boldsymbol{q}}\cdot\hat{\boldsymbol{u}})(\hat{\boldsymbol{c}}_s\cdot\widehat{\boldsymbol{\nabla}T}) = (\hat{\boldsymbol{q}}\cdot\widehat{\boldsymbol{\nabla}T})^2$. With this value of $|\boldsymbol{u}|$

equation (5.40) yields the following expression for ψ_{qs} in Callaway's approximation; let us call it ψ_{qs}^C

$$\psi_{qs}^C = -\hbar \frac{|\nabla T|}{k_B T^2}(\hat{q}\cdot\hat{u})\tau\left[\omega(qs)c_s(q) + \frac{1}{\tau_N}\frac{\langle\omega cq\tau\tau_N^{-1}\rangle}{\langle q^2\tau_N^{-1}(1-\tau\tau_N^{-1})\rangle}\right].\qquad(5.44)$$

Then from equation (5.22) the Callaway expression for the conductivity becomes

$$\begin{aligned}\mathscr{K}_C &= \frac{k_B T^2}{N_0\Omega|\nabla T|^2}\sum_{qs}\psi_{qs}^C X_{qs}\\ &= \frac{\hbar^2}{3N_0\Omega k_B T^2}\left[\langle\tau\omega^2 c^2\rangle + \frac{\langle\omega cq\tau\tau_N^{-1}\rangle^2}{\langle q^2\tau_N^{-1}(1-\tau\tau_N^{-1})\rangle}\right]\\ &= \mathscr{K}_{smrt} + \mathscr{K}_{N-drift},\end{aligned}\qquad(5.45)$$

where, again, the factor $\frac{1}{3}$ accounts for the average of $(\hat{q}\cdot\hat{u})^2 = (\hat{c}\cdot\hat{\nabla}T)^2$.

From equations (5.24) and (5.38) it can be seen that in Callaway's approximation we express

$$\begin{aligned}-\frac{\partial n_{qs}}{\partial t}\Big|_N &= \frac{\psi_{qs}}{\tau_N}\bar{n}_{qs}(\bar{n}_{qs}+1) - \frac{q\cdot u}{\tau_N}\bar{n}_{qs}(\bar{n}_{qs}+1)\\ &= P_{qsqs}\psi_{qs}\Big|_N + \sum_{q's'\neq qs}P_{qsq's'}\psi_{q's'}\Big|_N\\ &= \Gamma_{qs}\psi_{qs}\Big|_N + \sum_{q's'\neq qs}\Lambda_{qsq's'}\psi_{q's'}\Big|_N.\end{aligned}\qquad(5.46)$$

Clearly the second term with $u \neq 0$ gives a contribution from the off-diagonal N-processes operator Λ_N. Furthermore, writing equation (5.45) in the form

$$\mathscr{K}_C = \frac{\hbar^2}{3N_0\Omega k_B T^2}\langle\tau_C\omega^2 c^2\rangle\qquad(5.47)$$

we note that in Callaway's approximation the effective total relaxation time is given by

$$\tau_C = \tau(1+\beta/\tau_N)\qquad(5.48)$$

where

$$\beta = \frac{q}{\omega(qs)c_s(q)}\frac{\langle\omega cq\tau\tau_N^{-1}\rangle}{\langle q^2\tau_N^{-1}(1-\tau\tau_N^{-1})\rangle}.\qquad(5.49)$$

When the parameter β is set to zero, the effective relaxation time τ_C reduces to the *smrt* τ.

5.2.4 RELAXING ASSUMPTIONS IN ORIGINAL VERSION OF CALLAWAY'S MODEL

The derivation of the Callaway conductivity formula in the previous sub-section is based on two assumptions: (1) the assumption of isotropy, which allowed u to be taken parallel to ∇T, and (2) the continuum, or linear, dispersion relation for long wavelength acoustic phonons $\omega = cq$ which allows $c \parallel q$. Consideration of u in a direction unrelated to ∇T will make the formulation quite complicated. Allen (2013) has suggested how the Callaway formulation can be improved by choosing the orientation of u along a symmetry axis of a non-cubic crystal. However, Allen changed the condition of zero-rate-of-momentum to zero-momentum in N processes: *i.e.* rather than $\sum_{qs}\left(\frac{\partial n_{qs}}{\partial t}\right)_N q = 0$ he used the condition $\sum_{qs}(n_{qs})_N q = 0$. For details of changes to the second term in the conductivity expression in equation (5.45) we refer the reader to the paper by Allen.

Here we prefer to use the condition $\sum_{qs}\left(\frac{\partial n_{qs}}{\partial t}\right)_N q = 0$ and will assume that the constant vector u is oriented along the temperature gradient ∇T applied to the crystal (*i.e.* will retain the first assumption), but will remove the second assumption by keeping the terms $\hat{q} \cdot \hat{u}$ and $\hat{c} \cdot \vec{\nabla} T$ where these occur in the formulation.

We start by re-writing equation (5.40) as

$$-\frac{\hbar\omega(qs)}{k_B T^2} c_s(q) \cdot \nabla T = -\frac{q \cdot u}{\tau_N} + \psi_{qs}\left(\frac{1}{\tau_N} + \frac{1}{\tau_R}\right). \tag{5.50}$$

With this, equation (5.44) reads

$$\psi_{qs}^C = -\hbar\frac{|\nabla T|}{k_B T^2}\tau\left[\omega(qs)c_s(q)(\hat{q}\cdot\hat{u}) + \frac{1}{\tau_N}\frac{\langle\omega cq\tau\tau_N^{-1}(\hat{q}\cdot\hat{u})(\hat{c}\cdot\vec{\nabla}T)\rangle}{\langle q^2\tau_N^{-1}(1-\tau\tau_N^{-1})(\hat{q}\cdot\hat{u})^2\rangle}\right]. \tag{5.51}$$

The conductivity expression in equation (5.45) can thus be presented in the following tensor form, which can be confidently used for cubic as well as non-cubic crystal structures, using realistic dispersion relations for all acoustic and optical phonon branches

$$\mathcal{K}_C^{ij} = \frac{\hbar^2}{N_0\Omega k_B T^2}\left[\langle\tau\omega^2 c_i c_j\rangle + \frac{\langle\omega c_i q_j\tau\tau_N^{-1}\rangle^2}{\langle|q_i||q_j|\tau_N^{-1}(1-\tau\tau_N^{-1})\rangle}\right] \tag{5.52}$$

where q_α and c_α are the αth component of the phonon wave vector q and the group velocity c, respectively.

5.2.5 SRIVASTAVA'S MODEL

We discussed in the previous subsection that Callaway's treatment of the momentum conserving nature of N-processes results into a contribution from the off-diagonal N-processes operator—the so-called N-drift term. The effect of off-diagonal U-processes operator, within the relaxation time approach, has been studied by Simons (1975) and Srivastava (1976a, 1980). Here we describe Srivastava's approach which has the simplicity of a Callaway-like approach.

In the spirit of Callaway's approach we write

$$\begin{aligned}\frac{\partial n_{qs}}{\partial t}\bigg|_{\text{scatt}} &= \frac{q \cdot u - \psi_{qs}}{\tau_N}\bar{n}_{qs}(\bar{n}_{qs}+1) - \frac{\psi_{qs}}{\tau_R^{\text{eff}}}\bar{n}_{qs}(\bar{n}_{qs}+1) \\ &= \frac{\partial n_{qs}}{\partial t}\bigg|_N^{\text{Callaway}} - \frac{\psi_{qs}}{\tau_R^{\text{eff}}}\bar{n}_{qs}(\bar{n}_{qs}+1)\end{aligned} \tag{5.53}$$

with

$$\frac{1}{\tau_R^{\text{eff}}} = \frac{1}{\tau_r} + \frac{1}{\tau_U^{\text{eff}}}, \tag{5.54}$$

where τ_r is the relaxation time due to momentum non-conserving elastic processes and τ_U^{eff} is an effective relaxation time due to U-processes. (With $\tau_U^{\text{eff}} = \tau_U$ equation (5.53) becomes equivalent to the Callaway's expression). We define τ_U^{eff} by the relations

$$-\frac{\partial n_{qs}}{\partial t}\bigg|_U \simeq (P\phi)_{qs} = \frac{\phi_q}{\tau_U^{\text{eff}}}\bar{n}_{qs}(\bar{n}_{qs}+1), \tag{5.55}$$

which is obtained by approximating the phonon distribution to a drifting Planck's distribution (see equation (5.31)):

$$\psi_{qs} \simeq \phi_q = q \cdot u. \tag{5.56}$$

A detailed discussion on three-phonon processes is presented in section 6.4.1. Using equations (4.45), (5.56) and (6.52) we can express equation (5.55) for three-phonon U-processes as

$$\frac{1}{\tau_U^{eff}} = \frac{(P\phi)_{qs}}{\phi_q\bar{n}_{qs}(\bar{n}_{qs}+1)}$$

$$= \frac{\sum_{q's'q''s''}(\boldsymbol{G}\cdot\boldsymbol{u})(\bar{P}^{q''s''}_{qsq's'}+\frac{1}{2}\bar{P}^{q's'q''s''}_{qs})}{\boldsymbol{q}\cdot\boldsymbol{u}\,\bar{n}_{qs}(\bar{n}_{qs}+1)}, \tag{5.57}$$

where $\boldsymbol{G}$ is a reciprocal lattice vector and $\bar{P}$ denotes equilibrium transition rate. Using the decomposition $P = \Gamma + \Lambda$ we write

$$-\frac{\partial n_{qs}}{\partial t}\Big|_U \simeq (P\phi)_{qs} = \Gamma_{qs}\phi_{qs}\Big|_U + \sum_{q's'\neq qs}\Lambda_{qsq's'}\phi_{q's'}\Big|_U \tag{5.58}$$

so it becomes clear that for $\phi_{q's'}(\boldsymbol{q's'}\neq\boldsymbol{qs})\neq 0$ the second term gives a contribution from the off-diagonal U-processes operator Λ_U. The general idea behind equation (5.57) was first presented by Klemens (1958), as mentioned in section 5.2.2.

Now we follow the procedure in the previous subsection and calculate the deviation function ψ^S_{qs} in Srivastava's approximation

$$\psi^S_{qs} = -\frac{\hbar|\nabla T|}{k_B T^2}\tau_m\left[\omega(qs)c_s(\boldsymbol{q})\hat{\boldsymbol{q}}\cdot\hat{\boldsymbol{u}} + \frac{\boldsymbol{q}\cdot\hat{\boldsymbol{u}}}{\tau_N}\frac{\langle\omega c q\tau_m\tau_N^{-1}\rangle}{\langle q^2\tau_N^{-1}(1-\tau_m\tau_N^{-1})\rangle}\right], \tag{5.59}$$

where

$$\frac{1}{\tau_m} = \frac{1}{\tau_N} + \frac{1}{\tau_R^{eff}} = \frac{1}{\tau_N} + \frac{1}{\tau_r} + \frac{1}{\tau_U^{eff}}. \tag{5.60}$$

Thus the model conductivity reads

$$\begin{aligned}\mathscr{K}_S &= \frac{k_B T^2}{N_0\Omega|\nabla T|^2}\sum_{qs}\psi^S_{qs}X_{qs}\\ &= \frac{\hbar^2}{3N_0\Omega k_B T^2}\left[\langle\tau_m\omega^2 c^2\rangle\frac{\langle\omega c q\tau_m\tau_N^{-1}\rangle^2}{\langle q^2\tau_N^{-1}(1-\tau_m\tau_N^{-1})\rangle}\right].\end{aligned} \tag{5.61}$$

Writing this result as

$$\mathscr{K}_S = \frac{\hbar^2}{3N_0\Omega k_B T^2}\langle\tau_S\omega^2 c^2\rangle \tag{5.62}$$

we note that in this model

$$\tau_S = \tau_m(1 + \beta/\tau_N) \tag{5.63}$$

which reduces to τ_C when $\tau_U^{eff} = \tau_U$.

5.2.6 KINETIC THEORY EXPRESSION FOR $\mathscr{K}$

The results in equations (5.26), (5.36), (5.47) and (5.62) can all be expressed in the isotropic form of the kinetic theory expression

$$\mathscr{K} = \frac{1}{3}C_v^{sp}\bar{c}^2\bar{\tau}, \tag{5.64}$$

where C_v^{sp} is the phonon specific heat defined in equation (2.123), $\bar{c}$ is an average phonon speed and $\bar{\tau}$ is an average phonon relaxation time defined for a particular model (namely in Srivastava's model $\bar{\tau} = \tau_S$, in Callaway's model $\bar{\tau} = \tau_C$, and in the *smrt* $\bar{\tau} = \tau$).

In Chapter 6 we will derive expressions for τ_N and τ_U which will be used to express $\mathscr{K}_{smrt}$, $\mathscr{K}_C$ and $\mathscr{K}_S$ within the Debye scheme in Chapter 7.

5.2.7 AN ITERATIVE APPROACH

While the Klemens model, the Callaway model and the Srivastava model described in the previous three subsections are based on physical considerations, Omini and Sparavigna (1995) suggested an iterative (but non-variational) approach to solve the phonon Boltzmann equation beyond the *smrt* method. Broido *et al* (2005) provided further explanation of this approach. Here we describe this approach with a slightly different emphasis.

Let us express the linearised Boltzmann equation in the form [see equations (5.9), (5.12), (5.24) and (5.46)]

$$X_{qs} = (P\psi)_{qs} = \Gamma_{qs}\psi_{qs} + \sum_{q's' \neq qs} \Lambda_{qsq's'}\psi_{q's'}, \tag{5.65}$$

where Γ and Λ are, respectively, the diagonal and off-diagonal parts of the phonon collision operator. As expressed in equation (5.25), the diagonal part Γ is related to the *smrt* τ_{qs}, including all elastic and inelastic contributions. The contribution to the off-diagonal term Λ comes from inelastic scattering processes, such as three-phonon or four-phonon events. The iterative approach, similar to the other approaches discussed in previous subsections, attempts to incorporate the contribution to the phonon relaxation time beyond the *smrt* contribution from the off-diagonal part of the collison operator Λ.

Considering only three-phonon scattering events, we can express (also discussed in Chapter 6)

$$\Gamma_{qs}\psi_{qs} = \psi_{qs} \sum_{q's',q''s''} [\bar{P}^{q''s''}_{qs,q's'} \frac{1}{2}\bar{P}^{q's',q''s''}_{qs}]. \tag{5.66}$$

and

$$\sum_{q's' \neq qs} \Lambda_{qs,q's'}\psi_{q's'} = \sum_{q's' \neq qs,q''s'' \neq qs} [\bar{P}^{q''s''}_{qs,q's'}(\psi_{q's'} - \psi_{q''s''}) - \frac{1}{2}\bar{P}^{q's',q''s''}_{qs}(\psi_{q's'} + \psi_{q''s''})]. \tag{5.67}$$

Here $\bar{P}^{q''s''}_{qs,q's'}$ and $\bar{P}^{q's',q''s''}_{qs}$ represent, respectively, equilibrium transition rates for three-phonon Class 1 and Class 2 processes described in equations (4.48) and (4.49).

At the start of the iterative process, the contribution from the off-diagonal part of the operator is ignored. This is done by considering modes other than qs in equilibrium (*i.e.* by setting $\psi_{q's'} = \psi_{q''s''} = 0$). With this consideration we express the deviation function ψ_{qs} for the mode qs as

$$step\ 0: \quad \psi^{(0)}_{qs} = \Gamma^{-1}_{qs} X_{qs} \tag{5.68}$$

$$= -\frac{\hbar}{k_B T^2} \tau_{qs}\omega(qs)c_s \cdot \nabla T \tag{5.69}$$

$$step\ 1: \quad \psi^{(1)}_{qs} = \psi^{(0)}_{qs}$$
$$+ \Gamma^{-1}_{qs} \sum_{q's' \neq qs,q''s'' \neq qs} [\bar{P}^{q''s''}_{qs,q's'}(\psi^{(0)}_{q's'} - \psi^{(0)}_{q''s''})$$
$$- \frac{1}{2}\bar{P}^{q's',q''s''}_{qs}(\psi^{(0)}_{q's'} + \psi^{(0)}_{q''s''})] \tag{5.70}$$

$$step\ n: \quad \psi^{(n)}_{qs} = \psi^{(n-1)}_{qs}$$
$$+ \Gamma^{-1}_{qs} \sum_{q's' \neq qs,q''s'' \neq qs} [\bar{P}^{q''s''}_{qs,q's'}(\psi^{(n-1)}_{q's'} - \psi^{(n-1)}_{q''s''})$$
$$- \frac{1}{2}\bar{P}^{q's',q''s''}_{qs}(\psi^{(n-1)}_{q's'} + \psi^{(n-1)}_{q''s''})] \tag{5.71}$$

$$= \psi^{(0)}_{qs} + \Gamma^{-1}_{qs} \sum_{i=1}^{n} R^{(i)}, \tag{5.72}$$

where the term $R^{(i)}$ is defined from the previous step. The values of the deviation functions $\psi_{q's'}^{(i)}$ and $\psi_{q''s''}^{(i)}$ at ith iteration can be obtained by interpolation of ψ values taken from surrounding q-grid points.

Using equation (5.69), and assuming that the nth step provides a well converged result for the deviation function ψ_{qs}, we can express the iterative solution $\tau(qs)|^{\text{Iterative}}$ to the final relaxation time as

$$\tau(qs)|^{\text{Iterative}} = -\frac{k_B T^2}{\hbar} \frac{1}{\omega(qs)c_s \cdot \nabla T} \psi_{qs}^{(n)} \tag{5.73}$$

$$= \tau(qs)|^{\text{smrt}} + \Delta\tau(qs), \tag{5.74}$$

where $\tau(qs)|^{\text{smrt}} \equiv \tau(qs)$ is the *smrt* and

$$\Delta\tau(qs) = -\frac{k_B T^2}{\hbar} \frac{\Gamma_{qs}^{-1}}{\omega(qs)c_s \cdot \nabla T} \sum_{i=1}^{n} R^{(i)} \tag{5.75}$$

is an additional contribution. And the final conductivity expression becomes

$$\mathscr{K}|^{\text{Iterative}} = \frac{k_B T^2}{N_0 \Omega |\nabla T|^2} \sum_{qs} \psi_{qs}^{(n)} X_{qs}$$

$$= \frac{\hbar^2}{3N_0 \Omega k_B T^2} \langle \tau|^{\text{Iterative}} \omega^2 c^2 \rangle. \tag{5.76}$$

A couple of remarks are in order. Firstly, it may require many iterations for achieving a well converged result. It is reported (Broido *et al* 2005) that it typically takes approximately 50 iterations for bulk materials containing six phonon branches per q point. Secondly, this iterative procedure is non-variational in nature, *i.e.* it does not obey any variational principle. An iterative approach based on a firm variational principle will be described in section 5.3.6.

5.3 VARIATIONAL METHODS

A variational method gives an approximate estimate of a functional by using a first-order solution (called a trial function) of a given equation. In most early applications of the variational principle only one estimate (usually a lower bound) for the quantity of interest has been reported. In many other applications an upper bound has been obtained. But there are examples of applications where both upper and lower bounds have been obtained for the same quantity. These are known as *complementary variational principles* (CVPs). That is, CVPs provide *two-sided* (maximum and minimum) approximate solutions (or bounds) of a given equation.

There are many ways of deriving CVPs. We identify four approaches which can be applied in a systematic manner. These are:

(1) by involving the use of involutory transformations of the Euler equations, as discussed by Courant and Hilbert (1953). This method is especially useful for obtaining solutions of differential equations;

(2) from the hypercircle approach, as discussed by Synge (1957), applicable to a class of linear problems;

(3) from canonical Euler equations, as discussed extensively by Arthurs (1970). This method provides a systematic approach to many linear and non-linear problems involving differential, integral and matrix equations; and

(4) from direct use of the Schwarz inequality. As discussed by Diaz and Weinstein (1947), and Jensen *et al* (1969), this approach is applicable to many problems.

Subject to their applicability to a particular problem, all these methods can be shown to be connected, and sometimes equivalent.

The linearised phonon Boltzmann equation $X = P\psi$ in equation (5.18) for its solution (ψ, X) is a candidate for the variational approach. In the following section we demonstrate this by looking at the structure of the phonon collision operator P.

5.3.1 STRUCTURE OF PHONON COLLISION OPERATOR AND EXISTENCE OF A VARIATIONAL PROBLEM FOR THE LATTICE THERMAL CONDUCTIVITY

As discussed in section 5.1.1, elastic scattering processes can be described in terms of single-mode relaxation times and in such cases; therefore, the conductivity can be calculated exactly. On the other hand, anharmonic phonon processes, which are present at all temperatures, are inelastic in nature and can only be dealt with if the deviation function ψ is known. In the absence of direct knowledge on ψ one can write $\psi = P^{-1}X$, subject to the existence of P^{-1}, and calculate $\mathcal{K}$ exactly. In order to gain information about ψ and P^{-1} it is essential to investigate the structure of the phonon collision operator. Guyer and Krumhansl (1966) have made a thorough study and a few important observations are made here.

(i) The eigenvectors of P form a complete orthonormal set $\{|p_i\rangle\}$ and in principle we can express P and P^{-1} in terms of this set

$$P = \sum_i p_i |p_i\rangle\langle p_i| \tag{5.77}$$

$$P^{-1} = \sum_i \left(\frac{1}{p_i}\right) |p_i\rangle\langle p_i| \tag{5.78}$$

with

$$|p_i\rangle\langle p_i| = \delta_{ij}. \tag{5.79}$$

Clearly P^{-1} exists if $p_i \neq 0$ for all i.

(ii) The sets of eigenvalues $\{p_i\}$ and eigenvectors $\{|p_i\rangle\}$ are not known in any detail. Only partial information about the eigenvalues and eigenstates of the momentum conserving part N (where $P = N + R$) is known. In particular, only four distribution functions are known which are the eigenvectors $\{|\eta_i\rangle\}$ of N and correspond to the eigenvalue zero:

$$N|\eta_0\rangle = 0|\eta_0\rangle \tag{5.80}$$

$$\begin{aligned}
N|\eta_{1x}\rangle &= 0|\eta_{1x}\rangle \\
N|\eta_{1y}\rangle &= 0|\eta_{1y}\rangle \\
N|\eta_{1z}\rangle &= 0|\eta_{1z}\rangle,
\end{aligned} \tag{5.81}$$

where $|\eta_0\rangle$ and $|\eta_1\rangle$ are defined by

$$\begin{aligned}
n_{qs}(\delta T) &= \left[\exp\left(\frac{\hbar\omega}{k_B(T+\delta T)}\right) - 1\right]^{-1} \\
&\simeq \bar{n}_{qs} + \frac{\hbar\omega}{k_B T}\frac{\delta T}{T}\bar{n}_{qs}(\bar{n}_{qs}+1) \\
&= \bar{n}_{qs} + |\eta_0\rangle
\end{aligned} \tag{5.82}$$

and

$$n_{qs}(\mathbf{u}) = \left[\exp\left(\frac{\hbar\omega}{k_B T} - \mathbf{q}\cdot\mathbf{u}\right) - 1\right]^{-1}$$

$$\begin{aligned}
&\simeq \ \bar{n}_{qs} + \boldsymbol{u} \cdot \left(\frac{\partial n_{qs}}{\partial \boldsymbol{u}}\right)_{u=0} \\
&= \ \bar{n}_{qs} + \boldsymbol{q} \cdot \boldsymbol{u}\bar{n}_{qs}(\bar{n}_{qs}+1) \\
&= \ \bar{n}_{qs} + |\eta_1\rangle.
\end{aligned} \tag{5.83}$$

Equation (5.80) results from the conservation of energy and corresponds to an infinitesimal rise in the temperature of the phonon system. Since energy is conserved in general for an interaction process, we can express $P|p_0\rangle = 0|p_0\rangle$, where $|p_0\rangle$ is the eigenvector of P corresponding to the eigenvalue zero. Further, equation (5.81) results from the conservation of momentum and corresponds to the drifting distribution $|\eta_1\rangle$. The fact that $|\eta_i\rangle$ are orthogonal to one another expresses the time independence of the energy and momentum of the system. $|\eta_0\rangle$ is an even function of $\boldsymbol{q}$, and the remaining three, namely $|\eta_1\rangle$, are odd functions of $\boldsymbol{q}$.

(3) The eigenstates of N can be assumed to form a complete set of orthonormal subspace in $\boldsymbol{q}$ space: $|\eta_0\rangle;|\eta_1\rangle;|\delta\rangle$, where $|\delta\rangle = |\eta_2\rangle,|\eta_3\rangle,...$, etc, is the rest of the subspace and does not include zero-eigenvalue solutions.

Since $\{|\eta_i\rangle\}$ is assumed to be a complete orthonormal set, P can be expressed in the representation of N eigenstates. In this representation N has a diagonal structure

$$N = \begin{bmatrix} 0 & 0 & 0 \\ 0 & 0 & 0 \\ 0 & 0 & N_{22} \end{bmatrix}$$

$$= \begin{bmatrix}
0 & 0 & 0 & \cdots \\
0 & \begin{bmatrix} 0 & 0 & 0 \\ 0 & 0 & 0 \\ 0 & 0 & 0 \end{bmatrix} & 0 & \cdots \\
0 & 0 & \lambda_2 & \cdots \\
0 & 0 & 0 & \lambda_3 & \cdots \\
0 & 0 & 0 & 0 & \lambda_4 & \cdots \\
\vdots & \vdots & \vdots & \vdots & & \ddots
\end{bmatrix}. \tag{5.84}$$

The eigenvalues $\lambda_2,\lambda_3,\lambda_4,...$ are all positive, since the system will equilibrate. Thus we see that while the operator N does not have an inverse, N_{22} does (actually, diagonal with elements $\lambda_2^{-1},\lambda_3^{-1},...$, etc). In the same representation, for dielectrics, the momentum non-conserving operator R has the structure

$$R = \begin{bmatrix} 0 & 0 & 0 \\ 0 & R_{11} & R_{12} \\ 0 & R_{21} & R_{22} \end{bmatrix}. \tag{5.85}$$

(For metals, where electron–phonon processes transfer energy out of the phonon system, the above structure of R is not true.) R_{11} is a 3×3 diagonal matrix for isotropic or cubic systems, since the point-group symmetry operator for these systems commutes with R and the functions $|\eta_{1x}\rangle,|\eta_{1y}\rangle$ and $|\eta_{1z}\rangle$ form a basis for one of the three-fold degenerate irreducible representations. (For lower symmetry the structure of R_{11} may not be diagonal.) The matrix R as such is not diagonal and does not have an inverse. Therefore, $P = N + R$ is not diagonal in the above representation and has the structure

$$P = \begin{bmatrix} 0 & 0 & 0 \\ 0 & R_{11} & R_{12} \\ 0 & R_{21} & (N_{22}+R_{22}) \end{bmatrix}. \tag{5.86}$$

If we define a projection operator $\mathscr{P}_0 = |\eta_0\rangle\langle\eta_0|$ which projects out $|\eta_0\rangle$ from $\{|\eta_i\rangle\}$, then the projected part of P has the structure

$$P^* = (1-\mathscr{P}_0)P(1-\mathscr{P}_0)$$

$$= \begin{bmatrix} R_{11} & R_{12} \\ R_{21} & (N_{22} + R_{22}) \end{bmatrix}$$
$$= \begin{bmatrix} P_{11}^* & P_{12}^* \\ P_{21}^* & P_{22}^* \end{bmatrix}. \tag{5.87}$$

Obviously P^* is non-singular and possesses an inverse, though P itself does not. Further reduction of P^* is not possible for isotropic and cubic systems.

The deviation function $\boldsymbol{\psi}$ can be expressed in terms of the complete orthonormal set $\{|\eta_i\rangle\}$

$$\boldsymbol{\psi} = a_0|\eta_0\rangle + \sum_{\alpha=x,y,z} a_{1\alpha}|\eta_{1\alpha}\rangle + P^{-1}(1 - \mathscr{P}_0 - \mathscr{P}_1)\boldsymbol{X}, \tag{5.88}$$

where $\mathscr{P}_0$ and $\mathscr{P}_1 = |\eta_1\rangle\langle\eta_1|$ are projection operators introduced to orthogonalize to the null-space $\{|\eta_0\rangle, |\eta_{1x}\rangle, |\eta_{1y}\rangle, |\eta_{1z}\rangle\}$. The choice in equation (5.88) suggests that we can write

$$\boldsymbol{\psi} = \boldsymbol{\phi} - \delta\boldsymbol{\phi} \tag{5.89}$$

with

$$\begin{aligned} \boldsymbol{\phi} &= \sum_{\alpha} a_{1\alpha}|\eta_{1\alpha}\rangle \\ &= \boldsymbol{q} \cdot \boldsymbol{u} \end{aligned} \tag{5.90}$$

as a drifting distribution function discussed in equation (5.83), and the error function $\delta\boldsymbol{\phi}$ expressible in terms of the first and third terms of equation (5.88). As $\delta\boldsymbol{\phi}$ is an unknown part of $\boldsymbol{\psi}$, the distribution function $\boldsymbol{\phi}$ may be regarded as a first-order solution and can be used as a trial function to solve equation (5.18) variationally. Thus this is the origin of a variational problem for the lattice thermal conductivity.

5.3.2 PROPERTIES OF THE ANHARMONIC PHONON COLLISION OPERATOR

Before attempting to solve the linearised phonon Boltzmann equation $X = P\boldsymbol{\psi}$ variationally, we note that the collision operator P has the following properties:

(i) As P is a linear operator

$$((\boldsymbol{v}_1 + \boldsymbol{v}_2), P\boldsymbol{v}_3) = (\boldsymbol{v}_1, P\boldsymbol{v}_3) + (\boldsymbol{v}_2 + P\boldsymbol{v}_3) \tag{5.91}$$
$$P(a\boldsymbol{v}_1 + b\boldsymbol{v}_2) = aP\boldsymbol{v}_1 + bP\boldsymbol{v}_2, \tag{5.92}$$

where $\boldsymbol{v}_1, \boldsymbol{v}_2, \boldsymbol{v}_3$ are any vector functions, a and b are constants, and $(\, , \,)$ denotes an inner product in the Hilbert space which spans the field of the operator P.

(ii) P is symmetric and real (i.e. self-adjoint):

$$(\boldsymbol{v}_1, P\boldsymbol{v}_2) = (\boldsymbol{v}_2, P\boldsymbol{v}_1). \tag{5.93}$$

This can be proved by noting that the kernel $P_{qq'}$ is symmetric with respect to q and q'.

(iii) P is positive semi-definite:

$$(\boldsymbol{v}, P\boldsymbol{v}) \geq 0 \tag{5.94}$$

for any $\boldsymbol{v}$ in the domain of P. This can be proved by noting that the kernel $P_{qq'}$, which measures the transition probability for a phonon interaction process, is non-negative.

(iv) The operator P is semi-bounded. Its eigenvalues lie between zero and some finite positive value p_{max}:

$$0 \leq p_i \leq p_{max} < \infty. \tag{5.95}$$

The part of P projected onto the subspace of odd distribution functions (such as $\phi_q = q \cdot u$), namely, P^*, is irreducible, positive definite and bounded:

$$0 < p_{min} \leq p_i^* \leq p_{max} < \infty. \tag{5.96}$$

If we regard the operator P^* as a limiting case of a matrix operator, then according to the theorem of Perron and Frobenius (see, e.g., Gantmacher 1959) its shortest and largest eigenvalues satisfy

$$p_{min} \geq \left(\sum_{q'} P_{qq'} \right)_{min} \equiv \mu \tag{5.97}$$

$$\left(\sum_{q'} P_{qq'} \right)_{max} \equiv \lambda \geq p_{max}, \tag{5.98}$$

where $\sum_{q'} P_{qq'}$ represents a row sum (with $q = qs$). For a proof we refer the reader to the works of Benin (1970) and Srivastava (1975, 1976b, 1977a).

(v) It is possible to decompose P^* as

$$P^* = A - B \tag{5.99}$$

such that $A \geq P^*$ and A^{-1} exists. It is also possible to express

$$P^* = L + \mathscr{T}^* \mathscr{T} \tag{5.100}$$

such that L^{-1} exists and $\mathscr{T}^*$ is the conjugate of $\mathscr{T}$. It may be mentioned here that in deriving the theory of variational principles in this chapter we will consider $\mathscr{T}^* \mathscr{T} = \frac{1}{2}\mu\hat{\mathbf{I}}$. The decomposition in equation (5.100) also defines $(P^* L^{-1} - \hat{\mathbf{I}})$ as a positive definite operator.

(vi) The operator P^* is compact and spectrally completely continuous. In the decomposition $P^* = \Gamma + \Lambda$, the off-diagonal part Λ is Γ-compact and Γ-bound:

$$||\Gamma v|| \geq ||\Lambda v||; \quad (v, \Gamma v) \geq |(v, \Lambda v)|, \tag{5.101}$$

where $||...||$ represents the sup norm (see, e.g., Kato (1980), Griffel (1981) for a discussion on continuous operators, compactness, and sup ($\equiv$ supreme or maximum) norm). Thus the essential spectra of Γ and P^* are the same, and cover the range $[\mu, \lambda]$. These points are discussed in the works of Buot (1972) and Srivastava (1976b, 1977b).

Before dealing with the variational problem it should be made clear that when the collision operator P acts on ψ it acts as the full operator P, but when it acts on ϕ it acts as the projected operator P^*. In our discussion henceforth we will write P everywhere, but it should be considered as P^* in the sub-space of odd distributions.

5.3.3 COMPLEMENTARY VARIATIONAL PRINCIPLES USING CANONICAL EULER–LAGRANGE EQUATIONS

Let us first review the essence of the one-sided variational principle. Consider a functional $E(\varphi)$ of functions φ belonging to a real Hilbert space $\mathscr{H}_\varphi$ with inner product $(,)$. If $E(\varphi)$ is twice differentiable in the sense of Fréchet, then it is stationary at $\varphi = \psi$ where ψ is the solution of

$$\frac{\delta E(\varphi)}{\delta \varphi} = 0. \tag{5.102}$$

If $\delta^2 E / \delta \varphi^2$ is positive when $\varphi = \psi$, then $E(\varphi) \geq E(\psi)$ for all values of φ in a neighbourhood of ψ. If $\delta^2 E / \delta \varphi^2$ is negative when $\varphi = \psi$, then $E(\varphi) \leq E(\psi)$ in a neighbourhood of ψ. Equation (5.86) thus leads to either an upper or a lower bound (estimate) of $E(\psi)$.

Next, let us look at the essence of two-sided (or complementary) variational principles. Let $\mathscr{H}_\psi$ and $\mathscr{H}_\eta$ be two real Hilbert spaces with inner products $(\ ,\ )$ and $\langle\ ,\ \rangle$, respectively. Let $F(\boldsymbol{\psi},\boldsymbol{\eta})$ be a functional of the two functions $\boldsymbol{\psi}$ and $\boldsymbol{\eta}$. Let us also introduce a linear transformation $\mathscr{T}$ so that $\mathscr{T}\boldsymbol{\psi}$ belongs to $\mathscr{H}_\eta$, i.e. $\mathscr{T} : \mathscr{H}_\psi \to \mathscr{H}_\eta$, with $\mathscr{T}^* : \mathscr{H}_\eta \to \mathscr{H}_\psi$, defining $\mathscr{T}^*$, the conjugate of $\mathscr{T}$. Suppose $F(\boldsymbol{\psi},\boldsymbol{\eta})$ has the form

$$F(\boldsymbol{\psi},\boldsymbol{\eta}) \quad = \quad \langle\boldsymbol{\eta},\mathscr{T}\boldsymbol{\psi}\rangle - W(\boldsymbol{\psi},\boldsymbol{\eta}) \tag{5.103}$$

$$\equiv \quad (\mathscr{T}^*\boldsymbol{\eta},\boldsymbol{\psi}) - W(\boldsymbol{\psi},\boldsymbol{\eta}) \tag{5.104}$$

with $W(\boldsymbol{\psi},\boldsymbol{\eta})$ twice differentiable in the sense of Fréchet, or with linear first and second Gateaux derivates. Consider two other functions:

$$\boldsymbol{\varphi} = \boldsymbol{\psi} + \delta\boldsymbol{\varphi} \tag{5.105}$$

and

$$\boldsymbol{\xi} = \boldsymbol{\eta} + \delta\boldsymbol{\xi}. \tag{5.106}$$

Then we can write

$$F(\boldsymbol{\varphi},\boldsymbol{\xi}) = F(\boldsymbol{\psi},\boldsymbol{\eta}) + \delta F + \delta^2 F + 0(\delta^3) \tag{5.107}$$

with

$$\delta F \quad = \quad \left(\delta\boldsymbol{\varphi}, \left.\frac{\delta F(\boldsymbol{\varphi},\boldsymbol{\xi})}{\delta\boldsymbol{\varphi}}\right|_{\substack{\boldsymbol{\varphi}=\boldsymbol{\psi}\\\boldsymbol{\xi}=\boldsymbol{\eta}}}\right) + \langle\delta\boldsymbol{\xi}, \left.\frac{\delta F(\boldsymbol{\varphi},\boldsymbol{\xi})}{\delta\boldsymbol{\xi}}\right|_{\substack{\boldsymbol{\varphi}=\boldsymbol{\psi}\\\boldsymbol{\xi}=\boldsymbol{\eta}}}\rangle$$

$$= \quad \left(\delta\boldsymbol{\varphi}, \mathscr{T}^*\boldsymbol{\eta} - \left.\frac{\delta W(\boldsymbol{\varphi},\boldsymbol{\eta})}{\delta\boldsymbol{\varphi}}\right|_{\boldsymbol{\varphi}=\boldsymbol{\psi}}\right) + \langle\delta\boldsymbol{\xi}, \mathscr{T}\boldsymbol{\psi} - \left.\frac{\delta W(\boldsymbol{\psi},\boldsymbol{\xi})}{\delta\boldsymbol{\xi}}\right|_{\boldsymbol{\xi}=\boldsymbol{\eta}}\rangle.$$

$$\tag{5.108}$$

$F(\boldsymbol{\varphi},\boldsymbol{\eta})$ will be stationary at $\boldsymbol{\varphi} = \boldsymbol{\psi}, \boldsymbol{\xi} = \boldsymbol{\eta}$ if $\delta F = 0$. If $\delta\boldsymbol{\varphi}$ and $\delta\boldsymbol{\xi}$ are independent and arbitrary, then

$$\mathscr{T}^*\boldsymbol{\eta} - \frac{\delta W}{\delta\boldsymbol{\psi}} \quad = \quad 0 \quad\quad \text{in} \quad\quad \mathscr{H}_\psi \tag{5.109}$$

$$\mathscr{T}\boldsymbol{\psi} - \frac{\delta W}{\delta\boldsymbol{\eta}} \quad = \quad 0 \quad\quad \text{in} \quad\quad \mathscr{H}_\eta \tag{5.110}$$

must hold. These equations are a generalised form of the canonical Euler–Lagrange (or Hamilton) equations. From these, Noble (1964) and Arthurs (1970) derived complementary variational principles (CVPs) by showing that we can always find functionals $J(\boldsymbol{\phi})$ and $M(\boldsymbol{\xi})$ such that

$$J(\boldsymbol{\phi}) \leq F(\boldsymbol{\psi},\boldsymbol{\eta}) \leq M(\boldsymbol{\xi}) \tag{5.111}$$

if $\delta^2 J < 0$ and $\delta^2 M > 0$.

It should be appreciated that the functional $F(\boldsymbol{\psi},\boldsymbol{\eta})$ belongs to the product real Hilbert space $\mathscr{H} = \mathscr{H}_\eta \times \mathscr{H}_\psi$ with inner product $\{\ ,\ \}$ defined by

$$\{\boldsymbol{h}_1,\boldsymbol{h}_2\} = \langle\boldsymbol{\eta}_1,\boldsymbol{\eta}_2\rangle + (\boldsymbol{\psi}_1,\boldsymbol{\psi}_2), \tag{5.112}$$

where

$$\boldsymbol{h}_i = \begin{pmatrix} \boldsymbol{\eta}_i \\ \boldsymbol{\psi}_i \end{pmatrix} \quad\quad i = 1,2. \tag{5.113}$$

The functional $W(\boldsymbol{\psi},\boldsymbol{\eta})$ also belongs to the product space $\mathscr{H}$ and can be obtained by integrating equations (5.109) and (5.110).

Now we apply the complementary variational principles (CVPS) to the linearised phonon Boltzmann equation $X = P\psi$ which can be expressed with the help of equation (5.100) as

$$X = (L + \mathscr{T}^*\mathscr{T})\psi. \tag{5.114}$$

This equation can be written in canonical form, following equations (5.109) and (5.110), as

$$\mathscr{T}\psi \;=\; \eta = \frac{\delta W}{\delta \eta} \quad \text{in} \quad \mathscr{H}_\eta \tag{5.115}$$

$$\mathscr{T}^*\eta \;=\; X - L\psi = \frac{\delta W}{\delta \psi} \quad \text{in} \quad \mathscr{H}_\psi. \tag{5.116}$$

Equations (5.115) and (5.116) allow us to determine the functional W. Integrating equation (5.115) we get

$$W_\eta = \frac{1}{2}\langle \eta, \eta \rangle \quad \text{in} \quad \mathscr{H}_\eta \tag{5.117}$$

while integrating equation (5.116) we get

$$W_\psi = (\psi, X) - \frac{1}{2}(\psi, L\psi) \quad \text{in} \quad \mathscr{H}_\psi. \tag{5.118}$$

Therefore, a suitable functional W is

$$W(\eta, \psi) \equiv W_\eta + W_\psi = \frac{1}{2}\langle \eta, \eta \rangle + (\psi, X) - \frac{1}{2}(\psi, L\psi). \tag{5.119}$$

With this the functional $F(\psi, \eta)$ in equations (5.103)–(5.104) is now defined for the problem at hand.

Having defined the functional $F(\psi, \eta)$ we are now in a position to find functionals J and M which satisfy the inequalities in equation (5.111). First we assume that equation (5.115) can be solved for η in terms of ψ, i.e.

$$\eta_1 = \eta_1(\psi_1) = \mathscr{T}\psi_1 \quad \text{in} \quad \mathscr{H}_\psi \tag{5.120}$$

and then define

$$J(\psi_1) = -2F(\eta_1, \psi_1) \quad \text{via} \quad (5.103). \tag{5.121}$$

Secondly, we assume that equation (5.116) can be solved for ψ in terms of η, i.e.

$$\psi_2 = \psi_2(\eta_2) = L^{-1}(X - \mathscr{T}^*\eta_2) \quad \text{in} \quad \mathscr{H}_\eta \tag{5.122}$$

and define

$$M(\eta_2) = -2F(\eta_2, \psi_2) \quad \text{via} \quad (5.104). \tag{5.123}$$

These give

$$
\begin{aligned}
J(\psi_1) &= -2\langle \eta_1, \eta_1 \rangle + 2W(\eta_1, \psi_1) \\
&= 2(\psi_1, X) - (\psi_1, P\psi_1)
\end{aligned} \tag{5.124}
$$

and

$$
\begin{aligned}
M(\eta_2) &= -2(\mathscr{T}^*\eta_2, \psi_2) + 2W(\eta_2, \psi_2) \\
&= \langle \eta_2, \eta_2 \rangle + \langle X - \mathscr{T}^*\eta_2, L^{-1}(X - \mathscr{T}^*\eta_2) \rangle.
\end{aligned} \tag{5.125}
$$

J and M are stationary at ψ and η, respectively. At these values

$$\delta J/\delta \psi_1 = 0 = 2X - 2P\psi_1$$

and

$$\delta^2 J/\delta\boldsymbol{\psi}_1^2 = -2P$$

which is negative since P is positive definite operator (this is so because $\boldsymbol{\psi}_1$ which is a trial function can be recognised as $\boldsymbol{\phi}$ of equation (5.90) and hence P can be recognised as P^*, the projected operator). Similarly

$$\delta M/\delta\boldsymbol{\eta}_2 = 0 = 2\boldsymbol{\eta}_2 - 2\mathscr{T}^*L^{-1}X + 2L^{-1}\mathscr{T}^*\mathscr{T}\boldsymbol{\eta}_2$$

and

$$\delta^2 M/\delta\boldsymbol{\eta}^2 = 2(\hat{I} + L^{-1}\mathscr{T}^*\mathscr{T}) = 2PL^{-1}$$

which is positive if L is. Thus we have shown that the CVPS for the lattice thermal conductivity are given as

$$J(\boldsymbol{\psi}_1) \leq J(\boldsymbol{\psi}) \equiv (\boldsymbol{\psi}, X) \equiv M(\boldsymbol{\eta}) \leq M(\boldsymbol{\eta}_2). \tag{5.126}$$

The lower bound $J(\boldsymbol{\psi}_1)$ provides a maximum variational principle and the upper bound $M(\boldsymbol{\eta}_2)$ provides a minimum variational principle.

If we write $\boldsymbol{\psi}_1 \equiv \boldsymbol{\phi} = \boldsymbol{\psi} + \delta\boldsymbol{\phi}$, where $\delta\boldsymbol{\phi}$ is the error function defined in equation (5.89), then, as stated above, the functional

$$J(\boldsymbol{\phi}) = 2(\boldsymbol{\phi}, X) - (\boldsymbol{\phi}, P\boldsymbol{\phi}) \tag{5.127}$$

gives a maximum variational principle. In fact $J(\boldsymbol{\phi})$ is the functional used by Leibfried and Schlömann (1954) and by Roussopoulos (1953).

The functional $M(\boldsymbol{\eta}_2)$ is in an inconvenient form, since it contains the conjugate operator $\mathscr{T}^*$. However, we do not need to know either $\mathscr{T}$ or $\mathscr{T}^*$. The knowledge that they exist is sufficient, since we can solve for $\boldsymbol{\eta}_2$ in terms of $\boldsymbol{\psi}_2$, i.e.

$$\boldsymbol{\eta}_2 = \boldsymbol{\eta}_2(\boldsymbol{\psi}_2) = \mathscr{T}\boldsymbol{\psi}_2 = \boldsymbol{\psi}_3 \qquad \text{in} \qquad \mathscr{H}_{\boldsymbol{\psi}} \tag{5.128}$$

and define another functional $M(\mathscr{T}\boldsymbol{\psi}_2) = M(\boldsymbol{\psi}_3)$ in $\mathscr{H}_{\boldsymbol{\psi}}$:

$$\begin{aligned}
M(\boldsymbol{\psi}_3) &= (\mathscr{T}\boldsymbol{\psi}_2, \mathscr{T}\boldsymbol{\psi}_2) + (X - \mathscr{T}^*\mathscr{T}\boldsymbol{\psi}, L^{-1}[X - \mathscr{T}^*\mathscr{T}\boldsymbol{\psi}_2]) \\
&= 2(\boldsymbol{\psi}_2, X) - (\boldsymbol{\psi}_2, P\boldsymbol{\psi}_2) + ([X - P\boldsymbol{\psi}_2], L^{-1}[X - P\boldsymbol{\psi}_2]).
\end{aligned} \tag{5.129}$$

If we consider $\boldsymbol{\psi}_2 = \boldsymbol{\phi} = \boldsymbol{\psi} + \delta\boldsymbol{\phi}$, then M becomes

$$\begin{aligned}
M(\boldsymbol{\phi}) &= 2(\boldsymbol{\phi}, X) - (\boldsymbol{\phi}, P\boldsymbol{\phi}) + ([X - P\boldsymbol{\phi}], L^{-1}[X - P\boldsymbol{\phi}]) \\
&= J(\boldsymbol{\phi}) + (Y, L^{-1}Y)
\end{aligned} \tag{5.130}$$

with

$$Y = X - P\boldsymbol{\phi}. \tag{5.131}$$

The complementary variational bounds $J(\boldsymbol{\phi})$ and $M(\boldsymbol{\phi})$ can be obtained as normalisation-independent results. In $J(\boldsymbol{\phi})$ we include a variational one-parameter α and talk of $\alpha\boldsymbol{\phi}$ instead of $\boldsymbol{\phi}$ and maximise $J(\alpha\boldsymbol{\phi})$ with respect to α. The optimum value of α is

$$\tilde{\alpha} = \frac{(\boldsymbol{\phi}, X)}{(\boldsymbol{\phi}, P\boldsymbol{\phi})} \tag{5.132}$$

yielding

$$J(\tilde{\alpha}\boldsymbol{\phi}) \equiv \kappa^< = \frac{(\boldsymbol{\phi}, X)}{(\boldsymbol{\phi}, P\boldsymbol{\phi})^2}. \tag{5.133}$$

Similarly we include a variational one-parameter β with ϕ and minimize $M(\beta\phi)$ with respect to β. The optimum value of β is

$$\tilde{\beta} = \frac{(X,(PL^{-1}-\hat{I})\phi)}{(P\phi,(PL^{-1}-\hat{I})\phi)} \tag{5.134}$$

yielding[1]

$$M(\tilde{\beta}\phi) \equiv \kappa^> = (X,L^{-1}X) - \frac{(X,(PL^{-1}-\hat{I})\phi)^2}{(P\phi,(PL^{-1}-\hat{I})\phi)}. \tag{5.135}$$

The normalization-independent results $J(\tilde{\alpha}\phi)$ and $M(\tilde{\beta}\phi)$ are more convenient than $J(\phi)$ and $M(\phi)$, respectively. $J(\tilde{\alpha}\phi)$ has been used by many authors (Kohler 1948, Sondheimer 1950, Ziman 1960, Levine and Schwinger 1949) and in the context of thermal conductivity this result is generally known as Ziman's variational bound.

A more general variational trial function

$$\phi = \sum_{n=1}^{N} \alpha_n \phi_n \tag{5.136}$$

can be used to convert $J(\phi)$ to an improved normalization-independent lower bound by maximizing the result with respect to all the α_n. Similarly, an improved normalization-independent upper bound can be obtained by using equation (5.136) and minimizing $M(\phi)$ with respect to all the α_n. We will come to this issue in section 5.3.6.

5.3.4   COMPLEMENTARY VARIATIONAL PRINCIPLES USING SCHWARZ'S INEQUALITY

The CVPs derived in the previous subsection can also be derived from simple applications of the Schwarz inequality.

First we note that since P is positive semi-definite

$$(f,Pf) \geq = 0$$

for all vectors f in a Hilbert space $\mathscr{H}$ in which P is a correspondence. To derive the maximum variational principle consider $f = \psi - \phi = -\delta\phi$. Then

$$((\psi-\phi),P(\psi-\phi)) \geq = 0. \tag{5.137}$$

As P is symmetric, the above inequality leads to

$$(\psi,X) \equiv (\psi,P\psi) \geq 2(\phi,X) - (\phi,P\phi) \tag{5.138}$$

or

$$J(\psi) \geq J(\phi) \tag{5.139}$$

where $J(\phi)$ is the functional in equation (5.127). Introducing a variational parameter α and replacing ϕ by $\alpha\phi$ in equation (5.137) gives

$$((\psi,\alpha\phi),P(\psi-\alpha\phi)) \geq 0. \tag{5.140}$$

From this we get

$$(\psi,P\psi) - 2\alpha(\phi,P\psi) + \alpha^2(\phi,P\phi) \geq 0. \tag{5.141}$$

---

[1]Equation (5.194) at the end of section 5.3.6 shows how the coefficients $\kappa^<$ and $\kappa^>$ can be used to obtain the corresponding bounds on the lattice thermal conductivity $\mathscr{K}$ defined in equation (5.23).

The left-hand side of this inequality has a turning point with respect to α when

$$\alpha = \tilde{\alpha} = \frac{(\boldsymbol{\phi}, X)}{(\boldsymbol{\phi}, P\boldsymbol{\phi})}. \tag{5.142}$$

The second derivative of the left-hand side in equation (5.141) is $2(\boldsymbol{\phi}, P\boldsymbol{\phi})$, which is positive as P acting on $\boldsymbol{\phi}$ is positive definite. Substituting equation (5.126) into equation (5.141) we get

$$(\boldsymbol{\psi}, P\boldsymbol{\psi})(\boldsymbol{\phi}, P\boldsymbol{\phi}) \geq (\boldsymbol{\phi}, P\boldsymbol{\psi})^2 \tag{5.143}$$

which is the Schwarz inequality. From here we get

$$J(\boldsymbol{\psi}) \equiv (\boldsymbol{\psi}, X) \equiv (\boldsymbol{\psi}, P\boldsymbol{\psi}) \geq \frac{(\boldsymbol{\phi}, X)^2}{(\boldsymbol{\phi}, P\boldsymbol{\phi})} = \kappa^< \tag{5.144}$$

where $\kappa^<$ is the lower bound derived earlier in equation (5.133).

The Schwarz inequality in equation (5.143) is in general applicable to any operator P_j which has the formal properties of the operator P. To derive the minimum variational principle we consider an operator $(PL^{-1} - \hat{I})P$ which obviously has the formal properties of P. Then the Schwarz inequality in equation (5.143) can be extended to

$$(\boldsymbol{\psi}, (PL^{-1} - \hat{I})P\boldsymbol{\psi})(\boldsymbol{\phi}, (PL^{-1} - \hat{I})P\boldsymbol{\phi}) \geq (\boldsymbol{\phi}, (PL^{-1} - \hat{I})P\boldsymbol{\psi})^2. \tag{5.145}$$

Using $X = P\boldsymbol{\psi}$ and the symmetry property of P this gives

$$(X, L^{-1}X) - J(\boldsymbol{\psi}) \geq \frac{(\boldsymbol{\phi}, (PL^{-1} - \hat{I})PX)^2}{(\boldsymbol{\phi}, (PL^{-1} - \hat{I})P\boldsymbol{\phi})}. \tag{5.146}$$

From this we get

$$\begin{aligned} J(\boldsymbol{\psi}) & \leq & (X, L^{-1}X) - \frac{(X, (PL^{-1} - \hat{I})P\boldsymbol{\phi})^2}{(\boldsymbol{\phi}, (PL^{-1} - \hat{I})P\boldsymbol{\phi})} \\ & \equiv & \kappa^> \end{aligned} \tag{5.147}$$

where $\kappa^>$ is the upper bound derived earlier in equation (5.135). Notice that equation (5.146) gives a maximum principle for the functional $(X, L^{-1}X) - J(\boldsymbol{\psi}) \equiv (\boldsymbol{\psi}, (PL^{-1} - \hat{I})P\boldsymbol{\psi})$ which becomes a minimum principle for the functional $J(\boldsymbol{\psi}) \equiv (\boldsymbol{\psi}, X)$. Furthermore, it is easy to show that replacing P by $(PL^{-1} - \hat{I})P$ in equation (5.137) leads to the unnormalized upper bound $M(\boldsymbol{\phi})$ of equation (5.130). The upper bound $\kappa^>$ was originally derived by Jensen *et al* (1969).

5.3.5   SEQUENCES OF BOUNDS AND THEIR CONVERGENCE

In the above two subsections, we have essentially derived the CVPS, obtaining a lower bound $\kappa^<$ and an upper bound $\kappa^>$ for the conductivity coefficient $J(\boldsymbol{\psi}) : \kappa^< \leq J(\boldsymbol{\psi}) \leq \kappa^>$. In this and the following subsections we will demonstrate that the bounds $\kappa^<$ and $\kappa^>$ are actually the endpoint results for a sequence of lower bounds $\{\kappa_m^<\}$ $(m = 0, 1, 2, ...)$ and a sequence of upper bounds $\{\kappa_n^>\}$ $(n = 1, 3, 5, ...)$, respectively: $\kappa^< \equiv \kappa_0^<$ and $\kappa^> \equiv \kappa_1^>$. We will also show that the sequences $\{\kappa_m^<\}$ and $\{\kappa_n^>\}$ are monotonic and converge towards the exact solution $J(\boldsymbol{\psi})$ in the limits $m \to \infty$ and $n \to \infty$, respectively.

Let us define a set of equations

$$X_j = P_j\boldsymbol{\psi} \qquad j \geq 0, \tag{5.148}$$

where X_j and P_j bear the formal properties of X and P, respectively. As before, for a trial function $\alpha\phi$ we find an optimum value of α which makes $((\psi, \alpha\phi), P_j(\psi - \alpha\phi))$ an extremum, i.e.

$$\tilde{\alpha} = \frac{(\phi, P_j \psi)}{(\phi, P_j \phi)}. \tag{5.149}$$

Since P_j is positive semi-definite,

$$((\psi, \tilde{\alpha}\phi), P_j(\psi, \tilde{\alpha}\phi)) \geq 0 \tag{5.150}$$

which becomes the Schwarz inequality

$$(\psi, P_j \psi)(\phi, P_j \phi) \geq (\phi, P_j \psi)^2. \tag{5.151}$$

The sequences of normalized lower and upper bounds can now be derived.

5.3.5.1   Sequence of lower bounds

We define

$$\left. \begin{aligned} X_m &= (\hat{I} - PA^{-1})^m X \\ P_m &= (\hat{I} - PA^{-1})^m P \end{aligned} \right\} \qquad m = 0, 1, 2, \ldots \tag{5.152}$$

with $P = A - B$ and $A \geq P$ such that A^{-1} exists. Then from equations (5.148) and (5.151) we have

$$\kappa_m \equiv (\psi, X_m) = (\psi, P_m \psi) \geq \frac{(\phi, X_m)^2}{(\phi, P_m \phi)}. \tag{5.153}$$

For $m = 0$ this gives

$$\kappa_0 \equiv J(\psi) \geq \frac{(\phi, X)^2}{(\phi, P\phi)} = \kappa_0^< \tag{5.154}$$

thus showing that in fact $\kappa^<$ of the previous subsection is $\kappa_0^<$.

For $m = 1$ we get

$$\begin{aligned} \kappa_1 \equiv (\psi, X_1) &= (\psi, X) - (\psi, A^{-1}P\psi) \\ &= J(\psi) - (X, A^{-1}X) \\ &\geq \frac{(\psi, X_1)^2}{(\phi, P_1 \phi)}, \end{aligned}$$

where the last inequality is from equation (5.153). Thus

$$J(\psi) \geq (X, A^{-1}X) + \frac{(\phi, X_1)^2}{(\phi, P_1 \psi)} = \kappa_1^<. \tag{5.155}$$

And so on.

Writing explicitly, we have

$$m = 0 \qquad \kappa_0^< = \frac{(\phi, X)^2}{(\psi, P\phi)} \tag{5.156}$$

$$m = 1 \qquad \kappa_1^< = (X, A^{-1}X) + \frac{(\phi, (\hat{I} - PA^{-1})X)^2}{(\phi, (\hat{I} - PA^{-1})P\phi)} \tag{5.157}$$

etc.

This way we can obtain expressions for all the terms in the sequence of lower bounds $\{\kappa_m^<\}, m = 0, 1, 2, \ldots$.

To examine the nature of the sequence $\{\kappa_m^<\}$ we will find it convenient to use the unnormalized expression for $\kappa_m^<$. We define

$$
\begin{aligned}
J_m(\boldsymbol{\phi}) &= 2(\boldsymbol{\phi},X_m) - (\boldsymbol{\phi},P_m\boldsymbol{\phi}) \\
&= (\boldsymbol{\psi},X_m) - (\delta\boldsymbol{\phi},P_m\delta\boldsymbol{\phi}) \\
&= J_m(\boldsymbol{\psi}) - (\delta\boldsymbol{\phi},P_m\delta\boldsymbol{\phi}),
\end{aligned}
\tag{5.158}
$$

where equation (5.105) is used. Now we can express $J_m(\boldsymbol{\psi})$ as

$$
\begin{aligned}
J_m(\boldsymbol{\psi}) &= (\boldsymbol{\psi},X_m) \\
&= (\boldsymbol{\psi},X_{m-1}) - (X,A^{-1}X_{m-1}) \\
&= \ldots \\
&= (\boldsymbol{\psi},X) - \left(X,A^{-1}\sum_{l=0}^{m-1}X_l\right) \\
&= J(\boldsymbol{\psi}) - \left(X,A^{-1}\sum_{l=0}^{m-1}X_l\right).
\end{aligned}
\tag{5.159}
$$

Putting equation (5.159) into equation (5.158) we get

$$
J(\boldsymbol{\psi}) = J_m(\boldsymbol{\psi}) + \left(X,A^{-1}\sum_{l=0}^{m-1}X_l\right) + (\delta\boldsymbol{\phi},P_m\delta\boldsymbol{\phi})
\tag{5.160}
$$

or

$$
J(\boldsymbol{\psi}) = \kappa_m^<(\boldsymbol{\phi}) + (\delta\boldsymbol{\phi},P_m\delta\boldsymbol{\phi}),
\tag{5.161}
$$

where

$$
\kappa_m^<(\boldsymbol{\phi}) = 2(\boldsymbol{\phi},X_m) - (\boldsymbol{\phi},P_m\boldsymbol{\phi}) + \left(X,A^{-1}\sum_{l=0}^{m-1}X_l\right)
\tag{5.162}
$$

is the unnormalized lower bound on $J(\boldsymbol{\psi})$. The expression in equation (5.162) can be simplified by summing the geometric series

$$
\begin{aligned}
A^{-1}\sum_{l=0}^{m-1}X_l &= A^{-1}\sum_{l=0}^{m-1}(\hat{I}-PA^{-1})^l X \\
&= [\hat{I}-(\hat{I}-PA^{-1})^m]P^{-1}X.
\end{aligned}
\tag{5.163}
$$

(Notice that here we have an expression which has P^{-1}. This is no problem, as in this context P acts on the subspace of $\boldsymbol{\phi}$ and is positive definite so that its inverse exists.) Thus

$$
\kappa_m^<(\boldsymbol{\phi}) = 2(\boldsymbol{\phi},X_m) - (\boldsymbol{\phi},P_m\boldsymbol{\phi}) + (X,[\hat{I}-(\hat{I}-PA^{-1})^m]P^{-1}X).
\tag{5.164}
$$

Note that $\kappa_0^<(\boldsymbol{\phi}) \equiv J(\boldsymbol{\phi})$ of equation (5.127). It is easy to show from this expression that $\kappa_m^<(\boldsymbol{\phi})$ satisfies the following recursion relation:

$$
\kappa_{m+1}^<(\boldsymbol{\phi}) - \kappa_m^<(\boldsymbol{\phi}) = (Y,A^{-1}P_mY),
\tag{5.165}
$$

where $Y = X - P\boldsymbol{\phi}$ as in equation (5.131). Since $(Y,A^{-1}P_mY) \geq 0$, the sequence $\{\kappa_m^<(\boldsymbol{\phi})\}$ increases monotonically. It must therefore necessarily converge. To find the limit of its convergence, we expand $\delta\boldsymbol{\phi}$ in terms of the complete orthonormal set $\{|p_i\rangle\}$

$$
|\delta\phi\rangle = \sum_i c_i|p_i\rangle.
\tag{5.166}
$$

If $\{a_i\}$ are the eigenvalues of A, then in equation (5.145)

$$(\delta\phi, P_m \delta\phi) = \sum_i |c_i|^2 p_i \left(1 - \frac{p_i}{a_i}\right)^m$$
$$\rightarrow \quad 0 \qquad \text{for large } m.$$

Thus for $m \rightarrow \infty$, $J(\psi) = \kappa_\infty^<$, i.e. the sequence $\{\kappa_m^<\}$ converges to the exact coefficient $J(\psi)$ from below.

5.3.5.2 Sequence of upper bounds

We define

$$\begin{aligned}
X_n &= (PL^{-1} - \hat{I})^n X \\
P_n &= (PL^{-1} - \hat{I})^n P \\
X_n &= P_n \psi \qquad n = \text{integer}
\end{aligned} \tag{5.167}$$

with L defined in equation (5.100). From equations (5.148) and (5.151), we have

$$J_n(\psi) \equiv (\psi, X_n) = \frac{(\phi, X_n)^2}{(\phi, P_n \phi)}. \tag{5.168}$$

This gives

$$\begin{aligned}
\text{for } n = 1: \quad J(\psi) &\leq (X, L^{-1}X) - \frac{(\phi, (PL^{-1} - \hat{I})X)^2}{(\phi, (PL^{-1} - \hat{I})P\phi)} \\
&= \kappa_1^> \tag{5.169} \\
\text{for } n = 3: \quad J(\psi) &\leq (X, L^{-1}X) - (X, L^{-1}X_1) \\
&\quad + (X, L^{-1}X_2) - \frac{(\phi, X_3)^2}{(\phi, P_3 \phi)} \\
&= \kappa_3^> \tag{5.170}
\end{aligned}$$

and so on for $n = 5, 7, \ldots$. Thus with n as *odd integers* we get a sequence of upper bounds on $J(\psi) : \{\kappa_n^>\} n = 1, 3, 5, \ldots$. For clear identification we rewrite the first two terms of the sequence:

$$n = 1: \quad \kappa_1^> = (X, L^{-1}X) - \frac{(\phi, X_1)^2}{(\phi, P_1 \phi)} \tag{5.171}$$

$$n = 3: \quad \kappa_3^> = (X, L^{-1}X) - (X, L^{-1}X_1)$$
$$+ (X, L^{-1}X_2) - \frac{(\phi, X_3)^2}{(\phi, P_3 \phi)}. \tag{5.172}$$

It can be noticed that when n stands for an even integer we get a lower bound (which for $n = 0$ is the same as in equation (5.154) but yields new results for $n = 2, 4, 6, \ldots$).

To examine the nature of the sequence $\{\kappa_n^>\}$, $n = 1, 3, 5, \ldots$, we will again find it convenient to use the unnormalized expression for $\kappa_n^>$. We can express

$$J_n(\psi) \equiv (\psi, X_n) = 2(\phi, X_n) - (\phi, P_n \phi) + (\delta\phi, P_n \delta\phi). \tag{5.173}$$

Further, we express

$$J_n(\psi) = (\psi, X_n)$$

$$\begin{aligned}
&= (X, L^{-1} X_{n-1}) - (\psi, X_{n-1}) \\
&= \cdots \\
&= (X, L^{-1} X_{n-1}) - (X, L^{-1} X_{n-2}) + \cdots + (-1)^n (\psi, X) \\
&= (-1)^n \left\{ J(\psi) + \left(X, L^{-1} \sum_{l=0}^{n-1} (-1)^{-l-1} X_l \right) \right\} \\
&= (-1)^n \left\{ J(\psi) - (X, [\hat{I} - (\hat{I} - PL^{-1})^n] P^{-1} X) \right\}, \tag{5.174}
\end{aligned}$$

where the last step is written after summing the geometric series. When n is an odd integer, we can write from equations (5.173) and (5.174)

$$\begin{aligned}
J(\psi) &= -2(\phi, X_n) + (\phi, P_n \phi) + (X, [(PL^{-1} - \hat{I})^n + \hat{I}] P^{-1} X) \\
&\quad - (\delta\phi, P_n \delta\phi) \\
&= \kappa_n^{>}(\phi) - (\delta\phi, P_n \delta\phi). \tag{5.175}
\end{aligned}$$

Thus

$$\kappa_n^{>}(\phi) = -2(\phi, X_n) + (\phi, P_n \phi) + (X, [(PL^{-1} - \hat{I})^n + \hat{I}] P^{-1} X) \tag{5.176}$$

is an unnormalized upper bound for $\kappa \equiv J(\psi)$. Note that, for $n = 1$, $\kappa_1^{>}(\phi) \equiv M(\phi)$ of equation (5.130). As n has to be an odd integer, the required recursion relation is

$$\kappa_n^{>}(\phi) - \kappa_{n+2}^{>}(\phi) = (Y, L^{-1}(2\hat{I} - PL^{-1})(PL^{-1} - \hat{I})^n Y), \tag{5.177}$$

where, again, $Y = X - P\phi$ as in equation (5.131). The right-hand side of equation (5.177) is (semi)positive if $2\hat{I} - PL^{-1}$ is. Thus the sequence $\{\kappa_n^{>}(\phi)\}$ is monotonically increasing if $2L \geq P$. It must necessarily converge for this condition. To show that it converges to the exact coefficient $\kappa \equiv J(\psi)$, we expand $\delta\phi$ as in equation (5.166). If $\{l_i\}$ are the eigenvalues of L, then in equation (5.175)

$$\begin{aligned}
(\delta\phi, P_n \delta\phi) &= \sum_i |c_i|^2 p_i (p_i l_i^{-1} - 1)^n \\
&= \sum_i |c_i|^2 p_i \left(\frac{p_i}{l_i} \right)^n \left(1 - \frac{l_i}{p_i} \right)^n \\
&\to 0 \qquad \text{for large } n.
\end{aligned}$$

Thus for $n \to \infty$, $\kappa_\infty^{>} \to \kappa$, i.e. the sequence $\{\kappa_n^{>}\}$ converges to the exact coefficient $J(\psi) \equiv \kappa$ from above.

Before we close the discussion on the derivation of the CVPs, it should be noted that we have considered a common set of eigenfunctions for the operators P, A and L. This is equivalent to assuming that the operators P, A and L are mutually commutative. Mikhail (1985) has shown that it is possible to derive the sequences $\{\kappa_m^{<}\}$ and $\{\kappa_n^{>}\}$ without supposing a commutative nature for the two parts of the operator P.

In the evaluation of lower bounds $\{\kappa_m^{<}\}$ the operator A can either be considered as $\lambda\hat{I}$ (Benin 1970) with λ as the biggest eigenvalue (or the maximum of row sums) of the phonon collision operator P, or as Γ, the diagonal part of P (Srivastava 1976c). In the evaluation of upper bounds $\kappa_n^{>}$ the operator L can be considered as $L = P - \frac{1}{2}\mu\hat{I}$, with μ as the smallest eigenvalue (or the minimum of row sums) of the operator P (Srivastava 1975, 1976b).

Thus we conclude that the method of CVPs provides sequences of lower and upper bounds which bound the exact coefficient $J(\psi) \equiv \kappa$ from below and above, respectively. Calculation of a pair of bounds $\kappa_m^{<}$ and $\kappa_n^{>}$ will provide a width $\Delta_{mn} = \kappa_n^{>} - \kappa_m^{<}$ around the exact coefficient κ. The width Δ_{mn} can be made desirably small by making appropriate choice for m and n in the sequences of lower and upper bounds. This feature increases the applicability of the method of CVPs not only for the problem of lattice thermal conductivity but for any linear equation which can be expressed in the form of equation (5.18).

5.3.6   IMPROVEMENT OF VARIATIONAL BOUNDS BY SCALING AND RITZ PROCEDURES

In the preceding subsections, we have derived sequences of CVPS for the conductivity coefficient κ. Both unnormalized as well as normalized bounds were derived: $\kappa_m^<(\boldsymbol{\phi})$ and $\kappa_n^>(\boldsymbol{\phi})$ denoted unnormalized bounds and $\kappa_m^<$ and and $\kappa_n^>$ denoted normalized (or scaled) bounds.

In this section we first prove that normalized (or scaled) bounds are more convenient and improved (more accurate) results than unnormalized bounds. This is easily seen from the inequality

$$\kappa_0^<(\boldsymbol{\phi}) \equiv 2(\boldsymbol{\phi},X) - (\boldsymbol{\phi},P\boldsymbol{\phi}) \leq \frac{(\boldsymbol{\phi},X)^2}{(\boldsymbol{\phi},P\boldsymbol{\phi})} \equiv \kappa_0^< \tag{5.178}$$

i.e. $\kappa_0^< \geq \kappa_0^<(\boldsymbol{\phi})$, where $\kappa_0^<$ is the optimized version of $\kappa_0^<(\alpha\boldsymbol{\phi})$ with respect to the one-parameter α (a scale factor). This is what was done in section 5.3.4. The scaling procedure is common and is widely used in variational calculations (Robinson and Arthurs 1968, Pomraning 1967, Benin 1970, Jensen *et al* 1969).

Next we point out that the accuracy of improvement can be increased by adopting a Ritz procedure. Robinson and Arthurs (1968) and Arthurs (1970) have discussed the method of Rayleigh and Ritz. In this, one maximises $\kappa_m^<(\boldsymbol{\phi})$ or minimises $\kappa_n^>(\boldsymbol{\phi})$, not for the complete set of admissible functions in the domain D_P of P, but by using a smaller set of functions in D_R, a R-dimensional space of a linear combination of R independent admissible functions $\boldsymbol{\phi}_1, \boldsymbol{\phi}_2, \boldsymbol{\phi}_3, ..., \boldsymbol{\phi}_R$. Thus we write

$$(\boldsymbol{\psi},X) = \max_{\boldsymbol{\phi} \in D_P} \kappa_m^>(\boldsymbol{\phi}) \geq \max_{\boldsymbol{\phi} \in D_R} \kappa_m^<(\boldsymbol{\phi}) \tag{5.179}$$

and

$$(\boldsymbol{\psi},X) = \min_{\boldsymbol{\phi} \in D_P} \kappa_n^>(\boldsymbol{\phi}) \leq \min_{\boldsymbol{\phi} \in D_R} \kappa_n^>(\boldsymbol{\phi}). \tag{5.180}$$

For selected $\boldsymbol{\phi}$ optimised complementary bounds on $J(\boldsymbol{\psi}) \equiv (\boldsymbol{\psi},X)$ are obtained by evaluating the right-hand sides of equations (5.163) and (5.164).

To see explicitly how these optimised bounds are derived, we note that a selected $\boldsymbol{\phi}$ in the space D_R can be written as

$$\boldsymbol{\phi} = \sum_{r=1}^{R} \alpha_r \boldsymbol{\phi}_r, \tag{5.181}$$

where $\{\alpha_r\}$ are now R real constants (rather than a single constant as we considered earlier in scaling $\kappa_m^<(\boldsymbol{\phi})$ and $\kappa_n^>(\boldsymbol{\phi})$). The constants $\{\alpha_r\}$ are determined by maximising $\kappa_m^<(\sum_r \alpha_r \boldsymbol{\phi}_r)$, i.e. by solving

$$\frac{\partial \kappa_m^<(\sum_i \alpha_r \boldsymbol{\phi}_r)}{\partial \alpha_{r'}} = 0 \qquad r' = 1,2,...,R. \tag{5.182}$$

Similarly, we can write

$$\boldsymbol{\phi} = \sum_{r=1}^{R} \beta_r \boldsymbol{\phi}_r \tag{5.183}$$

and determine $\{\beta_r\}$ by minimising $\kappa_n^>(\sum_r \beta_r \boldsymbol{\phi}_r)$, i.e. by solving

$$\frac{\partial \kappa_n^>(\sum_r \beta_r \boldsymbol{\phi}_r)}{\partial \beta_{r'}} = 0 \qquad r' = 1,2,...,R. \tag{5.184}$$

With the optimum $\{\tilde{\alpha}_r\}$ and $\{\tilde{\beta}_r\}$, the optimum functionals thus obtained, say $[\kappa_m^<]_R$ and $[\kappa_n^>]_R$, are the improved bound results. Obviously $R = 1$ is the case of scaling by one-parameters α and β.

Let us explicitly write the optimised functional $[\kappa_0^<]$. We have

$$\kappa_0^< \left(\sum_r \alpha_r \boldsymbol{\phi}_r \right) = 2\sum_r \alpha_r (\boldsymbol{\phi}_r,X) - \sum_{rr'} \alpha_r \alpha_{r'} (\boldsymbol{\phi}_r,P\boldsymbol{\phi}_{r'})$$

$$= 2\sum_r \alpha_r X_r - \sum_{rr'} \alpha_r P_{rr'} \alpha_{r'}, \tag{5.185}$$

where X_r and $P_{rr'}$ are defined by

$$X_r = (\boldsymbol{\phi}_r, \boldsymbol{X}) \tag{5.186}$$

$$P_{rr'} = P_{r'r} = (\boldsymbol{\phi}_r, P\boldsymbol{\phi}_{r'}) \tag{5.187}$$

P being symmetric. From here we can write

$$\frac{\partial \kappa_0^<(\sum_r \alpha_r \boldsymbol{\phi}_r)}{\partial \alpha_{r'}} = 0 = 2X_{r'} - 2\sum_r \alpha_r P_{rr'}$$

or

$$X_{r'} = \sum_r \alpha_r P_{rr'} \tag{5.188}$$

and hence

$$\alpha_r \to \tilde{\alpha}_r = \sum_{r'} (P^{-1})_{rr'} X_{r'}. \tag{5.189}$$

With this optimum $\tilde{\alpha}_r$, the functional $\kappa_0^<(\boldsymbol{\phi})$ becomes the optimised version of the Ziman bound

$$[\kappa_0^<]_R = \sum_{rr'}^R X_r (P^{-1})_{rr'} X_{r'}. \tag{5.190}$$

Hamilton and Parrott (1969) used the method of Lagrange multipliers to derive this expression. In a similar way we can obtain all the optimised bounds $[\kappa_m^<]_R$ and $[\kappa_n^>]_R$.

An alternative, but equivalent, procedure for deriving the optimised bounds $[\kappa_m^<]_R$ and $[\kappa_n^>]_R$ is to follow the procedure described in section 5.3.5. The extremum condition for the left-hand side of

$$\left(\left(\boldsymbol{\psi} - \sum_r \alpha_r \boldsymbol{\phi}_r\right), P_j\left(\boldsymbol{\psi} - \sum_r \alpha_r \boldsymbol{\phi}_r\right)\right) \geq 0 \tag{5.191}$$

will directly result into equation (5.189) for the optimum $\tilde{\alpha}_r$. With that the left-hand side in equation (5.175) will yield the desired expressions for $[\kappa_m^>]_R$ and $[\kappa_n^>]_R$.

In equation (5.178) we proved that the scaled result $\kappa_0^<$ is an improved version of the unscaled result $\kappa_0^<(\boldsymbol{\phi})$. Now we show that the Ritz procedure yields a more improved result. For simplicity take $R = 1$. Then from equation (5.190)

$$[\kappa_0^<]_1 = X_1 (P^{-1})_{11} X_1 \tag{5.192}$$

while from equation (5.162) we have

$$\kappa_0^< = \frac{(\boldsymbol{\phi}, \boldsymbol{X})}{(\boldsymbol{\phi}, P\boldsymbol{\phi})} \equiv \frac{X_{11}^2}{P_{11}}. \tag{5.193}$$

From the Schwarz inequality we have

$$X_{11}(P^{-1})_{11} X_{11} \geq \frac{X_{11}^2}{P_{11}}$$

so that $[\kappa_0^<]_1 \geq \kappa_0^<$, which proves the point that a more general trial function yields a better variational result.

Details of numerical calculations of some bound results for the lattice thermal conductivity will be presented in chapter 7. More details can be found in Srivastava and Hamilton (1978). Before closing this section, it should be mentioned that the lattice thermal conductivity is related to the functional $J(\boldsymbol{\psi})$ (or, equivalently the coefficient κ) through the relation (see equation (5.23))

$$\mathcal{K} = \frac{k_B T^2}{N_0 \Omega |\nabla T|^2} J(\boldsymbol{\psi}). \tag{5.194}$$

5.3.7 THERMODYNAMIC INTERPRETATION OF THE VARIATIONAL PRINCIPLE

The mathematical description of the variational principle given in this section can be given a physical interpretation on the basis of steady-state thermodynamics. We will do this by expressing the variational integrals $(\boldsymbol{\psi}, X)$ and $(\boldsymbol{\psi}, P\boldsymbol{\psi})$ in terms of the rate of change of the entropy of the system.

We can associate with a heat flux $\boldsymbol{Q}$ an entropy flux $\boldsymbol{Q}/T$. The rate of change of the entropy due to the presence of a temperature gradient is given by

$$
\begin{aligned}
\dot{S}_{macro} &= \nabla \cdot \left(\frac{\boldsymbol{Q}}{T}\right) N_0\Omega \\
&= N_0\Omega \boldsymbol{Q} \cdot \nabla\left(\frac{1}{T}\right) \\
&= -\frac{N_0\Omega}{T^2}\boldsymbol{Q} \cdot \nabla T.
\end{aligned}
\tag{5.195}
$$

Using equation (5.20), we can express

$$
\begin{aligned}
\dot{S}_{macro} &= k_B \sum_{qs} \psi_{qs} \frac{\partial \bar{n}_{qs}}{\partial t}\bigg|_{diff} \\
&= k_B(\boldsymbol{\psi}, X).
\end{aligned}
\tag{5.196}
$$

Phonon scattering processes increase the entropy of the system, with the rate given by (Ziman 1960)

$$
\begin{aligned}
\dot{S}_{scatt} &\simeq -k_B \sum_{qs} \psi_{qs} \frac{\partial n_{qs}}{\partial t}\bigg|_{scatt} \\
&= k_B(\boldsymbol{\psi}, P\boldsymbol{\psi}).
\end{aligned}
\tag{5.197}
$$

Since $(\boldsymbol{\psi}, P\boldsymbol{\psi}) \geq 0$, $\dot{S}_{scatt}$ always tends to increase. In the steady state the increase in the entropy due to the microscopic scattering processes is balanced by the dissipation of existing temperature gradients

$$
\dot{S}_{macro} = -\dot{S}_{scatt}.
\tag{5.198}
$$

This balance relation is similar to that in equation (5.18) for the rate of change of the distribution function.

In the simplest form of the variational principle, we proved that the quantity $(\boldsymbol{\phi}, X)^2/(\boldsymbol{\phi}, P\boldsymbol{\phi})$ provides a minimum estimate for the coefficient $(\boldsymbol{\psi}, X)^2/(\boldsymbol{\psi}, P\boldsymbol{\psi}) \equiv (\boldsymbol{\psi}, X)$, where $\boldsymbol{\phi}$ is a trial function for the generally unknown function $\boldsymbol{\psi}$. In the language of the steady-state thermodynamics we then can say that the ratio $k_B(\boldsymbol{\phi}, X)^2/(\boldsymbol{\phi}, P\boldsymbol{\phi})$ provides a lower bound for the entropy production ratio $(\dot{S}_{macro})^2/\dot{S}_{scatt}$.

5.4 GREEN–KUBO LINEAR-RESPONSE THEORY

In this section, we derive an expression for lattice thermal conductivity based on the Green–Kubo linear-response theory. The Kubo linear-response formula for phonon conductivity is (Kubo 1957)

$$
\mathscr{K} = \frac{k_B T^2 N_0\Omega}{3}\Re\int_0^\infty \langle \boldsymbol{Q}(0) \cdot \boldsymbol{Q}(t)\rangle \, dt.
\tag{5.199}
$$

Here $\boldsymbol{Q}(t)$ is the heat current operator in the Heisenberg representation, and $\langle \ldots \rangle$ represents the canonical-ensemble average with respect to the total Hamiltonian H:

$$
\langle O \rangle = \frac{\mathrm{Tr}(e^{-\beta H}O)}{\mathrm{Tr}(e^{-\beta H})} \qquad \beta = 1/k_B T.
\tag{5.200}
$$

From equations (5.13) and (5.20), we write

$$Q(t) = \frac{1}{N_0\Omega} \sum_{qs} \hbar\omega(qs)\delta n_{qs}(t)c_s(q).$$ (5.201)

In writing this expression, we have recognised that it is the deviation function $\delta n_{qs} = n_{qs} - \bar{n}_{qs}$ which is responsible for heat current. Also, the distribution function n_{qs} is now regarded as the number-density operator for phonons in mode qs in the Heisenberg representation:

$$n_{qs}(t) = a^\dagger_{qs}(t)a_{qs}(t).$$ (5.202)

With this, equation (5.199) becomes

$$\mathscr{K} = \frac{\hbar^2 k_B \beta^2}{3N_0\Omega}\Re\int_0^\infty dt \sum_{qsq's'} \omega(qs)\omega(q's')c_s(q)\cdot c_{s'}(q')\mathscr{C}_{qsq's'}(t),$$ (5.203)

where

$$\mathscr{G}(t) = \langle\delta n_{qs}(0)\delta n_{q's'}(t)\rangle = \mathscr{C}_{qsq's'}(t)$$ (5.204)

is a *correlation function*. Calculation of $\mathscr{G}(t)$ can be made by several techniques, such as

   (i) the Zwanzig–Mori projection operator approach,
   (ii) double-time Green's function technique, and
   (iii) imaginary-time Green's function technique.
Here we will discuss the first two of these techniques.

5.4.1 EVALUATION OF THE CORRELATION FUNCTION $\mathscr{G}(T)$ BY THE ZWANZIG–MORI PROJECTION OPERATOR METHOD

5.4.1.1 Derivation of an integro-differential equation for $\mathscr{G}(t)$

We describe here the evaluation of $\mathscr{G}(t)$ by means of the Zwanzig–Mori projection method (see, e.g., Zwangig (1961), Mori (1965), Jancel (1969), Wilson and Kim (1973), Knauss and Wilson (1974) for a detailed description of the method). Let us define a projection operator $\mathscr{P}$ which picks out that part of $\delta n_{qs}(t)$ which contributes to $\mathscr{G}(t)$, i.e.

$$\langle\delta n_{qs}(0)\mathscr{P}\delta n_{q's'}(t)\rangle = \langle\delta n_{qs}(0)\delta n_{q's'}(t)\rangle = \mathscr{G}(t).$$ (5.205)

In doing so we consider that the operations on $\delta n_{qs}(t)$ by $\mathscr{P}$ and $(1-\mathscr{P})$ yield, respectively, the relevant and irrelevant parts of $\delta n_{qs}(t)$:

$$\delta n_{qs}(t) = \delta n_{qs}(t)|_r + \delta n_{qs}(t)|_{irr},$$ (5.206)

where

$$\delta n_{qs}(t)|_r = \mathscr{P}\delta n_{qs}(t)$$ (5.207)
$$\delta n_{qs}(t)|_{irr} = (1-\mathscr{P})\delta n_{qs}(t).$$ (5.208)

Let us introduce $\mathscr{P}$ as

$$\mathscr{P}O(t) = \delta n\frac{\langle\delta n O(t)\rangle}{\langle\delta n\delta n\rangle} \equiv \delta n\langle\langle\delta n O(t)\rangle\rangle$$ (5.209)

where δn stands for $\delta n(0)$. It is easy to prove that $\mathscr{P}$ thus introduced is a projection operator:

$$\mathscr{P}^2 O(t) = \mathscr{P}\delta n\frac{\langle\delta n O(t)\rangle}{\langle\delta n\delta n\rangle}$$

$$
\begin{aligned}
&= \delta n \frac{\langle \delta n \delta n \langle \delta n O(t) \rangle / \langle \delta n \delta n \rangle \rangle}{\langle \delta n \delta n \rangle} \\
&= \delta n \frac{\langle \delta n O(t) \rangle}{\langle \delta n \delta n \rangle} \\
&= \mathscr{P} O(t)
\end{aligned}
\tag{5.210}
$$

i.e. $\mathscr{P}^2 = \mathscr{P}$, thus proving the assertion.

To derive the equation of motion for $\mathscr{G}(t)$, we use Neumann's equation for $\delta n_{qs}(t)$, given by

$$
\frac{\mathrm{d}}{\mathrm{d}t} \delta n_{qs}(t) = \frac{\mathrm{i}}{\hbar}[H, \delta n_{qs}] = \mathrm{i}\mathscr{L} \delta n_{qs}(t).
\tag{5.211}
$$

Here $\mathscr{L}$ is the Liouville operator. In Zwanzig's method we look for an equation of evolution for the 'relevant part' $\delta n_{qs}|_r$ only. To eliminate the 'irrelevant part' $\delta n_{qs}(t)|_{irr}$, we calculate the time derivative of $\delta n_{qs}(t)|_r$ and $\delta n_{qs}(t)|_{irr}$:

$$
\begin{aligned}
\frac{\mathrm{d}}{\mathrm{d}t} \delta n_{qs}(t)\Big|_r &= \frac{\mathrm{d}}{\mathrm{d}t} \mathscr{P} \delta n_{qs}(t) \\
&= \mathrm{i}\mathscr{P}\mathscr{L}\delta n_{qs}(t)|_r + \mathrm{i}\mathscr{P}\mathscr{L}\delta n_{qs}(t)|_{irr} \\
&\quad - \mathrm{i}\mathscr{P}\mathscr{L}\delta n_{qs}(t)|_r + \mathrm{i}\mathscr{P}\mathscr{L}(1-\mathscr{P})\delta n_{qs}(t)|_{irr}
\end{aligned}
\tag{5.212}
$$

and

$$
\begin{aligned}
\frac{\mathrm{d}}{\mathrm{d}t} \delta n_{qs}(t)\Big|_{irr} &= \frac{\mathrm{d}}{\mathrm{d}t}(1-\mathscr{P})\delta n_{qs}(t) \\
&= \mathrm{i}(1-\mathscr{P})\mathscr{L}\delta n_{qs}(t)|_r + \mathrm{i}(1-\mathscr{P})\mathscr{L}\delta n_{qs}(t)|_{irr} \\
&= \mathrm{i}(1-\mathscr{P})\mathscr{L}\delta n_{qs}(t)|_r + \mathrm{i}(1-\mathscr{P})\mathscr{L}(1-\mathscr{P})\delta n_{qs}(t)|_{irr}
\end{aligned}
\tag{5.213}
$$

where we have used equations (5.206)–(5.208) and (5.211).

We first write the formal solution of equation (5.213) by considering it as a linear equation in $\delta n_{qs}(t)|_{irr}$ with the inhomogeneous term $(1-\mathscr{P})\mathscr{L}\delta n_{qs}(t)|_r$:

(i) *homogeneous solution*:

$$
\begin{aligned}
\frac{\mathrm{d}}{\mathrm{d}t} \delta n_{qs}(t)\Big|_{irr} &= \mathrm{i}(1-\mathscr{P})\mathscr{L}\delta n_{qs}(t)|_{irr} \\
\delta n_{qs}(t)\Big|_{irr}^{homog} &= \delta n_{qs}(0)\exp[\mathrm{i}(1-\mathscr{P})\mathscr{L}t].
\end{aligned}
\tag{5.214}
$$

(ii) *inhomogeneous solution*:

$$
\delta n_{qs}(t)|_{irr}^{inhomog} = \mathrm{i}\int_0^t \mathrm{d}t' \exp[\mathrm{i}(1-\mathscr{P})\mathscr{L}t'](1-\mathscr{P})\mathscr{L}\delta n_{qs}(t-t')|_r.
\tag{5.215}
$$

The homogeneous solution can be discarded by the 'molecular chaos hypothesis' $\delta n_{qs}|_{irr} = 0$. Therefore we have

$$
\delta n_{qs}(t)|_{irr} = \mathrm{i}\int_0^t \mathrm{d}t' \exp[\mathrm{i}(1-\mathscr{P})\mathscr{L}t'](1-\mathscr{P})\mathscr{L}\delta n_{qs}(t-t')|_r.
\tag{5.216}
$$

Substituting this result into equation (5.196) we get the equation of motion of the relevant part of δn_{qs}

$$
\frac{\mathrm{d}}{\mathrm{d}t} \delta n_{qs}(t)\Big|_r = \mathrm{i}\mathscr{P}\mathscr{L}\delta n_{qs}(t)|_r - \mathscr{P}\mathscr{L}\int_0^t \mathrm{d}t' \mathrm{e}^{[\mathrm{i}(1-\mathscr{P})\mathscr{L}t']}(1-\mathscr{P})\mathscr{L}\delta n_{qs}(t-t')|_r.
\tag{5.217}
$$

This form of Zwanzig's equation is equivalent to Liouville's equation.

Using Zwanzig's equation, we finally have the equation of motion of $\mathscr{G}(t)$

$$
\begin{aligned}
\frac{\mathrm{d}}{\mathrm{d}t}\mathscr{G}(t) &= \left\langle \delta n_{qs}\frac{\mathrm{d}}{\mathrm{d}t}\delta n_{q's'}(t)\right\rangle \\
&\equiv \left\langle \delta n_{qs}\frac{\mathrm{d}}{\mathrm{d}t}\delta n_{q's'}(t)\Big|_r\right\rangle \\
&= \mathrm{i}\langle \delta n_{qs}\mathscr{P}\mathscr{L}\delta n_{q's'}(t)|_r\rangle \\
&\quad - \int_0^t \mathrm{d}t'\langle \delta n_{qs}\mathscr{P}\mathscr{L}\exp[\mathrm{i}(1-\mathscr{P}\mathscr{L}t'](1-\mathscr{P})\mathscr{L}\delta n_{q's'}(t-t')|_r\rangle.
\end{aligned}
$$

(5.218)

Now from equation (5.209)

$$
\begin{aligned}
\langle \delta n_{qs}\mathscr{P}\mathscr{L}\delta n_{q's'}(t)|_r\rangle &= \langle \delta n_{qs}\delta n_{qs}\langle\langle \delta n_{qs}\mathscr{L}\delta n_{q's'}(t)|_r\rangle\rangle\rangle \\
&= \delta n_{qs}\delta n_{qs}\langle\langle \delta n_{qs}\mathscr{L}\mathscr{P}\delta n_{q's'}(t)|_r\rangle\rangle\rangle \\
&= \delta n_{qs}\delta n_{qs}\langle\langle \delta n_{qs}\mathscr{L}\delta n_{qs}\frac{\langle \delta n_{qs}\delta n_{q's'}(t)|_r\rangle}{\langle \delta n_{qs}\delta n_{qs}\rangle}\rangle\rangle \\
&\simeq \langle \delta n_{qs}\delta n_{q's'}(t)|_r\rangle\langle\langle \delta n_{qs}\mathscr{L}\delta n_{qs}\rangle\rangle \\
&= g\mathscr{G}(t)
\end{aligned}
$$

(5.219)

where the so-called *natural frequency* is given by

$$
g = \langle\langle \delta n_{qs}\mathscr{L}\delta n_{qs}\rangle\rangle.
$$

(5.220)

Also, from equation (5.209)

$$
\begin{aligned}
\langle \delta n_{qs}\mathscr{P}\mathscr{L}&\exp[\mathrm{i}(1-\mathscr{P})\mathscr{L}t'](1-\mathscr{P})\mathscr{L}\delta n_{q's'}(t-t')|_r\rangle \\
&= \langle \delta n_{qs}\delta n_{qs}\langle\langle \delta n_{qs}\mathscr{L}\exp[\mathrm{i}(1-\mathscr{P})\mathscr{L}t'](1-\mathscr{P})\mathscr{L}\delta n_{q's'}(t-t')|_r\rangle\rangle\rangle \\
&\simeq \langle \delta n_{qs}n_{q's'}(t-t')|_r\rangle\langle\langle \delta n_{qs}\mathscr{L}\exp[\mathrm{i}(1-\mathscr{P})\mathscr{L}t'](1-\mathscr{P})\mathscr{L}\delta n_{qs}\rangle\rangle \\
&= \mathscr{G}(t-t')f(t'),
\end{aligned}
$$

(5.221)

where

$$
f(t') = \langle\langle \delta n_{qs}\mathscr{L}\exp[\mathrm{i}(1-\mathscr{P})\mathscr{L}t'](1-\mathscr{P})\mathscr{L}\delta n_{qs}\rangle\rangle
$$

(5.222)

is called the *memory function*.

Zwanzig's equation of motion for $\mathscr{G}$, therefore, is

$$
\frac{\mathrm{d}}{\mathrm{d}t}\mathscr{G}(t) = \mathrm{i}g\mathscr{G}(t) - \int_0^t dt'\mathscr{G}(t-t')f(t').
$$

(5.223)

We must now calculate the natural frequency g and the memory function f.

5.4.1.2 Evaluation of $f(t)$ and g

We first write the Hamiltonian in the linear-coupling form

$$
H = H_0 + \lambda H',
$$

(5.224)

where H_0 is the harmonic or unperturbed part given by equation (4.37) and H' is the anharmonic part or perturbation given by equations (4.43) and (4.44). For simplicity we express H' in the form

$$
H' = \sum_{qs} S_{qs}(a_{qs}^\dagger - a_{qs})
$$

(5.225)

so that S_{qs} could be regarded as the linear coupling coefficient of the phonon mode qs to its surrounding or bath (namely the crystal). In equation (5.224) λ is a measure of anharmonicity, assumed to be small.

With equation (5.224), we have

$$
\begin{aligned}
\mathscr{L}\delta n_{qs} &= \frac{1}{\hbar}[H, \delta n_{qs}] \\
&= \frac{\lambda}{\hbar}[H', \delta n_{qs}] \\
&= \lambda\mathscr{L}'\delta n_{qs},
\end{aligned}
\tag{5.226}
$$

where

$$
\mathscr{L} = \mathscr{L}_0 + \lambda\mathscr{L}' \qquad \text{and} \qquad \mathscr{L}_0\delta n_{qs} = 0.
\tag{5.227}
$$

Therefore,

$$
\begin{aligned}
f(t') &= \lambda^2\langle\langle n_{qs}\mathscr{L}'\exp[i(1-\mathscr{P})\mathscr{L}t'](1-\mathscr{P})\mathscr{L}'n_{qs}\rangle\rangle \\
&= -\lambda^2\langle\langle(\mathscr{L}'n_{qs})\exp[i(1-\mathscr{P})\mathscr{L}t'](1-\mathscr{P})(\mathscr{L}'n_{qs})\rangle\rangle
\end{aligned}
\tag{5.228}
$$

as $\langle\langle A\mathscr{L}B\rangle\rangle = -\langle\langle(\mathscr{L}A)B\rangle\rangle$. It is sufficient to evaluate $f(t)$ up to λ^2 only. Therefore, we replace $\mathscr{P}$ by $\mathscr{P}_0$ and $\mathscr{L}$ by $\mathscr{L}_0$ in equation (5.228). This gives

$$
f(t') = -\lambda^2\langle\langle(\mathscr{L}'n_{qs})\exp[i(1-\mathscr{P}_0)\mathscr{L}_0t'](1-\mathscr{P}_0)(\mathscr{L}'n_{qs})\rangle\rangle_0,
\tag{5.229}
$$

where

$$
\begin{aligned}
\mathscr{P}_0 O &= \delta n_{qs}\langle\langle\delta n_{qs}O\rangle\rangle_0 \\
\langle O\rangle_0 &= \frac{\text{Tr}(e^{-\beta H_0}O)}{\text{Tr}(e^{-\beta H_0})}.
\end{aligned}
\tag{5.230}
$$

Next we write, using $\mathscr{P}_0(1-\mathscr{P}_0) = 0$,

$$
\begin{aligned}
\exp[i(1-\mathscr{P}_0)\mathscr{L}_0t'](1-\mathscr{P}_0)O &= \exp[i\mathscr{L}_0(1-\mathscr{P}_0)t'](1-\mathscr{P}_0)O \\
&\simeq e^{i\mathscr{L}_0t'}(1-i\mathscr{L}_0\mathscr{P}_0t')(1-\mathscr{P}_0)O \\
&= e^{i\mathscr{L}_0t'}(1-\mathscr{P}_0)O.
\end{aligned}
\tag{5.231}
$$

Therefore, from equations (5.209) and (5.229)–(5.231)

$$
\begin{aligned}
f(t') &= \frac{-\lambda^2}{\langle\delta n_{qs}\delta n_{qs}\rangle_0}\{\langle(\mathscr{L}'n_{qs})e^{i\mathscr{L}_0t'}\mathscr{P}_0(\mathscr{L}'n_{qs})\rangle_0 \\
&\quad -\langle(\mathscr{L}'n_{qs})e^{i\mathscr{L}_0t'}(\mathscr{L}'n_{qs})\rangle_0\}.
\end{aligned}
\tag{5.232}
$$

Now

$$
\begin{aligned}
\langle\delta n_{qs}\delta n_{qs}\rangle_0 &= \langle n_{qs}n_{qs}\rangle_0 - \bar{n}_{qs}^2 \\
&= \langle a_{qs}^\dagger a_{qs}a_{qs}^\dagger a_{qs}\rangle_0 - \bar{n}_{qs}^2 \\
&= \langle a_{qs}^\dagger a_{qs}\rangle_0\langle a_{qs}^\dagger a_{qs}\rangle_0 + \langle a_{qs}^\dagger a_{qs}^\dagger\rangle_0\langle a_{qs}a_{qs}\rangle_0 \\
&\quad + \langle a_{qs}^\dagger a_{qs}\rangle_0\langle a_{qs}a_{qs}^\dagger\rangle_0 - \bar{n}_{qs}^2 \\
&= \bar{n}_{qs}^2 + 0 + \bar{n}_{qs}(\bar{n}_{qs}+1) - \bar{n}_{qs}^2 \\
&= \bar{n}_{qs}(\bar{n}_{qs}+1),
\end{aligned}
\tag{5.233}
$$

where we have used Wick's factorisation scheme (Wick 1950), namely

$$\langle abcd \rangle = \langle ab \rangle \langle cd \rangle + \langle ac \rangle \langle bd \rangle + \langle ad \rangle \langle bc \rangle \tag{5.234}$$

and

$$\langle a_{qs}^\dagger a_{qs} \rangle_0 = \bar{n}_{qs} \qquad \langle a_{qs} a_{qs}^\dagger \rangle_0 = \bar{n}_{qs} + 1$$
$$\bar{n}_{qs} = (e^{\beta\hbar\omega} - 1)^{-1}. \tag{5.235}$$

Thus we have

$$
\begin{aligned}
f(t') &= \frac{-\lambda^2}{\bar{n}_{qs}(\bar{n}_{qs}+1)} \left(\langle (\mathscr{L}' n_{qs}) e^{i\mathscr{L}_0 t'} \mathscr{P}_0 (\mathscr{L}' n_{qs}) \rangle_0 \right. \\
&\quad \left. - \langle (\mathscr{L}' n_{qs}) e^{i\mathscr{L}_0 t'} (\mathscr{L}' n_{qs}) \rangle_0 \right).
\end{aligned} \tag{5.236}
$$

Let us consider the following in the first term:

$$
\begin{aligned}
\mathscr{P}_0 \mathscr{L}' n_{qs} &= n_{qs} \langle\langle n_{qs} \mathscr{L}' n_{qs} \rangle\rangle_0 \\
&= \frac{1}{\hbar} n_{qs} \langle\langle n_{qs} S_{qs} [a_{qs}^\dagger - a_{qs}, a_{qs}^\dagger a_{qs}] \rangle\rangle_0 \\
&= \frac{1}{\hbar} n_{qs} \langle\langle (a_{qs}^\dagger a_{qs} - \bar{n}_{qs}) S_{qs} [a_{qs}^\dagger - a_{qs}, a_{qs}^\dagger a_{qs}] \rangle\rangle_0 \\
&= 0
\end{aligned} \tag{5.237}
$$

as $\langle a_{qs}^\dagger a_{qs} a_{qs} a_{qs}^\dagger a_{qs} \rangle_0 = 0$, $\langle a_{qs} a_{qs}^\dagger a_{qs} \rangle_0 = 0$, etc. This reduces equation (5.236) to

$$
\begin{aligned}
f(t') &= \frac{\lambda^2}{\bar{n}_{qs}(\bar{n}_{qs}+1)} \langle (\mathscr{L}' n_{qs}) e^{i\mathscr{L}_0 t'} (\mathscr{L}' n_{qs}) \rangle_0 \\
&= \frac{\lambda^2}{\bar{n}_{qs}(\bar{n}_{qs}+1)} \langle S_{qs}[a_{qs}^\dagger - a_{qs}, n_{qs}] e^{i\mathscr{L}_0 t'} S_{qs}[a_{qs}^\dagger - a_{qs}, n_{qs}] \rangle_0 \\
&= \frac{\lambda^2}{\bar{n}_{qs}(\bar{n}_{qs}+1)} \langle S_{qs}[a_{qs}^\dagger - a_{qs}] S_{qs}(t')[a_{qs}^\dagger(t') - a_{qs}(t'), n_{qs}(t')] \rangle_0,
\end{aligned}
$$

where $e^{i\mathscr{L}t'} A = A(t')$ is used. We can further simplify terms so that

$$
\begin{aligned}
f(t') &= \frac{\lambda^2}{\hbar^2 \bar{n}_{qs}(\bar{n}_{qs}+1)} \langle S_{qs} S_{qs}(t') \rangle_0 \langle [a_{qs}^\dagger - a_{qs}, n_{qs}][a_{qs}^\dagger(t') - a_{qs}(t'), n_{qs}(t')] \rangle_0 \\
&= \frac{\lambda^2}{\hbar^2 \bar{n}_{qs}(\bar{n}_{qs}+1)} \langle S_{qs} S_{qs}(t') \rangle_0 \{ \langle a_{qs}^\dagger a_{qs}(t') \rangle_0 + \langle a_{qs} a_{qs}^\dagger(t') \rangle_0 \} \\
&= \frac{\lambda^2}{\hbar^2 \bar{n}_{qs}(\bar{n}_{qs}+1)} \langle S_{qs} S_{qs}(t') \rangle_0 \{ \bar{n}_{qs} e^{-i\omega(qs)t'} + (\bar{n}_{qs}+1) e^{i\omega(qs)t'} \},
\end{aligned} \tag{5.238}
$$

where we have used the following relations:

$$
\begin{aligned}
[a_{qs}, a_{qs}^\dagger] &= 1 \\
a_{qs}(t) &= a_{qs} e^{-i\omega(qs)t} \\
a_{qs}^\dagger(t) &= a_{qs}^\dagger e^{i\omega(qs)t}.
\end{aligned} \tag{5.239}
$$

Furthermore, using $\bar{n}_{qs} + 1 = \bar{n}_{qs} \exp(\beta\hbar\omega(q s))$ we can finally express

$$f(t') = \frac{\lambda^2}{\hbar^2 \bar{n}_{qs}} \langle S_{qs} S_{qs}(t') \rangle_0 \{ e^{i\omega(qs)t'} + e^{-i\omega(qs)t'} e^{-\beta\hbar\omega(qs)} \}. \tag{5.240}$$

We can similarly evaluate g:

$$
\begin{aligned}
g &= \langle \delta n_{qs} \mathscr{L} \delta n_{qs} \rangle / \langle \delta n_{qs} \delta n_{qs} \rangle \\
&= \lambda \langle n_{qs} \mathscr{L}' n_{qs} \rangle / \langle \delta n_{qs} \delta n_{qs} \rangle \\
&= \frac{\lambda}{\hbar} \langle n_{qs} [H', n_{qs}] \rangle / \langle \delta n_{qs} \delta n_{qs} \rangle \\
&= \frac{\lambda}{\hbar \langle \delta n_{qs} \delta n_{qs} \rangle} \frac{\mathrm{Tr}\{ \mathrm{e}^{-\beta H} a_{qs}^{\dagger} a_{qs} [H', a_{qs}^{\dagger} a_{qs}] \}}{\mathrm{Tr}(\mathrm{e}^{-\beta H})}.
\end{aligned}
\tag{5.241}
$$

To proceed we use the well-known perturbation expansion

$$
\mathrm{e}^{-\beta H} = \mathrm{e}^{-\beta H_0} \left(1 + \frac{\mathrm{i}\lambda}{\hbar} \int_0^{\mathrm{i}\hbar\beta} \mathrm{d}t\, H'(-t) + \dots \right).
\tag{5.242}
$$

The terms shown are sufficient to evaluate g up to λ^2. We can easily verify that the contribution to g from the term with λ vanishes, leaving

$$
\begin{aligned}
g &= \frac{\mathrm{i}\lambda^2}{\hbar^2 \langle \delta n_{qs} \delta n_{qs} \rangle} \\
&\quad \times \frac{\mathrm{Tr}\{ \mathrm{e}^{-\beta H_0} \int_0^{\mathrm{i}\hbar\beta} \mathrm{d}t\, S_{qs}(-t) \left(a_{qs}^{\dagger}(-t) - a_{qs}(-t) \right) a_{qs}^{\dagger} a_{qs} [H', a_{qs}^{\dagger} a_{qs}] \}}{\mathrm{Tr}(\mathrm{e}^{-\beta H_0})} \\
&= \frac{\mathrm{i}\lambda^2}{\hbar^2 \bar{n}_{qs}(\bar{n}_{qs}+1)} \\
&\quad \times \int_0^{\mathrm{i}\hbar\beta} \mathrm{d}t\, \langle S_{qs}(-t) S_{qs} \rangle_0 \langle \left(a_{qs}^{\dagger}(-t) - a_{qs}(-t) \right) a_{qs}^{\dagger} a_{qs} [a_{qs}^{\dagger} - a_{qs}, a_{qs}^{\dagger} a_{qs}] \rangle_0 \\
&= \frac{\mathrm{i}\lambda^2}{\hbar^2 \bar{n}_{qs}(\bar{n}_{qs}+1)} \\
&\quad \times \int_0^{\mathrm{i}\hbar\beta} \mathrm{d}t\, \langle S_{qs}(-t) S_{qs} \rangle_0 \{ \bar{n}_{qs}^2 \mathrm{e}^{-\mathrm{i}\omega(qs)t} - (\bar{n}_{qs}+1) \mathrm{e}^{\mathrm{i}\omega(qs)t} \},
\end{aligned}
\tag{5.243}
$$

where we have used Wick's factorisation scheme given in equation (5.234). The above result can be rewritten as

$$
\begin{aligned}
g &= \frac{\lambda^2}{\hbar^2 \bar{n}_{qs}(\bar{n}_{qs}+1)} \\
&\quad \times \int_0^{\beta} \mathrm{d}z \, \langle S_{qs} S_{qs}(\mathrm{i}\hbar z) \rangle_0 \{ (\bar{n}_{qs}+1)^2 \mathrm{e}^{-\hbar z\omega(qs)} - \bar{n}_{qs}^2 \mathrm{e}^{\hbar z\omega(qs)} \},
\end{aligned}
\tag{5.244}
$$

where $t = \mathrm{i}\hbar z$ and use is made of the cyclic invariance of trace

$$
\mathrm{Tr}(A_1 A_2 \dots A_m) = \mathrm{Tr}(A_2 \dots A_m A_1)
$$

so that

$$
\langle S_{qs}(t) \rangle = \langle S_{qs}(t-\beta) \rangle \qquad 0 < t < \beta.
\tag{5.245}
$$

Finally, we change β to t by using the identity

$$
\begin{aligned}
\mathrm{i}\hbar \int_0^{\beta} \mathrm{d}z \, & \exp(\pm\hbar z\omega(qs)) \langle S_{qs} S_{qs}(\mathrm{i}\hbar z) \rangle_0 \\
&= \int_0^{\infty} \exp(\mp\mathrm{i}\omega(qs)t) \langle S_{qs} S_{qs}(t) \rangle_0 - \exp(\pm\hbar z\omega(qs)) \\
&\quad \times \int_0^{\infty} \mathrm{d}t \exp(\mp\mathrm{i}\omega(qs)t) \langle S_{qs}(t) S_{qs} \rangle_0.
\end{aligned}
\tag{5.246}
$$

With this equation (5.244) becomes

$$
\begin{aligned}
g &= \frac{i\lambda^2}{\hbar^2 \bar{n}_{qs}(\bar{n}_{qs}+1)}\left[\bar{n}_{qs}^2 \int_0^\infty dt \exp(-i\omega(qs)t)\langle S_{qs}S_{qs}(t)\rangle_0 \right. \\
&\quad -(\bar{n}_{qs}+1)^2 \int_0^\infty dt \exp(i\omega(qs)t)\langle S_{qs}S_{qs}(t)\rangle_0 \\
&\quad +\bar{n}_{qs}(\bar{n}_{qs}+1)\left(\int_0^\infty dt\langle S_{qs}(t)S_{qs}\rangle_0 \exp(i\omega(qs)t)\right. \\
&\quad \left.\left. -\int_0^\infty dt \exp(-i\omega(qs)t)\langle S_{qs}(t)S_{qs}\rangle_0\right)\right].
\end{aligned}
\tag{5.247}
$$

5.4.1.3 Solution of Zwanzig's equation in the weak-coupling limit

In the van Hove weak-coupling limit, we let $\lambda \to 0$, $t \to \infty$, but $\lambda^2 t = y =$ finite. Zwanzig's equation in (5.223) then becomes

$$
\begin{aligned}
\lambda^2 \frac{d}{dy}\mathscr{G}\left(\frac{y}{\lambda^2}\right) &= \mathscr{G}\left(\frac{y}{\lambda^2}\right)\left(ig - \int_0^\infty dt f(t)\right) \\
&= \mathscr{G}\left(\frac{y}{\lambda^2}\right)\{ig - \hat{f}(0)\},
\end{aligned}
\tag{5.248}
$$

where

$$
\hat{f}(0) = \int_0^\infty dt\, f(t)e^{-igt}\Big|_{g=0}
\tag{5.249}
$$

is the Laplace transform of the memory function $f(t)$ at $g = 0$. Integrating equation (5.248) we get

$$
\ln\mathscr{G}\left(\frac{y}{\lambda^2}\right) = \frac{1}{\lambda^2}\{ig - \hat{f}(0)\}y + \text{constant}(= \ln\mathscr{G}(0))
$$

$$
\text{or}\qquad \mathscr{G}(t) = \mathscr{G}(0)e^{-t/\tau_{qs}},
\tag{5.250}
$$

where

$$
\tau_{qs}^{-1} = \frac{1}{\lambda^2}\{-ig + \hat{f}(0)\}
\tag{5.251}
$$

is the inverse relaxation time for phonons in mode qs. Also, note from equation (5.204) that

$$
\mathscr{G}(0) = C_{qsq's'}(0) = \langle \delta n_{qs}\delta n_{q's'}\rangle_0 = \delta_{qq'}\delta_{ss'}\bar{n}_{qs}(\bar{n}_{qs}+1)
\tag{5.252}
$$

so that

$$
C_{qsq's'}(t) = \delta_{qq'}\delta_{ss'}\bar{n}_{qs}(\bar{n}_{qs}+1)e^{-t/\tau_{qs}}
\tag{5.253}
$$

becomes the final solution for the correlation function. Furthermore, from equations (5.251), (5.238) and (5.247) we get, after some manipulation,

$$
\begin{aligned}
\tau_{qs}^{-1} &= \frac{1}{\hbar^2}\left(\int_0^\infty dt e^{i\omega(qs)t}\langle S_{qs}(t)S_{qs}\rangle_0 + \int_0^\infty dt\, e^{-i\omega(qs)t}\langle S_{qs}S_{qs}(t)\rangle_0\right) \\
&= \frac{1}{\hbar^2}\int_{-\infty}^\infty dt e^{i\omega(qs)t}\langle S_{qs}(t)S_{qs}\rangle_0 \\
&= \text{real and positive.}
\end{aligned}
\tag{5.254}
$$

In Chapter 6, we show that this result is identical to the *smrt* result in equation (5.24).

5.4.1.4 Conductivity expression

The expression for the conductivity in equation (5.203) now becomes

$$
\begin{aligned}
\mathscr{K} &= \frac{\hbar^2 k_B \beta^2}{3 N_0 \Omega} \sum_{q s q' s'} \omega(qs)\omega(q's')c_s(q)\cdot c_{s'}(q') \int_0^\infty dt\, C_{qsq's'}(0) e^{-t/\tau_{qs}} \\
&= \frac{\hbar^2 k_B \beta^2}{3 N_0 \Omega} \sum_{qs} \omega^2(qs) c_s^2(q) \bar{n}_{qs}(\bar{n}_{qs}+1) \int_0^\infty dt\, e^{-t/\tau_{qs}} \\
&= \frac{\hbar^2 k_B \beta^2}{3 N_0 \Omega} \sum_{qs} \omega^2(qs) c_s^2(q) \bar{n}_{qs}(\bar{n}_{qs}+1) \tau_{qs},
\end{aligned}
\tag{5.255}
$$

with τ_{qs} given in equation (5.254). This result for the conductivity is the Debye term or the 'smrt' result $\mathscr{K}_D$ (*i.e.* $\mathscr{K}_{\mathrm{smrt}}$) derived in section 5.2.1.

5.4.2 EVALUATION OF THE CORRELATION FUNCTION $\mathscr{G}(T)$ BY THE DOUBLE-TIME GREEN'S FUNCTION METHOD

In the previous subsection we evaluated the correlation function $\mathscr{G}(t)$ by using the Zwanzig–Mori projection operator method. In this subsection we evaluate this correlation function by using the double-time Green's function method. This technique is well explained by Zubarev (1960).

We first express $\mathscr{G}$ as

$$
\mathscr{G} = \frac{1}{\beta} \lim_{\varepsilon \to 0^+} \int_0^\infty dt\, e^{-\varepsilon t} \int_0^\beta d\beta' \langle \delta n_{qs} \delta n_{q's'}(t+i\hbar\beta') \rangle,
\tag{5.256}
$$

with $\langle ... \rangle$ as defined in equation (5.200). Here the factor $e^{-\varepsilon t}$ is introduced to deal with any divergence problem in the evaluation of $\mathscr{G}$ using the Green's function method. We next simplify $\mathscr{G}$ using the Wick's decoupling scheme, so that

$$
\begin{aligned}
\mathscr{G} &= \frac{1}{\beta} \lim_{\varepsilon \to 0^+} \int_0^\infty dt\, e^{-\varepsilon t} \int_0^\beta d\beta' \langle a_{qs}^\dagger a_{q's'}(t+i\hbar\beta') \rangle \langle a_{qs} a_{q's'}^\dagger(t+i\hbar\beta') \rangle \\
&= \frac{1}{\beta} \lim_{\varepsilon \to 0^+} \int_0^\infty dt\, e^{-\varepsilon t} \int_0^\beta d\beta' C_{qsq's'}^{(1)}(t+i\hbar\beta') C_{qsq's'}^{(2)}(t+i\hbar\beta').
\end{aligned}
\tag{5.257}
$$

The Fourier transforms of the single-particle correlation functions $C^{(1)}$ and $C^{(2)}$ are related to the following one-particle *retarded Green's functions*:

$$
C_{qsq's'}^{(1)}(\omega) = \frac{i}{e^{\beta\hbar\omega}-1}\left[G_{qsq's'}(\omega+i\varepsilon) - G_{qsq's'}(\omega-i\varepsilon)\right]
\tag{5.258}
$$

$$
C_{qsq's'}^{(2)}(\omega) = \frac{ie^{\beta\hbar\omega}}{e^{\beta\hbar\omega}-1}\left[G_{qsq's'}(\omega+i\varepsilon) - G_{qsq's'}(\omega-i\varepsilon)\right].
\tag{5.259}
$$

Using the Fourier transform

$$
C_{qsq's'}(t+i\hbar\beta') = \int_{-\infty}^\infty d\omega\, C_{qsq's'}(\omega) \exp\left[-i\omega(t+i\hbar\beta')\right]
\tag{5.260}
$$

we express

$$
\begin{aligned}
\mathscr{G} = &-\frac{1}{\beta} \lim_{\varepsilon \to 0^+} \int_0^\infty dt\, e^{-\varepsilon t} \int_{-\infty}^\infty d\omega_1 \\
&\times \int_{-\infty}^\infty d\omega_2 \frac{e^{\beta\hbar\omega_2}}{(e^{\beta\hbar\omega_1}-1)(e^{\beta\hbar\omega_2}-1)} \exp\left[-i(\omega_1-\omega_2)t\right]
\end{aligned}
$$

$$\times \int_0^\beta d\beta' \exp\left[\beta'\hbar(\omega_1 - \omega_2)\right] \left[G_{qsq's'}(\omega_1 + i\varepsilon) - G_{qsq's'}(\omega_1 - i\varepsilon)\right]$$
$$\times \left[G_{qsq's'}(\omega_2 + i\varepsilon) - G_{qsq's'}(\omega_2 - i\varepsilon)\right]. \tag{5.261}$$

Integrating over t and β' and then interchanging $\omega_1 \leftrightarrow \omega_2$ we get

$$\mathscr{G} = -\frac{i}{\hbar\beta} \lim_{\varepsilon \to 0^+} \int_{-\infty}^\infty d\omega_1 \int_{-\infty}^\infty d\omega_2 \frac{1}{\omega_1 - \omega_2} \frac{e^{\beta\hbar\omega_1} - e^{\beta\hbar\omega_2}}{(e^{\beta\hbar\omega_1} - 1)(e^{\beta\hbar\omega_2} - 1)}$$
$$\times \frac{1}{2}\left(\frac{1}{\omega_1 - \omega_2 + i\varepsilon} + \frac{1}{\omega_2 - \omega_1 + i\varepsilon}\right)$$
$$\times \left[G_{qsq's'}(\omega_1 + i\varepsilon) - G_{qsq's'}(\omega_1 - i\varepsilon)\right]\left[G_{qsq's'}(\omega_2 + i\varepsilon) - G_{qsq's'}(\omega_2 - i\varepsilon)\right]. \tag{5.262}$$

Using

$$\frac{1}{\omega_1 - \omega_2 + i\varepsilon} + \frac{1}{\omega_2 - \omega_1 + i\varepsilon} = -i2\pi\delta(\omega_1 - \omega_2) \tag{5.263}$$

we reduce the above expression to

$$\mathscr{G} = -\frac{\pi}{\hbar\beta} \lim_{\varepsilon \to 0^+} \int_{-\infty}^\infty d\omega_1 \left[G_{qsq's'}(\omega_1 + i\varepsilon) - G_{qsq's'}(\omega_1 - i\varepsilon)\right]$$
$$\times \int_{-\infty}^\infty d\omega_2 \left[G_{qsq's'}(\omega_2 + i\varepsilon) - G_{qsq's'}(\omega_2 - i\varepsilon)\right]$$
$$\times \frac{e^{\beta\hbar\omega_1} - e^{\beta\hbar\omega_2}}{(e^{\beta\hbar\omega_1} - 1)(e^{\beta\hbar\omega_2} - 1)} \frac{1}{\omega_1 - \omega_2}\delta(\omega_1 - \omega_2). \tag{5.264}$$

If we write $F(\omega_2) = G_{qsq's'}(\omega_2 + i\varepsilon) - G_{qsq's'}(\omega_2 - i\varepsilon)$ and let $\omega_1 - \omega_2 = x$, then the integration over ω_2 gives

$$\int_{-\infty}^\infty d\omega_2 \frac{e^{\beta\hbar\omega_1} - e^{\beta\hbar\omega_2}}{(e^{\beta\hbar\omega_1} - 1)(e^{\beta\hbar\omega_2} - 1)} \frac{F(\omega_2)}{(\omega_1 - \omega_2)}\delta(\omega_1 - \omega_2)$$
$$= \frac{e^{\beta\hbar\omega_1}}{(e^{\beta\hbar\omega_1} - 1)} \int_{-\infty}^\infty dx \frac{(1 - e^{-\beta\hbar x})F(\omega_1 - x)}{e^{\beta\hbar\omega_1}e^{-\beta\hbar x} - 1} \frac{\delta(x)}{x}.$$

Since the integrand peaks around $x = 0$, we can express $1 - e^{-\beta\hbar x} \simeq \beta\hbar x$, reducing the above integral to

$$\frac{\beta\hbar e^{\beta\hbar\omega_1}}{e^{\beta\hbar\omega_1} - 1} \int_{-\infty}^\infty dx \frac{F(\omega_1 - x)}{e^{\beta\hbar\omega_1}e^{\beta\hbar x} - 1}\delta(x)$$
$$= \beta\hbar \frac{e^{\beta\hbar\omega_1}}{(e^{\beta\hbar\omega} - 1)^2}F(\omega_1). \tag{5.265}$$

Thus equation (5.264) becomes

$$\mathscr{G} = -\pi \lim_{\varepsilon \to 0^+} \int_{-\infty}^\infty d\omega \frac{e^{\beta\hbar\omega}}{(e^{\beta\hbar\omega} - 1)^2} \left[G_{qsq's'}(\omega + i\varepsilon) - G_{qsq's'}(\omega - i\varepsilon)\right]^2. \tag{5.266}$$

Now we follow the *double-time Green's function* (or the equation-of-motion) method to evaluate $G_{qsq's'}(\omega)$. We write

$$G_{qsq's'}(\omega) = \int_{-\infty}^\infty dt\, e^{i\omega t} G_{qsq's'}(t) \tag{5.267}$$

and note that the one-particle retarded Green's function is defined as (Zubarev 1960)

$$G_{qsq's'}(t) = \langle\langle a_{q's'}(t); a_{qs}^\dagger\rangle\rangle$$

$$\equiv -i\Theta(t)\langle[a_{q's'}(t),a_{qs}^{\dagger}]\rangle \tag{5.268}$$

where $\Theta(t)$ is the step function (1 for $t > 0$ and 0 for $t < 0$). The equation of motion of $G_{qsq's'}$ is

$$-\hbar\frac{d}{dt}G_{qsq's'}(t) = [G_{qsq's'}(t),H(t)], \tag{5.269}$$

where the Hamiltonian of the phonon system is (see section 4.2)

$$H = \sum_{qs}\hbar\omega(qs)(a_{qs}^{\dagger}a_{qs}+\frac{1}{2}) + \sum_{\substack{q_1s_1q_2s_2\\q_3s_3}}\mathscr{V}_3(q_1s_1,q_2s_2,q_3s_3)A_{q_1s_1}A_{q_2s_2}A_{q_3s_3}$$

$$+\text{higher order anharmonic terms} + \text{other phonon scattering terms.} \tag{5.270}$$

In the following discussion we will only consider the third-order anharmonic term. The coefficient $\mathscr{V}_3$ is defined in equations (4.43) and (4.44) and we have used

$$A_{qs} = (a_{qs}^{\dagger} - a_{-qs}). \tag{5.271}$$

Using equation (5.268) for $G_{qsq's'}(t)$ and noting that $d\Theta(t)/dt = \delta(t)$ we can, after a little algebra, arrive at the following equation of motion:

$$i\frac{d}{dt}G_{qsq's'}(t) = \delta(t)\delta_{qq'}\delta_{ss'} + \omega(q's')G_{qsq's'}(t)$$

$$+\frac{3}{\hbar}\sum_{q_1s_1,q_2s_2}\mathscr{V}_3(q_1s_1,q_2s_2,-q's')D_1(t) \tag{5.272}$$

with

$$D_1(t) = \langle\langle A_{q_1s_1}(t)A_{q_2s_2}(t);a_{qs}^{\dagger}\rangle\rangle. \tag{5.273}$$

Taking the Fourier transform of equation (5.272) we get

$$(\omega-\omega(q's'))G_{qsq's'}(\omega) = \frac{1}{2\pi}\delta_{qq'}\delta_{ss'}$$

$$+\frac{3}{\hbar}\sum_{q_1s_1q_2s_2}\mathscr{V}_3(q_1s_1,q_2s_2,-q's')D_{q_1s_1,q_2s_2;qs}(\omega). \tag{5.274}$$

Notice that the equation of motion of the one-phonon Green's function leads to a higher-phonon Green's function. To evaluate the Green's function $D_1(t)$ we need the Green's functions

$$D_2(t) = \langle\langle B_{q_1s_1}(t)A_{q_2s_2}(t);a_{qs}^{\dagger}\rangle\rangle, \quad D_3(t) = \langle\langle A_{q_1s_1}(t)B_{q_2s_2}(t);a_{qs}^{\dagger}\rangle\rangle \text{ and}$$
$$D_4(t) = \langle\langle B_{q_1s_1}B_{q_2s_2}(t);a_{qs}^{\dagger}\rangle\rangle,$$

where $B_{qs} = (a_{qs}^{\dagger}+a_{qs})$.

These three-particle Green's functions are usually reduced to two-particle Green's functions by using Wick's decoupling scheme. Details of this are given in Pathak (1965). The final result can be written as

$$D_1(\omega) = F(q_1s_1,q_2s_2;\omega)\frac{1}{\hbar}\sum_{q's'}\mathscr{V}_3(-q_1s_1,-q_2s_2,q's')G_{q's'qs}(\omega), \tag{5.275}$$

where

$$F(q_1s_1,q_2s_2,\omega) = 6(N_1+N_2)\frac{\omega(q_1s_1)+\omega(q_2s_2)}{\omega^2-(\omega(q_1s_1)+\omega(q_2s_2))^2}$$

$$+6(N_2 - N_1) \frac{\omega(q_1 s_1) - \omega(q_2 s_2)}{\omega^2 - (\omega(q_1 s_1) - \omega(q_2 s_2))^2} \tag{5.276}$$

$$= 12\left((\bar{n}_{q_1 s_1} + \bar{n}_{q_2 s_2} + 1) \frac{\omega(q_1 s_1) + \omega(q_2 s_2)}{\omega^2 - (\omega(q_1 s_1) + \omega(q_2 s_2))^2} \right.$$

$$\left. +(\bar{n}_{q_2 s_2} - \bar{n}_{q_1 s_1}) \frac{\omega(q_1 s_1) - \omega(q_2 s_2)}{\omega^2 - (\omega(q_1 s_1) - \omega(q_2 s_2))^2} \right) \tag{5.277}$$

with

$$N_i = \langle A_{q_i s_i}^\dagger A_{q_i s_i} \rangle = \langle a_{q_i s_i}^\dagger a_{q_i s_i} \rangle + \langle a_{q_i s_i} a_{q_i s_i}^\dagger \rangle = 2\bar{n}_{q_i s_i} + 1. \tag{5.278}$$

From equations (5.274)–(5.277) we get

$$G_{qsq's'}(\omega) = \frac{\delta_{qq'} \delta_{ss'}}{2\pi[\omega - \omega(qs) - M_{qs}(\omega)]} \tag{5.279}$$

with

$$M_{qs}(\omega) = \frac{3}{\hbar^2} \sum_{q_1 s_1 q_2 s_2} \left| \mathscr{V}_3(q_1 s_1, q_2 s_2, -qs) \right|^2 F(q_1 s_1, q_2 s_2, \omega). \tag{5.280}$$

Writing

$$M_{qs}(\omega \pm i\varepsilon) = \Delta_{qs}(\omega) \mp i\Upsilon_{qs}(\omega) \tag{5.281}$$

we express

$$G_{qsq's'}(\omega + i\varepsilon) - G_{qsq's'}(\omega - i\varepsilon) = -\frac{\Upsilon_{qs}(\omega) + \varepsilon}{(\omega - \omega(qs) - \Delta_{qs}(\omega))^2 + (\Upsilon_{qs}(\omega) + \varepsilon)^2}$$

$$\times \frac{i}{\pi} \delta_{qq'} \delta_{ss'}. \tag{5.282}$$

Therefore, from equation (5.266) we get, after setting $\varepsilon = 0$

$$\mathscr{G} = \frac{1}{\pi} \int_{-\infty}^{\infty} d\omega \frac{e^{\beta\hbar\omega}}{(e^{\beta\hbar\omega} - 1)^2} \frac{\delta_{qq'} \delta_{ss'} \Upsilon_{qs}^2(\omega)}{\left[(\omega - \omega(qs) - \Delta_{qs}(\omega))^2 + \Upsilon_{qs}^2(\omega) \right]^2}. \tag{5.283}$$

Thus within the Green's function scheme the final result for the lattice thermal conductivity becomes

$$\mathscr{K} = \frac{\hbar^2 k_B \beta^2}{3N_0 \Omega} \sum_{qsq's'} \omega(qs)\omega(q's') c_s(q) \cdot c_{s'}(q') \mathscr{G}$$

$$= \frac{\hbar^2 k_B \beta^2}{3N_0 \Omega} \sum_{qs} \omega^2(qs) c_s^2(q)$$

$$\times \int_{-\infty}^{\infty} d\omega \frac{e^{\beta\hbar\omega}}{(e^{\beta\hbar\omega} - 1)^2} \frac{\Upsilon_{qs}^2(\omega)}{\left[(\omega - \omega(qs) - \Delta_{qs}(\omega))^2 + \Upsilon_{qs}^2(\omega) \right]^2}$$

$$\simeq \frac{\hbar^2 k_B \beta^2}{3\pi N_0 \Omega} \frac{\pi}{2} \sum_{qs} \omega^2(qs) c_s^2(q) \frac{e^{\beta\hbar\varepsilon(qs)}}{(e^{\beta\hbar\varepsilon(qs)} - 1)^2} \frac{1}{\Upsilon_{qs}(\varepsilon(qs))}$$

$$= \frac{\hbar^2 k_B \beta^2}{3N_0 \Omega} \sum_{qs} \omega^2(qs) c_s^2(q) \bar{n}_{qs}(\varepsilon(qs)) (\bar{n}_{qs}(\varepsilon(qs)) + 1) \tau_{qs}(\varepsilon(qs))$$

$$\tag{5.284}$$

with

$$\varepsilon(qs) = \omega(qs) + \Delta_{qs}(\omega)$$

$$\tau_{qs}(\varepsilon(qs)) = \frac{1}{2\Upsilon_{qs}(\varepsilon(qs))}$$

$$\bar{n}_{qs}(\varepsilon(qs)) = \frac{1}{\exp(\beta\hbar\varepsilon(qs)) - 1}. \tag{5.285}$$

In evaluating the integral in the equation above we have assumed small values of $\Upsilon_{qs}(\omega)$ for which the integrand peaks around $\omega = \varepsilon(qs)$. The result in equation (5.284) is the familiar single-mode relaxation-time expression (equations (5.26) and (5.255)), with the exception that the frequency shift Δ_{qs} appears in the arguments of $\bar{n}_{qs}$ and τ_{qs}. In this result $\varepsilon(qs)$ is the frequency of the perturbed normal modes (due to anharmonicity in our case). The term M_{qs} may be called 'self-energy' and the perturbed normal modes may be treated within a pseudoharmonic model. In this model phonons in mode qs are characterised by renormalised frequencies $\varepsilon(qs)$ and have a finite life-time given by $\tau_{qs}^{-1} = 2\Upsilon_{qs}$.

We can derive explicit expressions for τ_{qs}^{-1} and Δ_{qs}. Using the relation

$$\lim_{\varepsilon \to +0} \frac{1}{x \pm i\varepsilon} = \wp\left(\frac{1}{x}\right) \mp i\pi\delta(x), \tag{5.286}$$

where $\wp$ is the principal value, we write

$$\begin{aligned}
\tau_{qs}^{-1}(\varepsilon) &\equiv 2\Upsilon_{qs}(\varepsilon) = -2\Im M_{qs}(\varepsilon + i\varepsilon) \\
&= \frac{72\pi}{\hbar^2} \sum_{q_1 s_1, q_2 s_2} \left|\mathcal{V}_3(q_1 s_1, q_2 s_2, -qs)\right|^2 \\
&\quad \times \left[(\bar{n}_{q_1 s_1} + \bar{n}_{q_2 s_2} + 1)(\omega(q_1 s_1) + \omega(q_2 s_2))\right. \\
&\quad \times \delta\left(\varepsilon^2 - (\omega(q_1 s_1) + \omega(q_2 s_2))^2\right) \\
&\quad + (\bar{n}_{q_2 s_2} - \bar{n}_{q_1 s_1})(\omega(q_1 s_1) - \omega(q_2 s_2)) \\
&\quad \left. \times \delta\left(\varepsilon^2 - (\omega(q_1 s_1) - \omega(q_2 s_2))^2\right)\right]
\end{aligned} \tag{5.287}$$

and

$$\begin{aligned}
\Delta_{qs}(\omega) &= -\frac{36}{\hbar^2}\wp \sum_{q_1 s_1 q_2 s_2} \left|\mathcal{V}_3(q_1 s_1, q_2 s_2, -qs)\right|^2 \\
&\quad \times \left((1 + \bar{n}_{q_1 s_1} + \bar{n}_{q_2 s_2})\frac{\omega(q_1 s_1) + \omega(q_2 s_2)}{\omega^2 - (\omega(q_1 s_1) + \omega(q_2 s_2))^2}\right. \\
&\quad \left. + (\bar{n}_{q_2 s_2} - \bar{n}_{q_1 s_1})\frac{\omega(q_1 s_1) - \omega(q_2 s_2)}{\omega^2 - (\omega(q_1 s_1) - \omega(q_2 s_2))^2}\right).
\end{aligned} \tag{5.288}$$

[Note that the minus sign is missing in Pathak's article.] The relaxation-time result in equation (5.287) will be evaluated in Chapter 6 where it will shown that it is identical to the *smrt* result provided that the energy of the relaxing phonon in mode qs is normalised to include the shifted frequency $\varepsilon(qs)$.

5.5 SECOND SOUND AND POISEUILLE FLOW OF PHONONS

In the previous sections we have discussed the theory of lattice thermal conductivity in the steady state of heat current flow: i.e. Q was considered to depend explicitly on the direction of heat flow r but not on time t. If the heat current is time dependent (e.g. pulsed heat), then the Boltzmann equation (for slow spatial variation) is given by adding $\partial n/\partial t$ to equation (5.2)

$$\frac{\partial n}{\partial t} - c \cdot \nabla T \frac{\partial \bar{n}}{\partial t} + \frac{\partial n}{\partial t}\bigg|_{scatt} = 0. \tag{5.289}$$

If it is assumed that the pulsed heat has a wave-vector k and frequency φ dependence:

$$T = T_0 + \delta T \exp\left[i(k \cdot r - \varphi t)\right] \tag{5.290}$$

then the corresponding thermal conductivity depends on k and φ: $\mathscr{K} = \mathscr{K}(k, \varphi)$. In the steady state $\partial n/\partial t \to 0$ and $\mathscr{K} = \mathscr{K}(0,0)$. When the normal processes are very slow compared to the resistive processes, $N << R$, the expression for the conductivity $\mathscr{K}(k, \varphi)$ leads to strongly damped temperature oscillations.

An interesting phenomenon occurs in the limit when the normal processes are infinitely rapid ($\tau_N \to 0$). In this limit the temperature wave satisfies a dissipative wave equation (Guyer and Krumhansl 1966, Jackson and Walker 1971)

$$\frac{\partial^2 T}{\partial t^2} + \frac{1}{\tau_R}\frac{\partial T}{\partial t} + \frac{1}{3}c^2\nabla^2 T = 0. \tag{5.291}$$

If the resistive processes are very rare ($\tau_R \to \infty$), then the temperature disturbance propagates undamped with speed $c/\sqrt{3}$, where c is the phonon or 'first-sound' velocity. This phenomenon represents a collective propagation of phonons and is called 'second sound'. Detailed reviews of the theory of second sound in dielectric crystals have been given by Enz (1968), Hardy (1970) and Beck (1975).

The criteria for observation of second sound are

$$\begin{array}{c} \lambda_N << \lambda_{ss} << d << \lambda_R \\ \Delta t << d\sqrt{3}/c, \end{array} \tag{5.292}$$

where λ_N and λ_R are the phonon wavelengths for normal and resistive processes, respectively, and λ_{ss} is the second-sound wavelength. d is the sample thickness and Δt is the heat pulse duration. The condition $\lambda << \lambda_R$ can be met in pure crystals at low temperatures.

The possibility of observing second sound in solids was first suggested by Peshkov (1947). Successful observation of second sound has been made in solid helium, solid NaF and in Bi (see Beck (1975) for a list of successful experiments). Figure 5.1 shows the appearance of second sound in a pure NaF sample at 11 K, observed by Jackson and Walker (1971).

Under more stringent conditions

$$\begin{array}{c} \lambda_N << r \\ \lambda_N \lambda_R >> r^2, \end{array} \tag{5.293}$$

where r is the radius of (cylindrical) sample, Poiseuille flow of phonons may be observed. These conditions may be met in a small temperature range in a pure crystal in which the resistive processes have become negligible but the normal processes are still frequent in comparison with scattering by the boundaries. Poiseuille flow of phonons has been observed in solid helium (Thomlinson 1969) and in Bi (Kopylov and Meshov-Deglin 1971).

A recent theoretical study by Cepellotti et al (2015) suggests that criteria for the development of the hydrodynamic Poiseuille regime in two-dimensional materials with low isotopic disorder are less stringent than in conventional solids. This study is based on first-principles thermal conductivity calculations, employing the Callaway theory described in section 5.2.3, of five two-dimensional materials: graphene, graphane (hydrogenated graphene), boron nitride (BN), flurographene and molybdenum disulphide (MoS_2). It is found that N processes represent the dominant scattering mechanism in these two-dimensional materials at any temperature (with the exception of BN and MoS_2, where isotopic scattering is comparable to N-scattering because of the large isotopic disorder of Mo and B). This work emphasises that graphene, graphane and BN admit wave-like heat diffusion, with second sound present at room temperature and above.

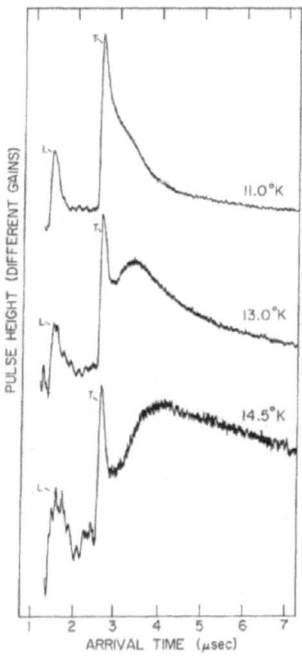

Figure 5.1 Heat pulse signals in a pure NaF sample in the $\langle 100 \rangle$ direction for several different temperatures. L and T are the longitudinal and transverse ballistic phonon pulses, and the third peak is the *second-sound* pulse. (From Jackson and Walker (1971).)

Cepellotti *et al* (2015) expressed the second-sound speed c_{ss}, the second-sound relaxation time τ_{ss} and the second-sound length λ_{ss} using the equations below

$$\left(c_{ss} \right)^2 = \frac{\langle \frac{c \cdot c}{2} \omega^2 \rangle}{\langle \omega^2 \rangle} \tag{5.294}$$

$$\tau_{ss}^{-1} = \frac{\langle \omega c q \tau_R^{-1} \rangle}{\langle \omega c q \rangle} \tag{5.295}$$

and

$$\lambda_{ss} = c_{ss} \tau_{ss}, \tag{5.296}$$

where c is the phonon (or first-sound) velocity, q is the phonon wave vector, and we have used the notation $\langle f \rangle = \sum_{qs} f_{qs} \bar{n}_{qs} (\bar{n}_{qs} + 1)$ as described in equation (5.27). Numerical evaluation of τ_{ss} was made for infinite materials characterised with only naturally occurring isotopic mass defects. Furthermore, in expressing equations (5.294) and (5.295), the authors restricted themselves to an isotropic two-dimensional material (so that c and q are parallel to ∇T, and there is no directional dependence of τ_R^{-1} (see Supplementray Note 3 in Cepellotti *et al* 2015). With these considertaions $\tau_{bs}^{-1} = 0$ (no boundary scattering) and $\tau_R^{-1} = \tau_{md}^{-1} + \tau_U^{-1}$. Figure 5.2 shows the second-sound propagation length λ_{ss} over a large temperature range for the five two-dimensional systems. As can be seen from this, the room-temperature propagation length λ_{ss} for the second-sound wave is of the order of microns for graphene and graphane, and about a tenth of a micron for BN.

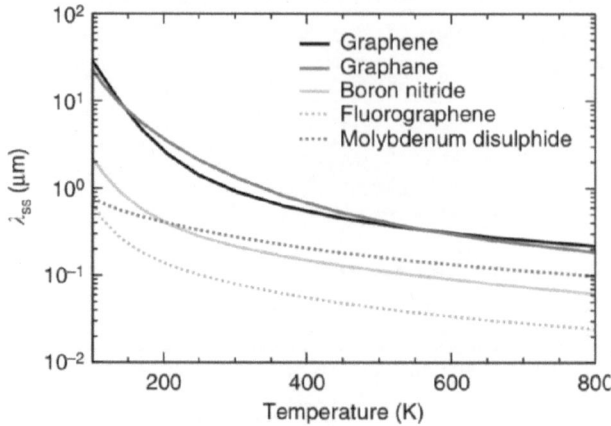

Figure 5.2   Second-sound scattering length λ_{ss} in two-dimensional materials. From Cepellotti *et al* (2015).

6 Phonon Scattering in Solids

In Chapter 4 we developed the concept of anharmonicity in crystals and its role in generating phonon–phonon interactions. That concept was used in Chapter 5 to develop the theory of lattice thermal conductivity. The results from the relaxation-time and linear-response approaches lead to the concept of phonon relaxation or damping. The formalism of complementary variational principles used the terminology of the phonon scattering operator P, whose diagonal elements are directly related to the single-mode phonon life-time (or relaxation, or damping). To make a quantitative calculation of lattice thermal conductivity it is, therefore, essential to obtain expressions for the phonon relaxation time τ_{qs}^{-1} and the kernel $P_{qsq's'}$ of the collision operator P. In this chapter we will derive expressions for both τ_{qs}^{-1} and $P_{qsq's'}$ by considering a few examples of phonon scattering processes in non-metallic crystals: such as boundary scattering, scattering with static imperfections, three-phonon scattering, four-phonon scattering, phonon–electron and phonon–hole scatterings in doped semiconductors, phonon scattering by magnetic impurities and phonon scattering by tunnelling states. We will present the theory of phonon–photon interaction related to the processes of infrared absorption and Raman scattering in crystals. Some applications of these scattering processes will be discussed in Chapters 7–12.

6.1 BOUNDARY SCATTERING

The boundaries of a crystal of finite size act as scattering regions and limit the effective mean free path of phonons. As boundary scattering does not occur uniformly throughout the crystal, it is essential to consider an explicit spatial dependence of the distribution function: $n_{qs} = n_{qs}(r)$. The problem of boundary scattering of phonons has been discussed by Casimir (1938), Berman (1953), Berman *et al* (1955), Ziman (1960) and Carruthers (1961). Here we follow the treatment given by Ziman (1960).

Let us express

$$n_{qs} = \bar{n}_{qs} + \psi_{qs}(r)\bar{n}_{qs}(\bar{n}_{qs} + 1), \tag{6.1}$$

where the explicit r-dependence of $n_{qs}(r)$ comes through the deviation function $\psi_{qs}(r)$, and the equilibrium distribution depends on the position only through the temperature: $\bar{n}_{qs} = \bar{n}_{qs}(T(r))$. In the steady state the distribution $n_{qs}(r)$ satisfies the Boltzmann equation given in equation (5.2)

$$-c_{qs} \cdot \nabla n_{qs}(r) = -\left.\frac{\partial n_{qs}}{\partial t}\right|_{\text{scatt}}. \tag{6.2}$$

Substituting equation (6.1) in equation (6.2), we obtain the following linearised form of the Boltzmann equation:

$$-c_{qs} \cdot \nabla T \frac{\partial \bar{n}_{qs}}{\partial T} = \frac{\psi_{qs}(r)\bar{n}_{qs}(\bar{n}_{qs} + 1)}{\tau_{bulk}(qs)} + c_{qs} \cdot \nabla \psi_{qs}(r)\bar{n}_{qs}(\bar{n}_{qs} + 1), \tag{6.3}$$

where we have expressed the right hand side of equation (6.2) in terms of a bulk relaxation time τ_{bulk}. Let us consider the situation at very low temperatures when the relaxation time for bulk processes is very long compared to boundary relaxation time, i.e. $\tau_{bulk} \to \infty$. In this limit equation (6.3) is reduced to

$$c_{qs} \cdot \nabla \psi_{qs}(r) = -c_{qs} \cdot \nabla T \frac{\partial \bar{n}_{qs}/\partial T}{\bar{n}_{qs}(\bar{n}_{qs} + 1)}. \tag{6.4}$$

DOI: 10.1201/9781003141273-6

The solution to this first-order differential equation is

$$\psi_{qs}(r) = \frac{\partial \bar{n}_{qs}/\partial T}{\bar{n}_{qs}(\bar{n}_{qs}+1)} \nabla T \cdot r + \alpha, \tag{6.5}$$

where α is a constant of integration. To determine α we express the following simple boundary condition:

$$\psi_{qs}(r_B)\Big|_{c_n} = p\psi_{q's}(r_B)\Big|_{-c_n} \tag{6.6}$$

which assumes that the fraction p of all phonons arriving at the surface r_B are reflected with their normal velocity c_n reversed. For the purely diffuse scattering (as in the theory of black body radiation) $p = 0$ and the boundary condition becomes

$$\psi_{qs}(r_B)\Big|_{c_n}^{\text{diff}} = 0. \tag{6.7}$$

This represents the Casimir limit in which all phonons, which have a positive normal velocity when they reach the boundary, lose the sense of their directionality and obey the equilibrium distribution. With equation (6.7), equation (6.5) becomes

$$\bar{n}_{qs}(\bar{n}_{qs}+1)\psi_{qs}(r) = \frac{\partial \bar{n}_{qs}}{\partial T}\nabla T \cdot (r - r_B). \tag{6.8}$$

The heat current over a cross-sectional area S_c is (see equation (5.20))

$$\begin{aligned}
QS_c &= \frac{1}{N_0\Omega}\sum_{qs}\int \hbar\omega_{qs}\psi_{qs}\bar{n}_{qs}(\bar{n}_{qs}+1)c_s(q)\cdot dS_c \\
&= \frac{1}{N_0\Omega}\sum_{qs}\int \hbar\omega_{qs}\frac{\partial \bar{n}_{qs}}{\partial T}\nabla T \cdot (r - r_B)c_s(q)\cdot dS_c \\
&= \frac{1}{N_0\Omega}\sum_{qs}\int\int \hbar\omega_{qs}\frac{\partial \bar{n}_{qs}}{\partial T}|\nabla T||r - r_B||c_s(q)|\cos^2\theta dS_c d\Omega, \tag{6.9}
\end{aligned}$$

where $(r - r_B)$ points in the solid angle $d\Omega$ and θ is the angle between $r - r_B$ and ∇T, and between $c_s(q)$ and dS_c (phonon velocity $c(q)$ is in the direction of $r - r_B$ and ∇T is in the direction of dS_c). We can further express equation (6.9) as

$$\begin{aligned}
Q &= \frac{1}{3}C_v^{sp}\bar{c}L_0|\nabla T| \tag{6.10} \\
&= \mathcal{K}|\nabla T|,
\end{aligned}$$

where C_v^{sp} is the phonon-specific heat defined in equation (2.123), $\bar{c}$ is an angularly averaged phonon velocity, and L_0 is an effective *boundary mean free path* in the Casimir limit

$$L_0 = \frac{3}{4\pi S_c}\int\int |r - r_B|\cos^2\theta d\Omega dS_c. \tag{6.11}$$

The mean free path L_0 can be worked out for a given geometry of the crystal. For a cylindrical shape with circular cross section of diameter D, it can shown that $L_0 = D$. For a square cross section of side d, the result is $L_0 \simeq 1.12d$.

In order to include the effect of specular reflection ($p \neq 0$) we use the boundary condition in equation (6.6). In this case equation (6.8) must be replaced by (Ziman 1960)

$$\bar{n}_{qs}(\bar{n}_{qs}+1)\psi_{qs}(r) = \frac{\partial \bar{n}_{qs}}{\partial T}\nabla T \cdot [(1-p)\{(r - r_B) + p(r - r'_B)\}$$

$$+p^2(\mathbf{r}-\mathbf{r}_B'')+\ldots\}], \tag{6.12}$$

where $\mathbf{r}_B',\mathbf{r}_B''$, etc, are the points on the surface where specular reflections would have taken place before the point $\mathbf{r}_B$ is reached. With this the expression for an effective boundary mean free path becomes

$$L = \frac{3}{4\pi S_c}\int\int d\Omega\,dS_c\cos^2\theta[(1-p)\{|\mathbf{r}-\mathbf{r}_B|$$
$$+p|\mathbf{r}-\mathbf{r}_B'|+p^2|\mathbf{r}-\mathbf{r}_B''|+\ldots\}]. \tag{6.13}$$

If we assume that the average position of $\mathbf{r}$ is in the middle of the circular cross section (for a cylindrical sample), then $|\mathbf{r}-\mathbf{r}'| = 3|\mathbf{r}-\mathbf{r}_B|, |\mathbf{r}-\mathbf{r}_B''| = 5|\mathbf{r}-\mathbf{r}_B|$, etc. Equation (6.13) then can be expressed, after summing the series, as

$$L = \frac{1+p}{1-p}L_0. \tag{6.14}$$

Thus the boundary mean free path becomes longer when specular reflections are present. However, the factor p varies with the surface condition and may also depend upon temperature. With decreasing temperature the average phonon wavelength increases and a surface of given roughness appears smoother.

As L does not depend on $\mathbf{q}$, we can define a constant relaxation time for a phonon with speed c_s in the polarisation mode s

$$\tau_{bs}^{-1}(s) = \frac{c_s}{L}. \tag{6.15}$$

This expression can be used to express, in the notation of equation (5.18), the elements of the phonon boundary operator:

$$P_{qq'}^{ss'}(bs) = \frac{c_s}{L}\bar{n}_{qs}(\bar{n}_{qs}+1)\delta_{ss'}\delta_{qq'}. \tag{6.16}$$

6.2 SCATTERING BY STATIC IMPERFECTIONS

Klemens (1955) has studied the scattering of phonons by static imperfections. By using perturbation theory, he has derived expressions for the single-mode phonon relaxation time due to scattering by a substitutional atom of different mass, by an atom of different binding force, by dislocations, and by grain boundaries. In this section we will derive an expression for phonon scattering due to mass difference (isotope mixing) and present a discussion of scattering by dislocations, stacking faults and grain boundaries.

6.2.1 MASS DIFFERENCE SCATTERING

Let

$$\bar{M} \equiv \sum_i f_i M_i = \frac{1}{N_0}\sum_n M_n \tag{6.17}$$

be the average mass in the solid, with f_i the fraction of unit cells having mass M_i and N_0 the number of unit cells in the solid. Then up to quadratic terms the crystal Hamiltonian is

$$H = \sum_n \frac{1}{2}M_n\dot{u}_n^2 + \mathscr{V}_2$$
$$= H_0 + H_{md}, \tag{6.18}$$

where

$$H_0 = \sum_n \frac{1}{2}\bar{M}\dot{u}_n^2 + \mathscr{V}_2 \tag{6.19}$$

is the unperturbed part and

$$H_{md} = \sum_n \frac{1}{2}(M_n - \bar{M})\dot{u}_n^2 = \sum_n \frac{1}{2}\Delta M_n \dot{u}_n^2 \tag{6.20}$$

is the perturbation due to mass difference.

From equation (4.79), we write

$$u_n(r) = -i\sqrt{\frac{\hbar}{2\rho N_0 \Omega}} \sum_{qs} \frac{1}{\sqrt{\omega(qs)}} e_{qs}(a_{qs}^\dagger - a_{-qs}) \exp\left[i(q \cdot r_n - \omega t)\right], \tag{6.21}$$

where ρ is the density of the solid. With this H_{md} becomes

$$
\begin{aligned}
H_{md} &= \frac{\hbar}{4\rho N_0 \Omega} \sum_{qsq's'} \sqrt{\omega(qs)\omega(q's')} \\
&\quad \times e_{qs}^* \cdot e_{q's'}(a_{qs}^\dagger - a_{-qs})(a_{q's'}^\dagger - a_{q's'})\mathscr{M}_{qq'} \\
&= \sum_{qsq's'} H_{md}(qs,q's'),
\end{aligned}
\tag{6.22}
$$
$$\tag{6.23}$$

where

$$\mathscr{M}_{qq'} = \sum_n \Delta M_n \exp\left[i(q-q')\cdot r_n\right]. \tag{6.24}$$

In a typical phonon scattering process due to isotope mixing we are interested in the transition probability $P_{qs}^{q's'}$ between an initial state $|i\rangle = |n_{qs}, n_{q's'}\rangle$ and a final state $|f\rangle = |n_{qs} - 1, n_{q's'} + 1\rangle$. From the golden rule formula we have

$$P_{qs}^{q's'} = \frac{2\pi}{\hbar} \left|\langle f|H_{md}(qs,q's')|i\rangle\right|^2 \delta(E_f - E_i). \tag{6.25}$$

Using equations (4.38) we obtain, with $E_i = \hbar\omega(qs), E_f = \hbar\omega(q's')$,

$$
\begin{aligned}
P_{qs}^{q's'} &= \frac{\pi}{2(\rho N_0 \Omega)^2} n_{qs}(n_{q's'} + 1)\omega(qs)\omega(q's')(e_{qs}^* \cdot e_{q's'})^2 \\
&\quad \times \left|\mathscr{M}_{qsq's'}\right|^2 \delta\left(\omega(qs) - \omega(q's')\right).
\end{aligned}
\tag{6.26}
$$

To evaluate $|\mathscr{M}_{qsq's'}|^2$ we assume that the isotopes are randomly distributed. Then

$$
\begin{aligned}
\left|\mathscr{M}_{qsq's'}\right|^2 &= \sum_{n,n'} \Delta M_n \Delta M_{n'} \exp\left[i(q-q')\cdot(r_n - r_{n'})\right] \\
&= \sum_n (\Delta M_n)^2 + \sum_{n'\neq n} \Delta M_n \Delta M_{n'} \exp\left[i(q-q')\cdot(r_n - r_{n'})\right] \\
&= \sum_n (\Delta M_n)^2 \\
&= N_0 \sum_i f_i(\Delta M_i)^2.
\end{aligned}
\tag{6.27}
$$
$$\tag{6.28}$$

The second term in equation (6.27) vanishes on average for a random distribution. Hence

$$
\begin{aligned}
P_{qs}^{q's'}(md) &= \frac{\pi}{2\rho^2 N_0 \Omega^2} n_{qs}(n_{q's'} + 1)\omega(qs)\omega(q's')\delta\left(\omega(qs) - \omega(q's')\right) \\
&\quad \times \left(\sum_{i,b} f_i(b)(\Delta M_i(b))^2 \left(e^*(b;qs) \cdot e(b;q's')\right)^2\right)
\end{aligned}
$$

$$
\begin{aligned}
= \quad & \frac{\pi}{2N_0} n_{qs}(n_{q's'}+1)\omega(qs)\omega(q's')\delta\left(\omega(qs)-\omega(q's')\right) \\
& \times \sum_b \Gamma_{md}(b)\left|e^*(b;qs)\cdot e(b;q's')\right|^2,
\end{aligned}
\tag{6.29}
$$

where the isotopic mass defect coefficient for the bth atom in the unit cell of average mass $\bar{M}(b)$ is

$$
\Gamma_{md}(b) = \sum_i f_i(b)\left(1-\frac{M_i(b)}{\bar{M}(b)}\right)^2 = \sum_i f_i(b)\left(\frac{\Delta M_i(b)}{\bar{M}(b)}\right)^2
\tag{6.30}
$$

with $f_i(b)$ as the fraction of the ith isotope of mass $M_i(b)$.

Following the discussion in equations (5.4)–(5.8) and (5.13)–(5.18) we can calculate the rate of change of n_{qs},

$$
\begin{aligned}
\left.\frac{\partial n_{qs}}{\partial t}\right|_{md} = \quad & \frac{\pi}{2N_0}\sum_{q's'}\omega(qs)\omega(q's')\bar{n}_{qs}(\bar{n}_{q's'}+1)\delta\left(\omega(qs)-\omega(q's')\right)(\psi_{q's'}-\psi_{qs}) \\
& \times \sum_b \Gamma_{md}(b)\left|e^*(b;qs)\cdot e(b;q's')\right|^2 \\
= \quad & \sum_{q''s''}\bar{P}_{qs}^{q''s''}\left(\psi_{q''s''}-\psi_{qs}\right) \\
= \quad & \sum_{q's'q''s''}\bar{P}_{qs}^{q''s''}\left\{\psi_{q's'}\delta_{s's''}\delta_{q'q''}-\psi_{q's'}\delta_{ss'}\delta_{qq'}\right\} \\
= \quad & -\sum_{q's'}P_{qq'}^{ss'}(md)\psi_{q'}^{s'}.
\end{aligned}
\tag{6.31}
$$
$$
\tag{6.32}
$$

(Note that ψ_{qs} will sometimes be expressed as ψ_q^s.) Thus the elements of the phonon mass-difference scattering operator are given by

$$
P_{qq'}^{ss'}(md) = \sum_{q''s''}\bar{P}_q^{q''s''}\left(\delta_{ss'}\delta_{qq'}-\delta_{s's''}\delta_{q'q''}\right)
\tag{6.33}
$$

with $\bar{P}_q^{q's'}$ given by equation (6.29) after replacing n_{qs} by $\bar{n}_{qs}$, etc.

The expression for the single-mode relaxation time is obtained by setting $\psi_{q's'}(q's'\neq qs)=0$ in equation (6.31):

$$
\begin{aligned}
\left.\frac{\partial n_{qs}}{\partial t}\right|_{md} = \quad & \frac{-\pi}{2N_0}\sum_{q's'}\omega(qs)\omega(q's')\bar{n}_{qs}(\bar{n}_{q's'}+1)\delta\left(\omega(qs)-\omega(q's')\right)\psi_{qs} \\
& \times \sum_b \Gamma_{md}(b)\left|e^*(b;qs)\cdot e(b;q's')\right|^2 \\
= \quad & -\bar{n}_{qs}(\bar{n}_{qs}+1)\psi_{qs}\tau_{qs}^{-1}(md).
\end{aligned}
$$

Thus

$$
\begin{aligned}
\tau_{qs}^{-1}(md) = \quad & \frac{\pi}{2N_0}\omega^2(qs)\sum_{q's'}\delta\left(\omega(qs)-\omega(q's')\right) \\
& \times \sum_b \Gamma_{md}(b)\left|e^*(b;qs)\cdot e(b;q's')\right|^2.
\end{aligned}
\tag{6.34}
$$

When using the isotropic continuum approximation, the summation over q' in the above expression can be changed to integration. Considering a single atomic species, we get

$$
\tau_{qs}^{-1}(md) = \frac{\Gamma_{md}\Omega}{4\pi}\omega^2(qs)\sum_{s'}\int dq'q'^2(e_{qs}^*\cdot e_{q's'})^2\delta\left(\omega(qs)-\omega(q's')\right)
$$

$$
\begin{aligned}
&= \frac{\Gamma_{md}\Omega}{4\pi}\omega^2(qs)\sum_{s'}\frac{1}{c_{s'}^3}\int d\omega(q's')\omega^2(q's')(e_{qs}^*\cdot e_{q's'})^2 \\
&\quad \times \delta(\omega(qs)-\omega(q's')) \\
&= \frac{\Gamma_{md}\Omega}{4\pi}\omega^4(qs)(e_{qs}^*\cdot e_{qs})^2\sum_{s'}\frac{1}{c_{s'}^3}.
\end{aligned}
\tag{6.35}
$$

Further, approximating $(e_{qs}^*\cdot e_{qs})^2 = 1/3$, we finally obtain

$$
\tau_{qs}^{-1}(md) = \frac{\Gamma_{md}\Omega}{4\pi\bar{c}^3}\omega^4(qs) = \Gamma_{md}\frac{\pi}{6N_0}\omega^2(qs)g(\omega)
\tag{6.36}
$$

with $3/\bar{c}^3 = \sum_s c_s^{-3}$ and the density of states $g(\omega)$ defined in equation (2.115).

For the general case of an impurity Klemens (1955) suggested the following form for Γ_{md}:

$$
\Gamma_{md} = \sum_i f_i\big[(\Delta M_i/M)^2 + 2(\Delta g_i/g - 6.4\gamma\Delta\delta_i/\delta)^2\big],
\tag{6.37}
$$

where δ_i is the radius of the impurity atom in the host lattice, δ is the radius of an atom in the virtual crystal, g_i is an average stiffness constant of the nearest-neighbour bonds from the impurity to host lattice, g is the average stiffness constant for the host atoms, $\Delta g_i = g_i - g, \Delta\delta_i = \delta_i - \delta$, and γ is an average anharmonicity of the bonds.

6.2.2 SCATTERING BY DISLOCATIONS, STACKING FAULTS, AND GRAIN BOUNDARIES

Time-dependent perturbation theory can also be applied to study scattering of phonons by other kinds of crystal imperfections such dislocations, stacking faults, and grain boundaries. However, numerically accurate results are not available, mainly due to uncertainty in the precise form of the strain field around such extended impurities. In this section we will, therefore, present a very simple qualitative picture of phonon scattering by such imperfections.

6.2.2.1 Dislocations

A dislocation is a line imperfection in a solid, with its strain field falling off inversely as the distance from the dislocation line. The region near the dislocation line where the crystal distortion is extremely large is called the *core* of the dislocation. The region outside the core, where the strain is small, is called the *elastic region*. The direction and the magnitude of the rigid displacement of the atoms causing a dislocation is given by the Burgers vector b. There are two simple types of dislocation: an *edge* dislocation is characterised by b being perpendicular to the dislocation line, and for a *screw* dislocation b lies along the dislocation line. A general dislocation may be expressed as a mixture of edge and screw types.

The elastic strain field around a dislocation is of the form (Cottrell 1953)

$$
Y = \frac{b}{2\pi r}\left\{\begin{array}{c}\sin\theta \\ \cos\theta\end{array}\right\},
\tag{6.38}
$$

where (r,θ) are cylindrical coordinates relative to the dislocation line as z axis. This strain gives rise to a change δv in the velocity of sound, where we can write (Ziman 1960)

$$
\delta v/v = \gamma Y,
\tag{6.39}
$$

where γ is Grüneisen's constant. An effective perturbation energy due to the dislocation strain field is

$$
\delta\mathcal{V}(r) = \hbar\delta\omega(q)
$$

$$
\begin{aligned}
&= \hbar q \delta v \\
&= \frac{\gamma b \hbar q v}{2\pi r} \begin{Bmatrix} \sin\theta \\ \cos\theta \end{Bmatrix}.
\end{aligned}
\tag{6.40}
$$

Using the golden rule formula of time-dependent theory, Klemens (1955) and Ziman (1960) arrived at the following result for the phonon relaxation time due to scattering by dislocations:

$$
\tau_{\text{dislocations}}^{-1} = C_1 N_d b^2 \omega,
\tag{6.41}
$$

where N_d is the number of dislocation lines per unit area and C_1 is a numerical constant which depends on the type of dislocation. The result in equation (6.41) should be regarded as crude, as no distinction has been made between dilatation and shear strains, or between transverse and longitudinal phonons.

For the effect of the dislocation core Stehle and Seeger (1956) proposed that there shold be a dilatation of amount

$$
\Delta = \frac{g b^2}{4\pi^2 r^2} \qquad r \le a,
\tag{6.42}
$$

where a is the core radius and g measures the strength of the anharmonic terms. Klemens (1958) estimated that this will cause the phonon relaxation as the square of the frequency

$$
\tau_{\text{core}}^{-1} = C_2 N_d \omega^2,
\tag{6.43}
$$

where C_2 is a constant.

6.2.2.2 Stacking faults

A two-dimensional imperfection may be a stacking fault or a twin boundary. A dislocation in a crystal may split into two partial dislocations, leaving a sheet of atoms in between, called a stacking fault. Klemens (1957) has estimated that an effective phonon relaxation time due to scattering by stacking faults is given by

$$
\tau_{\text{stacking fault}}^{-1} = C_3 N_s \omega^2,
\tag{6.44}
$$

where C_3 is a constant and N_s is the number of stacking faults crossing a line of unit length.

6.2.2.3 Grain boundaries

A grain boundary can be considered as an array of dislocations lying in the plane of the boundary. A grain boundary separates two regions of a crystal rotated relative to each other. The strain field of a grain boundary can be obtained by summing the strain field due to the individual dislocations. Neglecting the disorder in the immediate vicinity of each dislocation, Klemens (1955) showed that, if the spacing between dislocations is much less than the phonon wavelength, the phonon scattering by a grain boundary varies as ω^2. If the spacing between the dislocations is larger than the phonon wavelength, the dislocations scatter independently and the scattering varies as ω^n, with n between 0 and 1 (Klemens 1958).

6.3 PHONON SCATTERING IN ALLOYS

The high-temperature lattice thermal conductivity of a semiconducting or insulating single crystal alloy is usually lower than the average of the thermal conductivities of the constituent materials. To understand this we assume the alloy to be a random mixture of atoms, with different masses and volumes, arranged in a lattice. For example, for $Ga_{1-x}In_xAs$ alloys we can assume that the cations Ga and In are randomly distributed on the FCC sites, with each atom tetrahedrally bonded to its neighbours. To calculate the lattice thermal conductivity, we consider an ordered virtual crystal and treat disorder as a perturbation (Abeles 1963). The ordered virtual crystal has an average lattice constant (obeying the Vegard's law), and an average atomic weight $\bar{M} = \sum_i f_i M_i$, where M_i and f_i are, respectively, the atomic weight and fractional concentration of the ith component of the alloy. In such a virtual crystal phonon scattering from boundary, isotope mixing and anharmonicity can be treated as discussed in sections 6.1, 6.2 and 6.4, respectively.

When an atom of the virtual crystal is replaced by an atom of the alloy, we introduce a virtual impurity (of mass $\Delta M_i = M_i - \bar{M}$). This virtual impurity atom differs from the atoms of the virtual crystal in its mass, size and in the coupling forces to its neighbours. Thus phonons suffer scattering from the virtual impurities. This scattering can be treated using the theory of point-defect scattering as discussed in section 6.2. In highly disordered alloys high-frequency phonons are very strongly scattered by disorder scattering ($\propto \omega^4$) and thus most heat is transported by low-frequency phonons of long mean free path.

Sintered (hot pressed) alloys prepared from powder of small grain sizes (linear dimensions typically of the order of one micron) show marked relative increase in thermal resistance over their single crystal (zone-levelled) counterparts. Phonons with mean free path comparable to grain size are likely to suffer strong boundary scattering, over and above the disorder scattering due to point defects, and thus generate extra thermal resistance in sintered semiconductors or insulators. Both theoretical and experimental confirmations of this effect have been presented in the works of Parrott (1969b), Meddins and Parrott (1976), Bhandari and Rowe (1977), and Gaur (1978). Phonon scattering at boundaries of grains, characterised by grain size L, can be described by the analysis presented in section 6.1.

The influence of phonon–electron interaction on the lattice thermal conductivity of doped alloys (single crystals or sintered) has been studied by Steigmeir and Abeles (1964), Gaur *et al* (1966), Meddins and Parrott (1976) and Gaur (1978). The theory of phonon–electron scattering will be developed in section 6.5.

6.4   ANHARMONIC SCATTERING

The concept of non-interacting phonons (normal modes) in a crystal arises solely within the harmonic region of the crystal potential. In reality a crystal potential is anharmonic. This destroys the concept of non-interacting phonons. However, in most practical situations anharmonicity is only a small proportion of the whole crystal Hamiltonian. Therefore, in practice we retain the concept of phonons as quasiparticles and consider the picture of phonon–phonon interaction arising from anharmonic perturbation. This was assumed in developing the theory of lattice thermal conductivity in Chapter 5.

In this section we derive expressions for the single-mode relaxation time τ_{qs}^{-1} and the matrix elements $P_{qq'}^{ss'}$ for three- and four-phonon processes due to cubic anharmonicity in the crystal potential.

6.4.1   THREE-PHONON PROCESSES

The plan of this section is as follows. In section 6.4.1.1 we will derive the expression for the single-mode relaxation time τ_{qs} and the expression for the matrix elements $P_{qq'}^{ss'}$ for three-phonon processes from first-order time-dependent perturbation theory. The expression for τ_{qs} will also be derived from the projection operator method in section 6.4.1.2 and from the double-time Green's function method in section 6.4.1.3. It will be noticed that the three methods give identical expressions for τ_{qs}. The Debye method will be used to express $\tau^{-1}(smrt)$ and $\tau_{U,eff}^{-1}$ (the effective relaxation time for U processes in Srivastava's model, described in section 5.2.5) in integral form in section 6.4.1.4. In section 6.4.1.5 we will derive simple expressions for the relaxation rate for $ac - op$ three-phonon interaction.

6.4.1.1   Expressions for τ_{qs} and $P_{qq'}^{ss'}$ from time-dependent perturbation theory

Let us consider an initial state of the phonon system in which three phonons are identified with their modes: $|i\rangle \equiv |n_{qs}, n_{q's'}, n_{q''s''}\rangle$. Following the discussion in section 4.3, we consider the presence of anharmonic perturbation $\mathscr{V}_3$ to cause the system to change in time t to a final state: *for class 1* three-phonon events $|f\rangle = |n_{qs} - 1, n_{q's'} - 1, n_{q''s''} + 1\rangle$, and *for class 2 events* $|f\rangle = |n_{qs} - 1, n_{q's'} + 1, n_{q''s''} + 1\rangle$. The rate of occurrence of such a process per unit time, or the transition probability, is given by the golden rule formula

$$P_i^f(3\,ph) = \frac{2\pi}{\hbar}\left|\langle f|\mathscr{V}_3|i\rangle\right|^2 \delta(E_f - E_i), \tag{6.45}$$

where $E_i(E_f)$ is the initial (final) state energy of the three-phonon system: for class 1 events $E_f - E_i = \hbar(\omega(q''s'') - \omega(qs) - \omega(q's'))$, and for class 2 events $E_f - E_i = \hbar(\omega(qs) - \omega(q's') - \omega(q''s''))$.

For class 1 events equation (6.45) becomes

$$
\begin{aligned}
P_{qs,q's'}^{q''s''} &= \frac{2\pi}{\hbar^2}\left|\langle n_{qs} - 1, n_{q's'} - 1, n_{q''s''} + 1|\mathscr{V}_3|n_{qs}, n_{q's'}, n_{q''s''}\rangle\right|^2 \\
&\times \delta(\omega(qs) - \omega(q's') - \omega(q''s'')).
\end{aligned} \tag{6.46}
$$

For the isotropic continuum model, we have from equations (4.83)–(4.86)

$$
\begin{aligned}
\mathscr{V}_3 &= \frac{1}{3!}\sqrt{\frac{\hbar^3}{8\rho^3 N_0 \Omega}} \sum_{qsq's'q''s''} \sqrt{\frac{qq'q''}{c_s c_{s'} c_{s''}}} A_{qq'q''}^{ss's''} \delta_{q+q'+q'',G} \\
&\times (a_{qs}^\dagger - a_{-qs})(a_{-q's'}^\dagger - a_{q's'})(a_{q''s''}^\dagger - a_{-q''s''}),
\end{aligned} \tag{6.47}
$$

where $A_{qq'q''}^{ss's''}$ are the Fourier components of the phonon coupling constants. For a class 1 three-phonon process the matrix element in equation (6.46) reduces to

$$
\begin{aligned}
&\sqrt{\frac{\hbar^3}{8\rho^3 N_0 \Omega}}\sqrt{\frac{qq'q''}{c_s c_{s'} c_{s''}}} A_{qq'q''}^{ss's''} \delta_{q+q'+q'',G} \\
&\times \langle n_{qs} - 1, n_{q's'} + 1, n_{q''s''} + 1|a_{-qs}a_{-q's'}a_{q''s''}^\dagger|n_{qs}n_{q's'}n_{q''s''}\rangle.
\end{aligned} \tag{6.48}
$$

The factor $1/3!$ is cancelled out due to $3!$ equivalent terms from the summation in equation (6.47). Thus

$$
\begin{aligned}
P_{qs,q's'}^{q''s''} &= \frac{\pi\hbar}{4\rho^3 N_0 \Omega}\frac{qq'q''}{c_s c_{s'} c_{s''}}\left|A_{qq'q''}^{ss's''}\right|^2 n_{qs}n_{q's'}(n_{q''s''} + 1) \\
&\times \delta_{q+q'+q'',G}\, \delta(\omega(q''s'') - \omega(qs) - \omega(q's')).
\end{aligned} \tag{6.49}
$$

The expression for the transition probability for a class 2 event can be similarly worked out:

$$
\begin{aligned}
P_{qs}^{q's'q''s''} &= \frac{\pi}{4\rho^3 N_0 \Omega}\frac{qq'q''}{c_s c_{s'} c_{s''}}\left|A_{qq'q''}^{ss's''}\right|^2 (n_{qs} + 1)n_{q's'}n_{q''s''} \\
&\times \delta_{q+q'+q'',G}\, \delta(\omega(qs) - \omega(q's') - \omega(q''s'')).
\end{aligned} \tag{6.50}
$$

To obtain an expression for $P_{qq'}^{ss'}(3\,ph)$ we note from section 5.1.1 that

$$-\frac{\partial n_{qs}}{\partial t}\bigg|_{3\,ph} = \sum_{q's'} P_{qq'}^{ss'}(3\,ph)\psi_{q'}^{s'}$$

$$
= \sum_{\substack{q''s'' \\ q'''s'''}} \left[\left(P_{qs,q'''s'''}^{q''s''} - P_{q''s''}^{qs,q'''s'''} \right) \right.
$$

$$
\left. + \frac{1}{2} \left(P_{qs}^{q'''s''',q''s''} - P_{q'''s''',q''s''}^{qs} \right) \right].
$$

(6.51)

The first and second bracketed terms correspond to class 1 and class 2 events, respectively. The factor $1/2$ within the expression for class 2 events is to avoid double counting in the summation. Linearising the terms in equation (6.51) by writing $n_{qs} = \bar{n}_{qs} + \psi_{qs}\bar{n}_{qs}(\bar{n}_{qs} + 1)$, etc, we get

$$
-\frac{\partial n_{qs}}{\partial t}\Big|_{3\,ph} = \sum_{q's'} P_{qq'}^{ss'}(3\,ph)\,\psi_{q'}^{s'}
$$

$$
= \sum_{\substack{q''s'' \\ q'''s'''}} \left[\bar{P}_{qs,q'''s'''}^{q''s''} \left(\psi_q^s + \psi_{q'''}^{s'''} - \psi_{q''}^{s''} \right) \right.
$$

$$
\left. + \frac{1}{2} \bar{P}_{qs}^{q'''s''',q''s''} \left(\psi_q^s - \psi_{q'''}^{s'''} - \psi_{q''}^{s''} \right) \right].
$$

(6.52)

This yields

$$
P_{qq'}^{ss'}(3\,ph) = \sum_{\substack{q''s'' \\ q'''s'''}} \left[\frac{1}{2} \delta_{ss'}\delta_{qq'} \left\{ \bar{P}_{qs,q'''s'''}^{q''s''} + \bar{P}_{qs}^{q'''s''',q''s''} + \bar{P}_{qs,q''s''}^{q'''s'''} \right\} \right.
$$

$$
\left. - \delta_{s's'''}\delta_{q'q'''} \left\{ \bar{P}_{qs,q''s''}^{q'''s'''} + \bar{P}_{q''s'',q'''s'''}^{qs} - \bar{P}_{q'''s''',qs}^{q''s''} \right\} \right]
$$

(6.53)

$$
= \Gamma_{qs}\delta_{qq'}\delta_{ss'} + \Lambda_{qq'}^{ss'},
$$

(6.54)

where Γ and Λ are the diagonal and off-diagonal parts, respectively, of the phonon collision operator P.

In the single-mode relaxation time approximation it is assumed that only phonons in mode qs have a displaced distribution, and all other phonons have their equilibrium distribution (i.e. $\psi_{qs} \neq 0$, but $\psi_{q's'} = \psi_{q''s''} = 0$ in equation (6.52)). Thus, as in equation (5.24), we write

$$
-\frac{\partial n_{qs}}{\partial t}\Big|_{smrt} = \frac{n_{qs} - \bar{n}_{qs}}{\tau_{qs}} = \Gamma_{qs}\psi_{qs}
$$

(6.55)

giving the required expression for the *smrt* describing three-phonon processes,

$$
\tau_{qs}^{-1}(3\,ph) = \frac{\Gamma_{qs}}{\bar{n}_{qs}(\bar{n}_{qs} + 1)}
$$

$$
= \frac{1}{\bar{n}_{qs}(\bar{n}_{qs} + 1)} \sum_{\substack{q's' \\ q''s''}} \left(\bar{P}_{qs,q's'}^{q''s''} + \frac{1}{2} \bar{P}_{qs}^{q's',q''s''} \right)
$$

(6.56)

$$
= \frac{\pi\hbar}{4\rho^3 N_0 \Omega} \sum_{q's'q''s''} |A_{qq'q''}^{ss's''}|^2 \frac{qq'q''}{c_s c_{s'} c_{s''}} \delta_{q+q'+q'',G}
$$

$$
\times \left\{ \frac{\bar{n}_{q's'}(\bar{n}_{q''s''} + 1)}{(\bar{n}_{qs} + 1)} \delta\left(\omega(qs) + \omega(q's') - \omega(q''s'')\right) \right.
$$

$$
\left. + \frac{1}{2} \frac{\bar{n}_{q's'}\bar{n}_{q''s''}}{\bar{n}_{qs}} \delta\left(\omega(qs) - \omega(q's') - \omega(q''s'')\right) \right\}.
$$

(6.57)

6.4.1.2 Expression for τ_{qs} from the projection operator method

The Zwanzig–Mori projection operator method was described in section 5.4.1. Using this approach, the expression for the conductivity is given in equation (5.255). That expression contains the phonon relaxation time τ_{qs} which is defined in equation (5.254):

$$\tau_{qs}^{-1} = \frac{1}{\hbar^2} \int_{-\infty}^{\infty} dt e^{i\omega t} \langle [S_{qs}(t), S_{qs}] \rangle_0. \tag{6.58}$$

If the coupling coefficient S is derived from the cubic anharmonicity, then from equations (5.225), (5.271) and (6.47),

$$\begin{aligned} \mathscr{V}_3 &= \sum_{qs} S_{qs} A_{qs} \\ &= \sum_{qs,q's',q''s''} \mathscr{V}_3(qs,q's',q''s'') A_{qs} A_{q's'} A_{q''s''} \end{aligned} \tag{6.59}$$

with

$$\mathscr{V}_3(qs,q's',q''s'') = \frac{1}{3!} \sqrt{\frac{\hbar^3}{8\rho^3 N_0 \Omega}} \sqrt{\frac{qq'q''}{c_s c_{s'} c_{s''}}} A_{qq'q''}^{ss's''} \delta_{q+q'+q'',G}. \tag{6.60}$$

It should be remembered that the expression in equation (6.59) includes 3! equivalent terms from the summation over $qs,q's',q''s''$. From the above equations, we have

$$\begin{aligned} \langle [S_{qs}(t), S_{qs}] \rangle_0 &= \sum_{q's'q''s''} \left| f_3(qs,q's',q''s'') \right|^2 \\ &\quad \times \langle [A_{q's'}(t) A_{q''s''}(t), A_{q's'} A_{q''s''}] \rangle_0, \end{aligned} \tag{6.61}$$

where

$$\left| f_3(qs,q's',q''s'') \right|^2 = \frac{\hbar^3}{8\rho^3 N_0 \Omega} \frac{qq'q''}{c_s c_{s'} c_{s''}} \left| A_{qq'q''}^{ss's''} \right|^2 \delta_{q+q'+q'',G}. \tag{6.62}$$

The term $\langle \ldots \rangle_0$ in equation (6.61) can be evaluated by first using Wick's factorisation scheme and then using equations (5.235) and (5.239). For class 1 events

$$\begin{aligned} & \langle [a_{q's'}(t) A_{q''s''}(t), A_{q's'} A_{q''s''}] \rangle_0 |_{class\,1} \\ &= \langle a_{q's'}^{\dagger}(t), a_{-q's'} \rangle_0 \langle a_{-q''s''}(t) a_{q''s''}^{\dagger} \rangle_0 - \langle a_{q's'}^{\dagger} a_{-q's'}(t) \rangle_0 \langle a_{-q''s''} a_{q''s''}^{\dagger}(t) \rangle_0 \\ &= \exp[i(\omega(q's') - \omega(q''s''))t](\bar{n}_{q's'} - \bar{n}_{q''s''}). \end{aligned} \tag{6.63}$$

Similarly for class 2 events

$$\begin{aligned} & \langle [A_{q's'}(t) A_{q''s''}(t), A_{q's'} A_{q''s''}] \rangle_0 \Big|_{class\,2} \\ &= \exp[-i(\omega(q's') - \omega(q''s''))t](1 + \bar{n}_{q's'} + \bar{n}_{q''s''}). \end{aligned} \tag{6.64}$$

From equations (6.58) and (6.61)–(6.64) we get

$$\begin{aligned} \tau_{qs}^{-1} &= \frac{1}{\hbar^2} \int_{-\infty}^{\infty} dt e^{i\omega t} \sum_{q's'q''s''} \left| f_3(qs,q's',q''s'') \right|^2 \\ &\quad \times \Big\{ (\bar{n}_{q's'} - \bar{n}_{q''s''}) \exp[i(\omega(q's') - \omega(q''s''))t] \\ &\quad + \frac{1}{2}(1 + \bar{n}_{q's'} + \bar{n}_{q''s''}) \exp[-i(\omega(q's') + \omega(q''s''))t] \Big\}. \end{aligned} \tag{6.65}$$

The factor $1/2$ in the second term is introduced to avoid double counting of equivalent terms. Using the relation

$$\int_0^\infty dt\, e^{ixt} = \pi\delta(x) - i\wp\left(\frac{1}{x}\right) \tag{6.66}$$

we get

$$\int_{-\infty}^\infty dt \exp\left[i(\omega(qs) \pm \omega(q's') - \omega(q''s''))t\right] = 2\pi\delta\left(\omega(qs) \pm \omega(q's') - \omega(q''s'')\right). \tag{6.67}$$

Thus

$$\begin{aligned}
\tau_{qs}^{-1} &= \frac{2\pi}{\hbar^2} \sum_{q's'q''s''} |f_3(qs,q's',q''s'')|^2 \\
&\quad \times \Big\{ (\bar{n}_{q's'} - \bar{n}_{q''s''})\delta\left(\omega(qs) + \omega(q's') - \omega(q''s'')\right) \\
&\quad + \frac{1}{2}(1 + \bar{n}_{q's'} + \bar{n}_{q''s''})\delta\left(\omega(qs) - \omega(q's') - \omega(q''s'')\right) \Big\}.
\end{aligned} \tag{6.68}$$

Further, for class 1 events, we note the identity

$$\bar{n}_{qs}\bar{n}_{q's'}(\bar{n}_{q''s''} + 1) \equiv (\bar{n}_{qs} + 1)(\bar{n}_{q's'} + 1)\bar{n}_{q''s''}. \tag{6.69}$$

From this, we express

$$\bar{n}_{q's'} - \bar{n}_{q''s''} = \frac{(\bar{n}_{q's'} + 1)\bar{n}_{q''s''}}{\bar{n}_{qs}} = \frac{\bar{n}_{q's'}(\bar{n}_{q''s''} + 1)}{(\bar{n}_{qs} + 1)}. \tag{6.70}$$

For class 2 events, we have the identity

$$\bar{n}_{qs}(1 + \bar{n}_{q's'})(1 + \bar{n}_{q''s''}) = (\bar{n}_{qs} + 1)\bar{n}_{q's'}\bar{n}_{q''s''} \tag{6.71}$$

from which we express

$$1 + \bar{n}_{q's'} + \bar{n}_{q''s''} = \frac{\bar{n}_{q's'}\bar{n}_{q''s''}}{\bar{n}_{qs}}. \tag{6.72}$$

Therefore, combining equations (6.62) and (6.68)–(6.72) we get

$$\begin{aligned}
\tau_{qs}^{-1} &= \frac{\pi\hbar}{4\rho^3 N_0\Omega} \sum_{q's'q''s''} \left|A_{qq'q''}^{ss's''}\right|^2 \frac{qq'q''}{c_s c_{s'} c_{s''}} \delta_{q+q'+q'',G} \\
&\quad \times \Bigg(\frac{\bar{n}_{q's'}(\bar{n}_{q''s''} + 1)}{(\bar{n}_{qs} + 1)} \delta\left(\omega(qs) + \omega(q's') - \omega(q''s'')\right) \\
&\quad + \frac{1}{2}\frac{\bar{n}_{q's'}\bar{n}_{q''s''}}{\bar{n}_{qs}} \delta\left(\omega(qs) - \omega(q's') - \omega(q''s'')\right) \Bigg)
\end{aligned} \tag{6.73}$$

which is exactly the same result as derived in equation (6.57) from perturbation theory. This is so expected, as equation (5.254) holds in the weak coupling limit in which first-order perturbation is applicable.

6.4.1.3   Expressions for τ_{qs} and Δ_{qs} from the double-time Green's function method

In section 5.4.2 we used the double-time Green's function method and obtained equation (5.284) for the lattice thermal conductivity. That result contains the phonon life time τ_{qs} and the phonon frequency shift Δ_{qs} which are given in equations (5.287) and (5.288), respectively. Here we simplify those expressions for τ_{qs} and Δ_{qs}.

From equation (5.287) we have

$$
\begin{aligned}
\tau_{qs}^{-1} &= \frac{72\pi}{\hbar^2} \sum_{q's'q''s''} |\mathcal{V}_3(qs,q's',q''s'')|^2 \\
&\quad \times \big[(\bar{n}_{q's'} + \bar{n}_{q''s''} + 1)(\omega(q's') + \omega(q''s'')) \\
&\quad \times \delta(\varepsilon^2(qs) - (\omega(q's') + \omega(q''s''))^2) \\
&\quad + (\bar{n}_{q''s''} - \bar{n}_{q's'})(\omega(q's') - \omega(q''s'')) \\
&\quad \times \delta(\varepsilon^2(qs) - (\omega(q's') - \omega(q''s''))^2)\big].
\end{aligned}
\tag{6.74}
$$

Using equation (6.60) and the identity

$$
\delta((x^2 - a^2)) = \frac{1}{|2a|}[\delta(x-a) + \delta(x+a)], \quad \text{for } x > 0
\tag{6.75}
$$

we can write

$$
\begin{aligned}
\tau_{qs}^{-1}(\varepsilon) &= \frac{2\pi}{\hbar^2} \sum_{q's'q''s''} |f_3(qs,q's',q''s'')|^2 \Big[\frac{1}{2}(\bar{n}_{q's'} + \bar{n}_{q''s''} + 1) \\
&\quad \times \{\delta(\varepsilon(qs) - \omega(q's') - \omega(q''s'')) + \delta(\varepsilon(qs) + \omega(q's') + \omega(q''s''))\} \\
&\quad + \frac{1}{2}(\bar{n}_{q''s''} - \bar{n}_{q's'}) \\
&\quad \times \{\delta(\varepsilon(qs) - \omega(q's') + \omega(q''s'')) - \delta(\varepsilon(qs) + \omega(q's') - \omega(q''s''))\}\Big] \\
&= \frac{2\pi}{\hbar^2} \sum_{q's'q''s''} |f_3(qs,q's',q''s'')|^2 \\
&\quad \times \Big[\frac{1}{2}(\bar{n}_{q's'} + \bar{n}_{q''s''} + 1)\delta(\varepsilon(qs) - \omega(q's') - \omega(q''s'')) \\
&\quad + (\bar{n}_{q's'} - \bar{n}_{q''s''})\delta(\varepsilon(qs) + \omega(q's') - \omega(q''s''))\Big],
\end{aligned}
\tag{6.76}
$$

where we have neglected the term with $\delta(\varepsilon(qs) + \omega(q's') + \omega(q''s''))$ on the grounds of energy conservation. Using equations (6.62), (6.70) and (6.72), we can express equation (6.76) as

$$
\begin{aligned}
\tau_{qs}^{-1}(\varepsilon) &= \frac{\pi\hbar}{4\rho^3 N_0 \Omega} \sum_{q's'q''s''} \left|A_{qq'q''}^{ss's''}\right|^2 \frac{qq'q''}{c_s c_{s'} c_{s''}} \delta_{q+q'+q'',G} \\
&\quad \times \Big[\frac{\bar{n}_{q's'}(\bar{n}_{q''s''} + 1)}{(\bar{n}_{qs} + 1)}\delta(\varepsilon(qs) + \omega(q's') - \omega(q''s'')) \\
&\quad + \frac{1}{2}\frac{\bar{n}_{q's'}\bar{n}_{q''s''}}{\bar{n}_{qs}}\delta(\varepsilon(qs) - \omega(q's') - \omega(q''s''))\Big].
\end{aligned}
\tag{6.77}
$$

This is the same result as in equations (6.57) and (6.73) provided the renormalised phonon frequency $\varepsilon(qs)$ in the above result is considered equal to the harmonic phonon frequency $\omega(qs)$. Thus we see that all the three methods described in Chapter 5 produce the same result for the single-mode phonon relaxation time.

Similarly, from equation (5.288) the phonon frequency shift can be expressed as

$$
\begin{aligned}
\Delta_{qs} &= -\frac{1}{\hbar^2}\wp \sum_{q's'q''s''} |f_3(qs,q's',q''s'')|^2 \\
&\quad \times \Big[\frac{1}{2}\frac{(\bar{n}_{q's'} + \bar{n}_{q''s''} + 1)}{\varepsilon(qs) - \omega(q's') - \omega(q''s'')} + \frac{(\bar{n}_{q's'} - \bar{n}_{q''s''})}{\varepsilon(qs) + \omega(q's') - \omega(q''s'')}\Big] \\
&= -\frac{\hbar}{8\rho^3 N_0 \Omega} \sum_{q's'q''s''} \left|A_{qq'q''}^{ss's''}\right|^2 \frac{qq'q''}{c_s c_{s'} c_{s''}} \delta_{q+q'+q'',G}
\end{aligned}
\tag{6.78}
$$

$$\times \left[\frac{(\bar{n}_{q's'} - \bar{n}_{q''s''})}{(\varepsilon(qs) + \omega(q's') - \omega(q''s''))}_{\wp} \right.$$

$$\left. + \frac{1}{2} \frac{(\bar{n}_{q's'} + \bar{n}_{q''s''} + 1)}{(\varepsilon(qs) - \omega(q's') - \omega(q''s''))}_{\wp} \right]. \tag{6.79}$$

6.4.1.4 Acoustic–acoustic phonon interaction in the Debye model

The results for the phonon relaxation time and the frequency shift can in principle be evaluated by using an analytical procedure for Brillouin zone summation (e.g. the Gilat–Raubenheimer method described in section 2.6.2.1). However, it is much simpler to use the Debye model, described in sections 2.6.4 and 2.7.2. In this section we will treat three-phonon acoustic–acoustic interaction within the Debye model. A scheme for acoustic–optical interaction will be discussed in the next section.

Changing $\sum_{q's'} \to (N_0 \Omega/8\pi^3) \sum_{s'} \int d^3 q'$ and replacing the sum over q'' in the light of the Kronecker delta symbol, we express equation (6.57) as

$$\tau_{qs}^{-1} = \frac{\pi \hbar}{4\rho^3 N_0 \Omega} \frac{N_0 \Omega}{8\pi^3} 2\pi \sum_{\substack{s's'' \\ G}} \frac{1}{c_s c_{s'} c_{s''}} \int_0^\pi d\theta' \sin \theta' \int dq' q'^2 qq'q'' \left| A_{qq'q''}^{ss's''} \right|^2$$

$$\times \left(\frac{\bar{n}_{q's'}(\bar{n}_{q''s''} + 1)}{(\bar{n}_{qs} + 1)} \delta\left(\omega(qs) + \omega(q's') - \omega(q''s'') \right) \right.$$

$$\left. + \frac{1}{2} \bar{n}_{q's'} \bar{n}_{q''s''} \bar{n}_{qs} \delta\left(\omega(qs) - \omega(q's') - \omega(q''s'') \right) \right), \tag{6.80}$$

where

$$|q''| = |G - (q \pm q')| \tag{6.81}$$

with $+(-)$ sign for class 1 (2) events.

For N-processes $G = 0$. For U-processes we choose within the isotropic continuum model (Parrott 1963, Hamilton and Parrott 1969, Srivastava 1976a) (also see equations (4.51)–(4.52))

$$G = 2q_D \frac{q \pm q'}{|q \pm q'|}, \tag{6.82}$$

where again the $+$ and $-$ signs correspond to class 1 and class 2 events, respectively. With equation (6.82) we can express equation (6.81) as

$$|q''| = (1 - \varepsilon)q_D + \varepsilon|q \pm q'| \tag{6.83}$$

with $\varepsilon = +1(-1)$ for $N(U)$ processes, and q_D is the Debye radius. Expressions for $|A_{qq'q''}^{ss's''}|^2$ are given in equations (4.96)–(4.102) and (4.105)–(4.111). In general these three-phonon scattering strengths depend on angles between q, q' and q'', but one may consider angularly averaged values as presented in equations (4.105)–(4.111). Or, if preferred, these coefficients can be expressed in terms of the Grüneisen constant as in equation (4.135) or (4.138). With angularly averaged values, we can express equation (6.80) in the Debye model as

$$\tau_{qs}^{-1} = \frac{\hbar}{16\pi\rho^3} \sum_{\substack{s's'' \\ \varepsilon}} \frac{\overline{|A_{qq'q''}^{ss's''}|^2}}{c_s c_{s'} c_{s''}^2} \int dq' q'^2 qq'$$

$$\times \left(\frac{\bar{n}_{q's'}(\bar{n}_{q''s''} + 1)}{(\bar{n}_{qs} + 1)} \int_0^\pi d\theta' \sin \theta' q'' \delta(q'' - Cq - Dq') \right.$$

$$+\frac{1}{2}\frac{\bar{n}_{q's'}\bar{n}_{q''s''}}{\bar{n}_{qs}}\int_0^\pi d\theta' \sin\theta' q'' \delta(q''-Cq+Dq')\Bigg) \tag{6.84}$$

with $C=c_s/c_{s''}, D=c_{s'}/c_{s''}$. The angular integrals can be evaluated by using the Dirac delta function (see Appendix E), and thus

$$\tau_{qs}^{-1} = \frac{\hbar q_D^5}{16\pi\rho^3}\sum_{s's''}\underbrace{\frac{\overline{|A_{qq'q''}^{ss's''}|^2}}{c_s c_{s'} c_{s''}^2}}_{\varepsilon}\Bigg(\int dx' x'^2 x''_+\{1-\varepsilon+\varepsilon(Cx+Dx')\}\frac{\bar{n}_{q's'}(\bar{n}''_++1)}{(\bar{n}_{qs}+1)}$$

$$+\frac{1}{2}\int dx' x'^2 x''_-\{1-\varepsilon+\varepsilon(Cx-Dx')\}\frac{\bar{n}_{q's'}\bar{n}''_-}{\bar{n}_{qs}}\Bigg) \tag{6.85}$$

with $x=q/q_D, x'=q'/q_D, x''_\pm=Cx\pm Dx'$ and $\bar{n}''_\pm=\bar{n}(x''_\pm)$.

The expression derived in equation (6.85) is the single-mode relaxation time (smrt) defined in equation (6.57). The Debye model can also be used, with the help of equation (6.82), to evaluate the effective U-processes relaxation time τ_U^{eff} defined in equation (5.57). The result is

$$\tau_{U,eff}^{-1} = \frac{\hbar q_D^5}{8\pi\rho^3}\sum_{s's''}\frac{\overline{|A_{qq'q''}^{ss's''}|^2}}{c_s c_{s'} c_{s''}^2}$$

$$\times\Bigg[\int dx' x'^2\left(1+\frac{x'\mu'_+}{x}\right)\frac{\bar{n}_{q's'}(\bar{n}''_++1)}{(\bar{n}_{qs}+1)}$$

$$+\frac{1}{2}\int dx' x'^2\left(1-\frac{x'\mu'_-}{x}\right)\frac{\bar{n}_{q's'}\bar{n}''_-}{\bar{n}_{qs}}\Bigg] \tag{6.86}$$

with

$$\mu'_\pm = \frac{\{2-(Cx\pm Dx')\}^2 - x^2 - x'^2}{2xx'}. \tag{6.87}$$

The momentum and energy conservation conditions impose certain restrictions on the integration variable x'. The following inequalities can be worked out from equations (4.48)–(4.49) (Srivastava 1974, 1976a, Mikhail and Madkour 1985):

Class 1 events:

$$0 \leq x \leq 1$$
$$0, \frac{(1-C)x}{(1+D)} \leq x' \leq \frac{(1+C)x}{(1-D)}, \frac{(1-Cx)}{D}, 1 \qquad N\text{ processes}$$
$$0, \frac{(2-(1+C)x)}{(1+D)} \leq x' \leq \frac{(1-Cx)}{D}, 1 \qquad U\text{ processes} \tag{6.88}$$

Class 2 events:
N processes:

$$0 \leq x \leq 1$$
$$0, \frac{(C-1)x}{D+1}, \frac{(Cx-1)}{D} \leq x' \leq \frac{(C+1)x}{D+1}, \frac{(C-1)x}{D-1}, 1$$

U processes:

$$\frac{2}{1+C} \leq x \leq 1$$
$$0, \frac{2-(1+C)x}{1-D}, \frac{Cx-1}{D}, \frac{(C+1)x-2}{D+1} \leq x' \leq \frac{(C+1)x-2}{D-1}, 1. \tag{6.89}$$

It is evident from equation (6.85) that $\tau_{qs}^{-1} \propto g(\omega, T)$, where g is a rather complicated function of frequency and temperature. However, it would be interesting and might be useful to investigate the dependence of τ_{qs} in low- and high-temperature regions.

At high temperatures (HT) $\bar{n}_{qs} \simeq k_B T/\hbar\omega$, etc, and one gets from equation (6.85)

$$\tau^{-1}(HT) = (A_1\omega + A_2\omega^2)T \tag{6.90}$$

where A_1 and A_2 are constants with their relative strength being dependent on the range of ω, the combination $(ss's'')$ for a three-phonon process and temperature. (Notice that in this discussion we have dropped the argument qs from τ and ω for simplicity.)

At low temperatures (LT) $\bar{n}_{qs} \simeq \exp(-\hbar\omega/k_B T)$, $\bar{n}_{qs}+1 \simeq 1$ etc. Let us first discuss N-processes. For class 2 events we get

$$\tau_N^{-1}(s \rightarrow s' + s'') \propto \int_{ax}^{bx} dx' x'^2 (Cx - Dx')^2 \propto \omega^5 \tag{6.91}$$

where a and b are appropriately calculated from equation (6.89). For class 1 events, we get

$$\begin{aligned}
\tau_N^{-1}(s + s' \rightarrow s'') &\propto \int dx' x'^2 e^{-\alpha x'} (Cx + Dx')^2 \\
&= (c_1\omega^2 T^3 + c_2\omega T^4 + c_3 T^5)
\end{aligned} \tag{6.92}$$

where $\alpha \propto 1/T$ and we have used

$$\int dx' x'^n e^{-\alpha x'} \propto T^{n+1}.$$

Thus

$$\tau_N^{-1}(LT) \propto D_1 f_1(\omega) + D_2 f_2(\omega, T) \tag{6.93}$$

with the functions f_1 and f_2 appropriately obtained from equation (6.91). For U-processes, we note that $\omega' \geq \omega'_{min}$ in accordance with the energy and momentum conservation rules. Then for class 1 events

$$\tau_U^{-1}(s + s' \rightarrow s'') \propto \int_{x'_{min}} dx' x'^2 (Cx + Dx')(2 - Cx - Dx') e^{-\alpha x'}$$

$$\propto (a_1 T^5 + a_2\omega T^4 + a_3\omega^2 T^3) \exp(-\hbar\omega'_{min}/k_B T). \tag{6.94}$$

If $\omega'_{min} = 0$, a similar approach leads to

$$\tau_U^{-1}(s + s' \rightarrow s'') \propto T^3(b_1\omega + b_2\omega^2). \tag{6.95}$$

For class 2 events, decay of high-frequency phonons can be approximated to

$$\begin{aligned}
\tau_U^{-1}(s \rightarrow s' + s'') &\propto (c_1 g_1(\omega) + c_2 T^3 g_2(\omega)) \tag{6.96} \\
\text{or} &\propto (c_1 g_1(\omega) + c_3 g_3(\omega, T)) \exp(-\hbar\omega'_{min}/k_B T). \tag{6.97}
\end{aligned}$$

Thus

$$\tau_U^{-1}(LT) \propto Ag(\omega) + \underbrace{Bg(\omega)f(T)}_{\text{or} e^{-\sigma/T} g_3(\omega, T)} \tag{6.98}$$

where, again, A, B and σ are constants, and $f(T)$ may be T^3.

Some of the above results have been derived by various researchers using more simplified approaches. Herring (1954) derived the formula $\tau^{-1}(LT) \propto \omega^m T^{5-m}$, with $m = 1, 2, 3$ or 4, which is also seen in equations (6.92) and (6.96)–(6.97). The class 1 N process $T + L \rightarrow L$ in which a low-frequency transverse phonon ($\hbar\omega_T \ll k_B T \ll k_B\Theta_D$) participates, is known as the Landau–Rumer process (Landau and Rumer 1937). The main contribution to this process comes from the second term in equation (6.92), i.e. $\tau^{-1}(T+L \rightarrow L) \propto \omega T^4$ (see also Kwok (1967)). The relaxation rate of low-frequency (sub-thermal) longitudinal acoustic phonons via the processes $LA + TA \rightarrow LA$, $LA \rightarrow TA + TA$, and $LA \rightarrow TA + LA$ follows the frequency and temperature of the type (Pomeranchuk 1941, Kwok 1967): $\tau_{LA}^{-1}|_{\hbar\omega \ll k_B T} \propto \omega^4 T$.

As discussed in section 4.4 the allowed $ac + ac \rightleftharpoons ac$ processes, within the Debye model, are class 1 events:

$$L + L \rightarrow L \qquad N \text{ processes only}$$

$$L + T \rightarrow L; T + L \rightarrow L; T + T \rightarrow L \qquad N \text{ and } U \text{ processes}$$

class 2 events:

$$L \rightarrow L + L \qquad N \text{ processes only}$$

$$L \rightarrow L + T; L \rightarrow T + T; L \rightarrow T + L \qquad N \text{ and } U \text{ processes}$$

The process $T + T \rightleftharpoons T$ may be allowed but has zero strength. The areas of integrations in the $(x - x')$ space for the allowed processes are shown in figure 6.1. Although we have listed all possible

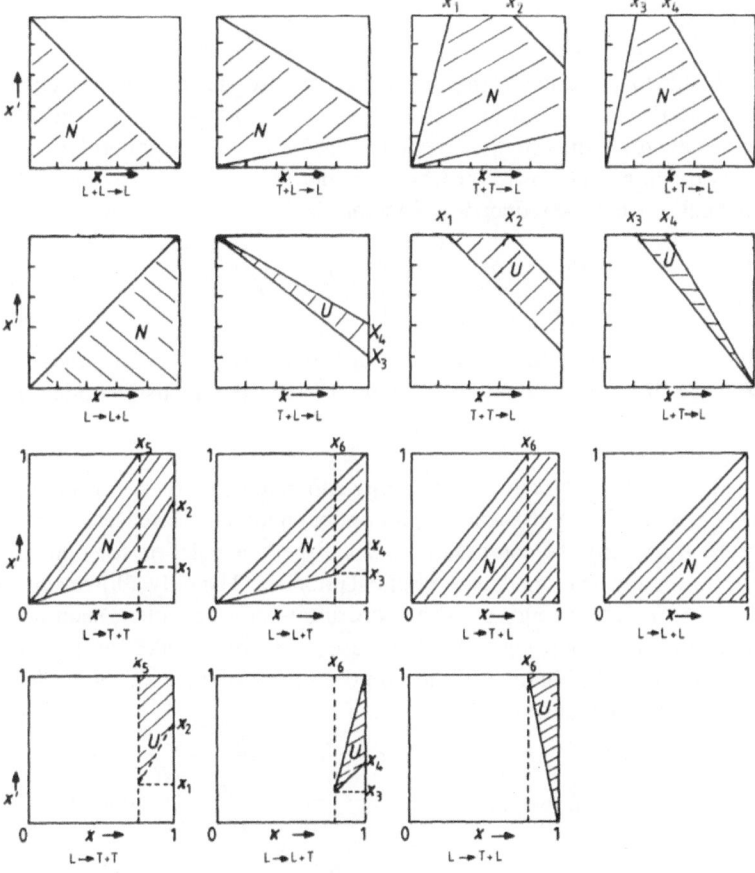

Figure 6.1 Areas of integration in the $(q - q')$ plane for the allowed three-phonon processes. The following symbols are used in the figure: $r = c_T/c_L$, $x_1 = (1-r)/(1+r)$, $x_2 = (1-r)/r$, $x_3 = (1-r)/2$, $x_4 = 1-r$, $x_5 = 2r/(1+r)$, $x_6 = (1+r)/2$.

processes, some of these become indistinguishable in a particular calculation. For example, in the second term in equation (6.56), the processes $L \rightarrow L + T$ and $L \rightarrow T + L$ can be represented by only one distinct process $L \rightarrow L + T$ (or $L \rightarrow T + L$). Similarly, when evaluating a matrix element of the type $(\phi, P\phi)$ it is sufficient to consider only one of the two processes $T + L \rightarrow L$ and $L + T \rightarrow L$.

Herring (1954) has used group theoretical considerations to study the role of energy and momentum conservation in the evaluation of anharmonic phonon relaxation times in different crystal systems (see also Ziman 1960).

6.4.1.5 Acoustic–optical phonon interaction

The interaction between acoustic and optical phonon branches is rather restrictive. Processes of the types $ac + op \rightleftharpoons ac$ and $op + op \rightleftharpoons ac$ are forbidden on the grounds of energy conservation. (Here ac and op mean acoustic and optical phonon branches, respectively.) Processes of the type $ac + op_1 \rightleftharpoons op_2$ are very restrictive in that op_2 must be a higher lying branch than op_1. Processes of the type $ac + ac \rightleftharpoons op$ are possible within the constraints of the energy selection rule. Blackman (1935), Ziman (1960), and Ecsedy and Klemens (1977) have all discussed these processes. The discussion here follows Srivastava (1980).

Assume that there are only two types of atomic masses present. Within the isotropic continuum approximation, we can study the interaction processes $ac + op \rightleftharpoons op$ and $ac + ac \rightleftharpoons op$ under two, rather extreme, limits.

(i) $m_1 / m_2 \simeq 1$:

For solids with the mass ratio $m_1 / m_2 \simeq 1$, we can use the continuum dispersion relation $\omega(qs) = c_s q$ for acoustic modes and $\omega_{op} = \omega_D + \omega_g + c_{op}(q_D - q)$ for optical modes. Here ω_D is the Debye frequency, ω_g is the phonon frequency gap at the Debye sphere and c_{op} is the optical phonon velocity. Following Klemens (1966), we consider the cubic anharmonic Hamiltonian in the presence of an optical mode to be reduced by a factor

$$\sqrt{R} = (\frac{2}{\sqrt{3}} \frac{\beta_1 - \beta_2}{\beta_1 + \beta_2})^{1/2},$$

where β_1 and β_2 are two effective force constants such that $\beta_1 / \beta_2 = (1 + \omega_g / \omega_D)^2$. The expression in equation (6.85) can then be modified accordingly to study the acoustic–optical phonon interactions.

(ii) $1 < m_1 / m_2 \leq 4$:

As a general result of lattice dynamics, the energy gap in the phonon spectrum increases with the mass ratio, and the group velocity associated with optical phonons decreases. Therefore for m_1 / m_2 significantly greater than unity we can assume a flat dispersion relation for optical phonons. This model does not allow for any heat transport by optical phonons. The only allowed $ac - op$ interaction process is $ac + ac \rightarrow op$. The energy conservation rule would not allow such an interaction for $m_1 / m_2 > 4$. Therefore for $1 < m_1 / m_2 \leq 4$ we can use the Debye model for acoustic phonons and an Einstein model for optical phonons. In the proposed Debye–Einstein model for $ac + ac \rightarrow op$ interaction equation (6.84) reduces to

$$\tau_{qs}^{-1}(ac + ac \rightarrow op)|_{\text{Debye–Einstein}} = \frac{\hbar}{8\pi\rho^3} \sum_{\substack{s's'' \\ \varepsilon}} \frac{\overline{|A_{qq'q''}^{ss's}|^2 R}}{c_s^2 c_{s'}^5 c_{s'',op}^2} \omega \omega_E''(\omega_E'' - \omega)^3$$

$$\times \frac{\bar{n}_{q's'}(\omega_E - \omega)(\bar{n}_{q''s''}(\omega_E) + 1)}{\bar{n}_{qs} + 1}, \qquad (6.99)$$

where $\omega \equiv \omega(qs)$ is restricted to

$$\omega_E'' - \omega_D' \leq \omega \leq \omega_D. \qquad (6.100)$$

Here we can allow for the possibility of three different Einstein-like optical modes with frequencies $\omega_E'' \equiv \omega_{q''s''}^E$. The result in equation (6.99) can be further simplified by assuming degenerate optical modes and by expressing $\overline{|A_{qq'q''}^{ss's''}|^2}$ in terms of an appropriate Grüneisen constant.

From equations (6.85) and (6.99) it can be seen that both at low and high temperatures the $ac + ac \rightarrow op$ interaction essentially has the same temperature dependence as the $ac + ac \rightarrow ac$ interactions. The frequency dependence of the $ac + ac \rightarrow op$ interaction, however, is not quite the same as for the $ac - ac$ interactions.

6.4.2 FOUR-PHONON PROCESSES

In the previous section we evaluated three-phonon relaxation times using first-order time-dependent perturbation theory and the cubic anharmonicity in the crystal Hamiltonian. The procedure can be extended to study four-phonon processes, within first-order perturbation theory, using the quartic anharmonicity. Four-phonon processes can also be contributed to, in second-order perturbation, by the cubic anharmonicity. The total transition probability of a four-phonon process can thus be written as

$$
P_i^f(4\,\text{ph}) = \frac{2\pi}{\hbar}\left(|\langle f|\mathcal{V}_4|i\rangle|^2 + \Big| \sum_{m \neq i,f} \frac{\langle f|\mathcal{V}_3|m\rangle\langle m|\mathcal{V}_3|i\rangle}{E_i - E_m} \Big|^2 \right)
$$
$$
\times \delta(E_f - E_i) \tag{6.101}
$$

where $\mathcal{V}_4$ is the quartic anharmonic term and E_m is the energy of an intermediate virtual phonon state $|m\rangle$.

The first term in equation (6.101) is contributed by four-phonon processes of the types $q + q' + q'' \rightleftharpoons q''' + G$, $q + q' \rightleftharpoons q'' + q''' + G$ and $q \rightleftharpoons q' + q'' + q''' + G$. The second term in equation (6.101) is contributed by processes of the type $q + q' \rightleftharpoons q'''' \rightleftharpoons q'' + q''' + G$, where q'''' is the wave vector of an intermediate virtual phonon state with energy $\omega'''' \equiv \omega(q''''s'''')$. However, it is necessary to consider only those states for which the matrix elements and the denominator are finite.

Following equation (6.46), we can write down the expression for the transition probability for four-phonon processes. For example, for the process $q + q' \rightarrow q'' + q''' + G$ we have

$$
P_{qsq's'}^{q''s''q'''s'''} = \frac{2\pi}{\hbar^2}\Bigg\{ \Big| \langle n-1, n'-1, n''+1, n'''+1 | \mathcal{V}_4 | n, n', n'', n''' \rangle \Big|^2
$$
$$
+ \Big| \sum_{q''''s''''} \langle n-1, n'-1, n''+1, n'''+1, n'''' | \mathcal{V}_3 | n-1, n'-1, n'', n''', n''''+1 \rangle
$$
$$
\times \langle n-1, n'-1, n'', n''', n''''+1 | \mathcal{V}_3 | n, n', n'', n''', n'''' \rangle
$$
$$
\times (\hbar(\omega + \omega' - \omega''''))^{-1} \Big|^2 \Bigg\} \delta(\omega'' + \omega''' - \omega - \omega') \tag{6.102}
$$

where $\omega'''' \neq (\omega + \omega')$ or $(\omega'' + \omega''')$. Notice that here we have used $\omega \equiv \omega(qs), n \equiv n_{qs}$, etc. For the process $q \rightarrow q' + q'' + q''' + G$ we have

$$
P_{qs}^{q's'q''s''q'''s'''} = \frac{2\pi}{\hbar^2}|\langle n-1, n'+1, n''+1, n'''+1 | \mathcal{V}_4 | n, n', n'', n''' \rangle|^2
$$
$$
\times \delta(\omega - \omega' - \omega'' - \omega'''). \tag{6.103}
$$

Following equations (6.52) and (6.56), the four-phonon single-mode relaxation time can now be expressed as

$$
\tau_{qs}^{-1}(4\,\text{ph}) = \frac{1}{\bar{n}_{qs}(\bar{n}_{qs}+1)} \sum_{\substack{q's',q''s'' \\ q'''s'''}} \left(\frac{1}{2}\bar{P}_{qsq's'q''s''}^{q'''s'''} + \frac{1}{2}\bar{P}_{qsq's'}^{q''s''q'''s'''} + \frac{1}{3!}\bar{P}_{qs}^{q's'q''s''q'''s'''} \right), \tag{6.104}
$$

where the factors $1/2$ and $1/3!$ are to compensate for over-counting of equivalent terms. Writing $\mathscr{V}_3$ as in equation (6.47) and $\mathscr{V}_4$ in an appropriate manner, and following the treatment presented in the previous section, we can finally obtain an expression for $\tau_{qs}^{-1}(4\,\text{ph})$ in the Debye model, at least in principle. However, the resulting expression will be very complicated, and the geometry of allowed processes very difficult to determine.

Four-phonon processes are expected to be important only at high temperatures. Ecsedy and Klemens (1977) have used a very simplified procedure to derive the interaction rate of four-phonon processes at high temperatures in terms of the Grüneisen constant and its dilational derivative. These authors express

$$\mathscr{V}_3(qs,q's',q''s'') = B_3\gamma\omega\omega'\omega'' \tag{6.105}$$

and

$$\mathscr{V}_4(qs,q's',q''s'',q'''s''') = B_4(\gamma^2 - \gamma')\omega\omega'\omega''\omega''' \tag{6.106}$$

where B_3 and B_4 are constants containing appropriate momentum conservation conditions, and γ and γ' are Grüneisen's constant and its dilational derivative, respectively. Using the high-temperature approximation $\bar{n} \simeq k_B T/\hbar\omega$, etc. and making some drastic approximations for integrations over q' and q'' these authors obtain the following expression for the single-mode relaxation rate for four-phonon processes:

$$\tau_{qs}^{-1}(4\,\text{ph})\Big|_{\text{high temp}} = B(5\gamma^4 + \gamma^2 - 2\gamma^2\gamma')\omega^2 T^2 \tag{6.107}$$

where B is a constant. Thus we see that the rate of four-phonon processes at high temperatures varies as ω^2 and T^2. This is in contrast with the linear T dependence for three-phonon processes. Ecsedy and Klemens estimate that the strength of four-phonon processes is *at least* two to three orders of magnitude *weaker* than for three-phonon processes.

6.5   PHONON–ELECTRON SCATTERING IN DOPED SEMICONDUCTORS

Addition of impurities to an insulator or a semiconductor reduces its thermal conductivity. Normally the maximum of the $\mathscr{K}$ against T curve is depressed. But below the temperature at which the conductivity is maximum a more striking phenomenon is observed: in certain materials the mean free path of phonons becomes nearly independent of temperature and yet very much smaller than the size of the material. This means that at these low temperatures boundary scattering of phonons is not very important. Nor can mass defect scattering explain such a drop in the conductivity at these temperatures. It is thought that the addition of impurities, i.e. doping, gives rise to extra scattering of phonons by electrons (holes) in the donor (acceptor) levels. In addition, when dopant atoms are completely ionised, due to compensation or high temperature, they scatter phonons via the mass-difference mechanism.

Actually one can consider phonon–electron scattering in a doped semiconductor to be of two types, depending on the donor (carrier) concentration and the extent of impurity wavefunction (and hence its effective Bohr radius). If the carrier concentration is typically less than 10^{17} cm^{-3}, the carriers form a semi-isolated or bound donor state lying inside the forbidden gap of the semiconductor, normally a few hundredths of an electron-volt below the conduction band minimum [1]. For higher concentration, the carriers form a well-defined impurity band which may become broad and overlap the conduction band of the host. In this case a few electrons become available at all temperatures and in fact form a degenerate assemblage. In this section we describe scattering of phonons from both types of electrons: free electrons (high concentration limit) and electrons in donor atoms (low concentration limit).

---

[1] Here we only consider 'shallow' impurities. Consideration of 'deep' impurities would be much more complicated. The interested reader may, however, like to follow the work of Sigmund and co-workers (Sigmund and Lassmann 1980, Mair and Sigmund 1981) on phonon scattering at deep centres.

First, we derive an expression for the matrix element for electron–phonon scattering in general. A doped semiconductor can be thought to be deformed in the neighbourhood of the impurity. If the lattice deforms by $\Delta(r)$, we could expect an energy change of the form

$$dU = C\Delta(r), \qquad (6.108)$$

where C is some parameter. The local deformation can be related to a local displacement vector field $u(r)$ by

$$\Delta(r) = \nabla \cdot u(r). \qquad (6.109)$$

Due to the perturbation dU, the electron–phonon interaction Hamiltonian is given by

$$H_{ep} = \int dr \Psi^\dagger(r) dU(r) \Psi(r), \qquad (6.110)$$

where $\Psi(r)$ are the second-quantised field operators defined by

$$\Psi(r) = \sum_k \psi_k(r) b_k \qquad \Psi^\dagger(r) = \sum_k \psi_k^*(r) b_k^\dagger. \qquad (6.111)$$

Here $\psi_k(r)$ is a one-electron wavefunction, and b_k and $b_k^\dagger$ are electron annihilation and creation operators defined by

$$\begin{aligned} b_k|f_k\rangle &= \sqrt{f_k}|f_k - 1\rangle \\ b_k^\dagger|f_k\rangle &= \sqrt{(1 - f_k)}|f_k + 1\rangle \end{aligned} \qquad (6.112)$$

with f_k being the electron occupation number. Thus the matrix element for an impurity electron to be scattered from a state k to another state k', by absorption of a phonon (qs) in the process $k + q \to k'$, is given by

$$\begin{aligned} \mathcal{M}_{k,k'} &= \langle f_{k'}, n_{qs} - 1|H_{ep}|f_k, n_{qs}\rangle \\ &= \langle n_{qs} - 1|\int dr \psi_{k'}^*(r) dU(r) \psi_k(r)|n_{qs}\rangle \sqrt{f_k(1 - f_{k'})}. \end{aligned} \qquad (6.113)$$

Using equations (4.38), (4.80) and (6.108)–(6.109), the matrix element can be expressed as

$$\mathcal{M}_{k,k'} = -\sum_{qs} \left(\frac{\hbar\omega(qs)}{2\rho N_0 \Omega c_s^2}\right)^{1/2} \Xi_{kk'}(qs)\{n_{qs}f_k(1 - f_{k'})\}^{1/2}, \qquad (6.114)$$

where

$$\Xi_{kk'}(qs) = \int dr \psi_{k'}^*(r) \psi_k(r) C\hat{q} \cdot e_{qs} e^{iq\cdot r} \qquad (6.115)$$

is a deformation potential matrix element whose evaluation will be discussed separately for the two limits (low and high carrier concentrations). For the process $k \to k' + q$ in which a phonon is emitted, the matrix element can be similarly obtained, with the factor n_{qs} replaced by $n_{qs} + 1$.

6.5.1 PHONON–ELECTRON SCATTERING IN MODERATELY AND HEAVILY DOPED SEMICONDUCTORS

In heavily doped semiconductors the carriers are considered to be degenerate and the deformation is expressed as a dilatation constant $E_d = C\hat{q} \cdot e_{qs}$. To evaluate the deformation potential matrix elements, we express the electron wave function $\psi_k(r)$ as a Bloch function:

$$\psi_k(r) = \frac{1}{\sqrt{N_0\Omega}} e^{ik\cdot r}\phi(r), \qquad (6.116)$$

where $\phi(r)$ is a periodic function of the host. Then equation (6.115) becomes

$$
\begin{aligned}
\Xi_{kk'}(qs) &= \frac{C}{N_0\Omega}\hat{q}\cdot e_{qs}\int dr\phi^*(r)\phi(r)\exp[i(k+q-k')\cdot r] \\
&= E_d\delta_{k+q,k'+G}.
\end{aligned}
\tag{6.117}
$$

It is clearly seen that the deformation potential matrix is zero for transverse phonons. In other words, transverse phonons do not couple with electrons of a degenerate gas. Further, in the temperature range where electron–phonon scattering is important only normal processes $(G = 0)$ are the most likely events. Thus we have

$$
\mathscr{M}_{k,k'} = -E_d\sum_q\left(\frac{\hbar\omega_L}{2\rho N_0\Omega c_L^2}\right)^{1/2}\{n_qf_k(1-f_{k'})\}^{1/2}\delta_{k+q,k'}.
\tag{6.118}
$$

From here the transition rate for the process $k+q\to k'$ can be written using first-order time-dependent perturbation theory:

$$
\begin{aligned}
P_{k+q}^{k'} &= \frac{2\pi}{\hbar}|\mathscr{M}_{kk'}|^2\delta(E_{k'}-E_k-\hbar\omega_q) \\
&= \frac{\pi\omega_qE_d^2}{\rho N_0\Omega c_L^2}n_qf_k(1-f_{k'})\delta(E_{k'}-E_k-\hbar\omega_q)\delta_{k+q,k'}.
\end{aligned}
\tag{6.119}
$$

The phonon relaxation time for the process $k+q\to k'$ is then

$$
\begin{aligned}
\tau_q^{-1}(k+q\to k') &= \frac{1}{\bar{n}_q(\bar{n}_q+1)}\sum_{kk'}\bar{P}_{k+q}^{k'} \\
&= \frac{\pi\omega_qE_d^2}{\rho N_0\Omega c_L^2}\frac{1}{(\bar{n}_q+1)}\sum_{kk'}\bar{f}_k(1-\bar{f}_{k'}) \\
&\quad\times\delta(E_{k'}-E_k-\hbar\omega_q)\delta_{k+q,k'},
\end{aligned}
\tag{6.120}
$$

where $\bar{f}$ represents the equilibrium electron distribution

$$
\bar{f}_k = [\exp(E_k-\zeta)+1]^{-1} = (e^\eta+1)^{-1}
\tag{6.121}
$$

with ζ the Fermi potential.

In equation (6.120) the sum over k' can be carried out using the Kronecker delta symbol. The sum over k can be changed to an integral over spherical polar coordinates (k,θ,ϕ), where $\theta = 0$ is the direction of q. Writing $z = \hbar\omega/k_BT$ we express

$$
\begin{aligned}
\mathscr{I} &= \sum_{kk'}\bar{f}_k(1-\bar{f}_{k'})\delta(E_{k'}-E_k-\hbar\omega_q)\delta_{k+q,k'} \\
&= \sum_k\frac{1}{(e^\eta+1)(1+e^{-(z+\eta)})}\delta(E_{k'}-E_k-\hbar\omega_q) \\
&= \frac{N_0\Omega}{8\pi^3}\int d\phi\int d\theta\sin\theta\int dk\,k^2\frac{1}{(e^\eta+1)(1+e^{-(z+\eta)})} \\
&\quad\times\delta(E_{k'}-E_k-\hbar\omega_q),
\end{aligned}
\tag{6.122}
$$

where the density of electron states is assumed not to have spin degeneracy, which is usually the case for electron scattering. The integration over ϕ gives 2π and we change the variable θ to Δ

$$
\begin{aligned}
\Delta &= E_{k'}-E_k-\hbar\omega_q \\
&= \frac{\hbar^2}{2m^\star}(k'^2-k^2)-\hbar\omega_q
\end{aligned}
$$

$$= \frac{\hbar^2}{2m^\star}(q^2 + 2kq\cos\theta) - \hbar\omega_q, \tag{6.123}$$

where we have used a parabolic electron band within the effective mass approximation. The Dirac delta function is then used to integrate over Δ to give

$$\mathcal{I} = \frac{N_0\Omega}{4\pi^2}\frac{m^\star}{q\hbar^2}\int_{k_0}^{k_F} dk\,k\frac{1}{(e^\eta + 1)(1 + e^{-(\eta+z)})}, \tag{6.124}$$

where k_F is the Fermi wave vector and $k_0 = |\frac{1}{2}q - m^\star c_L/\hbar|$ is the lowest allowed value of k. For a degenerate electron distribution we can stretch the limits of integration to $(-\infty, \infty)$. Now using the formula (Wilson 1953, p 335)

$$\int_{-\infty}^{\infty}\frac{F(\eta)d\eta}{(e^\eta + 1)(1 + e^{-(\eta+z)})} = -\frac{1}{(1 - e^{-z})}\int_{-\infty}^{\infty}\{G(\eta) - G(\eta - z)\}\frac{\partial\bar{f}_k}{\partial\eta}d\eta, \tag{6.125}$$

where

$$G(\eta) = \int^\eta F(\eta')d\eta' \tag{6.126}$$

equation (6.124) reduces to

$$\mathcal{I} = \frac{N_0\Omega}{4\pi^2}\frac{m^{\star 2}}{q\hbar^4}\frac{\hbar\omega_q}{k_B T}(\bar{n}_q + 1). \tag{6.127}$$

From equations (6.120) and (6.127), we finally obtain

$$\tau_q^{-1}(\boldsymbol{k} + \boldsymbol{q} \to \boldsymbol{k}') = \frac{m^{\star 2}E_d^2}{4\pi\rho\hbar^3}q, \tag{6.128}$$

where $\omega_q = c_L q$ is used.

A similar approach can be used to calculate $\tau_q^{-1}(\boldsymbol{k} \to \boldsymbol{k}' + \boldsymbol{q})$ for a process in which phonon emission takes place. The result is identical to equation (6.128). Adding the two contributions, the final result for phonon relaxation in a heavily doped, degenerate semiconductor becomes

$$\tau_q^{-1}(\text{ep}) = \frac{m^{\star 2}E_d^2}{2\pi\rho\hbar^3}q. \tag{6.129}$$

It can be noticed that this expression does not explicitly depend on carrier concentration. The electron effective mass $m^\star$ is, however, a function of carrier concentration and host temperature T.

If the lower limit of integration in equation (6.124) is maintained as k_0, then the integral may be evaluated in terms of elementary functions to give the following result which is valid for both low and high free-carrier concentrations (Ziman 1956, 1957):

$$\tau_q^{-1}(\text{ep}) = \frac{m^{\star 2}E_d^2 k_B T}{2\pi\rho c_L\hbar^4}\left[z - \ln\left(\frac{1 + \exp(\xi - \xi_0 + z^2/16\xi + z/2)}{1 + \exp(\xi - \xi_0 + z^2/16\xi - z/2)}\right)\right] \tag{6.130}$$

with $z = \hbar\omega/k_B T$, $\xi = m^\star c_L^2/2k_B T$ and $\xi_0 = \zeta/k_B T$. In the high carrier concentration limit with $\zeta > E_0$ and $\zeta - E_0 >> k_B T$, where $E_0 = \hbar^2\omega^2/8m^\star c_L^2 + \frac{1}{2}m^\star c_L^2 - \hbar\omega/2$, equation (6.130) reduces to equation (6.129).

For moderately doped semiconductors, Parrott (1979) reduced the expression in equation (6.130) to a simple form. Remember the relationship between n_c and the reduced Fermi energy ζ as (McKelvey, 1966)

$$n_c = 2(m^\star k_B T/2\pi\hbar^2)^{3/2}\exp(\zeta) = U_c e^\zeta. \tag{6.131}$$

For small carrier concentration n_c we can expand $\tau_q^{-1}(\text{ep})$ in a Taylor series about $n_c = 0$. We first obtain

$$\left(\frac{\mathrm{d}(\tau^{-1})}{\mathrm{d}n_c}\right)_{n_c=0} = \frac{E_d^2 m^{\star 2} k_B T}{\pi \hbar^4 \rho c_L} \frac{1}{2}\left(\frac{2\pi\hbar^2}{m^\star k_B T}\right)^{3/2} \\ \times \exp[-(\xi + z^2/16\xi)]\sinh(z/2). \tag{6.132}$$

Substituting for U_c one thus finds

$$\tau^{-1}(qs) = \frac{2n_c E_d^2 \sqrt{\pi\xi}}{\hbar\rho c_L^2}\exp[-(\xi + z^2/16\xi)]\sinh(z/2), \tag{6.133}$$

which for $z \to 0$ becomes

$$\tau^{-1}(qs) = \frac{n_c E_d^2 \omega}{\rho c_L^2 k_B T}\sqrt{\frac{\pi m^* c_L^2}{2 k_B T}}\exp\left(\frac{-m^* c_L^2}{2 k_B T}\right). \tag{6.134}$$

(Note that the root sign in the printed equation (2a) in Parrott (1979) should cover the 'ξ' sign.)

6.5.2 PHONON–ELECTRON SCATTERING IN LIGHTLY DOPED SEMICONDUCTORS

We now consider the case of lightly doped semiconductors where the impurity levels are 'bound states', lying within the band gap of the host. Since the impurity concentration is small, we may consider scattering of phonons by bound states in an atom at a time. To develop the theory of electron–phonon interaction in a such a system, it is necessary to have some knowledge of the band structure of both the host and the impurity donor electrons.

Assume that the electronic band structure of the semiconductor is many-valleyed. Then, because of the local energy change $\mathrm{d}U$, the electronic energy of the valleys will show a shift. This will result in writing a generalised form for the deformation potential constant C in equation (6.108) (Herring and Vogt 1956):

$$C\hat{q}\cdot e(qs) = \sum_{ij}(E_d\delta_{ij} + E_u\hat{k}_i\hat{k}_j)\hat{q}_i e_j(qs), \tag{6.135}$$

where E_d and E_u are constants due to dilatation and shear, respectively, and $\hat{k}_i$ is a unit vector along the ith conduction valley in the first Brillouin zone. The presence of shear gives rise to the possibility of coupling between transverse phonons and donor electrons.

Again, we exclusively deal with *shallow* impurity levels, whose binding energies usually lie between 0.01 and 0.1 eV below the conduction band minimum and orbits are of the order of 50 Å. The effective mass theory can be applied to study the electronic structure of such impurities. Neutral donor impurity electron wavefunctions satisfy the Schrödinger equation

$$(H_0 + U(r))\psi_k(r) = E_k\psi_k(r), \tag{6.136}$$

where H_0 represents the Hamiltonian of the host, and $U(r)$ is the impurity potential, usually considered in the hydrogenic form

$$U(r) = -e^2/\varepsilon_0 r \tag{6.137}$$

with ε_0 as the dielectric constant of the host.

The impurity wavefunction can be expressed as a linear combination of the host wavefunctions at the conduction band minima k_v:

$$\psi(r) = \sum_{v=1}^{N}\alpha^{(v)}\psi^{(v)}(r) \tag{6.138}$$

with the coefficient $\alpha^{(v)}$ obeying the requirements of the crystal point-group symmetry. $\psi^{(v)}(r)$ can be expressed as a linear combination of Bloch functions at k_v:

$$\psi^{(v)}(r) \quad = \quad \frac{1}{\sqrt{N_0\Omega}}\sum_k A_v(k)e^{ik\cdot r}\phi_k(r) \tag{6.139}$$

$$- \quad F_v(r)\psi^0_{k_v}(r), \tag{6.140}$$

where

$$\psi^0_{k_v}(r) = e^{ik_v\cdot r}\phi_{k_v}(r) \tag{6.141}$$

is the host Bloch function at k_v, and

$$F_v(r) = \frac{1}{\sqrt{N_0\Omega}}\sum_k A_v(k)\exp\left[i(k-k_v)\cdot r\right] \tag{6.142}$$

is an envelope function.

Substituting equations (6.135) and (6.138)–(6.142) into equation (6.115), we obtain the following result for the matrix element between the impurity electron levels n and n':

$$
\begin{aligned}
\Xi_{n'n}(qs) \quad &= \quad \frac{1}{N_0\Omega}\sum_{vv'}\sum_{ij}\alpha^{*(v')}_{n'}\alpha^{(v)}_n\left(E_d\delta_{ij}+E_u\hat{k}^{(v)}_i\hat{k}^{(v)}_j\right) \\
&\quad \times \hat{q}_i e_j \sum_{kk'}A^*_{v'}(k')A_v(k)\int dr\phi^*_{k'_{v'}}(r)\phi_{k_v}(r)\exp\left[i(k+q-k')\cdot r\right] \\
&= \quad \sum_{vv'}\sum_{ij}\alpha^{*(v')}_{n'}\alpha^{(v)}_n\left(E_d\delta_{ij}+E_u\hat{k}^{(v)}_i\hat{k}^{(v)}_j\right)\hat{q}_i e_j \\
&\quad \times \sum_{kk'}A^*_{v'}(k')A_v(k)\delta_{vv'}\delta_{k+q,k'} \\
&= \quad \sum_{ij}\hat{q}_i e_j\left(E_d\delta_{ij}\delta_{nn'}+\frac{1}{3}E_uD^{n'n}_{ij}\right)R_v(q) \tag{6.143} \\
&= \quad \tilde{\Xi}_{n'n}(qs)R(q) \tag{6.144}
\end{aligned}
$$

with

$$D^{n'n}_{ij} = 3\sum_{v=1}^N \hat{k}^{(v)}_i\hat{k}^{(v)}_j\alpha^{*(v)}_{n'}\alpha^{(v)}_n \tag{6.145}$$

and

$$
\begin{aligned}
R_v(q) \quad &= \quad \sum_k A^*_v(k+q)A_v(k) \\
&= \quad \int dr\left(F_v(r)\right)^2 e^{iq\cdot r}. \tag{6.146}
\end{aligned}
$$

In many cases, it is possible to use the isotropic approximation and express $R_v(q)$ as $R(q)$ and $F_v(r)$ as $F(r)$, with a hydrogenic form for $F(r)$

$$F(r) = \frac{1}{\sqrt{\pi a_B^3}}e^{-r/a_B}, \tag{6.147}$$

where a_B is an effective Bohr radius of the ground state of the donor. This gives

$$R(q) \quad = \quad \frac{1}{\pi a_B^3}\int dr e^{-2r/a_B}e^{-iq\cdot r}$$

$$\simeq \frac{4\pi}{\pi a_B^3} \int dr\, r^2 e^{-2r/a_B} \frac{\sin qr}{qr}$$

$$= [1 + (\frac{1}{2} a_B q)^2]^{-2}. \qquad (6.148)$$

For N valleys of the lowest conduction band, the Schrödinger equation (6.136) has N equivalent and degenerate solutions of the form given in equation (6.138), compatible with symmetry group of the host crystal. In the case of Ge, the conduction band shows four equivalent minima along

$$\text{Ge}: \quad \hat{k}^{(1)} = \frac{1}{\sqrt{3}}(1,1,1) \quad \hat{k}^{(2)} = \frac{1}{\sqrt{3}}(-1,1,1)$$

$$\hat{k}^{(3)} = \frac{1}{\sqrt{3}}(1,-1,1) \quad \hat{k}^{(4)} = \frac{1}{\sqrt{3}}(-1,-1,1). \qquad (6.149)$$

(Actually there are eight half-ellipsoids along the $\langle 111 \rangle$ directions. These may be considered to be equivalent to four ellipsoids.) In the case of Si, there are six valleys directed along

$$\text{Si}: \quad \hat{k}^{(1)} = (1,0,0) \quad \hat{k}^{(2)} = (-1,0,0) \quad \hat{k}^{(3)} = (0,1,0)$$

$$\hat{k}^{(4)} = (0,-1,0) \quad \hat{k}^{(5)} = (0,0,1) \quad \hat{k}^{(6)} = (0,0,-1). \qquad (6.150)$$

These N-fold degenerate solutions in Ge and Si form the representation T_d, appropriate to a substitutional impurity. When valley–orbit interactions and central cell corrections are taken into account, the degeneracy of the impurity state is split as indicated below:

$$\text{Ge}: \quad T_d \rightarrow \underset{\text{(singlet)}}{A_1} + \underset{\text{(triplet)}}{T_2}$$

$$\text{Si}: \quad T_d \rightarrow \underset{\text{(singlet)}}{A_1} + \underset{\text{(doublet)}}{E} + \underset{\text{(triplet)}}{T_1}.$$

The separation between the ground state A_1 and the next higher level is generally known as the chemical shift δE. This chemical shift is usually denoted as Δ in Si, but 4Δ in Ge.

The coefficients $\alpha^{(v)}$ can be evaluated by using group theory. The results are
For Ge:

$$(A_1): \quad \{\alpha_0^{(v)}\} = \frac{1}{2}(1,1,1,1)$$

$$(T_2): \quad \{\alpha_1^{(v)}\} = \frac{1}{\sqrt{2}}(1,0,0,-1)$$

$$\{\alpha_2^{(v)}\} = \frac{1}{\sqrt{2}}(0,1,-1,0)$$

$$\{\alpha_3^{(v)}\} = \frac{1}{\sqrt{2}}(1,-1,-1,1), \qquad (6.151)$$

where $n = 0$ denotes the singlet state, and $n = 1,2,3$ represent the triplet states.
For Si:

$$(A_1): \quad \{\alpha_1^{(v)}\} = \frac{1}{\sqrt{6}}(1,1,1,1,1,1)$$

$$(E): \quad \{\alpha_2^{(v)}\} = \frac{1}{2}(1,1,-1,-1,0,0)$$

$$\{\alpha_3^{(v)}\} = \frac{1}{2}(1,1,0,0,-1,-1)$$

$$(T_1): \quad \{\alpha_4^{(v)}\} = \frac{1}{\sqrt{2}}(1,-1,0,0,0,0)$$

$$\{\alpha_5^{(v)}\} = \frac{1}{\sqrt{2}}(0,0,1,-1,0,0)$$

$$\{\alpha_6^{(v)}\} = \frac{1}{\sqrt{2}}(0,0,0,0,1,-1). \tag{6.152}$$

With $\hat{k}^{(v)}$ and $\alpha_n^{(v)}$ known, the tensors $D_{ij}^{n'n}$ can now be evaluated easily. Here we evaluate these for Ge and the same procedure can be followed for Si.

Ge :

$$D_{11}^{00} = 3\sum_{v=1}^{4} \hat{k}_1^{(v)}\hat{k}_1^{(v)}\alpha_0^{*(v)}\alpha_0^{(v)} = 1$$

$$D_{11}^{10} = 3\sum_{v=1}^{4} \hat{k}_1^{(v)}\hat{k}_1^{(v)}\alpha_1^{*(v)}\alpha_0^{(v)} = 0$$

etc.

Thus

$$D^{00} = (D_{ij}^{00}) = \begin{pmatrix} 1 & 0 & 0 \\ 0 & 1 & 0 \\ 0 & 0 & 1 \end{pmatrix}$$

$$D^{01} = D^{10} = \begin{pmatrix} 0 & 0 & 1 \\ 0 & 0 & 1 \\ 1 & 1 & 0 \end{pmatrix}, D^{20} = \begin{pmatrix} 0 & 0 & -1 \\ 0 & 0 & 1 \\ -1 & 1 & 0 \end{pmatrix}, D^{30} = \begin{pmatrix} 0 & 1 & 0 \\ 1 & 0 & 0 \\ 0 & 0 & 0 \end{pmatrix}. \tag{6.153}$$

With the tensors $D_{ij}^{n'n}$ determined, we can now proceed with the calculation of phonon scattering from electrons bound to donor impurities. Here we discuss such a calculation for the host semiconductor Ge.

Consider scattering of phonons from electrons in the singlet ($n = 0$) or one of the triplet ($n = 1,2,3$) states. First-order perturbation theory cannot be applied to this case, as for $\hbar\omega(qs) \neq 4\Delta$ the energy is not conserved. However, for $|\hbar\omega(qs) - 4\Delta| \gg 0$ and $\hbar\omega(qs)$ much greater than level widths of the singlet and triplet states, perturbation theory in the second Born approximation can be applied. The near resonance condition, $\hbar\omega \simeq 4\Delta$, cannot be dealt with using perturbation theory at all and Green's function techniques must be used to calculate the attenuation of phonons in resonance with the impurity levels. Here we only discuss the application of second-order perturbation theory and refer the reader to the work of Kwok (1966) for the calculation of resonance attenuation by the thermodynamic Green's function technique.

We can think of three different types of phonon scattering processes.

(a) Elastic scattering:

$$\hbar\omega(qs) + \left(\begin{smallmatrix} singlet \\ or \\ triplet \end{smallmatrix}\right) \rightleftharpoons (int.) \rightleftharpoons \hbar\omega(q's') + \left(\begin{smallmatrix} singlet \\ or \\ triplet \end{smallmatrix}\right) \tag{6.154}$$

$$\omega(qs) = \omega(q's').$$

(b) Inelastic scattering:

$$\hbar\omega(qs) + (triplet) \rightleftharpoons (int.) \rightleftharpoons \hbar\omega(q's') + (singlet). \tag{6.155}$$

We can consider an acoustic phonon to be inelastically scattered into a high energy state ($\hbar\omega(q's') \simeq 4\Delta$) while the electron jumps down from the triplet to the singlet state.

(c) Phonon-absorption scattering:

$$\hbar\omega(qs) + \hbar\omega(q's') + (singlet) \rightleftharpoons (int.) \rightleftharpoons (triplet). \tag{6.156}$$

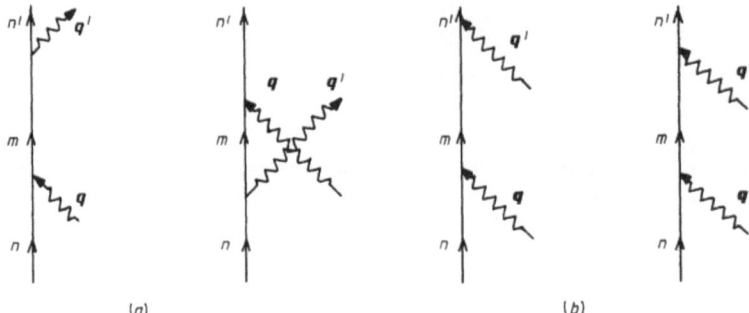

Figure 6.2 Phonon–electron scattering in lightly doped semiconductors by second-order processes. n, m and n' denote initial, intermediate and final electronic states, respectively. (a) represents elastic scattering for $E_n = E_{n'}$ and inelastic scattering for $E_n \neq E_{n'}$, and (b) represents 'thermally assisted' phonon absorption.

In such a process an acoustic phonon absorbs a low-frequency *thermal* phonon of energy $\hbar\omega(q's')$ and eventually promotes an electron from the singlet state to the triplet state.

In these equations (*int.*) denotes an appropriate intermediate (virtual) state. Phonon attenuation due to any of these processes can be calculated from the standard second-order perturbation theory. However, for the sake of brevity we will only present here the mechanics of such a calculation for the elastic scattering of phonons. For other cases we again refer the reader to the work of Kwok.

Let us represent the elastic process in equation (6.154) as

$$\hbar\omega(qs) + (n) \rightleftharpoons (m) \rightleftharpoons \hbar\omega(q's') + (n') \tag{6.157}$$

with the electronic energies in (n) and (n') being essentially the same: $E_n = E_{n'}$. Since there are only two different types of electronic states, namely singlet and triplet, if (n) is singlet then (m) has to be triplet and vice versa. There are basically two different kinds of virtual electronic transitions $(n) \to (n')$ which scatter a phonon from (qs) to $(q's')$. These are shown in figure 6.2.

The total scattering amplitude for the process described in equation (6.153) is given, in the second Born approximation, by

$$\sum_m \left\{ \frac{\langle q's', n'|H_{ep}|m\rangle\langle m|H_{ep}|qs, n\rangle}{E_m - (\hbar\omega(qs) + E_n)} + \frac{\langle q's', n'|H_{ep}|m\rangle\langle m|H_{ep}|qs, n\rangle}{E_m - (\hbar\omega(qs) + E_n)} \right\}. \tag{6.158}$$

Therefore, the total probability of phonon scattering $(qs) \to (q's')$ is

$$
\begin{aligned}
P_{qs}^{q's'} &= \frac{2\pi}{\hbar} \sum_{n,n'} |\text{Eq. (6.158)}|^2 \delta(\hbar\omega(qs) - \hbar\omega(q's')) \\
&= \frac{2\pi}{\hbar} \left(\frac{\hbar}{2\rho}\right)^2 \frac{1}{N_0\Omega} \frac{\omega(qs)\omega(q's')}{c_s^2 c_{s'}^2} n_{qs}(n_{q's'} + 1)\delta(\hbar\omega(qs) - \hbar\omega(q's')) \\
&\quad \times \sum_{n,n'} f_n(T)(1 - f_{n'}(T)) \\
&\quad \times \left| \sum_m \left(\frac{\Xi_{n'm}(q's')\Xi_{mn}(qs)}{E_m - E_n - \hbar\omega(qs)} + \frac{\Xi_{n'm}(qs)\Xi_{mn}(q's')}{E_m - E_n + \hbar\omega(qs)} \right) \right|^2,
\end{aligned} \tag{6.159}
$$

where we have used equations (6.108)–(6.109) and (4.79)–(4.80). Also, we have assumed that the electronic population is temperature dependent and satisfies

$$\sum_m f_m(T) = 1. \tag{6.160}$$

Now it is possible to write down the relaxation time for phonons in mode (qs):

$$\tau_{qs}^{-1}(\text{elastic process})$$

$$= \frac{1}{\bar{n}_{qs}(\bar{n}_{qs}+1)}\bar{P}_{qs}^{q's'}$$

$$= \frac{\pi\omega(qs)}{2\rho^2 N_0 \Omega c_s^2}\sum_{q's'}\frac{\omega(q's')}{c_{s'}^2}\sum_n \bar{f}_n(T)\delta\big(\omega(qs)-\omega(q's')\big)$$

$$\times \sum_{\substack{n'\\(E_n=E_{n'})}}\left|\sum_m\left(\frac{\Xi_{n'm}(q's')\Xi_{mn}(qs)}{E_m-E_n-\hbar\omega(qs)}+\frac{\Xi_{n'm}(qs)\Xi_{mn}(q's')}{E_m-E_n+\hbar\omega(qs)}\right)\right|^2, \tag{6.161}$$

where we have assumed the final state occupancy factor $\bar{f}_{n'}(T)$ to be negligible. The result obtained so far is for one donor impurity. If we assume a definite concentration of impurities, say $N_{ex}=$ number of donor electrons per cubic centimetre, and write $N_n(T)=N_{ex}\bar{f}_n(T)=$ number of electrons per unit volume in level n at temperature T, then in equation (6.161) the factor $\bar{f}_n(T)$ should be replaced by $N_n(T)$.

Using the continuum approximation, performing the sum over q' using the delta function, and using equation (6.144) we obtain the following result:

$$\tau_{qs}^{-1} = \frac{N_{ex}\omega^4(qs)}{4\pi\rho^2 c_s^2}\sum_{s'}\frac{1}{c_{s'}^5}\sum_n \bar{f}_n(T)R^2(q)R^2\left(\frac{c_s}{c_{s'}}q\right)$$

$$\times \sum_{\substack{n'\\(E_n=E_{n'})}}\left\langle\left|\sum_m\left(\frac{\tilde{\Xi}_{n'm}(q's')\tilde{\Xi}_{mn}(qs)}{E_m-E_n-\hbar\omega(qs)}+\frac{\tilde{\Xi}_{n'm}(qs)\tilde{\Xi}_{mn}(q's')}{E_m-E_n+\hbar\omega(qs)}\right)\right|^2\right\rangle_{\Omega'},$$

$$\tag{6.162}$$

where $\langle\ldots\rangle_{\Omega'}$ means angular average over $\hat{q}'$:

$$\langle\ldots\rangle_{\Omega'} = \frac{1}{4\pi}\int\int d\phi' d\theta' \sin\theta'(\ldots). \tag{6.163}$$

At this point, it is interesting to note from equation (6.153) that the matrices D^{10}, D^{20} and D^{30} all have zero diagonal matrix elements. Therefore, for the elastic process under consideration, we have

$$\tilde{\Xi}_{mn}(qs) = \frac{1}{3}E_u\sum_{ij}\hat{q}_i e_j(qs)D_{ij}^{mn} \tag{6.164}$$

and the electronic transitions are due entirely to the shear-strain produced by phonons. Equation (6.162) then becomes

$$\tau_{qs}^{-1} = \frac{N_{ex}\omega^4(qs)}{4\pi\rho^2 c_s^2}\frac{E_u^4}{81}\sum_{s'}\frac{1}{c_{s'}^5}\sum_n \bar{f}_n(T)R^2(q)R^2\left(\frac{c_s}{c_{s'}}q\right)$$

$$\times \sum_{n'}\left\langle\left|\sum_m\left\{\frac{\sum_{rp}\hat{q}'_r e_p(q's')D_{rp}^{n'm}\sum_{ij}\hat{q}_i e_j(qs)D_{ij}^{mn}}{E_m-E_n-\hbar\omega(qs)}\right.\right.\right.$$

$$\left.\left.\left.+\frac{\sum_{ij}\hat{q}_i e_j(qs)D_{ij}^{n'm}\sum_{rp}\hat{q}'_r e_j(q's')D_{rp}^{mn}}{E_m-E_n+\hbar\omega(qs)}\right\}\right|^2\right\rangle_{\Omega'}. \tag{6.165}$$

Notice that the result presented for τ_{qs} has been obtained by considering the angular average over $\hat{q}'$. It is common to further express this result by angularly averaging over $\hat{q}$

$$\bar{\tau}_{qs}^{-1} = \frac{1}{4\pi}\int\int d\phi d\theta \sin\theta\, \tau_{qs}^{-1}. \tag{6.166}$$

An explicit evaluation of the angular integrations over $\hat{q}$ and $\hat{q}'$ can be made by working withing the isotropic continuum approximation and using the spherical polar coordinate system. This means that the phonon dispersion can be characterised by pure longitudinal and transverse modes, and the three polarisation vectors e_{qs} form a suitable basis, with $e_L(q)$ along q. Thus we can choose $\hat{q}(= e_L), e_{T_1}$ and e_{T_2} as in equation (4.104).

Here we will only evaluate $\bar{\tau}_{qs}^{-1}$ for the transitions from the A_1 state as shown in figure 6.2(a). Then, in equation (6.161), $n = n' = 0$, and $m = 1, 2, 3$. For transitions almost to the degenerate triplet state T_2, we write $E_m - E_n = 4\Delta \pm \varepsilon, (\varepsilon \to 0^+)$. The angular term in equation (6.165) becomes

$$\langle \left| \sum_m \sum_{rp} \sum_{ij} (\hat{q}_i e_j)(\hat{q}'_r e'_p) D_{ij}^{m0} D_{rp}^{0m} \left\{ \frac{1}{(4\Delta \pm \varepsilon) - \hbar\omega(qs)} + \frac{1}{(4\Delta \pm \varepsilon) + \hbar\omega(qs)} \right\} \right| \rangle$$

$$= \frac{2(4\Delta)}{(4\Delta)^2 - (\hbar\omega(qs))^2} \langle \left| \sum_m \sum_{rp} \sum_{ij} (\hat{q}_i e_j)(\hat{q}'_r e'_p) D_{ij}^{m0} D_{rp}^{0m} \right| \rangle$$

as ε is set to zero. Thus in this particular case

$$\bar{\tau}_{qs}^{-1} = \frac{N_{ex}\omega^4(qs)}{4\pi\rho^2 c_s^2} \frac{E_u^4}{81} \frac{4(4\Delta)^2}{[(4\Delta)^2 - (\hbar\omega(qs))^2]^2} \bar{f}_0(T) R^2(q)$$

$$\times \sum_{s'} \frac{1}{c_{s'}^5} R^2\left(\frac{c_s}{c_{s'}}q\right)$$

$$\times \langle\langle \left| \sum_{m=1,2,3} \left(\sum_{ij} (\hat{q}_i e_j) D_{ij}^{m0} \right) \left(\sum_{rp} (\hat{q}'_r e'_p) D_{rp}^{0m} \right) \right|^2 \rangle\rangle_{\Omega\Omega'}, \qquad (6.167)$$

where

$$\langle\langle \ldots \rangle\rangle_{\Omega\Omega'} = \frac{1}{(4\pi)^2} \int\int d\phi d\theta \sin\theta \int\int d\phi' d\theta' \sin\theta' (\ldots) \qquad (6.168)$$

signifies averaging over both $\hat{q}$ and $\hat{q}'$. Making use of equations (6.153) and (4.104), we arrive at the following results:

$$\sum_{ij} \hat{q}_i e_j(q_L) D_{ij}^{10} = \frac{1}{\sqrt{2}} \sin 2\theta (\cos\phi + \sin\phi)$$

$$\sum_{ij} \hat{q}_i e_j(q_L) D_{ij}^{20} = \frac{1}{\sqrt{2}} \sin 2\theta (\sin\phi - \cos\phi)$$

$$\sum_{ij} \hat{q}_i e_j(q_L) D_{ij}^{30} = \sin^2\theta \sin 2\theta \qquad (6.169)$$

$$\sum_{ij} \hat{q}_i e_j(q_{T_1}) D_{ij}^{10} = \frac{1}{\sqrt{2}} \cos 2\theta (\cos\phi + \sin\phi)$$

$$\sum_{ij} \hat{q}_i e_j(q_{T_1}) D_{ij}^{20} = \frac{1}{\sqrt{2}} \cos 2\theta (\sin\phi - \cos\phi)$$

$$\sum_{ij} \hat{q}_i e_j(q_{T_1}) D_{ij}^{30} = \frac{1}{2} \sin 2\theta \sin 2\phi \qquad (6.170)$$

$$\sum_{ij} \hat{q}_i e_j(q_{T_2}) D_{ij}^{10} = \frac{1}{\sqrt{2}} \cos\theta (\cos\phi - \sin\phi)$$

$$\sum_{ij} \hat{q}_i e_j(q_{T_2}) D_{ij}^{20} = \frac{1}{\sqrt{2}} \cos\theta (\sin\phi + \cos\phi)$$

$$\sum_{ij} \hat{q}_i e_j(q_{T_2}) D_{ij}^{30} = \sin\theta \cos 2\phi. \qquad (6.171)$$

Denoting the double set of angular integrations in equation (6.167) by $\langle\langle qs, q's'\rangle\rangle_{\Omega\Omega'}$, it is easy to verify the following result:

$$\frac{3}{2}\langle qs, q'\mathsf{L}\rangle_{\Omega'} = \langle qs, q'\mathsf{T}_1\rangle_{\Omega'} + \langle qs, q'\mathsf{T}_2\rangle_{\Omega'}. \tag{6.172}$$

The result in equation (6.167) can therefore be written as

$$
\begin{aligned}
\bar{\tau}_{qs}^{-1} &= \frac{\omega^4(qs)N_{ex}\bar{f}_0(T)}{\pi\rho^2 c_s^2}\frac{E_u^4}{81}\frac{(4\Delta)^2}{[(4\Delta)^2 - (\hbar\omega(qs))^2]^2}R^2(q) \\
&\quad \times \langle\langle qs, q'\mathsf{L}\rangle\rangle_{\Omega\Omega'}\left[\frac{1}{c_\mathsf{L}^5}R^2\left(\frac{c_s}{c_\mathsf{L}}q\right) + \frac{3}{2}\frac{1}{c_\mathsf{T}^5}R^2\left(\frac{c_s}{c_\mathsf{T}}q\right)\right],
\end{aligned}
\tag{6.173}
$$

where $\langle\langle qs, q'\mathsf{L}\rangle\rangle_{\Omega\Omega'}$ can now be evaluated using equations (6.169)–(6.171). The results are

$$\langle\langle q\mathsf{L}, q'\mathsf{L}\rangle\rangle_{\Omega\Omega'} = \frac{48}{225}; \quad \langle\langle q\mathsf{T}_1, q'\mathsf{L}\rangle\rangle_{\Omega\Omega'} = \frac{32}{225}; \quad \langle\langle q\mathsf{T}_2, q'\mathsf{L}\rangle\rangle_{\Omega\Omega'} = \frac{40}{225}. \tag{6.174}$$

For a particular temperature, the phonon scattering off the singlet state shows three characteristic features: (i) for small phonon frequencies ($\hbar\omega \ll 4\Delta$) it shows the Rayleigh scattering law ($\tau^{-1} \propto \omega^4$), (ii) the Rayleigh type scattering shows a cut-off due to the factor $R(q)$, and (iii) at $\hbar\omega \simeq 4\Delta$ the denominator tends to zero and despite the cut-off factor we have a resonance. This is schematically shown in figure 6.3. With an increase in temperature, the electron has a larger probability of receiving enough thermal energy to shift off the singlet state. This results in a smaller number of electrons being available in the singlet state to scatter phonons and hence a lower phonon attenuation is expected with increased temperature.

Evaluation of $\bar{\tau}_{qs}^{-1}$ for elastic transition from the T_2 states can be done similarly. The same procedure can be followed to evaluate phonon relaxation due to inelastic and 'phonon assisted' scattering events. Suzuki and Mikoshiba (1971a) have made a complete calculation of the average phonon relaxation rates for all these processes. Their results are quoted below:

$$
\begin{aligned}
\bar{\tau}_{qs}^{-1} &= B\omega^4(4\Delta)^2\{[(4\Delta)^2 - \hbar^2\omega^2]^2 + 4\Gamma^2(4\Delta)^2\}^{-1} \\
&\quad \times F(\omega)\{N_{A_1}(T) + N_{T_2}(T)[2 + (4\Delta/\hbar\omega)^2]\}
\end{aligned}
\tag{6.175}
$$

$$
\begin{aligned}
\bar{\tau}_1^{-1}(qs) &= \frac{B}{2}\omega\left[1 - \exp\left(-\frac{\hbar\omega}{k_\mathrm{B}T}\right)\right]\left(\frac{1}{\hbar\omega} - \frac{1}{4\Delta + \hbar\omega}\right)^2 \\
&\quad \times N_{T_2}(T)\left(\frac{4\Delta}{\hbar} + \omega\right)^3 f\left(\frac{4\Delta}{\hbar} + \omega\right)\left[1 - \exp\left(\frac{-4\Delta - \hbar\omega}{k_\mathrm{B}T}\right)\right]^{-1}
\end{aligned}
\tag{6.176}
$$

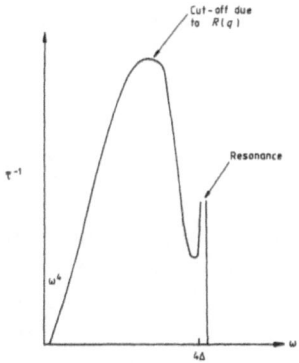

Figure 6.3 Schematic representation of the frequency dependence of phonon relaxation due to phonon scattering off the singlet state in elastic electron transitions.

$$\bar{\tau}_2^{-1}(qs) = \frac{B}{2}\omega\left[1-\exp\left(-\frac{\hbar\omega}{k_B T}\right)\right]\left(\frac{1}{\hbar\omega}+\frac{1}{4\Delta-\hbar\omega}\right)^2$$
$$\times N_{A_1}(T)\left|\frac{4\Delta}{\hbar}-\omega\right|^3 F\left(\frac{4\Delta}{\hbar}-\omega\right)\left|\left[\exp\left(\frac{4\Delta-\hbar\omega}{k_B T}\right)-1\right]^{-1}\right|, \quad (6.177)$$

where

$$B = (\pi\rho^2 c_s^2)^{-1}(E_u/3)^4 R^2(q)W_s$$

$$F(x) = c_L^{-5}R^2(x/c_L) + \frac{3}{2}c_T^{-5}R^2(x/c_T)$$

$$W_L = 48/225 \qquad E_{T_1} = 32/225 \qquad W_{T_2} = 40/225$$

$$N_{A_1}(T) + 3N_{T_2}(T) = N \qquad N_{T_2}/N_{A_1}(T) = \exp(-4\Delta/k_B T)$$

$$\Gamma = \Gamma_{A1} + \Gamma_{T_2} = \frac{1}{15\pi\rho}(E_u/3)^4\left(\frac{4\Delta}{\hbar}\right)^3 F\left(\frac{4\Delta}{\hbar}\right)\left\{1+2/\left[\exp\left(\frac{4\Delta}{k_B T}\right)-1\right]\right\} \quad (6.178)$$

with N_{A_1} and N_{T_2} as the number of electrons per unit volume in the A_1 and T_2 states, respectively. N is the total electron concentration, and Γ_{A_1} and Γ_{T_2} are the level widths of the A_1 and T_2 states, respectively. In equations (6.175)–(6.178) τ_{el}^{-1} represents phonon relaxation due to elastic scattering from the singlet (A_1) as well as triplet (T_2) states, τ_1^{-1} represents the relaxation rate due to inelastic scattering of phonons by electrons in the triplet state, and τ_2^{-1} represents the relaxation rate due to 'thermally assisted phonon absorption' for $\hbar\omega < 4\Delta$ and inelastic scattering by electrons in the singlet state for $\hbar\omega > 4\Delta$. Equation (6.175) includes the damping factors Γ_{A_1} and Γ_{T_2} to avoid the divergence at $4\Delta = \hbar\omega$. It can be easily verified that the result in equation (6.173) is included in equation (6.175) in the limit $\Gamma_{A_1} = 0$.

When $\hbar\omega \simeq 4\Delta$, the result in equation (6.177) is replaced by the formula for resonant phonon absorption (Kwok 1966),

$$\bar{\tau}_2(qs) \simeq \frac{\omega}{\rho c_s^2}\left(\frac{E_u}{3}\right)^2 R^2(q)[1-\exp\left(-\frac{\hbar\omega}{k_B T}\right)]$$
$$\times N_{A_1}(T)\bar{W}_s\left(\frac{\Gamma}{(4\Delta-\hbar\omega)^2+\Gamma^2}\right) \quad (6.179)$$

with $\bar{W}_L = 4/5, \bar{W}_{T_1} = 8/15$ and $\bar{W}_{T_2} = 2/3$.

6.5.3 PHONON–HOLE SCATTERING IN LIGHTLY DOPED SEMICONDUCTORS

The top of the valence band in the diamond/zincblende-structure semiconductors is six-fold degenerate, including spin degeneracy. When spin–orbit interaction is included, this level splits into a four-fold degenerate state Γ_8 and a two-fold degenerate state Γ_7. The ground state of a *shallow acceptor* is four-fold degenerate with the Γ_8 symmetry and can be described by (cf equations (6.138)–(6.142))

$$\psi(r) = \sum_{v=1}^{4}\alpha^{(v)}F_v(r)\psi_v^0(r), \quad (6.180)$$

where $\psi_v^0(r)$ is the Bloch function, at $k = 0$, of Γ_8 symmetry, and $F_v(r)$ are orbital envelope functions. The Γ_8 states correspond to the quantum number $M_J = 3/2$ for which the four spinors $\{\alpha^{(v)}\}$ are (Suzuki *et al* 1964)

$$\alpha_{3/2}^{(1)} = \begin{bmatrix} 1 \\ 0 \\ 0 \\ 0 \end{bmatrix} \qquad \alpha_{3/2}^{(2)} = \begin{bmatrix} 0 \\ iz(x+iy) \\ ixy \\ 0 \end{bmatrix}$$

$$\alpha_{3/2}^{(3)} = \begin{bmatrix} -\frac{1}{\sqrt{2}}(z^2 - \frac{x^2+y^2}{2}) \\ 0 \\ \frac{1}{2}\sqrt{\frac{3}{2}}(x^2 - y^2) \\ 0 \end{bmatrix} \frac{1}{r^2} \qquad \alpha_{3/2}^{(4)} = \begin{bmatrix} 0 \\ -\frac{i}{2}z(x+iy) \\ ixy \\ -\frac{\sqrt{3}}{2}iz(x-iy) \end{bmatrix} \frac{1}{r^2}. \qquad (6.181)$$

F_1 can be considered as an s-like function (as in equation (6.147)) and $F_i (i = 2, 3, 4)$ can be considered as a d-like function.

Internal static strain due to the impurity atoms themselves, or other defects, or externally applied uniaxial stress, or a magnetic field can split the Γ_8 quartet. A uniaxial stress splits the quartet into two Kramers doublets specified by the total magnetic quantum numbers $M_J = \pm 3/2$ and $\pm 1/2$. This splitting can be described by the spin Hamiltonian

$$H = \frac{2}{3}D_u^a[(J_x^2 - \frac{1}{3}J^2)e_{xx} + c.p.] + \frac{1}{3}D_{u'}^a[(J_xJ_y + J_yJ_x)e_{xy} + c.p.], \qquad (6.182)$$

where D_u^a and $D_{u'}^a$ are the deformation potential constants for the acceptor holes, $c.p.$ stands for cyclic permutation, J_α is the αth component of the total angular momentum operator ($J = 3/2$), and $e_{\alpha\beta}$ is the conventional strain component.

Phonon relaxation rates due to the scattering by holes can be calculated by expressing $e_{\alpha\beta}$ in terms of phonon creation and annihilation operators (using equation (6.21)) and following the discussion presented in section 6.5.2. Suzuki and Mikoshiba (1971b) chose all the F_ν's in equation (6.180) of the hydrogenic form as in equation (6.147) and calculated the results for the following processes:

$$\begin{array}{llll}
\text{elastic} & : & \hbar\omega + (M_J = \pm 3/2) \rightleftharpoons (\text{int.}) \rightleftharpoons \hbar\omega' + (M_J = \pm 3/2) \\
& & \hbar\omega + (M_J = \pm 1/2) \rightleftharpoons (\text{int.}) \rightleftharpoons \hbar\omega' + (M_j = \pm 1/2) \\
\text{inelastic} & : & \hbar\omega + (M_J = \pm 3/2) \rightleftharpoons (\text{int.}) \rightleftharpoons \hbar\omega' + (M_J = \pm 1/2) \\
& & \hbar\omega + (M_J = \pm 1/2) \rightleftharpoons (\text{int.}) \rightleftharpoons \hbar\omega' + (M_J = \pm 1/2)
\end{array}$$

$$\text{thermally assisted phonon absorption} :$$
$$\hbar\omega + \hbar\omega' + (M_J = \pm 1/2) \rightleftharpoons (\text{int.}) \rightleftharpoons (M_J = \pm 3/2).$$

The scattering of phonons by acceptors depends on the phonon frequency, the population of the states and the deformation potentials. Singh and Verma (1974) have shown that, in general, for $\hbar\omega > \Delta$ (energy difference between the Kramer's doublets) $\tau^{-1}(\omega) \propto \omega^2$, and for $\hbar\omega < \Delta, \tau^{-1}(\omega) \propto (c_1 + c_2\omega^4)$, where c_i are constants.

When a large enough magnetic field B is applied, the Γ_8 quartet splits into four levels specified by $M_J = 3/2, 1/2, -1/2, -3/2$. The spin Hamiltonian, up to the linear term in B, is given by

$$H = \beta g_1^a(\boldsymbol{B} \cdot \boldsymbol{J}) + \beta g_2^a \sum_{\alpha=x,y,z} B_\alpha J_\alpha^3, \qquad (6.183)$$

where β is the Bohr magneton, and g_1^a, g_2^a are the acceptor g-values which depend on the unperturbed eigenstates of the quartet. For $\boldsymbol{B} \parallel [001]$, g_1^a and g_2^a can be expressed in terms of the splitting coefficients of the quartet $g_{1/2}$ and $g_{3/2}$ (Suzuki and Mikoshiba 1971b)

$$\begin{aligned}
\frac{1}{2}g_{1/2} &= \frac{1}{2}g_1^a + \frac{1}{8}g_2^a \\
\frac{3}{2}g_{3/2} &= \frac{3}{2}g_1^a + \frac{27}{8}g_2^a.
\end{aligned} \qquad (6.184)$$

The four states with $M_J = 3/2, 1/2, -1/2, -3/2$ (which we refer to as 1, 2, 3, 4) have energies $3/2\beta Bg_{3/2}, 1/2\beta Bg_{1/2}, -1/2\beta Bg_{1/2}, -3/2\beta Bg_{3/2}$, respectively. The energy separations between these levels are

$$\Delta_{12} = \Delta_{34} = \frac{1}{2}(3g_{3/2} - g_{1/2})\beta B$$
$$\Delta_{13} = \Delta_{24} = \frac{1}{2}(3g_{3/2} + g_{1/2})\beta B$$
$$\Delta_{14} = 3g_{3/2}\beta B$$
$$\Delta_{23} = g_{1/2}\beta B. \tag{6.185}$$

Suzuki and Mikoshiba (1971b) have derived expressions for phonon–hole scattering rates including elastic, inelastic and 'thermally assisted absorption' processes between these levels. We do not reproduce those results here but refer the reader to their work.

6.6 PHONON SCATTERING DUE TO MAGNETIC IMPURITIES IN SEMICONDUCTORS

A number of experimental studies have shown that presence of magnetic impurities in non-metallic solids can drastically lower the thermal conductivity. The review article by Challis and de Goër (1984) gives an extensive report on this topic. One of the earliest investigations is due to Slack and co-workers (Slack 1972) who have studied the effect of substitutional Fe ions on the thermal conductivity of crystals of Ge, $ZnSO_4.7H_2O$, CdTe, $MgCr_2O_4$, $MgAl_2O_4$, ZnS, $KZnF_3$, and MgO. In these crystals the Fe^{2+} ions have either tetrahedral or nearly octahedral coordination with their nearest neighbours. The interaction of phonons with the Fe^{2+} ions appears to be strong for both types of coordination.

A substitutional site in the zincblende phase of the materials ZnS and CdTe has T_d symmetry. The crystal field of such a symmetry will split the lowest term, ^5D, of the free Fe^{2+} ion ($3d^6$) into an orbital doublet ^5E (ground state) and an orbital triplet $^5T_2$, with an energy separation $10Dq$ called the crystal-field splitting. Due to spin–orbit interaction, the ^5E level splits into five approximately equally spaced levels, and the $^5T_2$ level splits into six levels. The five energy levels of ^5E are, in increasing energy, Γ_1(singlet), Γ_4(triplet), Γ_3(doublet), Γ_5(triplet) and Γ_2(singlet). The interlevel spacing Δ of these states is $\simeq 15$ cm^{-1} for ZnS, and $\simeq 20.8$ cm^{-1} for CdTe. The ground state of d^4 ions in octahedral coordination has a very similar energy structure: the ground state is an orbital doublet ^5E which splits into five spin–orbit levels. The only significant difference in the energy levels for the tetrahedral and octahedral coordinations is in scale.

In this section we restrict our discussion to $3d^6$ ions (Fe^{2+}). Substitutional magnetic impurities in semiconductors, such as Fe^{2+} in ZnS or CdTe, can give rise to two types of phonon scattering: (i) Rayleigh scattering caused by local mass-difference and lattice distortion (as discussed in section 6.2) and (ii) resonant phonon scattering from the five low-lying ^5E levels of the Fe^{2+} ions.

Not all transitions between the five energy levels are allowed in a phonon scattering process. To understand this note that the normal modes of vibration of a tetrahedron (i.e. phonon states obeying T_d symmetry) have the irreducible representations $A_1 + E + T_2$. If we only consider the orbit–lattice interaction to be responsible for scattering the phonons, then the allowed one-phonon transitions correspond to only E-mode distortions of the tetrahedron. Only phonons of energy 2Δ can be scattered by this mechanism. But because of the spin–orbit coupling, there is some admixture of the orbital $^5T_2$ wavefunctions (at energy $10Dq$ above the ground state) into the ground state ^5E. This admixture allows additional scattering of phonons which produce T_2 distortion of the tetrahedron. Such a distortion allows phonons of energy 1Δ, 2Δ and 3Δ only between the levels shown by broken arrows in figure 6.4. However, these one-phonon transitions are much weaker than the 2Δ transitions of an E-mode distortion (shown by solid arrows) (see Slack (1972) for original references). The degenerate orbital levels ^5E and $^5T_2$ (before considering the spin–orbit interaction) are candidates for the Jahn–Teller distortion: such a degenerate orbital can couple with the lattice and split into a low-lying energy level. The Jahn–Teller distortion can be static or dynamic (phonon assisted).

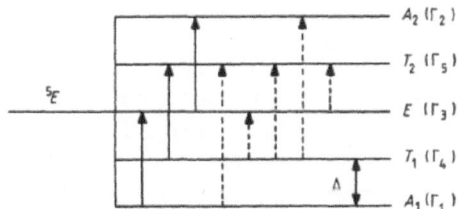

Figure 6.4 Energy-level scheme for Fe^{+2} ($3d^6$) impurities in tetrahedral coordination. The irreducible representations of the ground state and four low-lying excited states are given by Γ_i. The levels are equally spaced. The possible one-phonon absorption energies are Fe^{+2} ($3d^6$) tetrahedral coordination. The irreducible representations of the ground state and four low-lying excited states are given by Γ_i. The levels are equally spaced. The possible one-phonon absorption energies are 1Δ, 2Δ and 3Δ. The full arrows show the allowed transitions for E-mode distortions, and the broken arrows show the weakly allowed transitions for T_2-mode distortions.

The Jahn–Teller effect can lead to a change in the spin–orbit splitting Δ and to different inter-level spacings in figure 6.4. Although the Jahn–Teller effect is strong for various transition metal d^n ions, it is only a very weak effect for the Fe^{2+} ions in ZnS and CdTe and can be ignored in our discussion (Ham and Slack 1971, Challis and de Goër 1984).

The allowed one-phonon transitions between any two impurity levels in figure 6.4 can be described by resonant phonon interaction. A thermodynamic Green's function approach such as used by Kwok (1966), Klein (1969) or Joshi (1974) can be confidently used to calculate the frequency and temperature dependence of the resonance relaxation rate of phonons in such systems. Based on an application of the thermodynamic Green's functions, Joshi's work shows that for phonon frequencies close to $\Delta/\hbar = \omega_0$

$$\tau^{-1} \propto \frac{b^2(\omega)\tanh^2(\Delta/2k_BT)}{(\omega-\omega_0)^2+(\delta\omega)^2} \tag{6.186}$$

where $\delta\omega$ is a small quantity and $b(\omega)$ is an interaction constant which can be argued to be of the type $\sim \sqrt{\omega}$. The frequency dependence in the above expression is similar to that obtained in Kwok's work (see equation (6.179)).

Strictly speaking, a perturbation calculation is not valid for resonant scattering. However, using perturbation theory in the first- and second-order Born approximations, and by replacing the energy conserving Dirac delta function by a Lorentzian type line shape function, Morton and Lewis (1971) obtained the following expression for the resonant phonon relaxation rate between two spin–orbit levels:

$$\tau^{-1} = H_s N_s \frac{\omega}{\omega_0^3} F(\omega_0, T) \frac{\Gamma}{\Gamma^2 + (\omega - \omega_0)^2}, \tag{6.187}$$

where H_s represents the strength of the spin–phonon coupling, N_s is the spin density, Γ measures the level width and $F(\omega_0, T)$ is the electron population difference factor

$$F(\omega_0, T) = \frac{1 - \exp(-\hbar\omega_0/k_BT)}{3 + 5\exp(-\hbar\omega_0/k_BT)}. \tag{6.188}$$

It can be seen that the frequency dependence in both equations (6.186) and (6.187) is similar.

In order to explain the lattice thermal conductivity of non-metals doped with iron impurities, phenomenological expressions of the type $\tau^{-1} \propto \omega^4/(\omega^2 - \omega_0^2)^2$ (de Goër 1969), $\tau^{-1} \propto \omega^4 F(\omega_0, T)/(\omega^2 - \omega_0^2)^2$ (Neelmani and Verma 1972) and $\tau^{-1} \propto \omega F(\omega_0, T)/(\omega - \omega_0)^2$ (Srivastava and Verma 1974) have also been considered for resonant scattering of phonons.

6.7   PHONON SCATTERING FROM TUNNELLING STATES OF IMPURITIES

When a molecular impurity is introduced into an alkali halide lattice a barrier potential arises from the interactions of the lattice with the molecule. At energies above the barrier the molecule essentially has free *rotor states*. Below the barrier energy, the rotational motion of the molecule is perturbed and is described as states of angular oscillation, called *librational states*. At substitutional sites in alkali halides, dipolar molecular impurities experience an octahedral crystal potential with six minima corresponding to six $\langle 100 \rangle$ directions in the crystal. Due to the tunnelling of the impurity ions among the minima in the $\langle 100 \rangle$ directions, the librational ground state splits into levels known as *tunnelling* levels. Virtual excitation of the molecular impurity into a higher tunnelling energy level, or into a librational state, can resonantly absorb phonons of the lattice. The energy levels of the tunnelling states fall in the very-far-infrared and short-microwave spectral region. Thus it is possible to probe the tunnelling levels with low-temperature thermal conductivity measurements and also directly with very-far-infrared measurements.

In our discussion we will only consider the KCl:CN$^-$ and NaCl:CN^{-1} systems.

In KCl:CN^{-1} the librational ground state has six-fold degeneracy and splits into three tunnelling states: $A_{1g}(J=0)$, $T_{1u}(J=1, 1.4 \text{ cm}^{-1}$, triply degenerate) and $E_g(J=2, 2.4 \text{ cm}^{-1}$, doubly degenerate). The first excited librational state is 12-fold degenerate and splits into four triply degenerate states: $T_{2g}(J=2, 13.5 \text{ cm}^{-1})$, $T_{1u}(J=3, 16.4 \text{ cm}^{-1})$, $T_{2u}(J=3, 18.6 \text{ cm}^{-1})$ and $T_{1g}(J=4)$. (Here J is the rotational quantum number.) The energy values given in the parentheses were calculated by Seward and Narayanamurti (1966) who used a potential of octahedral symmetry (known as the Devonshire potential).

Above 60 K the CN$^-$ ion in the KCl crystal behaves like a freely rotating molecule, lending no possibility for resonant scattering. Below 10 K only the librational states are populated. The thermal conductivity of KCl:CN$^-$ shows two dips: at 0.6 K (1.6 cm^{-1}) and at 7 K (18 cm^{-1}) (see chapter 7). The lower dip is attributed to phonon scattering between the ground state $A_{1g}(J=0)$ and the tunnelling state $T_{1u}(J=1, 1.4 \text{ cm}^{-1})$ and/or $E_g(J=2, 2.4 \text{ cm}^{-1})$. The dip at the higher temperature corresponds to resonant phonon absorption due to the excitation of the CN$^-$ ion from the ground state $A_{1g}(J=0)$ to one or both of the librational states $T_{1u}(J=3, 16.4 \text{ cm}^{-1})$ and $T_{2u}(J=3, 18.6 \text{ cm}^{-1})$.

The theory of resonant phonon scattering from a two-level system can be applied to account for both the dips in the thermal conductivity curve of KCl:CN^{-1}. Kumar *et al* (1969) used an expression of the form $\tau^{-1} \propto \omega^2/(\omega^2 - \omega_r^2)^2$ for the resonant scattering of phonons to explain the observed dips in the phonon conductivity curves of the KCl:CN$^-$ and KBr:CN$^-$ systems. Singh and Verma (1971) used the theory of Klein (1969) to write the total resonant phonon scattering rate as

$$\tau^{-1}(\omega) = \frac{p\omega^2}{3\pi}\left(\frac{d_1 S_1(T)}{\rho(\omega_1)} \frac{\gamma_1(0)\gamma_1(T)\omega_1^2}{(\omega_1^2 - \omega^2)^2 + \gamma_1^2(T)\omega^4} \right.$$
$$\left. + \frac{d_2 S_2(T)}{\rho(\omega_2)} \frac{\gamma_2(0)\gamma_2(T)\omega_2^2}{(\omega_2^2 - \omega^2)^2 + \gamma_2^2(T)\omega^4} \right), \tag{6.189}$$

where p is the fractional impurity concentration, d is the degeneracy of the resonant mode, $\rho(\omega)$ is the (unperturbed) phonon density of states normalised to unity, ω_r is the resonance angular frequency, γ_r is the dimensionless half-width at ω_r and $S(T)$ is the temperature-dependent strength of the resonance. Singh and Verma took $S_1(T) = S_2(T) = 1, d_1 = d_2 = 3, \rho(\omega) = 0.75 \times 10^{-40}\omega^2$, and considered $\gamma_r(0)\gamma_r(T)$ as adjustable parameters. By evaluating the first term in equation (6.185) at $\omega_1 = 3 \times 10^{11} \text{ s}^{-1}(1.6 \text{ cm}^{-1})$ and the second term at $\omega_2 = 3.54 \times 10^{12} \text{ s}^{-1}(\simeq 18 \text{ cm}^{-1})$, these authors successfully explained both the dips in the thermal conductivity curve of KCl:CN$^-$.

In NaCl:CN$^-$ Callender and Pershan (1969) observed a librational splitting of 54 cm^{-1}, and a tunnel splitting of less than 0.1 cm^{-1}. Thus there is no possibility of resonant phonon scattering between the tunnelling states in this system. The thermal conductivity of NaCl:CN$^-$ shows only one dip (Narayanamurti and Pohl 1970): at a temperature above the peak in the conductivity against temperature curve. This dip corresponds to the energy of the librational state at $\sim 54\ \text{cm}^{-1}$. Thus in this case only one term in equation (6.189) would be needed to explain the resonant phonon scattering.

6.8   PHONON-SPIN INTERACTION

In section 6.6 we discussed phonon scattering due to magnetic impurities in semiconductors. Thermal conductivity of such systems is characterised by a dip at a temperature around or below the conductivity peak, as we will discuss in the next chapter. Anomalies in the thermal conductivity at the magnetic transition point have been observed in several magnetic crystals. The early work by Slack and Newman (1958) showed a decided minimum in the thermal conductivity of MnO at its antiferromagnetic–paramagnetic transition Nèel temperature T_N (120 K). Similarly, the work by Slack (1961) showed small cusps in the thermal conductivities of MnF_2 and CoF_2 at their antiferromagnetic–paramagnetic T_N values of 67 K and 38 K, respectively. Such anomalies are generally attributed to the effect of spin ordering, which causes coupling of phonons with spin waves (*i.e.* phonon-magnon coupling).

It is appealing to present Slack's qualitative explanation of the thermal conductivity mechanisms in diamagnetic CaF_2 and antiferromagnetic CoF_2 (Slack, 1961). In the diamagnetic insulator CaF_2 all of the heat is transported by phonons and any of the theories discussed in section 5.2–5.4 can be successfully applied to explain the thermal conductivity results over the entire temperature range. Both CaF_2 and CoF_2 crystals are similar in their lattice properties, and their thermal conductivities are nearly alike at high temperatures where anharmonic phonon scattering is dominant. There is no isotope scattering in CoF_2 as both Co and F have a single naturally occurring isotope. CaF_2 should have small isotope scattering. Thus, we should expect that a non-magnetic sample of CoF_2 would have thermal conductivity values quite similar to that of CaF_2. However, measured values of the conductivity of CoF_2 are much lower than that of CaF_2 below 100 K and exhibit a dip near its Nèel temperature.

In the paramagnetic state above the Nèel temperature T_N of 38 K, the magnetic lattice is disordered. Thus, above T_N phonons are scattered by the disordered lattice of magnetic moments. As the crystal is cooled to T_N and below, gradual ordering of the magnetic lattice is established. Thus, below T_N phonons are scattered by the ordered magnetic lattice as well as by the fraction of the magnetic lattice that remains disordered. The fraction of disordered magnetic moments varies as $\left(1 - \frac{M(T)}{M(0)}\right)$, where $M(T)$ is the magnetisation of one Co sublattice at temperature T. As a result, lattice thermal conductivity increases rapidly below T_N until it is limited by anharmonic, point defect impurities, or boundary scattering again. Below T_N, an additional contribution to thermal conductivity comes from magnons, quanta of spin orientational wave in the ordered lattice of the magnetic moments of the crystal. The total thermal conductivity is the sum of the phonon and magnon contributions: $\kappa = \kappa_{\text{ph}} + \kappa_{\text{magnon}}$. The magnon contribution κ_{magnon} can be calculated using the theory of lattice thermal conductivity, but using an appropriate magnon dispersion relation and density of states. Experimental studies performed by Slack (1961) suggest that $\kappa_{\text{magnon}} << \kappa_{\text{ph}}$. It is clear from the above discussion that scattering rates between phonons and magnons (*i.e.* phonon scattering rate due to phonon-spin interaction and magnon scattering rate due to magnon-phonon interaction) are required for explaining the thermal conductivity both sides of T_N. A first-principles treatment of phonon-spin interaction is a difficult task and has not yet been accomplished. Mattuck and Strandberg (1960), Sinha and Upadyaya (1962), Kawasaki (1963) and Stern (1965) have discussed the essential aspects of the phonon-spin interaction. Here we present a simplified treatment, as presented by Stern (1965).

Modulation of the Heisenberg exchange interaction under lattice vibrations can be expressed as

$$H^{\text{spin}} = \frac{1}{2} \sum_{l,\delta} J(\boldsymbol{\delta} + \boldsymbol{u}_{l+\delta} - \boldsymbol{u}_l) \boldsymbol{S}_l \cdot \boldsymbol{S}_{l+\delta}, \tag{6.190}$$

where S is a vector spin operator, J is the exchange coupling constant, the $\boldsymbol{u}_l$ is the displacement vector of the lattice site l and $\boldsymbol{\delta}$ is the distance vector between neighbours. The spin–lattice interaction can be expressed after expanding the above in the $\boldsymbol{u}_l$'s

$$H^{\text{spin-ph}} = \frac{1}{2} \sum_{l,\delta} \left[\nabla J \frac{\boldsymbol{\delta} \cdot (\boldsymbol{u}_{l+\delta} - \boldsymbol{u}_l)}{\delta} + \frac{1}{2} \nabla^2 J \left\{ \frac{\boldsymbol{\delta} \cdot (\boldsymbol{u}_{l+\delta} - \boldsymbol{u}_l)}{\delta} \right\}^2 \right] \boldsymbol{S}_l \cdot \boldsymbol{S}_{l+\delta}. \tag{6.191}$$

The atomic displacements $\boldsymbol{u}_l$'s can be expressed using the second quantised notation, as described in equation (4.79) or in equations (4.9) and (4.32). The spin vector operator S can also be expressed using the second quantised notation for magnons (see, *e.g.* Holstein and Primakoff (1940), and Sinha and Upadhyaya (1962)). The magnon energy spectrum can be taken as $E \propto k_{\text{sp}}^2$. And it would be sufficient to assume that a magnon state $|m>$ in the antiferromagnetic phase can be assumed to follow the Boltzmann distribution $\exp(-E_m/k_B T)$. Considering the Nèel temperature T_N much lower than the crystal's Debye temperature, it would be reasonable to use the long wavelength approximation $\omega = cq$ for phonon spectrum. First-order time dependent perturbation theory can be applied to calculate the probability of phonon-spin interaction processes of the types ph + magnon $\rightarrow$ magnon and magnon $\rightarrow$ magnon + ph. Formulation of all steps will be lengthy and will not be presented here. We simply quote Stern's final result for the phonon relaxation rate in antiferromagnets due to phonon-spin interaction (Stern, 1965)

$$\tau_q^{-1} \big|_{\text{ph-spin}} = B \omega^4 T^2, \tag{6.192}$$

where the constant B is a function of the exchange interaction term J, phonon velocity and crystal mass density.

6.9   PHONON–PHOTON INTERACTION

When a crystal is exposed to an electromagnetic field, the phonons of the crystal interact with the photons of the field. The subject of phonon–photon interaction is vast, but we will only present a brief discussion of infrared absorption and Raman scattering processes. Some applications of these processes will be discussed in chapters 8, 9 and 10.

6.9.1   INFRARED ABSORPTION

Infrared photons have energies comparable to the energies of long-wavelength optical phonons in most solids (between 0.01 and 0.1 eV). Therefore, when a solid is exposed to infrared light, phonon–photon interactions take place subject to energy and wave-vector conservation rules. Such an interaction is caused by electric dipoles which arise from the motion of charged ions in the optic mode vibrations (*cf* sections 2.2.2 and 2.4.4). As the photon represents a transverse electromagnetic wave, it interacts more strongly with optic phonons of tranverse polarisation than of longitudinal polarisation. Further, as the wavelengths of infrared photons are very large compared to the atomic dimensions in a crystal, the resultant wave vector of all phonons interacting with one photon must be nearly zero.

The interaction of electromagnetic radiation with phonons of a crystal can be studied by using standard time-dependent perturbation theory. Here we present a simple quantum mechanical description of one-phonon infrared absorption in ionic crystals. For the sake of simplicity, we use the rigid ion model and express the interaction Hamiltonian, to first order in the radiation field, as

$$H' = -\sum_{lb} \left(\frac{Z_b e}{m_b c} \right) \boldsymbol{p}(lb) \cdot \boldsymbol{A}(\boldsymbol{x}(lb)), \tag{6.193}$$

where $Z_b e$ is the charge on the bth ion of mass m_b, x and p are the ionic position and momentum vectors, respectively, and c is the speed of light. The vector potential A can be expressed in second quantised notation (similar to that described in section 4.2; also see Schiff (1968), and Donovan and Angress (1971)):

$$
\begin{aligned}
A(x(lb),t) &= -i\sum_{Q\lambda}\left(\frac{2\pi\hbar c^2}{\omega_{Q\lambda}\Omega_0}\right)^{1/2}f_{Q\lambda}\{a_{Q\lambda}^{\dagger}\exp[-i(Q\cdot x(lb)-\omega_{Q\lambda}t)]\\
&\quad -a_{Q\lambda}\exp[i(Q\cdot x(lb)-\omega_{Q\lambda}t)]\}\\
&= -i\sum_{Q\lambda}\left(\frac{2\pi\hbar c^2}{\omega_{Q\lambda}\Omega_0}\right)^{1/2}f_{Q\lambda}C_{Q\lambda},
\end{aligned}
\tag{6.194}
$$

where $a_{Q\lambda}\,(a_{Q\lambda}^{\dagger})$ is photon annihilation (creation) operator, $f_{Q\lambda}$ is a unit polarisation vector perpendicular to the photon wave vector Q, λ represents the two orthogonal transverse polarisations for a given Q, the summation takes all values of Q, positive and negative, and Ω_0 is a finite normalisation volume for the radiation field. Similarly, the momentum vector $p(lb)$ can be expressed, using equations (4.10) and (4.33), as

$$
\begin{aligned}
p(lb) &= \sum_{qs}\left(\frac{\hbar m_b\omega_{qs}}{2N_0\Omega}\right)^{1/2}e^*(b|qs)\,(a_{qs}\exp[i(q\cdot l-\omega_{qs}t)]\\
&\quad +a_{qs}^{\dagger}\exp[-i(q\cdot l-\omega_{qs}t)])\\
&= \sum_{qs}\left(\frac{\hbar\omega_{qs}}{2N_0\Omega}\right)^{1/2}e^*(b|qs)B_{qs}.
\end{aligned}
\tag{6.195}
$$

With equations (6.194) and (6.195), the interaction Hamiltonian becomes

$$
H' = \frac{i\hbar}{\sqrt{\Omega_0}}\sum_{lb}\sum_{qs}\sum_{Q\lambda}Z_b e\left(\frac{\pi\omega_{qs}}{N_0\Omega m_b\omega_{Q\lambda}}\right)^{1/2}e^*(b|qs)\cdot f_{Q\lambda}C_{Q\lambda}B_{qs}.
\tag{6.196}
$$

Consider the absorption of a photon in mode $Q\lambda$ and emission of a phonon in mode qs. This process is governed by the term $a_{Q\lambda}a_{qs}^{\dagger}$ in equation (6.196). The transition probability for this process is

$$
W = \frac{2\pi}{\hbar^2}|\langle f|H'|i\rangle|^2\delta(\omega_{Q\lambda}-\omega_{qs}).
\tag{6.197}
$$

The matrix element between the initial state $|i\rangle$ and the final state $|f\rangle$, $\langle f|H'|i\rangle$, contains the space-dependent term

$$
\exp[i(Q\cdot x(lb)-q\cdot l)] = \exp[i(Q-q)\cdot l]\exp[iQ\cdot x_b].
$$

After performing the summation over the Bravais lattice vector l, we can express

$$
\begin{aligned}
W &= \frac{2\pi^2 N_0\Omega}{\Omega_0}\left|\sum_b\frac{Z_b e}{\sqrt{m_b}}e^{iQ\cdot x_b}e^*(b|qs)\cdot f_{Q\lambda}\right|^2\\
&\quad\times\delta_{Q,q+G}\delta(\omega_{qs}-\omega_{Q\lambda}).
\end{aligned}
\tag{6.198}
$$

The Kronecker delta symbol $\delta_{Q,q+G}$ leads to the momentum conservation condition

$$
Q = q+G.
\tag{6.199}
$$

As in the optical region the electromagnetic wavelength is much bigger than the wavelength of phonons, we can consider $Q\simeq 0$. Thus the only appropriate reciprocal lattice vector is $G=0$ and we have

$$
Q = q\simeq 0
\tag{6.200}
$$

so that only an optical phonon can be emitted in the absorption process. In this limit the transition probability now reads

$$W = \frac{2\pi^2 N_0 \Omega}{\Omega_0} \left| \sum_b \frac{Z_b e}{\sqrt{m_b}} e^*(b|0s) \cdot f_0 \right|^2 \delta(\omega_{qs} - \omega_{Q\lambda}). \tag{6.201}$$

This is an essentially classical result as it does not contain $\hbar$. It is helpful to recall that $e(b|qs)/\sqrt{m_b}$ represents the time-independent normalised displacement of the bth atom in the long-wavelength optical mode (cf equations (2.48) and (2.56)). The summation under the modulus bars can be expressed as

$$
\begin{aligned}
\sum_b \frac{Z_b e}{\sqrt{m_b}} e^*(b|0s) &= Z_1 e \frac{e(1)}{\sqrt{m_1}} + Z_2 e \frac{e(2)}{\sqrt{m_2}} \\
&= Z_1 e u(1) + Z_2 e u(2) \\
&= M
\end{aligned}
\tag{6.202}
$$

where M is a dipole moment. Equation (6.201) can thus be expressed as

$$W = \frac{2\pi^2 N_0 \Omega}{\Omega_0} |M^* \cdot f_0|^2 \delta(\omega_{qs} - \omega_{Q\lambda}). \tag{6.203}$$

Let us consider zincblende materials as ionic crystals. For longitudinal phonons $e_L \parallel q \parallel Q \perp f$, so that $e_L \cdot f = 0$. Therefore, only transverse (optical) phonons can participate in the infrared absorption process. For the zincblende structure, the two TO branches are degenerate for small q. We may choose one of these to be parallel to the unit vector f, so that we can write $e_{T_1}^* \cdot f = +|e_{T_1}|$ and $e_{T_2}^* \cdot f = -|e_{T_2}|$. Equation (6.201) can also be expressed as

$$W(\omega) = \frac{2\pi^2 N_0 \Omega Q^*}{\Omega_0} \left| \left(\frac{|e_{T_1}|}{\sqrt{m_1}} + \frac{|e_{T_2}|}{\sqrt{m_2}} \right) \right|^2 \delta(\omega_{TO}(\Gamma) - \omega) \tag{6.204}$$

where Q^* is an effective charge on the ions. Thus the one-phonon infrared absorption process is governed by the conservation rules

$$\omega(Q) = \omega_{TO}(q) \qquad Q = q \simeq 0 \tag{6.205}$$

and the intensity is proportional to the square of the dipole moment.

In crystals with diamond structure (e.g. Si and Ge), there is a centre of inversion midway between the two identical atoms in the unit cell, and the dipole moment is identically zero. Such crystals are therefore one-phonon infrared *inactive*. However, such crystals can absorb a photon by a process which involves two phonons. This is caused by a second-order electric dipole moment produced by the two phonons created. Obviously the absorption is weaker in covalent crystals than in ionic crystals. The conservation rules for a single photon to generate two phonons are

$$
\begin{aligned}
\omega(Q) &= \omega_1(q_1) + \omega_2(q_2) \\
Q &= q_1 + q_2 \simeq 0.
\end{aligned}
\tag{6.206}
$$

Two-phonon absorption processes can involve optical as well as acoustic phonons of nearly equal and opposite wave vectors. Phonons near the zone edge are more likely to meet the requirements in equation (6.206). In addition, phonons at critical points, where $\nabla_q \omega(q) \simeq 0$, can also satisfy the conditions in equation (6.206). Thus two-phonon infrared absorption spectra can be used to obtain information on phonon density of states. A space-group analysis can be made to determine the two-phonon selection rules (Berman 1962, 1963, 1974, Spitzer 1967).

The net probability of optical absorption by a solid is proportional to the difference between stimulated phonon generation and absorption rates. Note that $\bar{n}(\omega)+1$ and $\bar{n}(\omega)$ represent, respectively, creation (emission) and destruction (absorption) of crystal phonons. An infrared absorption process with generation of a single phonon is thus proportional to $(\bar{n}(\omega)+1)-\bar{n}(\omega))$, i.e. independent of $\bar{n}(\omega)$ and therefore of temperature. Two-phonon infrared absorption processes are proportional to $(\bar{n}(\omega_1)+1)(\bar{n}(\omega_2)+1)-\bar{n}(\omega_1)\bar{n}(\omega_2) = 1+\bar{n}(\omega_1)+\bar{n}(\omega_2)$ and are thus temperature dependent.

For polar crystals, the long-wavelength TO modes, which are infrared active (one-phonon process), may be determined directly by absorption measurements on large thin crystals, or reflection measurements on bulk crystals. The long-wavelength LO modes are usually computed from the TO frequency using the Lyddane–Sachs–Teller relation (equation (2.87)). This requires the values of the ε_0 and ε_∞, the dielectric constant at very low and high frequencies, respectively. The LO frequency can also be determined via the Kramers–Kronig analysis or via a damped harmonic fit to infrared reflectance data (see, e.g., Burns (1985)).

6.9.2 RAMAN SCATTERING

When a crystal is exposed to an intense light source in the energy range 1 to 10 eV (e.g. the most intense lines of a mercury discharge tube, or a powerful continuous-wave laser), inelastic scattering of the light from the crystal is observed. This is the *Raman effect* (Raman 1928). The intensity of the scattered light is proportional to the intensity of the incident light.

Let $l = (a,n)$ represent a state of the solid, with a denoting an electronic level and n a vibrational level. The Raman scattering can be described as a two-stage process

$$\hbar\omega(\boldsymbol{Q})+E_i \rightarrow E_m \rightarrow \hbar\omega'(\boldsymbol{Q}')+E_f. \tag{6.207}$$

Here an absorbed photon of energy $\hbar\omega$ and wave vector $\boldsymbol{Q}$ excites the system from an initial state $i = (0,n)$ to an intermediate state $m = (a'',n'')$, and the system eventually relaxes to a final state $f = (0,n')$, releasing a photon of a different energy $\hbar\omega'$ and wave vector $\boldsymbol{Q}'$. (Note that we have considered the lowest electronic level for the initial and final states of the system, but the intermediate state is unrestricted.)

The interaction Hamiltonian responsible for the process in equation (6.207) is of the same form as in equation (6.193) but includes the electronic contribution as well,

$$H' = \sum_i \left(\frac{e}{mc}\right)\boldsymbol{p}_i \cdot \boldsymbol{A}(\boldsymbol{r}_i) - \sum_{lb} \left(\frac{Z_b e}{m_b c}\right)\boldsymbol{p}(lb) \cdot \boldsymbol{A}(\boldsymbol{x}(lb)), \tag{6.208}$$

where $\boldsymbol{r}_i$ is the position vector of an electron with mass m and, as before, m_b is the mass of the bth ion. The transition between the initial and final states can be expressed in terms of the *polarisability tensor* (Born and Huang 1954, Donovan and Angress 1971)

$$A_{\alpha\beta}(\omega)$$
$$= \sum_m \left[\frac{\langle f|M_\alpha|m\rangle\langle m|M_\beta|i\rangle}{E_m-E_f+\hbar\omega} + \langle f|M_\beta|m\rangle\langle m|M_\beta|i\rangle E_m - E_i - \hbar\omega\right], \tag{6.209}$$

where M is a vector dipole moment operator. Once again note that we have considered $i = (0,n), f = (0,n')$ and $m = (a'',n'')$. If we further consider $a'' = 0$, then the matrix elements of the operator M are defined by only the second term (ionic part) in equation (6.201). For the Raman scattering process, we consider $a'' \neq 0$ and approximately write $E_{m=(a'',n'')}-E_{f=(0,n'')} \simeq E_{(a'',0)}-E_{(0,0)}$, which is justified as vibrational energies are very much smaller than the energy difference between the electronic levels near the equilibrium nuclear configuration. We can now express the *electronic polarisability* in the form (for details see Born and Huang 1954, Donovan and Angress 1971)

$$
\begin{aligned}
A_{\alpha\beta}(\omega) &= \langle n'|A_{\alpha\beta}(\omega,x)|n\rangle \\
&= \sum_{a\neq 0}\left[\frac{\{M_\alpha(x)\}_{0a}\{M_\beta(x)\}_{a0}}{E_{(a,0)}-E_{(0,0)}+\hbar\omega}+\frac{\{M_\beta(x)\}_{0a}\{M_\alpha(x)\}_{a0}}{E_{(a,0)}-E_{(0,0)}-\hbar\omega}\right],
\end{aligned}
\tag{6.210}
$$

where $\{M_\alpha(x)\}_{0a}=\langle(0,n')|M_\alpha(x)|(a,n)\rangle$, and $x=\{x(lb)\}$ represents the nuclear configuration. (For cubic symmetry the polarisability is a scalar quantity.) We can expand $A_{\alpha\beta}(\omega,x)$ in Taylor's series involving nuclear displacements

$$
\begin{aligned}
A_{\alpha\beta}(\omega,x) &= A^{(0)}_{\alpha\beta}(\omega,x^0)+\sum_{qs}\mathscr{A}^{(1)}_{\alpha\beta}(qs)X(qs) \\
&\quad +\sum_{\substack{qq'\\ss'}}\mathscr{A}^{(2)}_{\alpha\beta}(qs,q's')X(qs)X(q's')+\cdots \\
&= A^{(0)}_{\alpha\beta}+A^{(1)}_{\alpha\beta}+A^{(2)}_{\alpha\beta}+\cdots,
\end{aligned}
\tag{6.211}
$$

where x^0 denotes the equilibrium configuration, and $X(qs)$ denotes a normal coordinate (cf equation (4.32)). The coefficient $\mathscr{A}^{(n)}_{\alpha\beta}$ is called the nth-order *Raman polarisability*.

It is useful to note a few more definitions which are often used in the liturature. *Electronic susceptibility* χ is defined as the dipole moment per unit volume induced by a unit field. *Raman susceptibility* is defined as $\partial\chi/\partial X(qs)$. The *Raman tensor* I_{ij} is defined as an entity proportional to $\partial\chi_{ij}/\partial X(qs)$.

The first term in equation (6.211) is independent of the nuclear displacements, and the only non-vanishing matrix elements correspond to $n'=n$. This elastic process is the *Rayleigh scattering*.

The second term in equation (6.211) corresponds to an inelastic scattering of a photon. This term contains a single phonon operator and non-vanishing matrix elements require $n'=n\pm 1$, with $\omega'(Q')=\omega(Q)\pm\omega(qs)$. Both longitudinal and transverse phonons can take part in the Raman effect. If the phonon involved comes from an optical branch, the process is the *first-order Raman scattering*. First-order Raman scattering can also be observed involving acoustic phonons, but with much smaller frequency shifts, and is sometimes called the Brillouin scattering. *Second-order* Raman scattering is due to the quadratic term in equation (6.211) and involves two phonons. If $\omega'(Q')<\omega(Q)$ one or more phonons are created in the crystal and the shift $\delta\omega=\omega-\omega'$ is called a *Stokes frequency*. If $\omega'(Q')>\omega(Q)$ one or more phonons are lost by the crystal and the shift $\delta\omega=\omega'-\omega$ is called an *anti-Stokes frequency*.

Selection rules for Raman-active phonons can be determined by standard group-theoretical methods (Heine 1960, Loudon 1964, Berman 1962, 1963, 1974). In general, a phonon can participate in a first-order Raman scattering if and only if its irreducible representation is the same as one of the irreducible representations of the polarisability tensor. Zincblende crystals are both Raman and infrared active in the first order. Also, while the LO modes in zincblende crystals are usually measured indirectly in infrared experiments, the selection rules allow these to be measured *directly* in the Raman scattering. In crystals with the diamond structure (which have a centre of inversion) only even-parity lattice vibrations can be Raman active and only odd-parity vibrations can be infrared active. For example, an optical phonon with the representation $\Gamma_{25'}$ (T_{2g}), which has even parity with respect to the inversion symmetry, is Raman active, but infrared inactive.

The energy and momentum conservation rules for the first-order Raman effect are

$$
\begin{aligned}
\Delta_1\omega &= \omega(Q)-\omega'(Q')=\pm\omega(q) \\
\Delta Q &= Q-Q'=\pm q\simeq 0.
\end{aligned}
\tag{6.212}
$$

Thus Raman shifts measure the frequencies of long-wavelength phonons. The intensity of Raman frequency shift is temperature dependent: proportional to $\bar{n}(\omega(q))$ for the anti-Stokes component and $\bar{n}(\omega(q))+1$ for the Stokes component.

The two phonons involved in the second-order Raman scattering can belong to either the acoustic or optical mode. The energy and momentum conservation rules are

$$
\begin{aligned}
\Delta_2 \omega &= \omega(\boldsymbol{Q}) - \omega'(\boldsymbol{Q}') = \pm \omega_1(\boldsymbol{q}_1) \pm \omega_2(\boldsymbol{q}_2) \\
\Delta \boldsymbol{Q} &= \boldsymbol{Q} - \boldsymbol{Q}' = \pm \boldsymbol{q}_1 \pm \boldsymbol{q}_2 \simeq 0.
\end{aligned}
\tag{6.213}
$$

The temperature dependence of the Stokes and anti-Stokes components can be easily worked out in terms of creation and destruction of phonons in the process. For example, the intensity of the process with the negative sign in equation (6.213) is proportional to $\bar{n}(\omega_1)\bar{n}(\omega_2)$.

In analogy with a second-order infrared spectrum, a second-order Raman spectrum contains information about the combined density of states of pairs of phonons with equal and opposite wave vectors governed by equation (6.206). As sharp features in a density-of-states spectrum are expressed in terms of a critical-point analysis, it is usual to follow this procedure for extracting information from a second-order Raman spectrum. A Kronecker square or product of the irreducible representations of the two participating phonons must be in common with the irreducible representations of the polarisability tensor.

For a detailed understanding of the Raman effect in crystals, the reader is referred to the review article by Loudon (1964) and to the series *Light Scattering in Solids I–V* edited by Cardona (1982) and Cardona and Güntherodt (1982a, b, 1984, 1989).

6.10   *AB INITIO* EVALUATION OF PHONON-PHONON INTERACTION

In sections 6.4 and 6.5 we presented expressions for phonon-phonon and phonon-electron scattering rates using simplified schemes, in particular Debye's isotropic continuum scheme for phonons and the effective mass scheme for electrons. With advances in numerical techniques and availability of computational power, these scattering rates can now be confidently and accurately evaluated using *ab initio* formulations presented in sections 3.2.3 and 3.3.2 for electron and phonon eigensolutions.

6.10.1   *AB INITIO* TREATMENT OF THREE-PHONON PROCESSES

The single mode relaxation time of phonon mode qs can be expressed, using equations (6.56) and (6.57), or equation (6.68), as

$$
\begin{aligned}
\tau_{qs}^{-1} &= \frac{2\pi}{\hbar^2} \sum_{q's'q''s''} \left| f_3(qs, q's', q''s'') \right|^2 \\
&\quad \times \left[(\bar{n}_{q's'} - \bar{n}_{q''s''}) \delta(\omega(qs) + \omega(q's') - \omega(q''s'')) \right. \\
&\quad \left. + \frac{1}{2}(1 + \bar{n}_{q's'} + \bar{n}_{q''s''}) \delta(\omega(qs) - \omega(q's') - \omega(q''s'')) \right] \\
&= \frac{\pi\hbar}{4N_0\Omega} \\
&\quad \times \sum_{\substack{q'q''G \\ s's''}} \left| \sum_{\substack{bb'b'' \\ \alpha\beta\gamma}} \frac{e_\alpha(b|qs)e_\beta(b'|q's')e_\gamma(b''|q''s'')}{\sqrt{m_b m_{b'} m_{b''}} \, \omega(qs)\omega(q's')\omega(q''s'')} \sum_{l',l''} \Psi_{\alpha\beta\gamma}(0b, l'b', l''b'') \right|^2 \\
&\quad \times \left[(\bar{n}' - \bar{n}'') \delta_{q+q',q''+G} \delta(\omega + \omega' - \omega'') \right. \\
&\quad \left. + \frac{1}{2}(1 + \bar{n}' + \bar{n}'') \delta_{q,q'+q''+G} \delta(\omega - \omega' - \omega'') \right].
\end{aligned}
\tag{6.214}
$$

Here we have used equations (4.43), (4.44), (6.60), (6.62) to express f_3 as a Fourier component of the cubic crystal potential $\mathscr{V}_3$

$$
\begin{aligned}
f_3(\boldsymbol{q}s,\boldsymbol{q}'s',\boldsymbol{q}''s'') &\equiv \Psi(\boldsymbol{q}s,\boldsymbol{q}'s',\boldsymbol{q}''s'') \\
&= \frac{i}{\sqrt{N_0\Omega}} \sum_{\substack{bb'b'' \\ \alpha\beta\gamma}} \left(\frac{\hbar^3}{8m_b m_{b'} m_{b''} \omega(\boldsymbol{q}s)\omega(\boldsymbol{q}'s')\omega(\boldsymbol{q}''s'')} \right)^{1/2} \\
&\quad \times e_\alpha(b|\boldsymbol{q}s)e_\beta(b'|\boldsymbol{q}'s')e_\gamma(b''|\boldsymbol{q}''s'') \\
&\quad \times \Psi_{\alpha\beta\gamma}(\boldsymbol{q}b,\boldsymbol{q}'b',\boldsymbol{q}''b'')
\end{aligned}
\tag{6.215}
$$

with

$$
\Psi_{\alpha\beta\gamma}(\boldsymbol{q}b,\boldsymbol{q}'b',\boldsymbol{q}''b'') = \sum_{l',l''} \Psi_{\alpha\beta\gamma}(\boldsymbol{0}b,l'b',l''b'') \exp(i\boldsymbol{q}'\cdot l') \exp(i\boldsymbol{q}''\cdot l''),
\tag{6.216}
$$

where l' and l'' are lattice vectors.

Ab initio evaluation of $\tau_{\boldsymbol{q}s}$ using the expression in equation (6.214) requires dealing with three aspects: (i) inputting phonon eigensolutions (*viz.* frequencies $\omega(\boldsymbol{q}s)$ and eigenvectors $e(b|\boldsymbol{q}s)$) obtained from the application of an established lattice dynamical theory within the harmonic approximation, (ii) inputting accurate numerical results for the Fourier components of the cubic anharmonic force constant $\Psi(\boldsymbol{q}s,\boldsymbol{q}'s',\boldsymbol{q}''s'')$ and (iii) performing realistically accurate Brillouin zone summation over the phonon wavevectors $\boldsymbol{q}'$ and $\boldsymbol{q}''$, subject to appropriate energy and momentum conservation conditions. We will cover these points in turn.

Ab initio schemes for obtaining phonon eigensolutions were discussed in sections 3.3.1.4 and 3.3.2.1. In section 4.9 we discussed two schemes for the *ab initio* numerical evaluation of the cubic force constant terms $\Psi_{\alpha\beta\gamma}(\boldsymbol{0}b,l'b',l''b'')$. And in section 4.8 we discussed a semi-*ab initio* scheme for a numerical evaluation of the third-order force constant tensor $\Psi(\boldsymbol{q}s,\boldsymbol{q}'s',\boldsymbol{q}''s'')$. This is based on the continuum approximation for $\mathscr{V}_3$ in terms of temperature-dependent Grüneisen constant $< \gamma(T) >$ calculated within the quasi-harmonic scheme and using phonon eigenvalues obtained from an *ab initio* scheme.

The Brillouin zone summation over the phonon wavevector $\boldsymbol{q}''$ may be performed in two different manners. In one approach, the momentum conservation condition can be removed by expressing $\omega(\boldsymbol{q}'') = \omega(\boldsymbol{q}\pm\boldsymbol{q}'-\boldsymbol{G}) \equiv \omega(\boldsymbol{q}\pm\boldsymbol{q}')$, where the $+(-)$ sign is required for term 1(2) in the square bracket in equation (6.214). This consideration will require obtaining phonon frequencies for wavevectors $\boldsymbol{q}$, $\boldsymbol{q}'$, $\boldsymbol{q}+\boldsymbol{q}'$ and $\boldsymbol{q}-\boldsymbol{q}'$. In an alternative approach, the summation over $\boldsymbol{q}''$ can be retained, subject to the consideration $\boldsymbol{q}'' = \boldsymbol{q}\pm\boldsymbol{q}'-\boldsymbol{G}$ but requiring the input $\omega(\boldsymbol{q}'')$ explicitly. This is what was described in equation (4.50). We will adopt the second approach in the following discussion.

Different choices for phonon wavevector grid can be made to perform numerical Brillouin zone summation over $\boldsymbol{q}'$ and $\boldsymbol{q}''$. For example, a Gaussian quadrature grid scheme, as employed by Broido *et al* (2005). Alternatively, the Monkhost-Pack scheme could be adopted, as discussed and employed in section 3.2.3.2. Care must be taken to ensure that the $\boldsymbol{q}$-grid is dense enough to adequately satisfy the energy conservation condition dictated by the Dirac delta functions in equation (6.214). A Gaussian smearing of the Dirac delta functions helps achieve the process. This is done by expressing

$$
\delta(x) = \lim_{\sigma\to 0} \frac{1}{\sqrt{\pi}\sigma} \exp(-x^2/\sigma^2)
\tag{6.217}
$$

with an appropriately chosen small value of the width σ. In the same spirit, the momentum conservation conditions may also be relaxed in the form

$$
\delta_{\boldsymbol{q}\pm\boldsymbol{q}',\boldsymbol{q}''+\boldsymbol{G}} = \begin{cases} 1 & \text{if } |q_\alpha \pm q'_\alpha - q''_\alpha - G_\alpha| \le q^0_\alpha, \quad \alpha = x,y,z \\ 0 & \text{otherwise,} \end{cases}
\tag{6.218}
$$

where q_α^0 is a vanishingly small positive quantity in wavenumber units. For the choice $\{q_i\} = \{q_i'\} = \{q_i''\}$ on a $N1 \times N_2 \times N_3$ Monkhorst-Pack grid, we can consider a value of σ slightly larger than the smallest frequency (assumed finite) on the grid, and q_α^0 slightly larger than the smallest $|q_\alpha|$ (assumed non-zero) on the grid. When making the momentum conservation test using equation (6.218) q'' should be treated as a member of the group $C(q)$, that is a symmetry-related member of the group of the wave vectors $\{q\}$ inside the Brillouin zone, (cf equation (1.38)).

6.10.2 *AB INITIO* TREATMENT OF FOUR-PHONON PROCESSES

In section 6.4.2 we discussed the essential aspect of four-phonon processes. Let us consider the single-mode relaxation rate for a phonon mode qs arising from four-phonon interactions within first-order perturbation theory. Following the procedure adopted for three-phonon processes, the four-phonon relaxation rate in equation (6.104) can be written explicitly as

$$
\begin{aligned}
\tau_{qs}^{-1}(4\,\mathrm{ph}) &= \frac{1}{\bar{n}_{qs}(\bar{n}_{qs}+1)} \sum_{\substack{q's',q''s'' \\ q'''s'''}} \left(\frac{1}{3!} \bar{P}_{qs}^{q's'q''s''q'''s'''} + \frac{1}{2} \bar{P}_{qsq's'}^{q''s''q'''s'''} + \frac{1}{2} \bar{P}_{qsq's'q''s''}^{q'''s'''} \right) \\
&= \frac{\pi\hbar^2}{8(N_0\Omega)^2} \\
&\quad \times \sum_{\substack{q'q''q''' \\ s's''s'''}} \left| \sum_{\substack{bb'b''b''' \\ \alpha\beta\gamma\delta}} \frac{e_\alpha(b|qs)e_\beta(b'|q's')e_\gamma(b''|q''s'')e_\delta(b'''|q'''s''')}{\sqrt{m_b m_{b'} m_{b''} m_{b'''}}\,\omega(qs)\omega(q's')\omega(q''s'')\omega(q'''s''')} \right. \\
&\quad \times \left. \sum_{l'l''l'''} \Xi_{\alpha\beta\gamma\delta}(0b,l'b',l''b'',l'''b''') \right|^2 \\
&\quad \times \left[\frac{1}{6} \frac{(\bar{n}'+1)(\bar{n}''+1)(\bar{n}'''+1)}{\bar{n}_{qs}+1} \delta_{q+q'+q''+q''',G}\,\delta(\omega-\omega'-\omega''-\omega''') \right. \\
&\quad + \frac{1}{2} \frac{\bar{n}'(\bar{n}''+1)(\bar{n}'''+1)}{\bar{n}_{qs}+1} \delta_{q+q',q''+q'''+G}\,\delta(\omega+\omega'-\omega''-\omega''') \\
&\quad \left. + \frac{1}{2} \frac{\bar{n}'\bar{n}''(\bar{n}'''+1)}{\bar{n}_{qs}+1} \delta_{q+q'+q'',q'''+G}\,\delta(\omega+\omega'+\omega''-\omega''') \right],
\end{aligned} \tag{6.219}
$$

where $\Xi_{\alpha\beta\gamma\delta}(0b,l'b',l''b'',l'''b''')$ are quartic force constant terms as defined in equation (4.159).

The relaxation rate $\tau_{qs}^{-1}(4\,\mathrm{ph})$ can be calculated by employing the special q-points scheme described in the previous section for $\tau_{qs}^{-1}(3\,\mathrm{ph})$. It should be pointed out that numerical evaluation of both the fourth-order force constants $\Xi(0b,l'b',l''b'',l'''b''')$ and the large number of summations in equation (6.219) will present huge computational challenges, as has been emphasized by Feng and Ruan (2016) and Feng *et al* (2017).

6.10.3 SEMI-*AB INITIO* TREATMENT OF THREE-PHONON PROCESSES

In section 4.8 we presented a semi-*ab initio* scheme for expressing the cubic anharmonic potential energy. Using equation (4.148) for $\mathcal{V}_3$, the single mode relaxation time of phonon mode qs can be expressed, where $\rho\Omega$ is the total atomic mass per unit cell, as

$$
\begin{aligned}
\tau_{qs}^{-1}(3\,\mathrm{ph}) &= \frac{\pi\hbar}{\rho N_0\Omega} \frac{<\gamma(T)>^2}{\bar{c}^2} \sum_{q's',q''s'',G} \omega\omega'\omega'' \\
&\quad \times \left[\frac{\bar{n}'(\bar{n}''+1)}{(\bar{n}+1)} \delta_{q+q',q''+G}\,\delta(\omega+\omega'-\omega'') \right. \\
&\quad \left. + \frac{1}{2} \frac{\bar{n}'\bar{n}''}{\bar{n}} \delta_{q,q'+q''+G}\,\delta(\omega-\omega'-\omega'') \right]
\end{aligned} \tag{6.220}
$$

with the temperature-dependent mode-averaged Grüneisen constant $< \gamma(T) >$ obtained as described in the sentence following equation (4.147). Note that the expressions in the square brackets in equations (6.214) and (6.220) are identical, on account of the identities in equations (6.69) and (6.71).

6.10.4   SEMI-*AB INITIO* TREATMENT OF FOUR-PHONON PROCESSES

Using equation (4.157) for $\mathcal{V}_4$, the single mode relaxation time of phonon mode qs can be expressed, where $\rho\Omega$ represents the total atomic mass per unit cell, as

$$
\tau_{qs}^{-1}(4\,\mathrm{ph}) = \frac{\pi\hbar^2}{6\rho^2(N_0\Omega)^2} \frac{< \gamma(T) >^2}{\bar{c}^4} \sum_{\substack{q's',q''s'',\\ q'''s''',G}} \omega\omega'\omega''\omega'''
$$
$$
\times \left[\frac{1}{3!} \frac{(\bar{n}'+1)(\bar{n}''+1)(\bar{n}'''+1)}{(\bar{n}+1)} \delta_{q,q'+q''+q'''+G} \delta(\omega-\omega'-\omega''-\omega''') \right.
$$
$$
+ \frac{1}{2} \frac{\bar{n}'(\bar{n}''+1)(\bar{n}'''+1)}{(\bar{n}+1)} \delta_{q+q',q''+q'''+G} \delta(\omega+\omega'-\omega''-\omega''')
$$
$$
\left. + \frac{1}{2} \frac{\bar{n}'\bar{n}''(\bar{n}'''+1)}{(\bar{n}+1)} \delta_{q+q'+q'',q'''+G} \delta(\omega+\omega'+\omega''-\omega''') \right]. \tag{6.221}
$$

In general $\tau_{qs}^{-1}(4\,\mathrm{ph})$ is a significantly weaker contribution compared to $\tau_{qs}^{-1}(3\,\mathrm{ph})$ at all temperatures. However, the contribution of four-phonon processes towards the total anharmonic relaxation rate becomes important above the Debye temperature. Numerical test results suggest that of the three classes of four-phonon processes, it is the second class of the type $q+q' \rightleftharpoons q''+q'''$ that makes the dominant contribution.

$$
\tau_{qs}^{-1}(4\,\mathrm{ph}) \approx \frac{\pi\hbar^2}{6\rho^2(N_0\Omega)^2} \frac{< \gamma(T) >^2}{\bar{c}^4} \omega^2 T^2
$$
$$
\times \sum_{\substack{q'q''q'''\\ s's''s'''}} \delta_{q+q',q''+q'''} \delta(\omega+\omega'-\omega''-\omega'''). \tag{6.222}
$$

The method of special points and the procedure described in equations (6.217) and (6.218) can be adopted to carry out the summations over the wavevectors q', q'' and q'''.

The expression in equation (6.221) can be converted to a very simplified analytical expression by using the high-temperature approximation for $\bar{n}$ and resorting to the isotropic continuum scheme, converting the summations over the wavevectors q', q'' and q''' to integrations over the Debye sphere, and summing over the polarisation indices s', s'' and s''' to account for allowed number of processes. Let us consider the diamond and zincblende cubic systems. Using the high-temperature approximation for $\bar{n} = k_B T/\hbar\omega$, the linear dispersion relation $\omega = cq$ and replacing the sum to integration $\sum_q \rightarrow \frac{N_0\Omega}{2\pi} \int_0^{q_D} dq q^2$ within the Debye sphere of radius q_D, Thomas and Srivastava (2017) obtained the following approximate analytical expression

$$
\tau_{qs}^{-1}(4\,\mathrm{ph})\Big|^{\mathrm{HT}} \simeq \frac{7}{216} \frac{k_B^2 < \gamma(T) >^2 q_D^5}{\pi\rho\bar{c}^5} \omega^2 T^2 \tag{6.223}
$$
$$
= B\omega^2 T^2. \tag{6.224}
$$

In estimating the numerical factor in this result, we have considered that on average there are six allowed combinations for each of the three classes of four-phonon scattering events and have used $q_D = (\frac{6p\pi^2}{\Omega})^{1/3}$, with $p = 2$ being the number of atoms per unit cell. The estimated value of B in equation (6.223) should provide a better alternative to the constant $B_4(5\gamma^4 + \gamma^2 - 2\gamma^2\gamma)$ in equation (6.107) which Ecsedy and Klemens (1977) obtained using a much more simplified procedure.

6.11 *AB INITIO* TREATMENT OF ELECTRON-PHONON INTERACTION

In section 6.5 we discussed electron-phonon scattering in doped semiconductors. The emphasis there was to derive relaxation rate of acoustic phonons due to their interaction with carriers forming either a semi-isolated (bound) state inside the semiconductor band gap or a broad band overlapping with the conduction band of the host. The interaction Hamiltonian was expressed using an acoustic deformation potential. Interaction between electrons and optical phonons involves consideration of optical deformation potential in non-polar semiconductors and the Frölich Hamiltonian in polar semiconductors. The topic of electron-phonon interaction (EPI) is much more general than that discussion. Many physical properties of solids get influenced by EPI. Different levels of semiempirical model Hamiltonians have been employed to study EPI. These have been thoroughly discussed in several review articles and monographs, and we refer the reader to the books by Ziman (1960), Grimvall (1981), Mahan (1993), and Alexandrov and Devreese (2010). *Ab initio* treatment of EPI, based on the density functional perturbation theory (DFPT), has been developed since the late 1980s (see, e.g. Baroni *et al* (1987), Baroni *et al* (2001)). Excellent description of these developments bas been presented in the book by Martin (2004) and the review article by Giustino (2017). Here we outline the *ab initio* DFPT treatment of EPI using the planewave pseudopotential scheme.

Following equation (6.110), we write the EPI Hamiltonian as

$$
\begin{aligned}
H_{\mathrm{ep}} &= \int \mathrm{d}r \boldsymbol{\Psi}^\dagger(r) \Delta V_{\mathrm{KS}}(\{\boldsymbol{x}(\boldsymbol{lb})\}) \boldsymbol{\Psi}(r) \\
&= \sum_{kk'} \int \mathrm{d}r \phi_{k'}^*(r) b_{k'}^\dagger \Delta V_{\mathrm{KS}}(\{\boldsymbol{x}(\boldsymbol{lb})\}) \phi_k(r) b_k,
\end{aligned} \tag{6.225}
$$

where $\boldsymbol{\Psi}(r)$ are the second-quantised field operators defined in equation (6.111), $\phi_k(r)$ is the pseudo wavefunction and ΔV_{KS} represents the change in the Kohn Sham potential due to displacements $\{u(\boldsymbol{lb})\} = \{\boldsymbol{x}(\boldsymbol{lb})\} - \{(\boldsymbol{l}+\boldsymbol{b})\}$ as defined in equation (4.1)

$$
\Delta V_{\mathrm{KS}}(\{\boldsymbol{x}(\boldsymbol{lb})\}) = \sum_{lb\alpha} \frac{\partial V_{\mathrm{KS}}}{\partial x_\alpha(\boldsymbol{lb})}\bigg|_0 u_\alpha(\boldsymbol{lb}). \tag{6.226}
$$

After expressing $u(\boldsymbol{lb})$ using equations (4.9) and (4.32), we can express ΔV_{KS} as

$$
\Delta V_{\mathrm{KS}}(\{\boldsymbol{x}(\boldsymbol{lb})\}) = \sum_{qs} \Delta V_{\mathrm{KS}}(qs)(a_{qs}^\dagger - a_{-qs}), \tag{6.227}
$$

where

$$
\begin{aligned}
\Delta V_{\mathrm{KS}}(qs) &= \mathrm{e}^{\mathrm{i}q \cdot r} \sum_{lb\alpha} \frac{-\mathrm{i}}{\sqrt{N_0 \Omega}} \sqrt{\frac{\hbar}{2m_b \omega(qs)}} \frac{\partial V_{\mathrm{KS}}}{\partial x_\alpha(\boldsymbol{lb})}\bigg|_0 e_\alpha(b;qs) \mathrm{e}^{\mathrm{i}q \cdot l} \tag{6.228} \\
&= \mathrm{e}^{\mathrm{i}q \cdot r} \Delta v_{\mathrm{KS}}(qs). \tag{6.229}
\end{aligned}
$$

Using equations (6.225), (6.227) and (6.229), we now write the EPI Hamiltonian in the form

$$
H_{\mathrm{ep}} = \sum_{kk'qs} \int \mathrm{d}r \phi_{k'}^\dagger(r) \mathrm{e}^{\mathrm{i}q \cdot r} \Delta v_{\mathrm{KS}}(qs) \phi_k(r) b_{k'}^* b_k (a_{qs}^\dagger - a_{-qs}). \tag{6.230}
$$

Adopting the reduced zone representation, we express $\phi_k(r) = \sum_n \phi_{kn}(r) = \mathrm{e}^{\pm \mathrm{i}q \cdot r}|nk>$, where $|nk>$ is the lattice-period part for electron wavector k and band index n. With this, equation (6.230) can be expressed as

$$
\begin{aligned}
H_{\mathrm{ep}} &= \sum_{nkmk'qs} \int \mathrm{d}r\, \mathrm{e}^{-\mathrm{i}k'\cdot r} \mathrm{e}^{-\mathrm{i}k\cdot r} \mathrm{e}^{\pm \mathrm{i}q\cdot r} < mk'|\Delta v_{\mathrm{KS}}(qs)|nk > \\
&\quad \times b_{k'}^{\dagger} b_k (a_{qs}^{\dagger} - a_{-qs}) & (6.231) \\
&= \sum_{nkmqs} g_{mn}(k, \pm qs) b_{k\pm q}^{\dagger} b_k (a_{qs}^{\dagger} - a_{-qs}), & (6.232)
\end{aligned}
$$

where, making use of the Fourier theorem in equation (4.82),

$$
g_{mn}(k, \pm qs) = < m(k \pm q)|\Delta v_{\mathrm{KS}}(qs)|nk > \tag{6.233}
$$

is the electron-phonon matrix element.

The term $\left.\frac{\partial V_{\mathrm{KS}}}{\partial x_\alpha(lb)}\right|_0$ appearing in scattering potential $\Delta v_{\mathrm{KS}}(qs)$ can be obtained either by employing the direct-space 'supercell technique' (*cf* section 3.3.1.4) or the momentum space DFPT formalism (*cf* section 3.3.2.1). In the 'supercell' approach, it is obtained as

$$
\left.\frac{\partial V_{\mathrm{KS}}}{\partial x_\alpha(lb)}\right|_0 \simeq \left[V_{\mathrm{KS}}(r; \tau_\alpha(lb) + \delta) - V_{\mathrm{KS}}(r; \tau_\alpha(lb)) \right] / \delta, \tag{6.234}
$$

where δ is a small displacement along the direction α of the bth atom in the lth unit cell. Within the pseudopotential DFPT formalism, $\Delta v_{\mathrm{KS}}(qs)$ is obtained as

$$
\Delta v_{\mathrm{KS}}(qs) = \Delta v_{\mathrm{ps}}^q(r) + \int \mathrm{d}r' \left[\frac{e^2}{|r - r'|} + f_{\mathrm{xc}}(r, r') \right] \Delta \rho^q(r), \tag{6.235}
$$

where f_{xc} is an exchange-correlation kernel and

$$
\Delta \rho^q(r) = 4 \sum_{nk}^{\mathrm{occ}} \phi_k^*(r) \Delta \phi_{n(k+q)}(r) \tag{6.236}
$$

(including a factor of 2 to account for spin degeneracy).

Let us denote by $|f_k n_{qs} >$ a state which has f_k electrons and n_{qs} phonons and remember the effects on this state of the creation and annihilation operators for electrons ($b_k^{\dagger}$ and b_k) and for phonons ($a_{qs}^{\dagger}$ and a_{qs}) as given in equations (6.112) and (4.38). The matrix element for the EPI process $k + q \to k'$ is then

$$
\begin{aligned}
\mathcal{M}_{mk+qs,nk'} &= \langle f_{nk'}, n_{qs} - 1|H_{ep}|f_{mk}, n_{qs} \rangle \\
&= g_{mn}(k, qs) \sqrt{n_{qs} f_k (1 - f_{k'})}. & (6.237)
\end{aligned}
$$

And the transition rate for the phonon absorption process $k + q \to k'$ is

$$
\begin{aligned}
P_{mk,qs}^{nk'} &= \frac{2\pi}{\hbar} \left| \mathcal{M}_{mk+qs,nk'} \right|^2 \delta(E_{nk'} - E_{mk} - \hbar \omega_q) \\
&= \frac{2\pi}{\hbar} \left| g_{mn}(k, qs) \right|^2 n_{qs} f_{mk} (1 - f_{nk'}) \delta(E_{nk'} - E_{mk} - \hbar \omega_q). & (6.238)
\end{aligned}
$$

Similarly, the transition rate for the phonon emission process $k \to q + k'$ is

$$
\begin{aligned}
P_{mk}^{qs,nk'} &= \frac{2\pi}{\hbar} \left| \mathcal{M}_{mk,qs+nk'} \right|^2 \delta(E_{nk'} - E_{mk} - \hbar \omega_q) \\
&= \frac{2\pi}{\hbar} \left| g_{mn}(k, -qs) \right|^2 (n_{qs} + 1) f_{mk} (1 - f_{nk'}) \\
&\quad \times \delta(E_{nk'} - E_{mk} + \hbar \omega_q). & (6.239)
\end{aligned}
$$

The expression in equations (6.238) and (6.239) can be used to calculate many electronic properties, including the relaxation rate of an electron in state $n\mathbf{k}$ (Grimvall, 1976). It can also be used to calculate the relaxation rate of a phonon in state $\mathbf{q}s$. The phonon relaxation rate is

$$
\begin{aligned}
\tau_{qs}^{-1} &= \frac{1}{\bar{n}_{qs}(\bar{n}_{qs}+1)} \sum_{mknk'} \left[\bar{P}_{mk,qs}^{nk'} + \bar{P}_{mk}^{qs+nk'} \right] \\
&= \frac{2\pi}{\hbar} \sum_{mknk'} \left[|g_{mn}(\mathbf{k},\mathbf{q}s)|^2 \frac{\bar{f}_{mk}(1-\bar{f}_{nk'})}{(\bar{n}_{qs}+1)} \delta(E_{nk'}-E_{mk}-\hbar\omega_q)\delta_{k+q,k'} \right. \\
&\quad \left. + |g_{mn}(\mathbf{k},-\mathbf{q}s)|^2 \frac{\bar{f}_{mk}(1-\bar{f}_{nk'})}{\bar{n}_{qs}} \delta(E_{nk'}-E_{mk}+\hbar\omega_q)\delta_{k,q+k'} \right] \\
&= 2\frac{2\pi}{\hbar} \sum_{mkn} \left[|g_{mn}(\mathbf{k},\mathbf{q}s)|^2 \frac{f_{mk}(1-\bar{f}_{nk+q})}{(\bar{n}_{qs}+1)} \delta(E_{nk+q}-E_{mk}-\hbar\omega_q) \right. \\
&\quad \left. + |g_{mn}(\mathbf{k},-\mathbf{q}s)|^2 \frac{f_{mk}(1-\bar{f}_{nk-q})}{\bar{n}_{qs}} \delta(E_{nk-q}-E_{mk}+\hbar\omega_q) \right]. \tag{6.240}
\end{aligned}
$$

Here $\bar{n}_{qs}$ is the Bose-Einstein distribution function, $\bar{f}_{mk}$ is the Fermi-Dirac distribution function and the additional factor of 2 in the last step accounts for the spin degeneracy of electron states. The Brillouin zone summation in equation (6.240) can be carried out by employing the special wavevector-points scheme discussed in section 6.10.1.

7 Phonon Relaxation and Thermal Conductivity in Bulk Solids

In this chapter we apply the theories developed in Chapters 4–6 to study phonon relaxation rates and thermal conductivity in bulk semiconductors and insulators and compare the results with experiment.

7.1 RELAXATION RATE DUE TO ISOTOPIC MASS DEFECTS

In figure 7.1(a) we present numerical results for phonon relaxation rate due to isotopic mass defects in enriched Ge. While the results obtained by using the isotropic continuum theory (see equation (6.36)) increase continuously as the fourth power of frequency, the results obtained from first-principles (see equation (6.34)) show the frequency variation of the type $\omega^2 g(\omega)$, where $g(\omega)$ is a realistic phonon density of states (see figure 2.19(b)). Only in the low-frequency regime does the result from the isotropic continuum model agree with the result from first-principles. This is because, as seen in figure 7.1(b), similar results for the funcion $\tau_{\text{iso}}^{-1}/\omega^2 \propto g(\omega)$ are only obtained from the two levels of theory in the low-frequency regime of up to 1.5 THz with linear dispersion relation for acoustic branches (see figure 3.4). Departure from realistic mass defect relaxation rate for high phonon frequencies is a general feature of the isotropic continuum theory.

7.2 SPECTRUM OF THREE-PHONON RELAXATION TIMES – *AB INITIO* RESULTS

As discussed in sections 4.3 and 6.4.1, anharmonic relaxation of a phonon mode qs due to three-phonon processes can be contributed by *class 1*, or coalescence, events $qs + q's' \rightarrow q''s''$ as well as *class 2*, or decay, events $qs \rightarrow q's' + q''s''$. Figure 7.2($a$) presents a spectrum of the total

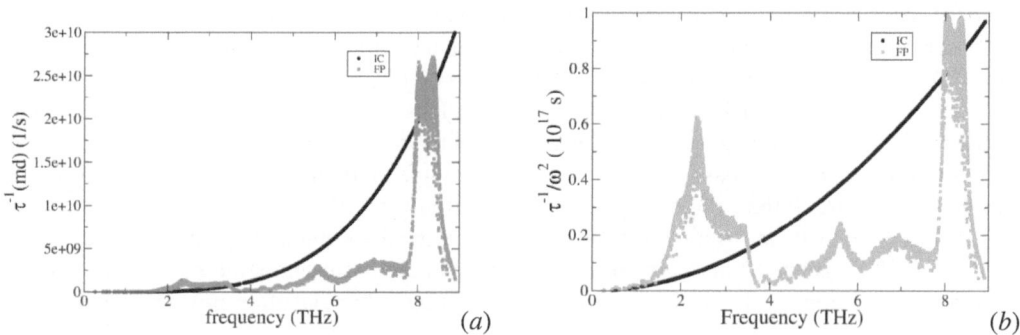

Figure 7.1 (a) Phonon relaxation rate, τ_{iso}^{-1}, due to isotopic mass defects in enriched Ge. (b) Plot of $\tau_{\text{iso}}^{-1}/\omega^2$ vs ω. Results from the isotropic continuum model are compared with results obtained from first-principles. Data points in black circles (IC) and coloured squares (FP) are the results from the applications of the isotropic continuum and first-principles methods, respectively.

DOI: 10.1201/9781003141273-7

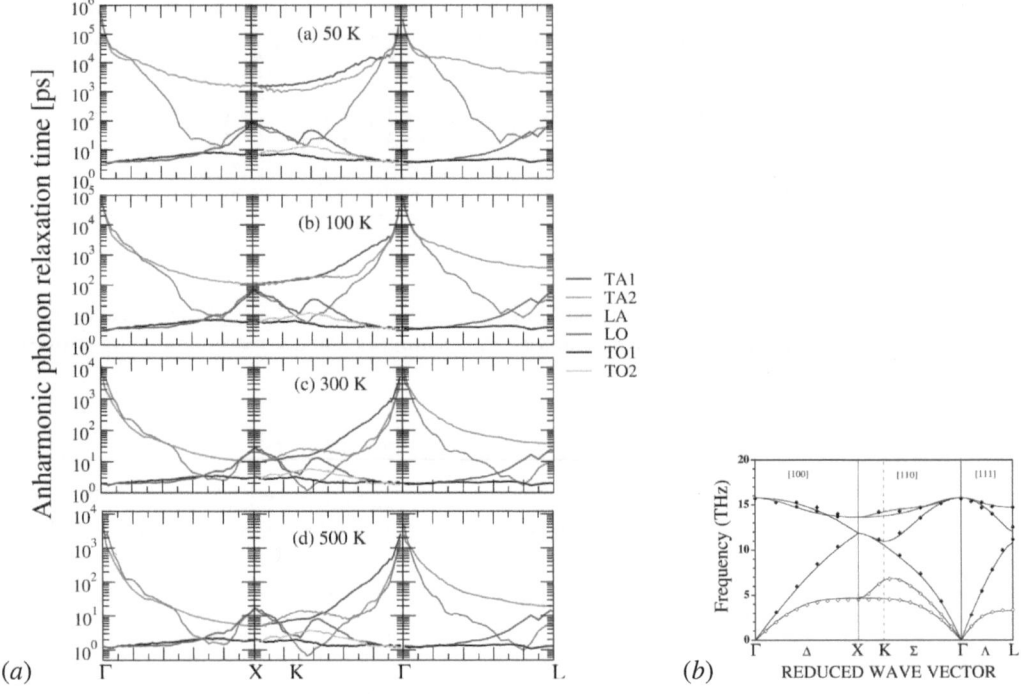

Figure 7.2 (*a*) Anharmonic phonon relaxation times, arising from three-phonon processes in Si, for wavevectors along high symmetry directions in the FCC Brillouin zone. Close to Γ along ΓK, from top to bottom are the curves for the TA1, TA2, LA, LO, TO1 and TO2 branches, respectively. [Original diagram in colour.] Reproduced from Garg *et al* (2014). (*b*) Phonon dispersion curves for Si.

three-phonon relaxation time of phonon modes in Si along the high symmetry directions in the FCC Brillouin zone, computed fully *ab initio* by Garg *et al* (2014) at 50, 100, 300 and 500 K. Many interesting and useful observations can be made by scrutinizing these results. To facilitate discussion we have presented the phonon dispersion curves in Si in figure 7.2(*b*) (which essentially has the same information as in figure 2.12(*b*)).

Firstly, the relaxation times of long wavelength modes, acoustic as well as optical, are isotropic. This observation, together with the linear dispersion relations for acoustic modes, can be considered to validate the frequency and temperature dependences of $\tau_{3\text{ph}}^{-1}$ as obtained using the elastic continuum model (see, section 6.4.1.4). Secondly, in general, long wavelength acoustic phonon modes are much longer lived compared to long wavelength optical modes. Thirdly, in contrast to optical modes, the relaxation time of an acoustic mode changes by a huge amount when its wave vector changes from the zone centre to the zone edge. In the next section we will provide a detailed discussion on the lifetimes of acoustic and optical modes governed by the three-phonon decay processes.

Ma *et al* (2014) analysed the frequency dependence of the room-temperature relaxation times of the TA and LA phonons in Si. For this purpose, they averaged the relaxation times by using

$$\bar{\tau}(\omega) = \frac{\sum_{qs} \tau(\omega(qs))\delta(\omega - \omega(qs))}{\sum_{qs} \delta(\omega - \omega(qs))} \tag{7.1}$$

with the delta function approximated with the Gaussian function of an appropriate broadening factor. Figure 7.3 shows the plots for the N and U processes for the LA TA1 and TA2 branches. As indicated in the figure, $\tau_{\text{U}}^{-1} \propto \omega^3$ for all three acoustic branches. For the LA branch τ_{N}^{-1} scales as ω^2. For

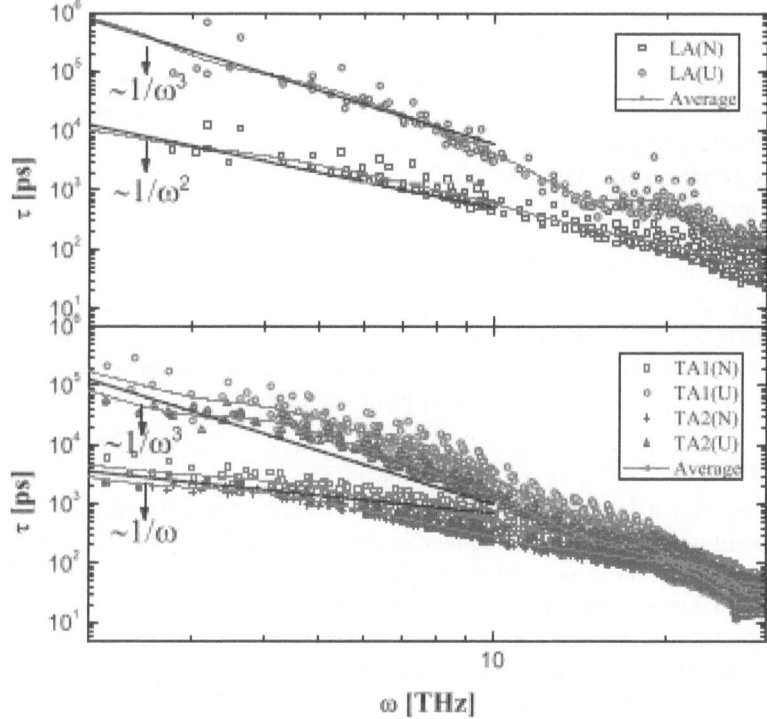

Figure 7.3 Frequency dependence of room-temperature anharmonic relaxation rates of LA and TA phonons in Si. [Original diagram in colour.] Reproduced from Ma *et al* (2014).

the TA branches τ_N^{-1} scales as ω at low frequencies and as ω^2 at higher frequencies. It was also pointed out that the anharmonic relaxation rate for the LA branch scales as ω^2 for high symmetry crystals (such as Si with point group O_h), while it is ω^3 and perhaps ω^4 for crystals with lower symmetry. These first-principles results are in general agreement with the discussion provided in section 6.4.1.4.

7.3 ANHARMONIC DECAY OF PHONONS

Because of the anharmonicity of lattice forces, a super-thermal or high-frequency phonon ($\hbar\omega \gg k_B T$), optical or acoustic, can decay into phonons of lower frequencies, subject to momentum and energy conservation rules (see equation (4.49)). Such a decay can be spontaneous as well as temperature dependent. The spontaneous decay is an example of a single-mode relaxation, as the decay process requires that only the high-frequency phonon be in excess population and that all other phonons remain in equilibrium. Several experimental methods can be applied to study anharmonic decay of phonons. These include Raman scattering, inelastic neutron scattering, time-resolved infrared absorption measurements and the method of vibronic sideband spectroscopy. Some of these methods are described in detail by Bron (1980).

7.3.1 HIGH-FREQUENCY ACOUSTIC PHONONS

High-frequency phonons usually have wavevectors close to the Brillouin zone boundary. Such phonons can decay into two phonons of lower frequencies and lower wavevectors, subject to the energy and momentum conservation rules.

TA *PHONONS*

Long life-times, of the order of microseconds or greater, have been observed for TA phonons in insulators and semiconductors (Grill and Weis 1975, Ulbrich *et al* 1980, Hu *et al* 1981, Lengfellner and Renk 1981, Baumgartner *et al* 1981). The experiments by Lengfellner and Renk (1981), performed between 3 and 20 K, also give evidence of strongly-temperature-dependent anharmonic lifetimes of zone-boundary TA phonons in thallium halide crystals.

The first-principles results in figure 7.3(*a*) confirm that in Si the near-zone-edge high frequency transverse acoustic (TA) phonons have, on average, the longest relaxation times of all phonon modes. This figure also shows strong temperature depence of the anharmonic lifetimes of TA phonons at low as well as high temperatures. The anharmonic lifetime of zone-edge TA phonons is due to their decay into two TA phonons of lower frequencies, such as $TA_i \rightarrow TA_i + TA_j$, with $i = 1, 2$ and $\omega(TA_i) > \omega(TA_j)$. Such a process is more likely to happen in anisotropic crystals, which are better characterised by non-degenerate transverse acoustic branches.

LA *PHONONS*

As discussed in section 6.4, anharmonic decay of high-frequency longitudinal acoustic (LA) phonons into two (or more) phonons of lower frequencies can take place both at low and high temperatures. The decay at low temperatures is spontaneous (temperature independent). It was shown in equation (6.91) that the spontaneous decay rate of a LA phonon via *N*-processes varies as the fifth power of the frequency of the decaying phonon: $\tau_N^{-1} \propto \omega^5$. This behaviour was first explained by Slonimskii (1937) and derived in several theoretical works, including Benin (1972) and Srivastava (1976d). Benin (1972) has argued that if the LA phonon has a large enough wave vector, the contribution of *U*-processes is similar to that of *N*-processes: $\tau_U \simeq \tau_N$.

In the theoretical work of Orbach and Vredevoe (1964), it was claimed that the process LA $\rightarrow$ LA + TA dominates over the process LA $\rightarrow$ TA + TA. In contrast, based on simple arguments within the contnuum model, Klemens (1967) concluded that the decay channel LA $\rightarrow$ TA + TA dominates over the channel LA $\rightarrow$ LA + TA. From numerical calculations, based on a phenomenological lattice dynamical model, of the decay rate of high-frequency LA modes in CaF_2 at low temperatures, Tua and Mahan (1982) confirmed the dominance of the LA $\rightarrow$ TA + TA channel. This was reconfirmed by Tamura (1985) who made numerical calculations, using the second and third-order elastic constants in the isotropic model. Further confirmation of this comes from the numerical estimates made by Mohamed *et al* (2021) who find that in quasi-isotropic III-nitrides the dominant anharmonic decay process is LA $\rightarrow$ TA + TA.

Experimental verification of the ω^5 decay rate of LA phonons with frequencies between 1.5 and 3 THz in CaF_2 is given in the work of Baumgartner *et al* (1981). Lengfellner and Renk (1981) used time-resolved far infrared laser spectroscopy to study anharmonic decay of zone-boundary phonons. Their experiment, performed at crystal temperatures between 3 and 20 K, gives evidence of strongly temperature-dependent anharmonic life-times of zone-boundary phonons. In fact their experiment indicates that the probed phonons decay at higher temperatures by decaying into 'thermal' phonons $(\hbar\omega \simeq k_B T)$.

Chou *et al* (2019) reported measurements of the lifetimes of LA phonons in wurtzite GaN. They also performed first-principles calculations of the anharmonic lifetimes of LA modes propagating along the c axis of GaN. A combined analysis of their experimental and theoretical works suggests that explanation of room-temperature long-lived THz LA modes requires their interaction with acoustic as well as optical phonons.

7.3.2   LONG WAVELENGTH OPTICAL PHONONS

Anharmonic forces in solids can also lead to decay of optical phonons. In principle, both *N* and *U* processes can contribute to decay of optical phonons. However, the most significant is the *N*-type decay of $q \approx 0$ optical phonons of frequency $\omega_{op}(q)$ into two acoustic phonons $\omega'(q')$ and $\omega''(q'')$, which satisfy

$$\begin{aligned} q &= q' + q'' = 0 \\ \omega_{op} &= \omega' + \omega''. \end{aligned} \tag{7.2}$$

The basic theoretical framework for studying the lifetime of long wavelength optical phonons was established decades ago (Maradudin and Fein (1962), Cowley (1963), Balkanski *et al* (1983), Menèndez and Cardona (1984)). Numerical calculations of the lifetime of the zone-centre LTO mode in Si, Ge and diamond were performed by Cowley (1965) by using parametrised forms of harmonic and cubic anharmonic models. Several phenomenologically simpler approaches have been developed for non-polar as well as polar semiconductors. Some of these are by Klemens (1966), Vallèe and Bogani (1991), Ridley (1996) and Barman and Srivastava (2004). In the so-called Klemens channel (Klemens, 1966) an optical phonon decays into two acoustic phonons with opposite wavevectors (op → ac + ac). The Ridley channel (Ridley, 1996) considers an optical phonon to decay into an optical phonon of lower branch and an acoustic phonon (op → op (lower branch) + ac). Vallèe and Bogani (1991) considered the decay of an optical mode into a lower mode of the same branch and an acoustic mode (op → op (same branch) + ac). Barman and Srivastava (2004) pointed out that in crystals with several atoms per unit cell, characterised with more than one set of LO and TO branches, further channels may also contribute in the decay of an optical mode (op (branch s) → op (branch s') + op (branch s''), where the sth branch is higher than the s'th and s''th branches. Simple schematic illustrations of these decay channels are presented in figure 7.4.

As we have briefly discussed in section 6.4, in the presence of anharmonicity an optical mode perturbs the force constants with alternating signs from linkage to linkage. This reduces the acoustic mode frequencies by a factor $\sqrt{R} = (2/\sqrt{3})(\beta_1 - \beta_2)/(\beta_1 + \beta_2)$. Using essentially the theory described in section 6.4, but a rather simplified expression for the anharmonic Hamiltonian, Klemens (1966, 1975) derived the following approximate result for the relaxation rate for the process in equation (7.2):

$$\tau^{-1} = \frac{2.5\gamma^2}{\pi} \frac{\hbar\omega_{op}^2}{Mv^2} \left[\frac{4}{3} \left(\frac{\beta_1 - \beta_2}{\beta_1 + \beta_2} \right)^2 \right] [1 + 2\bar{n}(\omega_{op}/2)], \tag{7.3}$$

where M is the atomic mass, v is the velocity of the acoustic phonons and γ is the Grüneisen constant.

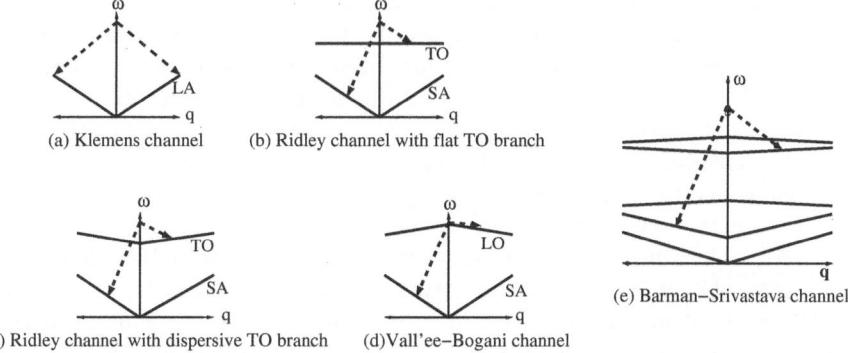

(a) Klemens channel   (b) Ridley channel with flat TO branch

(c) Ridley channel with dispersive TO branch   (d) Vall'ee–Bogani channel

(e) Barman–Srivastava channel

Figure 7.4 Schematic illustration of possible three-phonon decay channels for a zone-centre optical phonon. Klemens channel, Ridley channel, and Vallèe and Bogani channel for systems with two atoms per unit cell (*e.g.* diamond and zincblende structures) are illustrated in panels (*a*)–(*d*). Panel (*e*) shows an example of Barman-Srivastava channel for systems with more than two atoms per unit cell (*e.g.* wurtzite structure).

An accurate description and numerical evaluation of the decay rate of a long wavelength, or near zone-centre, optical phonon mode can be made by using the *ab initio* treatment of three-phonon processes in section 6.10.1. From equation (6.214), the decay rate of a phonon mode of polarisation s and $q = 0$ can be expressed as

$$
\begin{aligned}
\tau_{0s}^{-1} &= \frac{\pi\hbar}{8N_0\Omega} \\
&\times \sum_{\substack{q' \\ s's'' \\ \alpha\beta\gamma}} \left| \sum_{bb'b''} \frac{e_\alpha(b|0s)e_\beta(b'|q's')e_\gamma(b''|-q's'')}{\sqrt{m_b m_{b'} m_{b''}}\,\omega(0s)\omega(q's')\omega(q's'')} \sum_{l',l''} \Psi_{\alpha\beta\gamma}(0b, l'b', l''b'') \right|^2 \\
&\times (1 + \bar{n}(\omega(q's')) + \bar{n}(\omega(q's'')))\delta\left(\omega(0s) - \omega(q's') - \omega(q's'')\right).
\end{aligned} \tag{7.4}
$$

Using the semi-*ab-initio* treatment in section 6.10.3, a simpler expression for the decay of a zone-centre optical phonon can be obtained from equation (6.220)

$$
\begin{aligned}
\tau_{0s}^{-1} &= \frac{\pi\hbar}{2\rho N_0\Omega}\frac{<\gamma(T)>^2}{\bar{c}^2} \\
&\times \sum_{\substack{q' \\ s's''}} (1 + \bar{n}(\omega(q's')) + \bar{n}(\omega(q's'')))\delta\left(\omega(0s) - \omega(q's') - \omega(q's'')\right).
\end{aligned} \tag{7.5}
$$

Note that in writing both expressions above we have neglected the contribution from U processes by setting $\boldsymbol{G} = \boldsymbol{0}$. Both the *ab initio* and semi-*ab initio* expressions for the optical decay rate can be evaluated numerically by replacing the Dirac delta function by a Gaussian as expressed in equation (6.217) and by performing the Brillouin zone summation over $\boldsymbol{q}'$ by using the special wavevector scheme described in section 3.2.3.2.

Debernardi and co-workers used the *ab inito* expression in equation (7.4) to calculate the lifetimes of the zone-centre LTO modes in diamond structure semiconductors (Debernardi *et al*, 1995) and of the TO and LO modes in zincblende structure semiconductors (Debernardi, 1998). Barman and Srivastava (2004) used an expression similar to that in equation (7.5) but performed numerical calculations using the isotropic continuum scheme as explained in section 6.4.1.4. They investigated the relative contributions of allowed decay channels of LO and TO modes in semiconductors in cubic and hexagonal structures. Srivastava (2009) used a semi-*ab initio* expression similar to that in equation (7.5) and performed realistic Brillouin zone summation for cubic and hexagonal structures for III-N materials using phonon dispersion relations obtained from the adiabatic bond charge model.

Debernardi *et al* (1995) found that in diamond the Klemens channels LTO $\rightarrow$ TA + TA ($\approx 31\%$) and LTO $\rightarrow$ LA + LA ($\approx 15\%$) dominate. In contrast, in Si and Ge the Ridley channel LTO $\rightarrow$ TA + LA dominates ($> 94\%$). The relative importance of different channels for the decay rate of TO and LO phonons in zincblende materials depends on their phonon dispersion spectrum (Debernardi, 1998). The dominant mechanism for the decay of TO phonons is via the Klemens channel TO $\rightarrow$ TA + LA ($> 95\%$) in GaAs and GaP, but via the Klemens channel TO $\rightarrow$ LA + LA (100%) in AlAs and InP. The dominant mechanism for the decay of LO phonons is via the Klemens channel LO $\rightarrow$ TA + LA (96 %) in GaAs, and via the LO $\rightarrow$ LA + LA ($\approx 95\%$) in GaP and AlAs, but via the Ridley channel LO $\rightarrow$ TO + LTA (99%) in InP. Debernardi's work also finds that the decay rate in InP is an order of magnitude smaller than in GaAs, GaP and AlAs. This is because the Ridley process occurs in a small region of momentum space around the zone centre with vanishingly small contribution.

Barman and Srivastava (2004) examined trends in results. They predicted that in zincblende materials compared to the LO mode the anharmonic lifetime of the TO mode is shorter, similar and longer when cation/anion mass ratio is larger, similar and smaller than unity. They found that Klemens channel is forbidden in zincblende materials with cation-anion mass ratio larger than 3, such as InP, InN, GaN and AlSb. They also found that GaAs, with cation/anion mass ratio of 0.93, is the only III-V material for which the Vallèe-Bogani channel provides an important contribution in the decay of the LO mode. Furthermore, they found that in general for a given material the lifetime of the A_1(LO) mode in the wurtzite phase is smaller than the lifetime of the LO mode in the zincblende phase. This is due to an intricate balance between the reduction of the contributions from the Klemens and Ridley channels, and development of contribution from the Barman-Srivastava channel in the wurtzite phase. However, GaN is noted as an exception for which the Ridley channel provides the sole contribution for the decay of the LO mode in the zincblende phase and also for the A_1(LO) mode in the wurtzite phase. While in general agreeing with these predictions, further work on III-N materials by Srivastava (2009) pointed out that consideration of realistic Brillouin zone integration and use of realistic phonon dispersion relations play important roles in determining effective decay mechanisms.

Experimentally, anharmonic lifetimes of zone-centre phonons are usually extracted from their Raman linewidths. The Raman linewidth (full width at half maximum, or Γ_{FWHM}) of a phonon mode is related to its lifetime τ as[1]

$$\Gamma_{FWHM} = \hbar\tau^{-1}. \tag{7.6}$$

Menèndez and Cardona (1984) have summarised some experimental and theoretical determinations of the low-temperature FWHM of Raman phonons in diamond structure semiconductors. Parker *et al* (1967) observed the first-order Raman line in silicon, situated at 522 cm^{-1}, to have a half-width $\Delta\omega$ of 5 cm^{-1} at room temperature. Klemens' estimate of $\tau^{-1} \equiv \Delta\omega$ is in good agreement with experiment. Klemens' estimate for the decay rate of $q = 0$ LO phonons is also in good agreement with the width of the 1332 cm^{-1} first-order Raman line in diamond (Krishnan 1946, Solin and Ramdas 1970) in the temperature range 15 to 970 K. A detailed discussion of comparison between *ab initio* theory and experimental measurements can be found in Debernardi *et al* (1995) for C, Si and Ge, and in Debernardi (1998) for GaAs, GaP and InP. The work by Debernardi *et al* (1995) clearly suggests that the experimental values of FWHM for the LTO mode in C, Si and Ge are larger than the anharmonic contribution computed from the *ab initio* theory. Similarly, the work by Debernardi (1998) found that the anharmonic contribution to the FWHM for the LO and TO modes in GaAs, GaP and InP is smaller than the experimental values.

When comparing experimental and theoretical values of Raman linewidths, two factors must be kept under consideration: sample quality and sample temperature. As is well appreciated, real crystal samples do contain defects which also contribute to the broadening of Raman lines. In addition, at temperatures above the material Debye temperature anharmonic contributions higher than three-phonon decay must be included in theoretical calculations. It is fair, therefore, to express the FWHM as

$$\Gamma_{FWHM} = \hbar\tau_{defects}^{-1} + \hbar\tau_{3ph}^{-1} + \hbar\tau_{4ph}^{-1} \tag{7.7}$$

before attempting to compare theoretical and experimental values. The four-phonon contribution $\hbar\tau_{4ph}^{-1}$ to FWHM has been discussed in several works, including Maradudin and Fein (1962), Cowley (1963), Balkanski *et al* (1983), and Menèndez and Cardona (1984). Phonon relaxation rate due to four-phonon processes has also been discussed in sections 6.10.2 and 6.10.4. A word of caution may be appropriate here. As different samples usually contain unknown types and amounts of defects, the contribution $\hbar\tau_{defects}^{-1}$ may not be ascertained accurately, and hence comparison between experiment and theory may not always be made consistently.

7.4 LATTICE THERMAL CONDUCTIVITY OF UNDOPED SEMICONDUCTORS AND INSULATORS

In Chapter 5 we derived expressions for lattice thermal conductivity using two different theoretical routes. One route was to solve a linearized version of Boltzmann transport equation (BTE)

---

[1]If τ^{-1} is in units of THz, then $5.26\tau^{-1}$ gives Γ_{FWHM} in units of cm^{-1}.

and the other route was to follow the Kubo linear response formalism. In section 5.2 we described four different levels of the relaxation time approach for solving the BTE. In section 5.3 we described variational and complementary variational approaches for solving the BTE. Relaxation time formulations of solving the BTE have received most attention for numerical calculations of the conductivity. Relatively less effort has been made towards numerical calculations of the conductivity using variational methods. In this section we present and describe lattice thermal conductivity results obtained from applications of both the relaxation time and variational methods.

Before presenting and discussing thermal conductivity results, a brief mention of challenges involved in making numerical calculations is in order. To appreciate these issues, let us remind ourselves that thermal conductivity calculations for crystalline insulators require the following ingredients: phonon eigensolutions (phonon dispersion relations $\omega = \omega(qs)$ and phonon eigenvectors $e(b;qs)$) for all phonon branches; phonon density of states $g(\omega)$; atomic isotopic information; anharmonic (third- and forth-order) force constants; and a method of performing accurate Brillouin zone integration.

7.4.1 RELAXATION TIME RESULTS

Relaxation time theories can be grouped in five categories: simple phenomenological approach, complex phenomenological approach, non-phenomenological isotropic continuum approach, semi-*ab initio* approach and fully *ab initio* approach. A brief description of these follows.

7.4.1.1 Simple phenomenological approach

During 1960s, after the development of Callaway's theory in 1959 (see section 5.2.3), numerical evaluations of κ_{smrt} (the single-mode relaxation time expression in equation (5.26) and κ_C (Callway's expression in equation (5.47)) were performed mostly by considering only acoustic phonon branches, isotropic and linear dispersion relation $\omega = cq$, Debye's density of states expression (equation (2.115)), phenomenologically chosen simple forms of anharmonic relaxation times, and carrying out momentum space integration over Debye's sphere of a radius of an adjustable size. For example, the conductivity expression in Callaway's paper is

$$\mathscr{K}_C = \frac{k_B}{2\pi^2 c}(I_1 + \beta I_2), \tag{7.8}$$

where

$$I_1 = \int_0^{k_B\Theta_D/\hbar} \tau_C \left(\frac{\hbar\omega}{k_B T}\right)^2 \frac{e^{\hbar\omega/k_B T}}{\left(e^{\hbar\omega/k_B T} - 1\right)^2}\omega^2 d\omega \tag{7.9}$$

and

$$I_2 = \int_0^{k_B\Theta_D/\hbar} \frac{\tau_C}{\tau_N}\left(\frac{\hbar\omega}{k_B T}\right)^2 \frac{e^{\hbar\omega/k_B T}}{\left(e^{\hbar\omega/k_B T} - 1\right)^2}\omega^2 d\omega. \tag{7.10}$$

In the above equations $\omega = qc$, with c as an average (acoustic) phonon speed, and the Debye temperature θ_D is related to the Debye frequency ω_D or the Debye radius q_D as $k_B\theta_D/\hbar = \omega_D = cq_D$. The isotopic mass scattering rate is $A\omega^4$, with calculated from available isotopic masses (see equations 6.36 and 6.37). The low-temperature three-phonon scattering rates were taken as $\tau_N^{-1} = B_N\omega^2 T^3$ and $\tau_U^{-1} = B_U\omega^2 T^3\exp(-\Theta_D/\alpha T)$. This simple scheme, thus, used five adjustable parameters: c, B_N, B_U, α and Θ_D.

7.4.1.2 Detailed phenomenological approach

Holland (1963) introduced a couple of changes for evaluating the integrals I_1 and I_2 in the Callaway's conductivity expression in equation (7.8). These integrals were evaluated by taking into account explicit contributions from TA and LA phonon polarisations. Acknowledging the importance of using realistic phonon dispersion relations, he used different values of phonon speeds c_{TA} and c_{LA}, different values of Θ_D and different values of B_N and B_U in different frequency ranges. However, despite the use of several of these adjutable parameters, it was not possible to fit the high-temperature conductivity data satisfactorily. Guthrie (1966) argued that in expressing τ_U^{-1} as $\omega^n T^m$, the exponent m should be defined as m_I for *Class 1* and m_{II} it Class 2 type three-phonon events. He derived upper and lower bounds for m_I and m_{II}. Joshi *et al* (1970) recognised the importance of introducing the role of four-phonon processes at high temperatures. They expressed $\tau_{4ph}^{-1} = B_H \omega^2 T^2$, where B_H is an adjustable parameter. Sharma *et al* (1971) and Dubey and Verma (1973) incorporated the ideas of Holland (1963) and Guthrie (1966) and proposed to evaluate the Callaway conductivity expression by using three-phonon scattering rates in the form $B_I \omega^n T^{m_I}$ and $B_{II} \omega^n T^{m_{II}}$ for TA and LA phonons involved in N and U processes. In different temperature ranges different values of the exponents m_I and m_{II} were considered. Clearly, these considerations generate a plethora of adjustable parameters. Also, it should be noted that while consideration of parameters for anharmonic relaxation rates was based on partial consideration of phonon dispersion relations, the treatment of mass-defect scattering rate remained based on linear dispersion relation. Kaviany (2008) has compiled expressions and adjustable parameters for three-phonon relaxation rates used in different works for numerical evaluation of thermal conductivity.

7.4.1.3   Non-phenonomenological isotropic continuum approach

One of the first works which attempted a non-phenomenological approach for three-phonon scattering rates, though using the isotropic harmonic and anharmonic elastic continuum schemes and treating Grüneisen's constant as an adjustable parameter, is due to Srivastava (1974). Equation (6.85) presents the expression for the single-mode relaxation time and equation (6.86) presents an effective U-processes relaxation time. Detailed numerical calculations of the anharmonic relaxation rates and the thermal conductivity using this treatment appeared in further works (Srivastava 1976a, 1976d). An ad hoc treatment of acoustic-optical phonon interactions and the role of optical phonons in the conductivity was added to the formalism (Srivastava, 1980).

Using the Debye dispersion scheme described in section 2.6, an expression for the single-mode phonon relaxation time was derived in equation (6.85). Using this result for acoustic–acoustic phonon interaction and using the Debye scheme for evaluating the integral for the thermal conductivity in equation (5.26) we can express the *smrt* or the Debye result as

$$\mathscr{K}_{smrt} \equiv \mathscr{K}_D = \frac{\hbar^2 q_D^5}{6\pi^2 k_B T^2} \sum_s c_s^4 \int_0^1 dx\, x^4 \tau \bar{n}(\bar{n}+1), \tag{7.11}$$

where $x = q/q_D$, and τ and $\bar{n}$ are functions of x and the polarisation s. With the effective relaxation time defined in equations (5.54)–(5.60) and (6.85)–(6.87) the model conductivity given in equation (5.61) can be expressed within the Debye scheme as

$$\mathscr{K}_S = \frac{\hbar^2 q_D^5}{6\pi^2 k_B T^2} \left(\sum_s c_s^4 \int_0^1 dx\, x^4 \tau_m \bar{n}(\bar{n}+1) \right.$$
$$\left. + \frac{\left(\sum_s c_s^2 \int_0^1 dx\, x^4 \tau_m \tau_N^{-1} \bar{n}(\bar{n}+1) \right)^2}{\sum_s \int_0^1 dx\, x^4 \tau_N^{-1} (1 - \tau_m \tau_N^{-1}) \bar{n}(\bar{n}+1)} \right). \tag{7.12}$$

As explained in section 5.2, with $\tau_m = \tau$ equation (7.12) becomes the Callaway result $\mathscr{K}_C$.

The results in figure 7.5 show that the calculated thermal conductivity results for Si, GaAs and LiF using Srivastava's model expression κ_S (see equation 5.61) agree well with experimental measurements at low temperatures. Figure 7.6 shows the conductivity results using the $\mathscr{K}_{smrt}$, $\mathscr{K}_S$ and $\mathscr{K}_C$ formalisms for Geballe and Hull's natural sample of Ge. Clearly, $\mathscr{K}_S$ lies in between $\mathscr{K}_{smrt}$ and $\mathscr{K}_C$.

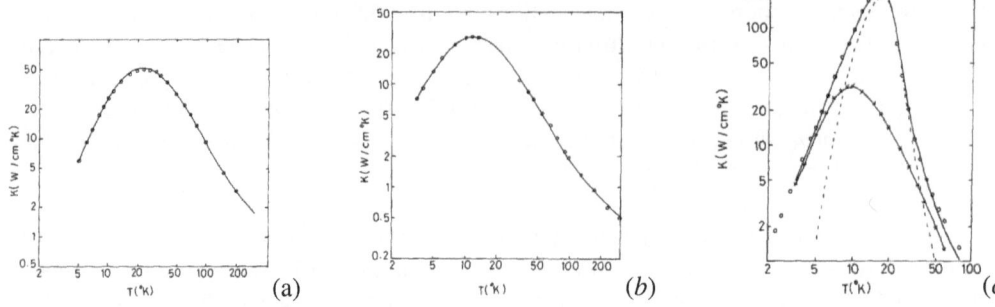

Figure 7.5 Lattice thermal conductivity of Si, GaAs and LiF. The solid curves are the results of the model conductivity $\mathscr{K}_S$ (see equation (5.61)). The open dots are experimental results (Holland and Neuringer, 1962 for Si; Holland, 1964 for GaAs; Thacher, 1967 for LiF). In (c) the curve with crosses represents $\mathscr{K}_{smrt}$ and the dotted curve is the N-drift contribution $\mathscr{K}_{N-drift}$ (see equation 5.37). Reproduced from Srivastava (1980).

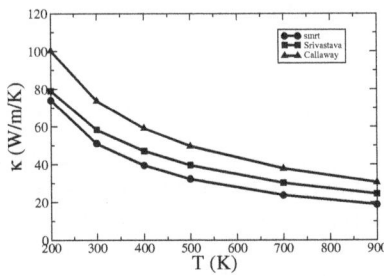

Figure 7.6 Lattice thermal conductivity results for the Geballe and Hull's natural sample of Ge using three different relaxation-time formalisms: $\mathscr{K}_{smrt}$ from the *smrt* theory, $\mathscr{K}_C$ from Callaway's theory and $\mathscr{K}_S$ from Srivastava's theory (see chapter 5). Symbols represent the numerical data from Table 4 in Srivastava (1976a) and the solid lines are fit to the data points.

7.4.1.4 Semi-*ab initio* approach

Using the semi-*ab initio* treatments of three-phonon and four-phonon scattering rates as described in equations (6.220) and (6.223), equation (6.34) for isotopic mass defect scattering rate and the special wavevector scheme for Brillouin zone summation, Thomas and Srivastava (2017) performed numerical calculations of $\mathscr{K}_{smrt}$ and $\mathscr{K}_C$ for Si and Ge. They also computed results using the Allen's variant (Allen, 2013) of Callaway's theory. The phonon eigensolutions were obtained using the DFPT method described in section 3.3.2.1.

Temperature variation of the mode-average Grüneisen constant, using equation (4.147), is shown in figure 7.7. The results suggest that, in the context of transport calculations in these materials, there is a strong temperature variation of the mode-average Grüneisen constant (and hence of the anharmonic force constants) up to about 100 K. The conductivity results over the wide temperature range 5–1700 K are presented in figure 7.8. Boundary, isotopic mass defect and three-phonon scattering rates are included at all temperatures. Four-phonon scattering rates were included close to and above the Debye temperature. It is pleasing to see that without the need for adjusting any variable or parameter, the computed theoretical results agree with carefully measured experimental results throughout the wide temperature range. The high-temperature conductivity variation in both Si and Ge is T^{-1} when only three-phonon processes are included. This changes to the experimentally deduced $T^{-1.1}$ variation when both three- and four-phonon processes are included. The Callaway's

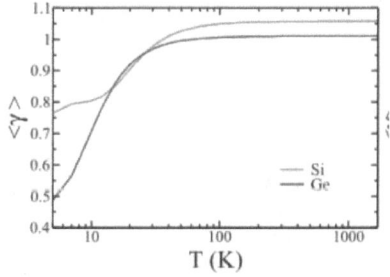

Figure 7.7 Temperature variation of $< \gamma >= \sqrt{\overline{\gamma^2}}$ (root-mode-square average Grüneisen constant) for Si (upper curve) and Ge (lower curve). [Original diagram in colour.] From Thomas and Srivastava (2017).

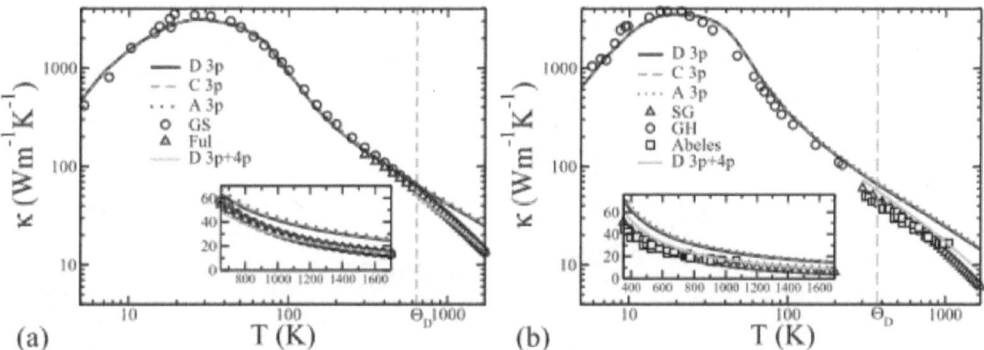

Figure 7.8 Lattice thermal conductivity of Si and Ge calculated within the semi-it ab initio scheme developed by Thomas and Srivastava (2017). The importance of the role of four-phonon processes in explaining the high-temperature results is highlighted. The solid (D 3p), dashed (C 3p) and dotted (A 3p) curves result from the Debye (*i.e.* smrt), Callaway and Allen conductivity expressions, respectively, with anharmonicity at the three-phonon level. Effective sample lengths used are $L_B = 4.5$ mm in the case of Si and $L_B = 2.5$ mm in the case of Ge, as reported in Glassbrenner and Slack (1964). Experimental measurements are shown by symbols: for Si open circles (expt GS) from Glassbrenner and Slack (1964) and filled triangles (expt Ful) from Fulkerson *et al* (1968); for Ge open circles (expt GH) from Geballe and Hull (1958), filled triangles (expt Abeles) from Abeles *et al* (1962) and open triangles (expt GS) from Glassbrenner and Slack (1964). For both Si and Ge, the golden colour solid curve (D 3 + 4p) results from the Debye (*i.e.* smrt) expression with four-phonon scattering included. [Original diamgram in colour.] From Thomas and Srivastava (2017).

conductivity result ($\mathcal{K}_C$) is only slightly higher than the *smrt* result ($\mathcal{K}_{smrt}$): less than 1% at 300 K and approximately 2.5% at 1000 K. These differences are much smaller than those in figure 7.6 obtained using the isotropic continuum theory.

Heat conduction by separate phonon branches

Figure 7.9 shows the contributions to the solid curve (D 3p) in figure 7.8 from the acoustic phonon branches. There are a few points to note from the results in panels (*a*) and (*b*). Throughout the temperature range, the contribution from the TA phonons is higher than that from the LA phonons. This difference becomes larger as temperature lowers below 70 K. The contribution from the TA phonons decreases and that from the LA phonons increases as the temperature increases up to about 100 K. The contribution of TA phonons is 70% at 70 K, which increases to 98% at 5 K. In contrast, the contribution of LA phonons is 30% at 70 K, which increases to 2% at 5 K. At and above room

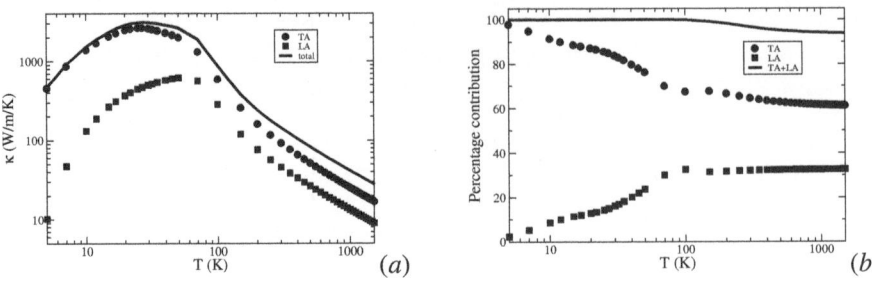

Figure 7.9 Contributions from acoustic phonons towards total thermal conductivity. Plotted are $\kappa_{smrt}(\text{TA})$, $\kappa_{smrt}(\text{LA})$ and κ_{smrt} (solid curve in figure 7.8). Calculated by the author, using exactly the same code which was employed for producing figure 7.8.

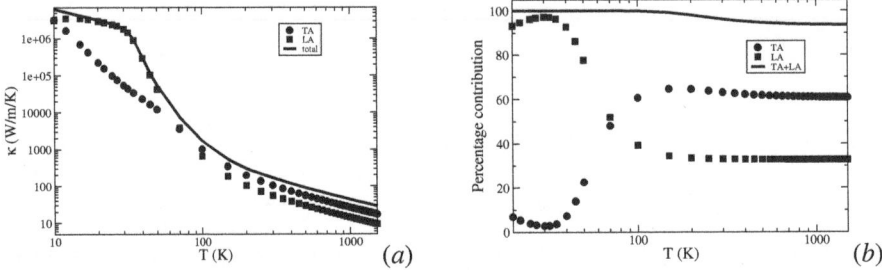

Figure 7.10 Same as figure 7.9 after setting $\tau^{-1}(\text{bs})\tau^{-1}(\text{md}) = 0$. Calculated by the author, using exactly the same code which was employed for producing figure 7.8.

temperature the TA and LA phonons make approximately 60% and 20% contribution, respectively. These findings for the relative contributions from TA and LA phonons are governed by the combined roles of specific dispersion spectrum for each phonon branch, occupation number for each phonon mode according to the Bose–Einstein distribution function and phonon lifetimes from boundary, mass-defect and three-phonon scatterings. In order to check the role played by the inclusion of boundary and mass-defect scatterings, we have presented results in figure 7.10 for a hypothetical Si sample for which three-phonon processes are considered as the only source of phonon scattering (*i.e.* when boundary and mass-defect scatterings are absent). Clearly, the relative contributions of the TA and LA phonons are interchanged at temperatures below approximately 70 K. At high temperatures, where anharmonic scattering dominates, the results are similar to that in figure 7.9.

Role of optical phonons

The uppermost curve in figure 7.9(*b*) shows the total percentage contribution from the acoustic branches. From this is it clear that up to about 100 K almost all conduction is contributed by acoustic phonons. The contribution from optical phonons (difference between the solid curve in figure 7.8 and the top curve in figure 7.9) becomes noticeable above 100 K, reaching approximately 3% at 300 K and 6% at 1500 K.

Logachev and Yur'ev (1973), Ecsedy and Klemens (1977), and Srivastava (1980) made phenomenological considerations to understand the role of optical phonons in thermal conduction. Let us consider a crystal structure with two atoms per unit cell. It would be convenient to consider two extreme cases of the mass ratio: $m_1/m_2 \simeq 1$ and $1 < m_1/m_2 \leq 4$. The Debye and Debye–Einstein dispersion relations can be used for the two cases, respectively (see section 6.4.1.5). Owing to small group velocity, optical phonons usually carry only a small percentage of total heat. On the other

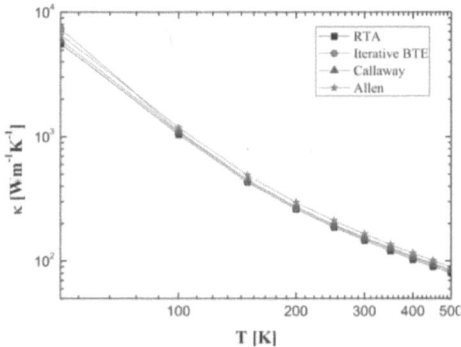

Figure 7.11 Lattice thermal conductivity of Si, obtained from a fully *ab-initio* approach using different relaxation-time methods of solving the linearized phonon Boltzmann equation. Here the legend 'RTA' stands for '*smrt*'. Reproduced from Ma *et al* (2014).

hand, acoustic–optic phonon scattering processes significantly reduce the contribution of acoustic phonons towards the thermal conductivity at high temperatures.

7.4.1.5 Fully-*ab initio* approach

Lattice thermal conductivity results using the fully *ab-intio* approach have recently been presented by several groups (Ward and Broido 2010, Chaput 2013, Garg *et al* 2014, Mingo *et al* 2014, Ma *et al* 2014, Feng *et al* 2017). The method for obtaining the basic ingredients for these calculations has been described in chapters 3–6. Figure 7.11 shows the results obtained by Ma *et al* (2014) for Si using the *smrt* theory, Callaway's model, Allen's model and the *iterative* approach. Their work considered only three-phonon events for phonon anharmonic interactions. In general, the results show the trend $\mathcal{K}_{\text{Allen}} > \mathcal{K}_{\text{C}} > \mathcal{K}_{\text{iterative}} > \mathcal{K}_{\text{smrt}}$. In agreement with the results obtained from the semi-ab initio theory in figure 7.10, this work shows that there is very little difference between the results obtained from these theoretical models at room temperature and above. There are, however, some differences below 100 K.

Feng *et al* (2017) performed detailed calculations of the thermal conductivity of naturally occurring boron arsenide (BAs), Si and diamond by employing the iterative solution of the phonon Boltzmann equation. The main focus of their work was to examine the role of four-phonon interactions in explaining the high-temperature thermal conductivity of these materials. Their results, shown in figure 7.12, are compared to available experimental data. Here $\mathcal{K}$ (3ph) and $\mathcal{K}$ (3ph + 4ph) show results using anharmonic interactions at three-phonon scattering alone and three- and four-phonon scatterings, respectively. The predicted $\mathcal{K}$ (3ph) results for Si and diamond agree with experimentally measured data below Debye temperature (*i.e.* below 600 K for Si and below 900 K for diamond). Above Debye temperature the predicted $\mathcal{K}$ (3ph + 4ph) results match well with the measured data. At 1000 K, $\mathcal{K}$ (3ph) overpredicts the measured data by 26% for Si and by 31% for diamond. Note that the application of the semi-*ab initio* theory also concluded the same level of the role of four-phonon scattering in explaining the thermal conductivity of Si and Ge (see figure 7.6).

Thermal conductivity of BAs is amongst the highest of known materials. Feng *et al* (2017) noted that inclusion of four-phonon interaction generates huge reduction in the conductivity of naturally occurring BAs. Even at room temperature the conductivity is reduced from 2241 to 1417 W/m K. At 1000 K the reduction grows to over 60%. It is found that for many frequencies the four-phonon scattering rate is comparable to, or even much stronger than, the three-phonon scattering rate. The significant role of four-phonon processes has specifically been identified for phonons with higher frequencies, as indicated in figure 7.13.

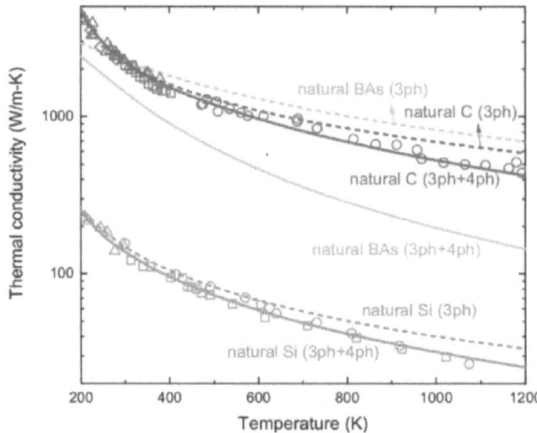

Figure 7.12 Lattice thermal conductivity of BAs, obtained from a fully *ab-initio* approach using the *iterative* relaxation-time method. Dashed lines give $\mathcal{K}$(3ph) and solid lines give $\mathcal{K}$(3ph + 4ph). Symbols represent measured data: for Si triangles (Ruf *et al*, 2000), squares (Abeles *et al*, 1962), circles (Glassbrenner and Slack, 1964); for diamond triangles (Wei *et al*, 1993), squares (Onn *et al*, 1992), circles (Olson *et al*, 1993) and diamonds (Berman *et al*, 1975). Reproduced from Feng *et al* (2017).

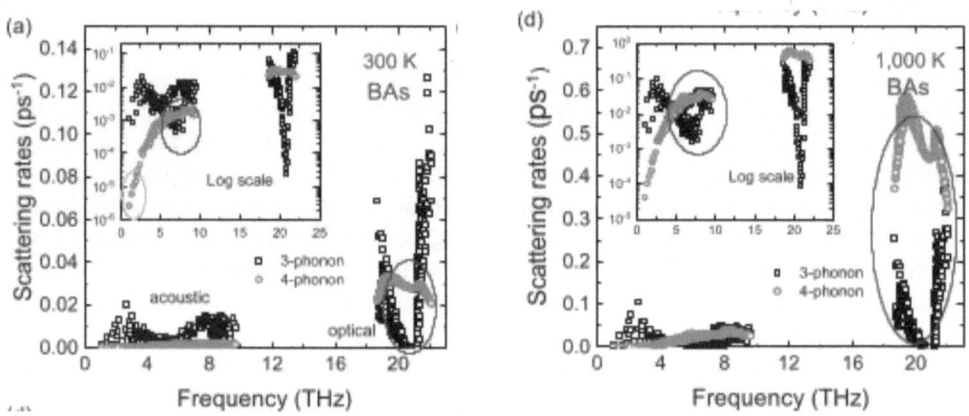

Figure 7.13 Three-phonon and four-phonon scattering rates in BAs. Blue ovals indicate the frequency region where four-phonon scattering plays an important role. Reproduced from Feng *et al* (2017).

7.4.1.6 High-temperature results: role of thermal expansion and four-phonon processes

It is generally thought that two mechanisms could be responsible for a stronger than T^{-1} dependence of $\mathcal{K}$: (i) higher-order phonon–phonon interactions (e.g. four-phonon processes, Pomeranchuk 1972) and (ii) temperature dependence of equilibrium parameters such as the phonon velocity and the anharmonic force constants or the Grüneisen constant γ. These parameters depend on volume and can in turn depend on temperature (Ranninger, 1965).

The parameter-free computational results obtained from both the semi-*ab initio* approach (section 7.4.1.4) and the fully-*ab initio* approach (section 7.4.1.5) demonstrate that the stronger than the T^{-1} high-temperature thermal conductivity variation can be successfully explained by adding four-phonon scattering rates to three-phonon scattering rates. Note that the semi-*ab initio* approach included a mode-average Grüneisen constant which, within the adopted quasi-harmonic approximation, shows only a slight increase with temperature (typically 0.1% – 0.2%) above Debye

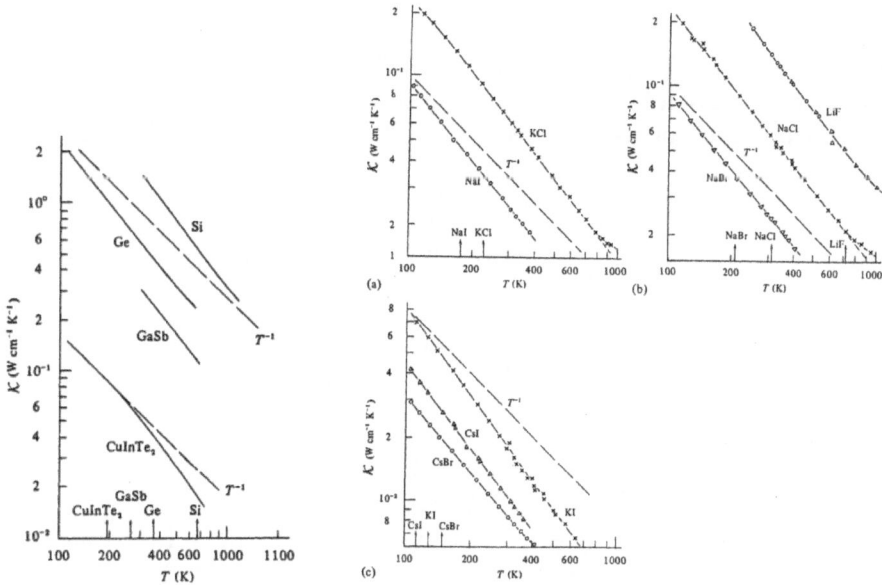

Figure 7.14 Temperature dependence of the thermal conductivity of a few semiconductors at high temperatures. For comparison, the T^{-1} curve is also shown. The Debye temperature for the materials is indicated by the arrows on the temperature scale. (Reproduced from Petrov *et al* (1976).)

temperature (*cf* figure 7.7). On the other hand, the fully-*ab initio* approach used cubic and quartic force constants evaluated at zero temperature. Ward and Broido (2010) investigated the effect of thermal expansion by evaluating the phonon relaxation rate τ_{qs}^{-1} using third-order force constants determined from a range of lattice constants. They found for both Si and Ge that τ_{qs}^{-1} is relatively insensitive to such changes, producing only small shift in the thermal conductivity ($\sim 1\% - 2\%$).

Figure 7.14 shows the temperature dependence of the thermal conductivity of several semiconductors and insulators at high temperatures. It is seen that the high temperature $\mathcal{K}$ varies as T^{-n}, with *n* in the range 1.2–1.5.

7.4.2 RESULTS FROM VARIATIONAL PRINCIPLES

In the previous section we have seen that relaxation time approaches can give information on the relative contributions of different polarisation branches to the thermal conductivity. The relative contribution of different scattering processes to the resistivity, however, cannot be directly obtained from the relaxation time approaches. Also, as the phonon–phonon ineteraction is inelastic in nature, it is unamenable to relaxation time treatments on philosophical grounds. The variational method, described in section 5.3, can treat all phonon interactions, elastic or inelastic, on equal ground and can provide additional information such as the relative contributions of different scattering processes to the thermal resistivity. Rigorous variational calculations of the thermal conductivity have been made by Hamilton and Parrott (1969), Benin (1970, 1972), and Srivastava (1976b, 1976c, 1977a, 1977c). Here we present some main points of their results.

7.4.2.1 Two-sided, or complementary, variational results

Using the method of complementary variational principles, described in section 5.3, we have derived sequences of lower and upper bounds on the lattice thermal conductivity: $\{\mathcal{K}_m^<\}, m = 0, 1, 2, \ldots$ and $\{\mathcal{K}_n^>\}, n = 1, 3, 5, \ldots$. The first two terms of the sequence of lower bounds are (from equations

Table 7.1

Two-sided variational bound results for Ge. Only three-phonon processes are considered and $\gamma = 2$ is used in the calculations (Srivastava 1976b, 1976c). Units: W cm^{-1} K^{-1}.

$T(K)$	$\mathcal{K}_0^<$	$\mathcal{K}_1^<(\Gamma)$	$\mathcal{K}_1^>(L)$
900	0.1416	0.2285	0.3195
300	0.4169	0.6926	1.010
100	1.245	2.604	4.589
60	2.776	3.849	14.17

(5.156)–(5.157) and (5.194))

$$\mathcal{K}_0^< = \frac{k_B T^2}{N_0 \Omega \mid \nabla T \mid^2} \frac{(\boldsymbol{\phi}, X)^2}{(\boldsymbol{\phi}, P\boldsymbol{\phi})} \tag{7.13}$$

$$\mathcal{K}_1^<(A) = \frac{k_B T^2}{N_0 \Omega \mid \nabla T \mid^2} \left[(X, \Gamma^{-1} X) + \frac{(\boldsymbol{\phi}, (\hat{I} - PA^{-1})X)^2}{(\boldsymbol{\phi}, (\hat{I} - PA^{-1})P\boldsymbol{\phi})} \right], \tag{7.14}$$

where $A \geq P$ and A^{-1} exists. The first term of the sequence of upper bounds is (from equations (5.171) and (5.194))

$$\mathcal{K}_1^>(L) = \frac{k_B T^2}{N_0 \Omega \mid \nabla T \mid^2} \left[(X, L^{-1} X) - \frac{(\boldsymbol{\phi}, (PL^{-1} - \hat{I})X)^2}{(\boldsymbol{\phi}, (PL^{-1} - \hat{I})P\boldsymbol{\phi})} \right], \tag{7.15}$$

where $0 < L \leq P \leq 2L$ and L^{-1} exists. In these results $\boldsymbol{\phi} = \{\phi_q\}$, with $\phi_q = \alpha q \cdot u$, where α is a one-parameter scaling factor and u is a constant vector in the direction of the temperature gradient.

As discussed in equation (5.100), one can construct $L = P - \frac{1}{2}\mu\hat{I}$, with μ as the minimum row sum of P: $\mu = (\sum_{q'} P_{qq'})_{min}$. Two different choices for A have been suggested. Benin (1970) constructed $A = \lambda\hat{I}$, with λ as the maximum row sum of P: $\lambda = (\sum_{q'} P_{qq'})_{max}$. With this choice, $A^{-1} = \lambda^{-1}\hat{I}$ is well defined and $A = \lambda\hat{I} \geq P$ is clearly satisfied. Srivastava (1976c) chose $A = \Gamma$, the diagonal part of P. For this choice, it remains to show that $\Gamma \geq P$. This, in turn, requires showing that in the decomposition $P = \Gamma + \Lambda$, the off-diagonal operator Λ is a negative definite operator. In Appendix F we show that this is the case when a trial function of the type $\phi_q = q \cdot u$ is used.

The isotropic continuum model described in section 6.4.1.4 was used to evaluate the matrix elements in equations (7.13)–(7.15). The results for the bounds $\mathcal{K}_0^<$, $\mathcal{K}_1^<(\Gamma)$ and $\mathcal{K}_1^>(L)$ for Ge are listed in Table 7.2. It can be seen that the lower bound $\mathcal{K}_1^<(\Gamma)$ is a big improvement over the Ziman limit $\mathcal{K}_0^<$. As $\mathcal{K}_1^>(L)$ and $\mathcal{K}_1^<(\Gamma)$ represent, respectively, an upper bound and a lower bound, the exact conductivity result should lie between these two estimates. With calculations of higher terms in the sequences $\{\mathcal{K}_m^<(\Gamma)\}$ and $\{\mathcal{K}_n^>(L)\}$, the exact conductivity result can be limited to a narrow band $\Delta_{mn} = \mathcal{K}_n^>(L) - \mathcal{K}_m^<(\Gamma)$.

7.4.2.2 Improvement of the lowest variational bound by the Rayleigh and Ritz procedure

It was proved in section 5.3.6 that the introduction of a variational one-parameter (single scale factor) produces a more convenient and more accurate, normalisation-independent, bound result. It was also shown there that the accuracy of improvement of a bound result can be increased by adopting a Ritz procedure.

It is clear from equation (5.90) that $q \cdot u$ is an admissible function for constructing the variational trial function ϕ_q. In general one can expect ϕ_q to be expressible as a power series in q. We thus generalise equation (5.181) and write

$$\phi_q = \hat{q} \cdot u \sum_{r=0}^{R} \sum_{s=1}^{3} a_r^s q^r = \sum_{rs} a_r^s \phi_r(q), \tag{7.16}$$

where s is the phonon polarisation index, and $\{a_r^s\}$ are $(R \times 3)$-parameters. With this general trial function the lowest variational bound $\mathcal{K}_0^<(\phi)$ turns into

$$[\mathcal{K}_0^<]_R = \frac{k_B T^2}{N_0 \Omega |\nabla T|^2} \sum_{rr'=0}^{R} \sum_{ss'=1}^{3} X_r^s (P^{-1})_{rr'}^{ss'} X_{r'}^{s'}, \tag{7.17}$$

where

$$X_r^s = (\phi_r, X_s) = \sum_q \phi_r(q) X_q^s \tag{7.18}$$

$$P_{rr'}^{ss'} = (\phi_r, P_{ss'} \phi_{r'}) = \sum_{qq'} \phi_r(q) P_{qq'}^{ss'} \phi_{r'}(q') \tag{7.19}$$

are generalisations of equations (5.186) and (5.187). X_q^s and $P_{qq'}^{ss'}$ are defined by equation (5.18), and $(P^{-1})_{rr'}^{ss'}$ is an element of the inverse matrix to $\{P_{rr'}^{ss'}\}$. The matrix elements in equation (7.18) can be evaluated using the procedure described in section 6.4.1.4. The areas in the q–q' space for various allowed three-phonon processes are obtained from equations (6.88)–(6.89) and are given in figure 6.1. Details of the evaluation of the matrix elements X_r^s and $P_{rr'}^{ss'}$ can be found in Hamilton and Parrott (1969) and Srivastava (1977c).

The variational parameters a_r^s which satisfy the phonon Boltzmann equation (5.18) are given by

$$X_r^s = \sum_{r's'} P_{rr'}^{ss'} a_{r'}^{s'}. \tag{7.20}$$

The conductivity is then

$$\mathcal{K} = \sum_s \mathcal{K}(s) = \frac{k_B T^2}{N_0 \Omega |\nabla T|^2} \sum_{rs} X_r^s a_r^s, \tag{7.21}$$

where $\mathcal{K}(s)$ is the contribution from polarisation s. The contribution of a scattering process 'y' to the total resistivity can be expressed as

$$\frac{W(y)}{W} = \frac{\sum P_{rr'}^{ss'}(y) a_r^s a_{r'}^{s'}}{\sum X_r^s a_r^s}. \tag{7.22}$$

The rate of improvement of the bound $[\mathcal{K}_0^<]_R$ as a function of R, the number of terms in equation (7.17), is illustrated in Table 7.3. It is clearly seen that the result for Ge has practically well converged for $R = 3$.

It is interesting to compare the improved version of the lowest bound, $[\mathcal{K}_0^<]_R$, with normalisation-independent (i.e. improved by using a single scale factor) bounds $\mathcal{K}_1^>(\Gamma)$ and $\mathcal{K}_1^>(L)$. At high temperatures $[\mathcal{K}_0^<]_3$ has reached the value predicted by $\mathcal{K}_1^<(\Gamma)$. At low temperatures $[\mathcal{K}_0^<]_3$ is even higher than $\mathcal{K}_1^<(\Gamma)$. Thus $[\mathcal{K}_0^<]_3$ indeed represents a very big improvement over the simple estimate $\mathcal{K}_0^<$ (namely, the original Ziman limit). It could be anticipated that only a few terms in the series in equation (7.17) would be required to arrive at a well improved limit $[\mathcal{K}_1^<(\Gamma)]_R$. Similarly, the bound $\mathcal{K}_1^>(L)$ could be improved *down* to a converged limit $[\mathcal{K}_1^>(L)]_R$. Thus the method of cvps has the potential of predicting the exact thermal conductivity to lie within a very narrow band $\Delta_{RR'} = [\mathcal{K}_1^>(L)]_R - [\mathcal{K}_1^<(\Gamma)]_{R'}$.

Table 7.2

Results of improvement of the lowest variational bound for the lattice thermal conductivity of Ge using the Rayleigh–Ritz procedure. Only three-phonon processes are considered, with $\gamma = 2$. Units: W cm^{-1} K^{-1}. (From Srivastava 1977c.)

T (K)	$[\mathscr{K}_0^<]_R$ R/0	1	2	3	$[\mathscr{K}_1^<(\Gamma)]$	$\mathscr{K}_1^>(L)$
900	0.1651	0.2165	0.2239	0.2285	0.2285	0.3195
300	0.5099	0.6822	0.7035	0.7129	0.6926	1.010
100	2.086	3.121	3.175	3.213	2.604	4.589
60	6.343	10.16	10.54	10.59	3.849	14.17
47	14.59	21.78	24.09	24.40		

7.4.2.3 Low-temperature results

Using the expression in equation (7.17) and the isotropic continuum model of section 6.4.1.4, Hamilton and Parrott (1969) calculated the thermal conductivity of Ge at low temperatures. They evaluated the three-phonon scattering strengths $|A_{qq'q''}^{ss's''}|^2$ in terms of the second- and third-order elastic constants as in equations (4.96)–(4.102). Figure 7.15 shows the results of their calculations for three samples of Ge. In all cases the results, which do not include any adjustable parameters, agree quite well with experiment at low temperatures. At high temperatures, there are some discrepancies between theory and experiment. This in great part can be explained to be due to four-phonon processes not being included and the use of temperature independent elastic constants in the calculations.

Figure 7.16 illustrates the percentage contribution of the transverse phonon modes to the conductivity. In all samples the transverse phonons carry between 80 and 90% of the total heat. In a similar, but simpler, calculation, Srivastava and Verma (1973) noticed that the contribution of

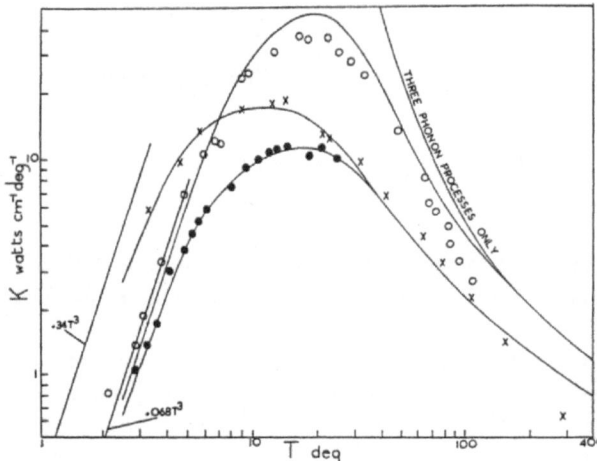

Figure 7.15 The thermal conductivity of Ge. The open and filled circles are the experimental results for the enriched and natural samples of Geballe and Hull (1958), while the crosses are the experimental results of Slack and Glassbrenner (1960). The solid curves are the variational calculations of Hamilton and Parrott (1969). (From Hamilton and Parrott (1969).)

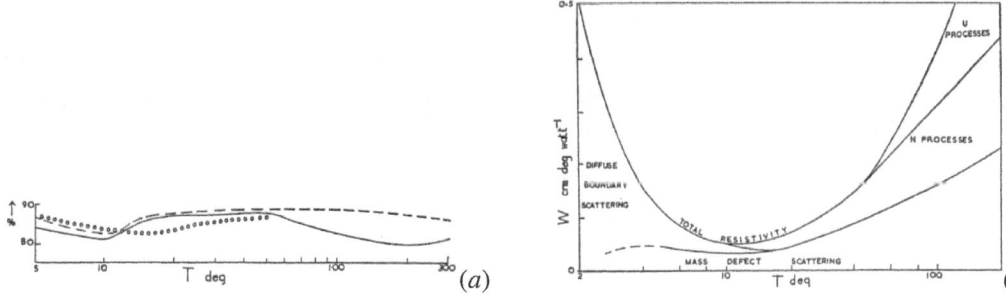

Figure 7.16   Results from the variational approach, as calculated by Hamilton and Parrott (1969). Panel (*a*) shows percentage contribution of transverse phonons towards the thermal conductivity. The various curves are for Ge samples with: circles for $L = 0.212$ cm, $\Gamma_{md} = 5.72 \times 10^{-4}$, solid curve for $L = 1.06$ cm, $\Gamma_{md} = 5.72 \times 10^{-4}$, broken curve for $L = 0.212$ cm, $\Gamma_{md} = 3.68 \times 10^{-5}$. Panel (*b*) shows contributions to the thermal conductivity of Ge from different scattering processes with $L = 1.06$ cm, $\Gamma_{md} = 5.72 \times 10^{-4}$. (From Hamilton and Parrott (1969).

transverse phonons is as high as 74% of the total conductivity. A similar conclusion was reached in section 7.4.1 from the relaxation-time methods.

7.4.2.4   Contributions from different scattering processes

One of the most interesting aspects of the variational method is that it can provide an analysis of the thermal conductivity in terms of contributions from various scattering processes. From their calculations, Hamilton and Parrott (1969) showed that the three-phonon processes produce an appreciable indirect effect on the thermal resistivity even when their direct contribution is negligible. From figure 7.16(b), it can be seen that at temperatures above the maximum of the $\mathscr{K}$ against T curve three-phonon processes are the major source of resistivity.

   Srivastava *et al* (1972) observed that the $\mathsf{T} + \mathsf{T} \rightleftharpoons \mathsf{L}$ (*N*) processes produce higher resistivity than the $\mathsf{T} + \mathsf{L} \rightleftharpoons \mathsf{L}$ (*N*) processes, the $\mathsf{L} + \mathsf{L} \rightleftharpoons \mathsf{L}$ (*N*) processes produce resistance similar in strength to $\mathsf{T} + \mathsf{L} \rightleftharpoons \mathsf{L}$ (*N*), and the processes $\mathsf{T} + \mathsf{T} \rightleftharpoons \mathsf{L}$ (*U*) generate higher resistivity than the processes $\mathsf{T} + \mathsf{L} \rightleftharpoons \mathsf{L}$ (*U*). That the $\mathsf{T} + \mathsf{T} \rightleftharpoons \mathsf{L}$ processes dominate the three-phonon thermal resistance can be understood on physical grounds: such processes can attenuate a transverse phonon with less expenditure of energy, and moreover, have the largest phase space in the q–q' space (see figure 6.1).

7.5   NON-METALLIC CRYSTALS WITH HIGH THERMAL CONDUCTIVITY

Using equations (7.11), (6.85) and (4.140), we can obtain the following approximate expression for lattice thermal conductivity at high temperatures when only three-phonon processes are expected to contribute significantly

$$\kappa(HT, 3ph) = B\bar{M}\Omega_{at}^{1/3}\Theta^3/(T\gamma^2), \tag{7.23}$$

where $\bar{M}$ is average atomic mass, Ω_{at} is average atomic volume, and B is an appropriately derived constant. This expression is of the form obtained by Leibfried and Sclömann (1954) and by Julian (1965). Consistent with this, Slack (1973) established four rules for finding a non-metallic crystal with high thermal conductivity: (i) low atomic mass, (ii) strong interatomic bonding, (iii) simple crystal structure, and (iv) low anharmonicity. Conditions (i) and (ii) mean a high Debye temperature Θ_D, condition (iii) means low number of atoms per primitive unit cell, or less number of optical branches, and condition (iv) means small Grüneisen constant γ.

   In metallic solids silver is the best heat conductor, with its room temperature thermal conductivity 430 W m^{-1} K^{-1}. Crystals for which the room temperature conductivity exceeds 100 W m^{-1} K^{-1}

Table 7.3

Adamantine (diamond-like) crystals with high room-temperature thermal conductivity (i.e. with $\mathscr{K} \geq 100$ W m^{-1} K^{-1}). Units are W m^{-1} K^{-1} (From Slack (1973).)

crystal	$\mathscr{K}$	crystal	$\mathscr{K}$	crystal	$\mathscr{K}$
Diamond	2000	BP	360	GaN	170
BN (cubic)	1300	AlN	320	Si	160
SiC	490	BeS	300	AlP	130
BeO	370	BAs	210	GaP	100

can be considered to possess a high thermal conductivity. There are at least 12 adamantine (diamond like) crystals with high thermal conductivity, with diamond possesing the highest value, 2000 W m^{-1} K^{-1} at room temperature. These are listed in Table 7.3. Several other crystals with α-boron, boron carbide, or graphite structure also possess high $\mathscr{K}$ values. There are no crystals with the rocksalt structure that possess high thermal conductivity.

Among the technologically promising nitride materials, AlN possesses the highest conductivity value, 320 W m^{-1} K^{-1}. The room-temperature conductivity of natural samples of diamond (2000 W m^{-1} K^{-1}) is 5 times that of copper. The room-temperature thermal conductivity of nearly isotopically pure diamond is reported to exceed 3320 W m^{-1} K^{-1} (Anthony *et al* 1990), Wei *et al* 1993). Similar to that of diamond, the in-plane thermal conductivity of graphite is also one of the largest, with the room-temperature value of approximately 2000 W m^{-1} K^{-1} (see Nihira and Iwata 1975 and references therein).

Carbon nanotubes (Iijima 1991, Dresselhaus *et al* 1996), consisting of self-supporting atomically perfect graphitic cylinders a few nanometres in diameter, also have an unusually high thermal conductivity. The experimental investigation by Hone *et al* (1999) indicates that the room-temperature thermal conductivity of a single-wall carbon nanotube may be comparable to that of diamond or in-plane graphite. Theoretical simulations carried out by Berber *et al* (2000) suggest that the room-temperature thermal conductivity of an isolated (10,10) carbon nanotube (consisting of a 400-atom large unit cell) is 6600 W m^{-1} K^{-1}, comparable to the thermal conductivity of an isolated mono-layer graphene.

The high thermal conductivity results for all forms of solid carbon are consistent with the rules presented above, in particular the low atomic mass (rule 1) and strong interatomic bonding (rule 2). The high thermal conductivity of single-crystal diamond results from the very stiff sp^3 covalent bonding between neighbouring atoms. The in-plane high conductivity of graphite is also due to the stiff sp^2 bonding between the neighbouring intra-layer carbon atoms. The very high thermal conductivity of carbon nanotubes results from stronger sp^2 bonds. In spite of the similarity between the acoustic phonon dispersion in a graphene sheet and a carbon nanotube (Jin *et al* 1995), the predicted temperature variation of the thermal conductivity of the carbon tubes is different from that of graphite (Hone *et al* 1999). Such a difference can be explained by noting that (i) graphite has additional phonon modes, corresponding to interplanar vibrations, (ii) the 'rolling up' of a graphene sheet to make a tube results in a significantly different phonon spectrum for the tube (Benedict *et al* 1996): the transverse component of the phonon wav vector ($q_\perp$) is quantized due to the periodic boundary conditions imposed by the cylindrical geometry, and (iii) due to (i) and (ii) the phonon scattering times may be significantly different for graphite and tubes.

There are several other non-admantine systems, such as bulk and mono-layer transition metal dichalogenides MoS$_2$and WS$_2$, that possess (in-plane) thermal conductivity larger than 100 W m^{-1} K^{-1}. We will discucss these some of these in the next chapter.

7.6   THERMAL CONDUCTIVOITY OF COMPLEX CRYSTALS

The main findings in section 7.4 can be generalised to the conductivity of all non-metallic crystals with simple structures: crystals with 1 or 2 atoms per unit cell, e.g. rare-gas, adamantine and rock-salt structures. The high temperature behaviour $\mathcal{K} \propto T^{-n}, n > 1$, is a general feature of crystals with simple structures.

For the discussion of thermal conductivity, we can regard all crystals which have three or more atoms in the unit cell as possessing complex structures. For example, in fluorite structure crystals with three atoms/cell there are three acoustic and six optical branches. The unit cell of β-boron contains 105 atoms and thus there are three acoustic and 312 optical branches.

In general the variation of thermal conductivity of complex crystals with temperature follows the same trend as in simple crystals, i.e. $\mathcal{K}$ reaches a maximum and then decreases with increasing temperature. However, (a) the magnitude of $\mathcal{K}$ both below and above the maximum is much reduced, and (b) the high-temperature dependence of $\mathcal{K}$ is much weaker than T^{-1} and in some cases virtually independent of T. Thus there is some similarity between the thermal conductivity of very complex solids and that of amorphous solids (see Chapter 12). The thermal conductivity of a dielectric crystal takes a minimum value when the wavelength of all the phonons becomes comparable to interatomic distance. Complex structure crystals (and amorphous solids) are candidates for the minimum $\mathcal{K}$ condition, which is important in designing thermoelectric regrigerators and generators.

Advances in the *ab initio* theoretical developments discussed in sections 3.3.2.1 and 7.4.1.5, together with the availability of computing power, have made parameter-free calculations of lattice thermal conductivity of complex crystals affordable. Additional interest in accurate prediction of the lattice thermal conductivity of complex crystals has been encouraged due to the possibility that such systems could behave like phonon-glass-electron-crystal (PGEC) and prove useful in the development of high figure of merit for thermoelectric applications (Slack, 1995). In this context we present and discuss here first-principles prediction of the lattice thermal conductivity of the type-II and type-VIII Si clatherates, which are three-dimensional open structures with 34 and 23 atoms per unit cell, respectively, as described in figure 3.13.

By employing the first-principles approaches described in sections 3.3.2.1 and 7.4.1.5, Härkönen and Karttunen (2016) computed the phonon dispersion relation and the lattice thermal conductivity of Si in three different structural forms: bulk (diamond structure) and clatherates of types II and VIII. The conductivity calculations did not include boundary scattering. The conductivity results are presented in figure 7.17. Due primarily to the neglect of boundary scattering of phonons, the theoretical results in panel (*a*) for diamond-structure Si are higher than experimental measurements

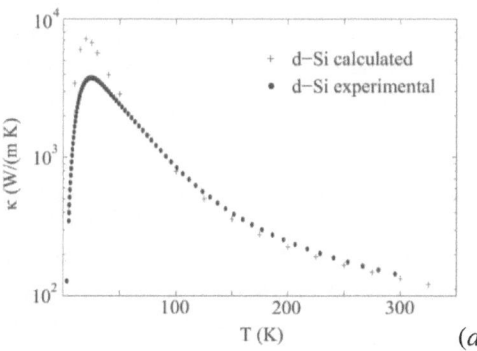
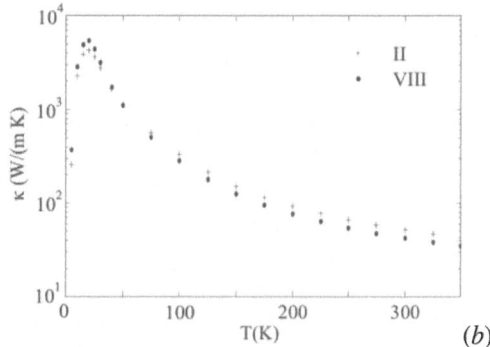

Figure 7.17   Thermal conductivity of (*a*) diamond-structured Si and (*b*) pristine Si clatherate structures II and VIII. Reproduced from Härkönen and Karttunen (2016).

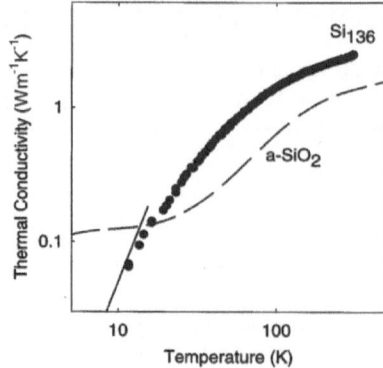

Figure 7.18   Lattice thermal conductivity of poly-crystalline Si clatherate structure II and vitrous silica (*a*-Si). Reproduced from Nolas *et al* (2003).

at temperatures below 100 K. Above 100 K, the theoretical results are comparable to first-principles results presented by other groups, see *e.g.* figure 7.11. The results in panel (*b*) predict marked lowering of the conductivity upon the formation of the complex open-structure clatherates of both type-II and type-VIII. At 300 K, the predicted conductivity values for type-II and type-VIII clatherates are approximately 52 and 43 W m^{-1} K^{-1}, respectively. These values are approximately one-third of the bulk (diamond-structure) result. In order to understand these results, we note from equation (5.64) that the conductivity is goverened by the product of three factors: phonon specific heat C_v^{sp}, square of average phonon speed $\bar{c}^2$ and average phonon relaxation time $\bar{\tau}$. As we have discussed in section 7.4.1, most heat is conducted by acoustic phonons. The range of acoustic phonons in the clatherate structures is nearly one-third of that in the diamond structure. This will make the value of C_v^{sp} much smaller for the clatherate structures. However, there is little change in the value of $\bar{c}^2$, as the slope of acoustic branches in the clatherate structures remains similar to to that in the diamond structure. The appearance of many flat optical branches has two effects. Low group velocity optical modes contribute very little towards conductivity. The presence of several optical branches generates strong reduction in the overall relaxation time of acoustic phonons due to acoustic-optical scatterings.

As mentioned above, the calculations by Härkönen and Karttunen (2016) did not include phonon boundary scattering, making it difficult to judge the magnitude and temperature variation of $\mathcal{K}$ in the low temperature range. It is valuable to study the experimentally measured conductivity results for the type-II Si clatherate. Figure 7.18 shows the results obtained by Nolas *et al* (2003) for poly-crystalline type-II Si clatherate (Si$_{136}$) and vitreous silica (*a*-Si). At room temperature, the conductivity of the clatherate is 2.5 W m^{-1} K^{-1}, almost 30 times lower than that of diamond-structures Si. The important observation to make is that the temperature dependence of the poly-crystalline type-II Si clatherate does not follow the glass-like behaviour characteristic of *a*-Si (see Chapter 12 for a detailed discussion of phonons in amorphous solids).

Roufosse and Klemens (1973) have reformulated the theory of the thermal conductivity of complex dielectric crystals in terms of anharmonic three-phonon interactions. Considering a statistical distribution of atoms within a unit cell as a simplification, these authors have shown that the three-phonon matrix element is inversely proportional to the square root of N, the number of atoms in the unit cell. However, a summation over all reciprocal lattice vectors increases the scattering rate. These two factors tend to cancel, so that the anharmonic relaxation rate is substantially independent of N.

It can be assumed that at low temperatures the thermal conductivity is practically determined by only acoustic phonons. Thus apart from the phonon scattering processes as expected in a simple crystal, two additional points should be considered. Firstly, for a complex crystal with N atoms per

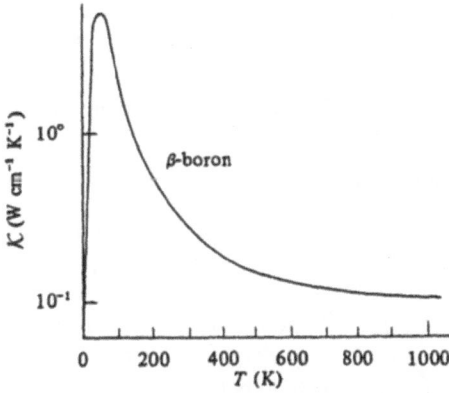

Figure 7.19 Temperature variation of the thermal conductivity of β-boron. (Reproduced from Petrov *et al* (1976).)

unit cell the upper limit in the conductivity integral (equation (7.11)) is a scaled frequency $\omega_R = \omega_D/N^{1/3}$, where ω_D is the Debye frequency of a simple crystal with 1 atom per unit cell. Secondly, because of a range of optical phonon frequencies available in a complex crystal, there will be an increased possibility for acoustic–optical interaction. Thus as compared to simple crystals, the low-temperature thermal conductivity of complex crystals can be significantly reduced in magnitude. If the anharmonic relaxation is assumed to be independent of N, then $\mathscr{K} \propto 1/N^{1/3}$ (Roufosse and Klemens 1973).

At high temperatures, the phonon mean free path decreases and may become comparable to the linear dimensions of the unit cell of the crystal lattice. The condition is more easily realised in complex crystals whose high-frequency vibrations, therefore, become localised and the phonon terminology tends to break down. In such a situation the energy is transmitted by activation or hopping of the localised modes. Thus the behaviour of $\mathscr{K}$ of complex crystals at high temperatures as well as its magnitude becomes characteristic of amorphous solids. Figure 7.19 shows the temperature dependence of $\mathscr{K}$ of β-boron.

7.7 LOW-TEMPERATURE THERMAL CONDUCTIVITY OF DOPED SEMICONDUCTORS

Reduction in the low-temperature thermal conductivity of doped semiconductors has been reported extensively. The theory described in section 6.5 is in general successful in explaining experimental observations.

7.7.1 HEAVILY DOPED SEMICONDUCTORS

It is evident from equation (6.129) that phonon scattering from a degenerate electron gas is species independent, but depends on the deformation potential E_d and (density-of-states equivalent) effective mass $m^\star$. A weak temperature dependence of $m^\star$ (of the form $m^\star \propto T^n, n \leq 0.5$) and some carrier concentration dependence of E_d can be expected. In the presence of strong electron–phonon scattering, the single-mode relaxation time result $\mathscr{K}_D$ should be expected to give a reasonably accurate account: the contribution of the N-drift term in Callaway's conductivity expression is reported to be less than 10% (Gaur and Verma 1967).

Using equation (6.130) for τ_{ep}^{-1}, and $\tau_{3\,ph}^{-1} = B\omega^2 T^3$ along with boundary and mass-defect scatterings, Gaur and Verma expressed $\tau^{-1} = \tau_{bs}^{-1} + \tau_{md}^{-1} + \tau_{ep}^{-1} + \tau_{3\,ph}^{-1}$ and were in general able to explain the low-temperature thermal conductivity of a number of heavily doped samples of n-Ge.

For P donors in Si with concentration higher than 4.7×10^{17} cm^{-3} Fortier and Suzuki (1976) reported a temperature dependence $\mathscr{K} \propto T^{2.2}$ for $1.1\text{K} \leq T \leq 4\text{K}$, and $\mathscr{K} \propto T^2$ for $4\text{K} \leq T \leq 12\text{K}$. This behaviour can presumably be explained by a phonon relaxation of the type in equation (6.129) with $m^* \propto T^n$, where n takes different values in the two temperature regions discussed above.

7.7.2 LIGHTLY DOPED SEMICONDUCTORS

When donor (or acceptor) impurity concentration is small (e.g. smaller than 4.7×10^{17}cm^{-3} in the n-Si samples of Fortier and Suzuki) scattering of phonons by bound donors (acceptors) is more likely to occur. The theory of such scattering was described in section 6.5.2.

Using equations (6.175)–(6.178) for phonon–electron scattering and $\tau_{3ph}^{-1} = B\omega^2 T^3$ for phonon–phonon scattering Suzuki and Mikoshiba (1971c) obtained good agreement between thereotical and experimental conductivity of Sb-doped Ge between 1 and 5 K. Low-temperature thermal conductivity measurements on samples of p-Ge have been made by Challis *et al* (1977). Using the Suzuki–Mikoshiba theory of phonon–hole interaction these authors were able to explain the $\mathscr{K}$ against T results down to 1 K. It was suggested that the experimental results below 1 K, down to 50 mK, can be explained by including resonant scattering of phonons at about half of the acceptors which are in distorted sites. The presence of random strain splits the $|M_J = \pm 3/2\rangle$ and $|M_J = \pm 1/2\rangle$ states. Resonant phonon scattering can take place between these acceptor levels, as described in section 6.5.3.

The ground state of a shallow donor in Si is partially split into a singlet (A_1), a doublet (E) and a triplet (T_2). The (A_1) level is the lowest energy state, the T_2 state is at 11.7 meV above A_1 and the E state is at 1.35 meV above T_2. The matrix elements of electron–phonon interaction between A_1 and T_2 states are zero (Hasegawa 1960). The relaxation rates of phonons due to elastic, inelastic and thermally assisted phonon absorption in n-Si can be calculated by following the prescription given in section 6.5.2. Fortier and Suzuki (1976) calculated the thermal conductivity of many samples of P doped Si and obtained reasonable agreement with experiment in low concentration limit ($\leq 4.7 \times 10^{17}$cm^{-3}). In figure 7.20 we show a comparison of their calculation with experiment for P-doped Si (concentration 2.5×10^{17}cm^{-3}). Also shown in the same figure is the conductivity of Li doped Si (the curve with the solid line) with concentration 2×10^{17}cm^{-3}. The weak scattering of phonons by P donors (also Sb and As donors) is due to very small contributions from the inelastic scattering and thermally assisted phonon absorption, which is a consequence of large valley–orbit splitting in such a case.

7.7.3 SEMICONDUCTORS WITH MAGNETIC IMPURITIES

The theory of phonon scattering by magnetic impurities in semiconductors was outlined in section 6.6. Srivastava amd Verma (1974) proposed a phenomenological model for the scattering of phonons by Fe^{2+} magnetic impurities in ZnS

$$\tau_{defect}^{-1} = A\omega^4 + \sum_i H_i \frac{\omega}{(\omega - \omega_{0i})^2} F_i(T). \tag{7.24}$$

Here the first term represents the Rayleigh scattering produced by the local mass difference and lattice distortion. In the second term H_i is an adjustable parameter which represents the strength of spin–phonon coupling, and $F_i(T)$ is the fractional electron population difference between the levels involved in the ith electronic transition (see figure 6.4). Using the above expression, Srivastava and Verma successfully explained the resonance structure of the $\mathscr{K}$ against T curve for cubic ZnS containing different concentrations of Fe^{2+}. This is shown in figure 7.21.

To explain the thermal conductivity of antiferromagnetic FeCl$_2$, (Nèel temperature $T_N = 23.5K$), Neelmani and Verma (1973) considered phonon-induced electronic excitations among the trigonal crystal and exchange-field-split energy levels of the Fe^{2+} ions in the crystalline lattice of FeCl$_2$. For

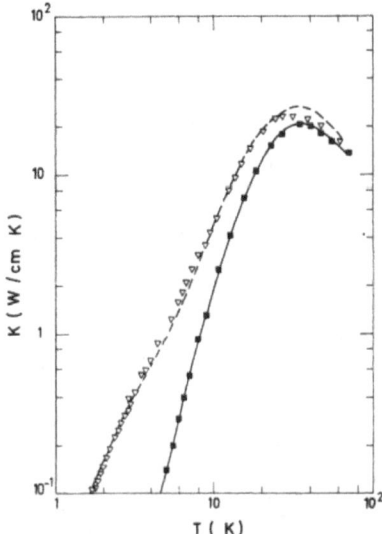

Figure 7.20 The low-temperature thermal conductivity of two n-Si samples. The dotted line represents the calculated $\mathcal{K}$ for P doped Si with impurity concentration $2.5 \times 10^{17} \mathrm{cm}^{-3}$ and boundary length $L = 0.302\mathrm{cm}$. The solid curve represents experimental conductivity for Li doped Si (with concentration $2 \times 10^{17}\mathrm{cm}^{-3}$ and boundary length $L = 0.158\mathrm{cm}$). (Reproduced from Fortier and Suzuki (1976).)

the ground state–triplet transitions, there are two groups of phonons with energies centred at 31 and 44 cm^{-1}. For each group of phonons, a resonant scattering of the form

$$\tau_r^{-1} = \sum_i H_i \frac{\omega^2 T^2}{(\omega_{0i}^2 - \omega^2)^2} \tag{7.25}$$

was considered. The 44 and 31 cm^{-1} transitions were considered to be active, respectively, below and above T_N. With these considerations, Neelmani and Verma successfully explained the 'two-maxima' structure of the $\mathcal{K}$ against T curve for $FeCl_2$, as shown in figure 7.22.

7.7.4 THERMAL CONDUCTIVITY OF DOPED ALKALI HALIDES

Phonon scattering from tunnelling states in doped alkali halides was discussed in section 6.7. Using equation (6.189), Singh and Verma (1971) explained the 'double-dip' structure in the $\mathcal{K}$ against T curve of KCl : CN$^-$ with different CN$^-$ concentrations (see figure 7.23). In this system the phonon scattering due to mass defect and force-constant changes was found not to play a significant role in causing the resonance dips in the $\mathcal{K}$ against T curves.

7.8 THERMAL CONDUCTIVITY OF DIFFERENT FORMS OF DIAMOND

Naturally abundant single crystal diamond

Thermal conductivity of diamond is higher than that of any other known three-dimensional materials. An important question is how is the thermal conductivity of diamond affected by the quality (e.g. crystallite size and defects) of a diamond sample? Even the purest form of natural diamond (e.g., a type IIa sample) is known to contain nitrogen in three different forms (see Morelli *et al* 1993a and references therein): isolated substitutional impurities, nitrogen-atom pairs (A aggregates) and larger

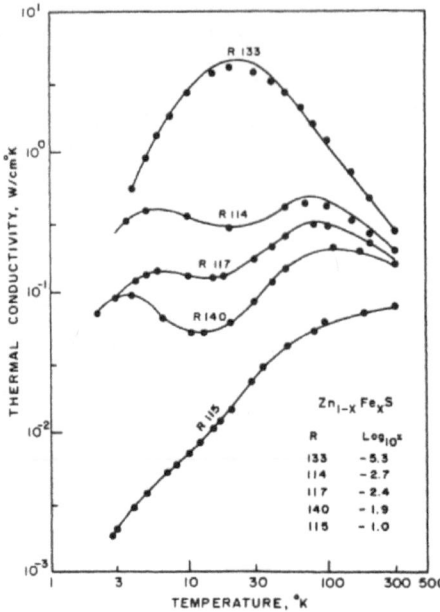

Figure 7.21 Thermal conductivity of cubic ZnS containing various concentrations of Fe impurity atoms. The solid curves represent the theoretical results within the Callaway model and the circles represent the experimental results of Slack (1972). (Reproduced from Srivastava and Verma (1974).)

nitrogen aggregates (B aggregates). However, the precise nature and concentration of such defects is not known. Morelli *et al* (1993b), and Morelli and Uher (1993) studied the influence of defects on the thermal conductivity of a single crystal of type-IIa natural diamond by irradiating the sample by fast neutrons and annealing it at different temperatures. The results, for the unirradiated sample and the sample irradiated to 4×10^{18} cm^{-2}, are shown by different symbols in figure 7.24(a).

Barman and Srivastava (2006) attempted to explain these results by employing the isotropic continuum version of Callaway's conductivity expression (see equations (5.56), (6.85) and (7.16–7.18)). The total phonon relaxation rate was expressed as

$$\tau^{-1} = \tau_{bs}^{-1} + \tau_{bs}^{-1} + \tau_{md}^{-1} + \tau_{3ph} + \tau_{aggregate}^{-1}, \tag{7.26}$$

where the first three terms refer to the boundary, (point) mass defect and three-phonon anharmonic scatterings and the last term represents the scattering from defect aggregates. Extended-sized defects can scatter small wavelength phonons geometrically (i.e., like boundaries) and long wavelength phonons like point defects (i.e., Rayleigh-like). Thus, we can express (Vandersande, 1977)

$$\tau_{aggregate}^{-1} = \begin{cases} nc\pi d^2/4 & \text{for} \quad qd \geq 1, \\ n\pi d^6 \omega^4/4c^3 & \text{for} \quad qd < 1, \end{cases} \tag{7.27}$$

where q is phonon wavenumber, c is phonon speed, d is the effective diameter of the defect region and n is the extended-defect concentration.

Theoretical matching of the experimentally measured results for the unirradiated sample required consideration of a small but finite amount of phonon scattering from extended-sized defects. This was taken as confirmation of the suggestion made by Morelli *et al* (1993a) that their natural type-IIa sample contains nitrogen defects in some mixture of the three possible configurations (isolated

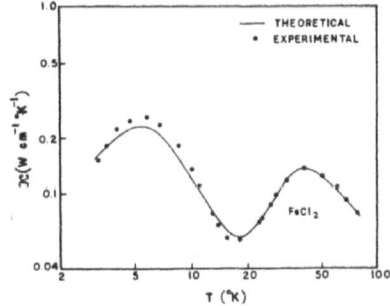

Figure 7.22 The thermal conductivity of antiferromagnetic $FeCl_2$. The solid curve is the theoretical result obtained by using the phonon resonant scattering (equation (7.24)) from the trigonal crystal and exchange-field-split energy levels of Fe^{2+} impurity ions and the Debye conductivity expression. The two-maxima structure in the experimental curve of Laurence (1971) is well explained. From Neelmani and Verma (1973).

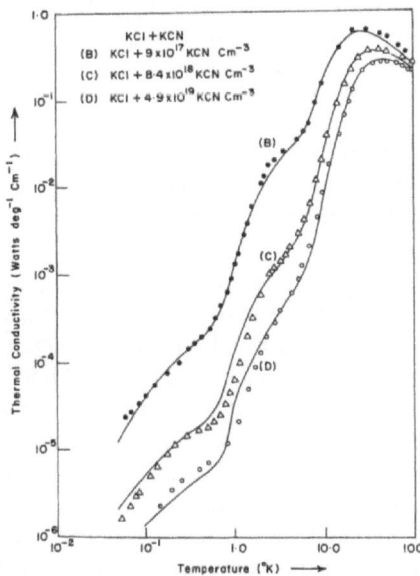

Figure 7.23 Thermal conductivity of KCl:CN. The experimental results are from Seward and Narayanamurti (1966). The solid curves are the theoretical results calculated on the basis of the Debye conductivity model and the resonant phonon scattering (equation (6.185)) from the tunnelling states of CN^- ions. Notice that the three-maxima structure is well reproduced by theory. (Reproduced from Singh and Verma (1971).)

substitutional impurities, A aggregates and B aggregates). The theoretical estimates for the concentration and size of the extended defect are: $n = 1 \times 10^{18}$ cm^{-3} and $d = 7$ Å. The calculations also revealed that there is significant contribution from the N-drift term in the Callaway's conductivity expression: $\mathscr{K}_C$ being approximately 50% higher than $\mathscr{K}_{smrt}$ at 300 K. This is consistent with the *ab initio* calculation by Ma *et al* (2014).

Exposure to the irradiation leads to two significant changes to the conductivity of the sample: (i) the magnitude is hugely reduced, e.g. by more than 20 times at room temperature, and (ii) there is a shallow dip in the conductivity at around 30 K. The experimental curve was reproduced by increasing the strength of the point mass defect scattering by approximately a hundred times, the

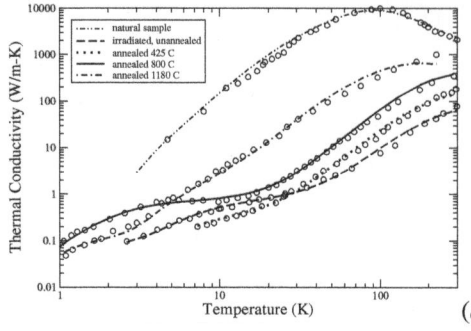
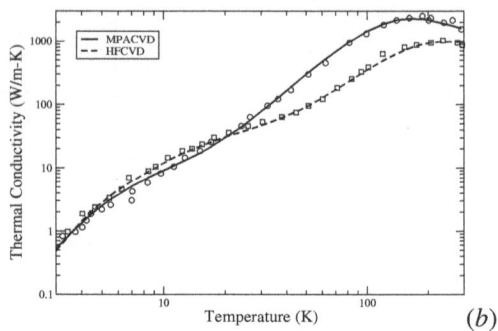

Figure 7.24 (*a*) Thermal conductivity of unirradiated and neutron-irradiated naturally abundant type IIa single crystal diamond. Lines show theoretical results and symbols represent experimental data from Morelli *et al* (1993a) and Morelli and Uher (1993). (*b*) Thermal conductivity of CVD diamond films. Lines show theoretical results and symbols represent experimental data from Morelli *et al* (1993b). Reproduced from Barman and Srivastava (2006).

extended-defect concentration by a thousand times and the extended-defect size by approximately twice. The dip in the conductivity results from an interplay between the scatterings of phonons in the two ranges of wavenumbers in equation (7.27). The observed dip in the conductivity at around 30 K is consistent with the estimate $T_{dip} \approx 50$ K obtained by equating the two regions in equation (7.27) for the 'dominant phonon' mode (Ziman, 1960): $\hbar\omega_{dom} = 1.6 k_B T_{dip} = \hbar \bar{c}/d$, where $\bar{c}$ is average phonon speed.

Upon annealing the sample to 425 °C the extended-defect size d is enlarged, but the densities of both point defects (A) and extended defects (n) are reduced. These changes result in a crossover in the conductivity curves for the annealed and unannealed samples at around 30 K. There is a shallow dip between 20 and 25 K for the annealed sample, which is partially masked by the point defect scattering. Annealing the sample to 800 °C further increases the size of the extended defect, and decreases the concentrations of both extended defects and point defects, in such a manner that the geometrical scattering parameter nd^2 reduces but the Rayleigh-like agglomerate scattering is maximized. As a result, the conductivity becomes higher than that of the unannealed sample but a pronounced dip develops at around 15 K. Annealing at 1180 °C further reduces both the geometrical and Rayleigh-like agglomerate scattering centres, resulting in increase in the conductivity. However, there is a cross-over at around 5 K between the conductivity curves for the 800 °C and 1180 °C anneals. In the temperature range below 5 K the lower conductivity for the 1180 °C anneal is suggested to be caused by an effect similar to resonant scattering of low-frequency phonons between two lowest energy levels of double-well potential in the disordered region (Pohl, 1962).

MPACVD and HFCVD grown diamond samples

Good quality diamond films can be produced in the laboratory by using chemical vapour deposition (CVD) techniques. Morelli *et al* (1993b) measured the thermal conductivity of two types of CVD diamond films: grown by a hot-filament process (HFCVD) and by a microwave plasma assisted process (MPACVD). The results are shown in figure 7.24(*b*). A comparison of results in panels (*a*) and (*b*) of the figure clearly indicates that the conductivity of the MPACVD film is higher that that of the 1180 °C anneal in the entire temperature range 3–300 K. On the other hand, the conductivity of the HFCVD film is comparable to that of the 1180 °C anneal in the temperature range 50–300 K, but becomes higher below 50 K. The conductivities of the MPACVD and HFCVD films are comparable in the temperature range 3–30 K. There are shallow dips in the conductivity curves for the MPACVD and HFCVD samples at around 15 K and 40 K, respectively.

Barman and Srivastava (2006) explained the temperature variation of the measured data by using the same method as they used for the single crystal type-IIa diamond. The theoretical modelling suggests that the average grain size in the CVD films is similar to that for the single crystal sample with the 1180 °C anneal. The point defect scattering parameter required for explaining the conductivity of the CVD films was much smaller than that for neutron irradiated (annealed and unannealed) single crystal samples, but larger than that for the natural sample. The strength of the geometrical scattering (nd^2) due to extended defects in the MPACVD gains was similar to that for the 1180 °C anneal, and approximately seven times larger for the HFCVD film. The Rayleigh-like scattering parameter (nd^6) for the MPACVD film was estimated to be an order of magnitude smaller than that for the 1180 °C anneal but an order of magnitude larger than that for the HFCVD film.

8 Phonons on Crystal Surfaces and Layered Crystals

8.1 INTRODUCTION

In Chapters 2–7 we have discussed lattice dynamics, phonon interactions and phonon conductivity in crystalline three-dimensional (bulk) solids. When a crystal surface is created, we can no longer expect to describe, even in the harmonic approximation, the surface vibrational modes as the normal modes (phonons) of the bulk. In general, we expect surface atoms to acquire relatively large displacement amplitudes. This is because by creating a free surface we have cut down the coupling of surface atoms with atoms which otherwise would have remained as a part of the bulk. This decrease in coupling constants, or equivalently, increase in displacement amplitudes, tends to lower the normal mode frequencies and produces a new class of vibrational modes called surface modes. Techniques now exist for fabrication of layered crystals, which can be described as surface systems. In this chapter we present both the continuum and lattice dynamical theories of surface vibrational modes.

8.2 CONTINUUM THEORY

In section 2.8 we briefly described the theory of elastic waves in three-dimensional crystals. Subject to appropriate boundary conditions, that theory can be applied to describe elastic waves on the surface of elastic media. Two different geometrical situations can be treated: (i) a medium occupying a half-space and (ii) a plate of finite-thickness. Solutions to equations for cases (i) and (ii) are known as Rayleigh waves (Rayleigh 1885) and Lamb waves (Lamb 1917), respectively. Rayleigh (Lamb) waves are produced in films of thickness much greater (less than or equal to) the acoustic wavelength. We mention in passing that waves with properties which are determined by boundaries are often referred to as *guided waves*. Here we present a brief discussion of Rayleigh waves.

Consider a cubic crystal with the free surface defined by $z = 0$. The x, y and z components of displacement field, u, v, w, satisfy the equations of motion given in equation (2.142), subject to the following boundary conditions:

$$\frac{\partial w}{\partial x} + \frac{\partial u}{\partial z} = 0$$
$$\frac{\partial w}{\partial y} + \frac{\partial v}{\partial z} = 0$$
$$c_{12}\left(\frac{\partial u}{\partial x} + \frac{\partial v}{\partial y}\right) + c_{11}\frac{\partial w}{\partial z} = 0 \qquad (8.1)$$

which arise due to vanishing of the stress components across the $z = 0$ surface. For the surface problem, we modify equation (2.138) and seek solutions of the form

$$(u, v, w) = (u_0, v_0, w_0)\exp\{q_{\parallel}[-\alpha_z z + \mathrm{i}(\alpha_x x + \alpha_y y - c_S t)]\}, \qquad (8.2)$$

where c_S is the velocity of surface (Rayleigh) waves, $q_{\parallel}$ is the magnitude of the two-dimensional wave vector parallel to the surface, α_z is a dimensionless attenuation constant, and α_x and α_y are the

DOI: 10.1201/9781003141273-8

direction cosines of the propagation direction. The term $\exp(-\alpha_z q_\parallel z)$ characterises an exponential decrease in the displacement amplitudes in the positive z direction (towards the bulk). Substitution of equation (8.2) into equation (2.142), subject to the boundary conditions in (8.1), leads to the following equation for surface waves on an isotropic continuum (see Wallis 1974):

$$g(p^6 - 8p^4 + 24p^2 - 16) - 16(p^2 - 1) = 0, \tag{8.3}$$

where $p = \rho c_S^2/\mu$, $g = (\lambda + 2\mu)/\mu$, and λ and μ are Lamé's elastic constants for an isotropic solid of density ρ. The Lamé constants are related to the second-order elastic constants c_{11}, c_{12} and c_{44} as shown in equation (4.93). For $p = 0$ and 1, the left-hand side of equation (4.93) takes values $16(1/g - 1)$ and 1, respectively, implying that there is a root between 0 and 1. As the left-hand side is essentially positive, we can deduce that a Rayleigh wave must always exist at the surface of an infinite solid (Stoneley 1924). For real values of α, the surface wave velocity c_S must be smaller than the velocity of bulk transverse and longitudinal waves. The quantity p, the ratio of c_S to bulk transverse wave velocity, ranges from $p = 0.96$ for $g = \infty$, the incompressible case, to 0.69 for $g = 4/3$, the smallest value of g consistent with crystal stability. The frequencies of Rayleigh waves lie in the acoustical branch of the phonon spectrum.

8.3 LATTICE DYNAMICAL THEORY

As described in Chapters 2 and 3, the vibrational properties of a crystal lattice are usually calculated by using the periodic (or cyclic) boundary conditions. When dealing with the dynamics of a surface one is faced with the problem of lack of periodicity in the direction normal to the surface. Furthermore, there may be large decreases in certain coupling constants at the surface compared to the bulk values which must be given due consideration. In this section we will consider surface solutions for the special case of a semi-infinite linear chain and then discuss the semi-infinite three-dimensional case.

8.3.1 MONATOMIC LINEAR CHAIN

Consider a semi-infinite monatomic linear chain with only nearest-neighbour interactions. Let atom 0 in figure 8.1 represent an adsorbed atom at the free end. Let Λ' be the coupling constant between the adatom and its neighbouring bulk atom, and Λ be the coupling constant between the bulk atoms. If u_n denotes the displacement of the nth atom from its equilibrium, then the equations of motion can be written as

$$m'\frac{d^2 u_0}{dt^2} = \Lambda'(u_1 - u_0) \tag{8.4}$$

$$m\frac{d^2 u_1}{dt^2} = \Lambda(u_2 - u_1) + \Lambda'(u_0 - u_1) \tag{8.5}$$

$$m\frac{d^2 u_n}{dt^2} = \Lambda(u_{n+1} + u_{n-1} - 2u_n) \qquad n \geq 2. \tag{8.6}$$

To obtain surface solutions, we write the displacements in the form (Wallis 1964, 1974)

$$u_0 = A' \exp(i\omega t) \tag{8.7}$$

$$u_n = A(-1)^n \exp(-qna + i\omega t) \qquad n \geq 1. \tag{8.8}$$

Equation (8.8) describes a displacement which decays exponentially from the surface atom (adatom) towards the bulk atoms. Substitution of (8.8) into (8.6) yields the result

$$\omega^2 = 2\frac{\Lambda}{m}(1 + \cosh qa). \tag{8.9}$$

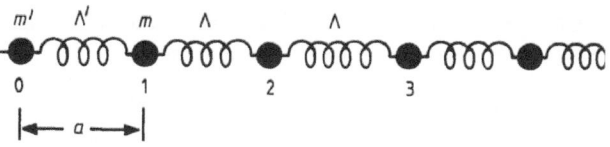

Figure 8.1 A semi-infinite monatomic linear chain. Atom 0 is an adatom. All bulk atoms have identical mass m, and the adatom has mass m'.

By substituting (8.7) and (8.8) into (8.4) and (8.5) and setting the determinant of the coefficients A' and A equal to zero, we get the following equation which must be satisfied by q:

$$[\Lambda' - \Lambda(1 + \exp(qa))]\left(\Lambda' - 2\frac{m'}{m}\Lambda(1 + \cosh qa)\right) - \Lambda'^2 = 0. \tag{8.10}$$

A solution of this equation has the form $q = q_0 + i\pi$, where q_0 is real and positive. This is the case when the ratio Λ'/Λ and m'/m satisfy the inequality (Wallis 1964, 1974)

$$\Lambda'/\Lambda > \frac{4(m'/m)}{[2(m'/m) + 1]}. \tag{8.11}$$

The surface mode frequency is then given by

$$\omega_S = \sqrt{(2\Lambda/m)(1 + \cosh q_0 a)} \tag{8.12}$$

and is larger than the maximum bulk-mode frequency $\sqrt{4\Lambda/m}$. If we consider $m' = m$ (no adsorbate), then equation (8.11) reduces to $\Lambda'/\Lambda > \frac{4}{3}$. In this situation a surface mode exists above the allowed band of frequencies of the infinite chain. If there is no perturbation in masses or coupling constants ($m' = m$, $\Lambda' = \Lambda$), then we are left with the surface atom for which one coupling constant is set to zero. By Rayleigh's theorem (Rayleigh 1885) this must give rise to a surface mode with frequency below the band of frequencies of the infinite chain. Such a frequency would be imaginary and therefore unphysical. Clearly, therefore, no surface modes exist if $m' = m$ and $\Lambda' = \Lambda$.

8.3.2 DIATOMIC LINEAR CHAIN

Consider a semi-infinite diatomic linear chain with nearest-neighbour interactions (figure 8.2). Let the surface atom have the smaller mass m ($m < M$). Let there be no change in the surface force constant. The equations of motion are

$$m\frac{d^2 u_1}{dt^2} = \Lambda(u_2 - u_1) \tag{8.13}$$

$$M\frac{d^2 u_{2n}}{dt^2} = \Lambda(u_{2n+1} + u_{2n-1} - 2u_{2n}) \quad n \geq 1 \tag{8.14}$$

$$m\frac{d^2 u_{2n+1}}{dt^2} = \Lambda(u_{2n+2} + u_{2n} - 2u_{2n+1}) \quad n \geq 2. \tag{8.15}$$

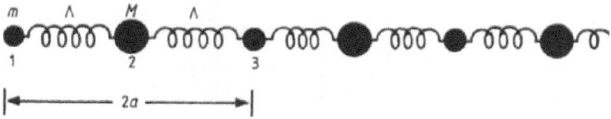

Figure 8.2 A semi-infinite diatomic linear chain with the lighter atom at the surface.

Wallis (1957, 1964) showed that there exists a surface solution to equations (8.13)–(8.15): the displacements are expressed as attenuated or damped oscillations

$$u_{2n} = A(-1)^n (m/M)^n \exp(i\omega t) \tag{8.16}$$

$$u_{2n+1} = A(-1)^{n+1} (m/M)^{n+1} \exp(i\omega t) \tag{8.17}$$

and the surface mode has the frequency

$$\omega_S = \sqrt{\Lambda(1/m + 1/M)}. \tag{8.18}$$

We know from section 2.2.2 that the bulk normal modes of a periodic diatomic linear chain ($M \geq m$) lie in two bands: $0 \leq \omega_{ac}^2 \leq 2\Lambda/M$ (acoustic branch), and $2\Lambda/m \leq \omega_{op}^2 \leq 2\Lambda(1/m + 1/M)$ (optical branch). In terms of squared frequencies, the surface mode given by equation (8.18) lies exactly at the centre of the forbidden gap between the optical and acoustical branches.

The existence of the surface mode is in accordance with Rayleigh's theorem. The creation of a free end reduces a coupling constant to zero, which leads to the drop of an optical normal-mode frequency into the forbidden gap.

8.3.3   A CRYSTAL WITH A SURFACE

For practical purposes, we can study surface phonons by maintaining an artificial periodicity normal to the surface. Consider a long elongated unit cell which in two dimensions is characterised by two primitive surface vectors a_1 and a_2, say. In the direction normal to the surface, the translation vector a_3 encompasses N_1 crystal layers plus a few (N_2 say) vacuum layers. The number N_2 (typically 4 or 5) is chosen such that two neighbouring crystal slabs are practically decoupled. The number N_1 is chosen such that there is no significant interaction between the two surfaces of a slab. Figure 8.3 presents an schematic illustration of such a unit cell. The lattice dynamics of such a periodic system can be studied by using any of the methods described in section 2.4, or by the first-principles methods described in Chapter 3.

To acquire a local minimum in free energy, the atomic geometry of a semiconductor surface may *relax* and *reconstruct*. A surface is called *unreconstructed* if the symmetry of the atomic arrangement on the surface remains the same as on parallel planes deeper into the bulk. If the separation of the surface plane from the next plane of atoms is different than the corresponding separation in the bulk crystal, and/or if the atomic positions in the surface plane are different than in the parallel planes in the bulk, then the surface is said to have *relaxed*. The surface translation vectors a_1 and a_2 must be appropriately chosen for a reconstructed surface geometry.

The normal modes of the periodic supercell system can be obtained by solving the following determinantal equation (see equation (2.50)):

$$|D_{\alpha\beta}(bb'|q) - \omega^2 \delta_{bb'} \delta_{\alpha\beta}| = 0. \tag{8.19}$$

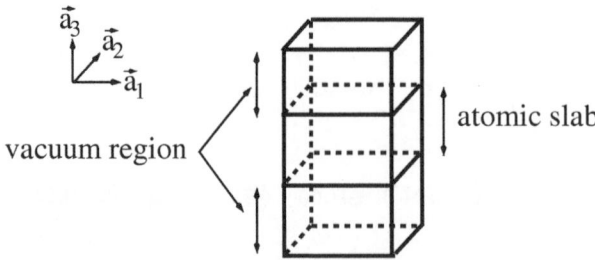

Figure 8.3   Schematic illustration of a supercell for surface calculation. The translation vectors a_1 and a_2 are the appropriately chosen surface vectors and a_3 is in the direction normal to the surface under study.

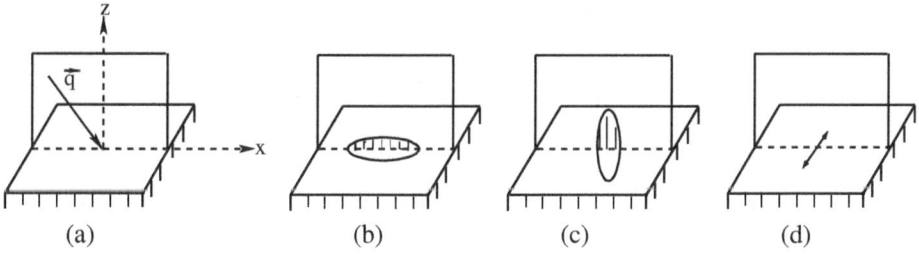

Figure 8.4   Schematic illustration of surface wave polarisation: (*a*) sagittal plane, (*b*) longitudinal polarisation, (*c*) sagittal polarisation and (*d*) shear horizontal polarisation.

The size of the dynamical matrix is $3p \times 3p$, where p is the number of atoms in the large unit cell. Although we have well defined q-vectors for this artificially defined periodic system, we are only interested in solving the dynamical matrix for phonon wave vectors $q_\parallel$ in the surface Brillouin zone (see section 1.9). (Also note that for the supercell geometry the extent of the $q_\perp$ component is much smaller than the extent of the $q_\parallel$ component.) The phonon spectrum obtained by solving equation (8.19) will contain frequencies of surface phonons along with folded bulk phonon features. Information about surface phonons can be extracted by investigating the magnitude and polarisation of the eigenvectors of the dynamical problem.

Accurate experimental measurements of surface phonon dispersion curves of clean and adsorbate covered surfaces are possible from the applications of helium atom scattering (HAS) and high-resolution electron energy loss spectroscopy (HREELS). We refer the reader to the interesting reviews back in 1990 and 1992 by Toennies who presented a historical survey of the field and expressed optimism for accurate experimental and theoretical surface phonon studies (Toennies 1990, 1992). It is clear from his review in 1990 (Toennies 1990) that by that time experimental studies of phonon dispersion curves had been accomplished for many of surfaces of metals, semiconductors, insulators and layered crystals.

As discussed in section 2.8, there are three mutually orthogonal polarisation directions of medium displacement for waves propagating in three-dimension isotropic continuum: a longitudinal and two (doubly-degenerate) transverse. No such clear distinction between longitudinal and transverse polarisations can be made for surface waves. There are three main types of polarisation of surface phonons, as illustrated in figure 8.4. We first note that the plane containing the phonon wave vector q and the surface normal is called the *sagittal plane*. *Longitudinal* surface polarisation refers to displacement along the surface wave vector. *Transverse* (or *shear vertical* or *sagittal*) surface polarisation refers to displacement along the surface normal (i.e. in the sagittal plane). *Shear horizontal* polarisation refers to displacement in the surface plane but normal to the phonon wave vector (i.e. normal to the sagittal plane). As indicated in figure 8.3, both the longitudinal and sagittal waves contain the x and z components of displacement, and the shear horizontal wave contains the y component of displacement.

8.4   PHONONS ON SEMICONDUCTOR SURFACES

Surface phonon dispersion curves have been calculated using several theoretical techniques at phenomenological , semi-phenomenological and *ab initio* levels. In this next section we present and discuss results for a few selected semiconductor surfaces and layered crystals, obtained using the *ab initio* DFT direct and DFPT techniques described in sections 3.3.1 and 3.3.2.1.

Surface phonon modes can be broadly classified as localised modes, gap modes, resonant modes or Fuchs-Kliewer modes. Modes outside the bulk region (*i.e.*, below or above the bulk continuum) are localised modes. Gap modes appear in 'stomach gaps' of surface-projected bulk continuum.

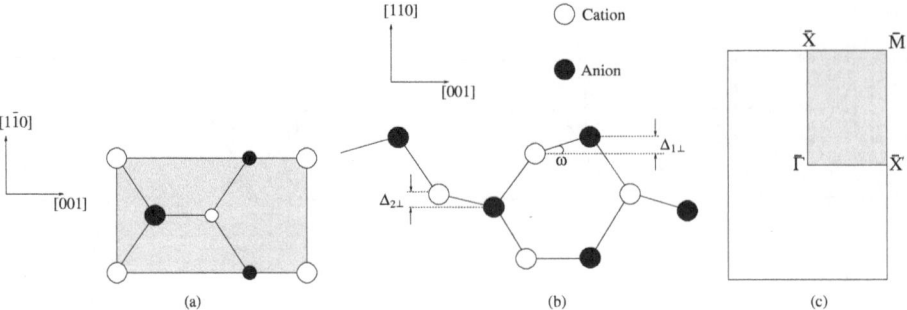

Figure 8.5 Schematic illustration of the relaxed atomic geometry of a III-V(110) surface: (*a*) surface unit cell; (*b*) side view with atomic relaxation parameters; (*c*) surface Brillouin zone.

Resonant modes are located inside the bulk continuum but have largest vibrational amplitude on surface atoms. Fuch-Kliewer modes are long wavelength optical phonons accompanied by long-range electric fields on polar surfaces (Fuchs and Kliewer 1965), with frequency ω_{FK} lying between the bulk TO and LO frequencies: $\omega_{TO} < \omega_{FK} < \omega_{LO}$.

8.4.1   UNRECONSTRUCTED III-V(110) SURFACES

The clean cleaved (110) surface of III-V and II-VI semiconductors is unreconstructed, *i.e.* it retains the bulk periodicity. However, atoms in the top few atomic layers relax to achieve the minimum energy configuration. In the top atomic layer, anions move away from the bulk in favour of s^2p^3 bonding with three neighbouring cations, and cations move into the bulk in favour of sp^2 bonding with three neighbouring anions. The resulting characteristic relaxation parameters are atomic layer relaxation parameters $\Delta_{1\perp}$ and $\Delta_{2\perp}$ of the top two atomic layers, and the rotational angle ω of the top layer atomic chain, as illustrated in figure 8.5.

As seen in figure 8.5, the (110) surface of a zincblende material is characterised by the presence of a mirror plane normal to the zig-zag chain direction. Thus, vibrational modes at the centre of the surface Brillouin zone ($\bar{\Gamma}$) as well as along the $\bar{\Gamma}\bar{X}'$ symmetry direction can be classified according to the irreducible representations A' and A'' of the point group C_s (or C_{1h} or m) of the surface unit cell. The A' modes correspond to displacements perpendicular to the surface atomic chain direction, and the A'' modes describe atomic displacements along the zig-zag chain direction (*i.e.* along $[1\bar{1}0]$). Such a clear classification is not possible along $\bar{\Gamma}\bar{X}$ and $\bar{\Gamma}\bar{M}$ directions. Along these directions modes show a mixture of shear-horizontal (SH) and sagittal (SG) polarisations. These contributions along $\bar{\Gamma}\bar{X}$ can be calculated as

$$\sum U_{SG}^2 = \sum_i U_i^2([\bar{1}10[]) + \sum_i U_i^2([110]) \qquad (8.20)$$

$$\sum U_{SH}^2 = \sum_i U_i^2([001]), \qquad (8.21)$$

where U is the magnitude of the eigenvector and the sum over the index i considers atoms in the surface layers in the supercell.

Fritsch *et al* (1995) used the DFT and DFPT methods to determine the relaxed atomic geometry and phonon dispersion curves for the InP(110) surface. The values for the structural parameters are $\Delta_{1\perp} = 0.64$ Å, $\Delta_{2\perp} = 0.11$ Å and $\omega = 28°$. The surface phonon dispersion curves are shown in figure 8.6. There is good agreement between theoretical and experimental results. Atomic vibrational patterns of the zone-edge surface acoustic modes are shown in figure 8.6.

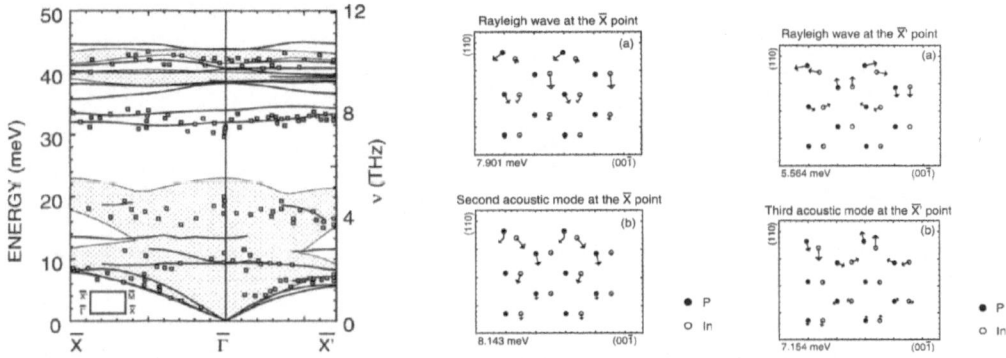

Figure 8.6 Phonon dispersion curves for InP(110). Thick solid lines are the theoretically obtained results, and the square symbols are the experimental results from HREELS measurements by Nienhaus and Mönch (1995). Also shown are the theoretically determined vibrational patterns of the zone-edge acoustic modes. Reproduced from Fritsch *et al* (1995).

Along the $\overline{\Gamma X}$ symmetry direction (atomic chain direction), there are two surface acoustic branches, the lower one being the Rayleigh wave. For short wavevectors (near $\overline{\Gamma}$), the Rayleigh wave is characterised by in-phase displacements along [110] and the upper mode has pure horizontal polarisation. At the $\overline{X}$ point, the lower (Rayleigh) wave turns into a vibrational pattern of the top-layer anions and the second-layer cations perpendicular to the chain direction. The second acoustic branch originates from vibrations of the first-layer cations and the second-layer anions mainly along surface normal, while the first-layer anions vibrate in the chain direction. Consistent with group theory, the lowest (Rayleigh) mode at the $\overline{X}'$ point is sagittally polarised (*i.e.* is a A' mode), with vibrations of the top-layer atoms mainly parallel to the surface and second-layer atoms along the surface normal. The second lowest mode is a pure shear horizontal mode (*i.e.* is a A'' mode), with vibrations along the chain direction.

The surface optical branch on InP(110) lies above the bulk continuum. It touches the bulk continuum near the zone centre. The highest surface mode at the zone centre is the Fuchs–Kliewer phonon mode. This mode is dominated by the motion of the second-layer P ions perpendicular to the zig-zag chain direction and parallel to the (110) plane. In other words, it is a A' mode.

Tütüncü (1997), and Tütüncü and Srivastava (1997) have presented a detailed account of dispersion curves and polarisation characteristics of surface phonons on III-V(110) by using the phenomenological adiabatic bond charge model.

8.4.2 RECONSTRUCTED SILICON SURFACES

The (111) surface is the natural cleavage plane for diamond structure crystals. The cleavage process usually occurs through the middle of the covalent bonds pointing straight along the [111] direction. As each half cut (dangling) sp^3 orbital is occupied by only one electron, the ideally terminated 1×1 structure is unstable and the surface structure prefers to reconstruct by rehybridisation of the dangling bonds. When cleaved in a very high vacuum at low and room temperatures, Si(111) exhibits a 2×1 reconstruction. Several higher-order reconstructions appear when the system is annealed to high temperatures. The most studied is the high-temperature 7×7 reconstruction.

The (001) surface of elemental semiconductors is usually grown by the molecular beam epitaxy (MBE) technique. For ideal termination, or 1×1 structure, each atom of the (001) surface is accompanied by two partially occupied dangling bonds. The room-temperature stable structure is the Si(001)-(2×1) reconstruction. This is achieved by the formation of surface dimers, which helps energy gain by saturation of one dangling bond per surface atom. However, for symmetric dimer

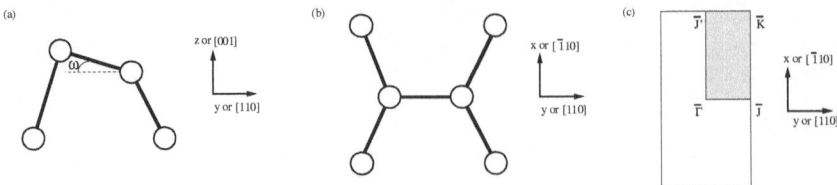

Figure 8.7 Atomic geometry of Si(001)-(2×1): (*a*) side view showing tilted dimer, (*b*) top view, (*c*) surface Brillouin zone.

models the p_z dangling bonds form degenerate π and π^* bands. As this is not a spin degeneracy, a Jahn-Teller-like distortion in the form of tilted dimers occurs: charge transfer from one dimer component to the other takes place, resulting in one fully occupied p_z dangling bond on the upper-lying dimer atom and an empty p_z dangling bond on the lower-lying dimer atom.

Here we discuss essential features of phonon modes on the Si(001)-(2×1), Si(111)-(2×1) and Si(111)-(7×7) surfaces.

8.4.2.1 Si(001)-(2×1)

Figure 8.7 shows the atomic geometry for the asymmetric dimer model for the Si(001)-(2×1) surface. The dimer bond length is approximately 2.25 Å (4% smaller than the bulk Si-Si bond length) and the bond tilt angle is approximately 16°. Calculations of surface phonon dispersion curves have been made by using *ab inito* (Fritsch and Pavone, 1995), semi-empirical (Allan and Mele 1984, Pollmann *et al* 1986) and phenomenological (Tütüncü *et al* 1997) schemes. The DFPT *ab initio* results (Fritsch and Pavone, 1995) are shown in figure 8.8(*a*). Several types of surface phonon modes have been predicted. These are indicated as RW, A_1, A_2, *r*, *s*, *ds* and *sb*, as explained further.

Figure 8.8(*a*) shows the phonon dispersion curves for the Si(001)-(2×1) surface with asymmetric dimer geometry. The RW (Rayleigh wave) mode lies below the bulk continuum along the symmetry directions $\bar{J}$–$\bar{K}$–$\bar{J}'$. The vibrations of the modes A_1 and A_2 have parity even (+) and odd (−)

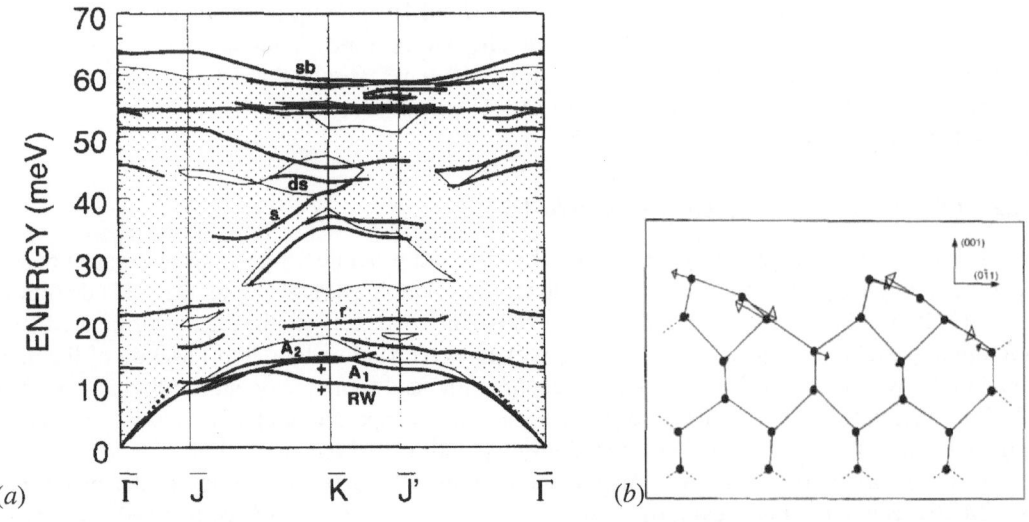

Figure 8.8 (*a*) Phonon dispersion curves for Si(001)-(2×1). Shaded region is projection of bulk results on the surface Brillouin zone. (*b*) Vibrational pattern for the dimer back bond (sb) mode at the J point of the surface Brillouin zone. Reproduced from Fritsch and Pavone (1995).

respectively with respect to the mirror plane perpendicular to the dimer row. The r (*rocking*) mode is resonant with the bulk continuum and is characterised by an opposing up and down displacements of dimer atoms. The s (*swinging*) mode results from the motion of the tilted dimer in the [001] direction (parallel to the dimer row). The mode ds is a *dimer stretch* mode and appears as a truly localised surface feature in the stomach gap around the $\bar{K}$ point. The mode labelled sb is a *bond stretch* mode, which is caused by a length oscillation between the dimer bond and the dimer back bonds and lies above the bulk optical phonon continuum. The vibrational pattern of this mode is sketched in figure 8.8(*b*). The calculated speed of the RW wave is in good agreement with the measured value 5020 ms^{-1} from the Brillouin light scattering experiment (Dutcher *et al* 1992).

From the application of the adiabatic bond charge model, Tütüncü *et al* (1997) made a few additional observations. The RW mode at the $\bar{K}$ point corresponds to the vibrations of the first layer atoms with components in both the surface normal and dimer bond directions, while the second layer atoms vibrate in the dimer row direction. They find that dimerization leads to the formation of new peaks in the phonon density of states. In particular, the peak in the stomach gap is a strong signature of dimer formation, since no stomach gap phonon modes are obtained for the unreconstructed surface geometry.

8.4.2.2 Si(111)-(2×1)

The lowest energy configuration of the Si(111)-(2×1) surface is the Pandey's π-bonded chain model (Pandey, 1981). The top and side views of the atomic structure are shown in figure 8.9. In this reconstruction, alternate [001] rows of atoms break their surface bonds and rebond with other atoms. A characteristic feature is the presence of five fold and seven fold rings of atoms involving the surface and sub-surface layers. Atoms in the rings buckle to allow for raised atoms (atom 1 in the figure) to be s^2p^3 coordinated and the lowered atoms (atom 2 in the figure) to be sp^2 coordinated.

Calculations of surface phonon dispersion curves have been made by using *ab initio* (Zitzlsperger *et al* 1997) and semi-empirical (Alerhand and Mele, 1988) schemes. Alderhand and Mele (1988) identified several types of surface phonon modes. These include *fivefold ring modes* above the bulk optical continuum, resonant *dimer mode* and *subsurface dimer mode*, and the Rayleigh wave (RW) mode below the bulk acoustic continuum. The phonon dispersion curves along the $\bar{\Gamma}\bar{J}$ symmetry direction, calculated by Zitslsperger *et al* (1997) using the DFPT *ab intio* scheme, are shown in figure 8.10. At the $\bar{J}$ point, the Rayleigh mode is found to be polarized perpendicular to the π-bonded chains, with large motion of the upper atom along the surface normal direction.

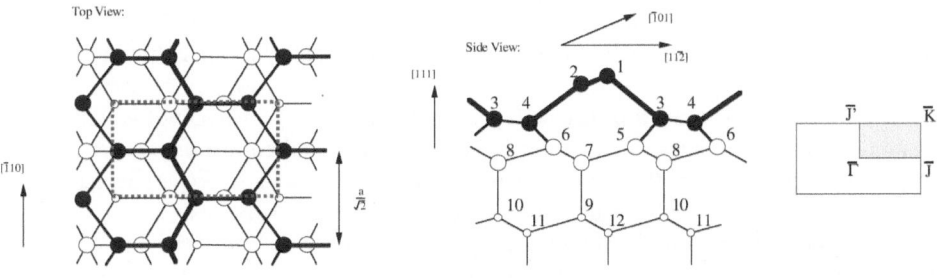

Figure 8.9 Atomic geometry of the Pandey's π-chain model for Si(111)-(2×1). The surface unit cell is shown by dotted lines. Also shown is the surface Brillouin zone.

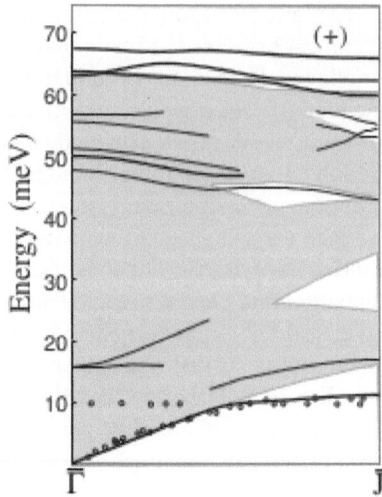

Figure 8.10 Phonon dispersion curves for Si(111)-(2×1), with Pandey's π-chain model with positive (+) chain tilt as shown in figure 8.9. The shaded region is projection of bulk results on the surface Brillouin zone. The open circles represent the He-atom inelastic scattering data from Horten *et al* (1986). Reproduced from Srivastava (1997), which was originally obtained from Zitslsperger *et al* (1997).

8.4.2.3   Si(111)-(7×7)

The agreed geometrical structure of the Si(111)-(7×7) surface is the dimer-adatom-stacking-fault (DAS) model, proposed by Takayanagi *et al* (1985). The top two (111) atomic layers are heavily reconstructed. The first reconstructed layer contains a stacking fault and the second reconstructed layer contains dimers. Above these two layers reside adatoms. The atomic configuration in a 7×7 unit cell is shown in figure 8.11.

The atomic layer sequence along [111] in the diamond structure is ...CcAaBb..., where capital and lowercase letters indicate layers of the basis atoms. The 7×7 unit cell in figure 8.11 can be viewed as two triangles. The triangle on the right shows adatoms on top of normal sequence CcAaB. The triangle on the left shows adatoms on top of the sequence CcAa/C, thus with the presence of a stacking fault at the slant.

- There are 12 adatoms in the *adatom layer* which lie in T_4 sites in a 2×2-like arrangement. Each triangle contains six adatoms, three at its corner sites and the other three at its edge centre sites.
- There are 42 atoms in the top surface layer *the stacking fault layer*, 36 of which are bonded to the adatoms and the remaining six are three-fold coordinated (called *restatoms*).
- There are nine dimers in the second layer (*the dimer layer*). These are connected by 8-member rings along the boundaries of the faulted and unfaulted triangular subunits.
- A 12-member atomic ring surrounds a large "hole" at each corner of the unit cell.

Phonon modes on the Si(111)-(7×7) surface have been studied both theoretically and experimentally. Theoretical studies include molecular dynamics simulations (Kim *et al* 1995, Štich *et al* 1996) and a semiempirical DFT approach (Liu L *et al* 2003). Experimental measurements have been attempted using HREELS (Daum *et al* 1987), HAS (Lange *et al* 1998) and Raman scattering (RS) (Liebhaber *et al* 2014) techniques. The most informative of these has been the RS work by Liebhaber *et al* (2014), and here we summarise those results.

The surface atomic arrangement belongs to the point group C_{3v} (3m), with a three fold axis of symmetry and reflections on three vertical mirror planes. The irreducible representations of the

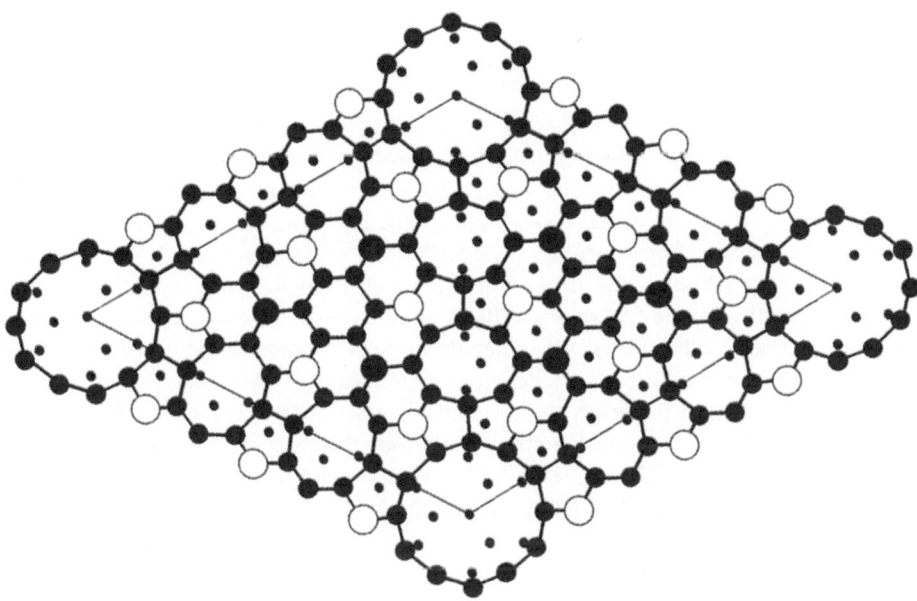

Figure 8.11 Top view of the DAS (dimer-adatom-stacking-fault) geometry of the Si(111)-(7×7) surface. Large open and large filled circles show adatoms and restatoms, respectively. The lower triangle unit contains a stacking fault. Increasing atomic sizes indicate increasing atomic layers along [111].

optical phonon modes corresponding to this point group are non-degenerate A_1 and doubly-degenerate E. Consider $x = [11\bar{2}]$ and $y = [\bar{1}10]$ as the surface in-plane axes, and $z = [111]$ as the surface normal axis. For Raman scattering in backscattering geometry, the Raman tensor diagonal elements xx and yy are relevant for A_1 modes and the off-diagonal elements xy and yx are relevant for E modes. In the Porto notation,[1] out-of-plane A_1 modes appear in the configurations $z(x,x)\bar{z}$ and $z(y,y)\bar{z}$, while the in-plane E modes can appear in the configurations $z(x,y)\bar{z}$ and $z(y,x)\bar{z}$.

Using the polarised Raman spectroscopy, Liebhaber *et al* (2014) observed six (zone-centre) phonon modes on the Si(111)-(7×7) surface. In agreement with HAS data, the backfolded acoustic Rayleigh wave (RW) mode is found at 62.5 cm^{-1} in the $z(y,y)\bar{z}$ configuration. Three modes at 115.3, 130.0 and 136.1 cm^{-1} were observed as in-plane wagging modes (adatom – atom underneath). In agreement with HREELS measurement, a mode at 250.9 cm^{-1} in polarisation configuration $z(y,x)\bar{z}$ was observed, which is localised at adatomic sites. A collective mode (involving adatoms, first- and second-layer atoms) was observed at 420.0 cm^{-1} in the polarisation configuration $z(y,x)\bar{z}$.

8.4.3 MOLECULAR ADSORPTION ON SURFACES

Many experimental and theoretical studies of equilibrium atomic geometry, electronic structure and phonon modes have been devoted to atomic and molecular adsorption, and overlayers on surfaces. A few examples of such studies for chemisorbed semiconductor surfaces can be found in Srivastava (1999). Here we present and discuss the zone-centre phonon results for the adsorption on the Si(001)

---

[1]This notation for describing a Raman spectrum by four symbols is due to Damen *et al* (1966). The symbols inside the parentheses are, from left to right, the polarisation of the incident and of the scattered light, while the symbols to the left and right of the parentheses are the propagation directions of the incident and of the scattered light, respectively.

surface of inorganic molecules XH_3 (X = N, P, and As) and the organic molecule C_2H_2–C_2O_3 (also expressed as $C_4H_2O_3$ or $C_2H_2(CO)_2O$).

Adsorption of trihydrides on Si(001)

Adsorbate covered semiconductor surfaces play important role in device manufacturing. In this respect, interaction of common chemical vapour deposition (CVD) gas sources with Si(001) and Si(111) surfaces have been studied quite widely, both experimentally and theoretically. Here we briefly discuss adsorption of ammonia (NH_3), phosphine (PH_3) and arsine (AsH_4) on the Si(001)-(2×1) surface. The down and up dimer atoms on Si(001)-(2×1) are electron deficient (electrophile) and electron rich (nucleaphile), respectively. Such molecules can adsorb both nondissociatively and dissociatively, depending on the coverage and flux during exposure. Being Lewis base, in its nondissociative form a XH_3 (X = N, P, and As) molecule initially adsorbs at the down (electrophilic) Si dimer atom. At room temperature, the molecule is more likely to dissociate as $XH_3 \rightarrow XH_2 + H$, with XH_2 and H adsorbing at the down and up dimer atoms, respectively. Characteristic features of the nondissociative and dissociative adsorption models are the stretch, bending and scissors modes. Fourier-transform infrared (FTIR) and HREELS measurements can identify these modes. Theoretical calculations of zone-centre optical phonons can be performed to identify these modes and support experimental studies if the molecule has been adsorbed in the dissociative form.

Based on the DFT-LDA scheme, Miotto *et al* (2001) employed the *ab initio* finite displacement method for zone-centre phonon, as briefly described in section 3.3.1.2. Total energy calculations favoured the gauche structural model for both the molecular and dissociative adsorptions, with trigonal-pyramidal geometry, as shown in figure 8.12. Considering the molecule and the top four atomic layers on Si(001)-(2×1), a 36×36 dynamical matrix was constructed by exerting finite displacements of atoms away from their equilibrium positions and extracting the linear part of the resultant Hellmann-Feynman forces. Results with pronounced surface character were analysed in terms of stretch, bend and scissors modes, and compared with HREELS and FTIR measurements (see Miotto *et al* 2001 for references).

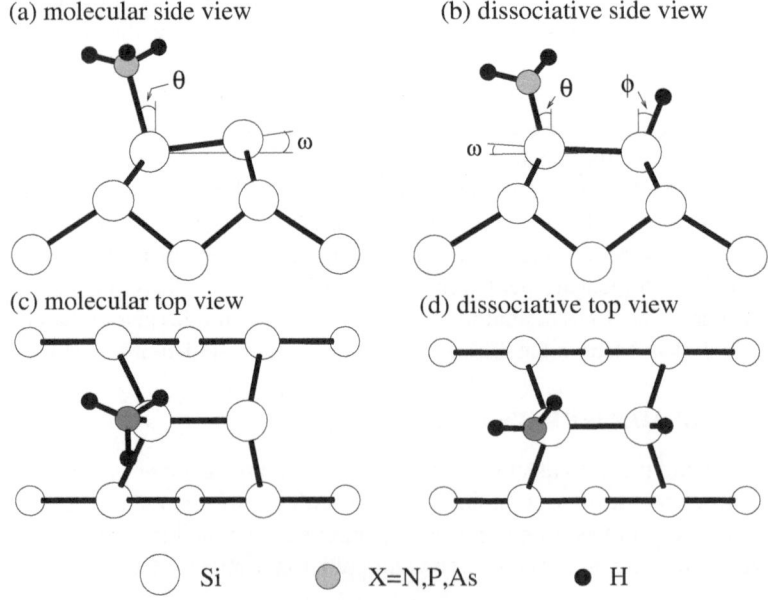

Figure 8.12 Gauch model of the molecular and dissociative adsorption of XH_3 (X = N, P, As) on Si(001)-(2×1). Reproduced from Miotto *et al* (2001).

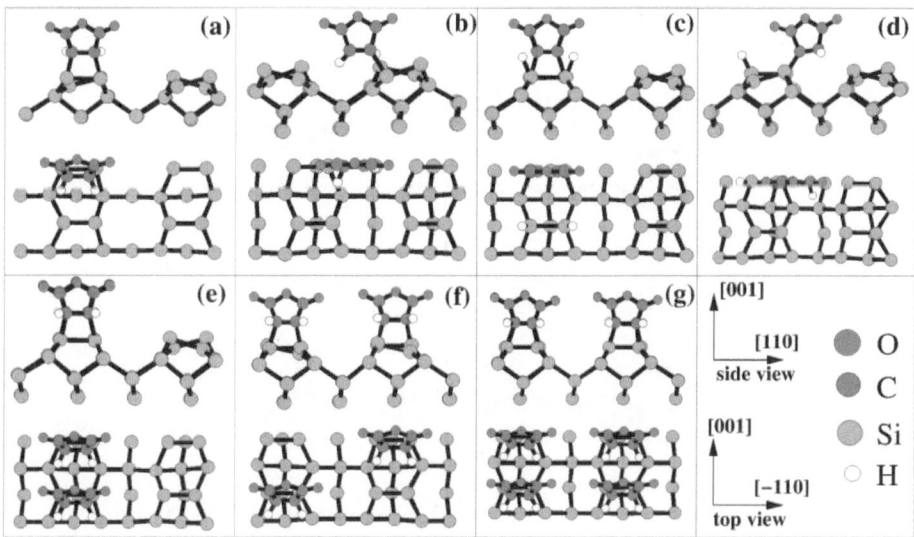

Figure 8.13 Schematic side and top views of possible models for the adsorption of maleic anhydride ($C_2H_2C_2O_3$) on the Si(001)-(4 × 2) surface. Panels (a)–(e) consider a 0.25 ML (monolayer) coverage and panels (f) and (g) consider a 0.5 ML coverage. (a) [2+2] cycloaddition via the C=C bond ([2+2]), (b) inter-dimer (the molecule is bonded to a single Si–Si dimer), (c) [2+2] cycloaddition via the C=C bond with hydrogen transfer from the molecule to the neighbouring Si–Si dimer ([2+2]$_h$), (d) inter-dimer-B (the molecule is bonded to neighbouring Si–Si dimers), (e) inter-dimer with hydrogen transfer from the molecule to the nucleophilic Si atom from the same dimer (inter-dimer$_h$), (f) [2+2] cycloaddition via the C=C bond at neighbouring dimers (neighbouring), (g) [2+2] cycloaddition via the C=C bond at Si–Si alternate dimers along two neighbouring dimer rows (zig-zag). Reproduced from Miotto *et al* (2005).

For the molecular adsorption, the theoretical (HREELS) results for the N–H stretch and N–H scissors modes were obtained as 3334 (3430) and 1501 (1630) cm^{-1}, respectively. For the dissociative adsorption the N–H stretch, N–H scissors, Si–H stretch and Si–H bend modes lie at 3447 (3330–3450), 1483 (1570), 2082 (2050–2075), 598 (603) cm^{-1}, respectively. For PH$_3$ adsorption, the theoretical (FTIR) results for the P–H stretch and P–H scissors modes were obtained as 2075 (2270) and 952 cm^{-1}, respectively. For the dissociative adsorption, the P–H stretch, P–H scissors, Si–H stretch and Si–H bend modes theoretical (HREELS, FTIR) results are 2162 (2300, 2245), 955 (1050, ...), 2049 (2100, 2097) and 628 (640, ...) cm^{-1}, respectively. For AsH$_3$ adsorption, the theoretical results for the As–H stretch and As–H scissors modes were obtained as 2856 and 776 cm^{-1}, respectively. For the dissociative adsorption, the As–H stretch, As–H scissors, Si–H stretch and Si–H bend modes lie at 2824, 820, 2108 and 610 cm^{-1}, respectively. The strongest experimental indication of the dissociative adsorption of ammonia (NH$_3$) and phosphine (PH$_3$) is the detection of the Si–H stretch mode (Shan *et al* 1996). The first-principles calculations provide support for the dissociative adsorption model.

Adsorption of maleic anhydride on Si(001)

Several possible models can be considered for the adsorption of maleic anhydride ($C_2H_2C_2O_3$) on the Si(001)-(4 × 2) surface. Schematic side and top views of some of these are shown in figure 8.13. Based on the DFT-LDA scheme, Miotto *et al* (2005) found that for 0.25 ML coverage (one molecule per 4 Si–Si dimer), the molecule adsorbs through a [2 + 2] cycloaddition reaction via the C==C functionality, shown in panel (a) in the figure. A slightly less energetically favourable is

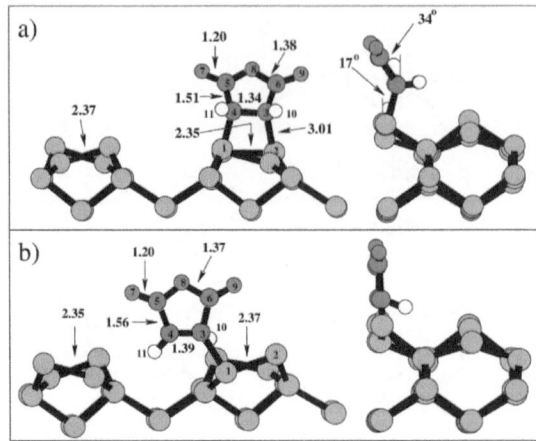

Figure 8.14 Relaxed atomic geometry for (*a*) [2+2] cycloaddition via the C=C bond ([2+2]) and (*b*) inter-dimer-B structures. Reproduced from Miotto *et al* (2005).

the interdimer B structure, shown in panel (*d*) in the figure. The equilibrium atomic geometries of these two structural models are shown in figure 8.14.

Employing the *ab initio* finite displacement method, these authors examined the frequency and vibrational pattern of some of the optical modes for these two structures. Overall, the calculated results for both the [2+2] cycloaddition and inter-dimer-B structures are in the same range as experimental measurement by Lopez *et al* (2001). Numerical results, in cm^{-1}, for some of the modes ([2+2] cycloaddition theoretical, inter-dimer-B theoretical, HREELS) are: C–H stretch mode (atoms 4–11): (3025–3062, 3032–3074, 2990–3070); C–O stretch mode (atoms 5–9): (1729, 1780, 1775); C–H bend mode (atoms 4–11 and atoms 3–10): (1096–1233, 1080–1210, 1080–1212); C–C stretch(ring) mode (atoms 4–5): (1014, 980, 960); C–O bend(ring) mode: (822-899, 818–897, 858). The similarity in the vibrational frequencies for the [2+2] cycloaddition and inter-dimer-B models is consistent with the interpretation of scanning tunnelling microscopic (STM) images by Bitzer *et al* (2001) and Hofer *et al* (2002) which indicate adsorption of the molecule in both structures. This led Miotto *et al* to propose a mixed domain structure for the adsorption of maleic anhydride on Si(001).

8.5 PHONONS ON MONOLAYER TRANSITION METAL DICHALCOGENIDES

Monolayer two-dimensional systems provide an extreme example of the thinnest possible surface system. Phonon dispersion curves for the monolayer graphene were presented in figure 3.9. Here we present a brief discussion of phonons on the monolayer transition metal dichalcogenides (ml-TMDs).

Figure 8.15 shows the atomic structure of ml-TMDs. Figure 8.16 shows the phonon dispersion curves and density of states for MoS_2, WS_2 and $MoTe_2$, obtained from the application of the DFPT method described in section 3.3.2.1. Some observations can be made from these results. Both the maximum optical and maximum acoustic frequencies show the trend $\omega(MoS_2) > \omega(WS_2) > \omega(MoTe_2)$. The minimum optical frequency shows the trend $\omega(WS_2) > \omega(MoS_2) > \omega(MoTe_2)$. For $MoTe_2$, the values of the maximum acoustic and the minimum of optical frequencies are very close to each other. These results are consistent with the relative atomic masses and bond strengths (or lattice constants) in these materials. As monolayer TMDs are periodic arrays of zig-zag tri-atomic chains, a rough understanding of these trends can be obtained from the analytical solutions for a linear triatomic mass chain (see, e.g. Kesavasamy and Krishnamurthy 1978).

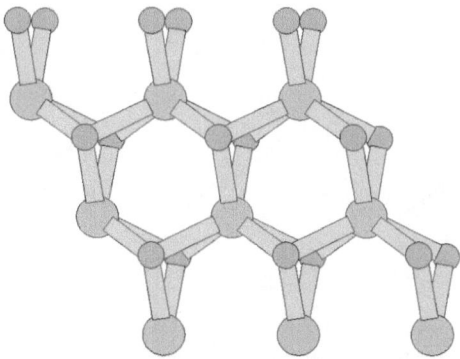

Figure 8.15   Atomic structure of monolayer transition metal dichalcogenides.

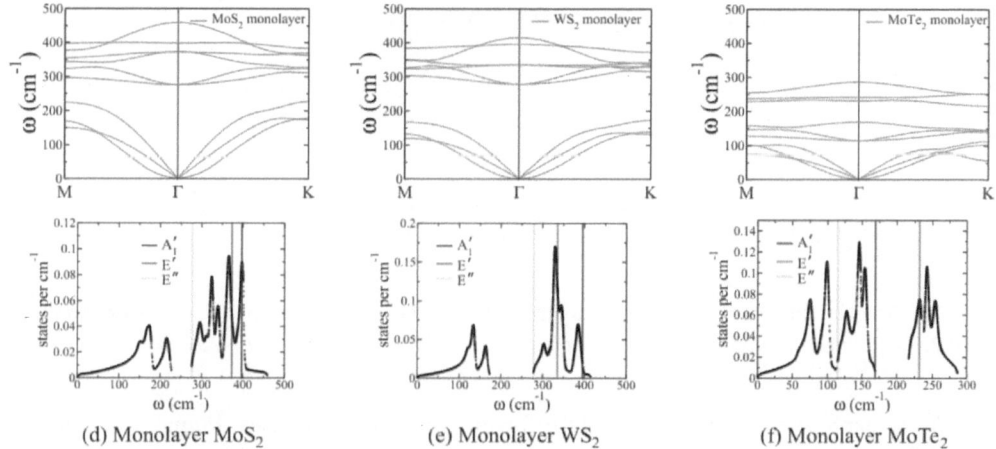

(d) Monolayer MoS₂   (e) Monolayer WS₂   (f) Monolayer MoTe₂

Figure 8.16   Phonon dispersion curves and density of states for monolayer MoS2, WS2 and MoTe2. Also indicated are the frequencies for the Raman modes A_1', E' and E''. Reproduced from Srivastava and Thomas (2018).

The point group symmetry of the monolayer TMD structure is D_{3h}. The eigenmodes of the optical atomic vibrations for ml-TMDs can therefore be decomposed from the zone-centre representation as

$$\Gamma_{\text{optical}} = A_1'(\text{R}) + A_2''(\text{IR}) + E'(\text{R} + \text{IR}) + E''(\text{R}), \tag{8.22}$$

where R and IR indicate Raman and infra-red active modes, respectively. Figure 8.17 illustrates the atomic displacement patters for the Raman modes.

For a pure and homogeneous sample of size larger than the intrinsic phonon mean free path, the full-width at half maximum FWHM(s) of a Raman active mode of polarisation s can be expressed as

$$\text{FWHM}(s) \equiv \hbar \tau_{0s}^{-1} = \hbar [\tau_{0s}^{-1}(\text{md}) + \tau_{0s}^{-1}(\text{anh})], \tag{8.23}$$

where $\tau_{0s}^{-1}(\text{md})$ and $\tau_{0s}^{-1}(\text{anh})$ are, respectively, the isotopic mass-defect and anharmonic scattering rates of the zone-centre phonon mode $\omega(q = 0, s)$. *Ab initio* calculation of $\tau_{0s}^{-1}(\text{md})$ can be made using equation (6.34). *Ab initio* calculations of anharmonic three-phonon and four-phonon scattering rates can be made using equations (6.214) and (6.219).

Srivastava and Thomas (2018) included all stable isotopic masses for Mo, W, S and Te to calculate $\tau_{0s}^{-1}(\text{md})$ using equation (6.34). They calculated the anharmonic scattering rate by using the

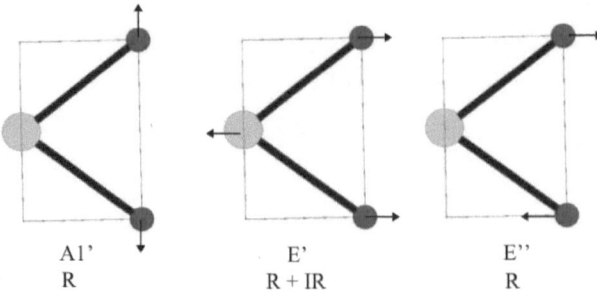

Figure 8.17 Vibrational pattern for Raman active modes on monolayer MoS2, WS2 and MoTe2. Reproduced from Srivastava and Thomas (2018).

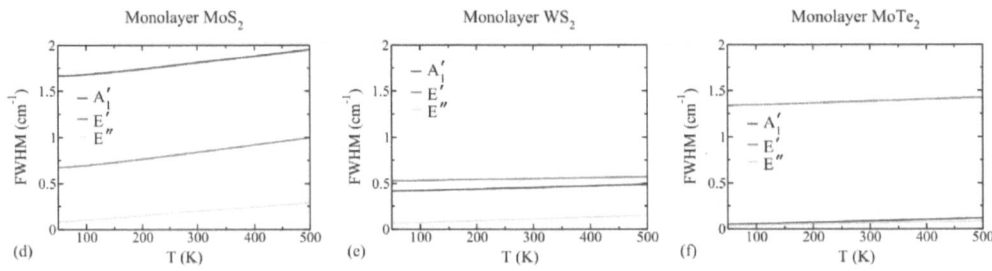

Figure 8.18 Temperature variation of anharmonically controlled Raman width for monolayer MoS2, WS2 and MoTe2. Reproduced from Srivastava and Thomas (2018).

semi-*ab initio* expression in equation (6.220) for three-phonon processes. Figure 8.18 shows the calculated Raman linewidth over a large temperature range. Note that $\tau_{0s}^{-1}(\text{md})$ is temperature independent. Also, note that the first of the three contributions in the second term for $\tau_{0s}^{-1}(\text{anh})$ in equation (6.214) is also temperature independent. With these considerations, we can express

$$\text{FWHM}(s) = \text{constant} + f(T), \qquad (8.24)$$

where $f(T)$ is a temperature-dependent anharmonic contribution. Figure 8.18 shows the computed results for the temperature variation of the Raman linewidths for monolayer MoS_2, WS_2 and $MoTe_2$. While in general the temperature variation of the results are similar to those extracted from Raman measurements for a few-layer TMDs (see, e.g. Sahoo *et al* 2013, Lanzillo *et al* 2013), it is difficult to make direct comparison between theory and experiment as real samples are characterised by inhomogeneity and unknown types and amounts of additional defects.

The first and second terms in equation (6.214) represent the scattering rate of a phonon mode (q,s) via coalescence (class 1) and decay (class 2) processes, respectively. Because of low-frequency locations, coalescence (decay) of the Raman modes A_1' and E' is more (less) probable in $MoTe_2$ than in MoS_2 and WS_2. For example, numerical results suggest that at room temperature the contribution towards the cubic anharmonic part of FWHM(A_1') from the decay process is 85% in MoS_2 and 35% in $MoTe_2$.

8.6 SURFACE SPECIFIC HEAT

We know from Chapter 2 that the density of vibrational modes of a continuum $g(\omega)$ varies as ω^2, and that the bulk phonon specific heat has the familiar T^3 form at low temperatures. As we have seen in the preceding section, when surfaces are present the normal mode frequencies are modified. The existence of surface modes gives rise to a surface contribution to the specific heat.

A proper calculation of surface specific heat is difficult (see Wallis 1974 and references therein). In a simple description we can write the total density of normal modes of an isotropic elastic slab as

$$g(\omega) = g_B(\omega) + g_S(\omega), \tag{8.25}$$

where B and S refer to bulk and surface (Rayleigh waves) contributions, respectively. While $g_B(\omega) \propto \omega^2$, it can be shown that $g_S(\omega) \propto \omega$. Now Debye's approach can be applied to determine a maximum cut-off ω_D by counting the total number of modes as $3N$, where N is the number of atoms in a slab: $3N = \int g(\omega)d\omega$. Using equation (8.25) one can express

$$\omega_D = \omega_D^B + \Delta\omega_D^S, \tag{8.26}$$

where ω_D^B is the bulk Debye frequency and it is found that

$$\Delta\omega_D^S \propto \mathscr{S}\omega_D^B, \tag{8.27}$$

$\mathscr{S}$ being the area of the slab surface.

Having determined ω_D, we can use the Debye method for calculating lattice heat capacity. At low temperatures, the surface contribution can be expressed as (Wallis 1974)

$$C_v^S \propto \mathscr{S}T^2. \tag{8.28}$$

Thus the low-temperature surface specific heat is proportional to the surface area and to the square of the absolute temperature: it becomes dominant as temperature decreases and the surface/volume ratio increases. Thus for fine powders the surface specific heat at sufficiently low temperatures may become comparable to the bulk T^3 contribution.

When one considers an adsorbate layer of atomic mass m_s, then the low-temperature specific heat contribution comes out to be

$$\Delta C_S^{adsorbate} \propto N_S \left(\frac{m_s - m}{m} \right) T^3 \tag{8.29}$$

where m is the mass of bulk atoms and N_S is the number of adsorbates.

8.7 ATTENUATION OF SURFACE PHONONS

Surface phonons can be attenuated by interaction with crystal impurities and imperfections, with surface or conduction electrons in metals and semiconductors, and with bulk phonons.

8.7.1 INTERACTION WITH IMPURITIES AND IMPERFECTIONS

Steg and Klemens (1970) have calculated the relaxation rate of the Rayleigh waves from impurity scattering using the isotropic continuum model. If there are mass defect impurities with additional mass Δm and concentration n per unit surface area, then the relaxation rate of a surface (Rayleigh) phonon of frequency ω is given by

$$\tau_{S,md}^{-1} = \frac{n(\Delta m)^2}{8\rho^2 B^2 c_S^4} \omega^5 [A_1(q_\parallel z) + A_2(q_\parallel z) + A_3(q_\parallel z)], \tag{8.30}$$

where c_S is the velocity of the Rayleigh phonon, B is a slowly varying function of Poisson's ratio, and the terms A_1, A_2 and A_3 are dimensionless parameters describing the scattering of the Rayleigh phonon into other Rayleigh phonons, into longitudinal bulk phonons, and into transverse bulk phonons, respectively. z is the distance of the impurity from the surface. If the wavelength λ_S of the surface phonon is greater than the distance z, then the first term in equation (8.30) dominates over the other two terms. In other words, then surface phonons are mostly scattered into other surface phonons. When $\lambda_S \ll z$, surface phonons are mostly scattered into bulk phonons. The frequency dependence in equation (8.30) suggests that undamped propagation of high-frequency surface waves ($\omega \geq 10$ GHz) may require very smooth surfaces.

8.7.2   INTERACTION WITH BULK PHONONS

The theory of anharmonic interaction between bulk phonons, described in sections 4.3, 4.4 and 6.4, can be appropriately modified to study anharmonic interaction between surface phonons and bulk phonons. The dominant interaction processes are S+B $\rightleftharpoons$ B, where S and B stand for ultrasonic surface phonons and thermal bulk phonons, respectively. Such an interaction process may be treated by perturbation theory, as described in section 6.4, provided $\omega_S \tau \gg 1$, where ω_S is the frequency of the Rayleigh (surface) wave and τ is the relaxation time of the thermal phonons. Maradudin and Mills (1968), and King and Sheard (1970) have studied these processes. Here we follow the approach of the latter authors.

Treating the surface wave as a normal mode of a semi-infinite solid we modify equation (4.79) to write the displacement field due to the surface modes q as

$$u(r) = \sum_J \sum_q \sqrt{\frac{\hbar}{2\rho S c_S \Sigma}} [B_J e^J(q) a_q e^{iq^J \cdot r} + B_J^* e^{*J}(q) a_q^\dagger e^{-iq^J \cdot r}], \qquad (8.31)$$

where $J = 1, 2, 3$ label the solutions with wave vector q^J, a_q and $a_q^\dagger$ are the annihilation and creation operators for surface phonons, c_S is the velocity of the surface wave, $\mathscr{S}$ is the surface area and B_J are numerical factors chosen to satisfy the boundary condition of zero stress at the surface (x, y) plane. We can write, analogous to equation (8.2),

$$q^J \cdot r = q_\parallel (\alpha_x x + \alpha_y y + i\alpha_z^J z) \qquad (8.32)$$

where $q_\parallel$ is the magnitude of q in the surface plane, with direction cosines α_x and α_y, and α_z is a positive and real attenuation constant. The factor Σ is an amplitude normalisation constant adjusted to ensure that each mode q contributes an elastic energy $\hbar\omega_q a_q^\dagger a_q$ to the Hamiltonian of the solid: the result is

$$\Sigma = \sum_{JJ'j} \frac{B_J^* B_{J'} e_j^{*J} e_j^{J'}}{i\alpha_z^J - i\alpha_z^{J'}} \qquad (j = x, y, z). \qquad (8.33)$$

Consider the process

$$q + (q', s') \rightleftharpoons (q'', s''), \qquad (8.34)$$

where (q', s') and (q'', s'') are bulk phonon states. Using equations (8.31), (4.61) and (6.45), King and Sheard (1970) obtained the following expression for the relaxation time τ_S of the surface phonon with frequency ω_S:

$$\tau_S^{-1} = \sum_{s's''} \int\int \frac{\hbar q_\parallel^2 q' q'' (\bar{n}' - \bar{n}'')}{64\pi^3 \rho^3 c_S c' c'' \Sigma}$$

$$\times \left| \sum_J \frac{F_J}{iq_\parallel \alpha_z^J + q_z' - q_z''} \right|^2 \delta(\omega_S + \omega' - \omega'') dq' dq_z'', \qquad (8.35)$$

where

$$F_J = \frac{1}{3!} \sum_{\substack{ijk \\ lmn}} \sum_P A_{ijk}^{lmn} B_J e_l^J \hat{q}_i^J e_m(q') \hat{q}_j' e_n'(q'') \hat{q}_k'', \qquad (8.36)$$

with P denoting the six permutations of the pairs (li), (mj) and (nk). The integrals in (8.35) can be simplified by introducing a new variable $\boldsymbol{p} = \boldsymbol{q''} - \boldsymbol{q'}$ and using polar coordinates (q', θ, ϕ) for $\boldsymbol{q'}$ with $\boldsymbol{p}$ as the axis. Only small $\boldsymbol{p}$ are important, so that we can regard $q'q'' \approx q'^2$, and $c'c'' \approx c'^2$. Integration over q' gives, for $s'' = s'$,

$$\tau_S^{-1} = \frac{\hbar \omega_S}{64\pi^3 \rho^3 c_S^2 \Sigma} \left(\frac{k_B T}{\hbar} \right)^4$$

$$\times \sum_{s'} D_4 \int \frac{1}{c'^7} \left| \sum_J \frac{F_J}{i\alpha_z^J - \lambda} \right|^2 \delta(\Delta) \sin\theta \, d\theta \, d\phi \, d\lambda \qquad (8.37)$$

where

$$\lambda = p_z / q_\| \qquad (8.38)$$

$$\Delta = 1 - \frac{c'p}{c_S q_\|} \cos\theta + \frac{p}{c_S q_\|} \sin\theta \frac{\partial c'}{\partial \theta} \qquad (8.39)$$

and

$$D_4 = \int_0^{\Theta_D / T} \frac{z^4 \exp(z)}{(\exp(z) - 1)^2} dz. \qquad (8.40)$$

The remaining integrals in (8.37) are difficult and must be evaluated in terms of line integrals along $p_z = \lambda q_\|$. Then equation (8.37) becomes

$$\tau_S^{-1} = \frac{\hbar \omega_S}{64\pi^3 \rho^3 c_S^2 \Sigma} \left(\frac{k_B T}{\hbar} \right)^4 D_4$$

$$\times \sum_{s'} \int \oint_{\Delta=0} \left| \sum_J \frac{F_J}{i\alpha_z^J - \lambda} \right|^2 \frac{\sin\theta}{c'^7 |\text{grad}\Delta|} d\ell \, d\lambda, \qquad (8.41)$$

where $d\ell$ is a line element in (θ, ϕ) space to which the gradient operator also refers.

The contours in the (θ, ϕ) space give the directions of the wave vectors of the thermal phonons which interact with the surface phonon for each value of λ. In the isotropic medium $\partial c'/\partial \theta = 0$ and $\Delta = 0$ defines a cone of semi-angle $\theta = \arccos(c_S q_\| / c'p)$ with the axis along $\boldsymbol{p}$. Since $q_\| \leq p$ and $c_S < c'$ the angle θ always exists. The wave vectors of the participating thermal phonons must lie on such a cone.

At low temperatures $D_4 = 4\pi^4/15$ and equation (8.41) reduces to

$$\tau_S^{-1} \propto \omega_S T^4. \qquad (8.42)$$

Therefore, the anharmonic damping of Rayleigh waves is governed by the same frequency and temperature dependence as observed in the Landau–Rumer (1937) process $T + L \rightleftharpoons L$ for bulk phonons (see also the second term in equation (6.92)).

9 Phonons and Thermal Transport in Nanocomposites

9.1 INTRODUCTION

In the previous chapter we discussed phonons on clean and covered crystal surfaces. Surfaces provide one class of low dimensional systems (LDS). Another class of LDS is a composite structure comprised of several intentionally arranged materials. Such a structure can be periodic in one, two or three dimensions. Also, periodicity in such systems can be in the nanometer or micrometer range. An essential aspect of physics in such systems is governed by the presence of interfaces between their constituent materials.

A periodic nanocomposite structure consists of identical inclusion of more than one material type in every unit cell. Such structures normally do not exist in nature but are fabricated in laboratory using crystal growth techniques such as molecular beam epitaxy (MBE) and chemical vapour deposition (CVD). We can classify such structures according to their newly adopted one-dimensional, two-dimensional and three-dimensional periodicities, respectively, as *planar superlattices* (PSLs), *nanowire superlattices* (NWSLs) and *nanodot superlattices* (NDSLs). A PSL, usually referred to as a superlattice (SL), in its simple form is a periodic system comprising alternating layers of two crystals. Superlattices (SLs) with thick alternating layers (typically 5 nm or more) are referred to as multi-quantum wells (MQWs). In its simple form a NWSL is a periodically embedded nanowire of a material in the matrix of another material. Similarly, in its simple form a NDSL is a periodically embedded nanodot of one material in the matrix of another material.

In this chapter we present some aspects of phonon physics in nanocomposite and microcomposite structures.

9.2 CONTINUUM THEORY OF PHONONS IN PLANAR SUPERLATTICES

The treatment given here is due to Rytov (1956), as reviewed by Jusserand and Paquet (1986a), and Jusserand and Cardona (1989).

For acoustic modes with wavelength much larger than the lattice parameter the atomic displacements can be considered as a continuum field u. Consider the two materials of a SL as two effective media characterised by densities ρ_1 and ρ_2, and elastic constants Λ_1 and Λ_2, respectively. Propagation of elastic waves along the SL direction, say the z direction, is governed by the Lagrangian density

$$\mathscr{L} = \frac{1}{2}\rho(z)\left(\frac{\partial u}{\partial t}\right)^2 - \frac{1}{2}\Lambda(z)\left(\frac{\partial u}{\partial z}\right)^2, \tag{9.1}$$

where $\rho = (\rho_1, \rho_2)$ and $\Lambda = (\Lambda_1, \Lambda_2)$ when z is in medium (1, 2).

The elastic wave is given by the one-dimensional equation

$$\rho\frac{\partial^2 u}{\partial t^2} = \frac{\partial}{\partial z}\left(\Lambda\frac{\partial u}{\partial z}\right) \tag{9.2}$$

DOI: 10.1201/9781003141273-9

subject to the following interface boundary conditions at $z = 0$:

$$\Lambda_1 \frac{\partial u_1}{\partial z} = \Lambda_2 \frac{\partial u_2}{\partial z} \qquad \text{(stress continuity)} \tag{9.3}$$

$$u_1 = u_2 \qquad \text{(displacement continuity).} \tag{9.4}$$

Further, we have due to the periodicity

$$u(z+d) = u(z)e^{iqd}, \tag{9.5}$$

where q is a wave vector of the SL, d_1 and d_2 are the thicknesses of layers 1 and 2, and d $(= d_1 + d_2)$ is the periodicity of the SL. Within each layer equation (9.2) reduces to the form given in equation (2.137)

$$\rho_{1,2} \frac{\partial^2 u_{1,2}}{\partial t^2} = \Lambda_{1,2} \frac{\partial^2 u_{1,2}}{\partial z^2} \tag{9.6}$$

which leads to the linear dispersion relation

$$\omega = \sqrt{\frac{\Lambda_{1,2}}{\rho_{1,2}}} q_{1,2}, \tag{9.7}$$

where $q_{1,2}$ is a phonon wave vector in the layer (1,2). The SL Brillouin zone edge along the z axis is related to the bulk Brillouin zone edge by $q^{BZ}/q_{1,2}^{BZ} = d_{1,2}/d$.

For the SL problem we try solutions

$$u_1(z) = (Ae^{iq_1 z} + Be^{-iq_1 z})e^{-i\omega t} \tag{9.8}$$

$$u_2(z) = (Ce^{iq_2 z} + De^{-iq_2 z})e^{-i\omega t}. \tag{9.9}$$

The continuity equations (9.3) and (9.4) become, using equations (9.8) and (9.9):
at $z = 0$

$$A + B = C + D \tag{9.10}$$

$$(A - B)\Lambda_1 q_1 = (C - D)\Lambda_2 q_2 \tag{9.11}$$

at $z = d_2$

$$Ce^{iq_2 d_2} + De^{-iq_2 d_2} = (Ae^{-iq_1 d_1} + Be^{iq_1 d_1})e^{iqd} \tag{9.12}$$

$$\Lambda_2 q_2(Ce^{iq_2 d_2} - De^{-iq_2 d_2}) = \Lambda_1 q_1(Ae^{-iq_1 d_1} - Be^{iq_1 d_1})e^{iqd}. \tag{9.13}$$

From equations (9.10)–(9.11) and (9.12)–(9.13) the following SL dispersion relation can be obtained:

$$\cos(qd) = \cos(q_1 d_1)\cos(q_2 d_2) - \frac{\Lambda_1 \rho_1 + \Lambda_2 \rho_2}{2\sqrt{\Lambda_1 \rho_1 \Lambda_2 \rho_2}} \sin(q_1 d_1)\sin(q_2 d_2). \tag{9.14}$$

Furthermore, the eigendisplacements are given as

$$\begin{aligned} u_1 = {} & (\Lambda_1 q_1 \sin(q_2 d_2))\cos(q_1 z) - (\Lambda_2 q_2 \cos(q_2 d_2))\sin(q_1 z) \\ & + (\Lambda_2 q_2 e^{iqd})\sin q_1(z + d_1) \qquad -d_1 \le z \le 0 \end{aligned} \tag{9.15}$$

$$\begin{aligned} u_2 = {} & (\Lambda_1 q_1 e^{iqd}\cos(q_1 d_1))\sin(q_2 z) + (\Lambda_2 q_2 e^{iqd}\sin(q_1 d_1))\cos(q_2 z) \\ & - \Lambda_1 d_1 \sin(q_2(z - d_2)) \qquad 0 \le z \le d_2. \end{aligned} \tag{9.16}$$

The dispersion relation in (9.14) can be re-expressed as

$$\cos(qd) = \cos\{\omega(d_1/v_1 + d_2/v_2)\} - \frac{\varepsilon^2}{2}\sin(\omega d_1/v_1)\sin(\omega d_2/v_2), \tag{9.17}$$

where

$$\varepsilon = (\Lambda_1\rho_1 - \Lambda_2\rho_2)/(\Lambda_1\rho_1 + \Lambda_2\rho_2) \tag{9.18}$$

is a measure of the mismatch of the acoustical impedances of the two layers, and v_1, v_2 are the sound velocities in bulk compounds 1, 2. In the limit of long wavelengths ($q \to 0, \omega \to 0$) equation (9.17) reduces to $\omega = qv$, where

$$v = d[(d_1/v_1 + d_2/v_2)^2 - \varepsilon^2(d_1/v_1)(d_2/v_2)]^{-1/2}. \tag{9.19}$$

When $\varepsilon = 0$, we get $\bar{\omega} = q\bar{v}$, where

$$\bar{v}^{-1} = (1-x)v_1^{-1} + xv_2^{-1} \tag{9.20}$$

with $x = d_2/d$ as the concentration of the compound 2 in the SL. The relation $\bar{\omega} = q\bar{v}$ can be interpreted as the folding of the continuum dispersion curve of an average compound of sound velocity $\bar{v}$. The relation

$$\cos(qd) = \cos(\bar{\omega}\bar{v}/d) \tag{9.21}$$

also yields doubly degenerate frequencies

$$\bar{\omega} \to \Omega_v = \begin{cases} 2\pi v\bar{v}/d & \text{at zone centre} \\ \pi(v+1)\bar{v}/d & \text{at zone edge} \end{cases} \tag{9.22}$$

with $v =$ integer. When ε is small the effect of the second term in (9.17) is to split the frequencies at the zone centre and zone edge. Thus this model predicts new zone-centre optical modes of acoustic nature. Figure 9.1 gives a schematic representation of folded acoustic curves of a SL using the continuum theory.

The zone-folding and the creation of minigaps in the acoustic phonon dispersion curves of SLS have been observed experimentally by acoustic transmission using superconducting tunnel junctions (Narayanamurti *et al* 1979) and by means of Raman scattering (Colvard *et al* 1980, Jusserand *et al* 1986, 1988).

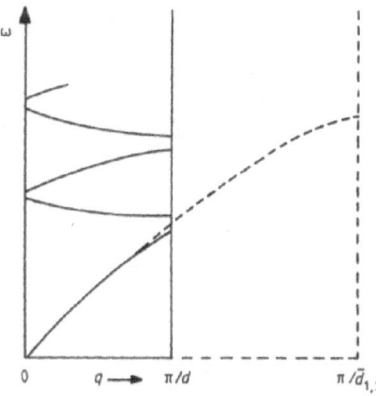

Figure 9.1 Schematic illustration of folded acoustic dispersion curves of a superlattice using the continuum theory. Note small splittings at the zone centre and zone edges. Also shown (dashed curve) is the corresponding curve for an average bulk of size $\bar{d}_{1,2}$.

9.3   ONE-DIMENSIONAL APPROACH FOR LATTICE DYNAMICAL THEORY OF PHONONS IN PLANAR SUPERLATTICES

A simple and useful model of longitudinal lattice dynamics along the SL growth direction can be developed by using the linear chain model discussed in section 2.2.2. Let us consider a superlattice $(XZ)_{N_1}(YZ)_{N_2}(001)$ whose repeat period along [001] contains N_1 layers of XZ and N_2 layers of YZ, both XZ and YZ being zincblende materials. We consider the [001] direction because most Raman scattering experiments have been performed along this direction. In this configuration only longitudinal phonon modes can be detected (see next section). Thus the one-dimensional model will be sufficient to analyse most of the experimental results.

For phonons propagating along the SL growth direction [001] planes of atoms move as a whole and the longitudinal and transverse vibrations are decoupled (Kunc 1973–74). Let us deal with the longitudinal case here. (The transverse case can be described analogously.) Let us further consider, for simplicity, only nearest-neighbour interactions. Moreover, we assume that both XZ and YZ have the same lattice constant a. The thickness of a monolayer is $b = a/2$.

Let masses, atomic displacements, and interplanar force constants be as shown in figure 9.2. Then, following the approach in section 2.2.2 but using a slightly different notation, we can write within XZ layers at $z = jb$

$$m_1 \frac{d^2 u_1(jb)}{dt^2} = \Lambda_1 [u_2(jb) + u_2((j-1)b) - 2u_1(jb)] \tag{9.23}$$

$$m_2 \frac{d^2 u_2(jb)}{dt^2} = \Lambda_1 [u_1((j+1)b) + u_1(jb) - 2u_2(jb)]. \tag{9.24}$$

For atomic vibrations in the XZ layers we try solutions

$$u_1(jb) = A_1 \exp(i(q_1 jb - \omega t)) \tag{9.25}$$

$$u_2(jb) = A_2 \exp(i(q_1 jb - \omega t)), \tag{9.26}$$

where $\pm q_1$ are the z components (along [001]) of the phonon wave vector q in XZ. From equations (9.23)–(9.24) and (9.25)–(9.26) we get

$$\cos(q_1 b) = \frac{(m_1 \omega^2 - 2\Lambda_1)(m_2 \omega^2 - 2\Lambda_1) - 2\Lambda_1^2}{2\Lambda_1^2} \tag{9.27}$$

and

$$\gamma_\pm = \frac{A_2}{A_1} = \frac{\Lambda_1(1 + \exp(\pm iq_1 b))}{(2\Lambda_1 - m_2 \omega^2)}. \tag{9.28}$$

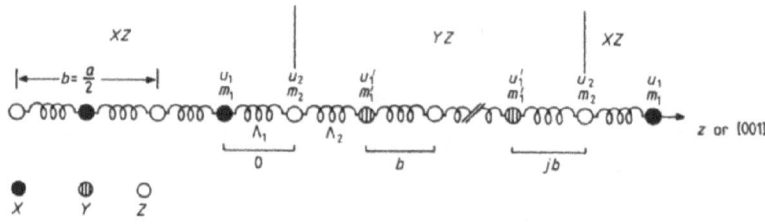

Figure 9.2   The linear chain model for a $(XZ)_{N_1}(YZ)_{N_2}(001)$ superlattice, where XZ and YZ represent zincblende materials. Thicknesses of layers XZ and YZ are $d_1 = N_1 b$ and $d_2 = N_2 b$, respectively. b is monolayer thickness, and $d = d_1 + d_2 = (N_1 + N_2)b$ is the superlattice periodicity.

Similarly, within YZ layers at $z = jb$

$$u_1'(jb) = C_1 \exp(i(q_2 jb - \omega t)) \tag{9.29}$$

$$u_2'(jb) = C_2 \exp(i(q_2 jb - \omega t)) \tag{9.30}$$

and

$$m_1' \frac{d^2 u_1'(jb)}{dt^2} = \Lambda_2[u_2'(jb) + u_2'((j-1)b) - 2u_1'(jb)] \tag{9.31}$$

$$m_2 \frac{d^2 u_2'(jb)}{dt^2} = \Lambda_2[u_1'((j+1)b) + u_1'(jb) - 2u_2'(jb)], \tag{9.32}$$

where $\pm q_2$ are the the phonon wave vector components q_z in YZ. From equations (9.29)–(9.30) and (9.31)–(9.32) we get

$$\cos(q_2 b) = \frac{(m_1'\omega^2 - 2\Lambda_2)(m_2\omega^2 - 2\Lambda_2) - 2\Lambda_2^2}{\Lambda_2^2} \tag{9.33}$$

and

$$\delta_\pm = \frac{C_2}{C_1} = \frac{\Lambda_2(1 + \exp(+iq_2 b))}{(2\Lambda_2 - m_2\omega^2)}. \tag{9.34}$$

For atomic vibrations in the SL we try solutions similar to those used in equations (9.8)–(9.9)

$$u_1(z) = (Ae^{iq_1 z} + Be^{-iq_1 z})e^{-i\omega t} \tag{9.35}$$

$$u_2(z) = (\gamma_+ Ae^{iq_1 z} + \gamma_- Be^{-iq_1 z})e^{-i\omega t} \tag{9.36}$$

$$u_1'(z) = (Ce^{iq_2 z} + De^{-iq_2 z})e^{-i\omega t} \tag{9.37}$$

$$u_2'(z) = (\delta_+ Ce^{iq_2 z} + \delta_- De^{-iq_2 z})e^{-i\omega t}, \tag{9.38}$$

where $u_2/u_1 = \gamma_\pm$ and $u_2'/u_1' = \delta_\pm$ are used. These solutions must obey the displacement and stress continuities at the interface between XZ and YZ. The discrete analogues of equations (9.3)–(9.4) are:

Displacement continuity:
at $z = 0$

$$u_2(0) = u_2'(0)$$

or

$$\gamma_+ A + \gamma_- B = \delta_+ C + \delta_- D \tag{9.39}$$

at $z = d_2 = N_2 b$

$$u_2(d_2) = u_2'(d_2)$$

or

$$\gamma_+ Ae^{-iq_1 d_1}e^{iqd} + \gamma_- Be^{iq_1 d_1}e^{-iqd} = \delta_+ Ce^{iq_2 d_2} + \delta_- De^{-iq_2 d_2}, \tag{9.40}$$

where equation (9.5) is used.
Stress continuity:
at $z = 0$

$$\Lambda_1[u_1(b) - u_1(0)] = \Lambda_2[u_1'(b) - u_1'(0)]$$

or

$$G_+ A + G_- B = H_+ C + H_- D \tag{9.41}$$

at $z = d_2 = N_2 b$

$$\Lambda_1[u_1(d_2 + b) - u_1(d_2)] = \Lambda_2[u_1'(d_2 + b) - u_1'(d_2)]$$

or

$$G_+Ae^{-iq_1d_1}e^{iqd} + G_-Be^{iq_1d_1}e^{iqd} = H_+Ce^{iq_2d_2} + H_-De^{-iq_2d_2}, \qquad (9.42)$$

where

$$G_\pm = \Lambda_1(e^{\pm iq_1b} - 1) \qquad H_\pm = \Lambda_2(e^{\pm iq_2b} - 1). \qquad (9.43)$$

Equations (9.39)–(9.43) define a 4×4 dynamical matrix equation whose solution represents the one-dimensional case of vibrations in the SL. The resulting eigenvalues are given by the expression

$$\begin{aligned}
\cos(qd) =& \cos(q_1d_1)\cos(q_2d_2) \\
&+ \left(\frac{2(\gamma_+\gamma_-H_+H_- + \delta_+\delta_-G_+G_-)}{(\gamma_-\delta_-G_+H_+ + \gamma_+\delta_+G_-H_- - \gamma_-\delta_+G_+H_- - \gamma_+\delta_-G_-H_+)} \right) \\
&\times \sin(q_1d_1)\sin(q_2d_2).
\end{aligned} \qquad (9.44)$$

When a common force constant $\Lambda_1 = \Lambda_2(= \Lambda)$ is assumed for both compounds XZ and YZ, then equation (9.44) reduces to

$$\begin{aligned}
\cos(qd) =& \cos(q_1d_1)\cos(q_2d_2) \\
&- \frac{1 - \cos(q_1b)\cos(q_2b)}{\sin(q_1b)\sin(q_2b)} \sin(q_1d_1)\sin(q_2d_2).
\end{aligned} \qquad (9.45)$$

This result is the discrete analogue of the continuum result in equation (9.14).

Before we present the results for a SL, let us discuss the terminology of complex phonon dispersion relation. The method of the complex phonon dispersion relations is similar to that of the complex electronic band structure (Heine 1963, Inkson 1980, Yip and Chang 1984). In general, due to symmetry properties of the dynamical matrix and the reality of ω^2, the complex phonon dispersion relation satisfies

$$\omega^2(\boldsymbol{q}) = \omega^2(-\boldsymbol{q}^*) = \omega^2(\boldsymbol{q}^*). \qquad (9.46)$$

The reflection symmetry $x, y \to -x, -y$ in the $(GaAs)_{N_1}(AlAs)_{N_2}(001)$ SL means that for real $\boldsymbol{q}_\parallel$ equation (9.46) can be expressed as

$$\begin{aligned}
\omega^2(\boldsymbol{q}_\parallel, q_r + iq_i) &= \omega^2(\boldsymbol{q}_\parallel, q_r - iq_i) \\
&= \omega^2(-\boldsymbol{q}_\parallel, q_r - iq_i) \\
&= \omega(\boldsymbol{q}_\parallel, -q_r + iq_i) \\
&= \omega(\boldsymbol{q}_\parallel, -q_r, -iq_i),
\end{aligned} \qquad (9.47)$$

where $q_z = q_r + iq_i$ is the z component of the phonon wave vector. As we are dealing with a linear chain model along the z direction, we will set $\boldsymbol{q}_\parallel = 0$, so that $\boldsymbol{q} = \boldsymbol{q}_z$. Thus for a given ω^2, the complex q solutions can be grouped in the form $\pm q_r \pm iq_i$. The solutions $q(\omega^2)$, for fixed ω^2, are known as the complex branches and can be classified as: (i) a real branch ($q_i = 0$), (ii) an imaginary branch of the first kind ($q_i \neq 0, q_r = 0$), (iii) an imaginary branch of the second kind ($q_i \neq 0, q_r = q_{max}$), and (iv) a complex branch ($q_r \neq 0, q_i \neq 0$). An imaginary or complex branch characterises an evanescent wave and indicates attenuation (i.e. decay length $\lambda = 1/q_i$) of a vibration.

Figure 9.3 shows the complex dispersion relations for the longitudinal modes in GaAs and AlAs from equations (9.27) and (9.33), respectively. The force constants Λ_1 and Λ_2 were fitted (Colvard et al 1985) to reproduce the zone-centre LO phonon frequencies in GaAs and AlAs, respectively. Imaginary branches of the first kind, real branches and imaginary branches of the second kind are plotted in the left-hand, middle and right-hand side panels, respectively. Because of the symmetry relation (9.47), only the phonon branches with $q_r \geq 0$ and $q_i \geq 0$ are plotted. The LA and LO real branches are connected by an imaginary branch of the second kind at the point q_{max}. With decreasing q_r the LO branch turns into an imaginary branch of the first kind at point $q_r = 0$.

Figure 9.4 shows the plot of equation (9.44) for the $(GaAs)_5(AlAs)_4(001)$ SL. The bulk LA modes of the GaAs and AlAs origins become the folded LA modes of the SL. The optical modes of the GaAs

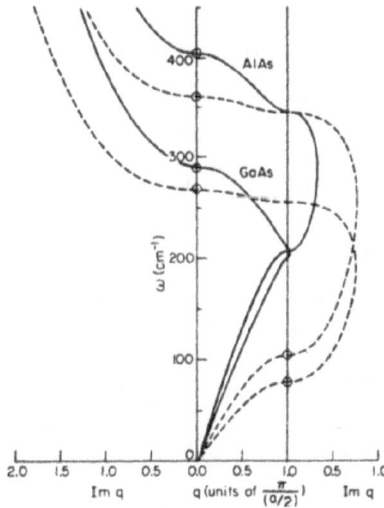

Figure 9.3 Complex phonon dispersion curves for longitudinal modes in GaAs and AlAs bulk materials. The left-hand, middle and right-hand panels show imaginary branches of the first kind, real branches and imaginary branches of the second kind, respectively. The force constants $\Lambda_1 = 0.907 \times 10^5$ dyn/cm (GaAs) and $\Lambda_2 = 0.954 \times 10^5$ dyn/cm (AlAs) are fitted to the frequencies shown in circles. (Reproduced from Colvard *et al* (1985).)

and AlAs origins are also folded into the SL Brillouin zone but now show quite flat dispersion. These folded LO modes in the SL appear in two separate groups, corresponding to the non-overlapping bulk LO modes of GaAs and AlAs as shown in figure 9.3. The flatness of the LO dispersion curves in the SL is in agreement with the confinement of the LO vibrations: the decay length $\lambda \equiv 1/q_i$ can be inferred from the complex dispersion curves in figure 9.3. At the GaAs LO mode frequencies the vibrations are mostly confined to the GaAs layers and extend less than one monolayer into the AlAs layers, where they are damped optical modes with $q = q_{max} = 2\pi/a$. At AlAs optical mode frequencies the vibrations are even more confined to the AlAs layers, decaying into the GaAs layers as damped acoustic modes with $q_r = 0$.

From the linear chain, nearest-neighbour interaction, model in equation (9.45) it can be shown that the jth confined LO mode in the GaAs (AlAs) layers of the SL corresponds to q vectors of the bulk given by (Jusserand and Paquet 1986b)

$$q_{1,2} = \frac{j\pi}{(N_{1,2}+1)b},\qquad(9.48)$$

where $j = 1, 2, \ldots, N_{1,2}$. Thus in the small-$q$ limit the jth folded frequency can be approximated as

$$
\begin{aligned}
\omega_j^2 &= \omega_{LO}^2 - v_L^2 q_{1,2}^2 \\
&= \omega_{LO}^2 - v_L^2 \left(\frac{j\pi}{(N_{1,2}+1)b} \right)^2,
\end{aligned}
\qquad(9.49)
$$

where ω_{LO} is the bulk LO phonon frequency, and v_L is the longitudinal sound velocity. This result suggests that the zone-centre optical phonon frequencies in general decrease with layer thickness in the SL.

Certain optical modes in $(GaAs)_{N_1}(AlAs)_{N_2}(001)$ superlattices have been reported to show an anisotropic behaviour (Merlin *et al* 1980, Ren *et al* 1987): the frequencies at the zone centre are different when $q \to 0$ from [001] and [100] directions (the second direction is an in-plane direction).

LONGITUDINAL

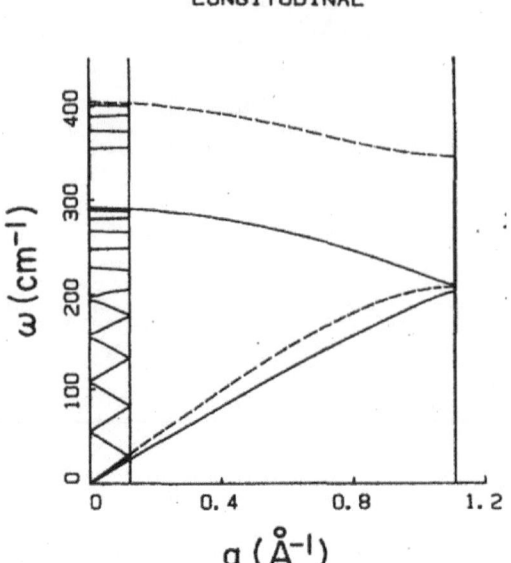

Figure 9.4 Dispersion of longitudinal phonon modes in $(GaAs)_5(AlAs)_4(001)$ using the linear chain model. The force constants are given in the caption for figure 9.3. The bulk phonon modes are also shown. (Reproduced from Colvard *et al* (1985).)

Ren *et al* (1987) have also studied the relation of the in-plane optical modes (due to anisotropy) observed by Merlin *et al* (1980) with the interface modes observed by Sood *et al* (1985).

9.4 RAMAN SCATTERING STUDIES OF PLANAR SUPERLATTICE PHONONS

A brief discussion of Raman scattering in crystals was presented in section 6.9.2. To study phonons in SLS made from III–V semiconductors, polarised monochromatic light is used and the scattered light is analysed in the so-called back scattering configuration: the incident and scattered photons propagate within the crystal in opposite directions perpendicular to the sample surface. The wave vector of the involved phonon is in general very small compared to the Brillouin zone extent, hence pronounced first-order Raman scattering is caused by zone-centre optical phonons. In the following we will deal only with the Stokes (phonon creation) process.

For a Raman active phonon its irreducible representation must coincide with one of the irreducible representations of the polarisability tensor. For a back scattering experiment on a (001) surface of a bulk III–V compound, the transverse optical (TO) phonon modes are forbidden for any polarisation configuration of the incident and scattered beams. The longitudinal optical (LO) phonon mode is allowed for incoming light (incident along the z direction) polarised along the x direction and scattered light (propagating along the $-z$ direction) polarised along the y direction (the so-called $z(x,y)\bar{z}$ configuration) and forbidden in the $z(x,x)\bar{z}$ configuration[1]. For a SL along [001], e.g. $(GaAs)_{N_1}(AlAs)_{N_2}(001)$, the point group symmetry is reduced from tetrahedral (T_d or $\bar{4}3m$ for bulk III–V) to quadratic (D_{2d} or $\bar{4}2m$). For this symmetry TO modes remain forbidden in the back scattering configuration. In the linear chain model for this SL the total number of zone-centre

---

[1] Also see section 8.4.2.3 for this notation

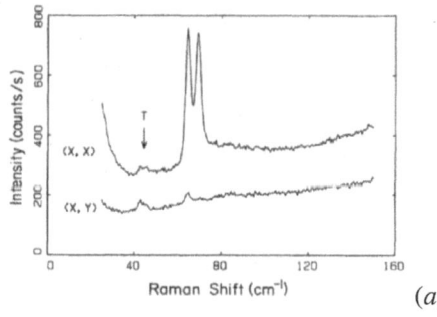
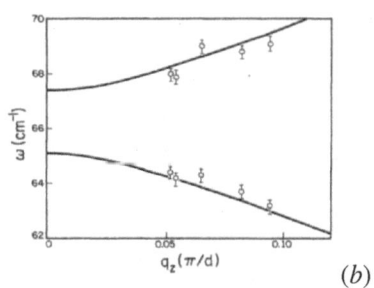

Figure 9.5 (a) Room-temperature Raman spectrum of $(GaAs)_5(AlAs)_4(001)$ near the first folded longitudinal acoustic gap. The sharp peaks are LA phonons, and T indicates transverse phonons. (b) Phonon dispersion near the first folded LA gap for $(GaAs)_5(AlAs)_4(001)$. The solid lines are calculated with the linear chain model. The Raman data are shown by vertically crossed large open circles. The superlattice corresponds to layer thicknesses $d_1 = 14$ Å, $d_2 = 12$ Å. (Reproduced from Colvard et al (1985).)

LO and (folded) LA modes is $2(N_1 + N_2) - 1$. These modes are Raman active in the two different configurations mentioned above.

Let us note that the D_{2d} group has the irreducible representations which are denoted as A_1, A_2, B_1, B_2 and E (see section 1.10). The irreducible representation of the polarisability tensor for the group D_{2d} must transform as xx or xy (Loudon 1964). Raman active phonons have symmetry A_1 (or Γ_1) for the $z(x,x)\bar{z}$ scattering, and B_2 (or Γ_4) for the $z(x,y)\bar{z}$ scattering.

In the following we discuss Raman scattering measurements of folded LA and confined LO phonons in $(GaAs)_{N_1}(AlAs)_{N_2}(001)$.

9.4.0.1 Folded LA phonons

The most striking features of the Raman spectrum from a SL are the folded acoustic phonons in the region $5 \text{ cm}^{-1} \leq \omega \leq 220 \text{ cm}^{-1}$. The folded LA phonons appear as doublets, with each doublet made of a $A_1(\Gamma_1)$ and a $B_2(\Gamma_4)$ mode. The Raman spectrum of $(GaAs)_5(AlAs)_4(001)$ at room temperature near the first folded LA gap is shown in figure 9.5(a). It can be seen that the intensity of the B_2 component ((x,y) spectrum) is much smaller than the intensity of the A_1 component ((x,x) spectrum). This is due to the fact that the B_2 modes are effectively of odd symmetry under inversion and since purely odd modes cannot ordinarily participate in Raman scattering, B_2 modes have a vanishing cross section. In figure 9.5(a) it can be seen that there also appears a folded TA phonon in both (x,x) and (x,y) spectra. This is due to a Brewster angle scattering geometry used in the experiment.

By collecting data for folded peaks at several laser wavelengths, the phonon dispersion near the first folded LA gap in $(GaAs)_5(AlAs)_4(001)$ is shown in figure 9.5(b). Also shown in the figure is the phonon dispersion from the linear chain model (equation (9.44)). It is very encouraging to note that the agreement between theory and experiment is good.

For large period SLs the actual extent of the Brillouin zone becomes of the same order of magnitude as the wave vector of the created phonon. Further, Raman shifts vary for different incident laser wavelengths. Consequently, one can take advantage of the dependence of the created phonon wave vector on the incident wave vector to determine entire phonon dispersion curves in thick SLs. Figure 9.6 shows experimental frequencies of folded LA modes at different wave vectors for a GaAs/AlAs SL with layer thicknesses $d_1 = 26$ Å, $d_2 = 14$ Å. The measured frequencies are in good agreement with theoretical dispersion curves obtained from Rytov's continuum model (equations (9.17) and (9.18)).

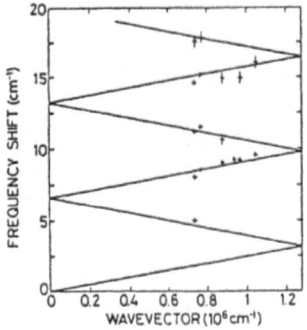

Figure 9.6 Experimental frequencies of folded LA modes in GaAs/AlAs superlattices (with $d_1 = 26$ Å, $d_2 = 14$ Å) compared with theoretical dispersion curve. (Reproduced from Jusserand and Paquet (1986a).)

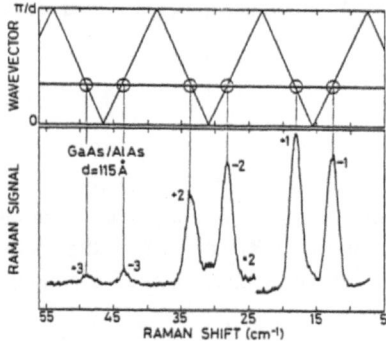

Figure 9.7 Low-frequency Raman spectrum for GaAs/AlAs superlattice with period $d = 115$ Å, and the corresponding theoretical dispersion curve. (Reproduced from Jusserand *et al* (1987).)

Apart from dispersion of folded LA modes, it is also possible to determine structural information from Raman scattering. Figure 9.7 shows a low-frequency Raman spectrum of a GaAs/AlAs SL of period $d = 115$ Å. The back scattering frequencies appear as doublets with the average frequency of the jth doublet as $\bar{\omega}_j = 2\pi j \bar{v}/d$, where $\bar{v}$ is the average phonon velocity in the SL. Thus a good estimate of d, the SL period, can be obtained from the measured doublet frequencies.

9.4.0.2 Confined LO phonons

The frequency difference of the optical phonon bands in GaAs and AlAs is very large (see figure 9.4). One therefore expects the GaAs (AlAs)-like optical phonons to be strictly confined in the GaAs (AlAs) layers of the SL, with very small penetration depth (typically less than a monolayer) within the barrier (well) layers. The non-dispersive frequency levels of LO confined modes are given by equation (9.49).

Figure 9.8 shows the Raman spectra, taken at liquid nitrogen temperature, in the GaAs-like LO phonon frequency range for bulk GaAs and GaAs/AlAs SL with $d_1 = d_2 = 20$ Å. Whereas the bulk LO phonons are active only in the $z(x,y)\bar{z}$ configuration, the SL LO phonons are active both in $z(x,x)\bar{z}$ (A_1 symmetry) and $z(x,y)\bar{z}$ (B_2 symmetry) configurations. The observed LO modes in the SL can be compared with the theoretical predictions from the linear chain model. However, when comparing experiment with theory it should be remembered that the treatment in section 9.3 does not account for electronic long-range forces which are produced by LO vibrations (cf section 2.4.4).

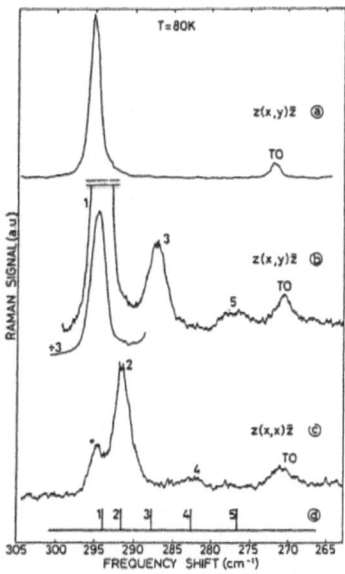

Figure 9.8 Raman scattering spectra on (*a*) pure GaAs; (*b*) and (*c*) GaAs/AlAs superlattice with $d_1 = d_2 = 20$ Å in the GaAs-like LO phonon frequency range. Modes 1, 3 and 5 are $B_2(\Gamma_4)$ type and modes 2 and 4 are $A_1(\Gamma_1)$ type. The line labelled by a star is a small forbidden contribution from mode 1. (Reproduced from Jusserand and Paquet (1986a).)

By fitting observed GaAs-like LO frequencies to the expression in equation (8.86), it is possible to obtain an estimate of the thickness of GaAs layers $d_1 = N_1 b$ (*viz.* the phonon quantum well width).

9.5 THREE-DIMENSIONAL TREATMENT OF PHONON DISPERSION RELATIONS IN PERIODIC NANOCOMPOSITE STRUCTURES

In section 9.3 we discussed the phonon dispersion characteristics of a superlattice using a one-dimensional spring-and-ball analytical model. Phonon dispersion curves for realistic superlattice structures can now be routinely obtained, subject to computing resources, by applying the *ab-initio* DFPT method described in sub-section 3.3.2.1. Here we present and discuss results for a PSL, an *embedded* NWSL and an *embedded* NDSL, each of ultrathin periodicity, using the Quantum Espresso package detailed by Giannozzi *et al* (2009). The general characteristics of *folding, confinement* and *gap openings* along an interface normal direction described in the sub-section 9.3 hold true for each of these constructs. Most notable is the possibility of opening of a gap (or gaps) both along and normal to the growth direction. Such a gap is usually referred to as the *true phononic gap* and the system is called a three-dimensional *phononic* system (or a phononic crystal).

Before discussing phonon dispersion curves for such periodic systems, it would be helpful to keep in mind that the phonon frequency spans continuously in the range 0–300 cm^{-1} for bulk Ge and and 0–500 cm^{-1} for Si (see, figures 2.12 and 2.15).

Si/Ge[001](4,4) PSL

Figure 9.9 shows the atomic structure and phonon dispersion curves for the Si/Ge[001](4,4) PSL, with a unit cell containing four bi-layers of Si and four bi-layers of Ge along the [001] direction. A few characteristic features can be noted. The maximum frequency in this superlattice is larger than the average of the maximum frequencies in bulk Si and bulk Ge. This is primarily due to the

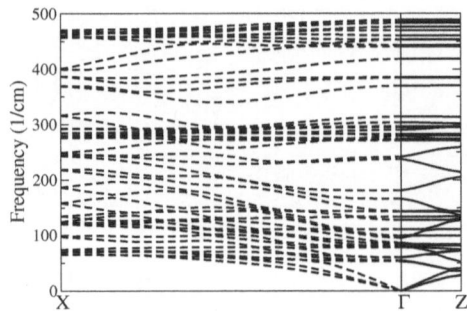

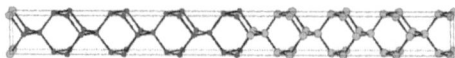

Figure 9.9 (left) Unit cell for Si/Ge[001](4,4) planar superlattice. (right) Phonon dispersion curves along the growth direction (ΓZ) and an in-plane direction (ΓX), calculated by the author using the Quantum Espresso package (Giannozzi *et al* 2009) and the parameters detailed in Srivastava and Thomas (2019).

superlattice lattice constant being close to the average of bulk Si and Ge lattice constants, which results in an increase in Si-Si bond lengths and a decrease in Ge-Ge bond lengths. The bulk acoustic branches get folded according to the superlattice periodicity. Lower lying bulk optical branches also get folded accordingly. However, higher lying bulk acoustic as well bulk optical branches tend to show confinement (*i.e.* flatness in dispersion). It is possible to 'extract' Si and Ge signatures in any of the branches. Modes above 300 cm^{-1} are Si-like. All these features are essentially similar to those discussed in the sub-section 9.3. In addition to zone-folding and confinement features, we notice splitting in the dispersion curves at the zone-edge point X, especially at 44, 300, 340, 400 and 460 cm^{-1}. These arise due to a combination of factors, including mass difference between Si and Ge, formation of Si-Ge interface bonds and relaxation of bond lengths close to the interface. Note a clear gap just above 300 cm^{-1} both along and normal to the growth direction.

Phononic gaps in Si/SiGe semiconductor superlattices have been measured using ultrafast pump and probe experiments (see, e.g. Ezzahri *et al* 2007) and discussed theoretically (see, e.g. Hepplestone and Srivastava, 2008). These will be discussed further in section 10.4.

Si/Ge NWSL

Figure 9.10 shows the atomic structure and phonon dispersion curves for a Si nanowire of $a/2 \times a/2$ square base embedded in a Ge unit cell of dimensions $a \times a \times a$. The equilibrium value of a is close to

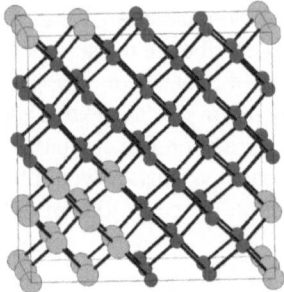

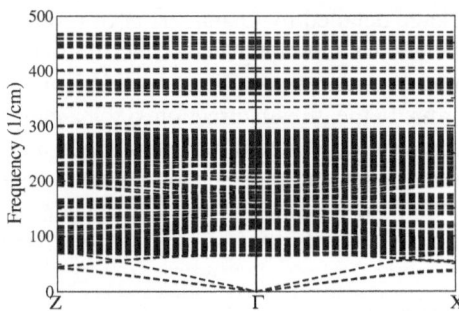

Figure 9.10 (left) Unit cell for an ultrathin Si nanowire embedded in an ultrathin Ge matrix. (right) Phonon dispersion curves along the nanowire growth direction (ΓZ) and along a direction across the nanowire (ΓX), calculated by the author using the Quantum Espresso package (Giannozzi *et al* 2009) and the parameters detailed in Srivastava and Thomas (2019).

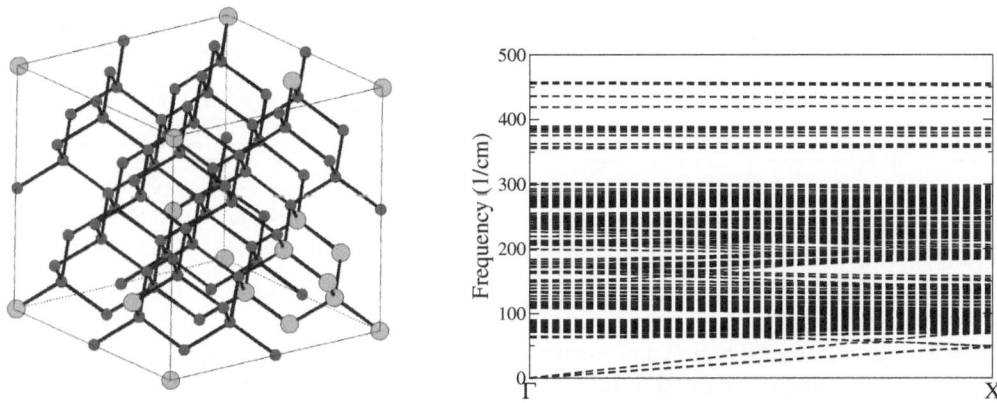

Figure 9.11 (left) Unit cell for an ultrathin Si nanodot embedded in an ultrathin Ge matrix. (right) Phonon dispersion curves along the nandot growth direction (ΓZ), calculated by the author using the Quantum Espresso package (Giannozzi *et al* 2009) and the parameters detailed in Srivastava and Thomas (2019).

$(a(Si) + 3a(Ge))/4$, as expected from Vegard's law. The maximum vibrational frequency is higher than the average of that in Si and Ge, but lower than that for the Si/Ge[001](4,4) PSL. This can be understood using the explanation given for the Si/Ge[001](4,4) PSL. The phonon dispersion curves along Γ–Z and Γ–X are, respectively, in the directions along and normal to the wire growth. There are clear signatures of zone folding and confinement along the growth direction. All the confined modes above 300 cm^{-1} are Si-like. The zone-edge degeneracies seen at the zone edge Z split at the zone edge X, due to vibrational waves crossing both Si and Ge regions. There are several phononic gaps, e.g. just above 300 cm^{-1}, around 350 cm^{-1}, just below and above 400 cm^{-1} and just below 450 cm^{-1}.

Si/Ge NDSL

Figure 9.11 shows the atomic structure and phonon dispersion curves for a cubic Si nanodot of dimensions $a/2 \times a/2 \times a/2$ embedded in a Ge unit cell of dimensions $a \times a \times a$. The equilibrium value of a is close to $(a(Si) + 7a(Ge))/8$, as expected from Vegard's law. A large number of phonon branches show confinement for this system. All the confined modes above 300 cm^{-1} are Si-like. Similar to the Si nanowire case, we can notice five phononic gaps. These are found above 300 cm$^{]1}$, around 375 cm^{-1}, around 400 cm^{-1}, around 425 cm^{-1} and around 450 cm^{-1}. However, as expected, compared to the nanowire case there is significantly less spreading of Si-like confined branches.

9.6 THERMAL CONDUCTIVITY OF PERIODIC NANOCOMPOSITE STRUCTURES

9.6.1 INTRODUCTION

Lattice thermal conductivity of bulk solids spans over four orders of magnitude, covering the range 10^{-1}–10^3 W m^{-1} K^{-1} at room temperature. Nanostructuring of solids can further increase this range (Kim *et al* 2007). Identification of key physical parameters of nanostructures, particularly nanocomposites, for tuning phonon transport is an important topic of both fundamental and practical importance. In this section we attempt to identify tuneable thermal transport characteristics of nanocomposites made of Si and Ge. We do this based on the application of a combination of the full-scale semi-*ab initio* approach (descibed in sub-sections 5.2 and 6.10.3) and a generalised and extended effective medium theory. The reason for this choice is made by realising that a direct application of full-scale *ab initio* or semi-*ab initio* methods is computationally limited to periodic systems containing a reasonably small number (typically less than 200) of atoms per unit cell.

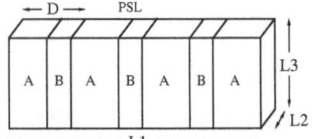 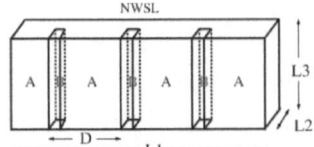 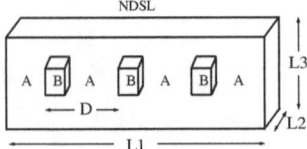

Figure 9.12 An schematic illustration of unit cell size D and sample dimensions for periodic nanocomposites in the form of planar superlattice (PSL), embedded nanowire (NWSL) and embedded nanodot (NDSL) made of two materials A and B.

Before we proceed, we mention that the application of the Boltzmann transport equation within a relaxation time scheme assumes the validity of the Landau-Peierls-Ziman condition (Ziman 1960): sample size (L) must be larger than the phonon mean free path (Λ), which in turn must be larger than the phonon wavelength (λ), i.e., $L > \Lambda > \lambda$. The energy at temperature T of the dominant phonon mode is $\hbar\omega_{\mathrm{dom}} = 1.6k_BT$ (Ziman 1960). Using the linear dispersion relation $\omega_{\mathrm{dom}} = 2\pi\bar{c}/\lambda$, the wavelength at temperature T of the dominant phonon mode of speed $\bar{c}$ can be estimated from $\lambda = h\bar{c}/1.6k_BT$. The choice $\bar{c} = 5000$ ms^{-1} gives $\lambda = 0.5$ nm. Typical values of λ and Λ at room temperature are: $\lambda = 1 - 10$ nm and $\Lambda = 10 - 100$ nm (Kim *et al* 2007).

Figure 9.12 provides a simple illustration of the unit cell size D and sample length L (taken as an effective value from the sample dimensions $L1$, $L2$ and $L3$) for PSL, NWSL and NDSL nanocomposites made of two materials A and B. Note that shapes of nanowire and nanodot are not restricted to this illustration. For discussing lattice thermal conductivity calculations of these, we consider three situations: $D << \Lambda$, $D \sim \Lambda$ and $D >> \Lambda$. We will refer to systems corresponding to these situations as ultrathin nanocomposites, nanocomposites and microcomposites, respectively.

9.6.2 DIRECT CALCULATION FOR ULTRATHIN NANOCOMPOSITES

As mentioned earlier, we consider ultrathin nanocomposites as periodic systems with unit cell size much shorter than average phonon mean free path ($D << \Lambda$). Such systems should be treated as new materials and their phonon properties such as dispersion relations, specific heat and thermal conductivity should be calculated rigourously employing a well established theoretical technique. Note that the volume of the Brillouin zone for a composite material will be smaller than the Brillouin zone for its constituent bulk materials. This will give rise to *mini Umklapp* phonon-phonon scattering processes (Ren and Dow 1982), consistent with the formation of smaller reciprocal lattice vectors. The state-of-the-art techniques discussed in sections 5.2, 6.2 and 6.10.1 – 6.10.4 can be straightforwardly applied to such systems, with one extra consideration. In addition to the intrinsic boundary, isotopic mass defect and anharmonic scatterings it would be important to include 'interface scattering' of phonons. This new type of scattering arises from the unavoidable possibility of mass smudging across boundaries between different chemical compounds making the nanocomposite structure.

Considering mass swap across interfaces as a small perturbation, we can employ an appropriately modified version of equation (6.34) to calculate the phonon interface scattering rate. Accordingly, the phonon scattering rate due to interface mass smudging (IMS) can be expressed as

$$\tau_{qs}^{-1}(\mathrm{IMS}) = \frac{\pi}{2N_0}\omega^2(qs)\sum_{q's'}\delta\big(\omega(qs) - \omega(q's')\big)$$
$$\times \sum_{b}^{\mathrm{IF}}\Gamma_{\mathrm{IMS}}(b)\big|e^*(b;qs)\cdot e(b;q's')\big|^2. \qquad (9.50)$$

Here $\Gamma_{\mathrm{IMS}}(b)$ is the mass-smudging disorder coefficient for the bth atom in the interface region. It is worth reemphasizing that it is difficult to control and quantify such type of mass disorder in real

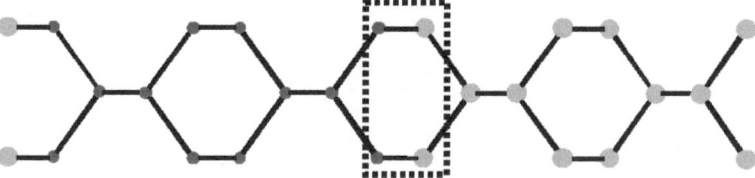

Figure 9.13 Interface region in the Si/Ge[001](4,4) planar superlattice structure.

nanocomposites fabricated in laboratory (Huxtable *et al* 2002a, 2002b). Theoretical calculations have employed different mass-mixing schemes across interfaces (Landry and McGaughey 2009, Hepplestone and Srivastava 2010, Thomas and Srivastava 2013, Garg and Chen 2013, Srivastava and Thomas 2019).

Figure 9.13 illustrates the interface region for the Si/Ge[001](4,4) PSL structure. Considering a simple Gaussian form (Srivastava and Thomas 2019) for mass swap across an interface, we can express

$$\Gamma_{\text{IMS}}(\boldsymbol{b}) = \sum_{b'j} \exp(-j^2\alpha)\left[1 - m_j(\boldsymbol{b}')/\bar{m}(\boldsymbol{b})\right]^2, \tag{9.51}$$

where $\bar{m}(\boldsymbol{b})$ is the average mass of the bth atom, $m_j(\boldsymbol{b}')$ is the mass of the b'th in the jth interface layer swapping with the bth atom, and α is a parameter determining the amount $j^2\alpha$ of atomic mass swapping in the jth interface layer. The choice $\alpha = 3.5$ amounts to mass swapping of approximately 3% in the first interface layer ($j = 1$), *viz* 3% of the first Ge atomic layer is occupied by Si atoms and 3% of the first Si atomic layer is occupied by Ge atoms.

Minimum conductivity in PSL

Garg and Chen (2013) carried out a fully *ab inito* investigation of the lattice thermal conductivity of Si/Ge[001] superlattices. They employed the single-mode relaxation time (*smrt*) conductivity expression (equation 5.26) and included the isotopic mass defect scattering rate (equation (6.34)) and three-phonon scattering rate (equation 6.214). For including the interface scattering rate, they evaluated $\Gamma_{\text{IMS}}(\boldsymbol{b})$ in a manner which is different from that described in equation (9.51). The interface region was considered to be four atomic layers (two bi-layers), and the masses of the atoms in that region were taken as the average of Si and Ge masses. Figure 9.14 shows the computed in-plane and cross-plane conductivity at room temperature as a function of superlattice period (D).

The result for the shortest period ($D = 11$ Å) is essentially that for the isotropic $Si_{0.5}Ge_{0.5}$ alloy. The conductivity along the in-plane direction increases monotonically with period. However, the cross-plane conductivity shows a minimum at a period of $D = 33$ Å. This minimum results from the cross-over around 1 THz between the intrinsic three-phonon scattering processes and the interface scattering. In short period superlattices most of the heat is conducted by low-frequency (below 1 THz) phonons due to the strong interface scattering of high-frequency phonons. These low-frequency phonons conduct heat coherently due to their long mean free paths. In thicker period superlattices higher frequency phonons begin to play a role, pointing to a transition from coherent to incoherent heat transport. Simkin and Mahan (2000) explained the minimum thermal conductivity of superlattices using a different language. According to their intuition, cross-plane thermal conductivity in PSLs should be explained as wave transport when $D < 2\Lambda$ and as particle transport when $D > 2\Lambda$.

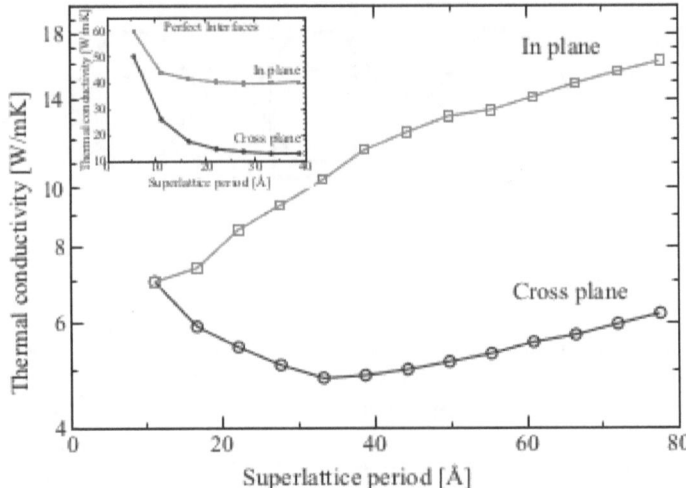

Figure 9.14 *Ab initio* prediction of the room-temperature lattice thermal conductivity of Si/Ge[001] super-lattices, with mass disorder in interface regions, as a function of superlattice periodicity (*D*). The inset shows results with perfect interfaces. Reproduced from Garg and Chen (2013).

Sample size dependence of conductivity in bulk, PSL, NWSL and NDSL

Sample size dependence of lattice thermal conductivity has been discussed in many experimental (Chang *et al* 2008, Xu *et al* 2014) and theoretical (Herring 1954, Mingo and Broido 2005, Nika *et al* 2012, Vermeersch *et al* 2016) studies. Using group theoretical arguments, Herring (1954) predicted that different size dependence is expected in different temperature ranges and for different sample dimensionalities. Some of his predictions have been verified from detailed numerical calculations (Nika *et al* 2012, Vermeersch *et al* 2016, Thomas and Srivastava 2017).

It was shown in figure 7.8 that, within Callaway's relaxation time scheme, the application of the semi-*ab initio* method (Thomas and Srivastava 2017) developed in section 7.4.1.4 reproduces the experimental measurements of the lattice thermal conductivity over a broad temperature range for bulk Si and Ge with sample lengths of mm size. Employing the same theory and numerical technique, Srivastava and Thomas (2019) examined the sample size dependence of the room-temperature con-ductivity results for the bulk and ultrathin Si/Ge nanocomposites (PSL, NWSL and NDSL). Presented in figure 9.15 are the results for the Si/Ge[001](4,4) PSL, a Si NW of $a/2 \times a/2$ square base embedded in a Ge unit cell of dimensions $a \times a \times a$, a Si ND of size $a/2 \times a/2 \times a/2$ embedded in a Ge unit cell of size $a \times a \times a$ and a Ge ND of size $a/2 \times a/2 \times a/2$ embedded in a Si unit cell of size $a \times a \times a$. It is clear that the conductivity of any of these nanocomposite (lower dimensional) systems shows weaker sample size dependence than the bulk (three-dimensional) materials. The in-plane nanocomposite conductivity shows a weaker sample size dependence than either of the two bulk materials does, and the cross-plane nanocomposite conductivity shows a weaker sample size dependence than the in-plane conductivity. Both in-plane and cross-plane conductivities show a quasi-ballistic variation of the form $\ln(L_{BS})$, where the sample size is expressed as an effective boundary length L_{BS}.

It is also interesting to note that for sample size of 1 micron the conductivity results show the trend: κ(bulk Si) $> \kappa$(bulk Ge) $> \kappa$(in-plane PSL) $> \kappa$(in-plane Si NWSL) $> \kappa$(cross-plane Si NWSL) $> \kappa$(Ge NDSL) $> \kappa$(cross-plane PSL) $> \kappa$(Si NDSL). The trend for sample size of 100 nm is: κ(bulk Si) $> \kappa$(bulk Ge) $> \kappa$(in-plane PSL) $> \kappa$(in-plane Si NWSL) $> \kappa$(Ge NDSL) $> \kappa$(cross-plane Si NWSL) $> \kappa$(cross-plane PSL) $> \kappa$(Si NDSL). For sample size smaller than 100 nm the results for κ(cross-plane PSL), κ(cross-plane Si NWSL), κ(Ge NDSL) and κ(Si NDSL) are very close to each other.

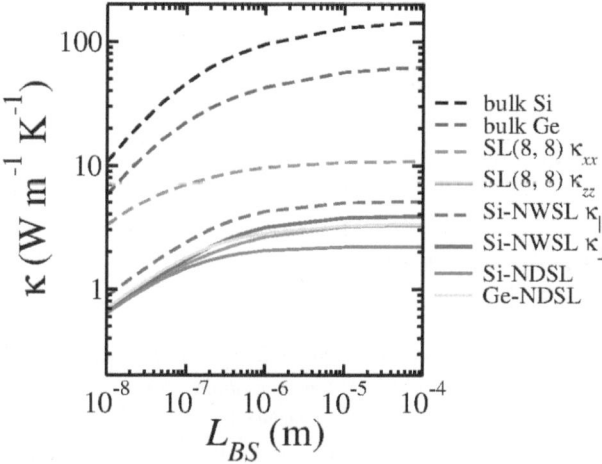

Figure 9.15 Sample size dependence of room-temperature thermal conductivity of ultrathin Si/Ge nanocomposites. For PSLs, κ_{xx} and κ_{zz} are, respectively, the in-plane and cross-plane conductivities. For NWSL $\kappa_{\parallel}$ and $\kappa_{\perp}$ are, respectively, the in-plane and cross-plane conductivities. See text for meanings of SL(8,8), Si–NWSL, Si–NDSL and Ge–NDSL. Sample size is presented as the boundary scattering length L_{BS}. For $L_{BS} = 1\mu$m, from top to bottom are the results for bulk Si, bulk Ge, SL(8,8) $\kappa_{\parallel}$, Si–NWSL $\kappa_{\parallel}$, Si–NWSL $\kappa_{\perp}$, Ge–NDSL, SL(8,8) $\kappa_{\perp}$ and Si–NDSL, respectively. [Original diagram in colour.] Reproduced from Srivastava and Thomas (2019).

Theory-Experiment Comparison

The thermal conductivity of nanocomposite structures is usually measured by using the 3ω method (Cahill *et al* 1994). Measurements for Si- and Ge-based nanocomposites have been reported by many groups, most notably by Lee *et al* (1997), Borca-Tasciuc *et al* (2000), Huxtable *et al* (2002a, 2002b), and Lee and Venkatasubramanian (2008). A direct and detailed comparison between theoretical predictions and experimental measurements is generally not possible. This is because first-principles theoretical calculations normally assume samples to be homogeneous with perfectly periodic structure at atomic level, except for some degree of mass smudging at interfaces, In contrast, even the best fabrication techniques produce samples which are inhomogeneous and contain uncontrollable amount of point and extended defects of unknown nature. Notwithstanding these realities, it is instructive to compare theoretical results with experimental measurements for a chosen system.

Figure 9.16(*a*) shows the results of calculation by Srivastava and Thomas (2020) for the Si/Ge PSL with Si and Ge layers each of thickness 2.2 nm and sample size $L \equiv L_{BS} = 400$ nm, n-doped with 10^{26} m^{-3} and using the IMS factor $\alpha = 2.3$ (which amounts to 10% intermixing of Si and Ge in the first interface layer). The electron-phonon scattering rate was included by using equation (6.134). The temperature variation of the conductivity is similar to what is obtained for crystalline bulk Si and Ge, except for a few expected differences. The maximum in the κ-T curve has shifted to a much higher temperature (around 100 K). The conductivity of the PSL is highly reduced compared to the bulk result for either of the constituent materials. These features arise from a combined effect of small sample size, superlattice-formation related characteristics of the phonon dispersion curves and the IMS scattering.

Figure 9.16(*b*) shows experimentally measured results for several p-type and n-type doped Si/Ge PSLs of similar periodicity and sample size grown by MOCVD (Borca-Tascius *et al* 2000). These results are hugely different from the theoretical predictions in figure 9.16(a). The experimental results reveal a plateau-like feature in the range 50 – 250 K, indicative of gross inhomogeneity or amorphousness. There is some indication of a peak in the κ-T curve above 250 K. These doped

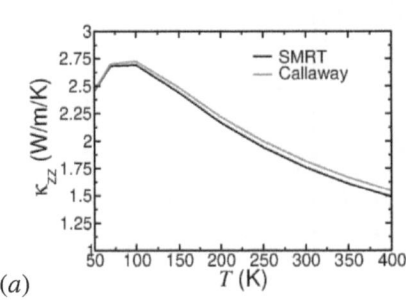
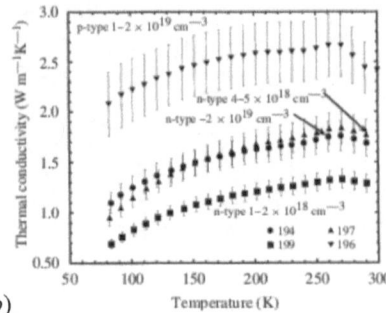

Figure 9.16 Cross-plane thermal conductivity of the Si/Ge[001] PSL of sample length $L = 400$ nm. (a) Theoretical results for Si(2.2 nm)/Ge(2.2 nm) PSL with n-type doping concentration $n = 10^{26}$ m^{-3} and consideration of 10% mass-smudging in the first interface layer (IMS factor $\alpha = 2.3$). The lower and upper curves represent results from the *smrt* and Callaway formalisms, respectively. Reproduced from Srivastava and Thomas (2020). (b) Experimentally measured results for several p-type and n-type doped Si(2 nm)/Ge(2 nm) PSL samples. Reproduced from Borca-Tasciuc *et al* (2000).

PSLs have smaller thermal conductivity values than the undoped sample, indicating the influence of high dislocation densities and doping on the thermal conductivity reduction. Accurate quantification of fabrication-related and doping-related atomic-level structural information, and further development in theoretical techniques to incorporate such features in first-principles calculations are required to understand the complex thermal conductivity behaviour of laboratory fabricated PSLs.

Experimental measurements by Lee and Venkatasubramanian (2008) show that κ in SiGe/Si NDSLs decreases monotonically with decreasing period. This is in good agreement with the theoretical prediction in figure 9.15. The experimental measurements suggest that at short periods room-temperature κ is as low as 2.0 – 2.7 W m^{-1} K^{-1}. This is consistent with the results in figure 9.15. This experimental work also indicates that low κ can be attained in SiGe/Si NDSLs either with a low superlattice period, a high nanodot areal density, or both.

9.6.3 EFFECTIVE MEDIUM THEORY FOR NANO- AND MICRO-COMPOSITES

As we mentioned earlier in this section, a full-scale first-principles calculation of the thermal conductivity of nanocomposites is limited to periodic systems containing no more than 200 atoms per unit cell. For nanocomposite structures made from diamond and zincblende materials, this would mean restricting to period size D in the region of 10–12 nm. For practical applications, nanostructures with larger periodicities may be required. Methods based upon effective medium approximations (EMAs) are widely used to calculate physical properties, including the thermal conductivity. Based on a general form of the EMA, Nan *et al* (1997) derived an expression for the thermal conductivity of a two-phase composite including thermal boundary resistance (TBR). To obtain physically meaningful results both the TBR and the size effects must be incorporated in the EMA theory. Incorporating size effect for a nanocomposite in the form of inclusion of one material in a homogeneous matrix of another, Minnich and Chen (2007) proposed a modified version of the EMA. This approach produces a closed-form expression for the effective thermal conductivity of a nanocomposite. This is achieved by including the effects of phonon scattering from the boundaries of spherical inserts in the calculation of the thermal conductivities of the insert and matrix. Note that, as mentioned at the beginning of this section, for the application of a relaxation-time theory of thermal conductivity the size of the insert must be greater than the phonon MFP ($L_{insert} > \Lambda$). Generalisations of the Minnich-Chen approach have been made to account for different insert shapes, where the effects of boundary scattering are not isotropic (Ordonez-Miranda 2011, Behrang *et al*

2013, 2014, 2015), and also to systems containing multiple insertions with different orientations (Siddiqui and Arif 2016). An extension of the Minnich-Chen approach that accounts for anisotropic TBR effects has been proposed by Thomas and Srivastava (2018a). Here we present the essential expressions for the thermal conductivity of a two-component composite structure, details of which can be found in the original citations in the references above.

Basic effective medium theory for isotropic systems

The general framework of the effective-medium theory for the thermal conductivity of an isotropic composite medium was developed by Nan (1993) and Nan *et al* (1997). The point-to-point thermal conductivity variation in a composite medium is expressed as

$$\kappa(\boldsymbol{r}) = \kappa^0 + \kappa'(\boldsymbol{r}), \tag{9.52}$$

where κ^0 denotes a constant part of a homogeneous medium and $\kappa'(\boldsymbol{r})$ is an arbitrary fluctuating part. By expressing the perturbed part in terms of the Green's function $\mathscr{G}$ for the homogeneous medium and the transition matrix $\mathscr{T}$ for the entire composite medium, the effective conductivity of the composite structure is obtained from the equation

$$\kappa_{\text{eff}} = \kappa^0 + < \mathbb{T} > \left(\mathbb{I} + < \mathbb{G}\mathbb{T} > \right)^{-1}, \tag{9.53}$$

where $\mathbb{I}$ is the unit tensor and $<>$ denotes spatial averaging. For a composite structure with n particles inserted in a host medium, and neglecting interparticle multiple scattering, $\mathbb{T}$ can be approximated as

$$
\begin{aligned}
\mathbb{T} &= \sum_n \mathbb{T}_n + \sum_{n,m \neq n} \mathbb{T}_n \mathbb{G} \mathbb{T}_m + \dots \\
&= \sum_n \mathbb{T}_n + \sum_n \kappa'_n \left(\mathbb{I} - \mathbb{G}\kappa'_n \right)^{-1}.
\end{aligned} \tag{9.54}
$$

Using this approach, Nan *et al* (1997) obtained explicit expressions for the thermal conductivity of isotropic composite structures with different particle sizes, shapes, volume fractions and topologies. The expressions given below are for a two-component composite system in which one component is *insert* and the other a *matrix* or host. Note that in figure 9.12 we have illustrated material B as insert and material A as matrix or host.

For heat flow parallel to the interface boundary (for PSLs and NWSLs), the effective thermal conductivity is simply the volumetrically weighted average of the two bulk components:

$$\kappa_{\parallel}^E = f\kappa^i + (1-f)\kappa^m, \tag{9.55}$$

where κ^i and κ^m are the conductivities of the insert and matrix (host), respectively, and f is the volume fraction of the insert within the unit cell of the periodic composite (insert+matrix) system.

The effective cross-plane effective thermal conductivity of the composite structure can be expressed as

$$\kappa_{\perp}^E = \eta \kappa^m, \tag{9.56}$$

where η is a structure-related dimensionless parameter which is defined in terms of an effective side d of the inserted particle and the Kapitza's thermal boundary resistance R_{TB} (Kapitza 1941). For PSLs, NWSLs and NDSLs of insert size d (layer thickness of one PSL component, and diameter for cylindrical wires and spherical dots), the explicit expressions for η are

$$\text{PSLs}: \quad \eta = \frac{\kappa^i}{\kappa^i - f[\kappa^i(1-\alpha) - \kappa^m]} \tag{9.57}$$

$$\text{NWSLs}: \quad \eta = \frac{\kappa^i(1+\alpha) + \kappa^m + f[\kappa^i(1-\alpha) - \kappa^m]}{\kappa^i(1+\alpha) + \kappa^m - f[\kappa^i(1-\alpha) - \kappa^m]} \tag{9.58}$$

$$\text{NDSLs}: \quad \eta = \frac{\kappa^i(1+2\alpha) + 2\kappa^m + 2f[\kappa^i(1-\alpha) - \kappa^m]}{\kappa^i(1+2\alpha) + 2\kappa^m - f[\kappa^i(1-\alpha) - \kappa^m]}, \tag{9.59}$$

where α is a dimensionless quantity defined as

$$\alpha = 2\kappa^m R_{TB}/d. \tag{9.60}$$

The IMS effect due to mass smudging can be accounted for *post priori*. Let W_{IMS}^{insert} and W_{IMS}^{matrix} represent the bulk thermal resistivities of the insert and matrix with only point-defects resulting from the matrix atom(s) replacing insert atom(s) and vice-versa, respectively. Then the IMS resistivity W_{IMS} can be taken as $W_{IMS} = W_{IMS}^{insert} + W_{IMS}^{matrix}$. With such consideration, we can obtain the modified effective cross-plane thermal conductivity from

$$1/\kappa_\perp^E (\text{with IMS}) = 1/\kappa_\perp^E + W_{IMS}. \tag{9.61}$$

Generalised effective medium theory for anisotropic systems[2]

The expression in equation (9.59) for the effective thermal conductivity of an isotropic matrix with periodic isotropic spherical insertions is similar to that for the effective polarizability in the electromagnetic literature where it is generally refereed to as the Garnett (or Maxwell Garnett) mixing formula (see, Sihvola 1999). Using that language, and ignoring the role of TBR (*i.e.* by setting $\alpha = 0$), we can rewrite the effective thermal conductivity expression in equations (9.53) and (9.59) as follows:

$$\begin{aligned}
\kappa^E &= \kappa^m + f(\kappa^i - \kappa^m)\mathbb{J}^{-1}\kappa^m \\
\mathbb{J} &= \kappa^m + (1-f)\frac{1}{3}(\kappa^i - \kappa^m),
\end{aligned} \tag{9.62}$$

where κ^i, κ^i and $\mathbb{J}$ are 3×3 matrices. Working with the Cartesian x, y, z coordinates, the cross-plane component $\kappa_\perp^E$ is any of the diagonal components κ_{ii}^E. With TBR included, we can express

$$\kappa^E = \kappa^m + f(\kappa^* - \kappa^m)\mathbb{J}^{-1}\kappa^m \tag{9.63}$$

$$\mathbb{J} = \kappa^m + (1-f)\frac{1}{3}(\kappa^* - \kappa^m) \tag{9.64}$$

$$\kappa^* = \kappa^i \left[\mathbb{I} + \frac{6}{d}\kappa^i R_{TB}\mathbb{I}\right]^{-1}, \tag{9.65}$$

where $\mathbb{I}$ is the 3×3 identity matrix.

The mixing formulae described above should be modified when either the matrix, the insert or both are anisotropic systems (see, *e.g.* Sihvola 1997a, Levy and Cherkev 2013). Sihvola (1997a, 1997b, 1999) has presented mixing formulae for the effective polarizability with isotropic spherical inserts in an anisotropic matrix, anisotropic spherical inserts in an isotropic matrix and anisotropic spherical inserts in an anisotropic matrix. Thomas and Srivastava (2018a) adopted that formalism to obtain a mixing formula for the effective thermal conductivity of a composite system comprised of anisotropic spherical inserts in an anisotropic matrix, including the treatment of the anisotropy of the Kapitza resistance between the insert and the matrix.

If either the matrix or insert (or both) exhibits an anisotropic thermal conductivity, then it is reasonable to assume that the surface region (thought of as a mixture of the matrix and insert regions) will also exhibit anisotropy. This behaviour will persist when we consider the Kapitza resistance of the surface region in the limit where it is thin and poorly conducting. For correctly calculating the effective thermal conductivity of the insert and its surface together, we must perform an affine transformation that restores the isotropy of the surface, resulting in an effective distortion of the insert shape (see, Sihvola 1997a and Appendix G).

---

[2]Here we consider spherical inserts, but the formalism is valid for cylindrical and disc inserts using only two relevant Cartesian coordinates.

The effective thermal conductivity matrix κ^E is obtained from the Maxwell Garnett mixing formula

$$\begin{aligned}
\kappa^E &= \kappa^m + f(\kappa^* - \kappa^m)\mathbb{J}^{-1}\kappa^m \\
\mathbb{J} &= \kappa^m + (1-f)\mathbb{N}_m(\kappa^* - \kappa^m),
\end{aligned} \tag{9.66}$$

where f is the volume fraction of the insert within the matrix and $\mathbb{N}_m$ is the matrix depolarization tensor (see Appendix G and Sihvola 1997a) accounting for any anisotropy in κ^m, and κ^* is the effective thermal conductivity tensor for the system comprising the insert and a thin boundary layer around it.

The effective thermal conductivity tensor for the insert and a thin boundary layer around it, κ^*, is obtained using Sihvola's generalization of the Maxwell Garnett mixing formula. We start by expressing κ^* using the Maxwell Garnett equation as

$$\begin{aligned}
\kappa^* &= \kappa^s + v(\kappa^i - \kappa^s)\mathbb{L}^{-1}\kappa^s \\
\mathbb{L} &= \kappa^s + (1-v)\mathbb{N}_s(\kappa^i - \kappa^s)[\kappa^s]^{-1},
\end{aligned} \tag{9.67}$$

where $v = a^3/(a+\delta)^3 \simeq 1 - 3\delta/a$ is the volume fraction of the spherical insert of radius a and δ is the thickness of the (thin) surface layer around the insert, and $\mathbb{N}_s$ is the depolarization tensor for the insert+surface system. With slight manipulation, the expression for κ^* can be re-written as

$$\kappa^* = \kappa^s + v(\kappa^i - \kappa^s)\mathbb{L}^{-1} \tag{9.68}$$

which is of the form presented in Nan *et al* (1997). With the assumed consideration $\delta \to 0$, we take $\kappa^s = 0$ but express $(\kappa^s)^{-1}\delta = \mathbb{R}^K$ as the Kapitza (surface, or interface, or thermal boundary) resistance, and express

$$\kappa^* = \kappa^i\left[\mathbb{I} + \frac{3}{a}\mathbb{N}_s\kappa^i\mathbb{R}^K\right]^{-1}. \tag{9.69}$$

The anisotropy in κ^* arises from the introduction of the intended direction of spherical distortion in constructing $\mathbb{N}_s$, and hence $\mathbb{R}^K$.

In Appendix G, we provide more discussion on the assumptions and procedure involved in obtaining $\mathbb{N}_s$, $\mathbb{N}_m$, κ^* and $\mathbb{R}^K$ for anisotropic spherical insertions in an anisotropic matrix.

For isotropic spherical inserts in an isotropic matrix, $\mathbb{N}_s = \mathbb{N}_m = \frac{1}{3}\mathbb{I}$, and $\mathbb{R}^K_{xx} = \mathbb{R}^K_{yy} = \mathbb{R}^K_{zz} = R_{\text{TB}}$, and the results in equations (9.63)–(9.65) can be recovered. The resulting expression for κ^E would then be identical to that proposed by Minnich and Chen (2007).

Thermal Boundary Resistance[3]

Accurate calculation of the thermal boundary resistance R_{TB} is a challenging problem and has not yet been satisfactorily formulated using first-principles approaches. Based on existing continuum-level approaches, we express R_{TBR} as the sum of weighted diffuse R^D_{TB} and specular R^S_{TB} contributions (Swartz and Pohl 1989, Chen 1998, Behrang *et al* 2013, Thomas and Srivastava 2018b, 2020). Following Thomas and Srivastava (2020), and write

$$\begin{aligned}
R_{\text{TB}} &= R^D_{\text{TB}} + R^S_{\text{TB}} \tag{9.70} \\
&= R^D_{\text{im}} + R^D_{\text{mi}} + R^S_{\text{im}} + R^S_{\text{mi}}, \tag{9.71}
\end{aligned}$$

where R^D_{ij} is a diffuse contribution for transport from material i to material j, and R^S_{ij} is the corresponding specular contribution.

---

[3] Also see Chapter 14

The diffuse contribution R_{ij}^{D} can be approximated as

$$R_{ij}^{D} \approx \frac{2(\langle C_i v_i \rangle_W + \langle C_j v_j \rangle_W)}{\langle C_i v_i \rangle \langle C_j v_j \rangle}. \tag{9.72}$$

Here

$$\langle C_j v_j \rangle = \sum_{qs} C_{j,qs} v_{j,qs} \tag{9.73}$$

$$\langle C_j v_j \rangle_W = \sum_{qs} C_{j,qs} v_{j,qs} \zeta_q, \tag{9.74}$$

where $C_{j,qs}$ is the specific heat for the phonon mode with momentum q and polarisation s and $v_{j,qs}$ is the mode speed (which in the x or y in-plane direction is $\sqrt{v_x^2 + v_y^2}$ and in the z cross-plane direction is $|v_z|$), and ζ_q is the fraction of diffusively scattered modes (see later for a description).

The specular contribution is expressed as (Thomas and Srivastava 2020)

$$R_{ij}^{S} \approx \frac{2}{\langle C_i v_i \rangle I_{ij}^{W}} (1 - I_{ij} - I_{ji}), \tag{9.75}$$

where

$$I_{ij} = \int_{\mu_{c,i}}^{1} \mathcal{T}_{ij}(\mu_i) \mu_i d\mu_i, \tag{9.76}$$

$$I_{ij}^{W} = \int_{\mu_{c,i}}^{1} \mathcal{T}_{ij}^{W}(\mu_i) \mu_i d\mu_i \tag{9.77}$$

with μ_i, $\mu_{c,i}$, $\mathcal{T}_{ij}$ and $\mathcal{T}_{ij}^{W}$ explained below.

Define impedances

$$\langle Z_j \rangle = \rho_j \sum_{qs} v_{j,qs} \bar{n}_{qs} / \sum_{qs} \bar{n}_{qs} \tag{9.78}$$

$$\langle Z_j \rangle_W = \rho_j \sum_{qs} v_{j,qs} |_q \bar{n}_{qs} / \sum_{qs} \bar{n}_{qs}, \tag{9.79}$$

where $\bar{n}_{qs}$ is the Bose-Einstein distribution function and the material density is ρ_j. Define

$$\mathcal{B} = \frac{4 \langle Z_i \rangle \langle Z_j \rangle}{(\langle Z_i \rangle \mu_i + \langle Z_j \rangle \mu_j)^2} \tag{9.80}$$

$$\mathcal{B}^{W} = \frac{4 \langle Z_i \rangle \langle Z_j \rangle}{(\langle Z_i \rangle_W \mu_i + \langle Z_j j \rangle_W \mu_j)^2} \tag{9.81}$$

$$\langle C_j v_j^3 \rangle = \sum_{qs} C_{j,qs} v_{j,qs}^3, \tag{9.82}$$

where $\mu_j = \cos\theta_j$, with θ_j representing the angle of incidence of the phonon wavevector to the interface. Transmission coefficients for different relative maximum frequencies in materials i, j are then given by the expressions in Table 9.1 (Chen 1998).

If only elastic scattering is considered, Snell's law relates the angles in materials i and j:

$$\frac{\sin\theta_i}{v_i} = \frac{\sin\theta_j}{v_j}. \tag{9.83}$$

Table 9.1

Transmission coefficients from material i to material j under different maximum frequency conditions. $\omega_{\text{max},i}$ and $\omega_{\text{max},j}$ are the highest frequency mode in materials i and j, respectively. From Thomas and Srivastava (2020).

	$\omega_{\text{max},i} = \omega_{\text{max},j}$	$\omega_{\text{max},i} < \omega_{\text{max},j}$	$\omega_{\text{max},i} > \omega_{\text{max},j}$
$\mathscr{T}_{ij}(\mu_i)$	$\mathscr{B}$	$\mathscr{B}$	$\frac{\langle C_j v_j^3\rangle}{\langle C_i v_i^3\rangle}\mathscr{T}_{ji}(\mu_j)$
$\mathscr{T}_{ij}^W(\mu_i)$	$\mathscr{B}^W$	$\mathscr{B}^W$	$\frac{\langle C_j v_j^3\rangle}{\langle C_i v_i^3\rangle}\mathscr{T}_{ji}^W(\mu_j)$
$\mathscr{T}_{ji}(\mu_j)$	$\mathscr{B}$	$\frac{\langle C_i v_i^3\rangle}{\langle C_j v_j^3\rangle}\mathscr{T}_{ij}(\mu_i)$	$\mathscr{B}$
$\mathscr{T}_{ji}^W(\mu_j)$	$\mathscr{B}^W$	$\frac{\langle C_i v_i^3\rangle}{\langle C_j v_j^3\rangle}\mathscr{T}_{ij}^W(\mu_i)$	$\mathscr{B}^W$

Effects of inelastic scattering can be included by using Chen's approximation (Chen 1998)

$$\frac{\sin\theta_i}{\sqrt{\langle C_j v_j\rangle}} = \frac{\sin\theta_j}{\sqrt{\langle C_i v_i\rangle}}. \tag{9.84}$$

Let us denote $h = v_i/v_j$ (elastic scattering) or $h = \sqrt{\langle C_j v_j\rangle/\langle C_i v_i\rangle}$ (inelastic scattering). When $h \leq 1$, the critical angle $\theta_{c,i}$ is given by $\arcsin(h)$. When $h > 1$, $\theta_{c,i} = \pi/2$.

The effects of specularity can be included by following the scheme presented by Koh *et al* (2009). Let us define a momentum dependent parameter ζ_q as the fraction of modes of momentum q that undergo diffusive scattering from the insert interfaces. This parameter can be expressed as

$$\zeta_q = 1 - \exp(-\varepsilon^2/\lambda_q^2), \tag{9.85}$$

where ε is the the average height of surface inhomogeneities and λ_q is the wavelength corresponding to momentum q. For numerical calculations, we can express $\varepsilon = \eta a_0$ where a_0 is the lattice spacing, and $\lambda_q = 2\pi/Q$, where $Q = 2\pi|q|/a_0$ and $|q| = \sqrt{q_x^2 + q_y^2 + q_z^2}$ is the norm of the lattice momentum vector in Cartesian co-ordinates. From these considerations

$$\zeta_q = (1 - e^{-\eta^2|q|^2}), \tag{9.86}$$

where value of η controls the extent of specular scattering. The limit $\eta \to 0$ refers to the specular limit $\zeta_q = 0$ for a smooth surface. The limit $\eta \to \infty$ refers to the diffuse limit $\zeta_q = 1$, corresponding to an infinitely rough surface.

Using equation (9.66) and the semi-*ab inito* theoretical scheme described in section 7.4.1.4, Thomas and Srivastava (2020) computed the anisotropic thermal boundary resistance between 2H WS$_2$ and 2H MoS$_2$, with a spherical volume fraction $f = 0.125$ of the former inserted in the latter host. The 2H structure of transition metal dichalcogenides (TMDs) is a three-dimensional version of the monolayer TMD structure shown in figure 8.15, where layers are repeated along the z-axis with a lattice constant much larger than the in-plane lattice constant. The in-plane and cross-plane components of the TBR, $R_{TB,x} = R_{TB,y}$ and $R_{TB,z}$ respectively, computed for different values of the specularity factor η discussed in equation (9.86) are shown in figure 9.17.

The modelling predicts a rapid decrease in both in-plane and cross-plane TBR as the temperature rises before saturating above 400 K. The choice $\eta = 5.0$ produces results of the diffuse limit and the specular limit is attained for the choice $\eta \leq 0.1$. The change in R_{TB} as η decreases is not monotonic. This is because in the modelling the diffuse contribution varies linearly with the factor

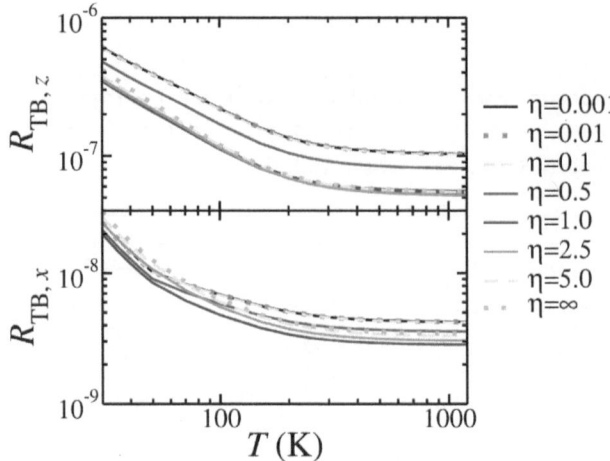

Figure 9.17 Thermal boundary resistance (in units of $m^2 KW^{-1}$), cross-plane $R_{TB,z}$ and in-plane $R_{TB,x}$, for a bulk 2H WS$_2$ spherical insert ($f = 0.125$) in bulk 2H MoS$_2$ matrix. Results for different values of the specularity factor η are shown. [Original diagram in colour.] Reproduced from Thomas and Srivastava (2020).

ζ_q but the specular contribution varies as $(1 - \zeta_q)^2$. For any combination of the diffuse and specularity contributions, it is found that $R_{TB,z} > R_{TB,x}$. This is consistent with the fact that the thermal conductivity across atomic planes in van der Waals systems, such as TMDs, is much smaller than the in-plane conductivity. At high temperatures, the DMM limit is much lower than the AMM limit for $R_{TB,z}$ than for $R_{TB,x}$. It is interesting to note that as temperature decreases (typically below 80 K), the AMM limit of $R_{TB,x}$ becomes lower than the DMM limit.

Effective boundary lengths

Let us discuss PSLs and NDSLs with spherical inserts. It is safe to consider the sample boundary itself as purely diffusive. Boundary scatterings for the insert and the matrix require consideration of specularity. Considering the insert, any phonon mode not scattered from the interface boundary due to specularity must be scattered from the sample boundary, and so the effective boundary length must be a weighted average of the sample length and the insert diameter. Considering the matrix, only the interface density contribution to the effective boundary length is affected by the specularity. The effective boundary length L_B may therefore be expressed as follows:

$$\text{PSLs}: \quad L_{B,j}^{-1} = \zeta_q L_j^{-1} + (1 - \zeta_q) L_{B,S}^{-1} \text{ for } j = \text{insert, matrix} \tag{9.87}$$

$$\text{NDSLs}: \quad L_B^{-1} = \begin{cases} \zeta_q L_{B,I}^{-1} + (1 - \zeta_q) L_{B,S}^{-1} \text{ for insert} \\ L_{B,S}^{-1} + \zeta_q L_{IS}^{-1} \text{ for matrix} \end{cases}, \tag{9.88}$$

where L_j^{-1} is the length of the jth segment in PSL, $L_{B,I} = d$ is the insert boundary length in NDSL, $L_{B,S}$ is the sample size (*i.e.* $L_{B,S} \equiv L_B$) and $L_{IS} = 4/\Sigma$ is the scattering length due to interface density $\Sigma = 6f/L_{B,I}$.

We note that the Minnich-Chen formulation of the EMA theory (Minnich and Chen 2007) requires use of the effective length $L_{B,I}$ of the insert and the effective length $L_{B,M}$ of the matrix. The Minnich-Chen thermal conductivity expressions reduce to Nan's expressions (Nan *et al* 1997) when $L_{B,I} = L_{B,M} = L_{B,S}$.

9.6.4 RESULTS FOR COMPOSITES WITH UNIT CELL SIZE IN THE RANGE LOW-nm TO MICRONS

In the previous two subsections, we presented the theoretical ingredients for thermal conductivity calculation of a composite system. In this respect it is helpful to consider three regions based on the size of the unit cell D in comparison with phonon mean free path Λ: Region R1 ($D << \Lambda$), region R2 ($D \sim \Lambda$) and region R3 ($D >> \Lambda$). Calculations for systems in region R1 can be made from a direct application of the phonon Boltzmann equation, as discussed in sections 7.4.1.4 and 7.4.1.5. For systems in regions R2 and R3, calculations are performed by applying the EMA theory, with input from bulk results.

Here we present and analyse results for the Si/Ge PSL and NDSL composites. For these isotropic systems, the cross-plane conductivity expressions are, from equations (9.56) to (9.60)

$$\text{Region R1 } (D << \Lambda): \quad \kappa_\perp = \quad \kappa_\perp (\text{composite system}), \tag{9.89}$$

$$\text{Region R2 } (D \sim \Lambda): \quad \kappa_\perp = \begin{cases} \kappa^m \frac{\kappa^i}{\kappa^i - f[\kappa^i(1-\alpha)-\kappa^m]}, & \text{PSL} \\ \kappa^m \frac{\kappa^i(1+2\alpha)+2\kappa^m+2f[\kappa^i(1-\alpha)-\kappa^m]}{\kappa^i(1+2\alpha)+2\kappa^m-f[\kappa^i(1-\alpha)-\kappa^m]}, & \text{NDSL} \end{cases} \tag{9.90}$$

$$\text{Region R3 } (D >> \Lambda): \quad \frac{1}{\kappa_\perp} = \begin{cases} \frac{f}{\kappa^i} + \frac{1-f}{\kappa^m}, & \text{PSL} \\ \frac{2+f}{h}\frac{1}{\kappa^i} + \frac{1-f}{h}\frac{1}{\kappa^m} & \text{NDSL} \end{cases}, \tag{9.91}$$

where $h = (1+2f)+2(1-f)\kappa^m/\kappa^i$. The expressions in equation (9.91) result from the expressions in equation (9.90) with the recognition that in region R3 the parameter $\alpha = \kappa^m R_{TB}/L_{BI}$ becomes negligibly small. Equation (9.91) helps explain that, for a given insertion factor f, a larger difference between the matrix and insert conductivities leads to a larger change in the conductivity of the composite compared to that of the matrix. This can be easily verified in the case of PSL, where a larger value of $(\kappa^m/\kappa^i - 1)$ results in a smaller value of $\kappa_\perp/\kappa^m$.

Figure 9.18 shows the cross-plane thermal conductivity results for Si/Ge composites (PSL and NDSL) of sample size 1 mm and unit cell size in the range low-nm to microns. The results have been obtained by considering a homogeneous interface between the constituent layers. We have set the boundary between the rgions R1 and R2 at the unit cell size $D = 10$ nm, the typical minimum for phonon mean free path (Kim *et al* 2007). And the boundary between regions R2 and R3 is arbitrarily set at 50 μm. Conductivity calculations for bulk Si, bulk Ge and ultrathin Si/Ge PSLs (*i.e.* in region R1) were performed using the Callaway's relaxation time scheme and the full-scale semi-*ab inito* formalism described in section 7.4.1.4. Note that the same method was used to obtain the results for Si/Ge PSLs presented in figure 9.15. The EMA expressions described in equations (9.90) were employed to obtain results for thicker unit cell PSLs and NDSLs with Ge inserted in Si matrix with a volume fraction $v = 0.125$.

Results in figure 9.18(*a*) of full-scale calculations for the Si($D/2$)Ge($D/2$)[001] PSL geometry in region R1 (for unit cell sizes smaller than 10 nm) are presented by blue colour diamond symbols. The variation of cross-plane conductivity, κ_{zz} with the unit cell size D is similar to that discussed in figure 9.14. The conductivity decreases as D increases, achieves a minimum at $D \sim 4$ nm and shows a slight increase for larger values of D. The EMA results for the PSL geometry with $D > 10$ nm smoothly 'extend' the full-scale results for $D < 10$ nm. The conductivity increases as D increases beyond 20 nm, saturating when $D \geq 20$ μm. The saturated value of the conductivity is the bulk weighted result obtained from equation (9.91) for region R3. Results in figure 9.18(*b*) predict that the thermal conductivity of the Si/Ge NDSL geometry also increases with unit cell size D and saturates when $D \geq 20$ μm at the bulk weighted result obtained from equation (9.91) for region R3.

Based upon the qualitative and quantitative discussions in this section, a few statements can be confidently made. Full-scale thermal conductivity calculations should be affordable for periodic composite structures with unit cell size in the range $D \leq 10$ nm. Generalised effective medium theories, which are valid for $D \geq 10$ nm, are capable of providing accurate results for systems of

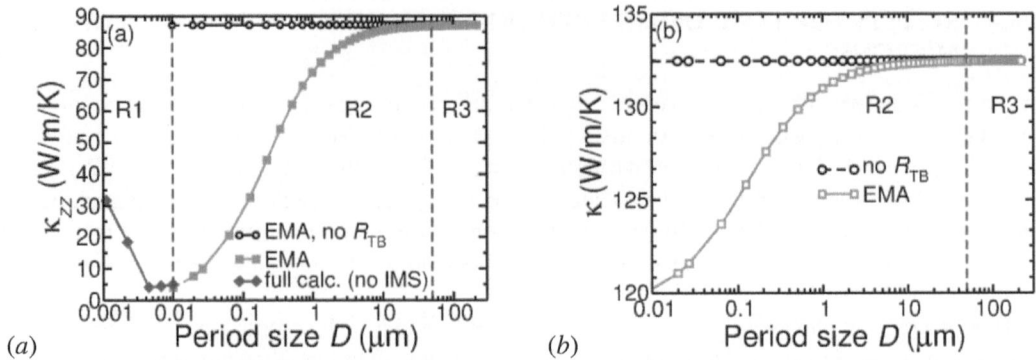

Figure 9.18 Variation of cross-plane room-temperature thermal conductivity of Si/Ge composites, of sample length 1 mm, with period size D for: (a) Si(D/2)Ge(D/2)001] PSL and (b) NDSL with $f = 0.125$ Ge insert in Si matrix. The regions marked $R1$, 2 and $R3$ correspond to $D << \Lambda$, $D \sim \Lambda$ and $D >> \Lambda$, respectively, where Λ is phonon mean free path. [Original diagram in colour.] Reproduced from Srivastava and Thomas (2020).

periodic sizes in the range 10 nm $< D \leq 50~\mu$m. The conductivity of composites with period sizes larger than 50 μm is simply the bulk weighted value.

9.7   DIMENSIONALITY DEPENDENCE OF THERMAL CONDUCTIVITY

Thermal conductivity of a material depends on several factors, such as sample length, average atomic mass, strengths of harmonic as well as anharmonic inter-atomic forces and dimensionality of crystal structure. We have earlier discussed the role of the first three factors. Here we provide a brief discussion on dimensionality dependence of thermal conductivity. This can best be done by examining the thermal conductivities of carbon solids in four different structural forms, *viz.* three-dimensional diamond, quasi two-dimensional graphite, two-dimensional monolayer graphene, and quasi one-dimensional carbon nanotubes (CNTs) and graphene nanoribbons (GNRs).

From the kinetic theory expression in equation (5.64), we note that $\kappa \propto C_v c^2 \tau$. At low temperatures, where τ is governed by boundary scattering, the temperature dependence of κ is that of the specific heat C_v. The low temperature specific heat scales as $C_v \sim T^{d/n}$, where d is the dimensionality of crystal structure and n refers to the acoustic phonon dispersion relation $\omega \sim q^n$. Experimental C_v results for diamond and graphite are presented in figure 9.19(a). Also presented in the inset is the theoretically predicted variation of C_v for monolayer graphene. In agreement with the discussion in equation (2.136), $C_v \propto T^3$ for the three-dimensional structure of diamond. The specific heat of graphite varies as $\sim T^3$ at very low temperatures (< 10 K) and as $\sim T^2$ in an intermediate temperature range (10–100 K). This behaviour is consistent with the quasi three-dimensional crystalline nature of graphite. Theoretical results suggest that for the two-dimensional structure of graphene $C_v \propto T$ in a wide low-temperature range (up to 50 K). Experimental data for the thermal conductivity of these forms of solid carbon are presented in figure 9.19(b). It is interesting to discuss both the room-temperature and low-temperature results.

At low temperatures, the thermal conductivity of diamond increases as T^3, as it does for other homogeneous three dimensional insulators. Theoretical calculations (Alofi and Srivastava 2013) estimate the variation (up to 40 K) of the conductivity of graphite as $\sim T^{2.4}$ in the basal plane and as $\sim T^{1.4}$ across the basal plane. The same theoretical scheme (Alofi and Srivastava 2013) estimates the variation of the conductivity of graphene as $\sim T^{1.6}$.

The room-temperature thermal conductivity can be affected by sample length and the level of sample homogeneity and chemical purity. For the samples studies in figure 9.19(b), the room-temperature thermal conductivity results show the ordering $\kappa_{CNT} > \kappa_{\text{suspended graphene}} >$

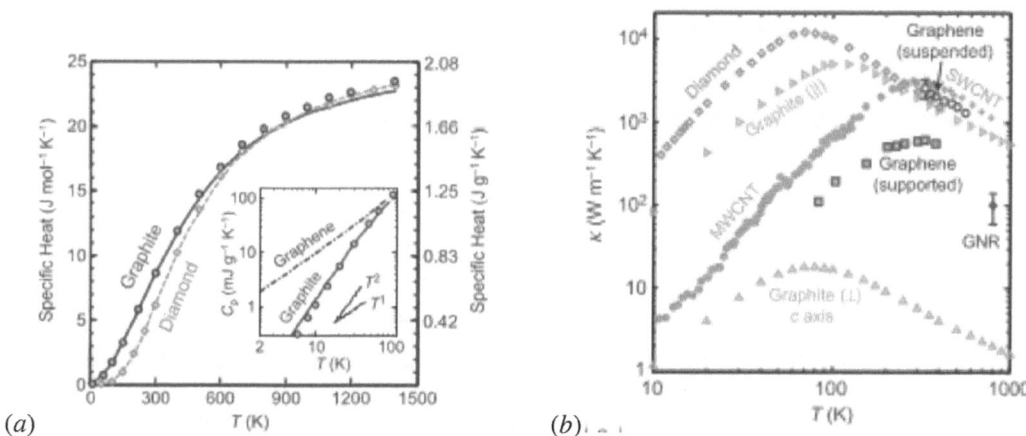

Figure 9.19 (a) Lattice specific heat and (b) lattice thermal conductivity of solid carbon with crystal structures of different dimensionalities. Symbols represent experimental data and lines show numerical calculations (as referenced in Pop *et al* (2012)). Reproduced from Pop *et al* (2012).

$\kappa_{\text{basal plane graphite}} > \kappa_{\text{supported graphene}} > \kappa_{\text{cross plane graphite}}$. The result for a GNR at ~ 800 K lies well below the conductivity of graphene or graphite (basal plane). The relatively lower conductivity of GNRs is due to surface or edge perturbations of phonon propagation.

9.8 NANOSTRUCTURING FOR ENHANCED THERMOELECTRIC PROPERTIES

A thermoelectric process, in which heat waste is converted to useful electrical power, is an important item for thermal management. Thermoelectric efficiency is usually discussed in terms of a dimensionless quantity called figure of merit ZT (Schmid 1960) defined as $ZT = S^2 \sigma T / (\kappa_{\text{carriers}} + \kappa_{\text{ph}})$, where S, σ, T, κ_{carriers} and κ_{ph} are Seebeck coefficient, electrical conductivity, temperature, thermal conductivity due to charge carriers and lattice thermal (phonon) conductivity, respectively. Conventionally, materials with heavy elements (e.g. Bi_2Te_3 and PbTe) have been used. Heavy cations and anions in such systems make their phonon spectrum quite narrow (and thus low C_v) and produce large anharmonic interactions. Low lattice thermal conductivity of these heavy cation and anion systems is the outcome of their narrow phonon spectrum and large anharmonicity. Similarly, as discussed further in sections 11.5.1 and 11.5.3, low thermal conductivity of sintered semiconductor alloys such as SiGe has been recognised useful in thermoelectric applications. Several other avenues have been explored. For achieving high ZT, use of materials exhibiting 'phonon glass electron crystal' (PGEC) behaviour has been advocated (Slack 1995). In practice, it is a big challenge to find or fabricate such materials. Use of materials containing large, complex unit cells, has also been explored. There is potential to reduce lattice thermal conductivity of materials containing large void spaces (e.g. skutterudites, clathrates and Zintl phases) through disorder within the unit cell or filling with heavy atoms (Synder and Toberer 2008). In early 1990s, theoretical predictions suggested that ZT could be greatly enhanced through nanostructuring (Hicks and Dresselhaus 1993). Following that, attention has turned to thin-film and nanostructuring designs. Venkatasubramanian *et al* (2001) demonstrated that Bi_2Te_3/Sb_2Te_3 thin-film superlattice structures (which can be classified as one-dimensional patterning) can offer enhanced ZT. Nakamura (2018) has outlined a three-dimensional patterning method for the independent control of phonon and electron transport in nanostructures containing connected epitaxial Si NDs and epitaxial Ge NDs embedded in epitaxial Si films. The proposed structures are shown in figure 9.20. Huge reduction in the thermal conductivity in the Si

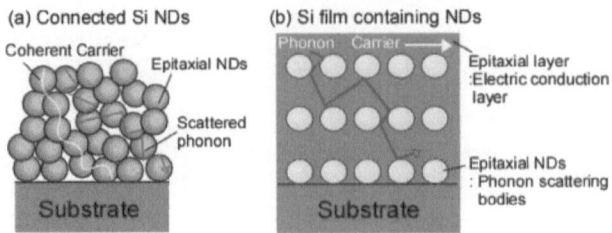

Figure 9.20 Nanostructures for independent control of thermal and electronic conductivities: (*a*) connected Si NDs, (*b*) Ge NDs embedded in Si film. Reproduced from Nakamura (2018).

film was observed, which was proposed to arise from efficient phonon scattering at the Ge NDs. On the other hand, electron scattering at the interface between the Ge NDs and Si did not significantly reduce the electron mobility. NDSL structures with Ge nanodots periodically embedded in Si matrix have also been fabricated and studied by other groups (Yang *et al* 2002, Liu J L *et al* 2003, Lee and Venkatasubramanian 2008). The work carried out by Lee and Venkatasubramanian showed that low thermal conductivity can be attained with a low superlattice period, a high ND areal density, or both.

Experimental investigations provide clear evidence that three-dimensional patterning can provide huge reduction in thermal conductivity. Established theoretical methods must be employed to verify the experimentally observed trends and numerical results for NDSLs. While it may not be possible to have atomic-level control of the dot-matrix interface region during fabrication of patterned structures, it is possible to make some predictions based on the EMA theory described in section 9.6. From computational point of view, there are four main factors that control the magnitude of thermal conductivity of periodically patterned nanostructures: sample length (L_B), insert side (d), repeat period size (D) and the amount of roughness/disorder (*i.e.* IMS) in the insert-matrix interface regions. Interface density is related to d, D and L_B, as mentioned after equation (9.88). Results in figures 9.15 and 9.18 clearly show that the conductivity becomes smaller both with decrease in sample length and in period size. Figure 9.21 shows how the thermal conductivity of Ge/Si NDSLs can change both with SL period size (D) and the amount of mass smudging (IMS) at the dot–matrix interface. While the phonon-interface scattering contribution can be controlled by changing the quality of the

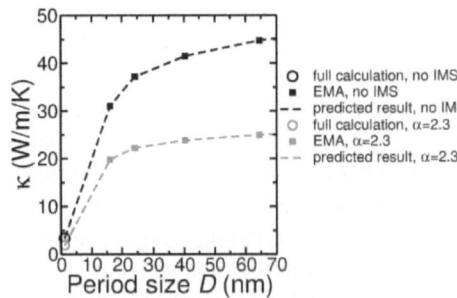

Figure 9.21 Predicted period size dependence of room-temperature thermal conductivity results for Ge nanodots of volume fraction $f = 0.125$ periodically embedded in Si matrix. Sample size is 500 nm and interface mass smudging (IMS) of different amounts is considered. Results for all period sizes are from the EMA theory, except for the smallest period size for which full-scale calculations are performed. The results with $\alpha = 2.3$ (lower curve) account for 10% IMS within one atomic layer across dot-matrix interface. Adopted from Srivastava and Thomas (2020).

interface regions, maintaining insert size (d) in low nm range will lead to strong phonon-boundary scattering for the insert material. Theory, therefore, provides qualitative support of the experimentally obtained trends.

9.9 BREAKDOWN OF FOURIER'S LAW

The kinetic theory expression for thermal conductivity (*cf.* equation (5.64) is based on the application of the Fourier law $\boldsymbol{Q} = -\mathcal{K}\nabla T$ (*cf.* equation (5.19). In section 9.6 we mentioned that the application of the Boltzmann transport equation to express the heat current Q within a relaxation time scheme assumes that the sample length is larger than phonon mean free path: $L > \Lambda$. Considering that typical values of Λ lie in the range $10 - 100$ nm (Kim *et al* 2007), it is reasonable to expect that the Fourier law of heat transfer will break down for solids of sizes in the low nm range. This indeed has been reported in several experimental works (see, *e.g.* Chang *et al* 2008, Yang *et al* 2010, Wilson and Cahill 2014). The failure of Fourier law has direct consequences for measurement and modelling of thermal boundary resistance R_{TB} and thermal conductivity κ in LDS, including nanocomposite structures of all dimensionalities, such as PSLs, NWSLs and NDSLs, discussed in the previous section. As yet, there is no firm criterion to predict when Fourier's law will break down in nanoscale heat transfer.

Various attempts have been made to model heat transport beyond the Fourier law. These include improvements of the kinetic theory (Struchtrup 2005, Péraud *et al* 2016, Sellitto *et al* 2017, Qu *et al* 2017), use of fluctuation-dissipation theorems (Banach and Larecki 2005, Larecki and Banach 2010, Pop 2010) and computer simulations (Hua and Cao 2014, 2016). Péraud *et al* (2016) extended the range of validity of Fourier's law into the kinetic transport regime. For this they derived the continuum equations and boundary conditions for heat transfer in the limit of small but finite mean free path from the asymptotic solution of the linearised phonon Boltzmann equation in the relaxation time approximation. Their asymptotic approach can also be used to calculate the thermal boundary resistance R_{TB} at the kinetic level, provided that the interface transmission and reflections coefficients are available.

Qu *et al* (2017) have presented a diffusive-nondiffusive two-parameter nanoscale heat conduction model. Following the Guyer–Krumhansl model (Guyer and Krumhansl 1966), Qu *et al* extended the Fourier law to the form

$$\boldsymbol{Q}^* = C_v \frac{\partial T^*}{\partial t} + \boldsymbol{\nabla} \cdot \boldsymbol{Q} \tag{9.92}$$

$$\boldsymbol{Q} = -\kappa \boldsymbol{\nabla} T^* + \frac{\Lambda_D \Lambda_B}{9} \left(\nabla^2 \boldsymbol{Q} + 2 \boldsymbol{\nabla}(\boldsymbol{\nabla} \cdot \boldsymbol{Q}) \right). \tag{9.93}$$

Here $\boldsymbol{Q}^*$ is the internal heat generation, T^* is a nonequilibrium temperature defined as $T^* = E/C_v$, where E is energy density and C_v is volumetric specific heat, and $\Lambda_B = c\tau_B$ characterises the nondiffusive heat transfer length with τ_B representing phonon relaxation time for nondiffisive heat transfer and c as the speed of sound. The two parameters of this model are κ, characterising diffusive heat transfer, and Λ_B, characterising nonlocal effects on the heat flux from nondiffusive heat transfer. Considering a heater-substrate system with heater temperature T_{heater}, an iterative procedure would be required to solve the equations (9.92) and (9.93) for T^* and $\boldsymbol{Q}$ simultaneously. Qu *et al* defined a nondimensional parameter ξ for spatially periodic heating with a period P

$$\xi = \frac{\Lambda_D \Lambda_B}{3(P/2\pi)^2}. \tag{9.94}$$

A small tolerance value for ξ can be set to predict the breakdown of Fourier's law.

9.10   PHONON INTERACTION WITH A TWO-DIMENSIONAL ELECTRON GAS

Potential device applications of semiconductor superlattices, quantum wells, metal-oxide-semiconductor (MOS) structures, and semiconductor–metal contacts are due mainly to formation of electron layers either at interfaces or in quantum wells. These electron layers show very distinct two-dimensional (2D) properties. A 2D electron gas (2DEG) can interact with phonons of both acoustic and optical branches. In 2D systems we have, in addition to bulk phonons (with nodes at interfaces), a series of 2D modes (so-called interface phonons). In this section we discuss phonon interaction with 2DEG.

9.10.1   ACOUSTIC PHONON INTERACTION WITH 2DEG

The interaction between a 2DEG and acoustic phonons can be studied by observing phonons emitted from a hot 2DEG or absorbed by a cold 2DEG. The participating phonons will be acoustic as long as $k_B T_{el} \ll \hbar\omega_{op} \simeq 50$ meV, where T_{el} = electron temperature, ω_{op} = optical phonon frequency. The phonons can be detected by using superconducting tunnel junctions (cf section 13.3). Here we discuss phonon emission and absorption from a 2DEG in a (001) plane of a Si MOSFET (MOS field-effect transistor). This approach is used by Hensel *et al* (1983a), Rothenfusser *et al* (1986) and Challis *et al* (1987).

Let us recall from section 6.5.2 that in bulk Si there are six degenerate conduction band minima along the $\langle 100 \rangle$ directions. The energy surface ellipsoid along [001] is described by

$$E_{3D} = \frac{\hbar^2 (k_x^2 + k_y^2)}{2m_t^\star} + \frac{\hbar^2 k_z^2}{2m_l^\star}, \tag{9.95}$$

where $m_t^\star$ and $m_l^\star$ are transverse and longitudinal electron effective masses. In the case of 2D systems, such as those discussed in this section, the z component of the energy is modified as discrete eigenvalues so that equation (9.95) becomes

$$E_{2D}^{(n)} = \frac{\hbar^2 (k_x^2 + k_y^2)}{2m_t^\star} + \left(E_0 + \frac{\pi^2 \hbar^2 n^2}{2m_l^\star L^2} \right), \tag{9.96}$$

where E_0 is the energy of the lowest subband, n is an integer and L is the width of the well containing the 2DEG. The ground-state energy ($n = 0$) is now two-fold degenerate, with a valley degeneracy factor $g_v = 2$ and a spin degeneracy factor $g_s = 2$. The Fermi wave vector k_F for the 2D case is given by

$$k_F = (2\pi N_s / g_v)^{1/2}, \tag{9.97}$$

where N_s is the carrier density in the 2DEG.

9.10.1.1   Phonon emission

Phonons can be emitted during the transition of an electron from state k to state k' within the lowest subband. We assume that the phonons emitted in the process are bulk Si phonons.

The emission rate of phonons qs is given by the golden rule formula

$$
\begin{aligned}
P_{emi}(qs) &= g_s g_v \sum_{kk'} \bar{P}_k^{k'+q} \\
&= g_s g_v \frac{2\pi}{\hbar} \sum_{kk'} \frac{|M(k,k')|^2}{\{\varepsilon(q,\omega)\}^2} \delta(E_k - \hbar\omega_q - E_{k'}),
\end{aligned} \tag{9.98}
$$

where the matrix element $M(k,k')$ can be calculated by following the scheme presented in section 6.5. Here $\varepsilon(q,\omega)$ represents a dynamic screening factor and can be evaluated using a reasonable scheme (also see section 3.3.2.3). Assuming equilibrium distribution for phonons and electrons, we can write, from equations (6.117) to (6.119),

$$
\begin{aligned}
\bar{P}_k^{k'+q} = \quad & g_s g_v \frac{\pi \omega_q [\Xi_{kk'}(qs)]^2}{\rho N_0 \Omega c_s^2 [\varepsilon(q,\omega)]^2} (\bar{n}_q + 1) \\
& \times \bar{f}_k (1 - \bar{f}_{k'}) \delta_{k,k'+q} \delta(E_k - E_{k'} - \hbar\omega_q),
\end{aligned}
\tag{9.99}
$$

where $\Xi(qs)$ are given by equation (6.115).

For the (001) plane equation (6.135) becomes

$$
C\hat{q} \cdot q_{qs} = E_d \hat{q} \cdot e(qs) + E_u \hat{q}_z e_z(qs)
\tag{9.100}
$$

with E_d and E_u as dilatation and shear deformation constants, respectively. For an isotropic (Debye) model, we can choose the two TA modes for each q to be polarised within and perpendicular to the plane (001) such that $e_s(qs) = \sin\theta$ and 0, respectively (with $\theta = 0$ along the z direction). Then equation (9.100) becomes

$$
\begin{aligned}
C\hat{q} \cdot e_{qs} \quad & = \quad E_d + E_u \cos^2\theta \qquad & \text{for LA modes} \\
& = \quad E_u \sin\theta \cos\theta \qquad & \text{for TA in} - \text{plane modes} \\
& = \quad 0 \qquad & \text{for TA out} - \text{of} - \text{plane modes.}
\end{aligned}
\tag{9.101}
$$

Further, for evaluating $\Xi_{k,k'}$ for the 2DEG problem, we modify equation (6.116) to write

$$
\psi_k(r) = \frac{1}{\sqrt{A}} \exp[i(k_x x + k_y y)]\phi(z),
\tag{9.102}
$$

where

$$
\phi(z) = \frac{1}{\sqrt{2z_0^3}} z e^{-z/2z_0}
\tag{9.103}
$$

is a variational functional with z_0 as a thickness parameter for the ground state (lowest subband 0), and A is the area of the 2DEG.

Using equations (9.99)–(9.103), we can finally express

$$
\begin{aligned}
P_{\text{emi}}(q,s) \quad = \quad & g_s g_v \frac{A}{4\pi^2} \frac{\pi\omega_q}{\rho N_0 \Omega c_s^2} \frac{|R(q_z)|^2}{[\varepsilon(q,\omega)]^2} \left(C\hat{q} \cdot e_{qs} \right)^2 \\
& \times (\bar{n}_q + 1) \int d^2 k \bar{f}_k (1 - \bar{f}_{k'}) \\
& \times \delta_{k,k'+q_{\parallel}} \delta(E_k - E_{k'} - \hbar\omega_q)
\end{aligned}
\tag{9.104}
$$

with $q_z = q\cos\theta, q_{\parallel} = q\sin\theta$ as the component of q in the (001) plane, and

$$
|R(q_z)|^2 = (1 + q_z^2 z_0^2)^{-3}.
\tag{9.105}
$$

The integral in equation (9.104) is similar to the integral I in equation (6.124) and can thus be evaluated similarly. Let us express

$$
\begin{aligned}
y \quad & = \quad E_k - E_{k'} - \hbar\omega_q \\
& = \quad \frac{\hbar^2}{2m^*}(k^2 - k'^2) - \hbar\omega_q \\
& = \quad \frac{\hbar^2 q_{\parallel}}{m^*}(k\cos\phi - k_0),
\end{aligned}
\tag{9.106}
$$

where ϕ is the angle between k and $q_\parallel$, and

$$k_0 = \frac{q_\parallel}{2} + \frac{m^* c_s}{\hbar \sin \theta}. \tag{9.107}$$

$y = 0$ requires $k \cos \phi_0 \equiv k_0 \leq k_F$, or $q_\parallel \leq 2k_F - 2m^* c_s / \hbar \sin \theta$. We next write

$$
\begin{aligned}
d^2 k &= k \, d\phi \, dk \\
&= -\frac{m^*}{\hbar^2 q_\parallel} dk \, dy \sin \phi
\end{aligned}
$$

so that the integral in equation (9.104) becomes

$$\frac{2m^*}{\hbar^2 q \sin \theta} \int dk \, \bar{f}_k (1 - \bar{f}_{k'}) \int dy \frac{\delta(y)}{\sin \phi} = \frac{2m^*}{\hbar^2 q \sin \theta} \int_{k_0}^\infty dk \frac{k \bar{f}_k (1 - \bar{f}_{k'})}{(k^2 - k_0^2)^{1/2}}$$

(the factor 2 counts for the two values of ϕ_0 satisfying $y = 0$, and we assume that the second integral can reasonably be approximated to $(\sin \phi_0)^{-1}$ throughout the range). Thus the emission from the 2DEG of phonons with polarisation s in a range of frequencies $d\omega$ in a solid angle $d\xi$ at an angle θ from surface normal is $P_{emi}(qs) d\xi \, d\omega$, where

$$
\begin{aligned}
P_{emi}(qs) &= g_s g_v A \frac{\pi m^*}{\rho \hbar^2 N_0 \Omega c_s \sin \theta} \frac{|R(q \cos \theta)|^2}{|\varepsilon(q, \omega)|^2} (\bar{n}_q + 1) \\
&\times (C \hat{q} \cdot e_{qs})^2 \int_{k_0}^\infty dk \frac{k \bar{f}_k (1 - \bar{f}_{k-q_\parallel})}{(k^2 - k_0^2)^{1/2}}. \tag{9.108}
\end{aligned}
$$

This integral can be evaluated numerically.

9.10.1.2 Phonon absorption

The absorption of a phonon of energy $\hbar \omega_q$ scatters an electron from an occupied state with $|k| \leq k_F$ to an unoccupied state with $|k'| > k_F$, subject to the conservation rules $k' = k + q_\parallel$ and $E_{k'} = E_k + \hbar \omega_q$. The procedure outlined above can be used to obtain an expression for phonon absorption rate $P_{abs}(qs)$. The result will be similar to equation (9.108), except that the factor $(\bar{n}_q + 1)$ will be replaced by $\bar{n}_q$, and k_0 will be given by

$$k_0^2 = \left(\frac{q_\parallel}{2} - \frac{m^* c_s}{\hbar \sin \theta} \right)^2$$

with

$$q_\parallel \leq 2k_F + \frac{2m^* c_s}{\hbar \sin \theta}.$$

The total emitted (absorbed) phonon power is

$$\mathscr{S} = \sum_{qs} \hbar \omega_q P(qs) \tag{9.109}$$

and emission (absorption) is the ratio

$$\frac{\Delta I(qs)}{I(qs)} = \frac{N_0 \Omega P(qs)}{A N_s c_s \cos \theta}. \tag{9.110}$$

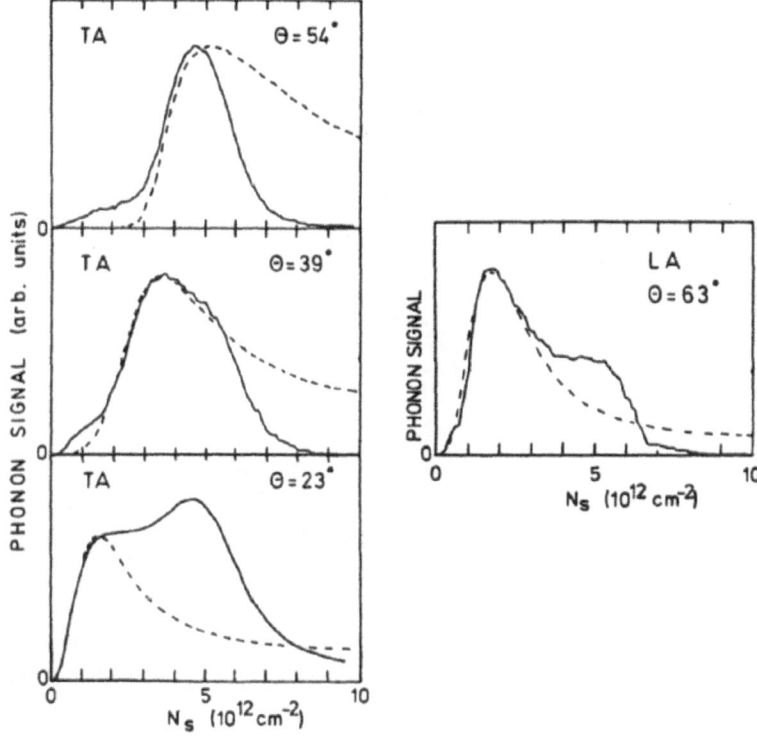

Figure 9.22   Phonon emission from a hot 2DEG formed in a Si-MOS structure. Dashed lines are theoretical results using $E_d = -6.0$ eV, $E_u = 9.0$ eV, $\rho = 2.33$ g cm^{-3}, $c_L = 9.4 \times 10^5$ cm s^{-1}. (Reproduced from Rothenfusser *et al* (1986).)

9.10.1.3   Measurements and theoretical results

Figure 9.22 shows phonon emission from a Si-MOS sample (Rothenfusser *et al* 1986). The figure also shows the result of the theoretical calculation by Rothenfusser *et al* (1986). The steep signal increase with at small values of N_s is very well reproduced by theory for emission of both TA and LA phonon modes. The dependence of the phonon signal on emission angle and polarisation is characteristic of the 2D nature of the electron gas. The theory could not explain the hump at $N_s = 5 \times 10^{12}$ cm^{-2} and the signal decrease beyond $N_s = 6 \times 10^{12}$ cm^{-2}. Rothenfusser *et al* (1986) suspected, however, that these anomalies are caused by the neglect in the theory of the population of higher subbands which lie close to E_F for larger values of N_s.

Figure 9.23 shows profiles of LA phonon absorption by a 2DEG in a Si MOSFET measured at lattice temperature 2.15 K. The agreement of theoretical calculations by Challis *et al* (1987), without allowing for screening, with experiment (Hensel *et al* 1983b) is convincing.

9.10.2   EMISSION OF ACOUSTIC PHONONS BY A 2DEG IN A QUANTISING MAGNETIC FIELD

When a large magnetic field B is applied to a system with a 2DEG at low temperatures, the electronic energy levels are quantised into the so-called Landau levels. This leads to the remarkable quantum Hall effect (see, e.g. von Klitzing 1986). When an electric pulse is applied to such a system, electrons can be excited to higher Landau levels. These electrons can 'relax' to their ground state by emitting phonons (Kent *et al* 1988). A brief discussion of this is also given in section 13.5.

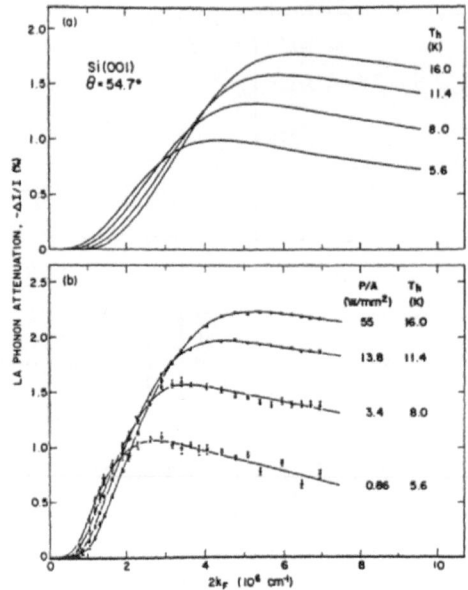

Figure 9.23 Absorption of LA phonons by a 2DEG in a Si MOSFET at lattice temperature 2.15 K and $\theta = 54.7°$ for different phonon temperatures: (*a*) theory (Challis *et al* 1987), (*b*) experiment (Hensel *et al* 1983b).

The theoretical method of the previous section can be extended to study the spectrum of acoustic phonons of a heated 2DEG in the presence of a quantising magnetic field. Challis *et al* (1987) and Toombs *et al* (1987) have made a thorough study of this point by suitably modifying the form of the electronic wavefunction in equation (8.94) in the presence of the quantising magnetic field.

9.10.3 MAGNETOPHONON EFFECT IN 2D SYSTEMS

In the previous section we have considered interaction between a 2DEG and acoustic phonons. Interaction between a 2DEG and longitudinal optic (LO) phonons also occurs. In the case of bulk (3D) polar semiconductors, the interaction between electrons and optical phonons is usually described by a model presented by Fröhlich (1954). The effective interaction of polar phonons with a 2DEG is strongly modified compared to the 3D case.

The dominant electron–phonon coupling is with LO phonons, due to the large electric polarisation associated with these modes. The Fröhlich Hamiltonian which describes such an interaction is

$$H_{\text{el}-\text{opph}} = 4\pi i \left[\frac{e^2 \hbar \omega_{\text{LO}}}{8\pi N_0 \Omega} \left(\frac{1}{\varepsilon_\infty} - \frac{1}{\varepsilon_0} \right) \right]^{1/2}$$
$$\times \sum_q \frac{1}{q} (a_q^\dagger e^{-iq \cdot r} - a_q e^{iq \cdot r}), \tag{9.111}$$

where ω_{LO} is the frequency of longitudinal optical phonon, and ε_0 and ε_∞ are, respectively, static and high-frequency dielectric constants.

In the 3D case the interacting electron can be treated as a 'polaron' (a quantum of the coupled electron–phonon system) with an effective mass given by

$$m_{pol}^{*3D} \cong m^*(1 + \alpha/6), \tag{9.112}$$

where m^* is the effective mass of the electron. The parameter α (the Fröhlich coupling constant), assumed to be very small ($\alpha << 1$), is defined as

$$\alpha = e^2 \left(\frac{1}{\varepsilon_\infty} - \frac{1}{\varepsilon_0} \right) \left(\frac{m^*}{2\omega_{LO}\hbar^3} \right)^{1/2}. \tag{9.113}$$

The energy of the lowest conduction band can be expressed as

$$E_{3D}(\boldsymbol{k}) = \frac{\hbar^2 k^2}{2m^*_{pol}} - \alpha\hbar\omega_{LO} + O(k^4), \tag{9.114}$$

where the last term is used to express non-parabolicity.

In 2D systems the effective polaron mass is given by (Das Sarma 1983)

$$m^{*2D}_{pol} = m^*(1 + \pi\alpha/8) \tag{9.115}$$

and the electronic dispersion relation is modified to

$$E(\boldsymbol{k}_\parallel) \cong \frac{\hbar^2 k_\parallel^2}{2m^{*2D}_{pol}} - \frac{\pi\alpha\hbar\omega_{LO}}{2} - \left(\frac{9\pi\alpha}{128} \right) \left(\frac{\hbar^2 k_\parallel^2}{2m^*} \right)^2 \frac{1}{\hbar\omega_{LO}}. \tag{9.116}$$

Absorption of LO phonons by electrons can result into a resonant inelastic scattering of electrons between Landau levels. This causes the magnetophonon effect: the magneto-conductivity exhibits oscillations which are periodic in $1/B$. The maxima in the conductivity occur at

$$\omega_{LO} = n\omega_c, \tag{9.117}$$

where $n = 1, 2, 3, ...$, and ω_c is the cyclotron resonance frequency. The magnetophonon effect can be observed in high purity and non-degenerate (3D) III–V semiconductors when electron mobility reaches the highest values. The effect can be observed in the degenerate case of 2D systems at high temperatures. The first observation of the magnetophonon effect in 2D systems was made in GaAs/GaAlAs heterostructures (Tsui et al 1980). The effect has also been observed in GaInAs/InP and GaInAs/AlInAs heterostructures.

10 Topological Nanophononics and Chiralphononics

10.1 INTRODUCTION

In this chapter we discuss some aspects of topology, phononics, chirality, topological nanophononics and topological chiralphononics. We begin by clarifying the meaning of the words *topology*, *nanophononics* and *chirality*. The word *topology* refers to a geometric property that remains unchanged upon any continuous change. For example, termination of a solid at an edge or a surface is a continuous (or first-order, or incremental) change. Relationship between the topological structure of bulk crystal to the presence of gapless boundary states comes from the bulk-boundary correspondence. The topic of *nanophononics* deals with the manipulation and engineering of phonon properties at the nanoscale. Thus *nanophononic topology* refers to the topological properties of nanophonic materials. An object which cannot be superimposed on its own mirror image is called a chiral object. *Chirality*, or handedness, is a geometric property which signifinies the mirror image asymmetry of an object or a physical variable. *Topological phononic chirality* refers to the topological properties of phonons in chiral materials.

10.2 TOPOLOGICAL SOLIDS

Solids whose properties are invariant under topological transformations are known as topological solids. Considering electronic properties, there are different types of topological solids. These include topological insulators (TIs) (Hasan and Kane 2010), topological semimetals (TSMs) (Yan and Felser 2017) and topological superconductors (TSCs) (Qi and Zhang 2011, Leijnse and Flensberg 2012). TI materials are characterised by electronic states with Dirac-like linear dispersion relation for edge or surface states that lie inside the band gap and can be protected by time reversal symmetry or by crystal lattice symmetry. There are two types of TSMs: Dirac semimetal (DSM) and Weyl semimetal (WSM). A three-dimensional gapless material with crossings of linear energy bands is a DSM if time reversal and inversion symmetry are preserved, and is a WSM if there is a lack of either time reversal or inversion symmetry. The crossing points are called Weyl nodes. There are two types of WSMs. In type-I linear crossing of two linear electronic bands occurs at the Fermi level with the energy cones collapsing at a pointlike Fermi surface. Type-II WSMs are characterised by tilted Weyl cones with Fermi surfaces consisting of electron-hole pockets which touch each other. A TSC has a superconducting gap in its bulk but has topologically protected metallic edge or surface states, which are made up of Majorana fermions (chargeless and spinless fermions which are their antiparticles (Majarona, 1937)).

Let us try to understand the concept of topological surface states at a more basic level. As discussed in section 2.2.1, when dealing with bulk properties, such as electronic band structure or phonon dispersion relation, we usually invoke periodic boundary condition to avoid unimportant end, or boundary, effects. This makes the electronic wavevector k and the phonon wavevector q good quantum numbers which ensure the translational symmetries $f(k+G) = f(k)$ and $f(q+G) = f(q)$, where G is a reciprocal lattice vector for the bulk crystal structure and f represents a property in reciprocal space. The termination of a crystal into a surface makes k and q complex vectors (*i.e.* these are no longer good quantum numbers). A brief explanation of the complex phonon dispersion

DOI: 10.1201/9781003141273-10

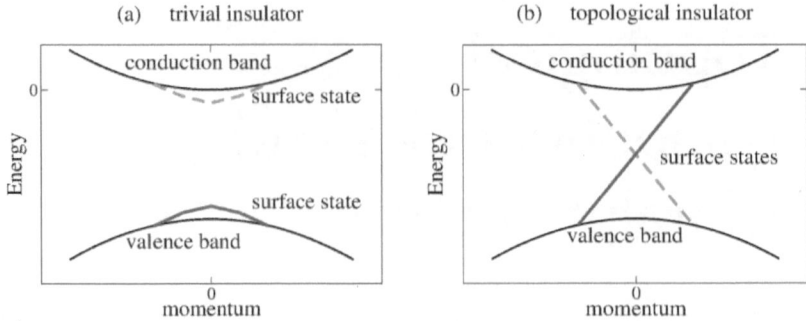

Figure 10.1 Schematic illustration of topologically trivial (*a*) and non-trivial (*b*) surface electronic states inside the bulk band gap of a semiconductor or insulator. For the trivial case, the lower(upper)-lying surface state is derived from a bulk valence(conduction) band of $p(s)$ orbital. For the non-trivial case, the valence and conduction band derived surface states cross each other inside the bulk band gap, with a linear dispersion relation in the region of the crossing point.

curves was presented in figure 9.3 and the text around that. The electronic wavefunction, or the atomic displacement, corresponding to a surface state is localised close to the surface, damps in an oscillatory fashion inside the crystal and decays monotonically in vacuum along the surface normal. In Chapter 8 we have presented and discussed several examples of surface phonon modes on semiconductor surfaces. Similarly, many examples of electronic states on semiconductor surfaces have been investigated, both theoretically and experimentally (see, e.g. Srivastava 1999). Such surface states normally originate from one bulk band and return to (or close at) the same bulk band. Figure 10.1(*a*) presents a simple illustration of two surface electronic states, one just above the bulk valence band and the other just below the conduction band minimum. For normal semiconductors, the lower(upper)-lying surface state is derived from an $p(s)$-orbital of the bulk (see, e.g. Srivastava *et al* 1983). A small perturbation, such as disorder or chemical adsorption, can easily destroy/change/modify such surface states. For this reason, such surface states are known as trivial surface states (or surface states with trivial topology). Electronic surface states in the bulk band gap of topological semiconductors or insulators obeying time reversal symmetry, continuously connect conduction and valence bands, as illustrated in figure 10.1(*b*), and are known as topological surface states (or surface states with non-trivial topology). Such special surface states are robust against any perturbation.

We now present clarification of the statements made above regarding trivial and non-trivial topological surface states by considering real systems. Let us consider the electronic structure of graphene (a two-dimensional system) and the (111) surface of the $Bi_{1-x}Sb_x$ alloy[1]. Graphene, characterised with inversion symmetry, has an even number of Dirac points (DPs) at the Fermi level. Following the Kramers theorem (Schiff 1968), a small perturbation (such as by changing the chemistry of the surface) can easily open a gap at the DPs (*i.e.* the DPs are destroyed). Thus, the surface states on graphene exhibit normal, or trivial, topology. In contrast, the $Bi_{1-x}Sb_x(111)$ surface is characterised with an odd number of DPs at the Fermi level (Teo *et al* 2008). These Dirac states are robust against any slight imperfection or impurities. Thus, $Bi_{1-x}Sb_x$ is a TI (*i.e.* its surface states exhibit non-trivial topology). The topological character of this alloy comes from Sb due to the inversion of its valence and conduction bands at the L point in the Brillouin zone.

What makes the surface states for $Bi_{1-x}Sb_x$ or $Sn_{1-x}In_xTe$ non-trivial? To understand this, let us consider bulk SnTe, which at room temperature has the simple rocksalt crystal structure. This

---

[1]This is so in a certain range of values for *x*. There are other bulk crystals that are predicted to be three-dimensional topological insulators, e.g. Bi_2Te_3, Bi_2Se_3, Sb_2Te_3, $Sn_{1-x}In_xTe$ etc.

IV-VI material is a narrow-gap ($E_g = 0.18$ eV) semiconductor with the valence band maximum (VBM) and conduction band minimum (CBM) at the L point of the Brillouin zone. In the absence of spin-orbit coupling, the VBM and CBM states at the L point are mainly contributed by (s-cation, p-anion) and (s-anion, p-cation), respectively (see, e.g. Tsang and Cohen 1971). [This is similar to the normal ordering of the band edge energies in III-V and II-VI semiconductors (see, e.g. Walter and Cohen 1971).] The presence of strong spin-orbit coupling inverts the cation/anion characters around the L point in these systems. The inverted orbital ordering produces a negative band gap in these systems. SnTe is a topological crystalline insulator (TCI), whose surface states are protected by crystal lattice symmetry. Considering the (001) surface, the presence of crystal mirror symmetry with respect to the family of {110} planes topologically protects pairs of surface states (Schmidt and Srivastava 2020). Angle-resolved photoemission measurements have confirmed the existence of non-trivial topological states on this surface (Sato *et al* 2013).

As topological considerations refer to surface states, it is obvious that surface electronic band structure should be calculated to examine the energy location and robustness of the Dirac states. However, it is not necessary to perform expensive surface state calculations, as bulk symmetry considerations can reveal information about the robustness of topological surface states. This requires examining winding of bands invoking the concepts of Chern number C and the topological invariant Z_2. The Z_2 index expresses whether the number of times the boundary state crosses the Fermi level between the zone centre and zone edge is even or odd. Let us consider the z surface plane for the bulk-boundary correspondence. The Chern number for a 2D system is defined as (see, e.g. Fukui *et al* 2005, Sato and Ando 2017)

$$
\begin{aligned}
C &= \frac{1}{2\pi} \sum_n^{occ} \int_{BZ} dk_x dk_y F_n \\
F_n &= (\nabla \times A_n)_z \\
A_n &= i < u_{nk} | \nabla_k | u_{nk} >,
\end{aligned}
\tag{10.1}
$$

where A_n is the Berry connection, u_{nk} is the cell-periodic part of the Bloch wave function for wavevector k and the nth band. $C = 0$ corresponds to a trivial state and $C = \pm 1, \pm 2, \pm 3, \ldots$ corresponds to a Chern insulator state. The Z_2 invariant[2] is defined in 2D as (see, e.g. Fukui *et al* 2005, Fu and Kane 2006, Bansil *et al* 2016, Sato and Ando 2017)

$$
Z_2 = \frac{1}{2\pi} \sum_n^{occ} \left[\oint_{half\,BZ} A_n \cdot dk - \int_{half\,BZ} dk_x dk_y F_n) \right] \mod 2.
\tag{10.2}
$$

The 3D extension requires computation of several 2D contributions using equation (10.2) by taking various pairs of time-reversal-invariant-momenta (TRIM) points in the 3D BZ. $Z_2 = 0$ corresponds to a trivial state and $Z_2 = 1$ corresponds to a TI state. The Chern number is not well defined in the presence of spin-orbit interaction. However, for a spin-preserving and time-reversal-invariant system, the Z_2 index coincides with the parity of the spin Chern number (Sato and Ando 2017). From these considerations for a crystal terminated on an inversion plane, Teo *et al* (2008) established a theorem which states that the surface fermion parity is determined by the parity invariants of the bulk band structure. The surface fermion parity determines which surface TRIM points are enclosed by an odd number of Fermi surface lines.

10.3 TOPOLOGICAL PHONONS

The concepts described in the previous section regarding electronic surface states in topological solids can be applied to discuss topological phonon modes. We will consider examples of topological phonons in 1D, 2D and 3D systems.

---

[2]Z is the group of integer numbers, and Z_2 is the quotient group of Z classifying even and odd numbers.

10.3.1 TOPOLOGICAL PHONONS IN 1D SYSTEMS

In section 2.2.2 we discussed phonon dispersion curves for a diatomic linear chain with alternating masses m_1 and m_2 connected with a single force constant Λ. Let us extend that exercise by considering a linear chain with alternating masses m_1 and m_2, alternating force constants Λ_1 and Λ_2, and unit cell length a, as illustrated in figure 10.2(a). The equations of motion and solutions can be straightforwardly obtained by following the approach in section 2.2.2. The dispersion relation can be obtained from the equation below:

$$\omega^2 = \frac{(\Lambda_1 + \Lambda_2)(m_1 + m_2)}{2m_1 m_2} \left[1 \pm \sqrt{1 - \frac{8\Lambda_1 \Lambda_2 m_1 m_2}{(\lambda_1 + \lambda_2)^2 (m_1 + m_2)^2} (1 - \cos qa)} \right]. \tag{10.3}$$

For the case when $m_1 = m_2 = m$, as shown in figure 10.2(b), equation (10.3) reduces to

$$\omega^2 = \frac{\Lambda_1 + \Lambda_2}{m} \left[1 \pm \sqrt{1 - \frac{2\Lambda_1 \Lambda_2}{(\lambda_1 + \lambda_2)^2} (1 - \cos qa)} \right]. \tag{10.4}$$

In the long wavelength limit $qa << \pi$, $\cos(qa) \approx 1 - (qa)^2/2$, and the lower (acoustic) and upper (optical) branches of the dispersion curves can be expressed as

$$\omega_{\text{ac}} = \sqrt{\frac{\Lambda_1 \Lambda_2}{2m(\Lambda_1 + \Lambda_2)}} \, (qa) \tag{10.5}$$

$$\omega_{op} = \sqrt{\frac{2(\Lambda_1 + \Lambda_2)}{m}} - O(qa)^2. \tag{10.6}$$

At the Brillouin zone boundary $qa = \pi$, and the frequencies are

$$\omega_{\text{BZ}} = \sqrt{\left(\frac{\Lambda_1 + \Lambda_2}{m} \right) \pm \left(\frac{\Lambda_1 - \Lambda_2}{m} \right)}. \tag{10.7}$$

For convenience sake, let us define the reduced frequency $\Omega = \sqrt{\frac{m}{\Lambda_1}} \omega$. The dispersion relation in equation (10.4) can then be expressed as

$$\Omega = (1 + \Lambda_2/\Lambda_1)^{1/2} \left[1 \pm \sqrt{1 - \frac{2\Lambda_2/\Lambda_1}{(1 + \Lambda_2/\Lambda_1)^2} (1 - \cos qa)} \right]^{1/2}. \tag{10.8}$$

Figure 10.2(c) shows the plot of Ω as a function of qa/π for the consideration $\Lambda_2 = 1.5\Lambda_1$. The reduced frequency Ω takes the values 0 and $\sqrt{5}$ at the zone centre, and $\sqrt{2}$ and $\sqrt{3}$ at the zone boundary.

Following the work of Pal and Ruzzene (2017), we now consider a superlattice structure, with a unit cell containing N contiguous primitive cells of type A ($\Lambda_1 < \Lambda_2$) and N contiguous primitive cells of type B ($\Lambda_1 > \Lambda_2$). As indicated in figure 10.3, this structure is characterised by the presence of two types of interfaces between the primitive cell types A and B. One interface (labelled A-B IF) connects the stronger springs between the A and B primitive cells and can be referred to as the $\Lambda_{\text{strong}} - \Lambda_{\text{strong}}$ IF. The other interface (labelled B-A IF) connects the weaker springs between the B and A primitive cells and can be referred to as the $\Lambda_{\text{weak}} - \Lambda_{\text{weak}}$ IF. Notice that the superlattice constructed in this manner is characterised by broken parity (or, space inversion) within the unit cell.

Numerically calculated phonon dispersion curves for such a superlattice, with $N = 20$ and $\Lambda_{\text{strong}} = 1.5 \Lambda_{\text{weak}}$, are shown in figure 10.3. While the BZ boundary for this superlattice is at

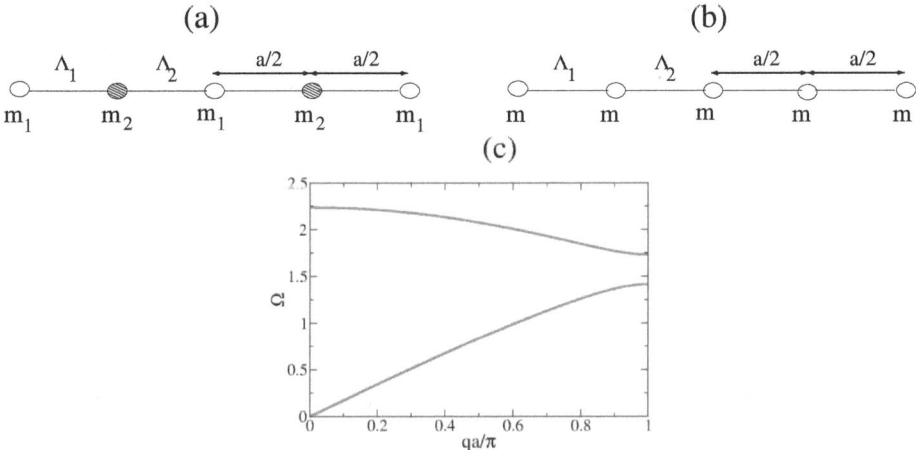

Figure 10.2 (*a*) Linear chain of alternating masses and force constants. (*b*) Linear chain of identical masses but alternating force constants. (*c*) Plot of the reduced phonon frequency $\Omega = \sqrt{m/\Lambda_1}\,\omega$ against the wavevector q for the chain in (*b*) with $\Lambda_2 = 1.5\,\Lambda_1$.

$q = \pi a/N$, the dispersion curves have been unfolded onto the first BZ for the bulk for a direct comparison with the results presented in figure 10.2. Apart from the appropriate number of folded acoustic and optical branches, there appear three additional interface modes. Two interface modes, featuring two $\Lambda_1 - \Lambda_1$ and $\Lambda_2 - \Lambda_2$ springs, lie within the bulk band gap. The lower mode (dashed line) is the B–A or $\Lambda_{\mathrm{weak}} - \Lambda_{\mathrm{weak}}$ IF mode, and the upper mode (dotted–dashed line) is the A–B or $\Lambda_{\mathrm{strong}} - \Lambda_{\mathrm{strong}}$ IF mode. A third mode, topologically trivial and localised at an interface having adjacent $\Lambda_{\mathrm{strong}} - \Lambda_{\mathrm{strong}}$ springs, lies above the bulk spectrum.

As discussed in the previous section in the context of electronic topological states, the existence of these superlattice interface phonon modes can be predicted by examining the Z_2 invariants associated with the A and B type bulk linear chains. The expression for the Z_2 invariant of the bulk chains is

$$Z_2 = \frac{\mathrm{i}}{2\pi} \int_{-\pi/a}^{\pi/a} \boldsymbol{u}^*(q) \frac{\partial}{\partial q} \boldsymbol{u}(q) \mathrm{d}q, \tag{10.9}$$

where $\boldsymbol{u}$ is the eigenvector associated with the dynamical problem $D(q)\boldsymbol{u}(q) = \omega^2 m\boldsymbol{u}(q)$. The two types of bulk chains (A and B) have distinct topological indices. Pal and Ruzzene (2017) showed that for both acoustic and optical branches $Z_2 = 0$ for the bulk chain B (*i.e.* when $\Lambda_1 > \Lambda_2$) and $Z_2 = 1$ for chain A (*i.e.* when $\Lambda_1 < \Lambda_2$). Thus, while both A and B chains have identical frequency spectrum, their branch topology is distinct as quantified by the different Z_2 values.

The difference in the Z_2 values for the two chain types can be further explained by considering the problem in the basis $\bar{\boldsymbol{u}} = [u^s, u^a]$, where u^s and u^a are, respectively, the symmetric and antisymmetric combinations of the displacements of the two atoms in a primitive unit cell. For chain B, at both the zone centre ($q = 0$) and zone edge ($q = \pi/a$), $\bar{\boldsymbol{u}}_{\mathrm{ac}} = [1,0]$ for the acoustic branch, and $\bar{\boldsymbol{u}}_{\mathrm{op}} = [0,1]$ for the optical branch. These modes do not flip in the range $[-\pi/a, \pi/a]$, resulting in $Z_2 = 0$. For chain A, for the acoutic branch $\bar{\boldsymbol{u}}_{\mathrm{ac}}$ takes values $[1,0]$ and $[0,1]$ at $q = 0$ and $q = \pi/a$, respectively, while for the optical branch $\bar{\boldsymbol{u}}_{\mathrm{op}}$ takes values $[0,1]$ and $[1,0]$ at $q = 0$ and $q = \pi/a$, respectively. This interchange between the acoustic and optical eigenvectors for chain A corresponds to $Z_2 = 1$ for both branches. Such a difference in the topology of the chains A and B generates a localised mode at an interface between the two chains in the superlattice structure. It should be pointed out that this topological behaviour is related to the two localised modes in the bulk band gap.

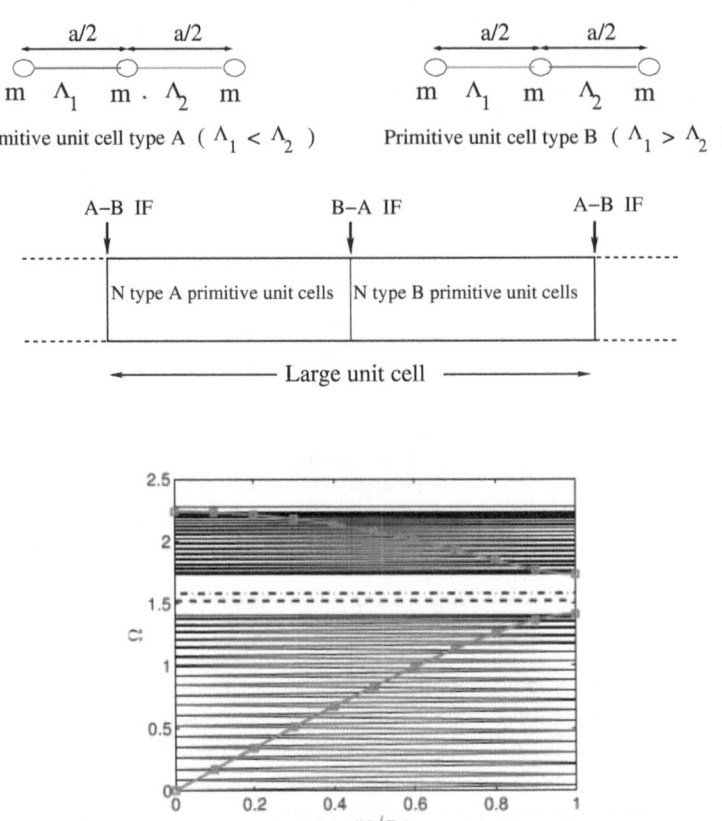

Figure 10.3 Phonon dispersion curves for a superlattice with a unit cell containing 20 contiguous primitive cells of type A (with $\Lambda_{\text{strong}} \equiv \Lambda_2$, $\Lambda_{\text{weak}} \equiv \Lambda_1$) and 20 contiguous primitive cells of type B (with $\Lambda_{\text{strong}} = \Lambda_1$, $\Lambda_{\text{weak}} = \Lambda_2$). Plotted along the y-axis is the reduced frequency $\Omega = \omega \sqrt{m/\Lambda_{\text{strong}}}$ with $\Lambda_{\text{strong}} = 1.5\,\Lambda_{\text{weak}}$. The two localised modes inside the acoustic–optical gap are topological in nature, while the mode above the optical continuum is topologically trivial. The dispersion diagram (original in colour) is reproduced from Pal and Ruzzene (2017).

Figure 10.4 shows the displacement amplitudes of the atoms for the two localised modes in the gap region. The localised displacement decays rapidly away from the interface in both cases. It is important to point out that while these localised modes are similar to the defect modes discussed in section 8.3, they cannot be eliminated by varying or changing the properties of the interfaces. Also, while these localised mode frequencies may vary with the interface properties, they cannot be moved into the bulk regions. This means that these are topologically protected interface states. In contrast, the localised mode above the bulk spectrum (*i.e.* above the bulk optical branch) is a trivial interface state, as this can be moved into the bulk bands by varying the interface type, and does not arise for low interface stiffness springs.

10.3.2   TOPOLOGICAL PHONONS IN 2D SYSTEMS

As an example of topological phonons in two-dimensional systems, Pal and Ruzzene (2017) considered a crystalline structure similar to graphene, but with dissimilar masses for the two basis atoms, so that inversion symmetry of the structure is broken. In order to follow their approach, we first discuss the lattice dynamics of graphene using the simple mass spring model.

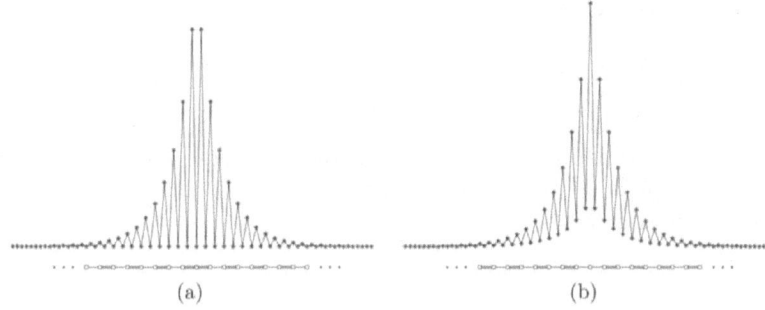

Figure 10.4 Amplitudes of (a) $\Lambda_{\text{strong}} - \Lambda_{\text{strong}}$ IF mode and (b) $\Lambda_{\text{weak}} - \Lambda_{\text{weak}}$ IF mode. For convenience, the adjacent *strong-strong* and *weak-weak* bonds are shown below the amplitude plots. [Original diagram in colour.] Reproduced from Pal and Ruzzene (2017).

10.3.2.1 Lattice dynamics of graphene

Figure 10.5 shows the atomic geometry of graphene, with a choice of the primitive unit cell and primitive translation vectors $\boldsymbol{a}_1$ and $\boldsymbol{a}_2$. Let (n_1, n_2) denote a hexagonal lattice point at $n_1\boldsymbol{a}_1 + n_2\boldsymbol{a}_2$, with n_1 and n_2 as integers. If one of the basis atoms is located at $(0,0)$, then the second basis atom is positioned at $\{\frac{1}{3}, \frac{1}{3}\}$, *i.e.* at $(\boldsymbol{a}_1 + \boldsymbol{a}_2)/3$. For the choice $\boldsymbol{a}_1 = a(1,0)$ and $\boldsymbol{a}_2 = a(\frac{1}{2}, \frac{\sqrt{3}}{2})$, the primitive translation vectors of the reciprocal lattice are $\boldsymbol{b}_1 = \frac{2\pi}{a}(1, -\frac{1}{\sqrt{3}})$ and $\boldsymbol{b}_2 = \frac{2\pi}{a}(0, \frac{2}{\sqrt{3}})$.

The equations of motion of the two basis atoms within the unit cell can be written down by extending the scheme discussed in section 2.3.2. With $\boldsymbol{u}$ and $\boldsymbol{v}$ as the two-dimensional displacement vectors of the two basis atoms, the equation of motion for the basis atoms can be written as

$$
\begin{aligned}
m\frac{\mathrm{d}^2}{\mathrm{d}t^2}\boldsymbol{u}(n_1, n_2) = {}& \Lambda_1[\boldsymbol{v}(n_1, n_2) - \boldsymbol{u}(n_1, n_2)] + \\
& \Lambda_2[\boldsymbol{v}(n_1, n_2 - 1) - \boldsymbol{u}(n_1, n_2)] + \Lambda_3[\boldsymbol{v}(n_1 - 1, n_2) - \boldsymbol{u}(n_1, n_2)]
\end{aligned}
\tag{10.10}
$$

$$
\begin{aligned}
m\frac{\mathrm{d}^2}{\mathrm{d}t^2}\boldsymbol{v}(n_1, n_2) = {}& \Lambda_1[\boldsymbol{u}(n_1, n_2) - \boldsymbol{v}(n_1, n_2)] + \\
& \Lambda_2[\boldsymbol{u}(n_1, n_2 + 1) - \boldsymbol{v}(n_1, n_2)] + \Lambda_3[\boldsymbol{u}(n_1 + 1, n_2) - \boldsymbol{v}(n_1, n_2)],
\end{aligned}
\tag{10.11}
$$

where m is the mass of a carbon atom, and Λ_1, Λ_2 and Λ_3 represent force constants as described

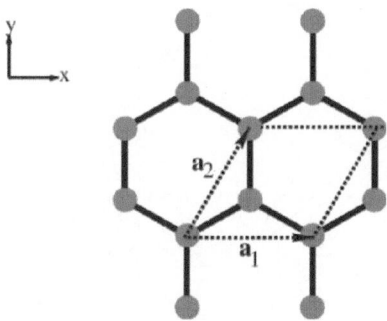

Figure 10.5 Atomic geometry of graphene. A choice of the primitive unit cell is indicated by dotted lines.

further on. The atomic displacement vectors can be written as

$$u(n_1, n_2) \;=\; A_1 \exp[i\boldsymbol{q} \cdot (n_1\boldsymbol{a}_1 + n_2\boldsymbol{a}_2) - \omega t] \tag{10.12}$$

$$v(n_1, n_2) \;=\; A_2 \exp[i\boldsymbol{q} \cdot (n_1\boldsymbol{a}_1 + n_2\boldsymbol{a}_2 + \boldsymbol{d}) - \omega t], \tag{10.13}$$

where $\boldsymbol{d} = \{\frac{1}{3}, \frac{1}{3}\}$ is the separation vector between the two basis atoms, ω is vibrational frequency, and $\boldsymbol{q} = k_1\boldsymbol{b}_1 + k_2\boldsymbol{b}_2$ is a phonon wavevector inside the Brillouin zone for $k_1, k_2 \leq 1$. The equations of motion can be be expressed as

$$\sum_j D_{ij}A_j = \omega^2 m A_i, \qquad i,j = 1,2 \tag{10.14}$$

with

$$D = \begin{bmatrix} \Lambda_1 + \Lambda_2 + \Lambda_3 & -e^{i\boldsymbol{q}\cdot\boldsymbol{d}}(\Lambda_1 + \Lambda_2 e^{-i\boldsymbol{q}\cdot\boldsymbol{a}_2} + \Lambda_3 e^{-i\boldsymbol{q}\cdot\boldsymbol{a}_1}) \\ -e^{-i\boldsymbol{q}\cdot\boldsymbol{d}}(\Lambda_1 + \Lambda_2 e^{i\boldsymbol{q}\cdot\boldsymbol{a}_2} + \Lambda_3 e^{i\boldsymbol{q}\cdot\boldsymbol{a}_1}) & \Lambda_1 + \Lambda_2 + \Lambda_3 \end{bmatrix}. \tag{10.15}$$

Note that the atomic displacements u and v, and the amplitudes A_1 and A_2, are all three-dimensional vectors, with two independent in-plane components and an out-of-plane component. And the force constants Λ_1, Λ_2 and Λ_3 are 3×3 matrices, containing radial, in-plane tangential and out-of-plane components, as detailed in Sahoo and Mishra (2012) and Saito *et al* (1998). With these considerations, the dynamical matrix in equation (10.15) is 6×6 in size, resulting in six phonon branches, as expected for a three-dimensional problem.

Sahoo and Mishra (2012) calculated phonon dispersion surves of graphene by taking into account force constant components between nearest neighbours and beyond. The results, with the choice of the force constant components given in their paper and in Saito *et al* (1998), are shown in figure 10.6. The three acoustic branches are labelled ZA, TA, LA and the three optical branches are labelled ZO, TO, LO, with ZA and ZO characterised as out-of-plane (or flexural) vibrational modes. The results obtained with consideration of interactions up to fourth nearest neighbours and presented in figure 10.6(*d*), compare quite well with *ab initio* results displayed in figure 10.7 (Dubay and Kress 2003). However, as seen from figure 10.6(*a*), inclusion of only nearest-neighbour interaction is not good enough to obtain good results for the dispersion curves. Indeed, Sahoo and Mishra's work demonstrates that even inclusion of interactions up to the third nearest neighbours does not reproduce the the quadratic dispersion of the ZA branch in the long wavelength region. The *ab inito* work by Dubay and Kress (2003) finds that neglecting the force constants before the 20th nearest-neighbours causes significant changes in either the low- or high-frequency regions.

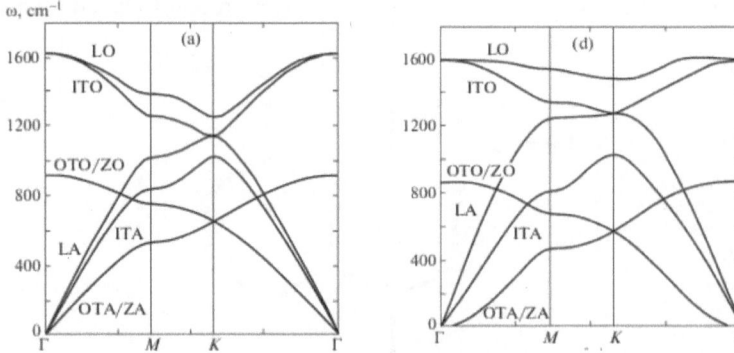

Figure 10.6 Phonon dispersion curves for graphene, obtained by using the mass and spring model and taking into account (*a*) nearest-neighbour interactions and (*d*) interactions up to fourth nearest neighbours. Reproduced from Sahoo and Mishra (2012).

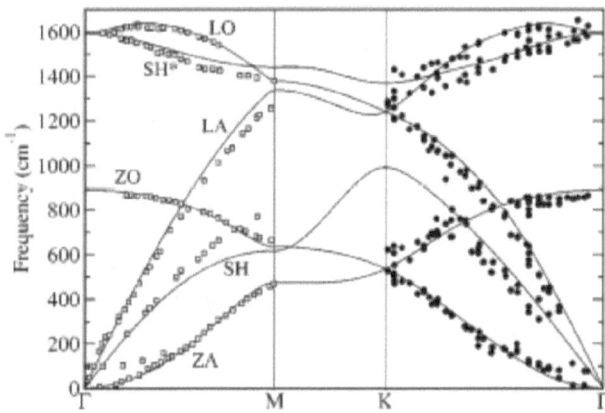

Figure 10.7 Phonon dispersion curves for graphene, obtained by using the *ab initio* method described in section 3.3.1.4. Symbols represent experimentally measured data points. Reproduced from Dubay and Kress (2003).

10.3.2.2 Topological phonons on graphene

Regardless of the interaction range, it is interesting to note that the ZO and ZA branches cross each other at the K point in the Brillouin zone. Also, these two branches are characterised with Dirac-like linear dispersion around the K point. The presence of Dirac-like linear dispersion of the ZA and ZO phonon branches at the K point on the graphene Brillouin zone prompts an investigation into the topological nature of these phonon modes. Note that there are two types of corner points on the Brillouin zone: K and K′. These are inequivalent, but time reversal exchanges K and K′. At the start of this chapter, we mentioned that crossings of linear dispersion curves in a gapless material with time reversal and inversion symmetries provides a signature for Dirac-like topology. As the graphene structure is characterised by inversion symmetry and the dispersion curves around the K point obey time reversal symmetry, we can confirm that the system is characterised with a Dirac-like phonon topology. With careful analysis of *ab initio* phonon dispersion curves for graphene, Li *et al* (2020) have suggested that there are four Dirac crossing points (DPs) of linearly dispersive phonon branches on the Brillouin zone, as shown in figure 10.8. Two of these are at the K point: one, labelled (DP1), is due to the ZO/ZA crossing and the other, labelled (DP2), is due to the LO/LA crossing. In addition, there are two more crossing points between the LO and TO branches along the $\Gamma - M$ and $\Gamma - K$ directions, labelled, respectively, as DP3 and DP4. The Berry phase

$$\gamma_n^{\text{Berry}} = \oint_C A_n(q) \cdot dl \tag{10.16}$$

was computed by defining a circle on the q_{xy} plane centred the K point. The phase for DP1 at K is $-\pi$ and at K′ is π. This indicates that the DP1 point at both K and K′ is topologically nontrivial, and also that their Berry phases are opposite to each other.

10.3.2.3 Topological phonons on graphene-like structure with broken inversion symmetry

Pal and Ruzzene (2017) investigated the topological properties of the ZA and ZO modes by performing lattice dynamical studies of a graphene-like structure with broken inversion symmetry. For this purpose, they considered the two basis carbon atoms in the primitive unit cell to have slightly different masses, $m_a = (1 + \beta)m$ and $m_b = (1 - \beta)m$ with $\beta \neq 0$, while maintaining the C_3 symmetry of the structure. Figure 10.9(*a*) shows the atomic structure.

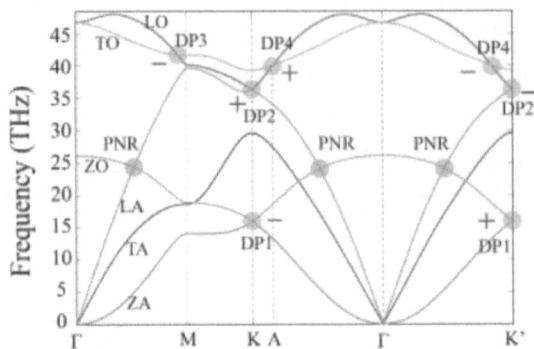

Figure 10.8 Topological Dirac points (DPs) and phonon nodal ring (PNR) in graphene. The Berry phases of the Dirac points DP1–DP4 are marked as '+' for π and '−' for $-\pi$. [Original diagram in colour.] Reproduced from Li *et al* (2020).

Considering only the nearest-neighbour out-of-plane interaction of spring constant k, the equations of motion for the masses in the primitive unit cell can be written, following equations (10.10)–(10.11), as

$$
m_a \frac{d^2}{dt^2} u_z(n_1, n_2) = k[v_z(n_1, n_2) - u_z(n_1, n_2)] +
$$
$$
k[v_z(n_1, n_2 - 1) - u_z(n_1, n_2)] + k[v_z(n_1 - 1, n_2) - u_z(n_1, n_2)] \tag{10.17}
$$

$$
m_b \frac{d^2}{dt^2} v_z(n_1, n_2) = k[u_z(n_1, n_2) - v_z(n_1, n_2)] +
$$
$$
k[u_z(n_1, n_2 + 1) - v_z(n_1, n_2)] + k[u_z(n_1 + 1, n_2) - v_z(n_1, n_2)], \tag{10.18}
$$

where u_z and v_z are the out-of-plane atomic displacement components. Following equations (10.14)–(10.15), these can be expressed as

$$
\begin{bmatrix} 3 - (1 + \beta)\Omega^2 & -e^{iq \cdot d}(1 + e^{-iq \cdot a_2} + e^{-iq \cdot a_1}) \\ -e^{-iq \cdot d}(1 + e^{iq \cdot a_2} + e^{iq \cdot a_1}) & 3 - (1 - \beta)\Omega^2 \end{bmatrix} \begin{bmatrix} A_1^z \\ A_2^z \end{bmatrix} = 0, \tag{10.19}
$$

where $\Omega = \omega\sqrt{m/k}$ is the dimensionless flexural frequency, and A_i^z is the amplitude of the ith flexural mode.

Figure 10.9(*b*) shows the dispersion curves for the ZA and ZO branches. For $\beta = 0$, the dashed lines show the crossing of the Dirac-like dispersion curves at the K point, in agreement with the result presented in figure 10.6(*a*). The results in solid curves, with $\beta = -0.2$, show that the DP is destroyed and a clear gap opens up at the K point. The same dispersion curves would appear with $\beta = +0.2$, but with the eigenvectors A_1^z and A_2^z flipped.

The Chern number C at the K and K' valleys can be computed by integrating the Berry curvature over a small region near the K and K' points (*cf* equation (10.1)). It is found that for $\beta > 0$, $C = -\frac{1}{2}$ for the ZA branch and $C = \frac{1}{2}$ for the ZO branch. For $\beta < 0$, the values are reversed. This means that breaking the inversion symmetry results in distinct valley Chern numbers for these branches. The bulk boundary correspondence, therefore, guarantees the presence of topologically protected localised modes inside the gap region between the the ZA and ZO branches at the interface between two strips, one having $\beta > 0$ (*i.e.* $m_a > m_b$) and the other having $\beta < 0$ (*i.e.* $m_b > m_a$).

The above approach can be extended to examine the phonon states around the K and K' valleys for a honeycomb structure composed of A/B basis atoms with broken inversion symmetry ($\mathscr{P}$), broken time-reversal symmetry ($\mathscr{T}$) and a combination of $\mathscr{P}$ and $\mathscr{T}$. Liu *et al* (2018) have summarized

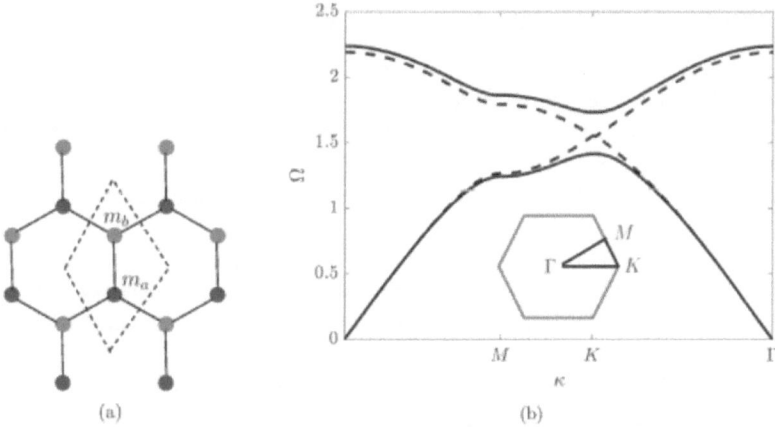

(a)   (b)

Figure 10.9 (*a*) Graphene-like structure with broken inversion symmetry. (*b*) Dispersion curves of the flexural branches ZA and ZO for the structure in (*a*). [Original diagram in colour.] Reproduced from Pal and Ruzzene (2017).

the results. Figure 10.10 shows the Dirac phonon cones, Barry curvature distribution and topological boundary phonon states for these situations. For the inversion and time-reversal symmetric case (*i.e.* when the two basis atoms have the same mass, as in graphene), identical Dirac cones are formed at the K and K′ valleys [figure 10.10(*a*)]. If only time-reversal symmetry is broken, $C = \pm 1$, and topologically protected phonon modes exist within the Dirac gap [figure 10.10(*b*)]. If only inversion symmetry is broken, $C = 0$, corresponding to a topologically trivial phase. However, for an interface structure with opposite basis atom masses on the two sides, there exist topologically protected and valley-momentum locked interface modes [figure 10.10(*c*) and the discussion presented above]. When both inversion and time-reversal symmetries are broken simultaneously, the K and K′ valleys are no longer frequency degenerate and their gaps can be tuned independently [figure 10.10(*d*)].

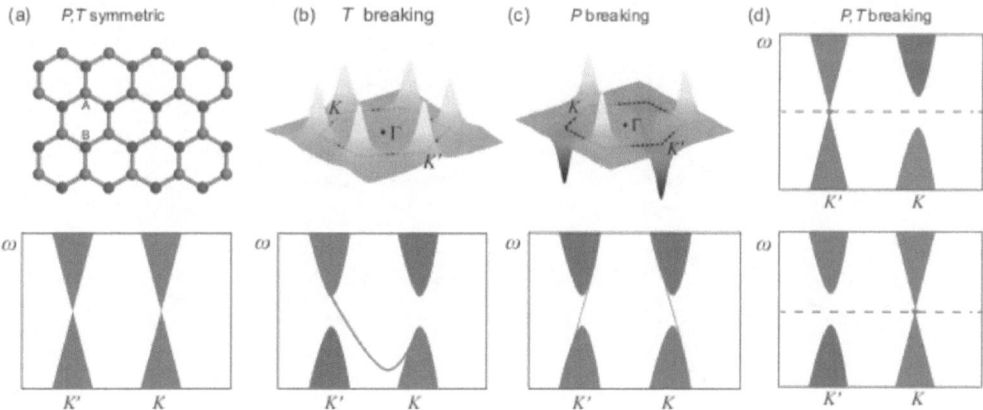

Figure 10.10 (*a*) Graphene-like honeycomb structure with basis atoms A and B (top), and the Dirac cones at the K and K′ valleys (bottom). (*b*)–(*d*) Berry curvature distribution (top) and Dirac cones (bottom) with broken time-reversal symmetry ($\mathcal{T}$), broken inversion symmetry ($\mathcal{P}$) and broken $\mathcal{P}\mathcal{T}$ symmetry, respectively. Red and blue colours represent phonon states localized at A and B basis atoms, respectively. [Original diagram in colour.] Reproduced from Liu *et al* (2018).

10.3.3 TOPOLOGICAL PHONONS IN 3D SYSTEMS

In the previous section we discussed the concept of Dirac phonons in 2D materials. Dirac phonons have also been predicted in three-dimensional (3D) materials. Similarly, several 3D TSMs have been identified. There are two distinct types of Weyl fermions in Weyl semimetals (WSMs). Type-I Weyl fermions are associated with a closed point like Fermi surface. Type-II Weyl fermions have an open Fermi surface. Similar to type-I and type-II Weyl fermions, there are also type-I and type-II Weyl phonons.

Here we present the results of theoretical predictions for the existence of Dirac phonons in Si, as discussed in Chen *et al* (2021). We also exemplify identification and analysis of the ideal type-II Weyl phonons in zinc-blende CdTe and wurtzite ZeSe, as presented by Xia *et al* (2019) and Liu *et al* (2021), respectively.

10.3.3.1 Dirac phonons in body-centred Si

In 3D semimetals, DPs in phonon spectrum, preserved by time reversal and inversion symmetry, are four fold degenerate with linear dispersion along all directions in momentum space. DPs can exist either at symmetry points (essential degeneracy) or along high symmetry lines (accidental degeneracy). Chen *et al* (2021) theoretically predicted the presence of Dirac phonons in Si (*cI*16), a metastable silicon allotrope with space group *Ia*-3, which crystallizes in a body-centred cubic structure with 16 atoms per primitive unit cell (Wosylus *et al* 2009).

Figure 10.11 shows the crystal structure, Brillouin zone and the phonon spectrum along the high symmetry directions of body centred Si (*cI*16). Also shown in that figure are phonon surface states and arcs projected on the (110) plane. The calculations were performed by employing the *ab inito* method described in section 3.3.1.4. The four-fold degenerate DP at around 8 THz is located at the symmetry point P. There is also a quadratic triple degenerate point (QTP) with frequency slightly above 8 THz at the symmetry point H. The topological nature of of these points was verified by

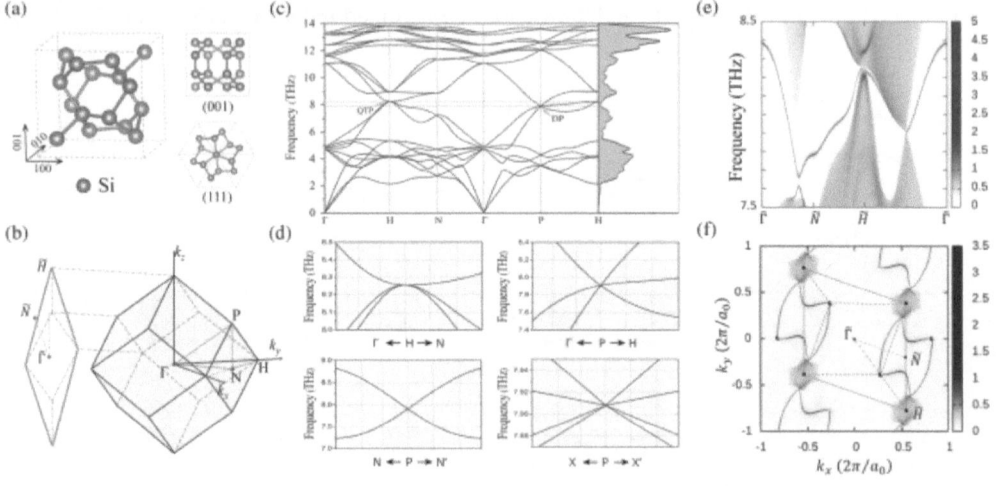

Figure 10.11 (*a*) The body-centred Si (*cI*16) structure viewed in the conventional cubic unit cell. (*b*) The Brillouin zone for the BCC lattice. (*c*) Phonon dispersion curves and density of states. (*d*) Details of phonon dispersion curves around the quadratic trip point (QTP) and the Dirac point (DP). (*e*) Phonon states on the (110) surface. Each of the two sets of surface states represents Weyl phonons. (*f*) Surface phonon arcs projected on the (110) surface. Phonon surface arcs cross over the BZ boundary and connect the projections of two nonequivalent DPs. [Original diagram in colour.] Reproduced from Chen *et al* (2021).

computing phonon local density of states and the corresponding isofrequency surface projected on the (110) surface. The surface states are composed of two sets, with each set viewed as the Weyl phonon states with opposite chirality. The phonon surface arcs cross over the Brillouin zone boundary and connect the projections of two nonequivalent DPs.

10.3.3.2 Weyl phonons in zinc-blende CdTe

Xia *et al* (2021) have investigated topological phonons in zinc-blende CdTe. Figure 10.12 shows the phonon spectrum of CdTe, obtained from the application of the *ab intio* method decribed in section 3.3.1.4, without the inclusion of spin-orbit interaction. There is a doubly-degenerate point in the X-W direction, originating from the crossing of the longitudinal acoustic (LA) and transverse optical (TO) branches at 3.5 THz. This crossing with linear dispersion along X-W is with respect to the two-fold rotational symmetry C_2 at the Brillouin zone boundary. The two crossing branches belong to opposite eigenvalues ± 1 of the symmetry operator C_2. A 3D Dirac phonon can be treated as the overlap of two Weyl phonons with opposite chirality. As the crossing of the two inverted phonon branches forms a tilted DP, it can be inferred that the crossing point belongs to type-II Weyl phonons. Detailed investigations, considering crystal symmetry, suggest that there are a total of 12 WPs along the symmetry direction X-W, where the C_2 symmetry coexists with the time-reversal symmetry. Xia *et al* determined the chirality $C = +1$ for phonon wavevectors $\left(\frac{2\pi}{a}, 0, \pm p\right)$, $\left(\pm p, \frac{2\pi}{a}, 0\right)$ and $\left(0, \frac{2\pi}{a}, \pm p\right)$, and the chirality $C = -1$ at $\left(\frac{2\pi}{a}, \pm p, 0\right)$, $\left(0, \pm p, \frac{2\pi}{a}, 0\right)$ and $\left(\pm p, 0, \frac{2\pi}{a}\right)$, with $p = 0.054$ Å^{-1}. Figure 10.13 shows the phonon isofrequency surface on the $q_x - q_y$ plane with $q_z = 2\pi/a$ (*a*) below, (*b*) at and (*c*) above the Weyl phonon frequency, and the locations of WPs (*d*) at the Brillouin zone {001} boundaries.

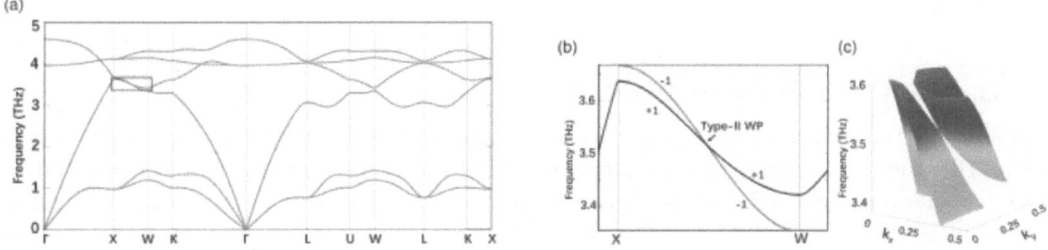

Figure 10.12 (*a*) Phonon dispersion curves for CdTe. (*b*) Enlarged version of the spectrum marked by the box in (*a*). (*c*) A tilted Weyl phonon formed by the inverted phonon branches on the $q_x - q_y$ plane with $q_z = 2\pi/a$, where *a* is the cubic lattice constant. [Original diagram in colour.] Reproduced from Xia *et al* (2019).

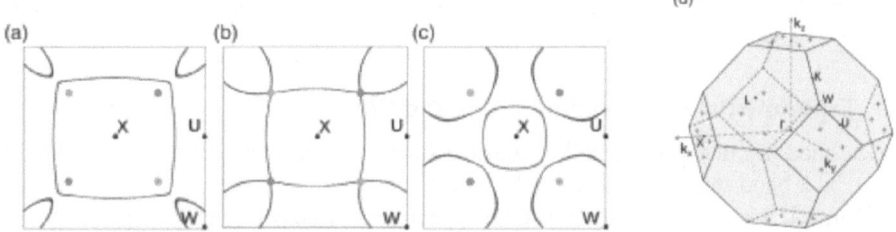

Figure 10.13 CdTe phonon isofrequency surface on the $q_x - q_y$ plane with $q_z = 2\pi/a$ (*a*) below, (*b*) at and (*c*) above the Weyl phonon frequency. (*d*) Locations of WPs at the Brillouin zone {001} boundaries. [Original diagram in colour.] Reproduced from Xia *et al* (2019).

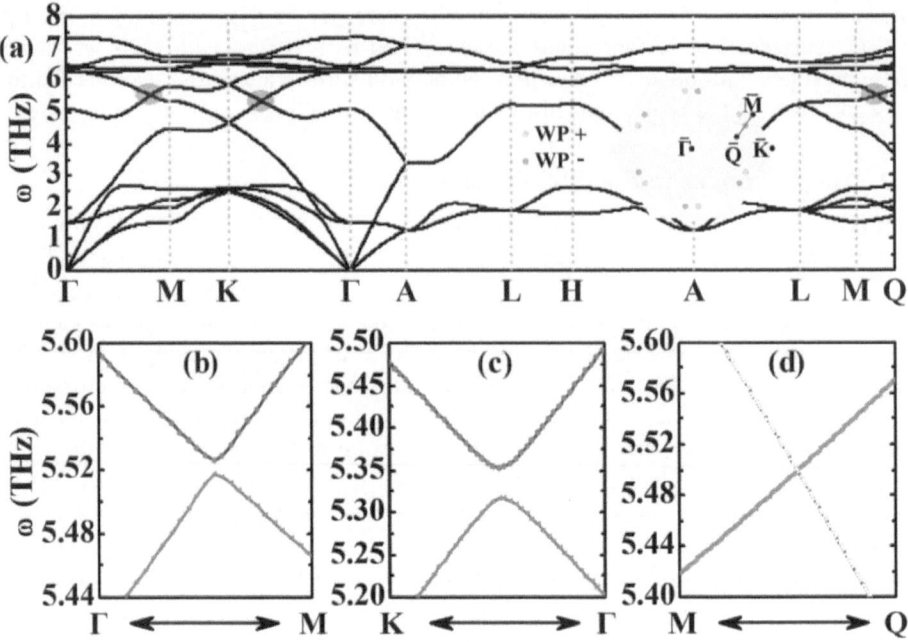

Figure 10.14 (*a*) Phonon dispersion curves for wurtzite ZnSe. (*b*)–(*d*) Enlarged version of dispersion curves in the range of optical branch crossings along $\Gamma - M$, $\Gamma - K$ and $M - Q$, respectively. [Original diagram in colour.] Reproduced from Liu *et al* (2021).

10.3.3.3 Weyl phonons in wurtzite ZnSe

Liu *et al* (2021) have made first-principles prediction of ideal type-II Weyl phonons in wurtzite ZnSe. The wurtzite structure is a noncentrosymmetric hexagonal structure with space group P6$_3$mc with four atoms per unit cell. Phonon eigensolution calculations were made by using the theoretical method described in section 3.3.1.4. The phonon dispersion curves are shown in panel (*a*) of figure 10.14. Enlarged versions of the dispersion curves along the symmetry directions $\Gamma - M$, $\Gamma - K$ and $M - Q$ in the frequency range 5.40 – 5.60 THz are presented in panels (*b*)–(*d*). The inversion of two optical branches gives rise to a WP along the M–Q symmetry line.

On the basis of time-reversal, mirror and sixfold rotational symmetries present in the system, 12 WPs have been predicted in the $q_z = 0$ plane. Six of these are characterised with the topological charge chirality $C = +1$ and the other six with the topological charge chirality $C = -1$. Clear verification of the type-II WPs was made from an examination of the topologically protected nontrivial surface phonon states and Fermi arcs.

10.3.3.4 Experimental investigation of DPs and WPs

Most optical measurements probe bulk electronic states in a material. As Dirac and Weyl fermions exist within bulk band structure, these can be investigated in optical measurements. Ultrafast pump-probe techniques, which are routinely used to generate and detect coherent optical phonons (see, chapter 13 for discussion), can therefore be used to investigate and control Dirac and WP in DSMs and WSMs.

Sie *et al* (2019) demonstrated that terahertz light pulses can be used to induce terahertz-frequency interlayer shear strain with large amplitude to produce a topologically distinct metastable phase in WSM WTe$_2$. Zhang *et al* (2019) have reported structural transition between the type-II Weyl semimetal phase and normal semimetal phase in bulk crystalline $WoTe_2$ by using ultrafast pump-probe and time-resolved second-harmonic-generation spectroscopy. Reflectivity changes, measured in pump-probe spectroscopy, are induced by coherently generated optical phonons of irreducible representation A_1. Weber (2021) has reviewed the topic of ultrafast investigation and control of Dirac and Weyl semimetals. It is pointed out that optical pulses may controllably and reversibly move, split, merge, or gap the material's Dirac and Weyl nodes. It is suggested that coherent phonons excited by an ultrafast optical pulse may offer mechanisms for similar control of the nodal structure. For example, using an ultrafast optical pulse, a DP may be split into a pair of Weyl nodes, or simply a gap at the DP may be opened. Advances in reversible control of the topological behaviour are expected to generate highly significant new technological applications, such as turning the proposal of a "topological field effect transistor" (Qian *et al* 2014) into reality.

10.4 PHONONIC SYSTEMS

Some naturally occurring crystals are characterised with absolute frequency gap(s) (clear frequency gap(s) in all directions). Some examples, with a clear optical–acoustic gap, are the III-V semiconductors AlN, AlAs, AlSb, GaN, GaP, GaSb, InN, InP, InAs, and II-VI semiconductors ZnS, ZnTe, CdS, CdSe, HgS, HgTe. An absolute frequency gap in a compound semiconductor arises from the combined effects of the mass ratio, crystal structure and the harmonic force constant matrix.

Phononic crystals are characterised by the presence of frequency gap(s) in their phonon spectrum in which vibrational energy cannot propagate and phonon wave cannot be focussed. Usually such crystals are artificial materials fabricated in laboratory with a periodic structure comprised of two or more constituents. The periodicity involved in phononic crystals is much larger (in the nm–m range) compared to that in natural crystals (in the sub-nm range). Nanophononic crystals have periodicities in the nm range. Acoustic metamaterials are also phononic crystals, but with some difference: periodicity in metamaterials may be similar to phononic crystals but the physical dimension of the structure is small compared to the wavelength of the acoustic wave.

Several experiments have revealed phononic band gaps in planar superlattice (PSL) structures. The first demonstration of such a gap was observed by Narayanamurti *et al* (1979) in the LA phonon branch for a $GaAs/Al_{0.5}Ga_{0.5}As[111]$ superlattice of equal individual layer thickness and period size 122 Å. Using superconducting tunnel junctions as the source and detector of quasimonochromatic phonons, the centre of the gap was determined at 0.93 meV (232 GHz) with a width of 0.16 meV (39 GHz).

Criteria for true phononic gap in two-component structures

Let us consider a ball-and-spring model of a periodic structure consisting of two atom types A and B with masses m_A and m_B and mass ratio $m = m_A/m_B$. The diatomic linear chain model discussed in section 2.2.2 clearly suggests the presence of a phononic gap in a 1D system when $m > 1$. For simplicity, assume square (cubic) unit cell for the 2D (3D) structure of unit cell size $L \times L$ ($L \times L \times L$), in which heavier atoms of type A are arranged in a square area (cubic volume) of size $\ell \times \ell$ ($\ell \times \ell \times \ell$). The packing fraction for the considered 2D (3D) structure is $P_f = L_f^2 (L_f^3)$, where $L_f = \ell/L$. As mentioned earlier, phononic gaps are controlled by a combination of factors including the mass ratio m, length fraction L_f, differences in force constant components and dimensionality of system. From model calculations, Hepplestone and Srivastava (2008) established that a true 2D (3D) phononic gap in square (cubic) structure will appear when $m > 10$ ($m > 15$) within the range $0.9 > L_f > 0.1$. These simple model criteria, however, generally prove insufficient in predicting phononic gap results for real systems.

Si/SiGe PSL as 1D nanophononic structure

Ezzahri *et al* (2007) presented generation and detection of coherent LA wave packet in a $Si(4\ nm)/Si_{0.4}Ge_{0.6}(8\ nm)[001]$ superlattice by femtosecond laser pulses. By analysing the 1D

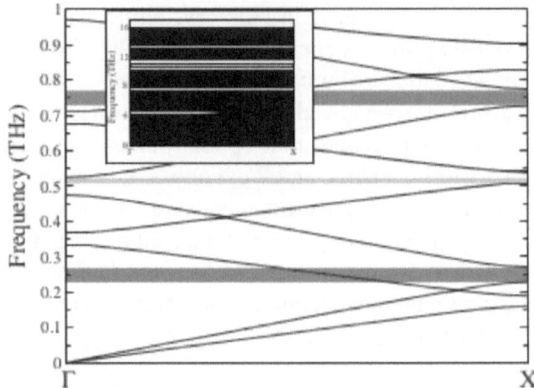

Figure 10.15 Phonon dispersion curves a Si(4 nm)/Si$_{0.4}$Ge$_{0.6}$(8 nm)[001] superlattice, obtained from the application of the bond charge model. Three lowest LA frequency gaps are highlighted, with the central region showing a true phononic gap. [Original diagram in colour.] Reproduced from Hepplestone and Srivastava (2008).

phonon dispersion relation using the continuum theory described in section 9.2, they concluded that the system exhibits three LA band gaps at 145 THz, 283 THz and 527 THz. Hepplestone and Srivastava (2008) made a full-scale (3D) theoretical investigation of the phonon spectrum of the same superlattice using the bond charge model (*cf* section 2.4.3.2). They obtained several frequency gaps for both LA and TA polarisations in the entire frequency range covering acoustic as well as optical regions), as shown in figure 10.15. This work predicts two lowest TA gaps at 175 THz and 350 THz, and two lowest LA gaps at 252 THz and 495 THz. The LA gaps are in agreement with the experimental results in Ezzahri *et al* (2007). In addition, the full-scale calculations reveal the presence of a true phononic gap between 507 and 515 THz, in which phonons of neither polarisation can propagate.

10.5 NANOPHONONIC TOPOLOGY

In section 10.4 we noted that PSL structures are characterised with the presence of phononic band gaps along their growth direction. The essential phononic physics of such systems is that of a 1D atomic chain with either alternating masses, or identical masses but alternating force constants, or both, as discussed in section 10.3.1. It was also shown in figure 10.3 that topologically robust interface states are generated in a phononic gap common to two 1D atomic chains with different topological indices. These considerations suggest that with fabrication of nanostructures containing interfaces between PSLs with periodicity in the nm range and different topological indices it would be possible to engineer topologically robust nanophononic devices. Esmann *et al* (2018) have successfully constructed a topologically robust interface state at 350 GHz at the junction of two concatenated GaAs/AlAs superlattices with different layer thicknesses. Their approach is described below.

Consider a GaAs/AlAs superlattice of period $d_0^{\text{GaAs}} + d_0^{\text{AlAs}}$ and layer thicknesses $d_{\text{GaAs}} = d_0^{\text{GaAs}}(1 + \delta)$ and $d_{\text{AlAs}} = d_0^{\text{AlAs}}(1 - \delta)$, with δ considered as a parameter in the range $-1 < \delta < 1$. This is shown in figure 10.16(*a*), with $d_{\text{GaAs}} = d_0^{\text{GaAs}}(1 + \delta) = a = \lambda/2$, where λ is a phonon wavelength. Theoretical calculations were performed by using the scheme described in section 9.2. Figure 10.16(*b*) shows the phonon dispersion curves for the structure with $d_0^{\text{GaAs}} = 6.7$ nm, $d_0^{\text{AlAs}} = 8.0$ nm, $d_0^{\text{GaAs}}\delta = 0.8$ nm and $d_0^{\text{AlAs}}\delta = 0.67$ nm. As δ is varied from -1 to $+1$, there is an exchange of the atomic displacement patterns, as well as of the Zak phases (*i.e.* the Berry phases for Bloch bands in one dimension), with respect to the centres of the material layers, as shown in figure 10.16(*c*).

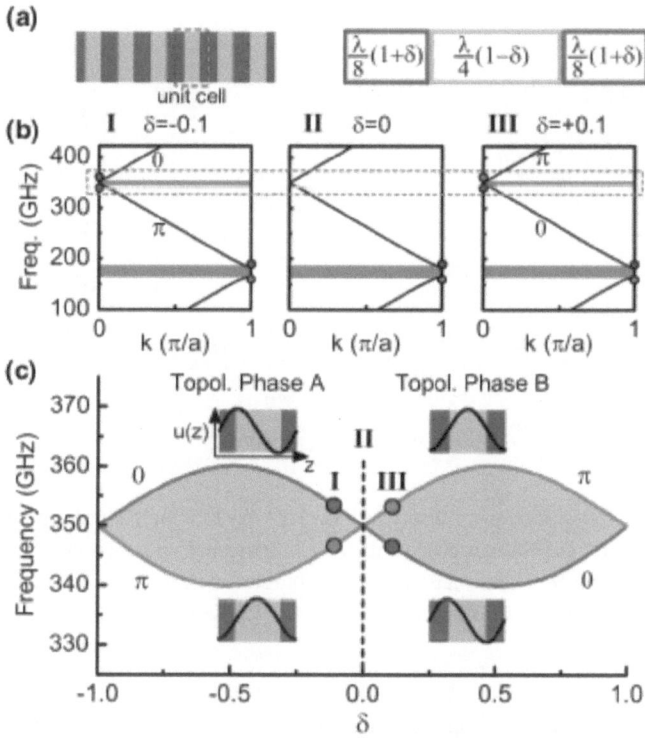

Figure 10.16 Phonon dispersion curves and topological phases of a nanophononic GaAs/AlAs superlattice containing 20 centrosymmetric unit cells: (*a*) Unit cell with layer thicknesses; (*b*) dispersion curves and band gaps; (*c*) topological phases. See text for details. [Original diagram in colour.] Reproduced from Esmann *et al* (2018).

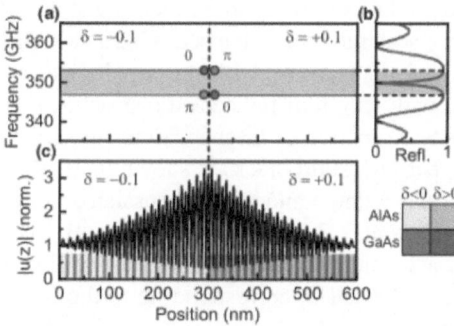

Figure 10.17 Topological interface state at 350 GHz at the junction of two concatenated nanophonic GaAs/AlAs superlattices shown in figure 10.16: (*a*) local band diagram; (*b*) phonon reflectivity; (*c*) displacement pattern, $|u_{n,k}|$, of the topological interface state together with that for the superlattice structure. The dotted vertical line separates different topological phases. [Original diagram in colour.] Reproduced from Esmann *et al* (2018).

Figure 10.17(*a*) shows the phonon dispersion cuves for a system containing two concatenated nanophonic GaAs/AlAs superlattices shown in figure 10.16: one with $\delta = -0.1$ and the other with $\delta = +0.1$. The dip in the phonon reflectivity in (*b*) corresponds to the topological mode confined to the interface between the two superlattices. The corresponding displacement pattern $|u(z)|$ is shown in (*c*).

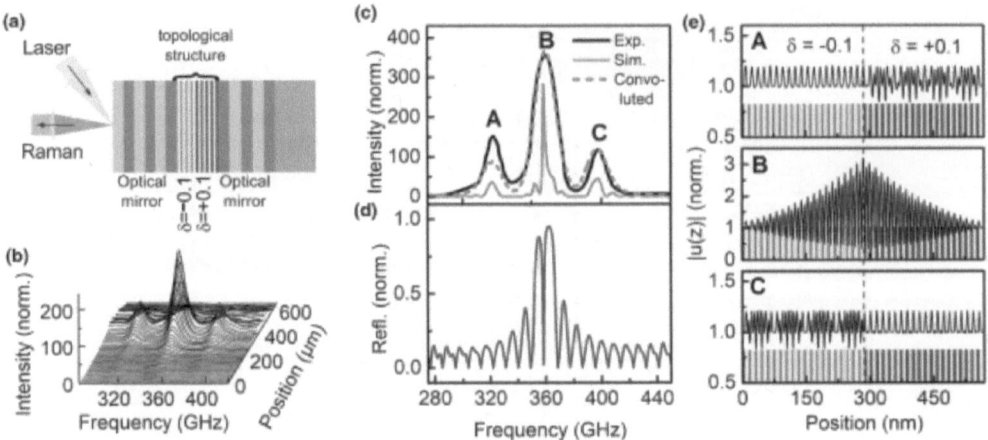

Figure 10.18 High-resolution Raman spectrum of a GaAs/AlAs topological nanophononic interface phonon mode. (*a*) Sample structure; (*b*) Raman spectrum; (*c*) Experimental and theoretical spectra; (*d*) Simulated acoustic reflectivity spectrum; (*e*) Confirmation of the topological nature of the interface state B, compared to the non-topological nature of the peaks A and C in panel (*c*). [Original diagram in colour.] Reproduced from Esmann *et al* (2018).

To confirm the esistence of topological states in real nanophononic systems, Esmann *et al* performed all-optical Raman scattering measurements on a system containing two concatenated nanophonic GaAs/AlAs superlattices shown in 10.18(*a*), which is similar to that in figure 10.16(*a*). The sample was grown by molecular-beam epitaxy on a GaAs(001) substrate and consisted of two parts in the central region: 20 superlattice unit cells with GaAs and AlAs bilayer thicknesses corresponding to $\delta = -0.1$ and 20 superlattice unit cells with GaAs and AlAs bilayer thicknesses corresponding to $\delta = +0.1$. This structure produces a common frequency gap at 354 GHz with inverted phonon branches for the two parts. This structure was enclosed within two GaAlAs-based optical mirrors, which serve as a resonant optical microcavity with the topological acoustic structure acting as a space of width 2λ (the optical path length of the acoustic structure). Double optical resonances in Raman spectra were recorded as a function of laser incidence position on the sample, as shown in figure 10.18(*b*). The Raman spectrum (solid black) is shown in 10.18(*c*). Also shown in this figure are photoelastic model calculations without and with a Gaussian convolution in solid red and dashed red, respectively. Three clear peaks at 323 GHz (A), 360 GHz (B) and 397 GHz (C) are observed. The clear dip in the simulated acoustic reflectivity of the sample shown in panel (*d*) is assigned to the topological interface mode at 360 GHz, corresponding to the Raman peak B in panel (*c*). The middle part in panel (*e*) confirms the topological nature of the interface mode B.

These concepts, of established direct connection between centrosymmetric unit cells and standard bilayers, bridging the fields of topology and nanophonics, are expected to be useful in optoelectronic, photonic and optomechanical applications.

10.6 PHONON CHIRALITY

In this section we will outline the concept of topologically chiral phonons in nanostructured systems.

10.6.1 PHONON ANGULAR MOMENTUM IN CRYSTALS

The angular momentum vector of lattice vibrations is defined as

$$J = \sum_{lb} u(lb) \times m_b \dot{u}(lb), \tag{10.20}$$

where $u(lb)$ is the displacement vector for the bth atom of mass m_b in the unit cell located at l. One of the components, say the z-component, can be expressed as

$$
\begin{aligned}
J_z &= \sum_b m_b \sum_l \left[u_x(lb)\dot{u}_y(lb) - u_y(lb)\dot{u}_x(lb) \right] \tag{10.21}\\
&= \sum_{lb} \begin{pmatrix} u_x(lb) \\ u_y(lb) \end{pmatrix}^T \begin{pmatrix} 0 & m_b \\ -m_b & 0 \end{pmatrix} \begin{pmatrix} \dot{u}_x(lb) \\ \dot{u}_y(lb) \end{pmatrix}. \tag{10.22}
\end{aligned}
$$

Using equations (2.48), (2.56), (4.9), (4.30) and (4.32), we can write

$$u(lb) = \frac{-i}{\sqrt{N_0 \Omega}} \sum_{qs} \sqrt{\frac{\hbar}{2m_b \omega_{qs}}} (a_{qs}^\dagger - a_{-qs}) e(b|qs) \exp[i(q \cdot l - \omega_{qs}t)], \tag{10.23}$$

where $e(b|qs)$ is a vibrational eigenvector. Using equations (10.3), (10.4), (2.58), (4.13), (4.26) and (4.29), the quantum mechanical operator $\hat{J}_z$ can be expressed as

$$\hat{J}_z = \hbar \sum_{qs} \sum_b e^T(b|qs) \begin{pmatrix} 0 & -m_b \\ m_b & 0 \end{pmatrix} e(b|qs)(a_{qs}^\dagger a_{qs} + \frac{1}{2}\hat{I}), \tag{10.24}$$

where

$$e(b|qs) = \begin{pmatrix} e_x(b|qs) \\ e_y(b|qs) \end{pmatrix}. \tag{10.25}$$

The operation of $\hat{J}_z$ on the phonon state $|n_{qs}\rangle$ produces, using equation (4.38),

$$\hat{J}_z|n_{qs}\rangle = \hbar j_z |n_{qs}\rangle, \tag{10.26}$$

where

$$j_z = \sum_{qs} \sum_b e^T(b|qs) \begin{pmatrix} 0 & -m_b \\ m_b & 0 \end{pmatrix} e(b|qs)(\bar{n}_{qs} + \frac{1}{2}). \tag{10.27}$$

At $T = 0$, $\bar{n}_{qs} = 0$ and the zero-point contribution to the z-component of the phonon angular momentum is

$$\frac{1}{2} \sum_{qs} \sum_b e^T(b|qs) \begin{pmatrix} 0 & -m_b \\ m_b & 0 \end{pmatrix} e(b|qs).$$

At high temperatures $\bar{n}_{qs} \approx k_B T / \hbar \omega_{qs}$ and all excited phonon modes contribute equally to the angular momentum component. Non-magnetic crystals are characterised by the presence of time-reversal symmetry. Using the symmetry properties of the dynamical eigenvalue problem, it can be shown that $j_z(qs) = -j_z(-qs)$, so that the total contribution to the each component of the angular momentum is zero.

10.6.2 PHONON PSEUDO ANGULAR MOMENTUM IN MONOLAYER THIN CRYSTALS

It is useful to note that the inversion (or, parity) symmetry ($\mathscr{P}$) and the time-reversal symmetry ($\mathscr{T}$) transform the linear momentum q and angular momentum j of a phonon in a crystal as follows:

$$
\begin{aligned}
\mathscr{P}(q,j) &\rightarrow (-q,j) \tag{10.28}\\
\mathscr{T}(q,j) &\rightarrow (-q,-j) \tag{10.29}\\
\mathscr{P}\mathscr{T}(q,j) &\rightarrow (q,-j). \tag{10.30}
\end{aligned}
$$

It follows that a phonon mode with wavevector q will have a zero angular momentum if it is characterised by the presence of the $\mathscr{P}\mathscr{T}$ symmetry. In order for a phonon to acquire a finite angular momentum, either the $\mathscr{P}$ or the $\mathscr{T}$ symmetry must be broken.

Atomically thin honeycomb structures AB lack the inversion symmetry ($\mathscr{P}$) within the plane and the time reversal symmetry ($\mathscr{T}$) is broken at the K and K' points in the hexagonal Brillouin zone. Due to these reasons, the phonon modes at K and K' are non-degenerate and acquire definite pseudo angular momenta (PAMs). The symmetry at these points is C_3 which produces a phase change to the eigenvector

$$\mathscr{R}\left(\frac{2\pi}{3},z\right)\boldsymbol{\eta}(b|qs) = e^{-i\frac{2\pi}{3}j_z(b,qs)}\boldsymbol{\eta}(b|qs), \tag{10.31}$$

where

$$\boldsymbol{\eta}(b|qs) = \boldsymbol{e}(b|qs)e^{i\boldsymbol{\tau}_b\cdot\boldsymbol{q}}, \tag{10.32}$$

with $\boldsymbol{\tau}_b$ as the coordinate of the bth basis atom in the unit cell. In above, $j_z(\boldsymbol{b},\boldsymbol{qs})$ is the PAM, or the z-component of the angular momentum, for the bth basis atom at $\boldsymbol{q}$ (K or K'). The value of the PAM of a phonon branch at K or K' can be -1, 0 or $+1$.

Using a spring and ball model, with $m_A = 1$, $m_B = 1.2$, longitudinal spring constant $\Lambda_L = 1$ and transverse spring constant $\Lambda_T = 0.25$, Zhang and Niu (2015) calculated the in-plane phonon dispersion curves for a two-dimensional honeycomb AB structure, and evaluated the PAM for different phonon polarisation branches at the BZ points K and K'. These are shown in figure 10.19.

10.6.3   PHONON CHIRALITY IN MONOLAYER THIN CRYSTALS

The chirality of the phonon modes at the K and K' points for a honeycomb AB structure is related to their PAMs. Let us expand the eigenvector e in right-handed and left-handed components

$$\boldsymbol{e} = e_R|R\rangle + e_L|L\rangle \tag{10.33}$$

and define the circular polarisation operator

$$\hat{\mathscr{C}}_z = |R\rangle\langle R| - |L\rangle\langle L|. \tag{10.34}$$

The eigenvalue of this operator produces the phonon circular polarisation

$$C_z = \hbar\boldsymbol{e}^\dagger\hat{\mathscr{C}}_z\boldsymbol{e} = \hbar\sum_b(|e_{R_b}|^2 - |e_{L_b}|^2) \tag{10.35}$$

$$= \sum_b C_z(\boldsymbol{b}) = \sum_b\sum_s C_z(\boldsymbol{b},s). \tag{10.36}$$

The left-hand ($C_z < 0$) and right-hand ($C_z > 0$) polarisations indicate the chirality of the phonon modes. The eigenvalue $C_z(\boldsymbol{b})$ represents the contribution of the bth basis atom to the phonon circular polarisation. At the K' point, as shown in figure 10.19, $C_z(A,2) = 0$, $C_z(B,2) = -\hbar$, while $C_z(A,3) = \hbar$ and $C_z(B,3) = 0$. In contrast, opposite circular vibrations of atoms A and B result in the PAM being zero for the phonon bands (polarisation indices) 1 and 4.

Helically-resolved Raman scattering experiments have been performed to investigate phonon chirality in monolayer transition-metal dichalcogenides (TMDs) (Chen *et al* 2015). In another experiment, Zhu *et al* (2018) identified phonons in monolayer WSe$_2$ by the intervalley transfer of holes through hole-phonon interactions during the indirect infrared absorption and confirmed their chirality by the infrared circular dichroism.

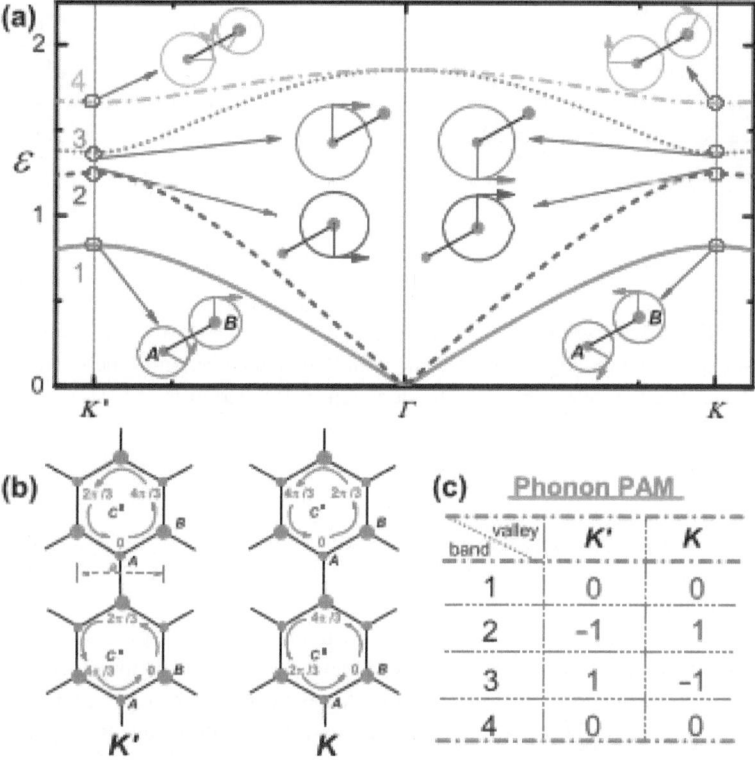

Figure 10.19 (*a*) Model in-plane phonon dispersion curves for a honeycomb AB structure, using the ratio 0.25 of the transverse and longitudinal force constants and the mass ratio $m_B/m_A = 1.2$. The phonon energy along the vertical axis is $\varepsilon = \hbar\omega$. (*b*) Phase correlation of the basis atoms A (upper two panels) and B (lower panels) at K′ (left panels) and K (right panels) points of the Brillouin zone. (*c*) Phonon pseudo angular momentum (PAM) for phonon polarisation branches 1–4 at K′ and K. [Original diagram in colour.] Reproduced from Zhang and Niu (2015).

10.6.4 TOPOLOGICALLY CHIRAL PHONONS

Topology and chirality can co-exit in crystals. Using a tight-binding model, Chang *et al* (2018) have investigated the topological electronic property of a class of chiral crystals. Angle resolved photoemission experiments have been performed to investigate surface electronic topology in the chiral topological crystals, some examples being CoSi, RhSi (Sanchez *et al* 2019) and PdGa (Schröter 2020). In analogy with the concept of topologically chiral electronic states, we can think of topologically chiral phonon modes.

As discussed in the previous sub-section, chiral phonons exist in asymmetric hexagonal two-dimensional systems. Xu *et al* (2018) have found nontrivial topology of chiral phonons in a bilayer system made of a monatomic hexagonal structure of atoms A (atomic mass m_A) centre-stacked with another monatomic hexagonal structure of atoms B (atomic mass m_B). Their *ab initio* theoretical work shows that circularly polarised phonons with nonzero Berry curvature are observed at the K and K′ points in the BZ, and that the chirality of those modes remains robust with change in the mass ratio of the A and B atoms and interlayer coupling.

In another *ab initio* investigation, Li *et al* (2021) find a chiral interface phonon mode at the line defect in the monolayer hexagonal boron nitride (h-BN) intralayer heterojunction, which is topologically protected. For this they modelled the line defect at the interface between the ordinary

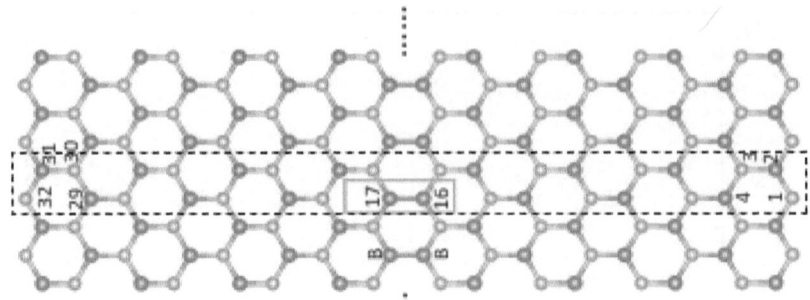

Figure 10.20 $\alpha\beta\beta\alpha$-type zigzag monolayer hexagonal boron nitride (h-BN) nanoribbon. The dashed rectangle indicates the 32-atom unit cell and the rectangle indicates the location of the interface (*i.e.* line defect) between the $\alpha\beta$ and $\beta\alpha$ type nanoribbons. The direction of the period is indicated by the dots. [Original diagram in colour.] Reproduced from Li *et al* (2021).

$\alpha\beta\alpha\beta$-type h-BN and $\beta\alpha\beta\alpha$-type h-BN zig-zag nanoribbons of a finite width, as shown in figure 10.20. Phonon eigensolutions were obtained by employing the *ab initio* DFPT method applied to a periodic supercell geometry.

The phonon dispersion curves, including all of in-plane and out-of-plane branches, are shown in figure 10.21. The results in panels (*a*) and (*b*) are for the N–N and B–B interfaces, respectively. The corresponding polarisation and vibrational amplitude results are shown in panels (*c*) and (*d*), and (*e*) and (*f*), respectively. Due to the breaking of the inversion symmetry arising from different atomic masses of the basis atoms in monolayer h-BN, there is a finite gap in the phonon spectrum at the K point in the BZ. Within that gap region, three interface phonon branches are obtained from the *ab*

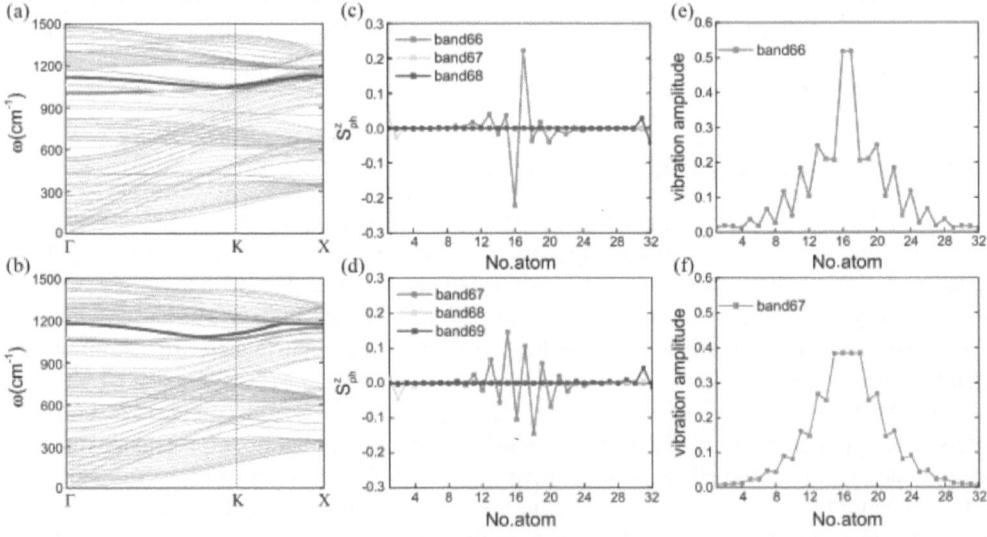

Figure 10.21 Phonon dispersion curves, interface polarisation at K and interface vibrational amplitude at K for $\alpha\beta\alpha\beta$-type zigzag monolayer h-BN nanoribbon with 32 atoms per unit cell. Thick curves show the interface branches. Panels (*a*), (*c*), and (*e*) show results for the N–N interface, and panels (*b*), (*d*), and (*f*) show results for the B–B interface. [Original diagram in colour.] Reproduced from Li *et al* (2021).

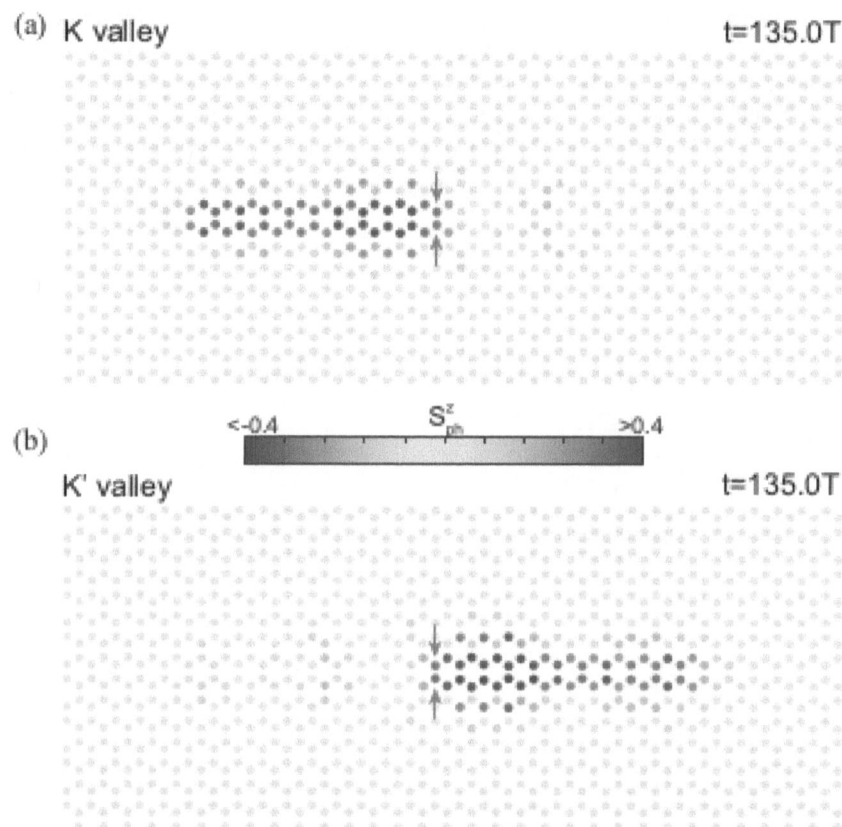

Figure 10.22 Prediction of chirality transmission of topological N–N interface phonons at K and K′ for a 32-atom zigzag monolayer h-BN nanoribbon. This is presented by taking a snapshot of the polarisation field at $t = 135T$. The arrow indicates the harmonic force exerted on the interface atoms. Dark and light dots, respectively, indicate the left-handed and right-handed vibrations of the atoms. [Original diagram in colour.] Reproduced from Li *et al* (2021).

initio calculations. Their dispersion, polarisation (chirality) and amplitude are shown by the blue, green and red curves.

The interface branch shown in red is topologically protected. In contrast, the interface branches in blue and green correspond to the boundary modes localised at the upper and lower boundaries, and are topologically trivial. For the topological interface mode at K, the atoms on the two sides of the interface have opposite left-handed or right-handed circularly polarised vibrations. As discussed in the previous sub-section, phonon polarisations at K′ are opposite to those at K.

Li *et al* (2021) further showed that chirality transmission of the topological phonons at the K and K′ valleys can be achieved by giving different harmonic force excitations to the interface atoms. To show that they chose the N–N interface, due to its localised polarisation distribution. They excited nitrogen atoms at the central interface with harmonic forces. Figure 10.22 shows a snapshot of the polarisation field at time $t = 135T$, where $T = 2\pi/\omega$ (ω being the eigenfrequency) is the characteristic time period of the system. Stable chirality transmission of the topological phonons is clearly evident. Corresponding to the opposite phonon group velocities, the transmission directions are opposite for the K and K′ valleys.

Phonon chirality can be considered as a new degree of freedom. Ordinary chiral phonons can easily be made unstable by their interaction with impurities and defects. Topologically chiral phonons, on the other hand, are robust against impurites and defects and can be useful in nondissipative thermal transport and thermal management in general. This finding paves the way for designing chiral phononic devices, based on the concept of topological valleyphononics, or topological chiralphononics.

11 Phonons and Thermal Transport in Impure and Mixed Crystals

11.1 INTRODUCTION

Defects in a crystal may be of a *chemical nature* such as a substitutional impurity atom, or of a *mechanical nature* such as a vacancy, or of an *extended nature* such as a dislocation or a stacking fault. In our discussion in this chapter, we will restrict ourselves to point defects such as substitutional impurities. The study of the effects of defects and disorder on the vibrational properties of crystals can be traced back to the works of Baden-Powell (1841), Hamilton (1940) and Rayleigh (1945). The modern theory of defect lattice dynamics dates from the work of Lifshitz (1943a, b). A good review of the modern theory is given by Maradudin (1965).

The presence of a small concentration of substitutional impurities gives rise to mass-defect and changes in force constants and may destroy the translational symmetry of the host crystal. These effects alter the frequencies of the normal modes of vibrations and the eigendisplacements in the crystal. In general, two types of vibrational modes may be created by defect atoms: *localised* modes and *resonance* modes. A localised mode is characterised by its frequency which lies in ranges forbidden to the normal modes of the perfect host crystal, and by its vibration amplitude which is large at the impurity site and decreases very rapidly with increasing distance. A localised mode is termed a *local mode* if its frequency lies above the maximum vibrational frequency of the perfect host lattice, and a *gap mode* if its frequency lies between the optical and acoustic bands of the host (i.e. in the reststrahlen band). A resonance mode is characterised by its frequency which lies in the ranges of frequencies allowed to the normal modes of the perfect host crystal, and by its vibration amplitude which is large at the impurity atom or at atoms with which the impurity atom interacts.

An impure crystal should be termed a disordered solid with increased defect concentration such that interaction between defect atoms begins to play an effective role. If the disorder is spatial, the solid is glass-like. (The vibrational properties of such a system will be discussed in Chapter 12). If the disorder is configurational, we have a disordered alloy (isotope mixture or mixed crystal); if the impurity content is not isotopic in nature but of a different chemical nature, then we have a *mixed crystal*. In general, two or more elements or compounds can be used to form a mixed system (single crystal or polycrystalline aggregate). The first experimental work on mixed crystals was made by Krueger *et al* (1928). The first theoretical study of a mixed linear chain was made by Matossi (1951). Developments in experimental techniques such as Raman scattering and infrared spectroscopy have stimulated a great deal of experimental and theoretical studies of long-wavelength optical phonons in mixed crystals.

Let us consider a mixed crystal of the type $A_{1-x}B_xC$. We can define three classes of mixed systems. One class exhibits the so-called *one-mode behaviour* in which one set of $q \simeq 0$ (long-wavelength) optical phonon frequencies (local or gap modes which are infrared or Raman active, or active under both investigations) are found to vary continuously with the concentration x from the mode frequencies of one end member (AC) to that of the other (BC), and all frequencies appear with approximately constant strengths. The other class exhibits the *two-mode behaviour* in which for a given concentration x there occur two sets of long-wavelength optical phonon frequencies (local and gap modes) close to those of the end members, with the strength of each mode being approximately proportional to the mole fraction of each component. Some mixed crystals show a

DOI: 10.1201/9781003141273-11

mixed-mode behaviour in which a *two-mode* behaviour is observed over a particular composition range and a *one-mode* behaviour over the remaining composition range.

In this chapter we discuss infrared and Raman scattering data on long-wavelength optical phonons in impure and mixed crystals of primarily covalent nature. Theoretical models are reviewed which explain experimental data and make some predictions as well. Finally, experimental and theoretical results for lattice thermal conductivity of mixed crystals are presented.

11.2   LOCALISED VIBRATIONAL MODES IN SEMICONDUCTORS

11.2.1   INFRARED ABSORPTION MEASUREMENTS

In section 6.9.1 we discussed the principle of infrared absorption in diamond and zincblende structure crystals. In zincblende crystals, which do not possess a centre of inversion, the TO modes give rise to an electric dipole moment and are infrared active, and may be determined directly by infrared absorption measurements on large thin crystals. Substitutional impurities in zincblende crystals have T_d symmetry and give rise to localised modes which can also be observed in infrared absorption experiments. Impurity complexes in diamond and zincblende crystals with defect symmetries $C_s, C_{2v}, C_{3v}, D_{3d}$ etc can also be infrared active.

Table 11.1 shows results of infrared measurements for some localised modes in tetrahedrally bonded semiconductors. Figure 11.1 shows the infrared absorption peaks due to Al and P substitutional impurities in GaAs. The absorption peak positions shift slightly to lower frequency when the temperature of the sample increases from liquid nitrogen to room temperature: the Al peak shifts from 362 cm^{-1} to 359 cm^{-1} and the P peak shifts from 355 cm^{-1} to 353 cm^{-1}. As Al and Ga have the same tetrahedral radii (1.26 Å), we expect the 362 cm^{-1} local mode frequency to be close to the optical mode frequency in AlAs. The P radius (1.10 Å) is also close to that of As (1.18 Å) and the local mode frequency of 353 cm^{-1} compares with the TO (Γ) value of 366 cm^{-1} for GaP.

It is quite interesting to study localised modes in GaAs:Si. At low concentration, Si is predominantly a donor, with a local mode $\omega(\text{Si}_{\text{Ga}}) = 384$ cm^{-1}. With increased concentrations, $> 10^{-18}$ cm^{-3}, Si dopant goes almost equally as a donor and an acceptor. The Si$_{\text{As}}$ local mode is observed at 399 cm^{-1}. Apart from local modes due to substitutional impurity of T_d symmetry, there are also observed modes of C_{3v} symmetry due to Si$_{\text{Ga}}$–Si$_{\text{As}}$ defects at 367, 393 and 464 cm^{-1}. Although there are only three local modes, a simple model for (Si$_{\text{Ga}}$–Si$_{\text{As}}$) pair defects in GaAs predicts four modes (figure 11.2) (Spitzer 1971).

When Zn (or Te) is introduced along with Si during growth of GaAs crystals, Zn goes as an acceptor and (Si$_{\text{Ga}}$–Zn$_{\text{Ga}}$)$_{pair}$ defects are formed. As a result the 399 cm^{-1} mode due to Si$_{\text{As}}$ becomes absent, the 393 cm^{-1} mode due to (Si$_{\text{Ga}}$–Si$_{\text{As}}$)$_{pair}$ is also virtually eliminated, and three new modes are observed which are attributed to the Si vibrations in (Si$_{\text{Ga}}$–Zn$_{\text{Ga}}$)$_{pair}$ defects with C_s symmetry: 378 cm^{-1}, 382 cm^{-1} and 395 cm^{-1}.

11.2.2   THEORY OF LOCALISED MODES

We describe here the theory of Dawber and Elliott (1963a, b) for the vibrational modes and infrared absorption due to point defects in a crystal. Consider a crystal with N_0 unit cells and p atoms per unit cell. Following the notation used in Chapters 2 and 4, the equation of motion of atom b of mass M_{lb} in unit cell l can be written as

$$M_{lb}\ddot{u}_\alpha(\boldsymbol{lb}) + \sum_{l'b'}\sum_\beta \Phi_{\alpha\beta}(\boldsymbol{lb};l'b')u_\beta(l'b') = 0, \tag{11.1}$$

where $\boldsymbol{u}$ is the displacement vector, $\alpha = x, y, z$, and Φ is the interatomic force constant tensor. In the presence of impurities in the crystal equation (11.1) is modified to

$$M_{lb}\ddot{u}_\alpha(\boldsymbol{lb}) + \sum_{l'b'}\sum_\beta \Phi_{\alpha\beta}(\boldsymbol{lb};l'b')u_\beta(l'b')$$
$$= \Delta M_{lb}\ddot{u}_\alpha(\boldsymbol{lb}) + \sum_{l'b'}\sum_\beta \Delta\Phi_{\alpha\beta}(\boldsymbol{lb};l'b')u_\beta(l'b'), \tag{11.2}$$

where ΔM and $\Delta\Phi$ are changes in the mass and force constants, respectively.

Table 11.1

Localised modes in semiconductors, observed in infrared absorption experiments. (1 THz = 33.3563 cm^{-1} = 4.1357 meV.)

Crystal	Impurity	Defect symmetry	Localised mode frequency (cm^{-1}) Temperature (K) is given in brackets
Si	^{10}B	T_d	644(300), 646(80)
	^{11}B	T_d	620(300), 622(80)
	^{10}B–^{6}Li	C_{3v}	683(300), 584(300), 534(300)
	^{11}B–^{6}Li	C_{3v}	657(300), 564(300), 534(300)
			659(80), 566(80), 536(80)
	^{10}B–^{7}Li	C_{3v}	681(300), 584(300), 522(300)
			683(80), 586(80), 523(80)
	^{11}B–^{7}Li	C_{3v}	655(300), 564(300), 522(300)
			656(80), 566(80), 523(80)
	^{10}B–^{10}B	D_{3d}	570(300)
	^{11}B–^{11}B	D_{3d}	615(80), 552(80)
	As	T_d	366(80) resonance
	P	T_d	441(80) resonance, 491(80) resonance
	^{14}C	T_d	570(300), 573(80)
	^{13}C	T_d	586(300), 589(80)
	^{12}C	T_d	605(300), 680(80)
Ge	Si	T_d	389(300)
	Si–Si	D_{3d}	476(300), 448(300)
GaSb	Al$_{Sb}$	T_d	316.7(77)
	P$_{Sb}$	T_d	324(77)
GaP	Al	T_d	444.7(77)
	C	T_d	606(20)
	O	T_d	464(77)
	As	T_d	272(300) reflection
GaAs	Al	T_d	362(80), 359(300)
	P	T_d	355(80), 353(300)
	Si$_{Ga}$	T_d	384(80)
	Si$_{As}$	T_d	399(80)
	Si$_{Ga}$–Si$_{As}$	C_{3v}	367(80), 393(80), 464(80)
	Si$_{Ga}$–Zn$_{Ga}$	C_s	378(80), 382(80), 395(80)
	^{10}B	T_d	540(80)
	^{11}B	T_d	517(80)
InP	^{10}B	T_d	543.5(77)
	^{11}B	T_d	522.8(77)
InSb	P$_{Sb}$	T_d	∼ 293(77)
ZnSe	Al	T_d	359(77)
	^{31}P	T_d	343(80)
	S	C_{3v}	266.5(300), 269(300)
	Ga–P	C_{3v}	∼ 343(80), 350(80)
CdTe	Se$_{Te}$	T_d	170(300)
CdS	Se	C_{3v}	182(300), 186(300)

The displacement vector $u(lb)$ can be expressed in terms of the polarisation vector $e(b \mid qs)$ and phonon creation and annihilation operators as in equations (2.48) and (4.79). Expressing

$$\chi(lb \mid qs) = \frac{1}{\sqrt{N_0 M_{lb}}} e(b \mid qs) e^{iq \cdot l} \qquad (11.3)$$

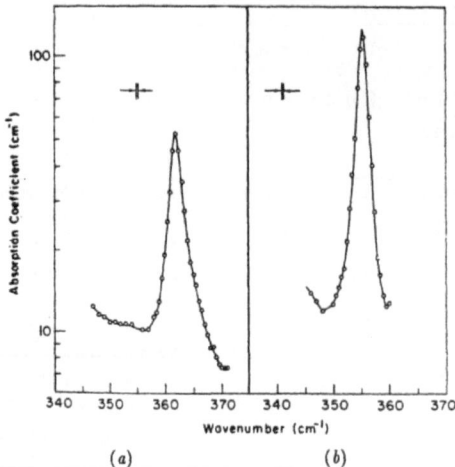

Figure 11.1 Liquid-nitrogen temperature absorption peaks due to substitutional impurities in GaAs: (*a*) Al content 69 parts per million, (*b*) P content 140 parts per million. (From Lorimor *et al* (1966).)

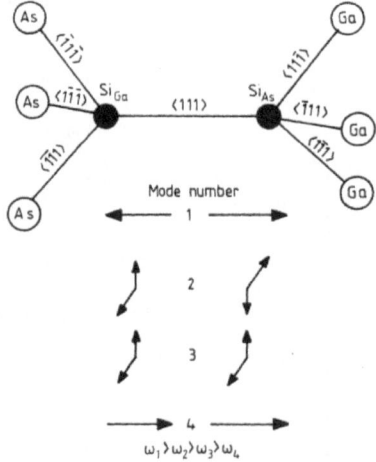

Figure 11.2 A simple model showing vibrational modes due to (Si$_{Ga}$–Si$_{As}$) pair defects with symmetry C_{3v} in GaAs.

the orthonormality condition in equation (4.23) reads

$$\sum_{lb} M_{lb} \sum_{\alpha} |\chi_\alpha(\boldsymbol{lb} \mid \boldsymbol{q}s)|^2 = 1. \tag{11.4}$$

With equations (4.79) and (11.3), equation (11.2) becomes for each normal mode $\boldsymbol{q}s$

$$-\omega^2 M_{lb}\chi_\alpha(\boldsymbol{lb} \mid \boldsymbol{q}s) + \sum_{l'b'}\sum_\beta \Phi_{\alpha\beta}(\boldsymbol{lb};\boldsymbol{l'b'})\chi_\beta(\boldsymbol{l'b'} \mid \boldsymbol{q}s)$$

$$= \sum_{l'b'}\sum_\beta D_{\alpha\beta}(\boldsymbol{lb};\boldsymbol{l'b'})\chi_\beta(\boldsymbol{l'b'} \mid \boldsymbol{q}s), \tag{11.5}$$

where

$$D_{\alpha\beta}(\boldsymbol{lb};\boldsymbol{l'b'}) = -\Delta M_{lb}\omega^2\delta_{\alpha\beta}\delta_{ll'}\delta_{bb'} + \Delta\Phi_{\alpha\beta}(\boldsymbol{lb};\boldsymbol{l'b'}). \tag{11.6}$$

The eigensolutions of the perturbed crystal are obtained by solving equation (11.5) which can be expressed as

$$\left| \sum_{l''b''} \mathcal{G}_{\alpha\beta}(lb;l''b'')D_{\beta\gamma}(l''b'';l'b') - \delta_{\alpha\gamma}\delta_{ll'}\delta_{bb'} \right| = 0, \qquad (11.7)$$

where the Green's function

$$\mathcal{G}_{\alpha\beta}(lb;l''b'';\omega) = \sum_{qs} \frac{\chi_\alpha(lb \mid qs)\chi_\beta^*(l''b'' \mid qs)}{\omega^2(qs) - \omega^2} \qquad (11.8)$$

is a solution of the equation

$$-\omega^2 M_{lb}\mathcal{G}_{\alpha\gamma}(lb;l''b'';\omega) + \sum_{l'b'}\sum_{\beta} \Phi_{\alpha\beta}(lb;l'b')\mathcal{G}_{\beta\gamma}(l'b';l''b'';\omega) = \delta_{\alpha\gamma}\delta_{ll'}\delta_{bb'}. \qquad (11.9)$$

Consider a single impurity of mass M_b' in the unit cell at the origin. For simplicity assume that there are no changes in force constants. Then

$$D_{\alpha\beta}(lb;l'b') = -M_{lb}\varepsilon_b\omega^2\delta_{\alpha\beta}\delta_{l,0}\delta_{l',0}\delta_{bb'} \qquad (11.10)$$

and from equations (11.3) and (11.8)

$$\mathcal{G}_{\alpha\beta}(0b;0b';\omega) = \frac{1}{N_0}\frac{1}{\sqrt{M_b M_{b'}}}\sum_{qs}\frac{e_\alpha(b \mid qs)e_\beta^*(b' \mid qs)}{\omega^2(qs) - \omega^2} \qquad (11.11)$$

and finally equation (11.7) reduces to

$$\left| M_b\varepsilon_b\omega^2\mathcal{G}_{\alpha\beta}(0b;0b';\omega) + \delta_{\alpha\beta}\delta_{bb'} \right| = 0. \qquad (11.12)$$

In the above equations, the mass defect is introduced through a dimensionless quantity

$$\varepsilon_b = (M_b - M_{b'})/M_b. \qquad (11.13)$$

Calculation of the Green's function $\mathcal{G}$ in equation (11.12) can be considerably simplified by using the symmetry of the defect (Maradudin 1965). For a substitutional defect in a cubic crystal equation (11.12) is reduced to

$$1 + \frac{\varepsilon_b\omega^2}{N_0}\sum_{qs}\frac{|e_\alpha(b \mid qs)|^2}{\omega^2(qs) - \omega^2} = 0 \qquad (11.14)$$

and has three-fold degenerate solutions. For the diamond structure, all M_b are identical and equation (11.14) can further be reduced to

$$1 + \frac{\varepsilon\omega^2}{6N_0}\sum_{qs}\frac{1}{\omega^2(qs) - \omega^2} = 0, \qquad (11.15)$$

where to satisfy the orthonormality relation in equation (4.23) all $|e_\alpha(b|qs)|^2$ is taken equal to $\frac{1}{3}p = 1/6$.

For local modes, the solutions of equation (11.15) yield $\omega^2 = \omega_{loc}^2 > \omega_{max}^2(qs)$. This allows the summation in equation (11.15) to be replaced by an integral so that

$$1 + \varepsilon\omega_{loc}^2\int_0^{\mu_m}d\mu\frac{\nu(\mu)}{\mu - \omega_{loc}^2} = 0, \qquad (11.16)$$

where $\mu = \omega^2(qs)$, $\int_0^{\mu_m} d\mu\, \nu(\mu) = 1$, and $\nu(\mu)$ is a density of unperturbed states per unit energy range

$$\nu(\mu) = \frac{\Omega}{3p} \sum_s \int \frac{dS_q}{\nabla_q \mu(qs)}, \tag{11.17}$$

with dS_q as an element of constant energy surface in q-space, and Ω as the volume of a unit cell.

The linear absorption coefficient $\alpha(\omega)$ due to the local mode from a defect at origin in a cubic crystal is given by

$$\alpha(\omega) = \frac{2\pi^2 D e^2}{\eta c} |\chi(0)|^2 f(\omega), \tag{11.18}$$

where η is the refractive index, e is the static charge on the impurity, D is the impurity density, and $f(\omega)$ is a shape function of frequency normalised to unity, $\int_0^\infty f(\omega)d\omega = 1$. The mean-square relative amplitude of the local mode at the defect atom is given by

$$M'|\chi(0)|^2 = \frac{M'}{M} \left(\varepsilon^2 \omega_{loc}^2 \int_0^{\mu_m} \frac{\nu(\mu)d\mu}{(\omega_{loc}^2 - \mu)^2} - \varepsilon \right)^{-1}. \tag{11.19}$$

Equations (11.16) and (11.19) can be computed provided the density of states $\nu(\mu)$ is available from a realistic calculation for the unperturbed crystal. If a Debye spectrum is used, then $\nu(\mu)$ takes the form

$$\nu_{Debye}(\mu) = 3\mu \frac{3\mu^{1/2}}{3\mu_m^{3/2}}, \tag{11.20}$$

with $\sqrt{\mu_m} = \omega_{max}$. In a study of Si Dawber and Elliott (1963a) used $\nu(\mu)$ which was computed by using Cochran's shell model (cf section 2.4.3). Their results for the relative frequency and relative amplitude of a localised mode at a defect atom in Si host are plotted in figure 11.3. This calculation predicts that the localised mode does not appear until $\varepsilon \geq 0.075$. A Debye spectrum for $\nu(\mu)$ would, however, give rise to a localised mode for all positive values of ε, i.e. for all light impurity masses $M' < M$.

The calculation of Dawber and Elliott predicts absorption for $D = 10^{19} \mathrm{cm}^{-3}$ impurities of various masses in Si. For B impurity in Si the predicted absorption peaks are at, without incorporating any changes in the force constants, 680 cm^{-1} (for ^{10}B) and 654 cm^{-1} (for ^{11}B). These values, although somewhat higher than those observed (table 11.1), show the correct relative peak positions for the two isotopes.

In general some mass criteria can be established for the occurrence of impurity modes. Consider a binary crystal. When the lighter mass is replaced, a lighter impurity ($\varepsilon > 0$) produces a local mode and a heavier impurity ($\varepsilon < 0$) causes a gap mode. When the heavier atom is replaced, two impurity modes are produced: a lighter impurity ($\varepsilon > 0$) gives rise to a local mode as well as a gap mode,

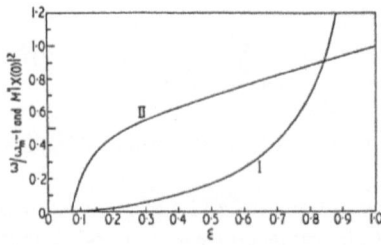

Figure 11.3 Relative frequency $\omega/\omega_m - 1$ (curve I) and relative amplitude (curve II) of the localised mode as a function of mass defect $\varepsilon = (M - M')/M$ at the origin in a Si crystal. (From Dawber and Elliott (1963a).)

whereas a heavier impurity mass ($\varepsilon < 0$) does not generate any new modes, local or gap. These criteria can often give a clue to the understanding of the one-mode or two-mode behaviour in mixed crystals, as discussed in section 11.4.2.

11.3 EXPERIMENTAL STUDIES OF LONG-WAVELENGTH OPTICAL PHONONS IN MIXED CRYSTALS

11.3.1 INFRARED REFLECTANCE MEASUREMENTS

Infrared reflectance measurements can be made on mixed crystals in the infrared reststrahlen region. Infrared reflectance spectra are commonly analysed by employing two procedures, a Kramers–Kronig dispersion analysis and a damped harmonic oscillator fit. The Kramers–Kronig analysis permits the determination of optical constants n (refractive index) and k (extinction coefficient), or, alternatively, the real ε_1 and imaginary ε_2 parts of the dielectric function ε from the reflectivity spectrum. In the analysis both ε_2 and $\text{Im}(1/\varepsilon)$ (the energy loss function) show two sharp maxima whose frequencies correspond to the long-wavelength ($q \simeq 0$) TO and LO phonon frequencies.

Figures 11.4(a) and 11.4(c) show long-wavelength TO and LO mode frequencies in $\text{Ga}_{1-x}\text{Al}_x\text{As}$ and $\text{Ga}_{1-x}\text{Al}_x\text{Sb}$, respectively, obtained from a Kramers–Kronig analysis. Typical two-mode behaviour is observed for both the alloys at all mixed crystal compositions.

In $\text{Ga}_{1-x}\text{Al}_x\text{As}$, the AlAs-like band is well defined for all compositions and shifts to slightly lower frequencies with decreasing values of x. Extrapolating the analysis to $x = 0$ one obtains the local mode of Al in GaAs: $\omega_{loc}(\underline{\text{Ga}}\text{As}:\text{Al}) = 356 \pm 2 \text{ cm}^{-1}$, which is in good agreement with the infrared absorption result listed in table 11.1. The GaAs-like band shifts to lower frequencies with decreasing values of $1 - x$. Extrapolation of the result to $x = 1$ gives the localised gap mode of Ga in AlAs at $\sim 252 \text{ cm}^{-1}$. The LO phonon frequency of the GaAs-like band and the TO phonon frequency of the AlAs-like band are essentially linear with concentration, whereas GaAs-like TO and AlAs-like LO modes show a non-linear dependence.

Tables 11.2 and 11.3 list several mixed semiconductor systems whose mode behaviours have been observed by infrared reflectance and Raman scattering spectroscopies.

In $\text{Ga}_{1-x}\text{Al}_x\text{Sb}$ the AlSb-like band grows in strength and shifts towards higher frequencies with increasing values of x. Extrapolation of LO and TO branches to $x = 0$ gives the local mode frequency for Al in GaSb: $\omega_{loc}(\underline{\text{Ga}}\text{Sb}:\text{Al}) = 312.3 \pm 0.9 \text{ cm}^{-1}$, which is in good agreement with 316.7 cm^{-1} listed in table 11.1. The GaSb-like band decreases in amplitude and shifts to lower frequencies with increasing values of x. For $x = 1$ one gets the Ga impurity mode in AlSb at a frequency $\omega_{loc}(\underline{\text{Al}}\text{Sb}:\text{Ga}) = 205.4 \pm 0.9 \text{ cm}^{-1}$). The variation with composition of mode frequencies, both AlSb-like and GaSb-like, has a definite non-linear dependence.

11.3.2 RAMAN SCATTERING MEASUREMENTS

As discussed in sections 6.9.2 and 9.4 Raman scattering is dependent on the polarisation of the incident and scattered light relative to the crystal axes. In the back scattering configuration for zincblende compounds and mixed crystals with zincblende structure, the LO phonon is observed for the (100) and (111) surfaces, while the TO phonon is observed for the (110) and (111) surfaces. Due to the microscopic nature of the compositional disorder in mixed crystals, there exist potential fluctuations which result in a q-vector relaxation process and hence in somewhat asymmetric and broader phonon lines.

The most studied mixed crystals are of the III–V family. In figure 11.4(b) is plotted the variation of AlAs- and GaAs-like LO and TO phonon modes as a function of composition x in $\text{Ga}_{1-x}\text{Al}_x\text{As}$. These measurements were made in the back scattering configuration on the (001) face, which only allows observation of LO modes. However, some deviation from strict back scattering geometry made it possible to measure weak TO modes as well. From a comparison of figures 11.4(a) and

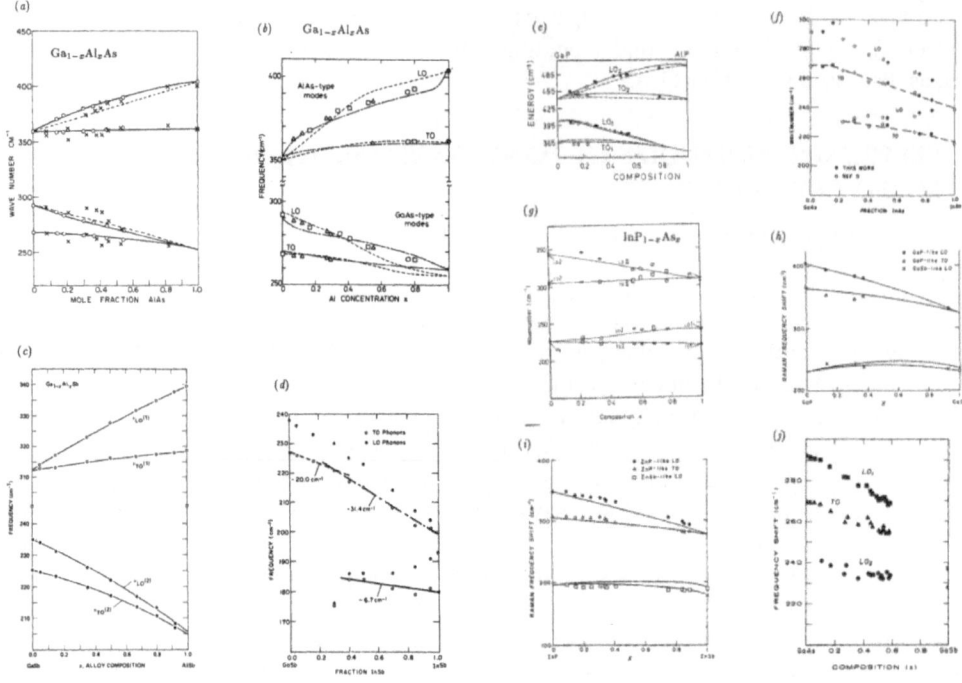

Figure 11.4 Long-wavelength LO and TO mode frequencies in III–V mixed crystals. IR = infrared reflectance measurements, RS = Raman scattering measurements. Solid curves show theoretical results from the MREI model. References are given in table 11.2. (Reproduced with permission from the original references.)

11.4(*b*), it can be seen that there is good agreement between the results from Raman scattering and infrared reflectance measurements.

A glance at table 11.2 reveals that almost all III–V mixed crystals show either two-mode or mixed-mode behaviour. Possibly $In_{1-x}Ga_xP$ is the only system which shows one-mode behaviour. However, more recent investigations (Galtier *et al* 1984, Jusserand and Slempkes 1984) suggest that this system can be considered to show a 'modified two-mode' behaviour. Abdelouhab *et al* (1989) have suggested that strain-free bulk $Ga_{1-x}In_xP$ materials show one-mode behaviour, but the presence of internal stress in epitaxially grown layers could induce a switchover from one-mode to two-mode behaviour.

II–VI mixed crystals grow either in the zincblende structure or wurtzite structure and can exhibit one-mode, mixed-mode or two-mode behaviour. Raman scattering and infrared reflectance experiments indicate that $Hg_{1-x}Cd_xTe$, $Cd_{1-x}Zn_xTe$, $ZnSe_{1-x}S_x$, $CdSe_{1-x}S_x$ and $Cd_{1-x}Mg_xTe$ show a two-mode behaviour (see figure 11.5 and table 11.3). On the other hand, $Cd_{1-x}Zn_xS$, $ZnTe_{1-x}Se_x$ and $Cd_{1-x}Zn_xSe$ show a one-mode behaviour with some structure and can therefore be regarded as crystals with mixed-mode behaviour.

Raman scattering measurements have also been reported for some metastable semiconducting $(III–V)_{1-x}IV_x$ alloys. The crystalline alloy $(GaAs)_{1-x}Ge_x$ exhibits a one-mode behaviour (Barnett *et al* 1982). The degenerate optic modes of Ge (for $x = 1$) split into GaAs-like LO and TO modes for $x < 1$. The crystalline alloy $(GaSb)_{1-x}Ge_x$ exhibits an intermediate mixed-mode behaviour, as shown in figure 11.6 (Krabach *et al* 1983).

The quaternary compound InGaAlP shows a partial *three-mode* behaviour (Asahi *et al* 1989), as is seen by the compositional dependencies of Raman shifts in figure 11.7. The compound $In_{1-x}Ga_xAs_yP_{1-y}$ is reported to show a *four-mode* behaviour (Jusserand and Slempkes 1984).

Table 11.2

Behaviour of long-wavelength optical phonons in mixed III–V semiconducting crystals in the zincblende structure. RS = Raman scattering; IR = Infrared reflectance.

System	Behaviour	Method	Reference
$Ga_{1-x}Al_xAs$	2 mode	RS, IR	Jusserand and Sapriel (1981),
			Kim and Spitzer (1979)
$Ga_{1-x}Al_xSb$	2 mode	IR	Lucovsky *et al* (1975)
$Ga_{1-x}Al_xP$	2 mode	IR	Lucovsky *et al* (1976),
		RS	Armelles *et al* (1988)
$Ga_{1-x}In_xAs$	mixed mode	IR	Lucovsky and Chen (1970),
	2 mode	RS	Emura *et al* (1988)
	(with strain)		
$Ga_{1-x}In_xSb$	mixed mode	IR	Brodsky *et al* (1970)
$In_{1-x}Ga_xP$	1 mode	RS	Besserman *et al* (1976),
	modified	RS	Jusserand and Slempkes (1984),
	2 mode		Abdelouhab *et al* (1989)
$In_{1-x}Al_xP$	2 mode	RS	Asahi *et al* (1989)
$In_{1-x}Al_xAs$	2 mode	RS	Emura *et al* (1987)
$GaP_{1-x}As_x$	2 mode	RS	Galtier *et al* (1984)
$InP_{1-x}As_x$	2 mode	RS, IR	Carles *et al* (1980),
			Talwar *et al* (1980),
			Kekelidze *et al* (1973)
$GaAs_{1-x}Sb_x$	mixed mode	IR	Cohen *et al* (1985),
	2 mode	RS	McGlinn *et al* (1986)
$InAs_{1-x}Sb_x$	mixed mode	IR	Lucovsky and Chen (1970)
$InP_{1-x}Sb_x$	2 mode	RS	Cherng *et al* (1989)
$GaP_{1-x}Sb_x$	2 mode	RS	Cherng *et al* (1989)
$(GaAs)_{1-x}Ge_x$	1 mode	RS	Barnett *et al* (1982)
$(GaSb)_{1-x}Ge_x$	mixed mode	RS	Krabach *et al* (1983)
$In_{0.49}Ga_{0.51-x}Al_xP$	3 mode	RS	Asahi *et al* (1989)
$In_{1-x}Ga_xAs_yP_{1-y}$	4 mode	RS	Jusserand and Slempkes (1984)

11.4 THEORETICAL MODELS FOR LONG-WAVELENGTH OPTICAL PHONONS IN MIXED CRYSTALS

Our discussion so far in this chapter can be summarised as follows: defect induced local or gap modes for small impurity concentration are broadened into vibrational bands (leading to one-, mixed- or two-mode behaviour) for high impurity concentrations in mixed crystals. A satisfactory theory of localised defect mode calculation was described in section 11.2.2. However, it will be a daunting task to develop a full-scale theory for the lattice dynamics of a disordered three-dimensional mixed crystal. In this section we discuss simple theoretical models which can be used to describe the mode behaviour of long-wavelength optical phonons in mixed crystals. A good review of various theoretical models has been given by Chang and Mitra (1971).

11.4.1 LINEAR CHAIN MODEL

Matossi (1951) and Lisitsa *et al* (1969) considered a periodic linear equidistant chain, with only nearest-neighbour force constants, to discuss the mode behaviour of a mixed crystal of the type $A_{1-x}B_xC$ with $x = 0.5$. The unit cell for this system is shown in figure 11.8. (For other values of x

Table 11.3

Behaviour of long-wavelength optical modes in mixed II–VI crystals. ZB = zincblende structure, W = wurtzite structure

System	Crystal structure	Behaviour	Method	Reference
HgTe–CdTe	ZB	2 mode	IR, RS	Carter *et al* (1971), Mooradian and Harman (1971)
MgTe–CdTe	ZB	2 mode	IR, RS	Nakashima *et al* (1973)
ZnTe–CdTe	ZB	2 mode	IR	Harada and Narita (1971)
			RS	Olego *et al* (1986)
ZnSe–ZnS	ZB	2 mode	IR, RS	Brafman *et al* (1968)
ZnSe–ZnTe	ZB	1 mode + weak structure	RS	Nakashima *et al* (1971)
CdSe–CdS	W	2 mode	IR, RS	Parrish *et al* (1967), Chang and Mitra (1968)
CdSe–ZnSe	W	1 mode (probably)	RS	Brafman (1972)
ZnS–CdS	W	1 mode + fine structure	IR	Lucovsky *et al* (1967)
CdSe–CdTe	W–ZB	2 mode	IR	Mityagin *et al* (1976) Górska and Nazarewicz (1973)
SnS–SnSe	CdI$_2$	2 mode or mixed mode	IR, RS	Garg (1986)

an appropriately big unit cell should be considered.)

The equations of motion for this chain are

$$
\begin{aligned}
M\ddot{u}_{4i} &= \Lambda(u_{4i+1} - u_{4i}) - \Lambda(u_{4i} - u_{4i+3}) \\
M'\ddot{u}_{4i+2} &= \Lambda'(u_{4i+3} - u_{4i+2}) - \Lambda'(u_{4i+2} - u_{4i+1}) \\
m\ddot{u}_{4i+1} &= \Lambda'(u_{4i+2} - u_{4i+1}) - \Lambda(u_{4i+1} - u_{4i}) \\
m\ddot{u}_{4i+3} &= \Lambda(u_{4i} - u_{4i+3}) - \Lambda'(u_{4i+3} - u_{4i+2}),
\end{aligned}
\tag{11.21}
$$

where u represents atomic displacement, and as shown in figure 11.8 M, M', m are atomic masses and Λ, Λ' are the nearest-neighbour interatomic force constants.

For long-wavelength modes $(q \simeq 0)$, we try solutions

$$
\begin{aligned}
u_{4i} &= U\mathrm{e}^{-\mathrm{i}\omega t} & u_{4i+2} &= V\mathrm{e}^{-\mathrm{i}\omega t} \\
u_{4i+1} &= W\mathrm{e}^{-\mathrm{i}\omega t} & u_{4i+3} &= W'\mathrm{e}^{-\mathrm{i}\omega t}.
\end{aligned}
\tag{11.22}
$$

With these displacement functions, the following vibrational frequencies can be derived:

$$
(1) \qquad \omega_1^2 = 0 \qquad \text{with} \quad U = V = W = W'. \tag{11.23}
$$

This corresponds to a pure translation and thus is of no importance for the present discussion.

$$
(2) \qquad \omega_2^2 = (\Lambda + \Lambda')/m \qquad \text{with} \quad U = V = 0, W = -W'. \tag{11.24}
$$

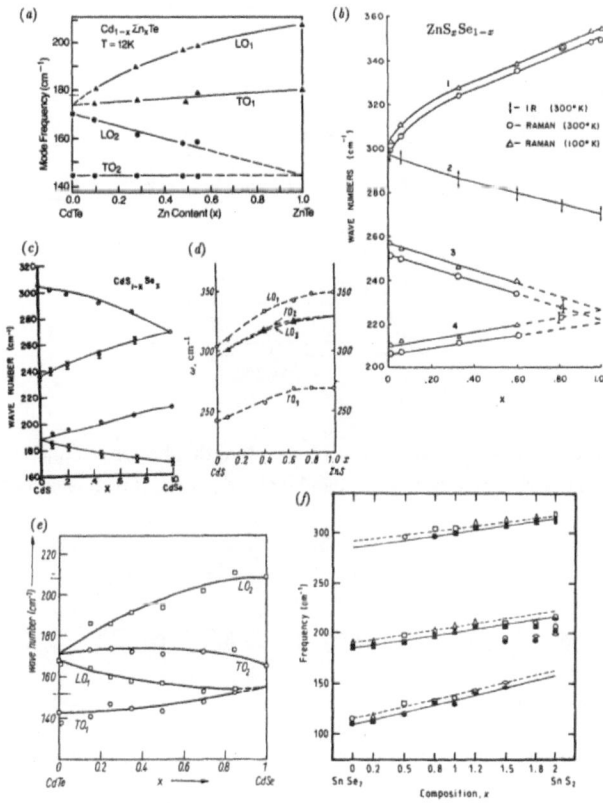

Figure 11.5 Same as figure 11.4 for II–VI mixed crystals. References are given in table 11.3. For ZnS_xSe_{1-x}, $1 = LO_1$, $2 = TO_1$, $3 = LO_2$, $4 = TO_2$. (Reproduced with permission from the original references.)

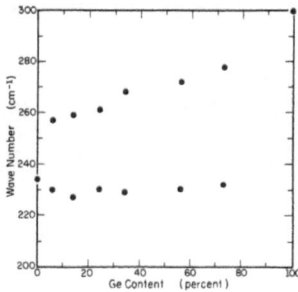

Figure 11.6 Long-wavelength opical phonons in the alloy $(GaSb)_{1-x}Ge_x$. (From Krabach *et al* (1983).)

This is an infrared inactive mode, but may be Raman active.

$$(3) \quad \omega_3^2 + \omega_4^2 = \frac{1}{2}\left[\omega_0^2\left(1 + \frac{m}{M+m}\right) + \omega_0'^2\left(1 + \frac{m}{M'+m}\right)\right] \tag{11.25}$$

$$\omega_3^2\omega_4^2 = \frac{1}{2}\omega_0^2\omega_0'^2\frac{Mm(1 + 2m/M + M'/M)}{(M+m)(M'+m)}, \tag{11.26}$$

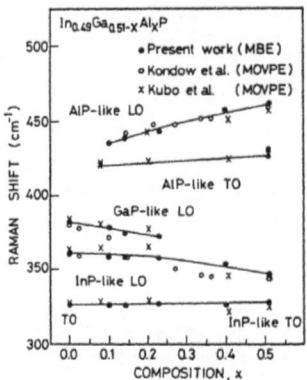

Figure 11.7 Compositional dependencies of long-wavelength phonon frequencies in epitaxially grown $In_{0.49}Ga_{0.51-x}Al_xP$. Partial three-mode behaviour can be clearly seen. (From Asahi *et al* (1989).)

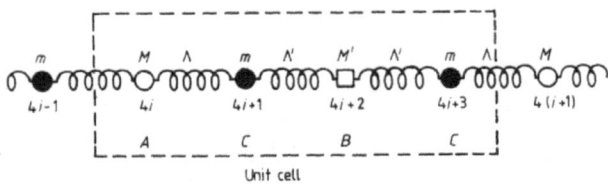

Figure 11.8 A linear chain with four atoms per unit cell to model a mixed-crystal $(AC)_{0.5}(BC)_{0.5}$. Only nearest-neighbour forces are considered, and the atoms are assumed equidistant. The chain is assumed to be periodic, so that the atom $4(i+1)$ is identical to the atom $4i$, etc.

where $\omega_0^2 = 2\Lambda(1/M + 1/m)$ and $\omega_0'^2 = 2\Lambda'(1/M' + 1/m)$ are the highest phonon frequencies in the pure compounds AC and BC, respectively. This case gives two infrared active modes. The amplitudes of ω_3 and ω_4 are given by

$$\frac{V}{U} = \omega_0'^2 \left[m\omega_0^2 - (M+m)\omega_{3,4}^2 \right] / \omega_0^2 \left[m\omega_0'^2 - (M'+m)\omega_{3,4}^2 \right]$$
$$\frac{W}{V} = 1 - (M-m)\omega_{3,4}^2 / m\omega_0^2 \qquad W' = W. \qquad (11.27)$$

It is interesting to consider two limiting cases:
(i) $m \ll M, M'$. This gives

$$\omega_3^2 + \omega_4^2 \simeq \frac{1}{2}(\omega_0^2 + \omega_0'^2)$$
$$\omega_3^2 \omega_4^2 \simeq \text{small} \qquad (11.28)$$

or, one particular solution,

$$\omega_3^2 \simeq 12(\omega_0^2 + \omega_0'^2) \qquad \omega_4^2 \simeq 0. \qquad (11.29)$$

This particular solution can be regarded as a one-mode behaviour.
(ii) $m \gg M, M'$. This gives

$$\omega_3^2 + \omega_4^2 \simeq \omega_0^2 + \omega_0'^2$$
$$\omega_3^2 \omega_4^2 \simeq \omega_0^2 \omega_0'^2. \qquad (11.30)$$

11.4.2   RANDOM ELEMENT ISODISPLACEMENT (REI) MODEL AND ITS MODIFICATIONS

A more appropriate scheme for long-wavelength optical phonons in mixed crystals was developed by Chen *et al* (1966) and modified by Chang and Mitra (1968), and Górska and Nazarewicz (1974). In this model, one considers a 'quasi-unit-cell' containing one unit of $AB_{1-x}C_x$ and makes two assumptions.

(i) *Isodisplacement.* Cations and anions of like species are considered to vibrate as rigid units, i.e. the units vibrate with the same phase and amplitude. This assumption is exactly true for the reststrahlen frequency in an ordered diatomic crystal, since there is no phase shift from unit cell to unit cell for $q \sim 0$ modes.

(ii) *Randomness.* It is assumed that the B and C ions are randomly distributed on the anion sublattice [1]. Thus each atom is subjected to forces produced by a statistical average of its neighbours. In the zincblende structure for $AB_{1-x}C_x$ each A is surrounded by exactly $4x$ ions of C and $4(1-x)$ ions of B. Furthermore, both B and C ions are assumed to have exactly $12(1-x)$ next-nearest neighbours of B and $12x$ next-nearest neighbours of C.

The equations of motion for the A,B and C atoms in $AB_{1-x}C_x$ within the modified random element isodisplacement (MREI) model are

$$
\begin{aligned}
m_A \ddot{u}_A &= -(1-x)\Lambda_{AB}(u_A - u_B) - x\Lambda_{AC}(u_A - u_C) + e_A E_{eff} \\
m_B \ddot{u}_B &= \Lambda_{AB}(u_B - u_A) - x\Lambda_{BC}(u_B - u_C) - e_B E_{eff} \\
m_C \ddot{u}_C &= -\Lambda_{AC}(u_C - u_A) - (1-x)\Lambda_{BC}(u_C - u_B),
\end{aligned}
\tag{11.31}
$$

where m_a, m_B, m_C are the masses of A, B, C, u_A, u_B, u_C are the displacement vectors, and $\Lambda_{AB}, \Lambda_{BC}, \Lambda_{AC}$ are nearest-neighbour interatomic force constants. E_{eff} is an effective field required to study long-wavelength modes and in a crystal of cubic symmetry is given by the Lorentz formula (see equation (2.84))

$$
E_{eff} = E + \frac{4\pi}{3}P,
\tag{11.32}
$$

where E is the macroscopic electric field and P is the polarisation. The condition of electrical neutrality of the crystal requires

$$
e_A - (1-x)e_B - xe_C = 0
\tag{11.33}
$$

in terms of two effective charges e_B and e_C which are considered as adjustable parameters.

For long-wavelength *transverse* optical phonons, one can approximate $E \simeq 0$. The polarisation is given by the equation

$$
P = N[e_A u_A - (1-x)e_B u_B - xe_C u_C] + N\alpha E_{eff},
\tag{11.34}
$$

where N is the ion-pair density. The electronic polarisability α is related to the high-frequency dielectric constant by the Clausius–Mossotti formula

$$
\frac{4\pi}{3}N\alpha = \frac{\varepsilon_\infty - 1}{\varepsilon_\infty + 2}.
\tag{11.35}
$$

Inserting equations (11.32)–(11.35) into equation (11.31) we get

$$
\begin{aligned}
m_A \ddot{u}_A &= -(1-x)\Lambda'_{AB}(u_A - u_B) - x\Lambda'_{AC}(u_A - u_C) \\
m_B \ddot{u}_B &= -\Lambda'_{AB}(u_B - u_A) - x\Lambda'_{BC}(u_B - u_C) \\
m_C \ddot{u}_C &= -\Lambda'_{AC}(u_C - u_A) - (1-x)\Lambda'_{BC}(u_C - u_B),
\end{aligned}
\tag{11.36}
$$

---

[1] Here we have considered A as cation and B and C as anions. For $A_x B_{1-x}C$ one should interchange 'cation' and 'anion', as well as 'A' and 'C', in the text.

where

$$\Lambda'_{AB} = \Lambda_{AB} - \frac{4\pi}{3} N e_A e_B \left(\frac{\varepsilon_\infty + 2}{3} \right)$$

$$\Lambda'_{AC} = \Lambda_{AC} - \frac{4\pi}{3} N e_A e_C \left(\frac{\varepsilon_\infty + 2}{3} \right)$$

$$\Lambda'_{BC} = \Lambda_{BC} + \frac{4\pi}{3} N e_B e_C \left(\frac{\varepsilon_\infty + 2}{3} \right). \qquad (11.37)$$

Writing $w_1 = u_A - u_B$ and $w_2 = u_A - u_C$, we can express the difference of equations (11.36) as

$$\ddot{w}_1 = \left(-(1-x) \frac{\Lambda'_{AB}}{m_A} - \frac{\Lambda'_{AB}}{mB} - x \frac{\Lambda'_{BC}}{m_B} \right) w_1 - \left(x \frac{\Lambda'_{AC}}{m_A} - x \frac{\Lambda'_{BC}}{m_B} \right) w_2$$

$$\ddot{w}_2 = \left(-x \frac{\Lambda'_{AC}}{m_A} - \frac{\Lambda'_{AC}}{m_C} - (1-x) \frac{\Lambda'_{BC}}{m_C} \right) w_2$$

$$- \left((1-x) \frac{\Lambda'_{AB}}{m_A} - (1-x) \frac{\Lambda'_{BC}}{m_C} \right) w_1. \qquad (11.38)$$

For $q \simeq 0$ modes, we try solutions

$$w_1 = A_1 e^{-i\omega t} \qquad w_2 = A_2 e^{-i\omega t}. \qquad (11.39)$$

Equation (11.38) then results into

$$\begin{bmatrix} -\omega^2 + K_1 & K_{12} \\ K_{12} & -\omega^2 + K_2 \end{bmatrix} \begin{bmatrix} A_1 \\ A_2 \end{bmatrix} = 0. \qquad (11.40)$$

Non-trivial eigenfrequencies are therefore determined from

$$\omega^4 - (K_1 + K_2)\omega^2 + (K_1 K_2 - K_{12} K_{21}) = 0, \qquad (11.41)$$

where

$$K_1 = (1-x) \frac{\Lambda'_{AB}}{m_A} + \frac{\Lambda'_{AB}}{m_B} + x \frac{\Lambda'_{BC}}{m_B}$$

$$K_2 = x \frac{\Lambda'_{AC}}{m_A} + \frac{\Lambda'_{AC}}{m_C} + (1-x) \frac{\Lambda'_{BC}}{m_C}$$

$$K_{12} = x \frac{\Lambda'_{AC}}{m_A} - x \frac{\Lambda'_{BC}}{m_B}$$

$$K_{21} = (1-x) \frac{\Lambda'_{AB}}{m_A} - (1-x) \frac{\Lambda'_{BC}}{m_C}. \qquad (11.42)$$

For long-wavelength *longitudinal* optical phonons, we have from Gauss' law

$$E = -4\pi P \qquad (11.43)$$

and therefore

$$E_{eff} = -\frac{8\pi}{3} P. \qquad (11.44)$$

With this, the longitudinal optical frequencies are determined from equation (11.41) after replacing Λ' by Λ'' and with

$$\Lambda''_{AB} = \Lambda_{AB} + \frac{8\pi}{3} N e_A e_B \left(\frac{\varepsilon_\infty + 2}{3\varepsilon_\infty} \right)$$

$$\Lambda''_{AC} = \Lambda_{AC} + \frac{8\pi}{3} N e_A e_C \left(\frac{\varepsilon_\infty + 2}{3\varepsilon_\infty} \right)$$

$$\Lambda''_{BC} = \Lambda_{BC} - \frac{8\pi}{3} N e_B e_C \left(\frac{\varepsilon_\infty + 2}{3\varepsilon_\infty} \right). \tag{11.45}$$

The quantities N and ε_∞ are assumed to be linearly dependent on composition

$$N = (1-x)N_{AB} + xN_{AC} \tag{11.46}$$

$$\varepsilon_\infty = (1-x)\varepsilon_{\infty,AB} + x\varepsilon_{\infty,AC}. \tag{11.47}$$

The force constants $\Lambda_{AB}, \Lambda_{AC}$ and Λ_{BC} are actually functions of the lattice constant. In mixed crystals, the lattice constant shows linear dependence on concentration x (known as Vegard's law). Therefore, one can assume

$$\frac{\Lambda_{AB}}{\Lambda_{AB_0}} = \frac{\Lambda_{AC}}{\Lambda_{AC_0}} = \frac{\Lambda_{BC}}{\Lambda_{BC_0}} = 1 - \theta x, \tag{11.48}$$

where the zero labels quantities for the pure compounds and θ is a constant dependent on the change in lattice constant $\Delta a / a$:

$$\theta x = \frac{\Delta \Lambda}{\Lambda_0} = \frac{2\Delta \omega}{\omega_0} = 6\gamma \frac{\Delta a}{a}, \tag{11.49}$$

where γ is the Grüneisen constant.

The MREI model contains five parameters: $e_B, e_C, \Lambda_{AB}, \Lambda_{AC}$ and Λ_{BC}. The force constants could be expected to be determined from observed quantities, thus leaving the effective charges e_B and e_C as the only two adjustable parameters. The force constants are subjected to boundary conditions, which are obtained by solving equation (11.41) for a gap mode at $x = 0$ and a local mode at $x = 1$.

TO *modes*

For $x = 0$

$$\omega^2 = \omega^2_{TO,AB} = \frac{\Lambda_{AB_0}}{\mu_{AB}} - \frac{4\pi}{3} \frac{N_{AB}e^2_{AB}}{\mu_{AB}} \left(\frac{\varepsilon_{\infty,AB} + 2}{3} \right) \tag{11.50}$$

$$\omega^2 = \omega^2_{gap,AB} = \frac{\Lambda_{AC_0} + \Lambda_{BC_0}}{m_C}. \tag{11.51}$$

For $x = 1$

$$\omega^2 = \omega^2_{TO,AC} = \frac{\Lambda_{AC_0}(1 - \theta)}{\mu_{AC}} - \frac{4\pi}{3} \frac{N_{AC}e^2_{AC}}{\mu_{AC}} \left(\frac{\varepsilon_{\infty,AC} + 2}{3} \right) \tag{11.52}$$

$$\omega^2 = \omega^2_{loc,AC} = \frac{\Lambda_{AB_0} + \Lambda_{BC_0}}{m_B}(1 - \theta), \tag{11.53}$$

where μ_{AB} and μ_{AC} are the reduced masses of the crystals AB and AC

$$\frac{1}{\mu_{AB}} = \frac{1}{m_A} + \frac{1}{m_B} \qquad \frac{1}{\mu_{AC}} = \frac{1}{m_A} + \frac{1}{m_C} \tag{11.54}$$

$e_{AB} \equiv e_B, e_{AC} \equiv e_C, \omega_{loc,AC}$ denotes the local mode frequency of impurity B in the crystal AC, and $\omega_{gap,AB}$ is the gap mode of impurity in the crystal AB.

LO *modes*

For $x = 0$

$$\omega^2 = \omega^2_{LO,AB} = \frac{\Lambda_{AB_0}}{\mu_{AB}} + \frac{8\pi}{3} \frac{N_{AB}e^2_{AB}}{\mu_{AB}} \left(\frac{\varepsilon_{\infty,AB} + 2}{3\varepsilon_{\infty,AB}} \right) \tag{11.55}$$

$$\omega^2 = \omega^2_{gap,AB} = \frac{\Lambda_{AC_0} + \Lambda_{BC_0}}{m_C}. \tag{11.56}$$

For $x = 1$

$$\omega^2 = \omega_{LO,AC}^2 \;=\; \frac{\Lambda_{AC_0}(1-\theta)}{\mu_{AC}} + \frac{8\pi}{3}\frac{N_{AC}e_{AC}^2}{\mu_{AC}}\left(\frac{\varepsilon_{\infty,AC}+2}{3\varepsilon_{\infty,AC}}\right) \tag{11.57}$$

$$\omega^2 = \omega_{loc,AC}^2 \;=\; \frac{\Lambda_{AB_0}+\Lambda_{BC_0}}{m_B}(1-\theta). \tag{11.58}$$

Notice that equations (11.51) and (11.56) are identical, and so are equations (11.53) and (11.58).

From the MREI model, one can derive conditions for mode behaviour in a mixed crystal. Let us remember that two-mode crystals exhibit local and gap modes (frequencies above and below, respectively, the optic band of the host lattice) when one component or the other is infinitely diluted in the mixed system. For $x = 1$ we can use either equations (11.52)–(11.53) or equations (11.56)–(11.57) to derive the condition for the existence of a local mode. Writing

$$\omega_{loc,AC}^2 \simeq \frac{\Lambda_{\text{eff}}}{m_B} > \omega_{max}^2 \simeq \frac{\Lambda_{\text{eff}}'}{\mu_{AC}} \tag{11.59}$$

and assuming that Λ_{eff} and Λ_{eff}' are of the same order of magnitude (which is the case for most zincblende structure mixed crystals), we conclude that for the existence of a local mode

$$m_B < \mu_{AC}. \tag{11.60}$$

Applying the above arguments for $x = 0$ and using equations (11.50)–(11.51) or (11.54)–(11.55), we derive the following condition for the existence of a gap mode:

$$m_C > \mu_{AB}. \tag{11.61}$$

Equation (11.60) is a stronger condition than equation (11.61). Thus those mixed systems which obey the inequality $m_B < \mu_{AC}$ exhibit two-mode type behaviour, whereas the opposite is true for the one-mode type behaviour. In other words, for a mixed crystal to exhibit two-mode (one-mode) behaviour it must (must not) have one substituting element whose mass is smaller than the reduced mass of the compound formed by the other two elements.

Thus, given the properties of the end members, the MREI theory can predict whether a mixed crystal will show one-mode or two-mode behaviour and also the dependence of the optic phonon frequencies on the mole fraction x.

It is interesting to apply Saxon–Hutner's theorem (Saxon and Hutner 1949) on energy gaps to the mixed system $AB_{1-x}C_x$: the frequencies which are forbidden in both end-member crystals AB and AC will be forbidden for any mixture of AB and AC (Jahne 1976). This means that the common gaps which are shared by pure AB and AC crystals persist in the $AB_{1-x}C_x$ mixed crystal over the entire composition range. An accurate picture for the common gap can be obtained from the optical phonon densities of states, and not necessarily from the reststrahlen bands, of the end-member crystals. Therefore, the criterion for one- (two-)mode behaviour is the non-existence (existence) of a common gap between the optical phonon densities of states of the end-member crystals. It may, however, be *sufficient* for one-mode behaviour to occur if the reststrahlen bands of the end-member crystals overlap.

11.4.3 APPLICATION OF THE MREI MODEL TO MIXED CRYSTALS

Let us consider two mixed crystals $GaP_{1-x}As_x$ and $Ga_{1-x}In_xP$. As seen from figure 11.4, the former shows two-mode behaviour, while the latter shows one-mode (or a very weak two-mode) behaviour. These behaviours are easily understood by noting that the optical phonon densities of states of GaAs and GaP do not overlap but those of InP and GaP do overlap (but in a narrow frequency range). The two-mode behaviour of $GaP_{1-x}As_x$ can also be explained from the inequality $m_P < \mu_{GaAs}$

which arises from equation (11.60). On the other hand, because $m_{Ga} > \mu_{InP}$ and $m_{In} > \mu_{GaP}$, the MREI predicts one-mode behaviour for $Ga_{1-x}In_xP$. For another example, consider $Ga_{1-x}Al_xAs$. This crystal, which exhibits a two-mode type bahaviour, meets the mass criterion $m_{Al} < \mu_{GaAs}$.

Calculations of TO modes (with $E_{eff} = 0$ in equations (11.31) and LO modes from the MREI model show good agreement with Raman scattering and infrared reflectance measurements for a number of mixed crystals. Such agreements can be seen in figures 11.4 and 11.5.

11.5   PHONON CONDUCTIVITY OF MIXED CRYSTALS

In this section we discuss the lattice thermal conductivity of mixed crystals. This could have been done in Chapter 7, but in view of the importance of the physics of phonons in mixed crystals, we have kept this topic for the present chapter.

An *ab initio* calculation of the lattice thermal conductivity of mixed crystals is a very difficult task, and indeed has not yet been achieved. The reason for this is clear. We remind ourselves that, using equation (5.64), calculation of κ requires determination of ω, c and τ for phonons of the crystal under investigation. Calculations of the first two of these properties require harmonic force constants, and calculations of τ require both harmonic and anharmonic force constants. For randomly disordered semiconductor alloys, estimates of changes in the harmonic force constant due to mass-disorder, changes in lattice constant and local bond length distribution would be required for calculation of phonon scattering rate generated by disorder, τ^{-1}(disorder). Mass-disorder, changes in lattice constant and local bond length distribution would also be required for a realistic calculation of anharmonic force constants, and hence τ^{-1}(anharmonicity). Considerations of these factors require thinking about possible crystal structures of such systems. In general, semiconductor alloys obey Vegard's law (i.e. have lattice constants very close to the fractional average of parent solids) but are characterised by distribution of bond lengths that are close to bond lengths of parent solids (Srivastava *et al* 1985). Considering a binary alloy A_xB_{1-x} made of parent solids A and B, it is possible to identify crystalline structures for certain concentration values x (Srivastava *et al* 1985). But modelling of random alloy structures requires a great deal of additional considerations (Srivastava *et al* 1985, Zunger *et al* 1990, Chouhan *et al* 2014). The simplest and most popular scheme is the *virtual crystal approximation* (VCA), in which lattice constant, bond lengths, harmonic force constant and anharmonic force are taken as the concentration average of parent compounds. The harmonic scattering rate due to mass disorder is calculated using the perturbation theory of isotopic mass defects discussed in section 6.2.1. Flicker and Leath (1973) developed a formulation within the *coherent potential approximation* (CPA) which calculates from the Kubo expression the lattice thermal conductivity of high-concentration mixed crystals in the harmonic approximation. Arrigoni *et al* (2018) made *ab initio* calculations of κ within the embedded-cluster approximation for alloy structure. In that they described an alloy by a small cluster of atoms embedded in an effective medium. The mass disorder and harmonic force constant disorder were treated by employing a theory similar to that discussed in section 11.2.2. However, they employed the VCA for calculating the third-order anharmonic force constants, *viz* concentration average of third-order force constants of parent compounds. Their work shows that it is necessary to include the contribution from interatomic force constant disorder in calculations of κ of III-V and II-VI semiconductor alloys. On the other hand, for alloys made of homopolar compounds such as Si_xGe_{1-x} mass disorder is the leading term and VCA is a reasonable approximation. Garg *et al* (2011) calculated the κ for bulk Si and Ge using the *ab-initio* method described in section 7.4.1.5. Using those results, they employed the VCA to calculate the room-temperature thermal conductivity of Si_xGe_{1-x} is shown in figure 11.9. It is pleasing to note good agreement with experimental results. Both theory and experiment suggest that very low conductivity values, close to $10~Wm^{-1}K^{-1}$, are obtained for the alloys with x values in the range $0.10 - 0.90$.

Within the VCA, discussion of phonon scattering in alloys was presented in section 6.3. To summarise, in a single-crystal undoped semiconductor alloy the main phonon scattering

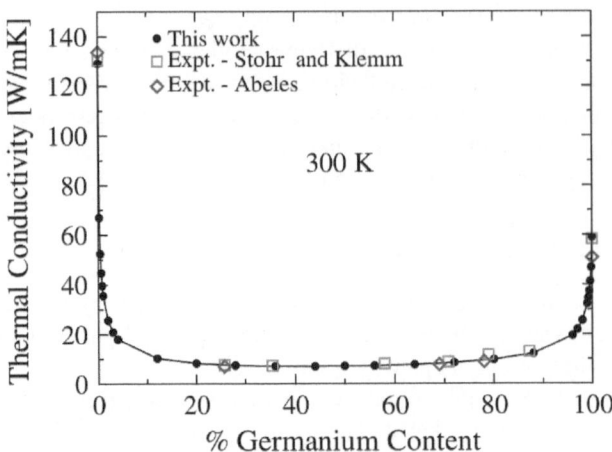

Figure 11.9   Room-temperature thermal conductivity of Si_xGe_{1-x}. Solid circles show predicted results from Garg *et al* (2011). Experimental values are shown by open square symbols (Stohr and Klemm 1954) and open diamond symbols (Abeles *et al*, 1962). [Original diagram in colour.] Reproduced from Garg *et al* (2011).

mechanisms include boundary scattering, isotope scattering, mass disorder scattering, strain scattering and anharmonic scattering. In a doped mixed crystal there will be, in addition, phonon–electron or phonon–hole scattering. The effect of sintering an alloy gives rise to strong scattering at grain boundaries. Expressions for phonon scattering due to these mechanisms were derived in chapter 6. Phonon scattering due to mass disorder and strain in alloys is characterised by the following relaxation time (*cf* equations (6.36) and (6.37)):

$$\tau^{-1}_{\text{mass disorder}} = \frac{\Omega}{4\pi\bar{c}^3}\Gamma_{md}\omega^4, \tag{11.62}$$

where $\bar{c}$ is an average phonon velocity, Ω is the unit cell volume and Γ_{md} is the mass-disorder parameter given by equation (6.37). Abeles (1963) has approximated equation (6.37) to

$$\Gamma_{md} = \sum_i f_i\left[\left(\frac{\Delta M_i}{M}\right)^2 + b\left(\frac{\delta-\delta_i'}{\delta}\right)^2\right], \tag{11.63}$$

where M_i is the atomic weight of the ith constituent with fractional concentration f_i, and b is regarded as an adjustable parameter. For mixed crystals of the type $A_{1-x}B_xC$, we can express

$$\Gamma_{md} \equiv \Gamma_{md}(x) = x(1-x)\left[\left(\frac{\Delta M}{M}\right)^2 + b\left(\frac{\Delta\delta}{\delta}\right)^2\right], \tag{11.64}$$

where

$$\Delta M = M_A - M_B \qquad M = (1-x)M_A + xM_B \qquad \Delta\delta = \delta_A' - \delta_B', \tag{11.65}$$

with δ_A', δ_B' and δ as the radii of atom A, atom B, and of the virtual crystal atom, respectively.

Abeles (1963) and Parrott (1963) independently approximated Callaway's expression (see equation (7.12)) for the lattice thermal conductivity of alloys in the limit of high temperatures. In the case of extreme (weak or strong) mass-disorder scattering their result for the thermal resistivity of the alloy can be reduced to

$$W(x) = W_P(x) + \frac{1}{3}\frac{(1+2\beta+25\beta^2/21)}{(1+5\beta/9)^2}A_1\Gamma_{md}(x) \tag{11.66}$$

$$= W_P(x) + A_2\Gamma_{md}(x), \tag{11.67}$$

where $\beta = \tau_N^{\ 1}/\tau_U^{\ 1}$ is taken as a constant ratio of N-process to U-process scattering, A_1 and A_2 are constants, and $W_P(x)$ is the thermal resistivity of the ordered virtual crystal. From equation (11.66), we can, therefore, express the thermal resistivity of the ternary system $A_{1-x}B_xC$ in the form

$$W(x) = (1-x)W_{AC} + xW_{BC} + x(1-x)W'_{A-B}, \tag{11.68}$$

where the first two terms give the virtual crystal result and the last term contains the contribution W'_{A-B} from the random distribution of the A and B atoms in the cation sublattice.

The result in equation (11.68) can be extended to a quaternary system $A_{1-x}B_xC_yD_{1-y}$:

$$
\begin{aligned}
W(x,y) = & \ (1-x)yW_{AC} + (1-x)(1-y)W_{AD} + xyW_{BC} \\
& + x(1-y)W_{BD} + x(1-x)W'_{A-B} + y(1-y)W'_{C-D},
\end{aligned} \tag{11.69}
$$

where W'_{C-D} is the $C-D$ disorder due to the random distribution of the C and D atoms in the anion sublattice. Thus in such a quaternary system alloy disorder gives rise to two bowing parameters, W'_{A-B} and W'_{C-D}.

11.5.1 THERMAL RESISTIVITY OF UNDOPED SINGLE CRYSTAL ALLOYS

Figure 11.10 shows the room temperature thermal resistivity of InAs–GaAs, InAs–InP, GaAs–GaP, InSb–GaSb, InP–GaP and GaAs–AlAs single crystal alloys. Also shown are the results of calculation using equation (11.68). There is reasonable agreement between theory and experiment. The interesting feature is that the thermal resistivity increases markedly with alloying and exhibits a maximum at composition near $x = 0.5$. This feature has been recognised to be very useful in high performance for thermoelectric power conversion since the figure of merit for such devices varies proportionally with the thermal resistivity (Parrott and Stuckes 1975).

The bowing in the thermal conductivity of mixed crystals is caused by mass-disorder scattering as well as strain scattering. The bowing in the conductivity as a function of x is predominantly due to mass-disorder scattering in $Ge_{1-x}Si_x$, whereas mainly due to strain scattering in $Ga_{1-x}Al_xAs$ (Abeles 1963). The effect of disorder and strain scattering is to reduce the sharpness of the fall in the conductivity at high temperatures: $\mathcal{K}(\text{alloy}) \propto T^{-n}, n < 1$. Near equimolar composition ($x \simeq 0.5$) where mass disorder is highest the phonon mean free path may be reduced to interatomic distance leading to $\mathcal{K} \propto T^0$. In such a situation heat transfer by energy quantum exchange between neighbouring atoms (i.e. due to phonon hopping) becomes possible.

$In_{1-x}Ga_xAs_yP_{1-y}$, a quaternary alloy which can be grown epitaxially on a lattice matched InP substrate, is a promising candidate for the light source (1.3–1.7 μm wavelength region) of an optical fibre communication system. The thermal resistivity of this alloy is an important parameter in the optimisation of such devices because it influences both device operation and operating life through the active-layer temperature. Figure 11.11 shows the thermal resistivity as a function of y for a $In_{1-x}Ga_xAs_yP_{1-y}$ lattice matched to InP. For the lattice matching condition, x and y compositions can be related by (Adachi 1983)

$$x = \frac{0.1894y}{0.4184 - 0.013y} \qquad (0 \le y \le 1). \tag{11.70}$$

It can be noticed that both theory and experiment reveal bowing in the resistivity, with a maximum bowing for y values between 0.6 and 0.75. These results suggest that the thermal resistivity of the

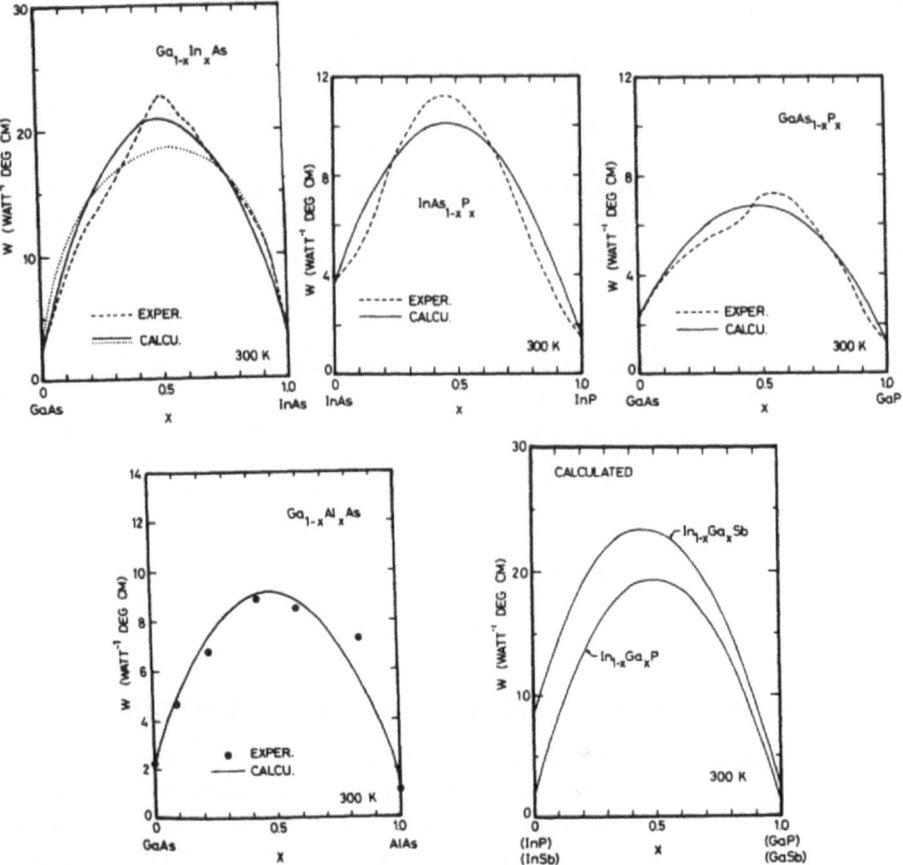

Figure 11.10 Room-temperature thermal resistivity of III–V mixed single crystals as a function of composition. The dashed curve is the experimental result. The solid curve is a theoretical fit using equations (11.64)–(11.66). The dotted curve is a theoretical calculation by Abeles (1963). (Reproduced from Adachi (1983).)

GaInAsP alloy in the composition range of fibre optic communication is about 20 times higher than that of InP. A three-dimensional view (Both *et al* 1986) of the thermal resistivity for this quaternary alloy is given in figure 11.12.

11.5.2 LOW-TEMPERATURE THERMAL CONDUCTIVITY OF DOPED SEMICONDUCTOR ALLOYS

As discussed in section 7.7, at low temperatures the phonon scattering by donors or acceptors due to impurities in semiconductors dominates over other scatterings. This can also be true for semiconductor alloys doped with donor or acceptor impurities. Here we consider one such case.

Briggs and Challis (1969) observed that at liquid helium temperatures the lattice thermal conductivity of $Ga_{1-x}In_xSb$ *increases* with increasing concentration x of InSb. This behaviour is explained as follows. Due to non-stoichiometry, there is a high concentration of acceptor impurities in pure GaSb. This is confirmed by its low-temperature lattice thermal conductivity which is about 100 times smaller than its expected value from the boundary scattering of phonons alone. On the other hand, InSb is free from such impurities. When InSb is mixed with GaSb to form $Ga_{1-x}In_xSb$, the

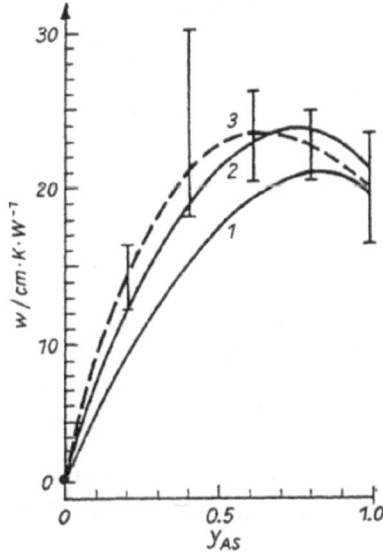

Figure 11.11 Room-temperature thermal resistivity of the GaInAsP alloy lattice matched to InP. Curve 1: after Both and Herrmann (1982), curve 2: theoretical result of Adachi (1983), curve 3: experimental results. (Reproduced from Both *et al* (1986).)

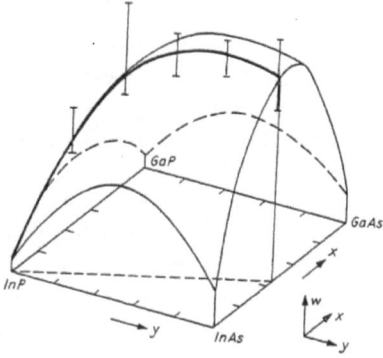

Figure 11.12 A three-dimensional representation of the room temperature thermal resistivity of $In_{1-x}Ga_xAs_yP_{1-y}$ alloys over the entire range of compositions. (Reproduced from Both *et al* (1986).)

impurity concentration in GaSb decreases and consequently the number of phonon-impurity scattering centres also decreases.

Sood and Verma (1973) investigated the contribution of phonon–hole scattering in the thermal conductivity of $Ga_{1-x}In_xSb$. These authors used the Callaway model for the conductivity. Phonon–hole scattering was calculated from Ziman's expression given in equation (6.130), and the mass-disorder scattering was calculated from equation (11.64). The deformation potential, m^*/m, and E_F were treated as adjustable parameters. It was concluded by these authors that the holes associated with the acceptor impurities in GaSb are largely responsible for the phonon scattering in the alloy in the temperature range 4–10 K. With increasing concentration of InSb, the density of acceptor centres as well as the hole effective mass decrease, and thus the phonon–hole coupling strength decreases. Consequently the low-temperature thermal conductivity increases. This is clearly seen in

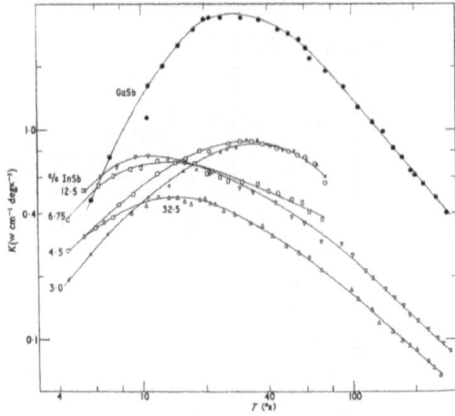

Figure 11.13 Variation of lattice thermal conductivity of GaSb–InSb alloys with temperature. InSb concentrations are given in mole %. (Reproduced from Briggs and Challis (1969).)

figure 11.13 for $x = 0.03, 0.045, 0.0675$ and 0.125. When the InSb concentration is increased to $x = 0.325$, the mass-disorder scattering of phonons becomes stronger than the phonon–hole scattering and the thermal conductivity is reduced.

11.5.3 THERMAL CONDUCTIVITY OF SINTERED SEMICONDUCTOR ALLOYS

In a mixed crystal, high-frequency phonons are very strongly scattered as ω^4, so that the heat transport is mainly due to low-frequency phonons of long mean free paths. Goldsmid and Penn (1968) and Parrott (1969b) pointed out that even at high temperatures boundary scattering of phonons can be important in mixed crystals. Smaller the crystal size, larger the additional boundary scattering resistance can be expected.

The realisation of the non-linear, indeed synergistic, combination of scattering processes of different phonon frequency dependence in determining the lattice thermal conductivity has led to some technological advancement. In particular, reduction in the thermal conductivity of semiconductor alloys makes them potential thermo-electric materials for high-temperature applications. The cost of preparing thermoelements is significantly reduced by using hot pressed (pressure sintered) materials. Apart from fabrication facility, hot-pressing substantially reduces thermal conductivity and thus improves the thermoelectric figure of merit compared to single crystal alloys (Parrott and Stuckes 1975).

Using theoretical calculations based on Callaway's model, equation (6.130) for phonon-free carrier scattering, and an effective grain boundary scattering, Meddins and Parrott (1976) predicted that for heavily doped Ge/Si alloys the effective grain size would be in the range $0.2\ \mu m < L < 2\ \mu m$. These limits may be altered somewhat by making a more detailed calculation. (Indeed, x-ray microscopic measurements suggest grain sizes in a range about an order of magnitude bigger than the theoretical prediction (Meddins and Parrott 1976).) Figure 11.14 shows a comparison of theoretical and experimental results for the lattice thermal resistivity of a few $Ge_{0.3}Si_{0.7}$ specimens. It is clearly seen that the effect of sintering is to increase the thermal resistivity by 20% or more. Further, the temperature dependence of the resistivity is less marked than T^{+1} in sintered materials than in single crystals.

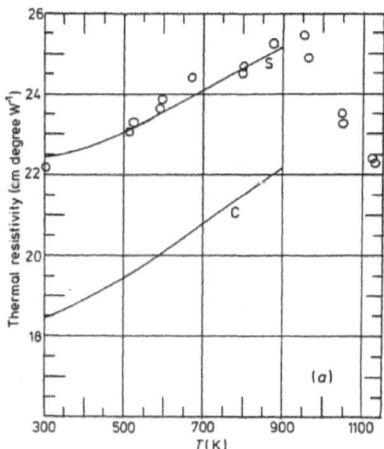

Figure 11.14 Lattice thermal resistivity of $Ge_{0.3}Si_{0.7}$ alloys. Open circles represent experimental results for sintered material with carrier concentration $n = 1.75 \times 10^{20} cm^{-3}$. The curves C and S are theoretical curves for crystalline and sintered materials, respectively. (Reproduced from Meddins and Parrott (1976).)

11.5.4   ROLE OF SINTERING ADDITIVES IN INCREASING THERMAL CONDUCTIVITY OF CERAMICS

Subject to favourable thermodynamic, kinetic and microstructural factors, sintering additivities can improve the thermal conductivity of ceramic powders. This has been demonstrated by Jackson *et al* (1997) for polycrystalline AlN that has been sintered with yttria (Y_2O_3). In general, additivities can enhance thermal conductivity of such ceramic samples by three mechanisms: (i) purifying the sample (*ie* reducing oxygen content), (ii) increasing sample density and (iii) increasing average grain size. Employing the isotropic continuum version of Callaway's conductivity expression, AlShaikhi and Srivastava (2008) examined the role of these mechanisms. In particular, for 8 wt% Y_2O_3 additive, the following features were established. At low temperatures (below 30 K) the dominant contribution to increase in the conductivity is exclusively due to the increase in grain size gained by sintering additive. In the intermediate temperature range (typically between 100 – 500 K), the increase in the conductivity is due to the combined effect of reduction in the oxygen-related impurity, reduction in grain size and increase in density (which was estimated by using the relation $\rho_{mix} = \alpha \rho_{additive} + (1 - \alpha)\rho_{powder}$, where α is the fraction of additive). The first of these factors was found to be most influential. At higher temperatures, anharmonic scattering becomes progressively more important, the defect scattering remains a dominant factor, and the role of increase in material density becomes a weak factor.

12 Phonons in Quasi-Crystalline and Amorphous Solids

12.1 INTRODUCTION

So far in this book we have discussed the physics of phonons in crystalline solids. In Chapter 11 where we discussed phonons in impure and mixed systems, the underlying structure was periodic: the disorder was compositional, not structural. Even in section 11.5 where we discussed sintered alloys, the grains were assumed to have crystalline structure. In this chapter we discuss the physics of phonons in quasi-crystals and in amorphous solids. Our discussion on quasi-crystals will be brief, and our discussion on amorphous solids will be sketchy as this subject is far from being well understood.

12.2 PHONONS IN QUASI-CRYSTALS

12.2.1 STRUCTURE OF QUASI-CRYSTALS

In 1984 Shechtman *et al* reported that rapidly solidified Al–Mn alloys exhibited electron diffraction patterns with *both* sharp peaks *and* with an icosahedral point-group symmetry (with six five fold, ten three fold and 15 two fold axes). Since this discovery, many other alloys have been found to exhibit similar structure. Such alloys known to-date can be classified as follows (Henley 1987).

(i) *i(Al–TM) class*: This is a class of alloys, made with Al and transition metals, which show icosahedral symmetry. Some examples are i($Al_{86}Mn_{14}$) (the original structure studied by Shechtman *et al* (1984)), $Al_{74}Mn_{20}Si_6$ and $Al_{79}Cr_{17}Ru_4$.

(ii) *The AlZnMg class*: Alloys such as $Al_{25}Zn_{38}Mg_{37}$, $Al_{44}Zn_{15}Cu_5Mg_{36}$ and $Al_{60}Cu_{10}Li_{30}$ also show icosahedral symmetry but have different atomic structure than the i(Al–TM) alloys.

(iii) *The d(Al–TM) class*: Al–TM alloys such as d($Al_{86}Mn_{14}$), $Al_{80}Pd_{20}$ and $Al_{80}Fe_{80}$ show quasi-crystalline structure of decogonal (ten-fold) symmetry in a plane, but periodic structure perpendicular to it.

(iv) $U_{20.6}Pd_{58.5}Si_{20.6}$ showing icosahedral symmetry.

(v) $(Ti_{0.9}V_{0.1})_2Ni$ showing icosahedral symmetry.

This list of alloys or their classes is not exhaustive.

The sharp diffraction peaks suggest presence of long-range order in these alloys, but the five-fold or ten-fold axes are incompatible with the classical concept of crystallography. The simultaneous presence of long-range positional order and icosahedral (or decagonal) symmetry can be resolved by constructing an *ordered aperiodic* or *quasi-periodic* structure (Levine and Steinhardt 1984). In this terminology, the alloys mentioned above are 'quasi-crystals'.

A quasi-crystalline structure, which falls between crystalline and amorphous structures, can be defined as follows (Levine and Steinhardt 1984).

(i) Every lattice point lies within some distance $R > 0$ of another lattice point, and the distance between any two lattice points is greater than some $R' > 0$.

(ii) If a subset of the lattice points is eliminated, another quasi-crystal lattice is obtained with nearest-neighbour distances increased by a constant factor. Thus a quasi-crystal is 'self-similar'.

(iii) The lattice has perfect long-range bond-orientational order.

DOI: 10.1201/9781003141273-12

(iv) The lattice has quasi-periodic translational order with $p(>1)$ linearly independent (incommensurate) lattice spacings along each lattice vector direction. For $p = 1$ a quasi-crystal reduces to a simple crystal.

The lattice points of a three-dimensional quasi-crystal are defined by a set of vectors x such that

$$x \cdot e_i = x_{in} \qquad x \cdot e_j = x_{jn'} \qquad x \cdot e_k = x_{kn''}, \tag{12.1}$$

where n, n' and n'' are integers, e_i are unit vectors along the axes of a regular polyhedron, $i > j > k$, and $i = 1, 2, \ldots, N$, with N as the number of axes of the polyhedron.

The lattice positions of a two-dimensional quasi-crystal are given by x_{in} and x_{jn} in equation (12.1), where e_i are unit vectors along the axes of a regular polygon, $i > j, i = 1, 2, \ldots, N$, and N is the number of axes of the polygon. Examples of two-dimensional quasi-periodic lattices are provided by Penrose's tiling patterns, which show mixture of order and unexpected deviations from order. Most Penrose two-dimensional patterns have local five-fold axes. An illustrative description of Penrose patterns is given in a fascinating article by Gardner (1977). A review of the remarkable properties and structure of quasi-crystals is given by Gratias (1987).

An example of a one-dimensional quasi-periodic lattice is the Fibonacci lattice whose points can be described in terms of a sequence of $p = 2$ linearly independent (incommensurate) intervals. Consider for simplicity a monatomic linear chain with two different interatomic lengths a and b. Define the pattern $r_0 = (a, b)$ as stage 0 of the sequence of intervals. The $(n + 1)$th stage of the sequence is obtained from the substitution rule

$$r_{n+1} = S r_n, \tag{12.2}$$

where S is a 2×2 matrix

$$S = \begin{pmatrix} 1 & 1 \\ 1 & 0 \end{pmatrix}. \tag{12.3}$$

It is easily seen that the Fibonacci chain has the sequence of intervals

$$(a, b) \to (a, b, a) \to (a, b, a, a, b) \to (a, b, a, a, b, a, b, a) \to \ldots. \tag{12.4}$$

The number of occurrences of the lengths a and b at stage n, $N_a^{(n)}$ and $N_b^{(n)}$, respectively, satisfy the following recurssion relation (Lu et al 1986):

$$\begin{aligned} N_a^{(n+1)} &= S_{11} N_a^{(n)} + S_{21} N_b^{(n)} \\ N_b^{(n+1)} &= S_{12} N_a^{(n)} + S_{22} N_b^{(n)}. \end{aligned} \tag{12.5}$$

The ratio $\tau = \lim\limits_{n \to \infty} (N_a^{(n)} / N_b^{(n)})$ is given by

$$\tau = \frac{S_{11} - S_{22} + [(S_{22} - S_{11})^2 + 4 S_{12} S_{21}]^{1/2}}{2 S_{12}} \qquad (S_{12} \neq 0) \tag{12.6}$$

$$= \frac{\sqrt{5} + 1}{2} = 1.61803\ldots \tag{12.7}$$

and is the 'golden mean'. The Fibonacci monatomic linear chain is shown in figure 12.1. It is obvious to note that for $a = b$ the chain becomes an ordinary monatomic linear chain.

You and Hu (1988) have developed a 'global cut-and-project' method to construct both generalised Fibonacci lattices and quasi-lattices.

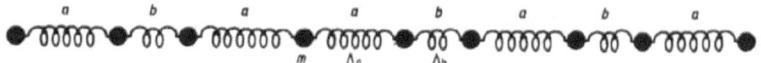

Figure 12.1 The Fibonacci monatomic linear chain (shown up to the third stage). m is the atomic mass, a and b are interatomic lengths, and Λ_a and Λ_b are the nearest-neighbour inter-atomic force constants. For $a = b$ the chain becomes an ordinary monatomic linear chain.

12.2.2 VIBRATIONAL PROPERTIES OF THE FIBONACCI MONATOMIC LINEAR CHAIN

Consider a finite chain of $N+2$ atoms, as shown in figure 12.1, to be connected by harmonic springs with nearest-neighbour force constants Λ_a and Λ_b. The equation of motion of the nth atom is given by

$$-m_n \omega^2 u_n = k_{n,n+1}(u_{n+1} - u_n) + k_{n-1,n}(u_{n-1} - u_n) \quad n = 1, 2, ..., N, \tag{12.8}$$

where m_n is the mass of the nth atom, $u_n e^{-i\omega t}$ is the displacement of the nth atom, and $k_{n,n+1}$ is the spring constant between atoms n and $n+1$. The set of N equations in equation (12.8) cannot be solved analytically, owing to the fact that the nearest neighbours of the nth atom cannot completely determine the nearest neighbours of the $(n+1)$th atom. Lu *et al* (1986) used a transfer matrix method to solve equation (12.8) numerically. A good description of the transfer matrix method can be found in Matsuda (1962).

We can express equation (12.8) as

$$u_{n-1} = c_n u_n + d_n u_{n+1}, \tag{12.9}$$

which can be written as

$$w_n = \begin{bmatrix} u_{n-1} \\ u_n \end{bmatrix} = \begin{bmatrix} c_n & d_n \\ 1 & 0 \end{bmatrix} \begin{bmatrix} u_n \\ u_{n+1} \end{bmatrix} = T_n w_{n+1}, \tag{12.10}$$

where

$$c_n = 1 + \frac{k_{n,n+1}}{k_{n-1,n}} - \frac{m_n \omega^2}{k_{n-1,n}} \qquad d_n = -\frac{k_{n,n+1}}{k_{n-1,n}}. \tag{12.11}$$

Following the structure of equation (12.10) we can write

$$w_1 = T_1 T_2 ... T_{N-1} T_N w_{N+1} = Q w_{N+1}. \tag{12.12}$$

If the fixed boundary condition $u_0 = u_{N+1} = 0$ is used, then equation (12.12) can be written as

$$\begin{bmatrix} 0 \\ u_1 \end{bmatrix} = \begin{bmatrix} Q_{11} & Q_{12} \\ Q_{21} & Q_{22} \end{bmatrix} \begin{bmatrix} u_n \\ 0 \end{bmatrix}. \tag{12.13}$$

For a non-trivial solution u_N, the vibrational frequency ω must satisfy the eigenvalue equation

$$Q_{11}(\omega) = 0. \tag{12.14}$$

This equation must be solved numerically.

Let us define an integrated (or cumulative) density of vibrational states as

$$g_{int}(\omega) = \int_0^\omega g(\omega') d\omega', \tag{12.15}$$

where $g(\omega)$ is the density of states as defined in section 2.5.

Consider all atoms to have the same mass, i.e. $m_n = m$ for all n. The integrated density of vibrational states for the Fibonacci monatomic chain of 2000 atoms as a function of a scaled frequency

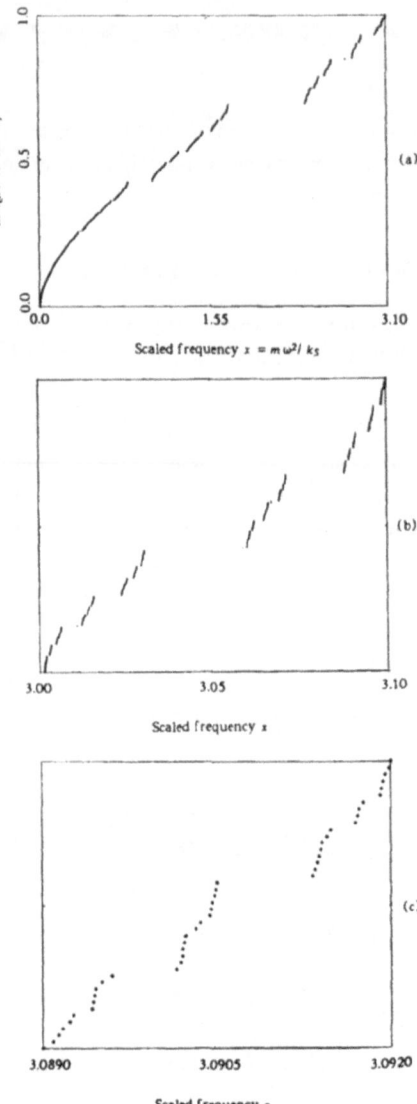

Figure 12.2 (a) Integrated vibrational density of states for the Fibonacci monatomic linear chain of 2000 atoms with $\Lambda_a/\Lambda_b = (\sqrt{5}-1)/2$. The horizontal axis is the scaled frequency $x = m\omega^2/\Lambda_b$. It is considered that $a > b$. (b) the same as (a), but the frequencies are enlarged around $x = 3.10$. (c) the same as (b), with further enlargement around $x = 3.0905$. The self-similarity in the density of states is clearly seen. (Reproduced from Lu *et al* (1986).)

$x = m\omega^2/\Lambda_b$, with $\Lambda_a/\Lambda_b = (\sqrt{5}-1)/2 \simeq 0.618$, is shown in figure 12.2. The following observations have been made by Lu *et al*.

(1) In the low-frequency region ($x << 0.1$), the spectrum behaves almost identically to that for a periodic monatomic linear chain.

(2) The spectrum is self-similar, which is a general feature of quasi-crystals. This behaviour is more explicitly seen in the high-frequency region as shown in figures 10.2(b) and (c). In the limit $N \to \infty$ the spectrum is 'Cantor-like', and there are an infinite number of gaps. Near the edges of bands in the spectrum the density of states exhibits van Hove singularities.

(3) Low-frequency vibrational wavefunctions are extended, while high-frequency waves tend to be localised and/or critical. This is a general feature of disordered solids (also see section 12.3.4).

Thus the Fibonacci chain processes the properties of both periodic and disordered chains. This conclusion can be extended to other types of quasi-crystals, including two- and three-dimensional cases.

12.2.3 FIBONACCI SUPERLATTICES

Merlin *et al* (1985) succeeded in growing, by molecular beam epitaxy, layers of GaAs and AlAs alternated in a Fibonacci sequence. This scheme of artificially fabricating one-dimensional quasi-crystalline superlattice structures has been followed by others (see, e.g. Dharma-wardana *et al* 1986). Raman scattering results from several Fibonacci superlattices have also been reported: these include GaAs/AlAs superlattices (Merlin *et al* 1985) and Si-Ge$_x$Si$_{1-x}$ strained-layer superlattices (Dharma-wardana *et al* 1987). An x-ray Bragg-driffraction analysis of the Fibonacci sequence of GaAs and AlAs layers has been presented by Tapfer and Horikoshi (1988).

It is expected that many more detailed theoretical and experimental studies of the structural, electronic and vibrational properties of Fibonacci superlattices will become available in the future.

12.3 STRUCTURE AND VIBRATIONAL EXCITATIONS OF AMORPHOUS SOLIDS

12.3.1 STRUCTURE

Amorphous solids (a-solids) lack long-range order, and also to some extent intermediate and short-range order. The lack of translational symmetry means that in a-solids we cannot use the concepts such as lattice and unit cell. This makes 'crystal structure' a rather ill defined term for a-solids. In fact one may consider an infinite number of possible structures for a-solids. In general, a-solids may show both macroscopic structural inhomogeities (e.g. voids, density fluctuations, etc) and microscopic structural defects (e.g. vacancies, broken bonds, etc). Experimentally, however, no single diffraction technique is wholly sufficient to determine the structure of a-solids. At best conventional neutron and x-ray scattering techniques can be used to determine the one-dimensional radial distribution function (RDF) which measures the probability of an atom at a distance from a given atom. This gives the description of the environment of an average atom.

The extended x-ray absorption fine structure (EXAFS) technique can be used to investigate the individual contribution to RDF from each type of atom present in a-solids. Infrared and Raman spectroscopies can be used to explore both microscopic and macroscopic details of structure. In particular, Raman spectroscopy has been used to correlate the width of the TO-like peak and its ratio to the TA-like band with the disorder in bond angles in a-Si and a-Ge (Lannin 1988).

Detailed studies using a combination of inelastic neutron scattering, EXAFS, infrared and Raman spectroscopies suggest that a-solids exhibit some order on the scale of several interatomic spacings. The range of this local ordering depends on the material and the condition of formation of the a-solid.

Amorphous solids may be metallic, semiconducting, or insulating. However, much more conclusive evidence exists for local structural ordering in a-semiconductors. A detailed discussion on the physics of amorphous materials can be found in Elliott (1983).

Theoretically, it is a formidable challenge to model real a-solids. A variety of random network models have been proposed for different types of a-solids such as Si, Ge, SiO$_2$, Se, etc. However, the best prescription available to date is the computer modelling of tetrahedrally bonded continuous random-networks with periodic boundary conditions (Wooten and Weaire 1987). In this scheme a periodic structure with a large cubic unit cell (say, with 216 atoms) is considered to model a-Si or a-Ge. The use of the periodic boundary conditions eliminates any surface states at the boundary of the unit cell. Inside the unit cell a highly random tetrahedral atomic structure is created and then relaxed following a Monte Carlo method. The model material is called *sillium* and is defined by the following rules.

(1) Each atom is tetrahedrally bonded to four neighbours.

(2) The total energy of the system is calculated by using the Keating potential with only bond-bending and bond-stretching terms.

(3) The only degrees of freedom in relaxing the system consists of *bond transpositions*, based on a few rules and technical considerations.

The general Keating potential is given in equation (2.69). In their model, Wooten and Weaire used only the first two terms in equation (2.69). The process of randomisation and annealing considers bond rearrangement or 'bond switch' involving the exchange of two parallel bonds. This is assisted by requiring that the bonds switched are not members of the same five, six, or seven fold ring. To avoid excessive computing time in the modelling no four fold rings are allowed, which involve large bond distortions.

The resulting annealed model for a-Si or a-Ge is shown in figure 12.3. In an earlier attempt Wooten *et al* (1985) modelled a slightly different version of the sillium for a-Si. For that earlier network, the calculated RDF (scaled to the bond length in Ge) shows an appealing degree of agreement with the experimental curve for a-Ge (figure 12.4). The correlation function shown in figure 12.4 is defined as $t(r) = g(r)/r$, where $g(r)$ is the RDF.

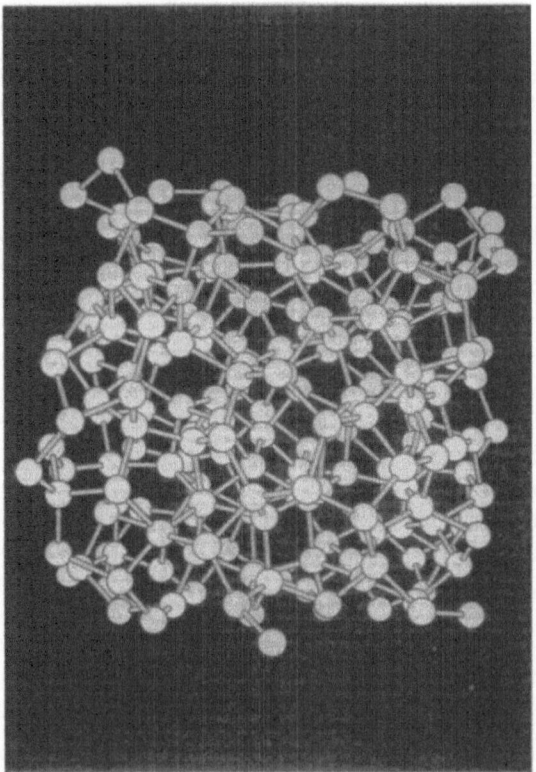

Figure 12.3 An annealed continuous random network model for tetrahedrally bonded 216-atom unit cell. For this model the correlation function $t(r) = g(r)/r$, where $g(r)$ is the radial distribution function, is found to be in agreement with experiment and the intensity of the (111) peak in the structure factor remains low. (Reproduced from Wooten and Weaire (1987).)

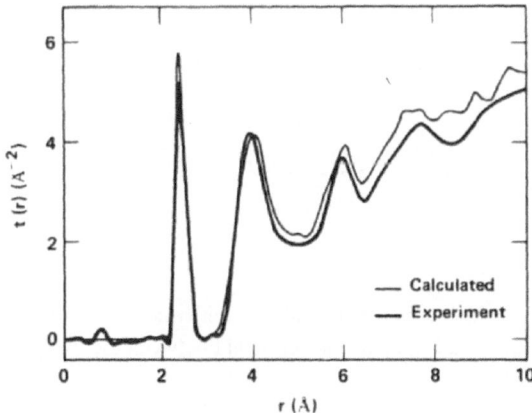

Figure 12.4 The correlation function of the continuous random network model of a-Si (scaled to model a-Ge) compared with experiment. (Reproduced from Wooten *et al* (1985).)

12.3.2 PHONONS

Because of the absence of long-range order, it is no longer possible to consider the vibrational modes in a-solids as plane waves with well defined q-vectors. In other words, phonons are not well defined quasi-particles in a-solids. However, inelastic neutron scattering experiments suggest that an approximate description of low-frequency vibrational excitations in a-solids can be given in terms of phonons with wavelengths longer than some characteristic structural unit of the material (see section 12.3.4). Even in that limit phonon dispersion relations $\omega(q)$ are not well defined and the phonon density of states $g(\omega)$ is regarded as the main quantity of interest. In our discussion of a-solids, however, we shall use the term phonons sometimes only for low-frequency vibrational modes, and sometimes for vibrational modes in general.

12.3.3 TUNNELLING STATES

It was proposed independently by Phillips (1972) and Anderson *et al* (1972) that many low-temperature thermal properties of certain types of a-solids, e.g. glasses, can be explained by considering scattering of Debye-like phonons by *additional excitations*. Such excitations can be represented most simply by two-level systems, or more generally by highly anharmonic oscillators. The two-level model leads to the concept of tunnelling states, similar to that discussed in section 6.7 in connection with impurity states in alkali halides. Consider a double-well potential with a barrier height V_0 and energy separation Δ between the two minima (figure 12.5). Certain atoms or groups of atoms can perform coupled rotation in such a double well via quantum mechanical tunnelling. The resulting states are called *tunnelling modes* and have frequencies in the range of low-frequency vibrational modes (phonons). The density of states of tunnelling modes is considered to be a constant. It should be emphasized that the model of tunnelling states in a-solids is phenomenological and lacks a firm theoretical or experimental foundation.

12.3.4 FRACTAL STRUCTURE AND FRACTONS

Certain aspects of the static geometrical, dynamical and transport properties of a-solids can be described in terms of the concept of fractal structures. There are different classes of fractal structures. One such class is the so-called self-similar geometry. As explained in section 12.2.2, a geometry is self-similar if it is indistinguishable as a function of length scale (or upon resolution). The 'fractal'

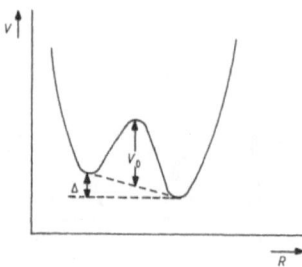

Figure 12.5 The concept of a double-well potential in open structure, with Δ being the energy separation (asymmetry) between the two minima. V_0 is the barrier height between the two potential wells.

concept, developed by Mandelbrot (1982), can be understood from a density argument. Consider a mass M inside a sphere of radius R. For a homogeneous (Euclidean) structure $M = AR^d$, where $d = 3$ is the *Euclidean dimension*, and A is a numeric constant. For a self-similar fractal structure $M = BR^{\bar{d}}$, where $\bar{d}$ is called the *fractal (or spectral, or Hausdorff) dimension*, and B is no longer a numeric constant but varies according to the lacunarity of the medium. In general, $\bar{d} \leq d$, because of the open structure or inhomogeneity of a-solids. A fractal geometry is usually realised within a space region $a \leq r \leq \xi$, where a and ξ are some appropriate short and large length scales, respectively. For $r > \xi$ a fractal geometry changes to the Euclidean geometry. Whereas Euclidean spaces posses the translational symmetry, fractal geometries show dilation symmetry (scale invariance). The concept of fractal geometry, therefore, is intermediate to the concepts of crystalline structures and disordered materials.

A few interesting examples of objects exhibiting fractal structure are gels, the track of a Brownian particle, diffusion-limited aggregates of metal particles of clusters and colloidal particles. An interesting discussion on fractals can be found in the book by Cusack (1987) where further references on the topic are also given. The self-similarity property of a two-dimensional fractal aggregate is illustrated in figure 12.6.

The vibrational quantum of a fractal network is called a *fracton* (or fractal phonon). Fractons are believed to be short wavelength ($\lambda < \xi$), localised vibrational modes of a-solids. With decreasing frequency fractons change into long wavelength ($\lambda > \xi$), propagating phonon-like modes. The crossover frequency ω_c is given by the relation

$$\omega_c \propto \xi^{-d/\bar{d}} \tag{12.16}$$

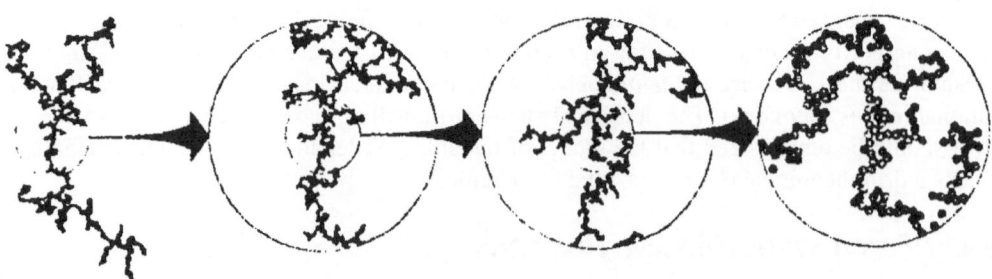

Figure 12.6 Schematic illustration of the self-similarity property of a two-dimensional disordered fractal aggregate. (From Jullien (1987).)

where $\bar{\bar{d}}$ is called *fracton dimension*. While low-frequency vibrational modes ($\omega < \omega_c$) have Debye-like phonon dispersion relation $\omega_{ph}(\lambda^{-1}) \propto \lambda^{-1}$, it has been suggested that fractons ($\omega > \omega_c$) have a dispersion relation of the type

$$\omega_{fr}(\lambda^{-1}) \propto \lambda^{-\bar{d}/\bar{\bar{d}}}. \tag{12.17}$$

We know from Chapter 2 that the density of Debye-like phonon states is given by

$$g_{ph}(\omega) \propto \omega^{d-1} \tag{12.18}$$

where $d = 3$ is the Euclidean dimensionality. Alexander and Orbach (1982) showed that the density of vibrational states in the fracton frequency range can be written as

$$g_{fr}(\omega) \propto \omega^{\bar{\bar{d}}-1}. \tag{12.19}$$

The precise shape of $g(\omega)$ in the vicinity of the crossover frequency ω_c and the relationship between g_{ph} and g_{fr} at ω_c are not fully established at present. Alexander *et al* (1983) have suggested the following relationship at the crossover:

$$g_{fr}(\omega)/g_{ph}(\omega) = \bar{\bar{d}}/d. \tag{12.20}$$

However, the discontinuity in the density of states at ω_c is somewhat controversial: other researchers believe that there should be a continuous transition between g_{ph} and g_{fr}.

For three-dimensional Euclidean spaces $d = \bar{d} = \bar{\bar{d}} = 3$. Alexander and Orbach (1982) 'conjectured' that $\bar{\bar{d}} = 4/3$ may be an exact relation for all site percolation networks [1] of fractal nature with dimensions greater than or equal to 2 ($d \geq 2$). However, this result is also somewhat controversial and should at best be considered as an excellent approximation.

It should be made clear that the concept of fractons is distinctly different from that of the tunnelling modes. Fractons are fundamental excitations above ω_c, whereas tunnelling modes are considered as low frequency excitations whose density of states is usually added to the extended phonon density of states.

The wavefunction of a fracton is localised (with $\lambda < \xi$). The extent of such localisation, however, is difficult to determine. It is sometimes useful to apply a scaling theory to study this problem. Assume that the length scale of a fractal geometry is scaled by a factor b. Then, for a self-similar structure, Rammal and Toulouse (1983) have suggested the following scaling behaviour for fracton frequency and density of states:

$$\omega(\xi/b) = b^a \omega(\xi) \tag{12.21}$$

$$g_{fr}(\xi; \omega) = b^{\bar{d}-a} g_{fr}(\xi; \omega b^{-a}). \tag{12.22}$$

If the scale factor b is chosen to be $b = \omega^{1/a}$, then equation (12.22) reduces to the power law

$$\begin{aligned} g_{fr}(\omega) &\propto \omega^{\bar{d}/a-1} \\ &\simeq \omega^{\tilde{d}-1}, \end{aligned} \tag{12.23}$$

where $\tilde{d} = \bar{d}/a$ defines a *scaled fractal (or Hausdorff, or spectral) dimensionality*. Notice that for $a = 1$, the frequency scaling becomes linear and there is no change in the frequency power law for the density of states. This theory can therefore be used to explain a scaling behaviour for the density of states in a-solids.

---

[1] We can simply understand a site percolating network as a regular lattice with some lattice sites randomly removed.

12.4 VIBRATIONAL PROPERTIES OF AMORPHOUS SOLIDS

The main features of the vibrational density of states in a-solids can be extracted from neutron scattering, infrared absorption (or reflection) and Raman scattering measurements.

In infrared absorption spectroscopy, the absorption coefficient $\alpha(\omega)$ is related to the vibrational infrared (IR) density of states $g_{IR}(\omega)$ of a-solids as

$$\alpha(\omega) \sim |M(\omega)|^2 g_{IR}(\omega), \tag{12.24}$$

where $M(\omega)$ is the dipole moment matrix element (cf equation (6.202)). In the low-frequency range it is assumed that $M(\omega) \propto \omega^2$ (Connell 1975) so that

$$g_{IR}(\omega) \propto \frac{\alpha(\omega)}{\omega^4}. \tag{12.25}$$

The intensity of light scattering is determined by the Raman tensor $I(\omega)$ (cf section 6.9.2). For a-solids, the Raman tensor is related to the Raman vibrational density of states $g_R(\omega)$ as (Shuker and Gamon 1971)

$$I(\omega) = C(\omega) \frac{[1 + \bar{n}(\omega, T)]}{\omega} g_R(\omega), \tag{12.26}$$

where $\bar{n}(\omega, T)$ is the Bose–Einstein distribution given in equation (2.2) and $C(\omega)$ measures the coupling of the vibrational modes of frequency ω with the light. For small ω it is assumed that $C(\omega) \propto \omega^2$ (Connell 1975) so that

$$g_R(\omega) \propto \frac{I(\omega)}{\omega[1 + \bar{n}(\omega, T)]}. \tag{12.27}$$

There is some uncertainty about the ω dependence of $C(\omega)$. Shuker and Gamon (1971) assume that $C(\omega)$ is independent of ω. In any case, the *reduced Raman tensor* can be defined as

$$I^R(\omega) = \frac{\omega^m I(\omega)}{C(\omega)[1 + \bar{n}(\omega, T)]} \propto g_R(\omega) \tag{12.28}$$

with $\omega^m / C(\omega) =$ constant.

The low-frequency vibrational density of states of a-Ge is shown in figure 12.7. There is fair agreement between $g_{IR}(\omega)$, $g_R(\omega)$ and the neutron scattering density of states.

A comparison of the Raman spectra of vitreous and crystalline silica is made in figure 12.8. The discrete Raman spectrum of the crystalline phase is replaced by a continuous spectrum for the glassy phase. This clearly demonstrates that the vibrational modes of the glass are not plane waves.

Theoretically it is much more difficult, at least in principle, to calculate the density of states of a-solids. In practice, however, various schemes have been adopted with different levels of simplification. These schemes can be grouped into two distinct classes: numerical methods and analytical methods. In such calculations both low- and high-frequency vibrational modes are treated on equal footing and are generally called phonons.

12.4.1 NUMERICAL METHODS

Numerical methods attempt to solve the dynamical matrix which is determined on the basis of a random-network cluster structural model and suitable interatomic forces. Most calculations reported to date have used the local structure of the corresponding crystalline situation, a large variety of intermediate range configurations (usually expressed in terms of rings of bonds), and a reasonable description of interatomic forces (such as the Born model, Keating model, or a more general

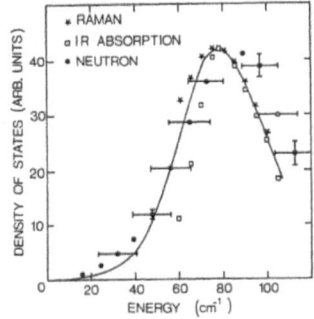

Figure 12.7 Low-frequency vibrational density of states of a-Ge as measured from Raman scattering, infrared absorption and neutron scattering data. (Reproduced from Brodsky (1983).)

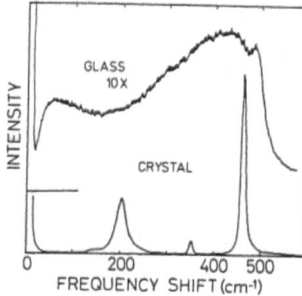

Figure 12.8 The Raman spectra of vitreous and crystalline silica. (From Shuker and Gamon (1971).)

valence-force-field model). The result for the vibrational density of states, $g(\omega)$, from such an approach, however, usually contains contributions from the surface of the cluster and suitable boundary conditions are required to eliminate such spurious modes from the calculated spectrum. For a better description of $g(\omega)$ a suitably large cluster is desired. This, however, results in a large dynamical matrix whose diagonalisation can be a genuine computational problem. Such a problem has been tackled by using efficient schemes such as the Lanczos method, or indirect methods such as the equation-of-motion method (Beeman and Alben 1977), or the recursion method (Meek 1976). (For a description of the recursion method, and the Lanczos method, and their applications, see Pettifor and Weaire (1985)).

12.4.2 ANALYTICAL METHODS AND OTHER IDEAS

Although it is important to use a proper numerical scheme employing a large cluster and a reasonable interatomic force model, it is nevertheless possible to extract some important features in the density of states spectrum of a large class of a-solids from a consideration of a small molecular unit cell with central forces only. For example, Sen and Thorpe (1977) have attempted to study the high-frequency features in the vibrational density of states of glassy AX_2 systems such as SiO_2 and GeS_2 in terms of the eigenmodes of an isolated AX_4 tetrahedron.

Weaire and Alben (1972) proved an important theorem by considering only central forces (actually from a bond-stretching potential of the Keating form) for a perfectly tetrahedral structural unit in an elemental solid: the spectral range of the density of states is bound between two limits (*cf* equation (2.67) and figure 2.13)

$$0 \le \omega \le \omega_{max} = \sqrt{\frac{8\alpha_K}{m}}, \tag{12.29}$$

where α_K is the Keating central force constant and m is the atomic mass. The spectrum is actually a continuum which is bound between two delta functions at the two ends. As the two limits are determined by the local geometry, they are expected to be valid for crystalline and a-solids alike. Local distortions may broaden the delta function at the upper limit of the spectrum.

It is also possible to use other simple structural models for which straightforward sotutions are possible, in particular if only short-range forces are considered. An example is an artificial pseudo lattice, called the *Bethe lattice* (an infinite Cayley tree) which maintains the same nearest-neighbour (e.g. tetrahedral) bonding for all atoms and generates an infinite non-periodic branching structure without any closed rings of bonds. (For a detailed description of Bethe lattices, see Thorpe (1981).) Another example is the cluster-Bethe-lattice model in which a finite-size cluster of atoms with precise local atomic arrangement is considered and a Bethe lattice is 'grafted' on atoms at the surface of the cluster. The grafting of a Bethe lattice is a mathematical device to eliminate cluster surface modes. Further, as the density of states of a Bethe lattice is rather smooth and featureless, this scheme does not introduce any additional structure into the density of states of the cluster. However, use of finite-size models coupled with short-range potentials in general leads to a rather artificial situation in that the density of states does not cover the very-long-wavelength region of the spectrum (Thorpe 1974).

12.4.3 COMPARISON WITH CRYSTALLINE DENSITY OF STATES

Figure 12.9 shows a comparison of the density of vibrational density of states for a continuous random network (CRN) model of a-Si with that for the diamond cubic structure (Meek 1976). Figure 12.9(a) was calculated with Born forces and the recursion method applied to a 344-atom microcrystal. Figure 12.9(b) shows the density of states obtained by a standard method of Brillouin zone summation. Clearly the calculation based on the finite-size microcrystallite shows the essential features of the density of states, albeit with somewhat broadened peaks. Figure 12.9(c) shows the results of calculation, using the same method as used for the crystalline case, for the CRN model of Steinhardt *et al* (1974) containing 201 atoms.

Strictly speaking, there are no analogues of acoustic and optical phonon modes in a-solids. However, it is instructive to study the spectrum of the density of states in a-solids in terms of features which are TA-, TO-, LA- and LO-like. The spectrum in figure 12.9(c) shows the same basic distribution of modes as in figure 12.9(a), but differs in detail. The theoretical curve also has reasonable similarity with the infrared absorption and Raman spectra for a-Si which are shown in figure 12.9(d). From his calculations using different CRN models, Meek showed that the shape of the TO-like peak of the spectrum broadens, which seems to depend on both the topology and bond-angle distortion of the network. The shape of the TA-like peak seems to depend more directly on the topology of the network and can give some information on the statistics of odd vs even numbered rings. The most significant difference between the crystalline and a CRN model is the virtual wash out of the LA and LO peaks due to the disorder present in the CRN. The dip between the LA and LO bands in the $g(\omega)$ of diamond cubic solids such as Si and Ge is considered to be a consequence of the presence of six-fold rings of bonds. An inspection of that part of the spectrum has been used to infer details of the ring statistics for a-Si or a-Ge.

However, questions like LO–TO splittings in a-solids remain unclear. Further, although a realistic calculation is likely to yield both extended (phonon-like) and localised (fracton-like) states in a a-solid, it is not easy to make a clear cut distinction between such states. These questions remain open for experimental studies as well.

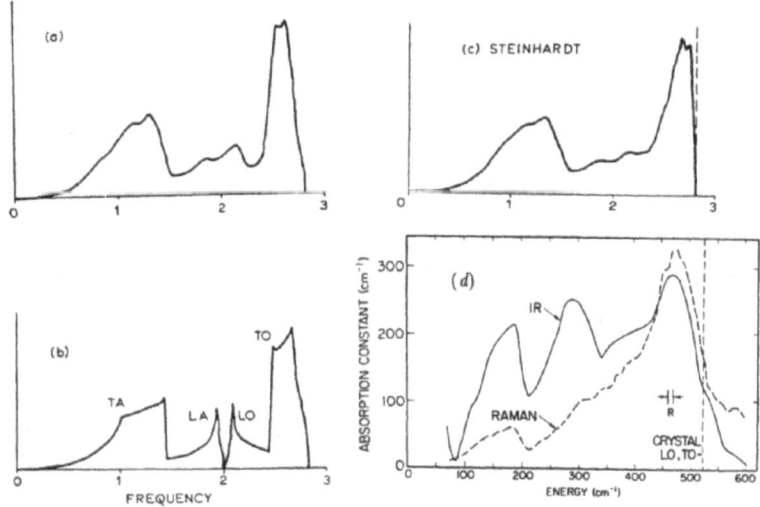

Figure 12.9 Comparison of the density of vibrational modes of a continuous random network model for a-Si with that for the diamond cubic structure: (*a*) result for a 344-atom microcrystal with Born forces using the recursion method, (*b*) result for crystalline Si using a standard Brillouin zone summation method, (*c*) result using the recursion method for the continuous random network model of Steinhardt containing 201 atoms, (*d*) Raman spectrum for a-Si. ((*a*)–(*c*) reproduced from Meek (1976), (*d*) reproduced from Weaire (1981), after Smith Jr *et al* (1971) and Brodsky and Lurio (1974).)

12.5 LOW-TEMPERATURE PROPERTIES OF AMORPHOUS SOLIDS

12.5.1 SPECIFIC HEAT

At high temperatures the specific heat of an a-solid is nearly the same as that of the crystalline phase of the same material. At low temperatures, however, the specific heat is in general larger for the amorphous phase. Also, whereas the low-temperature specific heat of crystalline (non-metallic) solids varies as $C_v = B_c T^3$ (see section 2.7), the specific heat of almost all known a-solids below about 1 K shows the behaviour $C_v = AT + B_a T^3$, with B_a being larger than B_c. The excess specific heat of a-solids is referred to as the anomalous specific heat. Another universal feature is that between 2 and 10 K there is a bump in the C_v/T^3 vs T curve. It should be emphasised that the linear term in C_v is not the electronic contribution, but is believed to be a purely vibrational effect. Figure 12.10 shows a comparison of the specific heat of SiO_2 in its crystalline and amorphous phases.

The origin of the general phenomenon of the temperature dependence of C_v in a-solids is considered to be a puzzle. Several different explanations have been put forward to explain the results. The linear behaviour of C_v vs T curve below about 1 K is usually explained in terms of contributions from tunnelling modes with a constant density of states. It has been suggested that the specific heat between 1 and 10 K can be explained in terms of contributions from both phonons and fractons. However, it is very difficult, if not impossible, to draw conclusions about details of phonon and fracton densities of states from experimental data on the specific heat of a-solids.

12.5.2 THERMAL CONDUCTIVITY

There is a marked difference in the thermal conductivity of an amorphous solid and its crystalline phase, both in magnitude and in temperature dependence. There are three main differences: (i) a-solids do not exhibit a peak in the $\mathscr{K}$ vs T curve which is characteristic of all crystalline solids, (ii)

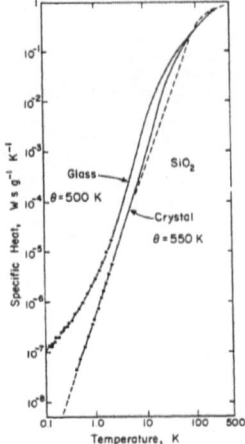

Figure 12.10 Variation of the specific heat of vitreous silica and crystalline silica (α-quartz) with temperature. Note that the specific heat of the glass decreases much more slowly with temperature. (Reproduced from Zeller and Pohl (1971).)

at low temperatures the thermal conductivity of all a-solids drops considerably and shows, roughly between 1–10 K, a near zero slope or a plateau, or even a dip, and (iii) below about 1 K the conductivity shows a temperature dependence T^n, with $n \simeq 1.8$ or $n \simeq 2$, thus dropping less strongly than the T^3 behaviour observed for crystalline solids (Zeller and Pohl 1971). These three differences are highlighted by regions A, B and C, respectively, in figure 12.11 where the thermal conductivity of fused-quartz is compared with that of crystalline quartz. The plateau in the $\mathcal{K}$ vs T curve of a-solids may be correlated with the bump in the C_v/T^3 vs T curve which also appears in the same temperature range. What is intriguing to note is that the magnitude and temperature dependence of the thermal conductivity of all a-solids is very similar. Figure 12.12 shows the results for a few a-solids. Freeman and Anderson (1986) have shown that the thermal conductivity of different a-solids can be scaled onto one universal curve both below and above the plateau.

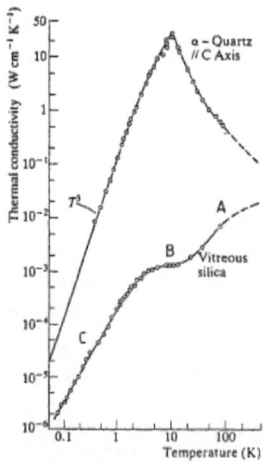

Figure 12.11 Thermal conductivity of crystalline and fused quartz. Regions A, B and C highlight the difference in the conductivity of the two phases. (Reproduced from Zeller and Pohl (1971).)

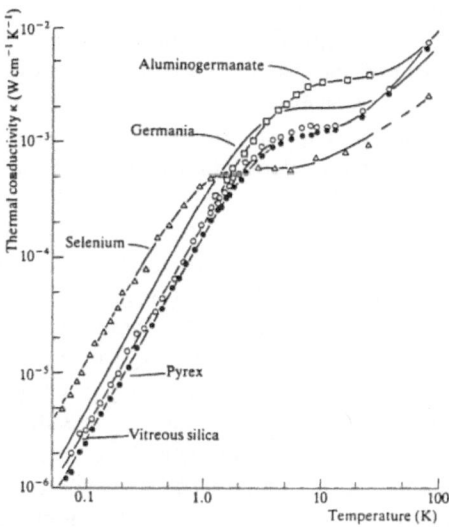

Figure 12.12 Thermal conductivity of a variety of glasses showing similar magnitude and temperature variation. (Reproduced from Zeller and Pohl (1971).)

Various mechanisms have been proposed to explain the temperature dependence of the conductivity of a-solids, but there is as yet no universal agreement on any particular model. Here we describe some recent viewpoints, which are open to scrutiny.

In order to explain the thermal conductivity of a-solids, let us consider the kinetic theory expression (see equation (5.64))

$$\mathscr{K} = \frac{1}{3}C_v^{sp}c\Lambda, \qquad (12.30)$$

where Λ is the mean free path of heat carriers. In keeping up with section 12.3, this expression must be contributed by the vibrational excitations with velocity c, namely low-frequency (long Λ) phonons and high-frequency (short Λ) fractons. In our discussion we will concentrate on the temperature dependence, rather than the magnitude of $\mathscr{K}$.

12.5.2.1 Below the plateau temperature (< 1 K, or $T/\Theta_D \leq 10^{-2}$)

At low temperatures, below 1 K, the main heat carriers are believed to be low-energy phonons. The main source of phonon scattering in this temperature range is the presence of tunnelling states. Resonant scattering of phonons takes place between two lowest energy levels of the anharmonic double-well potential of the type shown in figure 12.5. The mean free path of phonons participating in such scattering events has the temperature variation of the form (Jäckle 1972)

$$\Lambda \propto \coth\left(\frac{\hbar\omega}{2k_B T}\right), \qquad (12.31)$$

which for low frequencies, $\hbar\omega < k_B T$, reduces to $\Lambda \propto T$. This, together with the linear temperature variation of C_v of such phonons, leads to $\mathscr{K} \propto T^2$ which is in good agreement with experimental observation.

12.5.2.2 The plateau region ($10^{-2} \leq T/\Theta_D \leq 10^{-1}$)

Within the tunnelling state model, it can be shown (see Anderson (1981)) that apart from the resonant phonon scattering there can also be a non-resonant phonon scattering mechanism in a-solids characterised by a mean free path $\Lambda \propto T^{-3}$. This, together with the T^3 contribution for C_v in equation (12.30), would lead to a temperature independent $\mathscr{K}$ (plateau behaviour).

As an alternative treatment, Karpov and Parshin (1985) have suggested that the plateau behaviour can be explained on the basis of the resonant scattering of thermal phonons ($\hbar\omega \simeq k_B T$) by anharmonic oscillators, the density of states of which exhibits a van Hove singularity. Thermal phonons are also scattered resonantly and elastically by quasilocal harmonic oscillators which arise due to an atom or a group of atoms being weakly bound to the neighbouring atoms. (The quasilocalised harmonic oscillators are considered as additional excitations whose existence is related to the existence of the tunnelling states.) Such scatterings also lead to a temperature independent resistivity.

Orbach and co-workers (Alexander *et al* 1986) have suggested a different explanation. As discussed in the previous section, the picture of thermally excited vibrational excitations in a-solids changes from low-frequency propagating phonons ($\hbar\omega < k_B T$) to high-frequency spatially localised fractons at a temperature $T \simeq \hbar\omega_c/k_B$. Fractons cannot carry any appreciable amount of heat current, owing to their space localisation. Therefore, phonons are still the sole heat carriers, with a constant heat capacity (C_v = constant, the Dulong–Petit limit, since $k_B T > \hbar\omega$ in the plateau region). This means that the temperature variation of $\mathscr{K}(T)$ is controlled by the mean free path of phonons $\Lambda(T)$. In the plateau region the main phonon scattering process is the cubic anharmonic interaction between fractons and phonons of the type

$$\text{phonon} + \text{phonon} \rightleftharpoons \text{fracton}. \tag{12.32}$$

The inverse relaxation times of phonons and fractons taking part in such a scattering process can be calculated from the theory described in section 6.4.1. Orbach and co-workers used a localised fracton wavefunction, together with the phonon and fracton densities of states as described in the previous section, and argued that in the plateau region the phonon mean free path can be temperature independent. This would then explain $\mathscr{K}(T) = \mathscr{K}_{\text{plateau}}$ = constant.

12.5.2.3   Above the plateau temperature ($T/\Theta_D \geq 10^{-1}$)

Karpov and Parshin (1985) suggested that the thermal conductivity of a-solids above the plateau region is due to the low-frequency or prethermal phonons ($\hbar\omega < k_B T$) which are scattered off the two-level system (tunnelling states). At temperatures above the plateau region, the relative population of the two-level system decreases as T^{-1} with increasing temperatures. Thus the mean free path of prethermal phonons increases linearly with temperature: $\Lambda \propto T$. Since at these temperatures C_v = constant, it follows that $\mathscr{K} \propto T$. Any departure from the $\mathscr{K} \propto T$ behaviour at $T > 100$ K may be associated with the anharmonic interaction of prethermal phonons with thermal phonons.

A different prescription has been given by Orbach and co-workers (Alexander *et al* 1986), according to which heat conduction at temperatures above the plateau region is due to phonon-assisted fracton hopping. Consider the following anharmonic interaction:

$$\text{fracton} + \text{phonon} \rightleftharpoons \text{fracton}. \tag{12.33}$$

The energy conservation requirement can be satisfied by considering $\omega_{\text{ph}} << \omega_{\text{fr}}$. One can then consider the initial- and final-state fracton energies to be nearly the same, and refer to the interaction process as phonon-assisted fracton hopping. For this process to take place it is necessary that the final state fracton be within a distance $R(\omega)$, the fracton 'hop' distance, from the initial fracton position. It is estimated that the hopping distance is greater than the fracton localisation distance ξ

$$R(\omega)/\xi \sim (\omega/\omega_c)^{1/\bar{d}} > 1. \tag{12.34}$$

A lengthy calculation suggests that the fracton hopping lifetime increases linearly with temperature

$$\tau_{\text{fr}}^{-1} \propto T. \tag{12.35}$$

The fracton hopping conductivity is given by a slight modification of equation (12.30)

$$\mathcal{K}_{fr} = \int d\omega g_{fr}(\omega) C_{v,fr}^{sp}(\omega) D(\omega), \tag{12.36}$$

where g_{fr} is the fracton density of states, $C_{v,fr}^{sp}$ is the fracton specific heat, D is the diffusion constant for fracton hopping given by

$$D(\omega) \simeq \tau_{fr}^{-1}(\omega) R^2(\omega) \tag{12.37}$$

and the sum over the fracton polarisation modes has been carried out. As g_{fr} and $C_{v,fr}^{sp}$ are temperature independent, it is clearly demonstrated that $\mathcal{K}_{fr} \propto T$.

Considering fracton hopping as an extra mechanism for heat conduction we can express, at temperatures above the plateau region,

$$
\begin{aligned}
\mathcal{K} &= \mathcal{K}_{ph} + \mathcal{K}_{fr} \\
&= \mathcal{K}_{plateau} + AT \qquad T > T_{plateau}, \tag{12.38}
\end{aligned}
$$

where A is a constant for a particular material. The fractons which contribute most to $\mathcal{K}_{fr}$ are those which hop the greatest distance for a given fracton–fracton overlap and have energies much less than $k_B T$.

12.5.3 ACOUSTIC AND DIELECTRIC PROPERTIES

As with the specific heat and thermal conductivity, the low-temperature acoustic and dielectric properties of amorphous solids are also fundamentally different from those of their crystalline counterparts. Futhermore, in glasses the acoustic and dielectric absorption are qualitatively very similar. Good reviews of a wide range of acoustic and dielectric experiments are given in Hunklinger and Arnold (1976), Hunklinger and Schickfus (1981), and Golding and Graebner (1981). In brief the experimental results can be summarised as follows.

In the liquid nitrogen temperature range, a pronounced peak, or a broad shoulder, is observed in the ultrasonic absorption in amorphous solids. In many amorphous solids, a second absorption peak is observed at liquid helium temperatures. At temperatures below 1 K, the absorption rises again, but can be saturated at higher intensities. Figures 12.13 and 12.14 show these behaviours in vitreous silica.

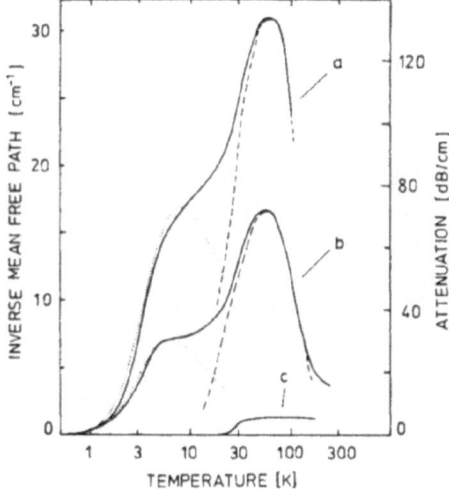

Figure 12.13 Temperature variation of the ultrasonic absorption in vitreous silica for longitudinal waves at (a) 930 MHz and (b) 507 MHz. Curve (c) represents the absorption in crystalline quartz at 1000 MHz for comparison. (Reproduced from Hunklinger and Arnold (1976).)

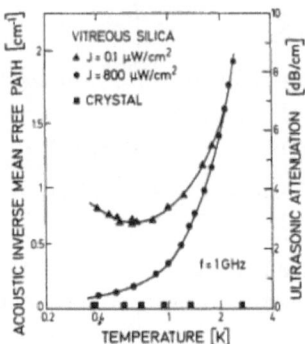

Figure 12.14 Same as in figure 12.13 below 2 K for longitudinal waves of 1 GHz. The absorption in a quartz crystal is shown for comparison. (Reproduced from Hunklinger (1977).)

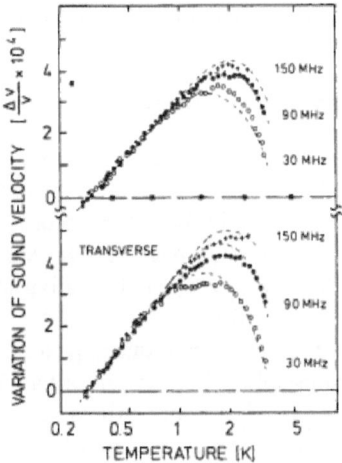

Figure 12.15 Relative variations of sound velocity with temperature in borosilicate glass. The upper and lower panels show the results for longitudinal and transverse polarisations, respectively. Below 1 K the temperature dependence becomes logarithmic. (Reproduced from Hunklinger and Arnold (1976).)

The temperature variations of sound velocity and dielectric constant in glasses are large compared with those in crystalline solids. Below 50 K the sound velocity increases with decreasing temperature until a frequency-dependent maximum is reached (typically between 1.5 and 6 K, see figure 12.15). At temperatures lower than 1 K, a logarithmic temperature dependence of the sound velocity is observed. When the temperature is reduced to $T \simeq \hbar\omega/2k_B$ (below 0.1 K) a shallow minimum is reached (see figure 12.16). An analogous behaviour is found for the temperature variation of the dielectric constant of glasses. With decreasing temperature, the dielectric constant decreases steeply, and below a few kelvin at GHz frequencies a logarithmic increase is observed (figure 12.17).

The tunnelling model offers a phenomenologically microscopic explanation for the acoustic and dielectric measurements in glasses.

12.5.4 AN INTERACTIVE DEFECT MODEL

In a critical review of the low-temperature properties of a-solids, Yu and Leggett (1988) have pointed out that while the observed effects are mostly consistent with the two-level system model, they do

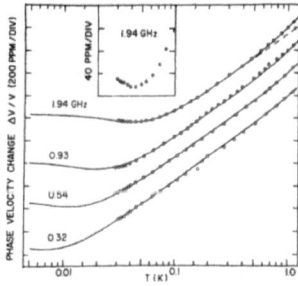

Figure 12.16 Same as figure 12.15 in vitreous silica at very low temperatures. The minimum is observed at $\hbar\omega \simeq 2k_B T$. (Reproduced from Golding *et al* (1976).)

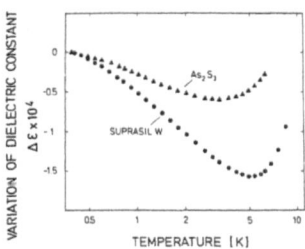

Figure 12.17 Temperature variation of dielectric constant in vitreous silica and vitreous As_2S_3 at 10 GHz. (Reproduced from Hunklinger and Schickfus (1981).)

not uniquely establish it. Furthermore, the model cannot explain the dramatic *quantitative* universality in the low-temperature behaviour of a-materials.

These authors have proposed to assume that glasses contain some sort of defects which have some low lying excitations for all energy range. Interaction between such defects via the elastic strain field is assumed to give rise to *collective modes* of vibration. These modes can be considered as many-body states representing a band of energies. The generalised dispersion relation for collective modes is of the form $\omega \propto qB(q)$, where $B(q)$ is a function of the wavevector q (March and Paranjape 1987). Bulk phonons can couple resonantly to induce transition between pairs of collective modes. Unlike the two-level system model, this model entails long-range interactions and results into a logarithmic density of states which can be helpful in explaining low-temperature specific heat, thermal conductivity and ultrasonic measurements.

In a further investigation of the model of strongly interacting defects via the elastic strain field, Yu (1989) noted that the dominant interactions are oscillatory at short distances and crossover to $1/r^3$ interactions at long length scales. She futher argued that this crossover can be associated with the plateau in the thermal conductivity and the bump in C/T^3.

Recent modelling work (Scalliet *et al* 2019) finds that the energy and temperature scales associated with defects in amorphous solids can be sensitively tuned by changing density.

13 Phonon Spectroscopy

13.1 INTRODUCTION

In Chapters 6–12 we have discussed various phonon interaction mechanisms in connection with the theory of lattice thermal conductivity. Indeed for many years, thermal conductivity was the only technique for studying phonon interactions in solids. In essence thermal conductivity studies provide an integrated picture for phonon interactions. However, many techniques have now become available which provide a wealth of new information regarding details of these interactions. Rather than establishing a steady-state heat current across the sample under study (as is done in thermal conductivity experiments), techniques of phonon spectroscopy are based on generation and detection of non-equilibrium phonon distributions. Such techniques involve generating beams of phonons in the energy range 10 GHz to 1 THz and monitoring their transport and interaction within crystals. Also, new techniques are capable of providing increased resolution in phonon energy, polarisation, or wave vector. The best energy resolutions reported to date lie in the range of a few MHz to a few GHz. Thus phonon spectroscopy has become a clearly defined and valuable tool in solid state physics. Of particular interest is phonon spectroscopy in the frequency range between the upper limit of microwaves and the lower end of far infrared for which there are almost no other suitable spectroscopic techniques. Phonon spectroscopy is also valuable for studying many neutral defects and excitations which strongly couple to phonons but not at all to photons. However, it should be remarked that although the new phonon spectroscopic techniques are analogous to the various methods of photon spectroscopy, so far it has not been possible to develop an analogue to the optical absorption or emission spectroscopy.

A very brief discussion of one of the phonon spectroscopies, namely, Raman scattering of light from phonons, was presented in section 6.9.2. In this chapter we provide a brief discussion on some selected topics in phonon spectroscopy. A detailed account of some of the earlier topics is given in review articles by Bron (1980) and Wybourne and Wigmore (1988) and references therein.

13.2 HEAT PULSE TECHNIQUES

In the heat pulse technique, originally introduced by von Gutfeld and Nethercot (1964), a short-duration (1 to 100 ns) electrical voltage or optical excitation is applied to a thin metallic film ($\sim$ 10 to 100 nm) evaporated on one face of the crystal under investigation. The heat pulse sequence generates a *non-equilibrium distribution* of phonons in the sample. A thin-film bolometer is evaporated on the opposite face of the sample to detect the pulses. Figure 13.1 shows a typical sample arrangement in a heat pulse experiment. If the pulse duration is shorter than the phonon transit time across the sample and if the phonon mean free path is longer than the distance between the pulse generator and detector, the phonon propagation takes place without suffering any scattering and is called 'ballistic'. Measurement of the 'time of flight' for phonons across the sample directly gives information on the group velocity c_g for longitudinal and transverse phonons. More important is the fact that since c_g is normally different for longitudinal and transverse phonons, different signals are detected for different phonon polarisation branches. This fact, therefore, permits measurements of the phonon dispersion relation. Furthermore, this technique also provides measurements of scattering of different polarisation branches of phonons.

This technique has been used to study the ratio of TA/LA scattering by magnetic ions (e.g. V^{3+} in Al_2O_3, Narayanamurti (1969)) and by free electrons/holes (e.g. n- and p-InSb, Maneval *et al* (1971), and Ladan and Maneval (1976)).

DOI: 10.1201/9781003141273-13

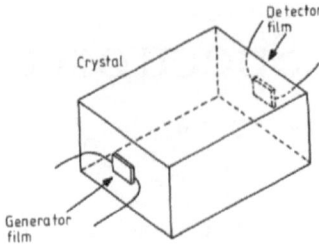

Figure 13.1   A typical set up for a heat pulse experiment.

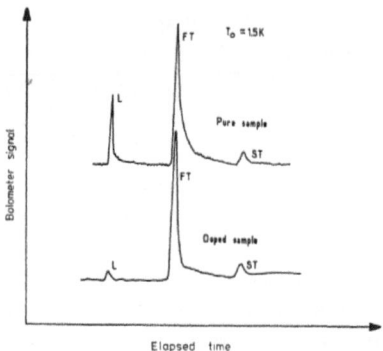

Figure 13.2   Detected heat pulses in InSb after propagation along the [110] direction at liquid helium temperatures. The horizontal and vertical scales are arbitrary. Notice that while longitudinal phonons are strongly scattered, transverse phonons are practically unattenuated from doping concentrations of the order of 10^{17} electrons cm^{-3}. (Reproduced from Maneval *et al* (1971).)

Figure 13.2 shows the bolometer signal corresponding to heat pulse propagation along the [110] direction in n-InSb samples (Maneval *et al* 1971). The upper trace is for a pure sample (with 1.9×10^{14} electrons cm^{-3}) and shows three distinct peaks: with increasing elapsed time these are for longitudinal (L), fast transverse (FT) and slow transverse (ST). In the [110] direction the FT and ST modes are non-degenerate. The difference in the amplitude of the three peaks results from the phonon focusing effect discussed in section 11.6. The lower trace is for a doped sample containing 6.5×10^{17} electrons cm^{-3}. (Both the upper and lower traces were taken at the same heater input power.) A comparison of the two traces reveals that while the FT/ST ratio is unaffected by doping, the L/FT ratio in the doped sample is smaller by a factor of eight compared to that in the pure sample. These observations can be explained by pointing out that in heavily doped semiconductors the coupling of electrons with transverse phonons is not permitted (cf section 6.5.1). Indeed, this experiment provides the first verification of the absence of interaction between purely transverse thermal phonons and free electrons in a material with an isotropic Fermi surface. However, in p-InSb in which the hole states are anisotropic, hole–phonon interaction has been observed to take place for both longitudinal and transverse phonon modes (Ladan and Maneval 1976).

13.3   SUPERCONDUCTING TUNNEL JUNCTION TECHNIQUE

Another phonon spectroscopic method is provided by the use of superconducting tunnel junctions. Such a technique is commonly used to generate and detect phonons in the frequency range

0.1–0.5 THz, with a resolution of several GHz. In exceptional applications (Josephson tunnelling, see below), resolutions of a few MHz have also been reported (Berberich *et al* 1982).

Simple tunnel junctions have the structure S-I-S, where S is a thin-film (≤ 1000 Å) superconductor (usually Al, Sn or Pb) and I is a thin insulating layer ($\simeq 10$ Å). In most phonon spectroscopic studies in the ^{4}He temperature range a Sn-I-Sn tunnel junction is mounted to generate non-equilibrium (or, non thermal) distribution of phonons on one face of the crystal, and a Al-I-Al junction is mounted to detect phonons on the opposite face of the crystal. The phonon generating junction is biased with a DC voltage and the current through it is modulated with a relatively small AC voltage. The detector junction is also voltage biased and observation of a modulated signal indicates the detection of the absorbed phonon.

It is useful to recall the Bardeen-Cooper-Schrieffer (BCS) theory of superconductivity (Bardeen *et al* 1957). According to this theory, an attractive interaction between a pair of electrons (called Cooper pairs), mediated by phonons, leads to an energy gap $E_g = 2\Delta$ between the superconducting ground state and the first excited state (the Fermi level E_F at $T = 0$ being at energy Δ). The energy levels close to E_F are occupied by Cooper pairs, while excited states can be occupied by *single particles* or *quasi-particles*. The energy gap 2Δ is 290 GHz for the Sn superconductor and 100 GHz for the Al superconductor.

At zero bias voltage ($V_0 = 0$), only Cooper pairs may tunnel through a tunnel junction. This is called *Josephson tunnelling*. Consider that a bias voltage V_0 lifts all electronic states on the left-hand side of the barrier by an amount eV_0 greater than the corresponding states in the superconductor on the right-hand side. If $V_0 < 2\Delta/e$ then a small current can flow due to the thermally excited quasi-particles. At a bias $V_0 > 2\Delta/e$ some Cooper pairs are broken into *single particles* or *quasi-particles*, which are promoted into the excited states. These single particles can tunnel across the insulating barrier into states above the energy gap in the second superconducting film. Both Cooper pair tunnelling and single-particle tunnelling give rise to an increase in the current through the junction. Josephson tunnelling has been used by Berberich *et al* (1982) in high-resolution spectroscopy. However, most tunnel spectrometers are only concerned with the current–voltage characteristics in single-particle tunnelling. The use of single-particle tunnel junctions for phonon generation and detection was first made by Eisenmenger and Dayem (1967).

Phonon generation and detection by single-particle tunnel junctions can be explained as follows. Within the generating tunnel junction, the excited single particles, which have tunnelled through the barrier, are in non-equilibrium distribution and decay back into Cooper pair equilibrium states by a two-stage process. Firstly, the excited single particles 'relax', by emitting phonons, into the first excited state (the upper edge of the energy gap) of the second superconducting film. Typical phonon energies in the 'relaxation' process lie in the range 0 and $eV_0 - 2\Delta_G$ (here the subscript G indicates the generating tunnel junction). In the second step the relaxed single particles 'recombine' to form de-excited Cooper pair states below the gap. The 'recombination' process emits phonons of energy $2\Delta_G$. If the bias energy eV_0 is larger than $2\Delta_G$, then a broad energy spectrum of 'relaxation' phonons and a narrow energy spectrum of 'recombination' phonons is generated. Figure 13.3 illustrates the origin of phonon generation from a single particle tunnel junction.

Phonons generated from the tunnel junction which travel ballistically through the crystal are detected by the second tunnel junction which is biased at $0 < eV_0 < 2\Delta_D$ (here the subscript D indicates the detecting tunnel junction). The arriving phonons with energy $\hbar\omega \geq 2\Delta_D$ can break Cooper pairs in the detector junction, thus causing an increase in the single-particle current. Because of the threshold $\hbar\omega = 2\Delta_D$ detector tunnel junctions can be used to spectrally separate phonons into two groups $\hbar\omega < 2\Delta_D$ and $\hbar\omega \geq 2\Delta_D$. If the generator junction bias lies in the range $2\Delta_G < eV_0 < 4\Delta_G$, only recombination phonons of energy $\hbar\omega = 2\Delta_G$ and relaxation phonons of energy $\hbar\omega < 2\Delta_G$ are produced in the generator. For this range of generator bias, therefore, only the flux of phonons generated in the recombination process in the generator junction and travelling to the detector junction is responsible for the current increase in the detector junction. By increasing the

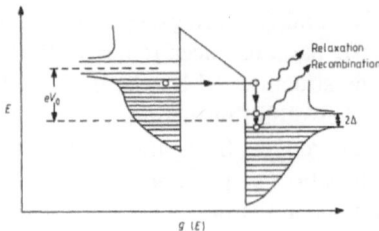

Figure 13.3 Schematic illustration of the origin of relaxation and recombination phonons generated from a single particle tunnel junction biased with V_0 at $T = 0$.

generator bias to $eV_0 > 4\Delta_G$, an additional increase in the detector current due to relaxation phonons (generated by the generator junction) with energies $\hbar\omega > 2\Delta_G$ can be observed. By increasing the bias voltage of the generator junction to $eV_0 > 6\Delta_G$ it is possible to detect relaxation phonons with $\hbar\omega > 4\Delta_G$, together with secondary relaxation phonons with $\hbar\omega > 2\Delta_G$.

By superimposing a small AC voltage or pulse δV_{AC} to the DC voltage V_0 it is possible to detect only those (nearly monochromatic) phonons which lie in the energy range $eV_0 < \hbar\omega < e(V_0 + \delta V_{AC})$. This is known as *modulated tunnel-junction spectroscopy*. This technique was first introduced by Kinder (1972) to study resonant phonon scattering at V^{3+} and V^{4+} impurity ions in sapphire (Al_2O_3).

There are many other applications of the tunnel junction technique including studies of tunnelling levels (e.g. OH^- in NaCl), phonon propagation in solid and liquid ^{4}He, and the dependence of the Kapitza conductance (cf section 14.5) on helium film thickness. Dietsche *et al* (1981) have used an extended tunnel junction method to study focusing of high-frequency phonons in Ge.

13.4 OPTICAL TECHNIQUES

Optical techniques can offer useful advantages over other phonon spectroscopies. For example, in optical techniques phonons are generated or detected completely within the sample, so that no influence of surface or any outside agent is encountered. Additionally, optical techniques can generate a wide range of phonon frequencies (up to many THz) including both optic and acoustic branches. Reviews of this topic are given by Narayanamurti (1981), Renk (1985) and Ulbrich (1985).

The principle of phonon generation by optical techniques can be explained with the help of figure 13.4. Consider a direct band gap semiconductor in which an electron has been photoexcited to a higher unoccupied energy level. This electron can lose its energy in several ways, some of which include phonon emission. Here we consider some of the phonon emission processes. First, the excited electron will quickly de-excite towards the bottom of the conduction band, emitting 'relaxation' phonons. The energy of such phonons will depend on the 'electron temperature' and can include the range of acoustic as well as optical branches. Secondly, electrons at the bottom of the conduction band can lose their energy non-radiatively, emitting 'recombination' phonons. Non-radiative recombinations of electrons and holes often take place at shallow or deep impurity and/or crystal defect levels between the conduction and valence bands of the host. Recombination at shallow donors or acceptors takes place via resonant phonon emission. In such a process the energy of the phonon emitted is equal to the ionisation energy of the impurity, as shown in figure 13.4. When an electron is captured by a deep level, a sequential relaxation of the lattice takes place near the impurity (or defect). As the lattice vibrates, the deep level moves up and down within the energy gap of the semiconductor and a number of phonons are emitted during the damping of lattice vibrations. This is thus a non-radiative, multiphonon emission process. The range of 'recombination' phonon energies can also include both acoustic and optical branches. There is a third mechanism

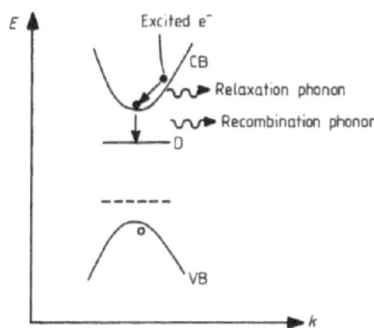

Figure 13.4 Principle of relaxation and recombination phonon generation in a direct band gap semiconductor.

for phonon emission in solids, the Auger process, although with rather small recombination cross section.

Optically generated LO phonons quickly decay via the processes LO → LA + TA and LO → TA + TA (see sections 6.4 and 7.3, although as discussed in section 7.3.2 other processes can also contribute). Similarly LA phonons can also decay into TA phonons. However, in the absence of impurity scattering (i.e. in pure sample) TA phonons can be quite long lived: these phonons can propagate through the sample and can be detected at the opposite face of the sample.

Photoexcitation experiments combined with tunnel junction detectors should provide some advantage over heat pulse and conventional tunnel junction spectroscopies in that generation, propagation and detection of TA phonons well up to the Brillouin zone boundary (~ 10 Å wavelength) can be studied.

13.5   PHONONS FROM LANDAU LEVELS IN 2DEG

As discussed in section 9.10, electrons can be confined to quasi two dimensions in suitably grown heterojunctions or multi-quantum wells. At low temperatures when a strong external magnetic field is applied perpendicular to the plane of the two-dimensional electron gas (2DEG), the electron motion in the plane is completely quantised: the energy spectrum consists of Landau levels $(v + \frac{1}{2})\hbar\omega_c$, where $v = 0, 1, 2, \ldots$ and $\hbar\omega_c$ is the cyclotron energy. (This leads to the quantum Hall effect (see von Klitzing (1986) for a discussion on this effect).) In the ideal case the Landau levels are sharp as a delta function. However, in a real system the Landau levels are broadened by impurities, phonons, or other scattering mechanisms. A schematic view of the density of states in a 2DEG is presented in figure 13.5.

Of particular interest is the modulation doped GaAs/Ga$_{1-x}$Al$_x$As system. The formation of the 2DEG is achieved by considering a geometry similar to that shown in figure 13.6. Typically a few micron (1–4 μm) thick undoped GaAs is deposited on a semi-insulating GaAs substrate, and a n-doped Ga$_{1-x}$Al$_x$As ($x < 0.4$ for direct band gap) thin film (10–100 nm) is grown on the top of the GaAs layers. If this heterostructure is grown by the MBE or the MOCVD method, an almost lattice-matched single crystal interface can be generated. The modulation doping technique minimises the effect of electron scattering by impurities. Furthermore, at low temperatures (e.g. liquid He temperature), there is very little scattering due to phonons. Thus the system GaAs/Ga$_{1-x}$Al$_x$As ($x < 0.4$) provides a high mobility interface whose Landau levels should be quite narrow.

When an electric current is applied, or an optical excitation is provided, electrons from the highest occupied Landau level in a GaAs/Ga$_{1-x}$Al$_x$As heterostructure will be excited to a higher unoccupied Landau level. These excited electrons will eventually be de-excited, giving rise to 'relaxation phonons' with energy equal to the Landau level separation $\hbar\omega_c$. The monochromatic phonons thus

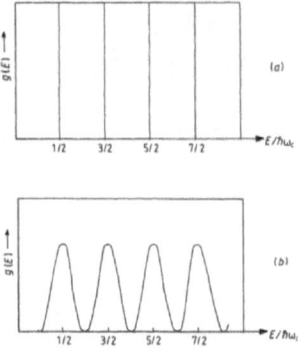

Figure 13.5 A schematic view of the density of states in a 2DEG in the presence of a strong magnetic field: (a) ideal case, (b) real case with the Landau levels broadened due to scattering mechanisms.

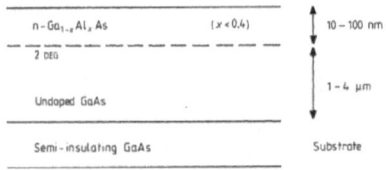

Figure 13.6 A typical GaAs–Ga$_{1-x}$Al$_x$As heterostructure used for the quantum Hall effect measurements.

generated evidently have potential in phonon spectroscopy (Kent *et al* 1988). Further, this system can also be used for monochromatic phonon detection.

13.6   PHONON FOCUSING AND IMAGING

Low-frequency sound waves propagate through a solid without scattering (i.e. ballistically) with the speed $v = (C/\rho)^{1/2}$, where ρ is the density and C is the elastic constant of the solid (*cf* section 2.8). In contrast, heat propagation in a crystal is a slow diffusive process which involves frequent scattering of phonons (*cf* chapters 5–7). However, high-frequency phonons generated by a heat pulse in single crystals at low temperatures (*e.g.* liquid He range) can propagate ballistically. In such a situation anharmonic phonon interactions will be negligible because at very low temperatures there will be very low ambient population of thermal phonons in the crystal. As the waves generated in a typical heat pulse experiment have an average wavelength which is much longer than the crystal lattice constant, the propagation of such waves can be described by the elastic wave equation as described in section 2.8.

The propagation of incoherent phonons (as generated in a heat pulse experiment) is in the direction of the group velocity $c_g = \nabla\omega(q)$, which for a given wave vector q is normal to the $\omega =$ constant energy surface (also called the slowness surface). As real crystals are elastically anisotropic, the group velocity c_g is not collinear with the wave vector q (and hence phase velocity) in all directions. In fact for regions with nearly zero curvature, $|\nabla\omega| \simeq 0$, a large number of q-states contribute to phonons with the same group velocity. Such a strong angular dependence of ballistic phonon energy flux in elastically anisotropic crystals is called 'phonon focusing' or 'phonon channeling'. And we see that regions of strong curvature in $\omega - q$ space lead to phonon 'defocusing'. Figure 13.7 demonstrates the phonon focusing effect for a transverse phonon mode along [100] in a cubic crystal.

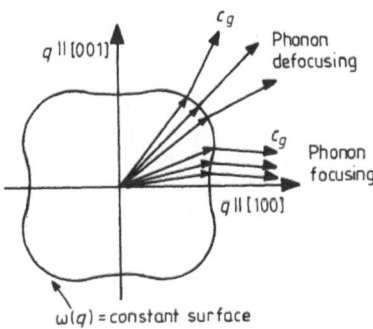

Figure 13.7 Illustration of a constant energy surface for a TA mode and the phonon focusing effect in a cubic crystal.

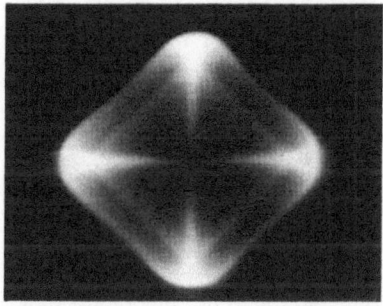

Figure 13.8 Phonon focusing image of the [100] slow transverse mode in Ge. Bright regions show high phonon flux. (From Northrup and Wolfe (1980).)

The intensity of ballistically propagating phonons in a heat pulse experiment can be recorded or 'imaged' by computer-aided techniques (Northrup and Wolfe 1980, Wolfe 1980). Figure 13.8 shows a high-resolution phonon focusing image of the $\langle 100 \rangle$ slow transverse mode in Ge.

A different technique for phonon imaging has been developed by Eisenmenger (1980). In this technique, ballistically propagating phonons in a Si crystal were absorbed in superfluid He film covering the crystal surface. Along the focusing directions, the rise in temperature leads to an increase in the He-film thickness by the fountain pressure in the superfluid ^{4}He. This effect is recorded or imaged by a camera.

The phonon focusing effect is caused by the elastic anisotropy of the crystal (Taylor *et al* 1969). A quantitative explanation of the effect can be provided by a focusing or energy-flux enhancement factor (Maris 1971)

$$A = \mathrm{d}\Omega_q / \mathrm{d}\Omega_{c_g} \tag{13.1}$$

which is the ratio of the solid angles in the $\boldsymbol{q}$-space and the group-velocity (or energy flux) space. For isotropic solids $A = 1$, while for anisotropic solids A varies over a wide range of values and indicates the phonon focusing effect in the limit $A \to \infty$.

Several alternative expressions for the focusing factor A can be derived (Maris 1986) and here we mention the approach of Philip and Viswanathan (1978). By expressing $\boldsymbol{q}$ and the group velocity $\boldsymbol{c}_g$ in spherical polar coordinates

$$
\begin{aligned}
\boldsymbol{q} &= q(\sin\theta_q\cos\phi_q, \sin\theta_q\sin\phi_q, \cos\theta_q) \\
\boldsymbol{c}_g &= |\boldsymbol{c}_g|(\sin\theta_v\cos\phi_v, \sin\theta_v\sin\phi_v, \cos\theta_v)
\end{aligned} \tag{13.2}
$$

we can express

$$
A^{-1} = \frac{\sin\theta_v \, \mathrm{d}\theta_v \, \mathrm{d}\phi_v}{\sin\theta_q \, \mathrm{d}\theta_q \, \mathrm{d}\phi_q}. \tag{13.3}
$$

From the Cartesian components in equation (13.2), we note that

$$
\begin{aligned}
\tan\theta_v &= (c_{g,x}^2 + c_{g,y}^2)^{1/2}/c_{g,z} \\
\tan\phi_v &= c_{g,y}/c_{g,x}
\end{aligned} \tag{13.4}
$$

which means that θ_v and ϕ_v are independent of the magnitude of the wave vector $\boldsymbol{q}$ and depend only on the angles θ_q and ϕ_q. With this in mind, we can express equation (13.3) as

$$
A^{-1} = \frac{\sin\theta_v}{\sin\theta_q} J, \tag{13.5}
$$

where

$$
\begin{aligned}
J &= \frac{\mathrm{d}\theta_v \, \mathrm{d}\phi_v}{\mathrm{d}\theta_q \, \mathrm{d}\phi_q} \\
&= \frac{\partial\theta_v}{\partial\theta_q}\frac{\partial\phi_v}{\partial\phi_q} - \frac{\partial\theta_v}{\partial\phi_q}\frac{\partial\phi_v}{\partial\theta_q}
\end{aligned} \tag{13.6}
$$

is the Jacobian of the transformation between the variables (θ_v, ϕ_v) and (θ_q, ϕ_q). Philip and Viswanathan have described the calculation of the Jacobian J in terms of the solutions of the Green–Christoffel determinant in equation (2.143) for cubic solids.

Numerical calculations of phonon focusing have also been presented by Northrup and Wolfe (1980) and McCurdy (1982). The effect of phonon dispersion on the phonon focusing factor has been studied by Dietsche *et al* (1981), Northrup (1982) and Tamura (1982, 1983a, 1983b).

13.7 FREQUENCY CROSSING PHONON SPECTROSCOPY

Although in general the thermal conductivity technique provides an average picture for phonon scattering in solids, in appropriate conditions it can be used to carry out phonon spectroscopy. Consider the thermal conductivity expression

$$
\mathscr{K} = \frac{1}{3}\int \mathrm{d}\omega C_v^{sp}(\omega)c\Lambda(\omega) = \int \mathrm{d}\omega \mathscr{K}(\omega), \tag{13.7}
$$

where the summation over polarisation indices is implicit. The quantity $\mathscr{K}(\omega)$ represents the conductivity due to phonons of frequency ω. At low temperatures where the mean free path Λ is independent of frequency ω, the spectrum of $\mathscr{K}(\omega)$ is essentially that of the specific heat $C_v^{sp}(\omega)$ and has the form shown in figure 13.9(a), with a maximum at the dominant phonon energy $\hbar\omega_{dom} \simeq 3.8k_BT$. If the crystal contains a two-level impurity or defect with the energy levels separated by $\hbar\omega_1$, then the spectrum $\mathscr{K}(\omega)$ will show a 'dip' or 'hole' due to resonant scattering of phonons with frequency $\omega = \omega_1$ as shown in figure 13.9(b). Consider that a second resonant frequency ω_2 is also present in the crystal (either due to a second pair of levels of the same impurity, or due to a second type of impurity). If ω_1 and ω_2 are well separated, the total scattering of a phonon beam is the sum of the two resonant scatterings. Suppose that the frequency ω_1 can be changed (keeping ω_2 fixed), or that both ω_1 and ω_2 can be changed, then when ω_1 becomes close to ω_2 (or, when 'frequency crossing' occurs) the two scattering events 'interfere' with each other and give rise to a 'dip' or 'minimum' in the thermal resistivity. Another interesting situation occurs when two *levels* either cross or anticross (repel). Phonon scattering between two anticrossing levels and a third level, referred to as 'anticrossing effect', gives rise to a Lorentzian maximum in the thermal resistivity. These effects can, thus, give spectroscopic information on the impurity ion(s) present in the crystal. A detailed discussion on the frequency crossing and level anticrossing effects is given by Challis and de Göer (1984) and Challis (1987) where original references can also be found.

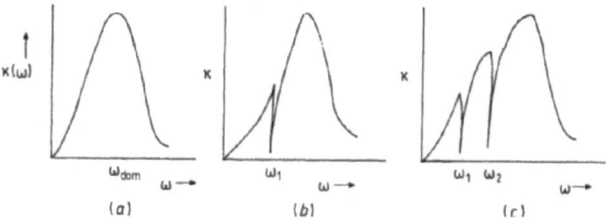

Figure 13.9 Illustration of the phonon conductivity spectrum $\mathscr{K}(\omega)$ at low temperatures: (*a*) a pure crystal, (*b*) a crystal containing a two-level impurity ion, (*c*) a crystal containing two two-level centres.

The situations described above can be realised in magnetothermal conductivity measurements of an insulator with paramagnetic impurities. An example is Fe^{2+} ions in Al_2O_3. The ground state has a singlet $|0\rangle$ and a doublet $|\pm 1\rangle$ energy level as shown in figure 13.10. When a magnetic field is applied parallel to the z axis, the doublet state splits and at a field $B = B_0 = 0.8$ T the transition energies ω_1 and ω_2 become equal. For this field, the frequency crossing effect produces a minimum in the low-temperature resistivity (figure 13.11(*a*)). With increasing magnetic field, there is a maximum in the resistivity at $B = 2.4$ T. This particular maximum (anti-crossing effect) has a frequency crossing effect (a minimum) superimposed onto its peak. In general the anti-crossing effect takes place because of coupling between two approaching levels due to off-axis fields, strains, etc. This is clearly seen in figure 13.11(*b*), where the truncated anti-crossing, or maximum, signal at $\theta = 0°$ changes into a clear Lorentzian shape at $\theta = 2°$.

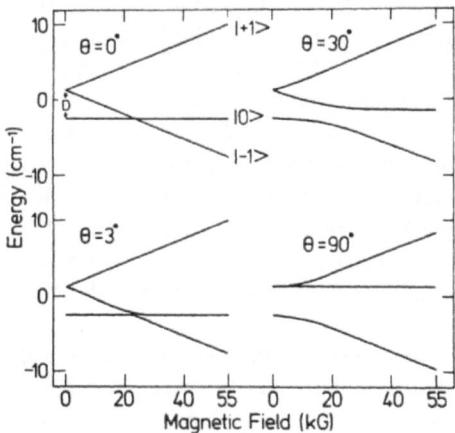

Figure 13.10 The energy levels of the ground state of Fe^{2+} in Al_2O_3 as a function of magnetic field at different angles θ relative to the z axis. When $\omega_1 = \omega_2$ a frequency crossing effect takes place. The levels $|-1\rangle$ and $|0\rangle$ anticross as θ increases from $\theta = 0^0$. (From Anderson and Challis (1973). Reproduced from Challis (1987).)

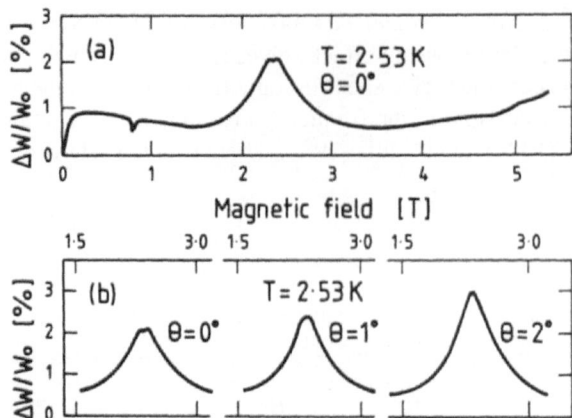

Figure 13.11 The reduced magnetothermal resistivity of Al_2O_3:Fe^{2+}. W_0 represents the resistivity of pure Al_2O_3. (*a*) A frequency crossing effect at $B = 0.8$ T, and a weak frequency crossing superimposed to a level anticrossing effect at $B = 2.4$ T; (*b*) the angular dependence of the anti-crossing effect. (From Challis (1987), after Anderson and Challis (1973).)

V^{3+} ions in Al_2O_3 also show a frequency crossing signal. In fact for V^{3+} the states $|-1\rangle$ and $|+1\rangle$ each split into eight levels due to hyperfine interaction. As a result, hyperfine structure is observed in the frequency crossing signal. So, in summary, this technique allows the determination of the parameters of a spin Hamiltonian with reasonably high precision.

13.8 PHONON ECHOES

Consider a phonon pulse of frequency ω and decay time constant T at time 0 propagating in a certain direction in a crystal. If a second non-propagating pulse of frequency ω or 2ω is applied at time τ ($\leq T$) which interacts non-linearly with the forward phonon pulse, then a backward, time-reversed, phonon pulse of frequency ω is generated at time 2τ. This phenomenon is called a two-pulse 'phonon echo' or 'polarisation echo'. The echo is labelled an (ω, ω)-echo ($(\omega, 2\omega)$-echo) if the frequency of the second pulse is ω (2ω). Higher-order echoes may also be observed at times $\tau = m\tau, m = 2, 3, 4, \ldots$. If a subsequent phonon pulse of frequency ω is applied at time t, then there results a 'stimulated phonon echo' at time $t + \tau$. When $t \gg T$ (the decay time of the first pulse) the decay time T_1 of the stimulated echo is called the storage time. The second, non-propagating, pulse can be an acoustic or electromagnetic (RF or microwave) pulse. Phonon echoes have been observed in many bulk solids such as piezoelectric crystals, paramagnets, glasses and doped semiconductors, and in powered samples of piezoelectric, magnetoelastic, normal metallic and superconducting materials. (For observing echoes in powders the wavelength of the phonon mode must be of the order of or smaller than the particle diameter.)

In general, the non-linear interaction of the forward phonon pulse can arise from the intrinsic anharmonicity of the crystal lattice potential, or from a crystal defect centre. For example, in doped semiconductors the non-linearity can arise through an electric field ionisation of trapped electrons, resulting into interaction of the forward phonon pulse with the space and time modulated free (or trapped) electron charge distribution. Examples of echo-active semiconducting materials include CdS, CdSe, CdTe and p-Si (e.g. Si:In).

Phonon echo studies have emerged as a powerful experimental technique for studying phase transitions and other phenomena in solids. Some useful references on phonon echoes are Golding and Graebner (1976, 1981), Fossheim *et al* (1978), Fossheim and Holt (1980, 1982), Kajimura (1982) and Melcher and Shiren (1982).

13.9 TIME- AND FREQUENCY-RESOLVED PHONON SPECTROSCOPIES

There are several variants of time- and frequency-resolved phonon spectroscopies. We provide a brief discussion on some of the methods.

13.9.1 TIME-RESOLVED COHERENT ANTI-STOKES RAMAN SCATTERING

The time-resolved coherent anti-Stokes Raman scattering (CARS) technique is used to study non-equilibrium phonon dynamics of long wavelength LO (Bron *et al* 1986) and TO modes (Ganikhanov and Vallee 1997). In section 7.3 we have presented some discussion on the role of crystal anharmonicity in controlling the lifetimes of these modes.

13.9.2 ULTRAFAST PUMP-PROBE TECHNIQUES

One of the first pump-probe techniques for thermal conductivity measurements was developed by Capinski *et al* (1999). Using a picosecond optical pump-probe technique, Capinski *et al* presented measurements of the cross-plane lattice thermal conductivity of short-period GaAs/AlAs planar superlattices (PSLs). A schematic illustration of their experiment and sample structre is shown in figure 13.12. A thin metallic film is deposited on the sample layer and a pump light pulse is focussed on a small spot on the surface of the metallic film. Change in the film temperature produces a proportional change in the optical reflectivity, which is measured by means of a time-delayed probe light pulse focussed to overlap with the pump pulse, Cross-plane thermal conductivity is determined by comparing the measured change in the reflectivity as a function of delay time to a change calculated from a numerical simulation of the heat flow. Capinski *et al* (1999) reported a typical error margin of ±10% in the measured value of the conductivity.

Coherent optical phonon spectroscopy

An ultrafast pump-probe technique has grown as coherent optical phonon spectroscopy (COPS). In this, the system is excited with an ultrafast pump laser pulse and the absorption of a second probe pulse is measured as a function of the pump frequency, the probe frequency, and their relative delay. The requirement in CARS measurements is that the pulse duration must be larger than the phonon period (Dekorsy *et al* 2000). This is to ensure that the detection of the Raman shifted intensity is not aggravated. In time-resolved Raman spectroscopies, therefore, resolution of the coherent phonon phase sensitivity is not guaranteed. Use of ultrafast Laser pulses, of duration shorter than the inverse of a fundamental phonon frequency, allows excitation of a large number of phonons in one mode

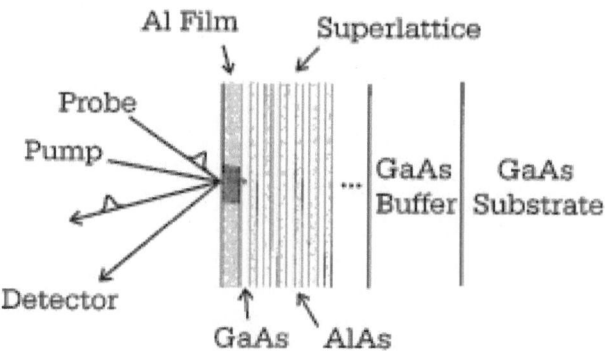

Figure 13.12 Pump-and-probe experimental set up for measuring cross-plane thermal conductivity of planar superlattices (PSLs). (From Capinski *et al* (1999).

with constant phase relation. Different from Raman spectroscopy, in the time-resolved COPS generation and detection of coherent optical phonons can be separated. A deeper understanding of the physics underlying generation and detection of coherent optical phonons can be found in Dekorsy *et al* (2000).

Several excitation mechanisms have been discussed, including impulsive stimulated Raman scattering (ISRS) (Yan and Nelson 1987), displacive excitation mechanisms (DECP) (Cheng *et al* 1991) and field screening mechanism (Cho *et al* 1990, Scholz and Stahl 1991). In the ISRS mechanism the pump impulsively generates coherent optical phonon modes through real or virtual electronic transitions (Hase *et al* 2009). In the DECP mechanism phonons are generated during the process of interband excitation from bonding to antibonding electron orbitals (Dekorsy *et al* 2000). In the field screening mechanism LO phonons are generated via ultrafast screening of longitudinal surface space-charge fields (Cho *et al* 1990, Scholz and Stahl 1991).

Generation and detection of coherent acoustic waves

Ultrafast ultrasonics pump-probe technique can be used to generate and detect coherent acoustic phonons. Irradiation of thin films by ultrafast laser pulses generates coherent acoustic waves. Such waves can be detected at the other side of the sample (transmission mode) or the same side of the sample (reflection mode). Bartels *et al* (1999) and Ezzahri *et al* (2007) generated and detected LA phonons in semiconductor superlattices.

Time- and frequency-domain thermoreflectance spectroscopies

Another ultrafast pump-probe spectroscopy uses time-domain thermoreflectance (TDTR) measurements (Jiang *et al* 2018). In this thermal properties of bulk as well as thin films can be measured through the reflectance change with temperature. Surface reflectance of a thin metal film, which is coated on samples, is measured under the exposure of pico- and femtosecond pulsed laser. The technique can be implemented as the TDTR method (Cahill 2004) as well as the frequency-domain thermoreflectance (FDTR) method (Schmidt *et al* 2009). Hangyo *et al* (2005) have reviewed the principle and variations of the terahertz time domain spectroscopy (THz-TDS) and its applications to solids.

14 Phonons in Liquid Helium

14.1 INTRODUCTION

So far in this book we have discussed the physics of phonons in crystalline and non-crystalline solids. The crystalline solids referred to in earlier chapters are classified as classical solids. Crystals with large zero-point energy, such as solid He, are called quantum solids. A study of the vibrational properties of quantum solids requires a knowledge of the concept of self-consistent harmonic approximation. We will not endeavour to discuss such a theory here, and the interested reader is referred to the articles by Horner (1974) and Eckstein *et al* (1970), and references therein. In this chapter our chief interest will be to provide an elementary discussion on some properties of *liquid helium*, although occassionally we shall refer to solid helium as well.

Helium exists in two stable isotopes: ^{4}He and ^{3}He. Natural helium contains mainly ^{4}He, with the lighter isotope ^{3}He in only about 1 part in 10^7.

At atmospheric pressure, ^{4}He is a liquid below 4.2 K and is called He I. When cooled further, it undergoes a second-order phase transition, becoming superfluid below 2.17 K (known as the λ-point) and is called He II. It remains a liquid down to 0 K unless a pressure of about 25 atm is applied when it solidifies into the hexagonal close packed (HCP) structure and at a much higher pressure into the face-centred cubic (FCC) structure. Below the λ-point liquid ^{4}He is considered to have undergone a Bose condensation, with its atoms having a spin of zero and therefore obeying Bose–Einstein statistics (equation (2.2)).

^{3}He is liquid below 3.2 K at atmospheric pressure and becomes superfluid at a much lower temperature than ^{4}He (below around 3 mK). Below its λ-point liquid ^{3}He becomes a degenerate Fermi liquid, with its atoms having a spin of $\frac{1}{2}$ and obeying Fermi–Dirac statistics. ^{3}He solidifies into the BCC structure above 33 atm.

Below the λ-point the properties of the quantum or superfluid He are interpreted in terms of elementary excitations. Above the λ-point the properties of the liquid can be studied by using the laws of ordinary classical hydrodynamics.

14.2 DISPERSION CURVE AND ELEMENTARY EXCITATIONS

Vibrational excitations in liquid He are considered in terms of density fluctuations and the dispersion curve for liquid He II, measured by using neutron inelastic scattering, is shown in figure 14.1. The dispersion curve, which has only one branch instead of three branches for solids, shows a prominant dip at $q \sim 2\,\text{Å}^{-1}$. It is interesting to mention that such a spectrum was originally proposed by Landau (1947) to account for the temperature dependence of the specific heat.

Excitations with wave vector up to about $1\,\text{Å}^{-1}$ (part A to B of the curve in figure 14.1) are regarded as longitudinal phonons with the dispersion curve approximately represented as

$$\omega = cq(1 + \gamma q^2), \tag{14.1}$$

where c is the long-wavelength limit of sound velocity and γ is a constant. In the long-wavelength limit, this relation reduces to the simple form $\omega = cq$. An interesting property of liquid He II is that its dispersion curve can be modified sustantially by the application of pressure (Narayanamurti *et al* 1973, Wyatt *et al* 1974). At the saturated vapour pressure (SVP), the phonon dispersion curve is anomalous in that it bends upwards (making γ positive in equation (14.1)) initially before bending downwards at point B. The upward dispersion at low pressures has a dramatic effect on phonon lifetimes. In particular, energy and wave vector conservation laws do not permit phonons above a critical frequency ω_c to take part in three-phonon processes. At pressures above about 19 Bar the

DOI: 10.1201/9781003141273-14

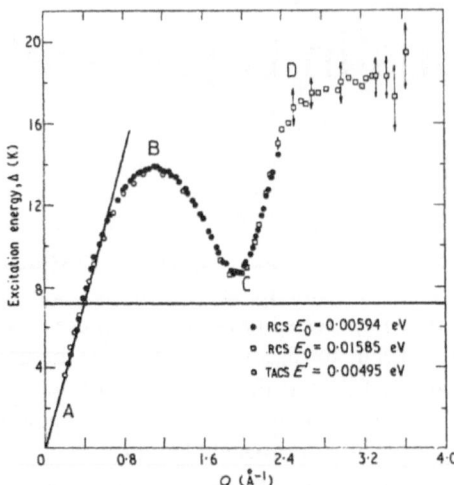

Figure 14.1 The experimental excitation spectrum of liquid He II (from Woods and Cowley (1973)). Quasi-particles in the regions A–B, B–C and C–D are phonons, R^- rotons and R^+ rotons, respectively. The horizontal line at $E = 7.16$ K represents the binding energy per atom in the liquid.

dispersion curve bends downwards from the origin (making γ negative in equation (14.1)). This dispersion curve does not allow for three-phonon interactions to take place at all.

Excitations representing the part B–C–D of the dispersion curve are called 'rotons'. Near the minimum at C the roton dispersion curve can be expressed by the relation

$$E = \Delta + \frac{(p - p_0)^2}{2\mu}, \tag{14.2}$$

where $p = \hbar q$, $E = \hbar\omega$, $\Delta/k_B \simeq 8.7$ K at C with $q_0 \cong 1.91$ Å^{-1} and $\mu \simeq 0.16 m_{He}$ is the effective mass of the roton. The part C–D of the curve is represented by R^+ rotons which possess positive group velocity c_g. The part B–C is represented by R^- rotons which possess negative group velocity with $c_g \parallel -q$.

The superfluidity of liquid He only occurs for flow velocities below a critical value. Landau showed that this corresponds to the minimum possible phase velocity which is given by the line joining the origin A to a point near the minimum C. This has been confirmed for ion motion. However, for flow in a capillary, the critical velocity is much smaller and is due to the generation of quantum vortices.

14.3 SPECIFIC HEAT

Thermodynamic properties of liquid He II are contributed by both phonons and rotons. The equilibrium distribution function for phonons with frequency ω is given by the Bose–Einstein form (see equation (2.2))

$$\bar{n}_{ph} = (e^{\hbar\omega/k_B T} - 1)^{-1}. \tag{14.3}$$

For rotons $\hbar\omega \gg k_B T$ and the distribution function reduces to the Maxwell–Boltzmann form

$$\bar{n}_{rot} \simeq e^{-\hbar\omega/k_B T}. \tag{14.4}$$

The following results for the specific heat per unit volume of liquid He II due to phonons and rotons can be evaluated (see, e.g., Khalatnikov 1965):

$$C_{ph} = \frac{2\pi^2 k_B^4 T^3}{15\hbar^3 c^3} \tag{14.5}$$

$$C_{rot} = \frac{2\mu^{1/2} q_0^2 \Delta^2}{(2\pi)^{3/2} \hbar k_B^{1/2}} \left[1 + \frac{k_B T}{\Delta} + \frac{3}{4} \left(\frac{k_B T}{\Delta} \right)^2 \right] T^{-3/2} e^{-\Delta/k_B T}. \tag{14.6}$$

In general above about 1 K, the roton contribution is higher than the phonon contribution. As the temperature decreases the roton contribution decreases exponentially with temperature, whereas the phonon contribution decreases as T^3. Thus at very low temperatures $C_{ph} > C_{rot}$. This is indeed confirmed by experimental measurements, reporting that the main contribution to the specific heat of liquid ^{4}He for $T \leq 600$ mK is due to phonons.

14.4 INTERACTIONS BETWEEN THE EXCITATIONS

At very low temperatures, phonons in liquid He II may be regarded as non-interacting quasi-particles characterised by a harmonic Hamiltonian H_0. In general, however, the Hamiltonian of liquid He can be expressed as (Landau 1941)

$$H = \int d^3 r \left(\frac{1}{2} v \cdot \rho v + E(\rho) \right), \tag{14.7}$$

where ρ and v are the liquid density and its velocity, respectively, and $E(\rho)$ is the internal energy density. It is instructive to express H as

$$H = H_0 + H_{PP} + H_{PR} + H_{RR}, \tag{14.8}$$

where

$$H_{PP} = \mathcal{V}_3 + \mathcal{V}_4 + \ldots \tag{14.9}$$

contains anharmonic terms (cubic, quartic, ...) of the phonon Hamiltonian, H_{PR} represents the phonon–roton interaction Hamiltonian, and H_{RR} represents the roton–roton interaction Hamiltonian.

Below about 600 mK only phonon–phonon interactions are important. Above about 600 mK the roton population starts to increase and so three types of interactions involving phonon and rotons must be considered. We will briefly describe these interactions.

14.4.1 ANHARMONIC INTERACTION BELOW ABOUT 600 mK

Because of the naturally pure nature of liquid ^{4}He, at very low temperatures phonons in general have very long lifetimes. The effect of the cubic anharmonicity $\mathcal{V}_3$ is to switch weak interaction between phonons, which can be treated by first-order perturbation theory. Because of liquid state $\mathcal{V}_3$ cannot be expressed in terms of elastic constants as was done in chapter 4. Instead $\mathcal{V}_3$ can be expressed as (Khalatnikov 1965)

$$\mathcal{V}_3 = \frac{v\rho' v}{2} + \frac{1}{3!} \frac{\partial}{\partial \rho} \left(\frac{c^2}{\rho} \right) \rho'^3, \tag{14.10}$$

where ρ' is the density variation with respect to its value in the liquid at rest: $\rho = \rho_0 + \rho'$. The first term represents the extra kinetic energy due to the perturbed density, and the second term is change in the potential due to the perturbed density. The quantities ρ' and v can be expressed in terms of second quantised notation introduced in chapter 4. However, we do not attempt to do this here.

Because of the nature of the dispersion relation in equation (14.1), three-phonon interaction above a critical frequency ω_c is not allowed by the energy conservation requirement. At low pressures ($P < 19$ Bar) and below the critical frequency ω_c we can expect three-phonon processes of the following types to take place:

$$q + q' = q'' \qquad \omega + \omega' = \omega'' \tag{14.11}$$

$$q = q' + q'' \qquad \omega = \omega' + \omega''. \tag{14.12}$$

Consider injecting a high-frequncy phonon into pure ^{4}He (liquid or solid phase) at $T = 0$. Such a phonon is expected to take part in the process (14.12), with its spontaneous decay rate following the frequency dependence $\tau^{-1} \propto \omega^5$ (see equation (6.91)). Next consider injecting an ultrasonic wave (a low-frequency phonon wave) in pure ^{4}He at a finite temperature ($T > 0$). Such an ultrasonic phonon (ω, q) will interact with thermal phonons (ω', q') within the system via the process (14.11). Consider ultrasonic waves with $\omega \leq 100$ MHz $<< \omega'$ so that $|q| << |q'|$. The frequency ω of an ultrasonic wave is expected to satisfy $\omega \tau_{th} > 1$, with τ_{th} as the lifetime of thermal phonons in liquid He below 600 mK. Thus first-order perturbation theory can be used to calculate the lifetime of the ultrasonic wave. The dominant process is the Landau–Rumer type process which has the behaviour (Abraham *et al* 1969, Wehner and Klein 1969) (also see equation (6.92))

$$\tau_{3\text{ph}}^{-1} \propto \omega T^4. \tag{14.13}$$

14.4.2   ANHARMONIC INTERACTION ABOVE ABOUT 600 mK

Interaction processes responsible for attenuation of quasi-particles in liquid He II above 0.6 K are quite complicated. In general three-phonon processes are not allowed and one must consider four-phonon, five-phonon, phonon–roton and roton–roton interactions. As described in section 14.2 there are two kinds of rotons: R^+ and R^-, and care must be taken to consider them separately. However, here we consider a few straight forward interaction mechanisms without distinguishing between R^+ and R^-.

14.4.2.1   Four-phonon processes

As discussed in section 6.4.2 four-phonon processes can arise from second-order terms in $\mathcal{V}_3$ and first-order terms in $\mathcal{V}_4$ of the Hamiltonian. Consider the four-phonon process

$$q + q_1' \rightleftharpoons q' + q_1'. \tag{14.14}$$

Using perturbation theory the transtition probability for such a process is expressed as

$$
\begin{aligned}
\langle q', q_1' | H_{pp} | q, q_1 \rangle &= \sum_{q''} \frac{\langle q', q_1' | \mathcal{V}_3 | q'' \rangle \langle q'' | \mathcal{V}_3 | q, q_1 \rangle}{\hbar \omega(q'') - \hbar \omega(q) - \hbar \omega(q_1)} \\
&\quad + \langle q', q_1' | \mathcal{V}_4 | q, q_1 \rangle.
\end{aligned}
\tag{14.15}
$$

Landau and Khalatnikov (1949) (also see Khalatnikov (1965)) considered the following fourth-order anharmonic potential:

$$\mathcal{V}_4 = \frac{1}{4!} \frac{\partial^2}{\partial \rho^2} \left(\frac{c^2}{\rho} \right) \rho'^4. \tag{14.16}$$

With equations (14.10) and (14.14)–(14.16) and by considering energy and momentum conservation laws, Landau and Khalatnikov derived the following results:

$$\tau_{4\text{ph}}^{-1} \propto \omega T^6 \qquad (\hbar \omega << k_B T) \tag{14.17}$$

$$\tau_{4\text{ph}}^{-1} \propto \omega^4 T^3 \qquad (\hbar \omega >> k_B T). \tag{14.18}$$

Eckstein showed (see Eckstein *et al* (1970)) that for those four-phonon processes which do not strictly conserve energy the attenuation varies as

$$\tau_{4\mathrm{ph}}^{-1} \propto \omega^3 T^4 \qquad (\hbar\omega \ll k_\mathrm{B}T) \qquad\qquad (14.19)$$

$$\tau_{4\mathrm{ph}}^{-1} \propto \omega^2 T^5 \qquad (\hbar\omega \gg k_\mathrm{B}T). \qquad\qquad (14.20)$$

From equations (14.17)–(14.20), we notice that the attenuation varies as $\omega^m T^n$, with $m+n=7$. Notice the contrast with the attenuation rates for three-phonon processes which vary as $\omega^m T^n$ with $m+n=5$ (*cf* section 6.4).

14.4.2.2 Roton–roton scattering

The momenta and energies of interacting rotons can be considered to be centred around (p_0, Δ). The energy and momentum conservation laws lead to the most effective roton–roton interaction as

$$R_1 + R_2 \rightleftharpoons R_3 + R_4. \qquad\qquad (14.21)$$

Such an interaction may be described by a δ-function of the distance between a pair of rotons at r_1 and r_2

$$H_{\mathrm{RR}} = V_0 \delta(r_1 - r_2), \qquad\qquad (14.22)$$

where V_0 is the coupling constant. The resulting roton relaxation time shows the following temperature dependence:

$$\tau_{\mathrm{RR}}^{-1} \propto T^{1/2} e^{-\Delta/k_\mathrm{B}T} \qquad\qquad (14.23)$$

which just comes from the thermal density of rotons.

14.4.2.3 Phonon–roton scattering

A low-energy phonon can collide with a roton to produce another phonon and another roton

$$P + R \rightleftharpoons P' + R'. \qquad\qquad (14.24)$$

Another possibility is that an energetic phonon can be absorbed (or emitted) in a four-particle interaction of the type

$$P + R \rightleftharpoons R' + R''. \qquad\qquad (14.25)$$

However, processes of the types $P + R \rightarrow R'$ and $R \rightarrow R' + P$, with roton energies close to the minimum Δ, are not allowed as they do not satisfy the energy and momentum conservation requirements simultaneously. (There is a small possibility for these processes to be allowed for high-energy R^+ rotons, provided the slope of the dispersion equals that of low-energy phonons.)

The Hamiltonian relevant for the phonon-roton interaction of the type in equation (14.24) can be written as (Khalatnikov 1965)

$$V_{\mathrm{ph-rot}} = -\frac{1}{2}(\boldsymbol{p}\cdot\boldsymbol{v} + \boldsymbol{v}\cdot\boldsymbol{p}) + \frac{1}{2}\left[\frac{\partial^2\Delta}{\partial\rho^2} + \frac{1}{\mu}\left(\frac{\partial p_0}{\partial\rho}\right)^2\right]\rho'^2. \qquad\qquad (14.26)$$

The process in equation (14.25) can be described by a δ-function interaction of the form in equation (14.22).

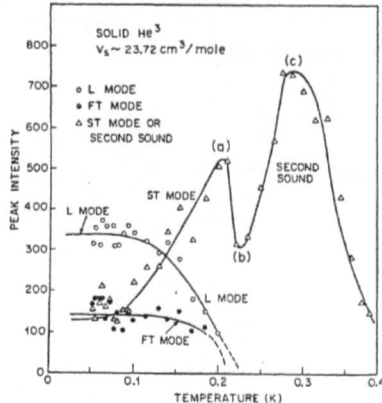

Figure 14.2 Mode decay and second sound measurements for heat pulses in solid ^{3}He. (From Narayanamurti and Dynes (1976).)

14.4.3 SECOND SOUND

Phonon–phonon interactions in liquid He II are strongest when the wave vectors of phonons involved are nearly parallel (N processes). As discussed in section 5.5, for $\omega\tau < 1$ the situation can be described by a temperature wave which propagates undamped with a second-sound velocity $c_{ss} = c/\sqrt{3}$, where c is the ordinary or first-sound velocity of phonons. Roton second sound is similar to phonon second sound but involves only excitations from the roton part of the dispersion curve.

Maris (1974) and Benin (1975) have calculated the dispersion relation and attenuation of phonon second-sound waves in superfluid ^{4}He at saturated vapour pressure (SVP).

Narayanamurti and Dynes (1976) have reported a study of a low-temperature (between 0.05 and 0.5 K) heat pulse propagation in solid ^{3}He (with molar volume between 24.5 and 23.7 cm^3 mole^{-1}) in the [110] direction. As seen in figure 14.2 the L and FT (fast transverse) modes decay rapidly between 0.15 and 0.2 K. However, the ST (slow transverse) mode picks up in intensity around 0.2 K, follows a rapid decay and then reaches a second peak at about 0.3 K. This second peak is interpreted as the collective second-sound mode with a velocity of 101 m s^{-1}.

14.5 KAPITZA RESISTANCE

When heat flows from one material to another, a small temperature difference δT develops across the interface and the ratio of δT and the heat current density $\dot{Q}/A$ is called thermal boundary resistance W_K[1]

$$W_K = \frac{\delta T}{\dot{Q}/A}. \tag{14.27}$$

Actually this effect was first observed by Kapitza (1941) for heat flow from a solid into superfluid He: historically, in this case the thermal boundary resistance is called the Kapitza resistance. In general, it is found that $W_K \propto T^{-n}$, where n lies between 3 and 4.

---

[1]Note that we have defined a thermal boundary resistance, and not a thermal boundary resistivity, in equation (14.27). The word 'resistance' is used for historical reasons, because resistivity usually describes a bulk (geometry-independent) property. Further, note that we consider only interfaces in which the thermal transport on at least one side is mainly due to phonons.

In section 9.6 we noted that for solid–solid interfaces Kapitza resistance is also known as thermal boundary resistance. In that section it was pointed out that there are two limiting models of the resistance: the acoustic mismatch model, and the diffuse mismatch model.

Acoustic mismatch model

Applying the laws of classical acoustics, the transmission coefficient for a longitudinal phonon beam incident *normal* to solid–liquid He interfaces is given by (Khalatnikov 1952, 1965)

$$\alpha = \frac{4z_1 z_2}{(z_1 + z_2)^2}, \tag{14.28}$$

where z_i are the acoustical impedances

$$z_i = \rho_i c_i \qquad i = 1, 2 \tag{14.29}$$

and ρ_i and c_i are the densities and sound velocities, respectively, in the two materials. Using Debye's isotropic continuum model of section 2.7 the following result for the Kapitza conductance W_K^{-1} can be derived (Wyatt 1981, Swartz and Pohl 1989):

$$W_K^{-1} = \frac{1}{4}C_v^{sp} c \alpha \tag{14.30}$$

$$= \frac{\pi^2 k_B^4 T^3 \alpha}{30\hbar^3 c^2}, \tag{14.31}$$

where C_v^{sp} is the low-temperature Debye specific heat (equation (14.5)) and c is the Debye phonon speed. (Note that when dealing with thermal boundary resistance between two solids the result in equation (14.31) should be multiplied by 3 to account for 3 phonon acoustic polarisation branches.)

The diffuse mismatch model

The acoustic mismatch model assumes complete specularity at the interface. The diffuse mismatch model considers the extreme opposite view and assumes that all the phonons are diffusely scattered at the interface (analogous to Casimir's model for boundary scattering, discussed in section 6.1). If the acoustic impedances of the two media in contact are very different, then the scattering at the interface will reduce the boundary resistance. This reduction can be typically over two orders of magnitude in the limiting case of diffuse mismatch (Swartz and Pohl 1989).

The acoustic mismatch model and the diffuse mismatch model provide, respectively, an upper limit and a lower limit to the Kapitza resistance. The observed Kapitza resistances have been found to lie between these two limits (Swartz and Pohl 1989).

For solid–liquid He interfaces the acoustic mismatch theory predicts $\alpha \simeq 0.01$ which is much smaller than experimental results which lie in the range 0.1–0.5. This disagreement can be qualitatively understood when account is taken of the sample temperature and the nature of the interface. At low temperatures (below 0.1 K), at smooth (polished, or defect-free) metal surfaces, phonons are not scattered and the acoustic mismatch model can explain the Kapitza resistance provided the effect of conduction electrons on phonon attenuation is taken into account. At temperatures above a few tenths of a kelvin, the effects of the helium boundary layer, and of imperfections at or near the interface, become important. At temperatures above 1 K and at sufficiently imperfect interfaces, phonons will scatter and the diffuse mismatch theory is likely to give a more realistic description of experimental measurements.

Making refined theoretical calculations, and by taking into account the nature of interface, Maris (1979) suggested that α is a function of phonon energy, and estimated that for low-frequency phonons across a solid–liquid He interface the transmission coefficient is of the order $\alpha \leq 0.1$, which is in much better agreement with experiment. Attempts to deal with different situations have been made by several theoreticians by considering the role of other excitations within the liquid He layers at the interface. These include surfons (surface phonons) in thin layers of He which eventually decay into He bulk phonons, vacancy waves in the He surface layer and their scattering from bulk phonons and tunnelling states which arise in the He layer due to an inhomogeneous crystal surface. However, mathematical modelling of these situations is extremely difficult, as very little is known about the surface in most experiments. Thus, at present it is generally recognised that our understanding, both theoretical and experimental, of the Kapitza resistance at solid–liquid He interfaces is far from being satisfactory. For detailed experimental and theoretical reviews of this subject the reader is referred to the works of Pollack (1969), Challis (1974), Wyatt (1981), and Swartz and Pohl (1989).

It was pointed out in section 9.6.3 that while an accurate description of thermal boundary resistance is a challenging problem, existing theories for solid-solid interfaces favour consideration of the sum of weighted diffuse (DMM) and specular (AMM) contributions. It was also shown that consideration of surface roughness and inelastic scattering may produce significant changes to the numerical estimate for Kapitza resistance. In the context of a solid/superfluid interface, important contributions of interface roughness (Adamenko and Fuks 1970), defects (Khalatnikov and Adamenko 1972) and inealstic scattering (Adamenko and Nemchenko 2013) have been theoretically proposed and explained. From their theoretical work Adamenko and Nemchenko (2013) showed that the largest inelastic contribution comes from the process in which a phonon from the solid is converted into two phonons in the superfluid. Adamenko and Fuks (1970) showed that a resonant phonon scattering mechanism is set up when the interface roughness becomes comparable to phonon wavelengths. Ramiere *et al* (2016) have corroborated Adamenko-Fuks's model of resonant phonon scattering due to nanoscale surface roughness by analysing their Si sample for the Si/liquid He interface.

14.6 QUANTUM EVAPORATION

Using the heat pulse technique at low temperatures, it has been shown (Wyatt 1984) that a phonon and/or a roton in liquid ^{4}He can be annihilated at the interface between the liquid and vacuum, resulting in the ejection of a free atom. To successfully observe this quantum evaporation process, it is necessary to work at low temperatures ($T \leq 0.1$ K) so that (i) the mean free path of the injected excitations (phonons or rotons) is long enough for scattering by thermal excitations of the liquid to be ignored and (ii) the evaporated atom travels ballistically to a bolometer detector. The energy of the injected excitation must be above the threshold of atomic binding energy in the liquid to meet the energy requirement for the process.

Anderson (1969) suggested that the simplest dynamical process which dominates the evaporation is the elastic single-atom emission across the liquid surface by the annihilation of a single elementary excitation in the liquid. This picture is analogous to thermionic or field emission. There is a similarity between the Bogoliubov theory (Bogoliubov 1947, 1958) for superconductivity and superfluidity. Thus one can see that the quantum evaporation is also analogous to the coupling between electrons at a superconducting and normal metal interface.

Widom (1969), Hyman *et al* (1969) and Cole (1972) have developed a simple theory of evaporation from liquid He ɪɪ. In this theory the elementary excitations of liquid He ɪɪ are represented by a gas of non-interacting quasi-particles. By considering the flux of quasi-particles incident upon the liquid interface with vacuum, the rate of evaporation of atoms from the surface is determined in terms of a phenomenological parameter. The theory assumes that the Hamiltonian of the liquid-vacuum system can be divided into three terms

$$H = H_L + H_g + \mathcal{V}_{int}, \tag{14.32}$$

where H_L is the Hamiltonian of the liquid, H_g is the free-particle Hamiltonian of the evaporated atoms, and $\mathcal{V}_{int}$ weakly couples the liquid with the vacuum via a transfer of atoms. In the second-quantised notation, one can express

$$\mathcal{V}_{int} = \sum_k \sum_q (T_{kq}^* c_k^\dagger b_q + T_{kq} c_k b_q^\dagger), \qquad (14.33)$$

where $c_k^\dagger$ (c_k) creates (annihilates) an atom of momentum $\hbar k$ in the vapour, and $b_q^\dagger$ (b_q) is an operator which acts only on the states of the liquid. In particular, we assume that b_q describes the removal of an atom from the liquid. The matrix element T_{kq} relates atomic wavefunctions in the liquid to those in the vacuum. This method of formulating the Hamiltonian of an interface system is similar to that of Cohen *et al* (1962) for the electron tunnelling between normal and superconducting metals.

The rate of evaporation of liquid atoms is given by

$$R = \sum_k W_{em}(k, 0), \qquad (14.34)$$

where $W_{em}(k, 0)$, the emission rate for an atom of momentum $\hbar k$, can be expressed as

$$W_{em}(k, n_k) = \sum_{i,f} P_i W_{i \to f}(k, n_k), \qquad (14.35)$$

with P_i as the occupation probability of the initial state $|i\rangle$ of the liquid. From the golden rule formula we can express

$$W_{i \to f}(k, n_k) = \frac{2\pi}{\hbar} \left| \langle n_k + 1, f | V_{int} | n_k, i \rangle \right|^2 \delta(E_k + \varepsilon_f - \varepsilon_i), \qquad (14.36)$$

where $|n_k, i\rangle$ and $|n_k + 1, f\rangle$ are, respectively, the initial and final states of the total system, $\varepsilon_i(\varepsilon_f)$ is the energy of the initial (final) state of the liquid, $E_k = \hbar^2 k^2 / 2m_{He}$ is the energy of a free (evaporated) atom, and n_k is the occupation factor of the free atom. An evaluation of the matrix element $\langle n_k + 1, f | \mathcal{V}_{int} | n_k, i \rangle$ is complicated, and has not been achieved so far. We will therefore not discuss it here.

The momentum $\hbar k$ and energy E_k of an atom evaporated from the liquid surface are uniquely determined by the momentum $\hbar q$ and energy $\hbar \omega_q$ of the incident quasi-particle. In order to match the quasi-particle wavefunction to the evaporated-atom wavefunction at the interface, the matrix element $|\langle n_k+, f | \mathcal{V}_{int} | n_k, i \rangle|^2$ must contain the parallel momentum conserving condition

$$q_\| = k_\|. \qquad (14.37)$$

Further, for one phonon (or roton) of energy $\hbar \omega_q$ evaporating one atom of energy E_k, the energy conservation in equation (14.36) reads

$$\hbar \omega_q = E_B + \frac{\hbar^2 k^2}{2m_{He}}, \qquad (14.38)$$

where E_B is the binding energy of an atom in the liquid.

Thus the evaporation as a single elastic emission process is subject to the momentum and energy conservation rules in equations (14.37) and (14.38), and its intensity is governed by the matrix element $|\langle n_k + 1, f | \mathcal{V}_{int} | n_k, i \rangle|^2$. Let us examine the effect of the selection rules in a little detail.

From bulk latent heat measurements it is established that the atomic binding energy in liquid ^{4}He is $E_B = 7.16$ K. This means that only rotons and high-frequency phonons can contribute to the evaporation process. The momentum conservation condition can be used to differentiate between the roton and phonon contributions. Equation (14.37) can be expressed as

$$q \sin \theta = k \sin \phi, \qquad (14.39)$$

where θ and ϕ are the angles the surface normal makes with the directions of the excitation in the liquid and the free atom, respectively. For a given q and θ the angle ϕ can be easily calculated

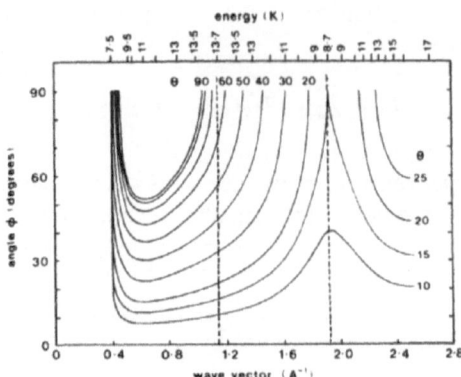

Figure 14.3 A plot of the angle of ejection ϕ of a free He atom as a function of wave vector of the liquid excitations for various angles of incidence θ. (From Wyatt (1984).)

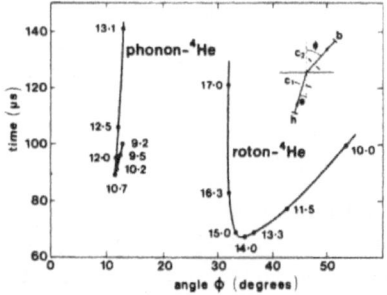

Figure 14.4 Calculated total signal times for phonon–atom and roton–atom as a function of the angle of atom evaporation ϕ. The angle of incidence for excitations is $\theta = 15^0$. (From Wyatt (1984).)

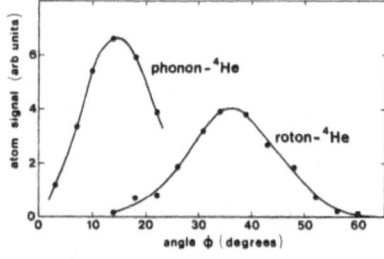

Figure 14.5 The signal intensity, integrated over the first 12 μs, of the evaporated helium atom as a function of the angle of ejection ϕ. The angle of excitation angle is $\theta = 13^0$. (From Wyatt (1984).)

with the help of equations (14.37)–(14.38) and the dispersion relations in equations (14.1)–(14.2). As shown in figure 14.3 for some q values there is only a limited range of θ which meets $\phi \leq 90^0$. There is a critical angle of incidence θ_c beyond which liquid excitations will be totally reflected from the surface and there will be no evaporation. The distinction between phonon–atom and roton–atom evaporation is clearly seen in figure 14.4 where the signal times are plotted against ϕ for excitation angle $\theta = 15^0$. Whereas phonons evaporate atoms at $\phi \sim 10^0$, roton–atom evaporation occurs over a wide range of ϕ centred around $35°$. Figure 14.5 shows time-integrated signal intensities of phonon–atom and roton–atom evaporations and the clear angular separation of the signals is apparent. The roton–atom signal has a larger angular width because of the greater range of q-values for the rotons compared to the phonons generated in a heat pulse. More conclusive measurements of phonon $\rightarrow$ atom and roton $\rightarrow$ atom evaporations, with conclusions similar to that shown in figure 14.5, have been made by Brown and Wyatt (1990).

A Density Functional Formalism

THOMAS-FERMI-DIRAC-WIGNER FORMALISM

The density functional formalism has been established for a long time. The first attempt was made in the late 1920s and early 1930s and is known as the Thomas-Fermi-Dirac-Wigner formalism. Good reviews of this scheme can be found in the book by March (1983) and the review article by Jones and Gunnarsson (1989).

In the Thomas-Fermi approximation, the electrons are treated as independent particles. Their kinetic energy is expressed in terms of the kinetic energy density of a system of non-interacting electrons with density n of the Fermi gas:

$$T^{TF}[n] = \int d\mathbf{r}\, t[n(\mathbf{r})] \tag{A.1}$$

with

$$t[n] = 2\frac{1}{(2\pi)^3} \int_{|\mathbf{k}| \leq k_F} d\mathbf{k} \frac{\hbar^2 k^2}{2m}, \tag{A.2}$$

where k_F is the Fermi wavenumber and the factor of 2 accounts for spin degeneracy of electron states. Combining equations (A.1) and (A.2), and using $2(4\pi/3)k_F^3/(2\pi)^3 = n$, we can express

$$T^{TF}[n] = \frac{3}{5} \int d\mathbf{r}[n(\mathbf{r})]^{5/3}. \tag{A.3}$$

Also, in the Thomas-Fermi approximation, the electron-electron interaction energy is considered as solely electrostatic

$$E_{es}[n] = \frac{e^2}{2} \int \int d\mathbf{r}\, d\mathbf{r}' \frac{n(\mathbf{r})n(\mathbf{r}')}{|\mathbf{r}-\mathbf{r}'|}. \tag{A.4}$$

The total electronic energy, in the presence of the ionic external potential $V_{ext}(\mathbf{r})$,

$$E_{el}^{TF}[n] = T^{TF}[n] + \int d\mathbf{r} V_{ext}(\mathbf{r})n(\mathbf{r}) + E_{es}[n] \tag{A.5}$$

is minimized subject to a fixed number of electrons

$$\int d\mathbf{r}\, n(\mathbf{r}) = N. \tag{A.6}$$

Application of the Euler-Lagrange variational principle $\delta[E_{el}^{TF}[n] - \lambda N] = 0$ results in the Thomas-Fermi equation

$$\frac{\hbar^2}{2m}(3\pi^2)^{2/3}n(\mathbf{r})^{2/3} + e^2 \int d\mathbf{r}' \frac{n(\mathbf{r})n(\mathbf{r}')}{|\mathbf{r}-\mathbf{r}'|} + V_{ext}(\mathbf{r}) + \lambda = 0, \tag{A.7}$$

where λ is a Lagrange multiplier.

The Thomas-Fermi model has several shortcomings. Firstly, this model ignores the shell structure of atoms and assumes that the external potential V_{ext} is slowly varying in space over a distance comparable to the de Broglie wavelengths of the electrons so that all the electrons can be localised

DOI: 10.1201/9781003141273-A

within the Fermi sphere in momentum space. With this consideration, the kinetic energy is expressed as the kinetic energy of a system of non-interacting electrons with a spatially slowly varying density $n(r)$. Also, the model does not accommodate the Fermi-Dirac statistics for electrons. However, improvements were made by Dirac (1930) and Wigner (1934). Dirac (1930) included an exchange term in the form

$$E_x^{\text{Dirac}} = \int dr \, \varepsilon_x[n(r)], \tag{A.8}$$

where $\varepsilon_x[n]$ is the exchange energy per particle of a homogeneous electron gas of density n. Wigner (1934) included a correlation term using $\varepsilon_c[n]$ as the correlation energy per particle of a homogeneous electron gas of density n. We will discuss $\varepsilon_x[n]$ and $\varepsilon_c[n]$ later in this appendix.

HOHENBERG-KOHN-SHAM FORMALISM

An exact formulation of the density functional theory (DFT) of many-body systems was laid down in mid 1960s by Kohn and co-workers in two seminal research papers (Hohenberg and Kohn (1964) and Kohn and Sham (1965)). This has evolved as a conceptually and practically useful method for studying the ground-state properties of many-electron systems. This theory has been described in a very large number of books and review articles. Here we will describe the DFT at a level that serves the purpose of developing the main subject of this book, the physics of phonons. We provide a brief discussion of the main points of the formalism for non-spin-polarised electronic systems. For a rigorous description of the theory we refer the reader to the book by Martin (2004).

The Hohenberg-Kohn-Sham density functional formalism is based on two theorems (Hohenberg and Kohn 1964, Kohn and Sham 1965) which are stated and proven below.

BASIC THEOREMS AND ELECTRONIC ENERGY FUNCTIONAL

Theorem 1: The density as basic variable The complete many-body ground-state wavefunction Ψ of an electronic system is a unique functional $\Psi[\rho(r)]$ of the electronic charge density $\rho(r)$.

Proof.
The proof proceeds by *reductio ad absurdum* (see Hohenberg and Kohn 1964, Kohn and Vashista 1983). Consider a system of N electrons which is subjected to some external potential $V_{\text{ext}}(r)$ and which has density $\rho(r)$ corresponding to a non-degenerate ground state $\Psi \equiv \Psi(r_1, r_2, \ldots, r_N)$. Let H be the Hamiltonian and E_{el} the total electronic energy of the system. Also, consider a different external potential $V'_{\text{ext}}(r)$, with ground state Ψ' which gives rise to the same density $\rho(r)$. Clearly (unless $V'_{\text{ext}}(r) - V_{\text{ext}}(r) = \text{constant}$) $\Psi' \neq \Psi$, since these states are eigenstates of different Hamiltonians. Let H' and E'_{el} be the Hamiltonian and the electronic energy associated with Ψ'. Then, since the ground-state energy is the minimum energy and as Ψ' is not an eigenstate of H, we have

$$\begin{aligned} E_{\text{el}} &= (\Psi, H\Psi) < (\Psi', H\Psi') \\ &= (\Psi', (H' + V_{\text{ext}} - V'_{\text{ext}})\Psi') \end{aligned}$$

or

$$E_{\text{el}} < E'_{\text{el}} + \int dr (V_{\text{ext}}(r) - V'_{\text{ext}}(r))\rho(r) \tag{A.9}$$

as Ψ' is assumed to have the same density as Ψ. Similarly, interchanging primed and unprimed quantities, we find

$$E'_{\text{el}} < E_{\text{el}} + \int dr (V'_{\text{ext}}(r) - V_{\text{ext}}(r))\rho(r). \tag{A.10}$$

Adding equations (A.9) and (A.10), we get

$$E_{\text{el}} + E'_{\text{el}} < E_{\text{el}} + E'_{\text{el}}. \tag{A.11}$$

Thus we conclude that $V_{\text{ext}}(r)$ is (to within a constant) a unique functional of $\rho(r)$. Now, since $V_{\text{ext}}(r)$ defines the Hamiltonian H, it can further be concluded that the full many-body ground state Ψ is also a unique functional of $\rho(r)$, i.e. $\Psi = \Psi[\rho(r)]$. This proves the theorem.

Before discussing the second theorem, we note that since $\rho(r)$ determines $V_{\text{ext}}(r)$ and Ψ, it also determines all other electronic properties of the system. In particular, we can express $E_{\text{el}} = E_{\text{el}}[V_{\text{ext}}, \rho] = E_{\text{el}}[\rho]$.

For a general density $n(r)$ the functional $\Psi[n(r)]$ is not known. Hohenberg and Kohn defined another functional $F[n]$ (also see Levy (1979)):

$$F[n] = (\Psi, (T + V_{ee})\Psi), \tag{A.12}$$

where $n(r)$ is a density which can be obtained from some antisymmetrised wavefunction $\Psi(r_1, r_2, \ldots, r_N)$, and T and V_{ee} are, respectively, the kinetic and electron–electron interaction energy operators of the many-body system. The functional $F[n]$ is universal in the sense that it refers neither to a specific system nor to the external potential $V_{\text{ext}}(r)$. With the help of $F[n]$ Hohenberg and Kohn further defined, for a given $V_{\text{ext}}(r)$, the electronic energy functional

$$E_{\text{el}}[V_{\text{ext}}, n] = \int dr V_{\text{ext}}(r)\rho(r) + F[n] \tag{A.13}$$

which must also be a unique functional of $n(r)$.

Theorem 2: The energy variational principle

The functional $E_{\text{el}}[V_{\text{ext}}, n]$ in equation (A.5) obeys a variational principle in the charge density n.

In other words, the functional $E_{\text{el}}[V_{\text{ext}}, n]$ assumes its minimum value (the ground-state energy) for the correct density $(n(r) = \rho(r))$ than for any other density distribution, $n(r)$, with the same total number of electrons N.

Proof.

Because of the assumed non-degeneracy of Ψ, it is well known that the conventional Rayleigh–Ritz functional of Ψ',

$$E_{\text{el}}[\Psi'] \equiv (\Psi', H\Psi') \tag{A.14}$$

has a lower value for the correct ground state Ψ than for any other Ψ' with the same number of particles (see also chapter 5). Let Ψ' be the ground state associated with some other admissible density $n(r)$. Then

$$\begin{aligned} \mathscr{E}_{\text{el}}[\Psi'] &= \int dr V_{\text{ext}}(r)n(r) + F[n] \\ &> \mathscr{E}[\Psi] = \int dr V_{\text{ext}}(r)\rho(r) + F[\rho]. \end{aligned} \tag{A.15}$$

This proves the minimum principle for $E_{\text{el}}[V_{\text{ext}}, n]$,

$$E_{\text{el}}[V_{\text{ext}}, \rho] < E_{\text{el}}[V_{\text{ext}}, n], \tag{A.16}$$

where $E_{\text{el}}[V_{\text{ext}}, \rho]$ is the correct ground-state energy associated with $V_{\text{ext}}(r)$ and N.

Having stated and proven the two basic theorems of the density functional formalism, we note that it is convenient to separate out from $F[\rho]$ the classical Coulomb energy and write

$$F[\rho] = \frac{e^2}{2} \int \int dr\, dr' \frac{\rho(r)\rho(r')}{|r - r'|} + \mathscr{G}[\rho] \tag{A.17}$$

so that the ground-state energy functional becomes

$$E_{\text{el}}[V_{\text{ext}}, \rho] = \int dr V_{ext}(r)\rho(r) + \frac{e^2}{2} \int \int dr\, dr' \frac{\rho(r)\rho(r')}{|r - r'|} + \mathscr{G}[\rho], \tag{A.18}$$

where $\mathscr{G}[\rho]$ is a universal functional of ρ, as $F[\rho]$ is.

Kohn and Sham (1965) proposed conveniently dividing $\mathscr{G}[\rho]$ into two parts,

$$\mathscr{G}[\rho] \equiv T_0[\rho] + E_{xc}[\rho] \tag{A.19}$$

where $T_0[\rho]$ is the kinetic energy of a system of *non-interacting particles* with density $\rho(r)$, and $E_{xc}[\rho]$ is another universal functional of ρ which represents all corrections to the independent-electron model, i.e. the non-classical *many-body effects* of exchange and correlation (xc), but which still remains unknown. With equations (A.18) and (A.19) we thus write

$$E_{el}[V_{ext}, \rho] = T_0[\rho] + \int dr V_{ext}(r)\rho(r) + \frac{e^2}{2} \int \int dr\,dr' \frac{\rho(r)\rho(r')}{|r-r'|} + E_{xc}[\rho]. \tag{A.20}$$

There are three major difficulties in evaluating the energy functional in equation (A.20): (i) one needs a method to determine the ground-state electronic charge density $\rho(r)$ which minimises E_{el}, (ii) evaluation of $T_0[\rho]$, given only $\rho(r)$, cannot be done straightforwardly as there is no information on wavefunctions, and (iii) the functional $E_{xc}[\rho]$ remains unknown, except for a few simple systems, and must therefore be represented in some approximate way.

The first two difficulties are resolved by following the proposal of Kohn and Sham (1965) which we describe below.

Self-consistent Kohn–Sham equations
Consider the minimisation of the energy functional

$$E_{el}[n] = T_0[n] + \int dr V_{ext}(r)n(r) + \frac{e^2}{2} \int \int dr\,dr' \frac{n(r)n(r')}{|r-r'|} + E_{xc}[n] \tag{A.21}$$

with respect to the density $n(r)$. Although this problem was originally considered by Kohn and Sham, here we follow the derivation of Kohn and Vashishta (1983). Let $V_{trial}(r)$ be a trial single-particle potential which gives rise to an electron density $n(r)$ defined as

$$n(r) \equiv \sum_{j=1}^{occ} |\phi_j(r)|^2, \tag{A.22}$$

where the summation is done over the *occupied states*, and $\phi_j(r)$ are wavefunctions of *fictitious non-interacting electrons* which satisfy the Schrödinger equation

$$\left(-\frac{\hbar^2}{2m}\nabla^2 + V_{trial}(r)\right)\phi_j(r) = \varepsilon_j\phi_j(r). \tag{A.23}$$

From a solution of this equation we can express

$$\begin{aligned} \sum_j \varepsilon_j &= \sum_j \left(\phi_j, \left(-\frac{\hbar^2}{2m}\nabla^2 + V_{trial}(r)\right)\phi_j\right) \\ &= T_0[n] + \int dr V_{trial}(r)n(r) \end{aligned} \tag{A.24}$$

so that equation (A.21) becomes

$$\begin{aligned} E_{el}[n] &\equiv \sum_j \varepsilon_j - \int dr V_{trial}(r)n(r) + \int dr V_{ext}(r)n(r) \\ &\quad + \frac{e^2}{2} \int \int dr\,dr' \frac{n(r)n(r')}{|r-r'|} + E_{xc}[n]. \end{aligned} \tag{A.25}$$

We need to minimise this quantity with respect to $V_{trial}(r)$, with $n(r)$ regarded as a functional of $V_{trial}(r)$, or equivalently with respect to $n(r)$, with $V_{trial}(r)$ regarded as a functional of $n(r)$.

Taking the variation with respect to $n(r)$ we see that the $V_{\text{trial}}(r)$ which minimises $E_{\text{el}}[n]$ must satisfy the condition

$$
\begin{aligned}
V_{\text{trial}}(r) &= V_{\text{ext}}(r) + e^2 \int dr' \frac{n(r')}{|r-r'|} + \frac{\delta E_{\text{xc}}[n]}{\delta n(r)} + \text{constant} \\
&= V_{\text{KS}}(r) + \text{constant}.
\end{aligned}
\tag{A.26}
$$

Here $V_{\text{KS}}(r)$ is an effective potential, known as the *Kohn–Sham potential*,

$$
\begin{aligned}
V_{\text{KS}}(r) &= V_{\text{ext}}(r) + e^2 \int dr' \frac{n(r')}{|r-r'|} + \frac{\delta E_{\text{xc}}[n]}{\delta n(r)} \\
&= V_{\text{ext}}(r) + V_{\text{H}}(r) + V_{\text{xc}}(r)
\end{aligned}
\tag{A.27}
$$

with V_{H} as the Coulomb potential and the functional derivative[1]

$$
V_{\text{xc}}(r) \equiv \frac{\delta E_{\text{xc}}[n]}{\delta n(r)}
\tag{A.28}
$$

defining an effective one-electron exchange-correlation potential.

In the *ground state* equations (A.23) and (A.22) become, respectively,

$$
\left(-\frac{\hbar^2}{2m} \nabla^2 + V_{\text{KS}}(r) \right) \phi_j(r) = \varepsilon_j \phi_j(r)
\tag{A.29}
$$

and

$$
\rho(r) = \sum_j^{\text{occ}} |\phi_j(r)|^2
\tag{A.30}
$$

and must be solved self-consistently. These are known as the self-consistent *Kohn–Sham equations*.

Having obtained the self-consistent charge density $\rho(r)$ from a solution of equations (A.29)–(A.30), the kinetic energy $T_0[\rho]$ is evaluated as

$$
\begin{aligned}
T_0[\rho] &= \sum_j (\phi_j, (\varepsilon_j - V_{\text{KS}}(r))\phi_j) \\
&= \sum_j \varepsilon_j - \int dr V_{\text{KS}}(r)\rho(r).
\end{aligned}
\tag{A.31}
$$

The local-density approximation

Kohn and Sham (1965) suggested circumventing the third difficulty, namely evaluation of $E_{\text{xc}}[\rho]$, by applying the *local density approximation* (LDA). In this approximation it is assumed that the electronic charge density in the system corresponds to that of a homogeneous electron gas, so that $\rho(r)$ is slowly varying and it is possible to approximate

$$
E_{\text{xc}}[\rho] \simeq \int dr \rho(r) \varepsilon_{\text{xc}}[\rho(r)]
\tag{A.32}
$$

with $\varepsilon_{\text{xc}}[\rho(r)]$ as the exchange-correlation energy per electron of a uniform electron gas of density $\rho(r)$. The corresponding exchange-correlation potential is given by

$$
V_{\text{xc}}(r) = \frac{d}{d\rho} \{ \varepsilon_{\text{xc}}[\rho(r)]\rho(r) \} \equiv \mu_{\text{xc}}[\rho(r)],
\tag{A.33}
$$

---

[1] The functional derivative $\frac{\delta E_{\text{xc}}}{\delta n(r)}$ of $E_{\text{xc}}[n(r)]$ with respect to $n(r)$ is defined using the equation $\delta E_{\text{xc}}[n] = E_{\text{xc}}[n + \delta n] - E_{\text{xc}}[n] = \int \frac{\delta E_{\text{xc}}}{\delta n(r)} \delta n(r) dr$.

where $\mu_{xc}[\rho]$ is the exchange-correlation contribution to the chemical potential of a uniform system. Expressing

$$\rho^{-1} = \frac{4\pi}{3}r_s^3 \tag{A.34}$$

with r_s as an average interelectron distance, we can write equation (A.25) as

$$V_{xc} \equiv \mu_{xc} = \varepsilon_{xc} - \frac{r_s}{3}\frac{d\varepsilon_{xc}}{dr_s}. \tag{A.35}$$

Using equations (A.20) (A.27), and (A.31)–(A.33), we can finally express the total ground-state electronic energy within the LDA as

$$\begin{aligned}
E_{el} &= \sum_j \varepsilon_j - \frac{e^2}{2}\int\int dr dr' \frac{\rho(r)\rho(r')}{|r-r'|} \\
&+ \int dr\{\varepsilon_{xc}(\rho(r)) - \mu_{xc}(\rho(r))\}\rho(r).
\end{aligned} \tag{A.36}$$

A number of prescriptions for ε_{xc} are available in the literature. Some of these are given below. Wigner (1938) (in Ryd units)

$$\begin{aligned}
\varepsilon_{xc} &= \varepsilon_x + \varepsilon_c \\
\varepsilon_x &= \frac{-0.9164}{r_s} \\
\varepsilon_c &= \frac{-0.88}{(7.8 + r_s)}.
\end{aligned} \tag{A.37}$$

Hedin and Lundqvist (1971) (in Ryd units)

$$\begin{aligned}
\varepsilon_{xc} &= \varepsilon_x + \varepsilon_c \\
\varepsilon_x &= \frac{-0.9164}{r_s} \\
\varepsilon_c &= -0.045\left[(1+x^3)\ln\left(1+\frac{1}{x}\right) + \frac{x}{2} - x^2 - \frac{1}{3}\right], \quad x = \frac{r_s}{21}.
\end{aligned} \tag{A.38}$$

Ceperley and Alder (1980), as parametrised by Perdew and Zunger (1981):
For an unpolarised electron gas (in Hartrees, 1 Hartree = 2 Ryd)

$$\begin{aligned}
\varepsilon_{xc} &= \varepsilon_x + \varepsilon_c \\
\varepsilon_x &= \frac{-0.4582}{r_s} \\
\varepsilon_c &= \begin{cases} -0.1423/(1+1.9529\sqrt{r_s}+0.3334r_s) & \text{for } r_s \geq 1 \\ -0.0480+0.0311\ln r_s - 0.0116r_s + 0.0020r_s\ln r_s & \text{for } r_s < 1. \end{cases}
\end{aligned} \tag{A.39}$$

X_α method (Slater 1974)

$$\begin{aligned}
\varepsilon_{xc} &= -\alpha\frac{9e^2}{8\pi}(3\pi^2)^{1/3}(\rho(r))^{1/3} \\
&= \frac{-0.68725e^2\alpha}{r_s} \qquad \frac{2}{3} \leq \alpha \leq 1
\end{aligned} \tag{A.40}$$

i.e. α is regarded as an adjustable parameter between 2/3 and 1. For $\alpha = 2/3$ equation (A.40) reduces to ε_x as given in equations (A.37)–(A.39). (Note that $e^2 = 2$ in Ryd units.)

Generalized gradient approximation

As mentioned above, application of the LDA is suitable for extended systems with slowly varying electron density. Several attepmts have been made to modify the LDA for incorporating inhomogeneous nature of electron density distribution in solids. One approach is to include gradient-type correction to the LDA exchange-correlation energy (see Hohenberg and Kohn (1964), Kohn and Sham (1965), Martin (2004)). In the so-called *generalized gradient approximation* (GGA) the change $\delta E_{xc}[\rho]$ is expressed to linear order in $\delta\rho$ and $\delta\nabla\rho = \nabla\delta\rho$:

$$\delta E_{xc} = \int d\boldsymbol{r}\left[\varepsilon_{xc} + \rho\frac{\partial\varepsilon_{xc}}{\partial\rho} + \rho\sum_{\alpha}\frac{\partial\varepsilon_{xc}}{\partial\nabla_{\alpha}\rho}\nabla_{\alpha}\right]\delta\rho(r). \tag{A.41}$$

The corresponding local potential can be expressed as (Martin 2004)

$$V_{xc}(\boldsymbol{r}) = \left[\varepsilon_{xc} + \rho\frac{\partial\varepsilon_{xc}}{\partial\rho} - \sum_{\alpha}\nabla_{\alpha}\left(\rho\frac{\partial\varepsilon_{xc}}{\partial\nabla_{\alpha}\rho}\right)\right]. \tag{A.42}$$

One of the most popular GGA schemes is due to Perdew, Burke and Ernzerhof (1996). Their expression, routinely referred to as the PBE expression, for the exchange part of the energy is

$$E_x^{GGA-PBE} = \int d\boldsymbol{r}\,\rho(r)\varepsilon_x[\rho(r)]F_x(s), \tag{A.43}$$

where, in the Hartree atomic units,

$$F_x = 1 + \kappa - \frac{\kappa}{1 + \frac{\eta s^2}{\kappa}}, \qquad s = \frac{|\nabla\rho(r)|}{2k_F\rho(r)} \tag{A.44}$$

with κ=0.804 and η=0.21951. The correlation energy expression is

$$E_c^{GGA-PBE} = \int d\boldsymbol{r}\,\rho[\varepsilon_c(r_s) + H(r_s,t)], \tag{A.45}$$

where, in the Hartree atomic units,

$$H = \gamma\ln\left\{1 + \frac{\beta}{\gamma}t^2\left[\frac{1 + At^2}{1 + At^2 + A^2t^4}\right]\right\},$$
$$A = \frac{\beta}{\gamma}\left[\exp\left\{-\frac{\varepsilon_c}{\gamma}\right\} - 1\right]^{-1}$$
$$t = \frac{|\nabla\rho(r)|}{2k_s\rho(r)}, \qquad k_s = \left(\frac{4k_F}{\pi}\right)^{\frac{1}{2}} \tag{A.46}$$

with γ=0.031091 and β=0.066725.

B The Pseudopotential Method

The fundamental concepts of the pseudopotential method are well described in the book by Harrison (1966) and in the research article by Cohen and Heine (1970). Here we present a brief discussion of some of the points of the method. Let us consider a solid as a collection of ion cores and valence electrons, with the valence electron wavefunctions orthogonal to core electron wavefunctions. In the pseudopotential approach it is assumed that ion cores do not play any role in determining electronic properties of molecules or solids. This means that both ionic core states and the strong potentials responsible for binding them can be dropped out of consideration in electronic structure calculations.

To understand this consider the one-electron Schrödinger equation

$$\hat{H}\boldsymbol{\psi} = \varepsilon\boldsymbol{\psi}, \tag{B.1}$$

where the Hamiltonian H is the sum of the kinetic energy T and an effective potential V_A. Let us expand the actual electronic wavefunction ψ as

$$\boldsymbol{\psi} = \boldsymbol{\phi} + \sum_c b_c \boldsymbol{\phi}_c, \tag{B.2}$$

where $\boldsymbol{\phi}$ is a smooth part, $\boldsymbol{\phi}_c$ is a core function corresponding to one of the bound states in the ion core, and b_c is determined from the condition that $\boldsymbol{\psi}$ and $\boldsymbol{\phi}_c$ are orthogonal to each other

$$\langle \psi \mid \phi_c \rangle = 0. \tag{B.3}$$

With the help of equations (B.2)–(B.3), equation (B.1) can be manipulated to take the form

$$\hat{H}\boldsymbol{\phi} + \sum_c (\varepsilon - E_c)|\phi_c\rangle\langle\phi_c|\boldsymbol{\phi} = \varepsilon\boldsymbol{\phi}, \tag{B.4}$$

where E_c is a core-state eigenvalue. This equation can be expressed as

$$(\hat{H} + V_R)\boldsymbol{\phi} = \varepsilon\boldsymbol{\phi} \tag{B.5}$$

or

$$(T + V_{\mathrm{ps}})\boldsymbol{\phi} = \varepsilon\boldsymbol{\phi}, \tag{B.6}$$

where V_R, defined by equation (B.4), is a repulsive potential operator. Phillips and Kleinman (1959), and Antoncik (1959) showed that the operator

$$V_{\mathrm{ps}} = V_A + V_R \tag{B.7}$$

represents a weakly attractive potential as a consequence of the cancellation between the attractive term V_A and the repulsive term V_R. Cohen and Heine (1970) showed that the cancellation between V_A and V_R can be almost complete in the core region. This is known as the 'cancellation theorem'. We term V_{ps} a pseudopotential[1] and the smooth wavefunction ϕ a pseudo-wavefunction.

---

[1] The use of the word 'pseudopotential' in the literature is rather unfortunate. It should be referred to as 'pseudopotential energy'.

DOI: 10.1201/9781003141273-B

There are many alternative prescriptions for defining a weak equivalent pseudopotential. Several categories of pseudopotentials have been applied to the study of electronic structure and ground-state properties of solids: (i) model pseudopotentials (Abarenkov and Heine 1965); (ii) empirical pseudopotential (Cohen and Bergstresser 1966, Chelikowsky and Cohen 1976, Schlüter *et al* 1975, Chelikowsky and Cohen 1979); and (iii) first-principles pseudopotentials (Hamann *et al* 1979, Kerker 1980, Bachelet *et al* 1982, Hamann 1989, Blöchl 1990, Vanderbilt 1990, Troullier and Martins 1991). The state-of-art first-principles pseudopotentials can be constructed to maintain 'norm-conservation' and 'transferability' from one atomic configuration to another, and thus can be used to calculate electronic and ground-state properties of atoms, molecules, or solids (Bachelet *et al* 1982, Troullier and Martins 1991). First-principles schemes of generating 'ultra-soft psudopotentials' (Blöchl 1990, Vanderbilt 1990) do not maintain the requirement of norm conservation but generate pseudofunctions that are much 'smoother' than the norm-conserving schemes provide.

Here we will discuss some aspects of the pseudopotential theory in relation to the Kohn-Sham equation in momentum space, using a basis set of plane waves, as discussed in section 3.2.3. In general a pseudopotential is 'non-local', i.e. it is not simply a function $|r|$ but can be represented as an angular-momentum-dependent potential. From equation (3.23) we write, for an electron wavevector k within the Brillouin zone and any two reciprocal translation vectors G and G',

$$
\begin{aligned}
V_{KS}(k+G, k+G') &= V_{ps}(k+G, k+G') + V_H(G'-G) + V_{xc}(G'-G) \\
&= V_{ps}(k+G, k+G') + V scr(G'-G),
\end{aligned} \tag{B.8}
$$

where $V_{ps}(k+G, k+G')$ and $V_{scr}(G)$ are Fourier components of the crystalline ionic pseudopotential and the screening potential, respectively. Following equation (3.18) we then write a Fourier component of the 'screend pseudopotential' $V(k+G, k+G')$ as

$$
\begin{aligned}
V(k+G, k+G') &= \frac{1}{M} \sum_b S_b(G-G')[v_b(k+G, k+G') + v_{scr,b}(G'-G)] \\
&= \frac{1}{M} \sum_b S_b(G-G') v_b^{at}(k+G, k+G'),
\end{aligned} \tag{B.9}
$$

where M is the number of atoms per unit cell, the G are reciprocal-lattice vectors, $S_b(G)$ are the structure factors, and $v_b(k+G, k+G')$, $v_{scr,b}(G)$ and $v_b^{at}(k+G, k+G')$ are the form factors of the ionic (i.e. unscreened) pseudopotential, the screening potential and the atomic (i.e. screened) pseudopotential, respectively, for the bth atom.

Model pseudopotential method

The concept of model pseudopotentials is well described by Abarenkov and Heine (1965). A simple model pseudopotential is the *empty core pseudopotential*. In real space the empty-core ionic pseudopotential is considered in the form

$$
v_{ion}(r) = \begin{cases} 0 & \text{for } r \le r_c \\ -\frac{ze^2}{r} & \text{for } r > r_c \end{cases}, \tag{B.10}
$$

where r_c is some radius around the nucleus of charge ze. Using the Thomas-Fermi screening model, appropriate for simple metals, the atomic (*i.e.* screened) potential can be expressed as

$$
v_{at}(r) = \begin{cases} 0 & \text{for } r \le r_c \\ -ze^2 \frac{e^{-k_s r}}{r} & \text{for } r > r_c \end{cases}, \tag{B.11}
$$

where k_s is the Thomas-Fermi screening length

$$
k_s = 6\pi z e^2 / E_F \tag{B.12}
$$

and E_F is the Fermi energy. The Fourier transform of the atomic potential is

$$v_{at}(G) = \frac{1}{\Omega_{at}} \frac{-4\pi z e^2}{G^2 + k_s^2} \cos(Gr_c), \tag{B.13}$$

where Ω_{at} is the the atomic volume. Note that at $G = 0$ the atomic pseudopotential approaches the screening limit $-\frac{2}{3}E_F$ for metals.

Empirical pseudopotential method
The earliest application of the pseudopotential method for the electronic band structure of semi-conductors was made by Cohen and co-workers in 1960s and 1970s (Cohen and Bergstresser 1966, Chelikowsky and Cohen 1976). In their approach, known as the 'empirical pseuodpoten-tial method' (EPM) the atomic (*i.e.* screened) pseudopoential form factors $v_b^{at}(\boldsymbol{k} + \boldsymbol{G}, \boldsymbol{k} + \boldsymbol{G}')$ are chosen as adjustable parameters to obtain experimentally measured electronic band gap values for semiconductors of diamond and zincblend structures. Both 'local' and 'non-local' types of form factors, respectively $v_b^{at}(|\boldsymbol{G}|)$ (Cohen and Bergstresser 1966) and angular momentum dependent $v_{b,\ell}^{at}(|\boldsymbol{k} + \boldsymbol{G}, \boldsymbol{k} + \boldsymbol{G}'|)$ (Chelikowsky and Cohen 1976) were employed. Values of such form factors are only required for $\boldsymbol{G}$ vectors for which the structure factor $S_b(\boldsymbol{G})$ is non-zero.

For extending the electronic structure calculations using large unit cells, such as an artifically periodic structure to model surfaces, it was found convenient to express the atomic and ionic pseu-dopotentials in analytic forms (Schlüter *et al* 1975, Chelikowsky and Cohen 1979). The atomic form factors were expressed as

$$v_{at}(G) = a_a \frac{G^2 - a_2}{1 + e^{a_3(G^2 - a_4)}}, \tag{B.14}$$

where a_i are adjustable parameters. The ionic form factors were expressed as

$$v_{ion}(G) = \frac{b_1}{G^2}[\cos(b_2 G) + b_3]e^{-b_4 G^4}, \tag{B.15}$$

where b_i are adjustable parameters. With $v_{at}(G)$ and v_{ion} being available, a self-consistent band structure calculation can be performed by starting with $v_{at}(G)$ at iteration zero and combining it with v_{ion} and v_{scr} made from the resulting electronic wavefunction at higher iteration steps. The first example of this procedure appeared in the work of Schlüter *et al* (1975) for the electronic band calculations of the unconstructed (1×1) and reconstructed (2×1) model structures of Si(111) surfaces.

Ab-initio pseudopotential method
Angular-momentum-dependent pseudopotential for the bth ion in the primitive unit cell of a solid can be expressed as

$$v_b(\boldsymbol{r}) = \sum_\ell \mathscr{P}_\ell v_{b,\ell}(|\boldsymbol{r}|)\mathscr{P}_\ell, \tag{B.16}$$

where $\mathscr{P}_\ell$ is the projection operator for angular momentum ℓ. The ionic pseudopotential $v_b(\boldsymbol{r})$ can be decomposed into a local part and a non-local correction part:

$$\begin{aligned} v_b(\boldsymbol{r}) &= v_b(|\boldsymbol{r}|) + \sum_\ell \mathscr{P}_\ell \big(v_{b,\ell}(|\boldsymbol{r}|) - v_b(|\boldsymbol{r}|)\big)\mathscr{P}_\ell \\ &= v_b^L(|\boldsymbol{r}|) + \Delta v_b^{NL}(\boldsymbol{r}). \end{aligned} \tag{B.17}$$

The local part v^L has the $-z_b e^2/r$ Coulomb tail but does not have the $1/r$ singularity at the origin, and the non-local correction part Δv is a short-ranged function of $|\boldsymbol{r}|$. It is usual to consider only $\ell = 0, 1$ and 2 in the definition of $v_b(\boldsymbol{r})$.

The total crystal pseudopotential can be constructed as a sum of non-overlapping ionic pseudopotentials

$$V_{ps}(\boldsymbol{r}) = \sum_{\boldsymbol{p}}\sum_{b} v_b(\boldsymbol{r}-\boldsymbol{p}-\boldsymbol{\tau}_b),\tag{B.18}$$

where $\boldsymbol{p}$ is a Bravais lattice vector, and $\boldsymbol{\tau}_b$ is the position vector of the bth ion in the primitive unit cell. A Fourier component of $V_{ps}(\boldsymbol{r})$ is given by

$$
\begin{aligned}
V_{ps}(\boldsymbol{k}+\boldsymbol{G},\boldsymbol{k}+\boldsymbol{G}') &= \frac{1}{N_0\Omega}\int d\boldsymbol{r}\, e^{-i(\boldsymbol{k}+\boldsymbol{G}')\cdot\boldsymbol{r}} V_{ps}(\boldsymbol{r}) e^{i(\boldsymbol{k}+\boldsymbol{G})\cdot\boldsymbol{r}}\\
&= \frac{1}{N_0\Omega}\sum_{\boldsymbol{p}}\sum_{b}\int d\boldsymbol{r}'\, e^{-i(\boldsymbol{k}+\boldsymbol{G}')\cdot(\boldsymbol{r}'+\boldsymbol{p}+\boldsymbol{\tau}_b)} v_b(\boldsymbol{r}')\\
&\quad \times e^{i(\boldsymbol{k}+\boldsymbol{G})\cdot(\boldsymbol{r}'+\boldsymbol{p}+\boldsymbol{\tau}_b)}\\
&= \left(\frac{1}{N_0}\sum_{\boldsymbol{p}} e^{i(\boldsymbol{G}-\boldsymbol{G}')\cdot\boldsymbol{p}}\right)\frac{1}{\Omega}e^{i(\boldsymbol{G}-\boldsymbol{G}')\cdot\boldsymbol{\tau}_b}\\
&\quad \times \int d\boldsymbol{r}\, e^{-i(\boldsymbol{k}+\boldsymbol{G}')\cdot\boldsymbol{r}} v_b(\boldsymbol{r}) e^{i(\boldsymbol{k}+\boldsymbol{G})\cdot\boldsymbol{r}}\\
&= \frac{1}{M}\sum_{b} e^{i(\boldsymbol{G}-\boldsymbol{G}')\cdot\boldsymbol{\tau}_b}\\
&\quad \times \left(\frac{1}{\Omega_{at}}\int d\boldsymbol{r}\, e^{-i(\boldsymbol{k}+\boldsymbol{G}')\cdot\boldsymbol{r}} v_b(\boldsymbol{r}) e^{i(\boldsymbol{k}+\boldsymbol{G})\cdot\boldsymbol{r}}\right)\\
&= \frac{1}{M}\sum_{b} S_b(\boldsymbol{G}-\boldsymbol{G}') v_b(\boldsymbol{k}+\boldsymbol{G},\boldsymbol{k}+\boldsymbol{G}').\tag{B.19}
\end{aligned}
$$

In deriving the above steps we have substituted $\boldsymbol{r}' = \boldsymbol{r}-\boldsymbol{p}-\boldsymbol{\tau}_b$. Also, here M is the number of basis ions in the primitive unit cell, N_0 is the number of unit cells, Ω_{at} is the atomic volume, and $S_b(\boldsymbol{q})$ is the structure factor for ion b. The ionic form factor $w_b(\boldsymbol{q},\boldsymbol{q}')$ can be expressed, following equation (B.17), as

$$v_b(\boldsymbol{q},\boldsymbol{q}') = v_b^L(|\boldsymbol{q}-\boldsymbol{q}'|) + \sum_{\ell}\Delta v_{b,\ell}^{NL}(\boldsymbol{q},\boldsymbol{q}')\tag{B.20}$$

with the local part as

$$v_b^L(|\boldsymbol{G}|) = \frac{1}{\Omega_{at}}\int d\boldsymbol{r}\, e^{-i\boldsymbol{G}\cdot\boldsymbol{r}} v_b^L(|\boldsymbol{r}|).\tag{B.21}$$

The Fourier transform of a non-local correction term is

$$\Delta v_{b,\ell}^{NL}(\boldsymbol{q},\boldsymbol{q}') = \frac{1}{\Omega_{at}}\int d\boldsymbol{r}\, e^{-i\boldsymbol{q}'\cdot\boldsymbol{r}}\mathscr{P}_\ell \Delta v_{b,\ell}(|\boldsymbol{r}|)\mathscr{P}_\ell e^{i\boldsymbol{q}\cdot\boldsymbol{r}}\tag{B.22}$$

Using the identity

$$e^{i\boldsymbol{q}\cdot\boldsymbol{r}} = 4\pi\sum_{\ell_1 m_1}(i)^{\ell_1} j_{\ell_1}(qr) Y_{\ell_1 m_1}(\hat{\boldsymbol{r}}) Y_{\ell_1 m_1}^*(\hat{\boldsymbol{q}})\tag{B.23}$$

in terms of spherical Bessel functions $j_\ell(x)$ and spherical harmonics $Y_{\ell m}$, and noting that the projection operator $\mathscr{P}_\ell$ picks out only the $\ell_1 = \ell$ component, we can express

$$\Delta v_b^{NL}(\boldsymbol{q},\boldsymbol{q}') = \frac{4\pi}{\Omega_{at}}\sum_{\ell}(2\ell+1)\Delta v_{b,\ell}(|\boldsymbol{q}|,|\boldsymbol{q}'|) P_\ell(\hat{\boldsymbol{q}}\cdot\hat{\boldsymbol{q}}'),\tag{B.24}$$

where the addition theorem

$$\sum_{m} Y_{\ell m}(\hat{\boldsymbol{q}}') Y_{\ell m}^*(\hat{\boldsymbol{q}}) = \frac{2\ell+1}{4\pi} P_\ell(\hat{\boldsymbol{q}}\cdot\hat{\boldsymbol{q}}')\tag{B.25}$$

has been used. The P_ℓ are the Legendre polynomials, and $\Delta w_{b,\ell}(q,q')$ are the matrix elements between the spherical Bessel functions

$$\Delta v_{b,\ell}(q,q') = \int_0^\infty dr \ r^2 \Delta v_{b,\ell}(|\mathbf{r}|) j_\ell(qr) j_\ell(q'r) \tag{B.26}$$

which, depending on the form of $\Delta v_{b,\ell}(|\mathbf{r}|)$, can be evaluated either analytically or numerically.

If the decomposition in equation (B.17) is not preferred, then a Fourier component $v_b(\mathbf{q},\mathbf{q}')$ of the total non-local pseudopotential can be directly expressed in the form of equations (B.24)–(B.26) after replacing $\Delta v_{b,\ell}$ by $v_{b,\ell}$.

Bachelet et al (1982) have generated a consistent set of 'norm-conserving' non-local ionic pseudopotentials for the entire periodic table. They have fitted their pseudopotentials to an analytical form in r-space which can conveniently be used in calculations. In particular, they have decomposed $w_{b,\ell}(r)$ into a long-range Coulomb part (ℓ-independent) and a short-range ℓ-dependent part:

$$v_{b,\ell}(r) = v_{b,core}(r) + \Delta v_{b,\ell}(r). \tag{B.27}$$

The core potential is considered as originating from Gaussian-type effective charges

$$v_{b,core}(r) = -\frac{z_b e^2}{r} \sum_{i=1}^2 c_i^{core} \text{erf}\left(\frac{\sqrt{\alpha_i^{core}}}{r}\right), \tag{B.28}$$

where $z_b e$ denotes the valence charge. The remaining term, $\Delta v_{b,\ell}(r)$, is expanded in Gaussian-type functions

$$\Delta v_{b,\ell}(r) = \sum_{i=1}^3 (A_i + r^2 A_{i+3}) e^{-\alpha_i r^2}. \tag{B.29}$$

For easy tabulation and reproduction, Bachelet et al transformed the coefficients A_i into a set of coefficients C_i. There are various ways of obtaining A_i from the set $\{C_i\}$. One approach is described in Srivastava (1999).

Hartwigsen et al (1998) have presented analytic expressions for norm-conserving relativistic LDA pseudopotentials for all elements from H to Rn. Their pseudopotentials require significantly fewer parameters that those tabulated by Bachelet et al (1982).

Another popular method of generating norm-conserving pseudopotentials is due to Troullier and Martins (1991). This method generates softer, smoother and fully non-local (both in radial and angular coordinates) pseudopotentials.

For further discussion of the pseudopotential concept and methods, including ultrasoft pseudopotentials and projected augmented waves (PAWs), we refer the reader to the book by Martin (2004).

C The $2n+1$ Theorem of Perturbation Theory

In section 4.9.2 of Chapter 4, the *ab initio* third-order density functional perturbation theory (third-order DFPT) was discussed for calculating third-order anharmonic force constants. That formalism is based on the use of the so-called $2n+1$ theorem of the perturbation theory. In that section we provided several useful references for the $2n+1$ theorem. In this Appendix we provide a proof of this theorem for a single particle system. Our approach follows Dalgarno and Stewart (1956), and Bransden and Joachain (2000).

Let us consider the one-electron Schrödinger equation for an unperturbed and non-degenerate system

$$\hat{H}_0 \phi_j^{(0)} = \varepsilon_j^{(0)} \phi_j^{(0)}, \tag{C.1}$$

where the Hamiltonian H_0 is the sum of the kinetic energy T and an effective potential V, $\varepsilon_n^{(0)}$ is the eigenvalue and $\phi_j^{(0)}$ is the orthonormal eigenfunction of the jth state.

Let us now consider the eigenvalue problem

$$(\hat{H}_0 + \lambda V')\phi_j = \varepsilon_j \phi_j \tag{C.2}$$

for the system when it has been perturbed by a small amount of potential $\lambda V'$. Here λ is a real parameter which will be used to distinguish between different orders of the perturbation.

Following the time-independent non-degerate perturbation theory, we expand the eigenvalues ε_j and the eigenfunction $\boldsymbol{\phi}_j$ in powers of the parameter λ:

$$\varepsilon_j = \sum_{r=0}^{\infty} \lambda^r \varepsilon_j^{(r)} \tag{C.3}$$

$$\phi_j = \sum_{r=0}^{\infty} \lambda^r \phi_j^{(r)}. \tag{C.4}$$

Equation (C.2), with these substitutions, now reads

$$(\hat{H}_0 + \lambda V') \sum_{r=0}^{\infty} \lambda^r \varepsilon_j^{(r)} = \sum_{r=0}^{\infty} \lambda^r \varepsilon_j^{(r)} \sum_{s=0}^{\infty} \lambda^s \varepsilon_j^{(s)}. \tag{C.5}$$

Equating terms of equal powers of λ from both sides of this equation, we obtain the following set of equations

$$\hat{H}_0 \phi_j^{(0)} = \varepsilon_j^{(0)} \phi_j^{(0)} \tag{C.6}$$

$$\hat{H}_0 \phi_j^{(1)} + V' \phi_j^{(0)} = \varepsilon_j^{(0)} \phi_j^{(1)} + \varepsilon_j^{(1)} \phi_j^{(0)} \tag{C.7}$$

$$\hat{H}_0 \phi_j^{(2)} + V' \phi_j^{(1)} = \varepsilon_j^{(0)} \phi_j^{(2)} + \varepsilon_j^{(1)} \phi_j^{(1)} + \varepsilon_j^{(2)} \phi_j^{(0)} \tag{C.8}$$

$$\cdots\cdots\cdots\cdots = \cdots\cdots\cdots\cdots$$
$$\hat{H}_0 \phi_j^{(r)} + V' \phi_j^{(r-1)} = \varepsilon_j^{(0)} \phi_j^{(r)} + \varepsilon_j^{(1)} \phi_j^{(r-1)} + \varepsilon_j^{(2)} \phi_j^{(r-2)} + \ldots + \varepsilon_j^{(r)} \phi_j^{(0)}. \tag{C.9}$$

DOI: 10.1201/9781003141273-C

Premultiplying each of the equations (C.6)–(C.8) and making use of the orthonormality relation $< \phi_i | \phi_j >= \delta_{ij}$ we obtain expressions for the orders $r = 1, 2, 3$ of the energy correction

$$\varepsilon_j^{(1)} = < \phi_j^{(0)} | H_0 | \phi_j^{(0)} > \tag{C.10}$$

$$\varepsilon_j^{(2)} = < \phi_j^{(0)} | V - \varepsilon_j^{(1)} | \phi_j^{(1)} > . \tag{C.11}$$

$$\varepsilon_j^{(3)} = < \phi_j^{(0)} | V - \varepsilon_j^{(1)} | \phi_j^{(2)} > - \varepsilon_j^{(2)} < \phi_i^{(0)} | \phi_j^{(1)} > . \tag{C.12}$$

Taking the inner product of equation (C.7) with $\phi_j^{(2)}$ and the inner product of equation (C.8) with $\phi_j^{(1)}$, and taking the difference of these two we obtain

$$< \phi_j^{(0)} | V - \varepsilon_j^{(1)} | \phi_j^{(2)} > = < \phi_j^{(1)} | V - \varepsilon_j^{(1)} | \phi_j^{(1)} > - \varepsilon_j^{(2)} < \phi_j^{(0)} | \phi_j^{(1)} > . \tag{C.13}$$

Substitute equation (C.13) in equation (C.12) to obtain

$$\varepsilon_j^{(3)} = < \phi_j^{(1)} | V - \varepsilon_j^{(1)} | \phi_j^{(1)} > - 2\varepsilon_j^{(2)} < \phi_i^{(0)} | \phi_j^{(1)} > . \tag{C.14}$$

From equations (C.10), (C.11) and (C.14), we note that while the first-order correction energy $\varepsilon_j^{(1)}$ requires knowledge of the unperturbed wave-function $\phi_j^{(0)}$, the second- and third-order energy correction terms $\varepsilon_j^{(2)}$ and $\varepsilon_j^{(3)}$ require knowledge of the unperturbed wave-function $\phi_j^{(0)}$ and the first-order correction to the wave-function $\phi_j^{(1)}$. Proceeding in a similar manner we can show that the fourth-order and fifth-order energy corrections $\varepsilon_j^{(4)}$ and $\varepsilon_j^{(5)}$ require knowledge of $\phi_j^{(0)}$, $\phi_j^{(1)}$ and $\phi_j^{(2)}$.

A generalisation of equations (C.12) and (C.14) is

$$\varepsilon_j^{(n)} = < \phi_j^{(0)} | (V - \varepsilon_j^{(1)}) | \phi_j^{(n-1)} > - \sum_{r=1}^{n-2} \varepsilon_j^{(r+1)} < \phi_i^{(0)} | \phi_j^{(n-r-1)} > \quad (n > 1). \tag{C.15}$$

Dalgarno and Stewart (1956) derived a further generalisation of the nth-order energy correction, as follows

$$\begin{aligned}
\varepsilon_j^{(n)} = {} & < \phi_j^{(2)} | (V - \varepsilon_j^{(1)}) | \phi_j^{(n-3)} > \\
& - \sum_{r=3}^{n-1} \varepsilon_j^{(r)} \{ \{ < \phi_i^{(0)} | \phi_j^{(n-r)} > + < \phi_i^{(1)} | \phi_j^{(n-r-1)} > + < \phi_i^{(2)} | \phi_j^{(n-r-2)} > \} \\
& - \varepsilon_j^{(2)} \{ < \phi_i^{(1)} | \phi_j^{(n-3)} > + < \phi_i^{(2)} | \phi_j^{(n-4)} > \}.
\end{aligned} \tag{C.16}$$

The above process can be continued until the expression for $\varepsilon_j^{(n)}$ contains the highest order wave-function of the $(\frac{1}{2}n)$th order if n is even or of the $\frac{1}{2}(n-1)$th order if n is odd.

From the results presented above we can make the statement that if the wavefunctions $\phi_j^{(0)}, \phi_j^{(1)}, ... \phi_j^{(n)}$ are known, then the energy $\varepsilon_j^{(2n+1)}$ can be obtained. In other words, evaluation of the $(2n+1)$th order energy only requires the wavefunction up to the nth order. This is the essence of the $(2n+1)$ order perturbation theory.

D Derivation of Tensor Expression for Thermal Conductivity

The Fourier heat current equation in (5.19) is

$$\boldsymbol{Q} = -\mathscr{K}\boldsymbol{\nabla}T. \tag{D.1}$$

Regarding the vectors $\boldsymbol{Q}$ and $\boldsymbol{\nabla}T$ as (3×1) matrices, and the transpose vector $(\boldsymbol{\nabla}T)^T$ as a (1×3) matrix, we express the above equation as

$$
\begin{aligned}
\boldsymbol{Q}(\boldsymbol{\nabla}T)^T &= -\mathscr{K}\boldsymbol{\nabla}T(\boldsymbol{\nabla}T)^T \\
\boldsymbol{Q}(\boldsymbol{\nabla}T)^T[\boldsymbol{\nabla}T(\boldsymbol{\nabla}T)^T]^{-1} &= -\mathscr{K}\boldsymbol{\nabla}T(\boldsymbol{\nabla}T)^T[\boldsymbol{\nabla}T(\boldsymbol{\nabla}T)^T]^{-1} \\
\boldsymbol{Q}(\boldsymbol{\nabla}T)^T(\boldsymbol{\nabla}T\cdot\boldsymbol{\nabla}T)^{-1} &= -\mathscr{K}[\boldsymbol{\nabla}T(\boldsymbol{\nabla}T)^T][\boldsymbol{\nabla}T(\boldsymbol{\nabla}T)^T]^{-1} \\
\frac{\boldsymbol{Q}(\boldsymbol{\nabla}T)^T}{|\boldsymbol{\nabla}T|^2} &= -\mathscr{K}\,\mathbb{I} \\
\frac{\boldsymbol{Q}(\boldsymbol{\nabla}T)^T}{|\boldsymbol{\nabla}T|^2} &= -\mathscr{K},
\end{aligned}
\tag{D.2}
$$

where $\mathbb{I}$ is the (3×3) unit matrix, and we have used the results $\boldsymbol{A}^T\boldsymbol{B} = \boldsymbol{A}\cdot\boldsymbol{B}$ and $(\boldsymbol{\nabla}T)(\boldsymbol{\nabla}T)^T = ((\boldsymbol{\nabla}T)^T)^T(\boldsymbol{\nabla}T)^T = (\boldsymbol{\nabla}T)^T\cdot(\boldsymbol{\nabla}T)^T = |\nabla T|^2$. Equation (D.2) can be expressed in terms of the conductivity tensor elements as

$$\mathscr{K}_{ij} = -\frac{1}{|\boldsymbol{\nabla}T|^2}Q_i\nabla_j T \qquad i,j = x,y,z, \tag{D.3}$$

which is equation (5.21). Here Q_i is the ith row element of the column matrix $\boldsymbol{Q}$ and ∇_j is the jth column element of the row matrix $(\boldsymbol{\nabla}T)^T$. The isotropic reduction of the above equation is

$$\mathscr{K} = -\frac{1}{|\boldsymbol{\nabla}T|^2}\boldsymbol{Q}\cdot\boldsymbol{\nabla}T, \tag{D.4}$$

which is equation (5.23).

E Evaluation of Integrals in Section 6.4.1.4

Consider the integral

$$I = \int_0^\pi d\theta' \sin \theta' x x' x'' \delta\{x'' - (Cx \pm Dx')\}. \tag{E.1}$$

Let

$$\Delta_\pm = x'' - (Cx \pm Dx') \tag{E.2}$$

$$= (1 - \varepsilon) + \varepsilon|x \pm x'| - (Cx \pm Dx'). \tag{E.3}$$

The $+(-)$ sign corresponds to three-phonon class 1(2) events. From this

$$\varepsilon|x \pm x'| \, d\Delta_+ = \mp \varepsilon x x' \sin \theta' d\theta', \tag{E.4}$$

where $\cos \theta' = \hat{x}.\hat{x}'$. The integral I can now be expressed as

$$I = \int_{\Delta_1}^{\Delta_2} d\Delta_\pm \, x''(\mp \varepsilon|x \pm x'|) \delta(\Delta_\pm), \tag{E.5}$$

where

$$\Delta_1 \equiv \Delta_\pm(\theta' = 0) = (1 - \varepsilon) + \varepsilon|x - x'| - (Cx \pm Dx')$$
$$\Delta_2 \equiv \Delta_\pm(\theta' = \pi) = (1 - \varepsilon) + \varepsilon|x \mp x'| - (Cx \pm Dx').$$

It can be verified that $\Delta_2 > \Delta_1$ for class 1 U-processes and class 2 N-processes, and $\Delta_1 > \Delta_2$ for class 1 N-processes and class 2 U-processes. Using the property of the Dirac delta function the integral I can be evaluated: the result is

$$I = (Cx \pm Dx')\{1 - \varepsilon + \varepsilon(Cx \pm Dx')\}, \tag{E.6}$$

where the first factor comes from setting $\Delta_\pm = 0$ in equation (E.2) and the second factor comes from setting $\Delta_\pm = 0$ in equation (E.3).

F Negative-definiteness of the Phonon Off-diagonal Operator Λ

In the case of three-phonon interaction, we can express, from equations (6.52)–(6.54),

$$(P\boldsymbol{\phi})_q = \Gamma_q \phi_q + \sum_{q' \neq q} \Lambda_{qq'} \phi_{q'}, \tag{F.1}$$

where $\boldsymbol{\phi}$ belongs to a subspace of odd distribution functions and $q \equiv qs$. The diagonal part Γ of P is positive-definite and possesses an inverse. The off-diagonal part can be expressed as

$$(\Lambda\boldsymbol{\phi})_q = \sum_{q' \neq q, q'' \neq q} \left[\bar{P}_{qq'}^{q''} (\boldsymbol{\phi}_{q'} - \boldsymbol{\phi}_{q''}) - \frac{1}{2} \bar{P}_q^{q'q''} (\boldsymbol{\phi}_{q'} + \boldsymbol{\phi}_{q''}) \right]. \tag{F.2}$$

In general it is not easy to prove the positive- or negative-definite property of Λ (Simons 1972). However, in the case of our assumed trial function $\phi_q = \boldsymbol{q} \cdot \boldsymbol{u}$ (equation (5.90)), it can be shown that

$$(\boldsymbol{\phi}, \Lambda\boldsymbol{\phi}) = -\sum_{ss's''} \left[J_{\substack{101\\101\\ss''s'}} + J_{\substack{011\\011\\s''s's}} - J_{\substack{110\\110\\s'ss''}} \right], \tag{F.3}$$

where we can express, following the scheme and notation in section 6.4.1.4 (also see Hamilton and Parrott (1969), Srivastava (1977d) and Mikhail (1985)),

$$J_{\substack{tt't''\\mm'm''\\ss's''}} = \text{constant} \sum_{\varepsilon} \int dx \int dx' x^{t+2} x'^{t''+2} (Cx + Dx')^{t''+1}$$
$$\times \bar{n}\bar{n}'(\bar{n}_1'' + 1) \varepsilon^{m''} G^{mm'm''} \tag{F.4}$$

with

$$
\begin{aligned}
G^{110} &= \mu'[1 - \varepsilon + \varepsilon(Cx + Dx')] \\
G^{101} &= (x + x'\mu') \\
G^{011} &= (x\mu' + x')
\end{aligned}
$$

$$x = q/q_D \quad C = c_s/c_{s''} \quad D = c_{s'}/c_{s''} \quad \mu' = \cos\theta' = \hat{q} \cdot \hat{q}' \tag{F.5}$$

and $\varepsilon = +1(-1)$ for N (U) processes.

With the help of equation (F.4), we can express equation (F.3) as

$$(\boldsymbol{\phi}, \Lambda\boldsymbol{\phi}) = \sum_{\varepsilon} (\boldsymbol{\phi}, \Lambda\boldsymbol{\phi})_{\varepsilon} = \sum_{\varepsilon} \int \int D_{\varepsilon} Q_{\varepsilon} dx dx', \tag{F.6}$$

where D_{ε} is a positive quantity for both N and U processes. The quantity Q_{ε} which determines the sign of $(\boldsymbol{\phi}, \Lambda\boldsymbol{\phi})_{\varepsilon}$ can be worked out to be

$$
\begin{aligned}
Q_{\varepsilon} &= -\varepsilon x x'' G^{101} - \varepsilon x' x'' G^{011} + x x' G^{110} \\
&= -\varepsilon x''[x(x + x'\mu') + x'(x' + x\mu')] + x x' \mu'[1 - \varepsilon + \varepsilon x''], \tag{F.7}
\end{aligned}
$$

DOI: 10.1201/9781003141273-F

where $x'' = Cx + Dx'$.

For N processes with finite x and x'

$$Q_N = Q_{\varepsilon=1} = -x''[(x-x')^2 + (2+\mu')xx'] < 0. \tag{F.8}$$

Thus the operator Λ_N is *negative-definite*. For U processes

$$Q_U = Q_{\varepsilon=-1} = x''[(x-x')^2 + xx'(2+\mu')] + 2xx'\mu'. \tag{F.9}$$

Clearly Q_U is positive when μ' is positive. However, both positive and negative values of μ' are possible. The most negative value of μ' comes for the critical case in which $\boldsymbol{q}, \boldsymbol{q}'$, and $\boldsymbol{q} + \boldsymbol{q}'$ form an equilateral triangle (Mikhail 1985). In this case $\mu' = -1/2$. Furthermore, from equation (6.88) we note that for U processes $x' \geq [2 - (1+C)x]/(1+D)$, so that $x'' = Cx + Dx' \geq [(C-D)x + 2D]/(1+D)$. Assuming that $c_L \leq 2c_T$ for most solids, it can be shown that for all allowed U processes within the isotropic continuum model $x'' \geq 2r/(1+r) \geq 2/3$, where $r = c_T/c_L$. For $\mu' = -1/2$ and $x'' = 2/3$, $Q_U = \frac{2}{3}(x-x')^2 \geq 0$. Thus the operator Λ_U is *non-negative*. Therefore, the sign of Λ will be negative (positive) if N (U) processes dominate the three-phonon interaction.

The use of the trial function $\phi_q = \boldsymbol{q} \cdot \boldsymbol{u}$ is only justified when N processes dominate over U processes. Therefore, we conclude that in the development of the theory of the complementary variational principles in Chapters 5 and 7, where we have used the decomposition $P = P^* = \Gamma + \Lambda$, the off-diagonal part Λ is a negative-definite operator.

G Geometry-dependent Depolarization Tensor

In Chapter 9 we derived an expression for the effective thermal conductivity of a structure comprised of periodically positioned inserts in a matrix, with either or both of these being an anisotropic thermal conductivity material. In order to correctly evaluate the effective thermal conductivity of the insert and its surface together an affine transformation must be performed that restores the isotropy of the surface (Sihvola 1997a, 1997b, 1999, Levy and Cherkaev 2013, Kushch *et al* 2017, Thomas and Srivastava 2018a).

Consider an isotropic sphere inserted in an anisotropic medium. The thermal conductivity of the anisotropic medium can be diagonalized by fixing the Cartesian x, y, z coordinate system along the eigenaxes of the conductivity. The corresponding scalar potential obeys the equation $\nabla \cdot \kappa^m \cdot \nabla \phi(r) = 0$. This equation can be transformed into the familiar Laplace equation $\nabla'^2 \phi(r') = 0$ by the affine coordinate transformation $r' = \mathbb{N}_m r$, where $\mathbb{N}_m$ is an appropriate transformation matrix (a scale factor tensor or a structure-dependent depolarization tensor). The medium can now be treated as isotropic. But in the primed coordinate system the sphere will be transformed into an ellipsoid. However, the Laplace equation can be solved for the ellipsoid with the help of an 'depolarization' matrix $\mathbb{N}_s$ containing three depolarization factors as eigenvalues (Sihvola 1997a, 1997b, 1999). A similar consideration can be made even when the inserted sphere is anisotropic.

Consider the volume insertion factor $v = a^3/(a + \delta)^3 \simeq 1 - 3\delta/a$, where a is the radius of a spherical insert and δ is the thickness of the (thin) surface layer. Further consider a prolate spheroid with the z-semiaxis being larger than the x- and y-semiaxes: $a_{l,z} > a_{l,x} = a_{l,y}$, where $l = s, m$ for the surface and matrix, respectively. For the thermal conductivity problem, we have to define two affine transformations. Following Sihvola (1997a), Thomas and Srivastava (2018a) defined the insert+surface-related affine transformations as

$$a_{s,x} = \sqrt{\frac{\mathbb{R}^K_{xx}}{\mathbb{R}^K_{zz}}} a, \qquad a_{s,y} = \sqrt{\frac{\mathbb{R}^K_{yy}}{\mathbb{R}^K_{zz}}} a, \qquad a_{s,z} = a, \tag{G.1}$$

where $\mathbb{R}^K$ is the thermal boundary resistance tensor. The matrix related affine transformations are considered as

$$a_{m,x} = \sqrt{\frac{\kappa^s_{zz}}{\kappa^s_{xx}}} a, \qquad a_{m,y} = \sqrt{\frac{\kappa^s_{zz}}{\kappa^s_{yy}}} a, \qquad a_{m,z} = a, \tag{G.2}$$

where κ^s is the thermal conductivity tensor for the surface region. Assuming that both κ^s and κ^m are diagonal for the respective mixing steps, the depolarization tensors for the surface and matrix, $\mathbb{N}_s$ and $\mathbb{N}_m$, will be diagonal matrices with components (Sihvola 1999)

$$\mathbb{N}_{l,zz} = \frac{1 - e^2}{2e^3} \left(\ln \frac{1 + e}{1 - e} - 2e \right), \text{ where } e = \sqrt{1 - \frac{a^2_{l,x}}{a^2_{l,z}}}, \tag{G.3}$$

$$\mathbb{N}_{l,xx} = \mathbb{N}_{l,yy} = \frac{1}{2} \left(1 - \mathbb{N}_{l,zz} \right)$$

for $l = s, m$.

DOI: 10.1201/9781003141273-G

The anisotropic Kapitza resistance matrix is defined as

$$
\mathbb{R}^K =
\begin{pmatrix}
\lim\limits_{\substack{\delta \to 0 \\ \kappa^s_{xx} \to 0}} \left(\dfrac{\delta}{\kappa^s_{xx}} \right) & 0 & 0 \\[1em]
0 & \lim\limits_{\substack{\delta \to 0 \\ \kappa^s_{yy} \to 0}} \left(\dfrac{\delta}{\kappa^s_{yy}} \right) & 0 \\[1em]
0 & 0 & \lim\limits_{\substack{\delta \to 0 \\ \kappa^s_{zz} \to 0}} \left(\dfrac{\delta}{\kappa^s_{zz}} \right)
\end{pmatrix},
\tag{G.4}
$$

where we assume that the system is oriented such that κ^s is diagonal and take the limits $\delta \to 0$ and $\kappa^s \to 0$ such that

$$
\frac{\mathbb{R}^K_{xx}}{\mathbb{R}^K_{zz}} = \lim_{\substack{\kappa^s_{zz} \to 0 \\ \kappa^s_{xx} \to 0}} \left(\frac{\kappa^s_{zz}}{\kappa^s_{xx}} \right), \qquad
\frac{\mathbb{R}^K_{yy}}{\mathbb{R}^K_{zz}} = \lim_{\substack{\kappa^s_{zz} \to 0 \\ \kappa^s_{yy} \to 0}} \left(\frac{\kappa^s_{zz}}{\kappa^s_{yy}} \right).
\tag{G.5}
$$

To ensure that the effective medium approximation (EMA) remains valid, in the above we have invoked the reasonable assumption that the relative scale of boundary resistance in different directions resembles that of the different thermal conductivities in the boundary shell.

With these considerations, κ^* in equation (9.67) in Chapter 9 can be expressed as

$$
\begin{aligned}
\kappa^* &= \kappa^s + v(\kappa^i - \kappa^s)\mathbb{L}^{-1}, \\[0.5em]
&\simeq \kappa^i \left[\lim_{\substack{\delta \to 0 \\ \kappa^s \to 0}} \mathbb{L} \right]^{-1} \\[0.5em]
&= \kappa^i \left[\mathbb{I} + \frac{3}{a} \mathbb{N}_s \kappa^i \mathbb{R}^K \right]^{-1},
\end{aligned}
\tag{G.6}
$$

which is equation (9.69).

For isotropic spherical inserts in an isotropic matrix, $\mathbb{R}^K_{xx} = \mathbb{R}^K_{yy} = \mathbb{R}^K_{zz} = R_{TB}$, $\mathbb{N}_s = \mathbb{N}_m = \frac{1}{3}\mathbb{I}$, and $\kappa^l_{xx} = \kappa^l_{yy} = \kappa^l_{zz}$, with $l = i, s, m$.

References

Abraham B M *et al* 1969 *Phys. Rev.* **181** 347
Abarenkov I V and Heine V 1965 *Phil. Mag.* **12** 529
Abeles B 1963 *Phys. Rev.* **131** 1906
Abeles B *et al* 1962 *Phys. Rev.* **125** 44
Abdelouhab R M *et al* 1989 *J. Appl. Phys.* **66** 787
Adachi S 1983 *J. Appl. Phys.* **54** 1844
Adamenko I N and Fuks I M 1970 *Zh. Eksp. Teor. Fiz.* **59** 2071
Adamenko I N and Nemchenko E K 2013 *Fiz. Nizk. Temp.* bf 39 975
Alexander S and Orbach R 1982 *J. Physique Lett.* **43** L625
Alexander S *et al* 1983 *Phys. Rev.* B **28** 4615
——1986 *Phys. Rev.* B **34** 2726
Alexandrov A S and Devreese J T 2010 *Advances in Polaron Physics* (Springer, New York)
Anderson A C 1981 *Amorphous Solids: Low Temperature Properties* ed W A Phillips (Berlin: Springer) p 65
Anderson B R and Challis L J 1973 *J. Phys. C: Solid State Phys.* **6** 266
Anderson P W 1960 *Fizika Dielektrikov* ed G I Skanavi (Moscow: Akademia Nauk) p 290
——1969 *Phys. Lett.* **29A** 563
Anderson P W *et al* 1972 *Phil. Mag.* **25** 1
Allan D C and Mele E J 1984 *Phys. Rev. Lett.* **53** 826
Allen P B 2013 *Phys. Rev.* B **88** 144302
Alofi A and Srivastava G P 2013 *Phys. Rev.* B **87** 115421
AlShaikhi A and Srivastava G P 2008 *J. Phys. D: Appl. Phys.* **41** 185407
Anthony R *et al* 1990 *Phys. Rev.* B **42** 1104
Antoncik E 1959 *J. Phys. Chem. Solids* **10** 314
Armelles G *et al* 1988 *Solid State Commun.* **65** 779
Arrigoni M *et al* 2018 *Phys. Rev.* B **98** 115205
Arthurs A M 1970 *Complementary Variational Principles* (Oxford: Clarendon)
Asahi H *et al* 1989 *J. Appl. Phys.* **65** 5007
Asaumi K and Minomura S 1978 *J. Phys. Soc. Japan* **45** 1061
Ashcroft N W and Mermin N D 1976 *Solid State Physics* (Philadelphia: Saunders)
Bachelet G B *et al* 1982 *Phys. Rev.* B **26** 4199
Baden-Powell J 1841 *View of the Undulatory Theory as Applied to the Dispersion of Light* (Cambridge: Cambridge University Press)
Baldereschi A 1973 *Phys. Rev.* B **7** 5212
Balkanski M *et al* 1983 *Phys. Rev.* B **28** 1928
Banach Z and Larecki W 2005 *J. Phys. A: Math. Gen.* **38** 8781
Banerjee R and Varshni Y P 1969 *Can. J. Phys.* **47** 451
Bansil A *et al* 2016 *Rev. Mod. Phys.* **88** 021004
Bardeen J *et al* 1957 *Phys. Rev.* **108** 1175
Barman S and Srivastava G P 2004 *Phys. Rev.* B **69** 235208
——2006 *Phys Rev* B **73** 073301
Barnett S A *et al* 1982 *Electron. Lett.* **18** 891
Baroni S *et al* 1987 *Phys. Rev. Lett.* **58** 1861
Baroni S *et al* 2001 *Rev. Mod. Phys.* **73** 515

Barron T H K and Klein M L 1974 *Dynamical Properties of Solids* (edited by G. K. Horton and A. A. Maradudin, North-Holland, Amsterdam) Vol. I, p. 391

Bartels A *et al* 1999 *Phys. Rev. Lett.* **82** 1044

Baumgartner R *et al* 1981 *Phys. Rev. Lett.* **47** 1403

Beck H 1975 *Dynamical Properties of Solids* vol II, ed G K Horton and A A Maradudin (Amsterdam: North-Holland) ch 4, p 205

Beeman D and Alben R 1977 *Adv. Phys.* **26** 339

Behrang A *et al* 2013 *J. Appl. Phys.* **114** 014305

——2014 *Appl. Phys. Lett.* **104** 063106

——2015 *RSC. Adv.* **5** 2768

Bendt P and Zunger A 1983 *Phys. Rev. Lett.* **50** 1684

Benedict L X *et al* 1996 *Sol State Commun.* **100** 177

Benin D 1970 *Phys. Rev.* B **1** 2777

——1972 *Phys. Rev.* B **5** 2344

——1975 *Phys. Rev.* B **13** 1105

Berber S *et al* 2000 *Phys. Rev. Lett.* **84** 4613

Berberich P *et al* 1982 *Phys. Rev. Lett.* **49** 1500

Berman J L 1962 *Phys. Rev.* **127** 1093

——1963 *Phys. Rev.* **131** 1489

——1974 *Encyclopedia of Physics* vol XXV/2b, ed S Flügge (Berlin: Springer)

——1953 *Proc. R. Soc.* A **220** 171

Berman R *et al* 1955 *Proc. R. Soc.* A **231** 130

——1975 *J. Phys.* C **8** L430

Besserman R *et al* 1976 *Solid State Commun.* **20** 485

Bethe H 1929 *Ann. Phys., Lpz.* **3** 133

Bhagavantam S 1966 *Crystal Symmetry and Physical Properties* (New York: Academic)

Bhandari C M and Rowe D M 1977 *J. Phys. D: Appl. Phys.* **10** L59

Bilz H and Kress W 1979 *Phonon Dispersion Relations in Insulators* (Berlin: Springer)

Birch F 1947 *Phys. Rev.* **71** 809

Biswas R and Hamann D R 1987 *Phys. Rev.* B **36** 6434

Bitzer T *et al* 2001 *J. Chem. Phys.* B **105** 4535

Blackman M 1935 *Phil. Mag.* **19** 989

Blöchl P E 1990 *Phys. Rev.* B **41** 5414

Blöchl P E *et al* 1994 *Phys. Rev.* B **49** 16223

Boardman A D, O'Connor D E and Young P A 1973 *Symmetry and its Applications in Science* (New York: McGraw-Hill)

Bogoliubov N N 1947 *J. Phys. USSR* **11** 23

——1958 *J. Phys. USSR* **34** 41

Borca-Tasciuc T *et al* 2000 *Superlattices and Microstructures* **28** 199

Born M 1914 *Ann. Phys., Lpz.* **44** 605

Born M and Huang K 1954 *Dynamical Theory of Crystal Lattices* (Oxford: Oxford University Press)

Both W and Herrmann F P 1982 *Cryst. Res. Technol.* **17** K117

Both W *et al* 1986 *Cryst. Res. Technol.* **21** K85

Bouckaert L, Smoluchowski P and Wigner E P 1936 *Phys. Rev.* **50** 58

Brafman O 1972 *Solid State Commun.* **11** 447

Brafman O *et al* 1968 *Localized Excitations in Solids* ed R F Wallis (New York: Plenum) p 602

Bransden B H and Joachain C J 2000 *Quantum Mechanics* second edition (Prentice Hall)

Briggs A G and Challis L J 1969 *J. Phys. C: Solid State Phys.* **2** 1353

Brodsky M H 1983 *Light Scatering in Solids* vol I, ed M Cardona (Berlin: Springer) p 205

Brodsky M H and Lurio A 1974 *Phys. Rev.* B **9** 1646

Brodsky M H *et al* 1970 *Phys. Rev.* B **2** 3303

Broido D A *et al* 2005 *Phys. Rev.* B **72** 014308

Bron W E 1980 *Rep. Prog. Phys.* **43** 303

Bron W E *et al* 1986 *Phys. Rev.* B **34** 6961

Brown M and Wyatt A F G 1990 *J. Phys. Condensed Matter* **2** 5025

Broyden C G 1965 *Math. Comput.* **19** 577

Brüesch P 1982 *Phonons: Theory and Experiments* vol I (Berlin: Springer)

Brugger K 1964 *Phys. Rev.* **133** A1611

Buot F A 1972 *J. Phys. C: Solid State Phys.* **5** 5

Burke K *et al* 2016 *Phys. Rev.* B **93** 195132

Burns G 1985 *Solid State Physics* (New York: Academic)

Cahill D G 2004 *Rev. Sci. Instrum.* **75** 5119

Cahill D G *et al* 1994 *Phys Rev* B **50** 6077

Callaway J 1959 *Phys. Rev.* **113** 1046

——1974 *Quantum Theory of the Solid State* (New York: Academic)

Callender R H and Pershan P S 1969 *Phys. Rev. Lett.* **23** 947

Capinski W S *et al* 1999 *Phys. Rev.* B **59** 8105

Carles R *et al* 1980 *J. Phys. C: Solid State Phys.* **13** 899

Cardona M (ed) 1982 *Light Scattering in Solids* vol I (*Topics in Appl. Phys.* **8**) (Berlin: Springer)

Cardona M and Güntherodt G (eds) 1982a *Light Scattering in Solids* vol II (*Topics in Appl. Phys.* **50**) (Berlin: Springer)

——1982b *Light Scattering in Solids* vol III (*Topics in Appl. Phys.* **51**) (Berlin: Springer)

——1984 *Light Scattering in Solids* vol IV (*Topics in Appl. Phys.* **54**) (Berlin: Springer)

——1989 *Light Scattering in Solids* vol V (*Topics in Appl. Phys.* **66**) (Berlin: Springer)

Carruthers P 1961 *Rev. Mod. Phys.* **32** 92

Carter D L *et al* 1971 *Physics of Semimetals and Narrow Gap Semiconductors* ed D L Carter and R J Bate (Oxford: Pergamon) p 275

Casimir H B G 1938 *Physica* **5** 495

Cepellotti A *et al* 2015 *Nature Commun.* **6** 6400

Ceperley D M and Alder B J 1980 *Phys. Rev. Lett.* **45** 566

Chadi D J and Cohen M L 1973 *Phys. Rev.* B **8** 5747

Challis L J 1974 *J. Phys. C: Solid State Phys.* **7** 481

Challis L J 1987 *Proc. Physics of Phonons* ed T Paszkiewicz *Lecture Notes in Physics* (Berlin: Springer) p 264

Challis L J and de Göer A M 1984 *The Dynamical Jahn–Teller Effect in Localized Systems* ed Yu E Perlin and M Wagner (Amsterdam: Elsevier) p 533

Challis L J *et al* 1977 *Phys. Rev. Lett.* **39** 558

——1987 *Physics of Phonons* ed T Paskiewicz (Berlin: Springer) p 348

Chang C W *et al* 2008 *Phys. Rev. Lett.* **101** 075903

Chang G *et al* 2018 *Nature Materials* **17** 878

Chang I F and Mitra S S 1968 *Phys. Rev.* **172** 924

——1971 *Adv. Phys.* **20** 359

Chaput L 2013 *Phys. Rev. Lett.* **110** 265506

Chaput L *et al* 2011 *Phys. Rev.* B **84** 094302

Chelikowski J R and Cohen M L 1976 *Phys. Rev.* B **13** 826

——1979 *Phys. Rev.* B **20** 4150

Chen G 1998 *Phys. Rev.* B **57** 14958

Chen S-Y *et al* 2015 *Nano. Lett.* **15** 2526

Cheng T K *et al* 1991 *Appl. Phys. Lett.* **59** 1923

Chen Y S *et al* 1966 *Phys. Rev.* **151** 648

Chen Z J *et al* 2021 *Phys. Rev. Lett.* **126** 185301
Cherng Y T *et al* 1989 *J. Appl. Phys.* **65** 3285
Cho G C *et al* 1990 *Phys. Rev. Lett.* **65** 764
Chou T-H *et al* 2019 *Phys. Rev.* B **100** 094302
Chouhan R K *et al* 2014 *Phys. Rev.* B **89** 060201
Cochran W 1959a *Proc. R. Soc.* A **253** 260
Cochran W 1959b *Phys. Rev. Lett.* **3** 412
Cochran W and Cowley R A 1962 *J. Phys. Chem. Solids* **23** 447
Cohen M H *et al* 1962 *Phys. Rev. Lett.* **8** 316
Cohen M L and Bergstresser T 1966 *Phys. Rev.* **141** 789
Cohen M L and Heine V 1970 *Solid State Physics* vol 24, ed F Seitz and D Turnbull (New York: Academic) p 37
Cohen R M *et al* 1985 *J. Appl. Phys.* **57** 4817
Cole M W 1972 *Phys. Rev. Lett.* **28** 1622
Colvard C *et al* 1980 *Phys. Rev. Lett.* **45** 298
——*et al* 1985 *Phys. Rev.* B **31** 2080
Connell G A N 1975 *Phys. Status Solidi* B **69** 9
Cottrell A H 1953 *Dislocations and Plastic Flow in Crystals* (Oxford: Oxford University Press)
Courant R and Hilbert D 1953 *Methods of Mathematical Physics* vol 1 (New York: Inter-Science)
Cowley R A 1963 *Adv. Phys.* **12** 421
——1965 *J. Phys.* (Paris) **26** 659
Cuffari D and Bongiorno A 2020 *Phys. Rev. Lett.* **124** 215501
Cunningham S L 1974 *Phys. Rev.* B **10** 4988
Cusack K N E 1987 *The Physics of Structurally Disordered Matter: An Introduction* (Bristol: Adam Hilger)
Dalgarno A and Stewart A L 1956 *Proc. Roy. Soc. London, Ser. A* **238** 269
Damen T C, Porto S P S and Tell B 1966 *Phys. Rev.* **142** 570
Das Sarma S 1983 *Phys. Rev.* B **27** 2590
Daum *et al* 1987 *Phys. Rev. Lett.* **59** 1593
Dawber P G and Elliott R J 1963a *Proc. Phys. Soc.* **81** 453
——1963b *Proc. R. Soc.* A **273** 222
de Gironcoli S 1995 *Phys. Rev.* B **51** 6773
de Goer A M 1969 *J. Physique* **30** 389
de Launay J 1956 *Solid State Physics* vol 2, ed F Seitz and D Turnbull (New York: Academic) p 219
Debernardi A 1998 *Phys. Rev.* B **57** 12847
Debernardi A and Baroni S 1994 *Sol. State Commun.* **91** 813
Debernardi A *et al* 1995 *Phys. Rev. Lett.* **75** 1819
Debye P 1912 *Ann. Phys., Lpz.* **39** 789
——1914 *Vortraege über die kinetische Theorie der Materie und der Elektrizitaet* (Berlin: Teubner) p 19
Deinzer G *et al* 2003 *Phys. Rev.* B **67** 144304
——*et al* 2004 *Phys. Rev.* B **69** 014304
Dietsche W *et al* 1981 *Phys. Rev. Lett.* **47** 660
Dekorsy T *et al* 2000 *Light Scattering in Solids* VIII (eds. M Cardona and M Güntherodt, Springer, Berlin) **76** 169
Dharma-wardana M W C *et al* 1986 *Phys. Rev.* B **34** 3034
——1987 *Phys. Rev. Lett.* **58** 1761
Diaz J B and Weinstein A 1947 *J. Math. Phys.* **26** 133
Dick B G and Overhauser A W 1958 *Phys. Rev.* **112** 90
Dirac P A M 1930 *Proc. Cambridge Phil. Soc.* **26** 376

Dolling G 1963 *Inelastic Scattering of Neutrons. Chalk River Conf.* (Vienna: International Atomic Energy Agency) p 41

Dolling G and Waugh J L T 1965 *Lattice Dynamics* ed R F Wallis (Oxford: Pergamon) p 19

Donovan B and Angress J F 1971 *Lattice Vibrations* (London: Chapman and Hall)

Dovesi R *et al* 2020 *J. Chem. Phys.* **152** 204111

Drabble J R 1966 *Semiconductors and Semimetals* vol 2, eds R K Willardson and A C Beer (New York: Academic) p 75

Dresselhaus M S *et al* 1996 *Science of Fullerenes and Carbon Nanotubes* (Academic Press, San Diego)

Dubay O and Kresse G 2003 *Phys. Rev.* B **67** 035401

Dubey K S and Verma G S 1973 *Phys. Rev.* B **7** 2879

Duffy J W and Trickey S B 2011 *Phys. Rev.* B **84** 125118

Dutcher J R *et al* 1992 *Phys. Rev. Lett.* **68** 2464

Eckstein S G *et al* 1970 *Physical Acoustics: Principles and Methods* ed W P Mason and R N Thurston (New York: Academic) p 243

Ecsedy D J and Klemens P G 1977 *Phys. Rev.* B **15** 5957

Eisenmenger W 1980 *Proc. 3rd Int. Conf. on Phonon Physics in Condensed Matter* ed H J Maris (New York: Plenum) p 303

Eisenmenger W and Dayem A H 1967 *Phys. Rev. Lett.* **18** 125

Enz C P 1968 *Ann. Phys., NY* **46** 114

Esfarjani K and Stokes H T 2008 *Phys. Rev.* B **77** 144112

Esfarjani K *et al* 2011 *Phys. Rev.* B **84** 085204

Einstein A 1907 *Ann. Phys., Lpz.* **22** 180

Elliott S R 1983 *Physics of Amorphous Materials* (London: Longman)

Elliott R J and Gibson A F 1982 *An Introduction to Solid State Physics and its Applications* corrected reprint (Hong Kong: Macmillan)

Emura S *et al* 1987 *J. Appl. Phys.* **62** 4632

——*et al* 1988 *Phys. Rev.* B **38** 3280

Epstein S 1974 *The Variation Methods in Quantum Chemistry* (Academic, New York), Chap. 5

Esmann M *et al* 2018 *Phys. Rev.* B **97** 155422

Evarestov R A and Smirnov V P 1983 *Phys. Status Solidi* B **119** 9

Ewald P P 1921 *Ann. der Physik* **64** 253

Eyring H *et al* 1940 *Quantum Chemistry* (New York: Wiley)

Ezzahri Y *et al* 2007 *Phys. Rev.* B **75** 195309

Fabian J and Allen P B 1997 *Phys. Rev. Lett.* **79** 1885

Fehlner W R and Vosko S H 1977 *Can. J. Phys.* **55** 2041

Feng T and Ruan X 2016 *Phys. Rev.* B **93** 045202

Feng T *et al* 2017 *Phys. Rev.* B **96** 161201(R)

Fermi E 1928 Rend Accad Naz Lincei **6** 602

Feynman R P 1939 *Phys. Rev.* **56** 340

Flicker J K and Leath P L 1973 *Phys. Rev.* B **7** 2296

Fortier D and Suzuki K 1976 *J. Physique* **37** 143

Fossheim K and Holt R M 1980 *Phys. Rev. Lett.* **45** 730

——1982 *Physical Acoustics* vol 16, ed W P Mason and R N Thurston (New York: Academic) p 217

Fossheim K *et al* 1978 *Phys. Rev.* B **17** 964

Freeman J J and Anderson A C 1986 *Phys. Rev.* B **34** 5684

Fritsch J and Pavone P 1995 *Surf. Sci.* **344** 159

Fritsch J *et al* 1995 *Phys. Rev.* B **52** 11326

Froyen S and Cohen M L 1982 *Solid State Commun.* **43** 447

——1984 *Phys. Rev.* B **29** 3770

Fröhlich H 1954 *Adv. Phys.* **3** 325

Fu L and Kane C L 2006 *Phys. Rev.* B **74** 195312

Fuchs K 1935 *Proc. R. Soc.* A **151** 585

Fuchs R and Kliewer K L 1965 *Phys. Rev.* **140** 2076

Fukui T *et al* 2005 *J. Phys. Soc. Jpm.* **74** 1674

Fulkerson W *et al* 1968 *Phys. Rev.* **167** 765

Galtier P *et al* 1984 *Phys. Rev.* B **30** 726

Ganikhanov F and Vallèe F 1997 *Phys. Rev.* B **55** 15614

Gantmacher F R 1959 *Applications of the Theory of Matrices* (New York: Inter-Science) ch 3

Gardner M 1977 *Sci. Am.* **236** 110

Garg A K 1986 *J. Phys. C: Solid State Phys.* **19** 3949

Garg J and Chen G 2013 *Phys. Rev.* B **87** 140302

Garg J *et al* 2011 *Phys. Rev. Lett.* **106** 045901

Garg J *et al* 2014 *Length-scale Dependent Phonon Interactions* (eds. S L Shind'e and G P Srivastava, Springer, New York) Ch 4

Gaur N K S 1978 *Physica* **93B** 212

Gaur N K S *et al* 1966 *Phys. Rev.* **144** 628

Gaur N K S and Verma G S 1967 *Phys. Rev.* **159** 610

Geballe T H and Hull G W 1958 *Phys. Rev.* **110** 773

Giannozzi P *et al* 1991 *Phys. Rev.* B **43** 7231

Giannozzi P *et al* 2009 *J. Phys: Condens Matter* **21** 395502

Gilat G 1976 *Methods in Comput. Phys.* **15** 317

Gilat G and Raubenheimer L J 1966 *Phys. Rev.* **144** 390

Giustino F 2017 *Rev. Mod. Phys.* **89** 015003

Glassbrenner C J and Slack G A 1964 *Phys. Rev.* **134** A1058

Golding B and Graebner J E 1976 *Phys. Rev. Lett.* **37** 852

——1981 *Amorphous Solids: Low Temperature Properties* ed W A Phillips (Berlin: Springer) p 107

Golding B *et al* 1976 *Phys. Rev. Lett.* **37** 1248

Goldsmid H J and Penn A W 1968 *Phys. Lett.* **27A** 523

Gonze X and Vigneron J P 1989 *Phys. Rev.* B **39** 1312

Górska M L and Nazarewicz W 1973 *Phys. Status Solidi* B **57** K65

——1974 *Phys. Status Solidi* B **65** 193

Gratias D 1987 *Contemp. Phys.* **28** 219

Griffel D H 1981 *Applied Functional Analysis* (Chichester: Ellis Horwood)

Grimvall G 1976 *Phys. Scr.* **14** 63

——1981 *The Electron-phonon Interaction in Metals* (North-Holland, Amsterdam)

Grill W and Weis O 1975 *Phys. Rev. Lett.* **35** 588

Guthrie G L 1966 *Phys. Rev.* **152** 801

Grüneisen E 1908 *Ann. Phys., Lpz.* **26** 393

Guyer R A and Krumhansl 1966 *Phys. Rev.* **148** 766

Ham F S and Slack G A 1971 *Phys. Rev.* B **4** 777

Hamann D R 1989 *Phys. Rev.* B **40** 2980

Hamann D R *et al* 1979 *Phys. Rev. Lett.* **43** 1494

Hamilton W R 1940 *Mathematical Papers* (Cambridge: Cambridge University Press)

Hamilton R A H and Parrott J E 1969 *Phys. Rev.* **178** 1284

Hangyo M *et al* 2005 *Int. J. Infrared and Millimeter Waves* **26** 1661

Harada H and Narita S 1971 *J. Phys. Soc. Japan* **30** 1628

Hardy R J 1970 *Phys. Rev.* B **2** 1193

Härkönen V J and Karttunen A J 2016 *Phys. Rev.* B **93** 024307

Harrison W A 1966 *Pseudopotentials in the Theory of Metals* (New York: Benjamin)
Hartwigsen C *et al* 1998 *Phys. Rev.* B **58** 3641
Hasan M Z and Kane C L 2010 *Rev. Mod. Phys.* **82** 3045
Hase M *et al* 2009 *Phys. Rev.* B **79** 174112
Hasegawa H 1960 *Phys. Rev.* **118** 1513
Hedin L and Lundqvist B I 1971 *J. Phys. C: Solid State Phys.* **4** 2064
Heine V 1960 *Group Theory in Quantum Mechanics* (Oxford: Pergamon)
Heine V 1963 *Proc. Phys. Soc.* **81** 300
Heine V and Jones R O 1969 *J. Phys. C: Solid State Phys.* **2** 71
Henley C L 1987 *Comments Condensed Matter Phys.* **13** 59
Hensel J C *et al* 1983a *Phys. Rev.* B **28** 1124
——1983b *Phys. Rev. Lett.* **51** 2302
Hepplestone S P and Srivastava G P 2008 *Phys. Rev. Lett.* **101** 105502
——2010 *Phys. Rev.* B **82** 144303
Hermann C 1930 *Z. Kristall* **75** 159
Herman F J 1959 *J. Phys. Chem. Solids* **8** 405
Herring C 1954 *Phys. Rev.* **95** 954
Herring C and Vogt E 1956 *Phys. Rev.* **101** 944
Hicks L D and Dresselhaus M S 1993 *Phys. Rev.* B **47** 12727
Hofer W A *et al* 2002 *Chem. Phys. Lett.* **355** 347
Hohenberg P and Kohn W 1964 *Phys. Rev.* **136** B864
Holstein T and Primakoff H 1940 *Phys. Rev.* **58** 1098
Hone J *et al* 1999 *Phys. Rev.* B **59** R2514
Holland M G 1963 *Phys. Rev.* **132** 2461
——1964 *Phys. Rev.* **134** A471
Holland M G and Neuringer L J 1962 *Proc. Int. Congr. on the Physics of Semiconductors* Exeter (London: Institute of Physics) p 475
Horner 1974 *Dynamical Properties of Solids* vol 1, ed G K Horton and A A Maradudin (Amsterdam: North-Holland) p 451
Horten U *et al* 1986 *Phys. Rev. Lett.* **57** 2947
Hu P *et al* 1981 *Phys. Rev. Lett.* **46** 192
Hua B and Cao Y-C 2014 *Heat Mass Transf.* **78** 755
——2016 *Proc. R. Soc.* A **472** 2015081
Hunklinger S 1977 *Festkörperprobleme* **XVII** 1
Hunklinger S and Arnold W 1976 *Physical Acoustics* vol 12, ed W P Mason and R N Thurston (New York: Academic) p 155
Hunklinger S and Schickfus M v 1981 *Amorphous Solids: Low-Temperature Properties* ed W A Phillips (Berlin: Springer) p 81
Huxtable S T *et al* 2002a *Appl. Phys. Lett.* **80** 1737
——2002b *Proc. of 2002 ASME Int. Mechanical Engineering Congress and Exposition vol IMECE2002-34239* pp 1–5
Hyman D S *et al* 1969 *Phys. Rev.* **186** 231
Ihm J, Zunger A and Cohen M L 1979 *J Phys C: Solid State Phys.* **12** 4409
Iijima S 1991 *Nature* (London) **354** 56
Inkson J C 1980 *J. Phys. C: Solid State Phys.* **13** 369
Jackson H E and Walker C 1971 *Phys. Rev.* B **3** 1428
Jackson T B *et al* 1997 *J. Am. Ceram. Soc.* **80** 1421
Jäckle J 1972 *Z. Phys.* **257** 212
Jancel R 1969 *Foundations of Classical and Quantum Statistical Mechanics* (Oxford: Pergamon)
Jahne E 1976 *Phys. Status Solidi* B **75** 221

Jensen H H *et al* 1969 *Phys. Rev.* **185** 323

Jepsen O and Anderson O K 1971 *Solid State Commun.* **9** 1763

Jiang P *et al* 2018 *J. Appl. Phys.* **124** 161103

Jin Y *et al* 1995 *J. Chem. Phys.* **103** 6697

Johnson D D 1988 *Phys. Rev.* B **38** 12807

Jones R O and Gunnarsson O 1989 *Rev. Mod. Phys.* **61** 689

Joshi A W 1982 *Elements of Group Theory for Physicists* (New Delhi: Wiley Eastern)

Joshi S K and Rajagopal A K 1968 *Solid State Physics* vol 22 (New York: Academic)

Joshi Y P 1974 *Phys. Status Solidi* B **65** 823

Joshi Y P *et al* 1970 *Phys. Rev.* **139** 642

Julian L 1965 *Phys. Rev.* **137** 128

Jullien R 1987 *Comments Condensed Matter Phys.* **13** 177

Jusserand B and Cardona M 1989 *Light Scattering in Solids* vol V, ed M Cardona and G Güntherodt (Berlin: Springer) p 49

Jusserand B and Paquet D 1986a *Heterojunctions and Semiconductor Superlattices* ed G Allen, G Bastard, N Boccara, M Lannoo and M Voos (Berlin: Springer) p 108

——1986b *Phys. Rev. Lett.* **56** 1752

Jusserand B and Sapriel J 1981 *Phys. Rev.* B **24** 7194

Jusserand B and Slempkes S 1984 *Solid State Commun.* **49** 95

Jusserand B *et al* 1986 *Phys. Rev.* B **33** 2897

——*et al* 1987 *Phys. Rev.* B **35** 2808

——*et al* 1988 *Gallium Arsenide and Related Compounds 1987* ed A Christou and H S Rupprecht (Inst. Phys. Conf. Ser. 91) (Bristol: Institute of Physics) p 47

Kajimura K 1982 *Physical Acoustics* vol 16, ed W P Mason and R N Thurston (New York: Academic) p 295

Kapitza P L 1941 *J. Phys. USSR* **4** 181

Karpov V G and Parshin D A 1985 *Sov. Phys.–JETP* **61** 1308

Karttunen A A *et al* 2011 *Inorg. Chem.* **50** 1733

Kato T 1980 *Perturbation Theory of Linear Operators* (Springer-Verlag, Berlin)

Kaviany M 2008 *Heat Transfer Physics* (Cambridge University Press, New York)

Kawamura M *et al* 2014 *Phys. Rev.* B **89** 094515

Kawasaki K 1963 *Prog. Theor. Phys.* **29** 801

Keating P N 1966 *Phys. Rev.* **145** 637

Kekelidze N P *et al* 1973 *J. Phys. Chem. Solids* **34** 2117

Kellermann E W 1940 *Phil. Trans. R. Soc.* A **238** 513

Kent A J *et al* 1988 *Surf. Sci.* **196** 410

Kerker G P 1980 *J. Phys. C: Solid State Phys.* **13** L189

Kesavasamy K and Krishnamurthy N 1978 *Am. J. Phys.* **46** 815

Khalatnikov I M 1952 *Zh. Eksp. Teor. Fiz.* **22** 687

——1965 *An Introduction to the Theory of Superfluidity* (New York: Benjamin)

Khalatnikov M and Adamenko I N 1972 *Zh. Eksp. Teor. Fiz.* **63** 745

Kim O K and Spitzer W G 1979 *J. Appl. Phys.* **50** 4362

Kim J *et al* 1995 *Phys. Rev.* B **52** 14709

Kim W *et al* 2007 *Nanotoday* **2** 40

Kinder H 1972 *Phys. Rev. Lett.* **28** 1564

King P J and Sheard F W 1970 *Proc. R. Soc.* A **320** 175

Kittel C 1986 *Introduction to Solid State Physics* 6th edn (New York: Wiley)

Klein M V 1969 *Phys. Rev.* **186** 839

Klemens P G 1951 *Proc. R. Soc.* A **208** 108

——1955 *Proc. Phys. Soc.* A **68** 1113

——1957 *Can. J. Phys.* **35** 441
——1958 *Solid State Physics* vol 7, ed F Seitz and D Turnbull (New York: Academic) p 1
——1966 *Phys. Rev.* **148** 845
——1967 *J. Appl. Phys.* **38** 4573
——1975 *Phys. Rev.* B **11** 320
Knauss D C and Wilson R S 1974 *Phys. Rev.* B **10** 4383
Koh Y K *et al* 2009 *Advanced Functional Materials* **19** 610
Kohler M 1948 *Z. Phys.* **124** 679
Kohn W and Sham L J 1965 *Phys. Rev.* **140** A1133
Kohn W and Vashishta P 1983 *Theory of the Inhomogeneous Electron Gas* ed S Lundqvist and N H March (New York: Plenum) p 79
Koster G F 1957 *Solid State Physics* vol 5, ed F Seitz and D Turnbull (New York: Academic) p 174
Kopylov V N and Meshov-Deglin L P 1971 *JETP Lett.* **14** 21
Krabach T N *et al* 1983 *Solid State Commun.* **45** 895
Kress G *et al* 1995 *Europhys. Lett.* **32** 729
Kress W 1972 *Phys. Status Solidi* B **49** 239
Krishnan K S 1946 *Proc. Ind. Acad. Sci.* A **24** 45
Krueger F *et al* 1928 *Ann. Phys., Lpz.* **85** 110
Krumhansl J A 1965 *Proc. Phys. Soc.* **85** 921
Kubo R 1957 *J. Phys. Soc. Japan* **12** 570
Kumar A *et al* 1969 *Phys. Rev.* **178** 1480
Kunc K 1973–74 *Ann. Phys., Paris* **8** 22
Kunc K 1985 *Electronic Structure, Dynamics, and Quantum Structural Properties of Condensed Matter* ed J T Devereese *et al* (New York: Plenum) p 227
Kunc K and Gomes Dacosta P 1985 *Phys. Rev.* B **32** 2010
Kunc K and Martin R M 1981 *Phys. Rev.* B **24** 2311
——1982 *Phys. Rev. Lett.* **48** 406
——1983 *Ab Initio Calculation of Phonon Spectra* ed J T Devereese, V E van Doren and P E van Camp (New York: Plenum) p 65
Kushch *et al* 2017 *Proc. R. Soc.* A **473** 20170472
Kwok P C 1966 *Phys. Rev.* **149** 666
——1967 *Solid State Physics* vol 20, ed F Seitz and D Turnbull (New York: Academic) p 213
Ladan F R and Maneval J P 1976 *Phonon Scattering in Solids* ed L J Challis, V W Rampton and A F G Wyatt (New York: Plenum) p 331
Lam P K and Cohen M L 1982 *Phys. Rev.* B **25** 6139
Lamb H 1917 *Proc. Roy. Soc. Lond.* A **93** 114
Landau L 1941 *J. Phys. USSR* **5** 71
——1947 *J. Phys. USSR* **11** 91
Landau L and Khalatnikov I M 1949 *Sov. Phys.–JETP* **19** 637
Landau L and Rumer G 1937 *Phys. Z. Sowjetunion* **11** 18
Landau L D and Lifshitz E M 1959 *Theory of Elasticity* (Oxford: Pergamon)
Landry E S and McGaughey A J H 2009 *Phys. Rev.* B **79** 075316
Lange G *et al* 1998 *Europhys. Lett.* **41** 647
Lannin J S 1988 *Physics Today* July 28
Lanzillo N A *et al* 2013 *Appl. Phys. Lett.* **103** 093102
Larecki W and Banach Z 2010 *J. Phys. A: Math. Gen.* **43** 385501
Laurence G 1971 *Phys. Lett.* **34A** 308
Leake J A *et al* 1969 *Phys. Rev.* **181** 1251
Lee M L and Venkatasubramanian R 2008 *Appl. Phys. Lett.* **92** 053112
Lee S-M *et al* 1997 *Appl. Phys. Lett.* **70** 2957

Lehman G and Taut M 1972 *Phys. Status Solidi* B **54** 469

Leibfried G and Ludwig W 1961 *Solid State Physics* vol 12, ed F Seitz and D Turnbull (New York: Academic) p 275

Leibfried G and Schlömann E 1954 *Nachr. Akad. Wiss. Göttingen Math. Phys.* Kl II(a) **4** 71

Leijnse M and Flensberg K 2012 *Semicond. Sci. Technol.* **27** 124003

Lengfellner H and Renk K F 1981 *Phys. Rev. Lett.* **46** 1210

Levine H and Schwinger J 1949 *Phys. Rev.* **75** 1423

Levine D and Steinhardt P J 1984 *Phys. Rev. Lett.* **53** 2477

Levy M 1979 *Proc. Natl Acad. Sci., USA* **76** 6062

Levy O and Cherkaev E 2013 *J. Appl. Phys.* **114** 164102

Li J *et al* 2020 *Phys. Rev.* B **101** 081403(R)

Liebhaber M *et al* 2014 *Phys. Rev.* B **89** 045313

Lifshitz E M 1943a *J. Phys. USSR* **7** 215

——1943b *J. Phys. USSR* **8** 89

Lisitsa M P *et al* 1969 *Phys. Status Solidi* **34** 269

Liu J L *et al* 2003 *Phys. Rev.* B **67** 165333

Liu L *et al* 2003 *Phys. Rev.* B **68** 201301

Li X *et al* 2021 *Phys. Rev.* B **104** 054103

Liu P-F *et al* 2021 *Phys. Rev.* B **103** 094306

Liu Y *et al* 2018 *National Sci. Rev.* **5** 314

Logachev Yu A and Yur'ev M S 1973 *Sov. Phys. Solid State* **14** 2826

Lopez A *et al* 2001 *Surf. Sci.* **477** 219

Lorimor O G *et al* 1966 *J. Appl. Phys.* **37** 2509

Loudon R 1964 *Adv. Phys.* **13** 423

Löwdin P-O 1951 *J. Chem. Phys.* **19** 1396

Lu J P *et al* 1986 *Phys. Rev.* B **33** 4809

Lucovsky G and Chen M F 1970 *Solid State Commun.* **8** 1397

Lucovsky G *et al* 1967 *Solid State Commun.* **5** 113

——1975 *Phys. Rev.* B **12** 4135

——1976 *Phys. Rev.* B **14** 2503

Ma J *et al* 2014 *Phys. Rev.* B **90** 035203

Madsen G K H *et al* 2016 *phys. stat. sol.* a **213** 802

Mahan G D 1993 *Many-Particle Physics* (Plenum, New York), 2nd Ed.

Mair J and Sigmund E 1981 *J. Physique* **42** C6-232

Majorana E 1937 *Il Nuovo Cimento* **14** 171

Mandelbrot B B 1982 *Fractal Geometry of Nature* (Freeman: San Francisco)

Maneval J P *et al* 1971 *Phys. Rev. Lett.* **27** 1375

Maradudin A A 1965 *Rep. Prog. Phys.* **38** 331

Maradudin A A and Fein A E 1962 *Phys. Rev.* **128** 2589

Maradudin A A and Mills D L 1968 *Phys. Rev.* **173** 881

Maradudin A A and Vosko S H 1968 *Rev. Mod. Phys.* **40** 1

Maradudin A A *et al* 1971 *Solid State Physics* Supplement 3 (New York: Academic)

March N H 1983 *Theory of the Inhomogeneous Electron Gas* ed S Lundqvist and N H March (New York: Plenum)

March N H and Paranjape B V 1987 *Phys. Rev.* A **35** 5285

Mariot L 1972 *Group Theory and Solid State Physics* (Englewood Cliffs, NJ: Prentice-Hall)

Maris H J 1971 *J. Acoust. Soc. Am.* **50** 812

——1974 *Phys. Rev.* A **9** 1412

——1979 *Phys. Rev.* B **19** 1443

——1986 *Nonequilibrium Phonons in Nonmetallic Crystals* ed W Eisenmenger and A A Kaplyanskii (Amsterdam: North-Holland) p 51

Martin R M 1969 *Phys. Rev.* **186** 871

——1970 *Phys. Rev.* B **1** 4005

——2004 *Electronic Structure - Basic Theory and Practical Methods* (Cambridge: Cambridge University Press)

Martin R M and Kunc K 1983 *Ab Initio Calculation of Phonon Spectra* ed J T Devereese, V E van Doren and P E van Camp (New York: Plenum) p 49

Matossi F 1951 *J. Chem. Phys.* **19** 161

Matsuda H 1962 *Prog. Theor. Phys. Suppl.* **23** 23

Mattuck R D and Strandberg M W P 1960 *Phys. Rev.* **119** 1204

Mauguin Ch 1931 *Z. Kristall.* **76** 542

McCurdy A K 1982 *Phys. Rev.* B **26** 6971

McGlinn T C *et al* 1986 *Phys. Rev.* B **33** 8396

McKelvey J P 1966 *Solid State and Semiconductor Physics* (International Edition) (Harper and Row, New York, Evanston, London, and John Weatherhill Inc., Tokyo)

McWhan D B and Marezio M J 1966 *J. Chem. Phys.* **45** 2508

Meddins H R and Parrott J E 1976 *J. Phys. C: Solid State Phys.* **9** 1263

Meek P E 1976 *Phil. Mag.* **33** 897

Melcher R L and Shiren N S 1982 *Physical Acoustics* vol 16, ed W P Mason and R N Thurston (New York: Academic) p 341

Menèndez J and Cardona M 1984 *Phys. Rev.* B **29** 2051

Merlin R *et al* 1980 *Appl. Phys. Lett.* **36** 43

——1985 *Phys. Rev. Lett.* **55** 1768

Mie G 1903 *Ann. Phys., Lpz.* **11** 657

Mikhail I F I 1985 *J. Phys. C: Solid State Phys.* **18** 5801

Mikhail I F I and Madkour S S R 1985 *J. Phys. C: Solid State Phys.* **18** 3427

Mingo N and Broido D A 2005 *Nano. Lett.* **5** 1221

Mingo N *et al* 2014 *Length-scale Dependent phonon Interactions* (eds. S L Shind'e and G P Srivastava, Springer, New York) Ch 5

Minnich A and Chen G 2007 *Appl. Phys. Lett.* **91** 073105

Miotto R *et al* 2001 *J. Chem. Phys.* **114** 9549

——2005 *J. Chem. Phys.* **123** 074708

Mityagin Yu A *et al* 1976 *Sov. Phys.–Solid State* **17** 1341

Mohamed A *et al* 2021 *J Phys: Condens Matter* **33** 075501

Monkhorst H J and Pack J D 1976 *Phys. Rev.* B **13** 5189

Mooradian A and Harman T C 1971 *Physics of Semimetals and Narrow Gap Semiconductors* ed D L Carter and R J Bate (Oxford: Pergamon) p 297

Morelli D T and Uher C 1993 *Appl. Phys. Lett.* **63** 165

Morelli D T *et al* 1993a *Phys. Rev.* B **47** 131

——1993b *Appl. Phys. Lett.* **62** 1085

Morgan D J 1969 *Solid State Theory: Methods and Applications* ed P T Landsberg (New York: Wiley) p 155

Mori H 1965 *Prog. of Theor. Phys.* **33** 423

Morita A *et al* 1972 *J. Phys. Soc. Japan* **32** 29

Morse P M and Feshbach H 1953 *Methods of Theoretical Physics* (McGraw-Hill, New York)

Morton I P and Lewis M F 1971 *Phys. Rev.* B **3** 552

Musgrave M J P and Pople J A 1962 *Proc. R. Soc.* A **268** 474

Nakamura Y 2018 *Sci. Technol. Adv. Mater.* **19** 31

Nakashima T *et al* 1971 *J. Phys. Soc. Japan* **30** 1508

Nakashima S *et al* 1973 *J. Phys. Soc. Japan* **35** 1437

Nan C-W 1993 *Prog. Mater. Sci.* **37** 1

Nan C-W *et al* 1997 *J. Appl. Phys.* **81** 6692

Narayanamurti V 1969 *Phys. Lett.* **30A** 521

——1981 *J. Physique* **42** C6-221

Narayanamurti V and Dynes R C 1976 *Phonon Scattering in Solids* ed L J Challis, V W Rampton and A F G Wyatt (New York: Plenum) p 93

Narayanamurti V and Pohl R O 1970 *Rev. Mod. Phys.* **42** 201

Narayanamurti V *et al* 1973 *Phys. Rev. Lett.* **31** 687

——1979 *Phys. Rev. Lett.* **43** 2012

Neelmani and Verma G S 1972 *Phys. Rev.* B **6** 2026

——1973 *Phys. Rev.* B **7** 1650

Nelin G and Nilsson G 1972 *Phys. Rev.* B **5** 3151

Nettleton R E 1963 *Phys. Rev.* **132** 2032

Nienhaus H and Mönch W 1995 *Surf. Sci.* **328** L561

Nihira T and Iwata T 1975 *Jpn. J. Appl. Phys.* **14** 1099

Nika D L *et al* 2012 *Nano. Lett.* **12** 3238

Nilsson G and Nelin G 1971 *Phys. Rev.* B **3** 364

Noble B 1964 *Math. Res. Centre Report* no 473, Wisconsin University, Madison, USA

Nolas G S *et al* 2003 *Appl. Phys. Lett.* **82** 910

Northrup G A 1982 *Phys. Rev.* B **26** 903

Northrup G A and Wolfe J P 1980 *Phys. Rev.* B **22** 6196

Olego D J *et al* 1986 *Phys. Rev.* B **33** 3819

Olson J R *et al* 1993 *Phys. Rev.* B **47** 14850

Omini M and Sparavigna A 1995 *Physica* B **212** 101

Onn D G *et al* 1992 *Phys. Rev. Lett.* **68** 2806

Orbach R and Vredevoe L A 1964 *Physics (Long Island City, N. Y.)* **1** 91

Ordonez-Miranda J *et al* 2011 *Appl. Phys. Lett.* **98** 233111

Pal R K and Ruzzene M 2017 *New J. Phys.* **19** 025001

Pandey K C 1981 *Phys. Rev. Lett.* **47** 1913

Parker J H *et al* 1967 *Phys. Rev.* **155** 712

Parlinski K *et al* 2000 *J. Phys. Chem. Solids* **61** 87

Parrish J F *et al* 1967 *II–VI Semiconducting Compounds* ed D G Thomas (New York: Benjamin) p 1164

Parrott J E 1963 *Proc. Phys. Soc.* **81** 726

——1969a *Solid State Theory - methods and applications* (Ed. P T Landsberg) (London: Wiley - Interscience)

——1969b *J. Phys. C: Solid State Phys.* **2** 147

——1971 *Phys. Status Solidi* B **48** K159

——1979 *Rev. Int. Hautes. Temp. Refract.* **16** 393

Parrott J E and Stuckes A D 1975 *Thermal Conductivity of Solids* (Pion Ltd)

Pathak K N 1965 *Phys. Rev.* **139** A1569

Peierls R 1929 *Ann. Phys., Lpz.* **3** 1055

——1935 *Ann. Inst. H Poincaré* **5** 177

Péraud J-P M *et al* 2016 *Phys Rev* B **93** 045424

Perdew J P and Zunger A 1981 *Phys. Rev.* B **23** 5048

Perdew J P *et al* 1996 *Phys. Rev. Lett.* **77** 3865

Peshkov V 1947 *Report on the Cambridge Low Temperature Conference* (London: Physical Society) p 19

Petrov A V *et al* 1976 *High Temp. High Pressures* **8** 537

Pettifor D G and Weaire D L (eds) 1985 *The Recursion Method and Its Applications* (Berlin: Springer)

Philip J and Viswanathan K S 1978 *Phys. Rev.* B **17** 4969

Phillips J C 1956 *Phys. Rev.* **104** 1263

——1968 *Phys. Rev.* **166** 832

Phillips J C and Kleinman L 1959 *Phys. Rev.* **116** 287

Phillips W A 1972 *J. Low. Temp. Phys.* **7** 351

Pickett W E *et al* 1978 *Phys. Rev.* B **17** 815

Pittalis S *et al* 2011 *Phys. Rev. Lett.* **107** 163001

Pohl R O 1962 *Phys. Rev. Lett.* **8** 481

——1981 *Amorphous Solids: Low-Temperature Properties* ed W A Phillips (Berlin: Springer) p 27

Pollack G L 1969 *Rev. Mod. Phys.* **41** 48

Pollmann J *et al* 1986 *Appl. Phys.* A **41** 21

Pomeranchuk I 1941 *J. Phys. (USSR)* **4** 259

——1972 *Sobranie Nauchseykh Trudov* (Collected scientific works) Vol 1 (Moscow: Nauka)

Pomraning G C 1967 *J. Math. Phys.* **8** 2096

Pop E 2010 *Nano Res.* **3** 147

Pop E *et al* 2012 *MRS Bull.* **37** 1273

Powell B M 1970 *Solid State Commun.* **8** 2157

Press W H *et al* 1986 *Numerical Recipes* (Cambridge University Press, Cambridge)

Pulay P 1969 *Mol. Phys.* **17** 197

Qi X-L and Zhang S-C 2011 *Rev. Mod. Phys.* **83** 1057

Qian X *et al* 2014 *Science* **346** 1344

Qu Z *et al* 2017 *AIP Advances* **7** 015108

Raman C V 1928 *Ind. J. Phys.* **2** 387

Ramiere A *et al* 2016 *Nature Materials* **15** 512

Rammal R and Toulouse G 1983 *J. Physique Lett.* **44** L13

Ranninger J 1965 *Phys. Rev.* A **140** 2031

Rayleigh L 1945 *The Theory of Sound* (New York: Dover)

——1885 *Proc. Lond. Math. Soc.* **17** 4

Ren S-F *et al* 1987 *Phys. Rev. Lett.* **59** 1841

Ren S Y and Dow J D 1982 *Phys. Rev.* B **25** 3750

Renk K F 1985 *Nonequilibrium Phonon Dynamics* ed W E Bron (New York: Plenum) p 59

Ridley B K 1996 *J. Phys.: Condens. Matter* **8** L511

Reissland J A 1973 *The Physics of Phonons* (New York: Wiley)

Robinson P D and Arthurs A M 1968 *J. Math. Phys.* **9** 1364

Rothenfusser M *et al* 1986 *Phys. Rev.* B **34** 5518

Roufosse M and Klemens P G 1973 *Phys. Rev.* B **7** 5379

Roussopoulos P 1953 *CR Acad. Sci. Paris* **236** 1858

Ruf T *et al* 2000 *Solid State Commun.* **115** 243

Rustagi K C and Weber W 1976 *Solid State Commun.* **18** 673

Rytov S M 1956 *Akust. Zh.* **2** 71 (*Sov. Phys. Acoust.* **2** 68 (1956))

Sahoo R and Mishra R R 2012 *J. Expt. and Theor. Phys.* **114** 805

Sahoo S *et al* 2013 *J. Chem. Phys.* **117** 9042

Saito R *et al* 1998 *Physical Properties of Carbon Nanotubes* (Imperial College Press, London), pp. 35–72

Sanchez D S *et al* 2019 *Nature* **567** 500

Sato M and Ando Y 2017 *Rep. Prog. Phys.* **80** 076501

Sato T *et al* 2013 *Phys. Rev. Lett.* **110** 206804

Saxon S D and Hutner R W 1949 *Philips Res. Rep.* **4** 81

Scalliet C *et al* 2019 *Nat. Materials* **10** 5102

Scheffler M *et al* 1985 *Phys. Rev.* B **31** 6541

Schiff L I 1968 *Quantum Mechanics* 3rd edn (New York: McGraw-Hill)

Schlüter M *et al* 1975 *Phys. Rev.* B **12** 4200

Schmid H J 1960 *Applications of Thermoelectricity* (Methuen, London)

Schmidt A J *et al* 2009 *Rev. Sci. Instrum.* **80** 09490

Schmidt T M and Srivastava G P 2020 *Comp. Mater. Sci.* **182** 109777

Schoenflies A 1923 *Theorie der Kristallstruktur* (Berlin: Borntraeger)

Scholz R and Stahl A 1991 *phys. stat. sol.* b **168** 123

Schröder U 1966 *Solid State Commun.* **4** 347

Schröter N B M *et al* 2020 *Science* **369** 179

Sellitto A *et al* 2017 *Continuum Mech.* **29** 411

Sen P N and Thorpe M F 1977 *Phys. Rev.* B **15** 4030

Seward W D and Narayanamurti V 1966 *Phys. Rev.* **148** 463

Sham L J 1974 *Dynamical Properties of Solids* vol 1, ed G K Horton and A A Maradudin (Amsterdam: North-Holland) p 301

——1969 *Phys. Rev.* **188** 1431

Sham L J and Kohn W 1966 *Phys. Rev.* **145** 561

Shan J *et al* 1996 *J. Chem. Phys.* **100** 4961

Shang H *et al* 2017 *Comp. Phys. Commun.* **215** 26

Sharma P C *et al* 1971 *Phys. Rev.* B **4** 1306

Shechtman D *et al* 1984 *Phys. Rev. Lett.* **53** 1951

Shuker R and Gamon R W 1971 *Light Scattering in Solids* (Paris: Flammarion) p 334

Siddiqui M U and Arif A F M 2016 *Materials* **9** 694

Sie E J *et al* (2019) *Nature* **565** 61

Sigmund E and Lassmann K 1980 *Phonon Scattering in Condensed Matter* ed H J Maris (New York: Plenum) p 417

Sihvola A 1997a *Electromagnetics* **17** 69

——1997b *Electromagnetics* **17** 269

——1999 *Electromagnetic mixing formulas and applications* The Institue of Electrical Engineers, London

Simkin M V and Mahan G D 2000 *Phys. Rev. Lett.* **84** 927

Simons S 1972 *Lett. Nuovo Cimento* **5** 423

——1975 *J. Phys. C: Solid State Phys.* **8** 1147

Singh D P and Verma G S 1971 *Phys. Rev.* B **4** 4647

Singh M and Verma G S 1974 *Phys. Status Solidi* B **65** 813

Sinha K P and Upadhyaya U N 1962 *Phys. Rev.* **127** 432

Slack G A 1961 *Phys. Rev.* **122** 1451

——1972 *Phys. Rev.* B **6** 3791

——1973 *J. Phys. Chem. Solids* **34** 321

——1995 *CRC Handbook of Thermoelectrics* ed D M Rowe (Boca Raton, FL: CRC) p 407

Slack G A and Bartram S F 1975 *J. Appl. Phys.* **46** 89

Slack G A and Glassbrenner C J 1960 *Phys. Rev.* B **120** 782

Slack G A and Newman R 1958 *Phys. Rev. Lett.* **1** 359

Slater J C 1974 *The Self-consistent Field for Molecules and Solids* vol 4 (New York: McGraw-Hill)

Slonimskii G L 1937 *Zh. Eksp. Teor. Fiz.* **7** 1457

Skriver H L 1984 *The LMTO Method—Muffin Tin Orbitals and Electronic Structure* (Berlin: Springer)

Smith Jr J E *et al* 1971 *Phys. Rev. Lett.* **26** 642

Solin S A and Ramdas A K 1970 *Phys. Rev.* B **1** 1687

Soma T 1976 *Phys. Status Solidi* B **76** 753

——1977 *Phys. Status Solidi* B **82** 319

——1978a *Phys. Status Solidi* B **87** 345

——1978b *J. Phys. Soc. Japan* **44** 469

——1980 *J. Phys. Soc. Japan* **48** 115

Soma T and Kagaya H-M 1983a *Phys. Status Solidi* B **118** 245

——1983b *Solid State Commun.* **46** 773

Soma T and Morita A 1972 *J. Phys. Soc. Japan* **32** 38

Soma T *et al* 1981 *Solid State Commun.* **39** 1193

Sondheimer E H 1950 *Proc. R. Soc.* A **203** 75

Sood K C and Verma G S 1973 *Phys. Rev.* B **7** 5316

Sood A K *et al* 1985 *Phys. Rev. Lett.* **54** 2111, 2115

Spitzer W G 1967 *Semiconductors and Semimetals* vol 3, ed R K Willardson and A C Beer (New York: Academic)

Spitzer W G 1971 *Festkörperprobleme* **XI** 1

Srivastava G P 1974 *Pramana* **3** 209

——1975 *Phys. Status Solidi* B **68** 213

——1976a *Phil. Mag.* **34** 795

——1976b *J. Phys. C: Solid State Phys.* **9** 3037

——1976c *Phys. Status Solidi* B **77** 131

——1976d *Pramana* **6** 1

——1977a *Phys. Status Solidi* B **80** 657

——1977b *J. Phys. C: Solid State Phys.* **10** L63

——1977c *J. Phys. C: Solid State Phys.* **10** 1843

——1980 *J. Phys. Chem. Solids* **41** 357

——1984 *J. Phys. A: Math. Gen.* **17** L316 and **17** 2737

——1997 *Rep. Prog. Phys.* **60** 561

——1999 *Theoretical Modelling of Semiconductor Surfaces* (Singapore: World Scientific)

——2009 *J. Phys.: Condens. Matter* **21** 174205

——2015 *Rep. Prog. Phys.* **78** 026501

Srivastava G P and Hamilton R A H 1978 *Phys. Rep.* **38** 1

Srivastava G P and Kunc K 1988 *J. Phys. C: Solid State Phys.* **21** 5087

Srivastava G P and Thomas I O 2018 *Phys. Rev.* B **98** 035430

——2019 *J. Phys. Condens. Matter* **31** 055303

——2020 *Nanomaterials* **10** 00673

Srivastava G P and Verma G S 1973 *Phys. Rev.* B **7** 897

Srivastava G P and Weaire D L 1987 *Adv. Phys.* **36** 463

Srivastava G P *et al* 1972 *Phys. Rev.* B **6** 3053

——1983 *J. Phys. C: Solid State Phys.* **16** 3627

——1985 *Phys. Rev.* B **31** 2561

Srivastava V P and Verma G S 1974 *Phys. Rev.* B **10** 219

Steg R G and Klemens P G 1970 *Phys. Rev. Lett.* **24** 381

Stehle H and Seeger A 1956 *Z. Phys.* **146** 242

Steigmeir E F and Abeles B 1964 *Phys. Rev.* **136** 1149

Steinhardt P *et al* 1974 *J. Non-crystal. Solids* **15** 199

Stern H 1965 *J. Phys. Chem. Solids* **26** 153

Štich *et al* 1996 *Phys. Rev.* B **54** 2642

Stillinger F H and Weber T A 1985 *Phys. Rev.* **31** 5262

Stohr H and Klemm W 1954 *Z Anorg Allg Chem* **241** 304

Stoneley R 1924 *Proc. Roy. Soc. Lond.* A **106** 416

Struchtrup S 2005 *Macroscopic Transport Equations for Rarefied Gas Flows: Approximation Methods in Kinetic Theory – Interaction of Mechanics and Mathematics* (Springer, New YorK)

Suzuki K and Mikoshiba N 1971a *J. Phys. Soc. Japan* **31** 186

——1971b *J. Phys. Soc. Japan* **31** 44

——1971c *Phys. Rev.* B **3** 2550

Suzuki K *et al* 1964 *J. Phys. Soc. Japan* **19** 930

Swartz E T and Pohl R O 1989 *Rev. Mod. Phys.* **61** 605

Synder G J and Toberer E S 2008 *Nature Materials* **7** 105

Synge J L 1957 *The Hypercircle in Mathematical Physics* (Cambridge: Cambridge University Press)

Takayanagi K *et al* 1985 *J. Vac. Sci. Technol.* A **3** 1502

Talwar D N *et al* 1980 *J. Phys. C: Solid State Phys.* **13** 3775

Tamura S 1982 *Phys. Rev.* B **25** 1415

——1983a *Phys. Rev.* B **27** 858

——1983b *Phys. Rev.* B **28** 897

——1985 *Phys. Rev.* B **31** 2574

Tapfer L and Horikoshi Y 1988 *Gallium Arsenide and Related Compounds 1987* ed A Christou and H S Rupprecht (Inst. Phys. Conf. Ser. 91) (Bristol: Institute of Physics) p 533

Taylor B *et al* 1969 *Phys. Rev. Lett.* **23** 416

Teo J C Y *et al* 2008 *Phys. Rev.* B **78** 045426

Tersoff J 1988 *Phys. Rev.* B **37** 6991

——1989 *Phys. Rev.* B **39** 5566

Thacher P D 1967 *Phys. Rev.* **156** 975

Thomas L H 1927 *Proc. Cambridge Phil. Soc.* **23** 542

Thomas I O and Srivastava G P 2013 *Phys. Rev.* B **88** 115207

——2017 *J. Phys.: Condens. Matter* **29** 505703

——2018a *Phys. Rev.* B **98** 094201

——2018b *Nanomaterials* **8** 01054

——2020 *J. Appl. Phys.* **127** 024304

Thomlinson W C 1969 *Phys. Rev. Lett.* **23** 1330

Thorpe M F 1974 *Amorphous and Liquid Semiconductors* ed J Stuke and W Brenig (London: Taylor and Francis) p 835

——1981 *Excitations in Disordered Systems* ed M F Thorpe (New York: Plenum) p 85

Toennies J P 1990 *Superlattices and Microstructures* **7** 193

——1992 *Europhys. News* **23** 63

Togo A *et al* 2015 *Phys. Rev.* B **91** 094306

Toombs G A *et al* 1987 *Solid State Commun.* **64** 577

Torres V J B and Stoneham A M 1985 *Handbook of Interatomic Potentials. III. Semiconductors* (Harwell Laboratory, Theoretical Physics Division)

Trommer R *et al* 1980 *Phys. Rev.* B **21** 4869

Troullier N and Martins J L 1991 *Phys. Rev.* B **43** 1993

Tsang Y W and Cohen M L 1971 *Phys. Rev.* B **3** 1254

Tsui D C *et al* 1980 *Phys. Rev. Lett.* **44** 341

Tua P F and Mahan G D 1982 *Phys. Rev.* B **26** 2208

Tütüncü H M 1997 *Ph. D. thesis, University of Exeter, UK*

Tütüncü H M and Srivastava G P 1997 *J. Phys. Chem. Solids* **58** 685

Tütüncü H M *et al* 1997 *Phys. Rev.* B **56** 4656

——2017 *Phys. Rev.* B **95** 214514

Ulbrich R G 1985 *Nonequilibrium Phonon Dynamics* ed W E Bron (New York: Plenum) p 101

Ulbrich R G *et al* 1980 *Phys. Rev. Lett.* **45** 1432

Vallèe F and Bogani F 1991 *Phys. Rev.* B **43** 12049

van Camp P E, van Doren V E and Devereese J T 1983 *Ab Initio Calculation of Phonon Spectra* ed J T Devereese, V E van Doren and P E van Camp (New York: Plenum) p 25

Vanderbilt D 1990 *Phys. Rev.* B **41** 7892

Vanderbilt D and Louie S G 1984 *Phys. Rev.* B **30** 6118

Vandersande J W 1977 *Phys. Rev.* B **15** 2355

Venkatasubramanian R *et al* 2001 *Nature* **413** 597

Vermeersch B *et al* 2016 *Appl. Phys. Lett.* **108** 193104

Verstraete M J *et al* 2008 *Phys. Rev.* B **78** 045119

Vetelino J F and Mitra S S 1969 *Phys. Rev.* **178** 1349

von Gutfeld R J and Nethercot A H 1964 *Phys. Rev. Lett.* **12** 641

von Klitzing K 1986 *Rev. Mod. Phys.* **58** 519

Wallis R F 1957 *Phys. Rev.* **105** 540

——1964 *Surf. Sci.* **2** 146

——1974 *Prog. Surf. Sci.* **4** 233

Walter J P and Cohen M L 1971 *Phys. Rev.* B **4** 1877

Ward A and Broido D A 2010 *Phys. Rev.* B **81** 085205

Ward A *et al* 2009 *Phys. Rev.* B **80** 125203

Warren J L 1968 *Rev. Mod. Phys.* **40** 38

Waugh J L T and Dolling G 1963 *Phys. Rev.* **132** 2410

Weaire D 1981 *Amorphous Solids: Low-Temperature Properties* ed W A Phillips (Berlin: Springer) p 13

Weaire D and Alben R 1972 *Phys. Rev. Lett.* **29** 1505

Weber C P 2021 *J. Appl. Phys.* **129** 070901

Weber W 1974 *Phys. Rev. Lett.* **33** 371

——1977 *Phys. Rev.* B **15** 4789

Wehner R K and Klein R 1969 *Phys. Rev. Lett.* **23** 1372

Wei L *et al* 1993 *Phys. Rev. Lett.* **70** 3764

Weinstein B A 1977 *Solid State Commun.* **24** 595

Weinstein B A and Piermarini G J 1975 *Phys. Rev.* B **12** 1172

Wick G C 1950 *Phys. Rev.* **80** 268

Widom A 1969 *Phys. Lett.* **29A** 96

Wigner E P 1934 *Phys. Rev.* **46** 1002

——1938 *Trans. Faraday Soc.* **34** 678

Wigner E and Seitz F 1933 *Phys. Rev.* **43** 804

——1934 *Phys. Rev.* **46** 509

Wilson A H 1953 *The Theory of Metals* 2nd edn (Cambridge: Cambridge University Press)

Wilson R B and Cahill D G 2014 *Nature Commun.* **5** 5075

Wilson R S and Kim S K 1974 *Phys. Rev.* B **7** 4674

Wiser N 1963 *Phys. Rev.* **129** 62

Wolfe J P 1980 *Physics Today* December 44

Woods A D B and Cowley R A 1973 *Rep. Prog. Phys.* **36** 1135

Wooten F and Weaire D 1987 *Solid State Physics* vol 40, ed F Seitz and D Turnbull (New York: Academic) p 1

Wooten F *et al* 1985 *Phys. Rev. Lett.* **54** 1392

Wosylus A *et al* 2009 *Z. Anorg. Allg. Chem.* **635** 700

Wyatt A F G 1981 *Nonequilibrium Superconductivity, Phonons, and Kapitza boundaries* ed K E Gray (New York: Plenum) p 31

——1984 *Physica* **126B** 392

Wyatt A F G *et al* 1974 *Phys. Rev. Lett.* **33** 1425

Wybourne M N and Wigmore J K 1988 *Rep. Prog. Phys.* **51** 923

Xia B W *et al* 2019 *Phys. Rev. Lett.* **123** 065501

Xu X *et al* 2014 *Nat. Commun.* **5** 3689

——2018 *J. Phys.: Condens. Matter* **30** 225401

Yan B and Felser C 2017 *Annu. Rev. Condens. Matter Phys.* **8** 337

Yan Y-X and Nelson K A 1987 *J. Chem. Phys.* **87** 6240

Yang B *et al* 2002 *Appl. Phys. Lett.* **80** 1758

Yang N *et al* 2010 *Nano Today* **5** 85

Yates B 1972 *Thermal Expansion* (New York: Plenum)

Yin M T and Cohen M L 1980 *Phys. Rev. Lett.* **45** 1004

——1982 *Phys. Rev.* B **26** 3259

Yip S-K and Chang Y-C 1984 *Phys. Rev.* B **30** 7037

You J Q and Hu T B 1988 *Phys. Status Solidi* B **147** 471

Yu C C 1989 *Phys. Rev. Lett.* **63** 1160

Yu C C and Leggett A J 1988 *Comments Condensed Matter Phys.* **14** 231

Yu S C *et al* 1978 *Solid State Commun.* **25** 49

Zeller R C and Pohl R O 1971 *Phys. Rev.* B **4** 2029

Zhang L and Niu Q 2015 *Phys. Rev. Lett.* **115** 115502

Zhang M Y *et al* (2019) *Phys. Rev. X* **9** 021036

Zhu H *et al* 2018 *Science* **359** 579

Ziman J M 1956 *Phil. Mag.* **1** 191

——1957 *Phil. Mag.* **2** 292

——1960 *Electrons and Phonons* (Oxford: Clarendon)

Zitslsperger M *et al* 1997 *Surf. Sci.* **377-379** 108

Zubarev D N 1960 *Usp. Fiz. Nauk* **71** 71 (*Sov. Phys. Usp.* **3** 320)

Zunger A *et al* 1990 *Phys. Rev. Lett.* **65** 353

Zwangiz R 1961 *Phys. Rev.* **124** 983

Index